Low Frequency Waves and Turbulence in Magnetized Laboratory Plasmas and in the Ionosphere

IOP Plasma Physics Series

Series Editors

Richard Dendy
Culham Centre for Fusion Energy and the University of Warwick, UK

Uwe Czarnetzki
Ruhr-University Bochum, Germany

About the Series

The IOP Plasma Physics ebook series aims at comprehensive coverage of the physics and applications of natural and laboratory plasmas, across all temperature regimes. Books in the series range from graduate and upper-level undergraduate textbooks, research monographs and reviews.

The conceptual areas of plasma physics addressed in the series include:

- Equilibrium, stability and control
- Waves: fundamental properties, emission, and absorption
- Nonlinear phenomena and turbulence
- Transport theory and phenomenology
- Laser-plasma interactions
- Non-thermal and suprathermal particle populations
- Beams and non-neutral plasmas
- High energy density physics
- Plasma-solid interactions, dusty, complex and non-ideal plasmas
- Diagnostic measurements and techniques for data analysis

The fields of application include:

- Nuclear fusion through magnetic and inertial confinement
- Solar-terrestrial and astrophysical plasma environments and phenomena
- Advanced radiation sources
- Materials processing and functionalisation
- Propulsion, combustion and bulk materials management
- Interaction of plasma with living matter and liquids
- Biological, medical and environmental systems
- Low temperature plasmas, glow discharges and vacuum arcs
- Plasma chemistry and reaction mechanisms
- Plasma production by novel means

Low Frequency Waves and Turbulence in Magnetized Laboratory Plasmas and in the Ionosphere

Hans Pécseli
University of Oslo, Physics Department, Oslo, Norway

IOP Publishing, Bristol, UK

ISBN 978-0-7503-1251-6 (ebook)
ISBN 978-0-7503-1252-3 (print)
ISBN 978-0-7503-1253-0 (mobi)

DOI 10.1088/978-0-7503-1251-6

Version: 20161101

IOP Expanding Physics
ISSN 2053-2563 (online)
ISSN 2054-7315 (print)

British Library Cataloguing-in-Publication Data: A catalogue record for this book is available from the British Library.

Published by IOP Publishing, wholly owned by The Institute of Physics, London

IOP Publishing, Temple Circus, Temple Way, Bristol, BS1 6HG, UK

US Office: IOP Publishing, Inc., 190 North Independence Mall West, Suite 601, Philadelphia, PA 19106, USA

Alice said to the cat 'Would you tell me, please, which way I ought to go from here?'
'That depends a good deal on where you want to get to,' said the Cat.
'I don't much care where ...' said Alice
'Then it doesn't much matter which way you walk,' said the Cat.
'...so long as I get somewhere,' Alice added as an explanation
'Oh, you're sure to do that,' said the Cat, 'if only you walk long enough.'

Lewis Carroll, *Alice's Adventures in Wonderland*

Contents

Preface

It is often argued that the studies of turbulent flows had their significant break-through with Reynolds' experiments where he studied flow of water through long pipes. It was eventually realised that an enhanced mixing rate is one (maybe even the most) important property of such flows. Since that time, turbulence studies have found applications relevant for industry as well as for the environment and are considered to be of utmost importance.

When plasma media, magnetized plasmas in particular, became an active topic of research it was realised that also in this case turbulent fluctuations could develop. Turbulent or anomalous transport also became relevant here. This transport mechanism determines for instance the steepness of the plasma density gradients that can be maintained. A problem of particular importance concerns confinement of hot plasma in fusion experiments, but as a principle the question has relevance also for plasmas found in nature such as the Earth's near and more distant space environments.

Fully developed turbulence requires free energy to be available. In Reynolds' experiment it was the energy needed to force the flow through the pipe. In general we need an instability to mediate the transfer of free energy to oscillations, and then some nonlinear interaction of many modes being excited in order to distribute the energy over a turbulent wavenumber spectrum. These basic ideas seem applicable to turbulence in general, not only in neutral flows. The first part of this book is devoted to discussions of some relevant plasma instabilities and the free energy that drives them. In the second part we discuss nonlinear models and the interactions of many modes. Examples of low frequency plasma turbulence in magnetized plasmas are presented and illustrated by observational results. Some theoretical tools available for turbulence modeling are also outlined. Fluid models for plasma turbulence and classical fluid turbulence have many similarities, and for later reference it is found reasonable to include a summary of some basic ideas for modeling turbulence in neutral flows. This book also attempts to summarize a number of relevant statistical tools and methods that can be applied to give an insight into the nature of the fluctuations. Descriptions of some basic phenomena, such as single particle motion etc., have also been included in order to make the presentation self-contained. The book is a result of discussions with colleagues over the years, and parts of the text has been used in lectures and seminars for Masters and PhD students.

Hans L Pécseli
Oslo, 2016

Acknowledgments

The enthusiasm and interest of students in Tromsø and Oslo who followed the plasma physics courses where these lecture notes were used, was most stimulating. Parts of the material were also used at the Max-Planck Institute PhD School in Greifswald, where the author sometimes lectured. Solveig Kjus read large parts of the notes concerning electrostatic drift waves and offered many valuable comments. My colleagues Mitsuo Kono, Bård Krane and Jan Trulsen also offered many valuable comments on the presentation and contents. Communications by Odd Erik Garcia are also greatly appreciated. Liv Larssen from the Auroral Observatory in Tromsø helped, as so often before, in typing and editing parts of the manuscript and in preparing many of the figures in the first stages of the writing. Her patience and help are gratefully acknowledged. Bjørn Lybekk contributed several figures and also his help and enthusiasm is greatly appreciated. The present notes borrow the notation and formulations from several papers on electrostatic drift waves by Francis F Chen (Chen 1964, 1965, 1966, 1970) and others (Hendel *et al* 1968). Concerning the part on turbulence I am indebted to Dwight Nicholson for many discussions. Writing the last section on analytical tools in turbulence, Part VII, made me think of him again. He was a good friend and a very fine scientist.

The book is dedicated to my family.

References

Chen F F 1964 Normal modes for electrostatic ion waves in an inhomogeneous plasma *Phys. Fluids* **7** 949–55

Chen F F 1965 Excitation of drift instabilities in thermionic plasmas *Plasma Phys. (J. Nucl. Energy Part C)* **7** 399–417

Chen F F 1966 Microinstability and shear stabilization of a low-β, rotating, resistive plasma *Phys. Fluids* **9** 965–81

Chen F F 1970 Linear drift waves UCLA Lecture Notes unpublished

Hendel H W, Chu T K and Politzer P A 1968 Collisional drift waves—identification, stabilization, and enhanced plasma transport *Phys. Fluids* **11** 2426–39

Author biography

Hans László Pécseli

The author, Hans László Pécseli, graduated as a civil engineer, solid state electronics, from the Technical University of Denmark in Lyngby, north of Copenhagen. He obtained his PhD in plasma physics at the Risø National Laboratory, in Roskilde, Denmark, and a Dr. Scient. degree from the University of Århus, Denmark. After a sabbatical year at the University of Iowa, Iowa City, he became head of the basic research unit in the Physics department, Risø National Laboratory working at the Q-machine there. He also had a position as adjoined professor at the University of Tromsø in Norway, now the Arctic University of Norway, where he participated in the experimental work on the magnetized toroidal plasma device, Blaamann. He was called to a position as full professor in plasma-space physics at the Department of Physics, University of Oslo, Norway. He has contributed to the literature on plasma and space physics, meteorology and also marine biology.

Professor Pécseli served as member of sub-committees of the national science foundations in Norway and Denmark, and as consultant for the national science foundations in Sweden and Finland. Together with Otto Kofoed-Hansen he received the bi-annual physics prize of the Danish Physical Society in 1989 for studies of plasma turbulence. He is a fellow of the American Physical Society, and member of the Royal Danish Academy of Sciences and Letters, and the Norwegian Academy of Sciences and Letters.

IOP Publishing

Low Frequency Waves and Turbulence in Magnetized Laboratory Plasmas and in the Ionosphere

Hans Pécseli

Chapter 1

Introduction

Turbulence in neutral fluids is a classical problem addressed by an overwhelming number of books and journal publications. In spite of all this, it is not so easy to find a definition of what turbulence is, it is much easier to find what it is *not*. A somewhat restrictive discussion can start with the classical experiments of Reynolds who observed the transition from laminar to turbulent flow in pipes as he increased the flow velocity. The transition occurred at some critical value of the *Reynolds number*, $\mathrm{Re} \equiv U_0 L/\nu$, expressed in terms of the pipe diameter L, the flow velocity U_0 in the central part of the pipe, and the kinematic viscosity ν. Detailed laboratory experiments (carried out much later with equipment and an accuracy not available for Reynolds) demonstrated that the onset of turbulence was associated with a transition from stable to unstable flows, i.e. at very carefully tailored conditions the growth of an externally introduced small perturbation could be observed. These instabilities did not saturate in some large amplitude coherent modes but rather in messy flow conditions that could be described only by statistical methods. If a numerical simulation is carried out in terms of some properly chosen orthogonal set of basis functions, it is generally found that the turbulent state of a flow is associated with a state where many modes are interacting. The surprising observation was that while it is so to speak impossible to make any predictions concerning individual realizations of the randomly varying velocity fields in neutral turbulence, accurate and surprisingly simple results can be obtained for some statistical averages, the Kolmogorov–Oubokhov structure functions, for instance.

We can generalize the discussion slightly by allowing for free energy being introduced in the system for a finite time only, and study decaying turbulence, but these details will not be addressed here, although they also hold historical interest. Some general observations can be made: turbulence requires free energy to be available (in Reynolds' experiment the energy to drive the flow). We need an instability mediating the transfer of free energy to oscillations and fluctuations, and

doi:10.1088/978-0-7503-1251-6ch1

the nonlinear interaction of many modes being excited in order to distribute energy over the wavenumber spectrum. These basic ideas seem applicable to turbulence in general, not only in neutral flows. The first part of this book is devoted to discussing some relevant plasma instabilities and the free energy that drives them. In the second part we discuss nonlinear models and the interactions of many modes. We present examples of low frequency plasma turbulence in magnetized plasmas, illustrated by some experimental and observational results. Some theoretical tools available for turbulence modeling are also outlined. Fluid models for plasma turbulence and classical fluid turbulence have many similarities and for later reference it is found reasonable to include a summary of some ideas for modeling turbulence in neutral flows. If it was so that turbulence saturated in a stage of a Gaussian random process it would be relatively easy to give a statistical description. It is thus found that such a Gaussian random process is completely described by its space-time varying correlation function which can be relatively easy to obtain. To lowest order it is actually so that fully developed strong turbulence in neutral flows is relatively well described by a model assuming a Gaussian random process. Only higher order structure functions that account for instance for intermittency effects require more detailed, and as it turns out much more complicated models. Turbulence in magnetized plasmas is usually not well described by Gaussian models, and the correlation functions contain only a limited part of the relevant statistical information. The description on inherently non-Gaussian processes is, however, complicated and no universal approach can be advocated. The analysis will often depend on the relevant application. The book attempts to summarize a number of relevant statistical tools and methods that can be applied to give an insight into the nature of the fluctuations.

A magnetized inhomogeneous plasma can sustain particular low frequency electrostatic waves, called ‘drift-waves’ because they propagate in the direction of the electron diamagnetic drift. They are often called ‘universal instabilities’ since they are associated with plasma density gradients (Chen 1984), or more generally plasma pressure gradients, which are unavoidable in laboratory as well as naturally occurring plasmas, and hence ‘universal’. An inhomogeneous plasma is not in perfect thermodynamic equilibrium and the pressure tends to expand the plasma. The expansion can provide free energy for driving an instability (Chen 1965a, 1965c, 1984). Alternatively, though more formally, it can be argued that the source of free energy is the gradient drift current associated with the pressure gradient. These drift waves will be given particular attention in part because they are considered to be important for the turbulent transport (and thereby the plasma confinement) in magnetized plasmas in laboratory and in space.

Historically, drift waves were identified as low frequency fluctuations (well below the ion cyclotron frequency), located at the edge of the plasma column in linear plasma devices (Buchel’nikova 1964, Buchel’nikova *et al* 1967, Hendel *et al* 1968, Rowberg and Wong 1970, Rogers and Chen 1970, Politzer 1971) as reviewed by for instance Timofeev and Shvilkin (1976) and also by Lehnert (1967), where an extensive reference list containing many of the early experimental studies can be found. Detailed experimental studies of drift waves were carried out in Q-machines (Motley 1975). Extensive theoretical investigations of the stability of drift waves were made by, for instance, Chen (1964, 1965b, 1965a, 1965c, 1966), and Schlitt and Hendel (1972), for resistive drift waves in particular. Parts of the presentations in Part II are based on these and related works.

Apart from being 'universal', drift waves are also interesting from a more 'pedagogical' point of view by offering a simple, yet illustrative, analysis involving basic plasma phenomena, wave interactions, stationary solutions to nonlinear dynamic equations, etc. It is also interesting that the theoretical description of electrostatic drift waves, linear as well as nonlinear, to a very large extent is similar to that describing Rossby waves in stratified media in rotating systems. The analogy is rooted in the fact that the Lorentz force and the Coriolis force have identical analytical forms, differing only in the physical meaning of the variables.

The present notes contain an introduction to drift waves, with an emphasis on a fluid description of the problem, which is then generalized to include the basics elements of a kinetic description. In gyro-kinetic models the plasma dynamics are studied on perpendicular spatial scales comparable to the gyro-radius and frequencies much lower than the particle cyclotron frequencies. The helical trajectories of the magnetized plasma particles can be decomposed into a relatively slow motion of the guiding center along the field line and a fast circular motion, the gyro-motion. For fusion plasma studies, in particular for plasma turbulence in these conditions, these gyro-kinetic models have been most valuable, and great progress has been made in describing collisionless plasma dynamics, collisionless drift waves in particular. Most of the present analysis is kept in simpler terms using fluid models.

Most of the analysis deals with electrostatic fluctuations, with a brief outline of low frequency electromagnetic effects. We study highly conductive plasma media where electrons move almost freely with respect to the ions, and a finite conductivity mediates the transfer of free energy to the fluctuations. The relative electron-ion motion can be inhibited by anomalous effects, often called anomalous resistivity, where a complete description is likely to involve kinetic plasma effects, the formation of phase-space structures in particular (Kruer and Dawson 1972). Although these phenomena are important, they are not considered here: any summary has its limitations. There are some repetitions in the text, which hopefully makes it easier to read; in principle you *can* skip some sections, and concentrate on the part you need here and now.

The present treatise contains elements from two previous books by the author (Pécseli 2000, 2012). These parts are used with kind permission from Cambridge University Press© and Taylor and Francis/CRC-Press, respectively. Many basic characteristics of plasma physics are assumed to be known, Debye shielding, fluid models of plasmas, etc. Descriptions of a few basic features (concerning single particle motions and collisions, for instance) are presented mostly for completeness and for later reference in the text. The aim of the work is to describe low frequency waves (mostly electrostatic) as relevant for plasma laboratory experiments and for ionospheric conditions. The topic is also important for fusion plasma conditions, for instance, but these will not be touched at all, although many of the discussions of statistical methods, etc. can also be useful in these general contexts. Magnetic shear has great importance for low frequency waves in fusion plasma applications, but is mentioned only in passing in the present work, since several other aspects of fusion plasma physics are left out, such as zonal flows.

References

Buchel'nikova N S 1964 Universal instability in a potassium plasma *Nucl. Fusion* **4** 165–8
Buchel'nikova N S, Salimov R and Eidel'man Y I 1967 Plasma turbulence in the presence of a drift instability *Sov. Phys. JETP* **25** 548–56
Chen F F 1964 Normal modes for electrostatic ion waves in an inhomogeneous plasma *Phys. Fluids* **7** 949–55
Chen F F 1965a Resistive overstabilities and anomalous diffusion *Phys. Fluids* **8** 912–9
Chen F F 1965b Spectrum of low-β plasma turbulence *Phys. Rev. Lett.* **15** 381–3
Chen F F 1965c Universal overstability of a resistive, inhomogeneous plasma *Phys. Fluids* **8** 1323–33
Chen F F 1966 Microinstability and shear stabilization of a low-β, rotating, resistive plasma *Phys. Fluids* **9** 965–81
Chen F F 1984 *Introduction to Plasma Physics and Controlled Fusion* vol 1 2nd edn (New York: Plenum)
Hendel H W, Chu T K and Politzer P A 1968 Collisional drift waves – identification, stabilization, and enhanced plasma transport *Phys. Fluids* **11** 2426–39
Kruer W L and Dawson J M 1972 Anomalous high-frequency resistivity of a plasma *Phys. Fluids* **15** 446–53
Lehnert B 1967 Experimental evidence of plasma instabilities *Plasma Phys.* **9** 301–37
Motley R W 1975 *Q Machines* (New York: Academic)
Pécseli H L 2000 *Fluctuations in Physical Systems* (Cambridge: Cambridge University Press)
Pécseli H L 2012 *Waves and Oscillations in Plasmas* (London: Taylor and Francis)
Politzer P A 1971 Drift instability in collisionless alkali metal plasmas *Phys. Fluids* **14** 2410–25
Rogers K C and Chen F F 1970 Direct measurements of drift wave growth rates *Phys. Fluids* **13** 513–6
Rowberg R E and Wong A Y 1970 Collisional drift waves in the linear regime *Phys. Fluids* **13** 661–71
Schlitt L G and Hendel H W 1972 Effects of parallel wavelength on the collisional drift instability *Phys. Fluids* **15** 1578–89
Timofeev A V and Shvilkin B N 1976 Drift-dissipative instability of an inhomogeneous plasma in a magnetic field *Sov. Phys. Usp.* **19** 149–68

Part I

Plasma basics

IOP Publishing

Low Frequency Waves and Turbulence in Magnetized Laboratory Plasmas and in the Ionosphere

Hans Pécseli

Chapter 2

Basic concepts of the motion of charged particles

Many of the basic properties of a plasma can be illustrated by single particle motion. In this chapter we summarize some of the basic dynamics of charged particles in electric and magnetic fields. The equations of motion are basically Newton's second law, in terms of the Lorentz force, for a particle with velocity **U** at a position **R**, all quantities being functions of time

$$
\begin{aligned}
M\frac{\mathrm{d}}{\mathrm{d}t}\mathbf{U}(t) &= q\mathbf{E}(\mathbf{R}(t),\, t) + q\mathbf{U}(t)\times\mathbf{B}(\mathbf{R}(t),\, t),\\
\frac{\mathrm{d}}{\mathrm{d}t}\mathbf{R}(t) &= \mathbf{U}(t),
\end{aligned}
\tag{2.1}
$$

for particles with charge q and mass M. Positions and velocities of individual particles will in the following be denoted by capital letters. A thermal velocity will also be given a capital letter. In the present case we assume the electric and magnetic fields to be *a priori* given. The fields have to be obtained at the actual position of the particle at time t, i.e. along the particle orbit. The analysis in the following is included for completeness and reference. The results as such can be found in many other works (Boyd and Sanderson 1969, Chen 1984), although the formulations can be different: the same results can often be reached in different ways.

2.1 Single particle orbits

The general problem of determining the orbits of charged particles in arbitrary given electric and magnetic fields as in (2.1) is quite complicated, even for the case where these fields are time stationary. To get an overview of the sort of trajectories one can expect, it is an advantage to consider some individual and relatively simple cases. Although we rather seldomly experience fields being constant in space and time, we include also this somewhat over-idealized case. A more general study of the motion

doi:10.1088/978-0-7503-1251-6ch2

of charged particles in electric and magnetic fields is presented by, for instance, Trubnikov (1965).

2.1.1 $\mathbf{B} \neq 0$ and $\mathbf{E} = 0$

Assume first that there is no electric field component in the direction $\perp\mathbf{B}$. In this case a particle will gyrate in a circular orbit with a radius determined by the initial velocity. The equation of motion is

$$\frac{\mathrm{d}}{\mathrm{d}t}\mathbf{U}_\perp = \frac{q}{M}\mathbf{U}_\perp \times \mathbf{B}. \tag{2.2}$$

The acceleration is $\perp\mathbf{U}_\perp$, implying that no work is done on the particle, giving $|\mathbf{U}_\perp|$ to be constant. We see this by taking the scalar product with $\mathbf{U}$ on both sides of (2.2) and find $\mathbf{U} \cdot \mathrm{d}\mathbf{U}/\mathrm{d}t = \frac{1}{2}\mathrm{d}U^2/\mathrm{d}t = 0$ implying the stated result. A time-stationary magnetic field, be it homogeneous or inhomogeneous, cannot do any work on a particle.

We can take the 'cross-product' with $\mathbf{B}$ on both sides of (2.2) and find

$$\frac{\mathrm{d}}{\mathrm{d}t}\mathbf{U}_\perp \times \mathbf{B} = \frac{M}{q}\frac{\mathrm{d}^2}{\mathrm{d}t^2}\mathbf{U}_\perp = -\frac{q}{M}\mathbf{U}_\perp B^2,$$

using (2.2). We have $\mathrm{d}^2\mathbf{U}_\perp/\mathrm{d}t^2 = -(qB/M)^2\mathbf{U}_\perp$ which is in reality two equations, one for each of the vector components in the plane perpendicular to $\mathbf{B}$. Solutions are e.g. $U_x = a_1(qB/M)\cos(t\,qB/M) + a_2(qB/M)\sin(t\,qB/M)$ and $U_y = a_3(qB/M)\cos(t\,qB/M) + a_4(qB/M)\sin(t\,qB/M)$, taking $\mathbf{B}$ along the $\hat{\mathbf{z}}$-direction. The constants a_1, a_2, a_3, and a_4, all with dimension *length*, are determined through the initial conditions specifying positions and velocities at $t = 0$. We can introduce qB/M as a part of the coefficient for convenience. For determining the particle position we solve $\mathrm{d}X(t)/\mathrm{d}t = U_x$, etc. The position vector follows the same differential equation as the velocity vector. The resulting particle orbit is a circle with radius r_L, the Larmor radius and an angular frequency Ω_c, where

$$\Omega_c = \frac{qB}{M} \qquad \text{and} \qquad r_L = \frac{MU_\perp}{qB}. \tag{2.3}$$

These parameters describe the basic elements of the motion of charged particles in a magnetic field, see also figure 2.1.

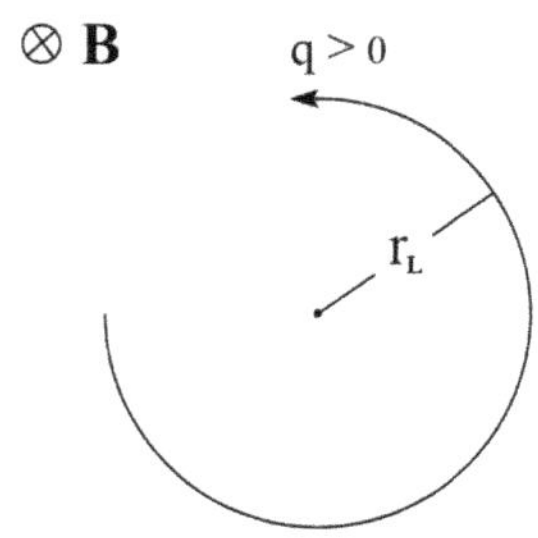

Figure 2.1. The basic gyration orbit of a charged particle in a magnetic field pointing into the plane as indicated.

If you change into a frame of reference rotating with the gyro-frequency (this is not an inertial frame) the particle is seen to be at rest but experiencing a fictitious centrifugal force $M\Omega_c^2 r$ taken at $r = r_L$ pointing in the outward radial direction. Since $\mathbf{U}_\perp \times \mathbf{B} = 0$ in this frame, there must here be another force balancing the centrifugal force: this is $q\mathbf{E}_\perp = -q\mathbf{V} \times \mathbf{B}$ where the velocity $V = r\Omega_c$ taken at $r = r_L$ is assumed to be non-relativistic. We find $\mathbf{E}_\perp \| \mathbf{r}$ with $|E_\perp| = r\Omega_c B$. This electric field is induced in your new frame of reference when crossing magnetic field lines by moving with the velocity $\mathbf{V}$. Just as the centrifugal force is fictitious and originating from the frame being non-inertial, the electric field balancing is fictitious as well. Maxwell's equations are also valid in the rotating frame. Through Poisson's equation the electric field is related to a spatially distributed (fictitious) charge density ρ' so that $2\pi r \ell E_\perp = (2\pi/\varepsilon_0)\ell \int_0^r \rho'(\xi)\xi \, \mathrm{d}\xi$, where ℓ is some arbitrary length scale along the homogeneous magnetic field lines. Due to the present symmetry of the problem, ℓ cancels in the final result. We find that $\rho' = \frac{1}{2}\Omega_c B \, \varepsilon_0 =$ constant. The reader may verify that the physical dimension of ρ' is indeed charge per volume.

There are many intriguing aspects of electrodynamics in rotating systems, but these fall outside the present discussion. Some of these problems are analyzed by, for instance, Schiff (1939), Trocheris (1949), Irvine (1964), Corum (1980), with a summary by Parks (2004).

2.1.2 E || B

We now include also a static electric field **E**. If the field has a component $E_\|$ along the magnetic field, we have a steady acceleration in this direction, implying

$$U_\| = \frac{q}{M} E_\| t + U_{0\|} \qquad \text{and} \qquad R_\| = \frac{q}{2M} E_\| t^2 + U_{0\|} t + R_{0\|} \tag{2.4}$$

are unaffected by the magnetic field.

2.1.3 E ⊥ B

If we now in addition to the magnetic field have a constant electric field in the direction $\perp\mathbf{B}$, we know that by a suitable change of reference this electric field can be made to vanish (Feynman *et al* 1963, Pécseli 2012). The appropriate frame is the one moving with a drift velocity $\mathbf{E}_\perp \times \mathbf{B}/B^2$. In this frame the particle is subject only to the magnetic field force and it will be gyrating with the gyro-frequency which is the same in both frames as long as we deal with non-relativistic velocities. In the rest frame the orbit is therefore a combination of rotation and drift, with a trajectory known as a 'cycloid'. This result can be obtained by other arguments, e.g. by writing out the components of the equation of motion and making the appropriate change in frame of reference (Boyd and Sanderson 1969, Chen 1984, Pécseli 2012)

If the electric field is replaced by another force field, say gravity with acceleration **g**, we can replace the electric force $q\mathbf{E}_\perp$ by the gravitational force $M\mathbf{g}$; the motion of the particles does not depend on the physical origin of the force, only its direction and magnitude. The orbit is determined by Newton's second law. By this argument we find that the particle has a drift $M\mathbf{g} \times \mathbf{B}/(qB^2)$ in the rest frame, in addition to its gyrating motion. Generally, we have for an arbitrary constant force, $\mathbf{F}\perp\mathbf{B}$ a particle drift velocity

$$\mathbf{U}_D = \frac{\mathbf{F} \times \mathbf{B}}{qB^2}. \tag{2.5}$$

In figure 2.2 we show three examples of cycloid motion as followed by a particle with drifting gyro-center. A trivial example (not shown) is a straight line, found when the initial velocity of the particle is exactly $\mathbf{U}_D$, as given in (2.5).

2.1.4 Curvature drifts

A special case of the gravitational drift is found when a charged particle moves along curved field lines with velocity $U_{\|}$, where we here let the **B**-perpendicular velocity be negligible, corresponding to negligible Larmor radius. If we let R be the radius of curvature of the field line, we find that the particle experiences a centrifugal force $MU_{\|}^2/R$ in the direction pointing away from the center of the curvature. Seen from the particle's frame of reference this centrifugal force cannot be distinguished from a gravitational force, and it gives rise to a drift as in (2.5). Consequently, the particle experiences a curvature drift $U_{DC} \equiv MU_{\|}^2/(qBR)$ in the direction perpendicular to the magnetic field lines. The centrifugal force, correct in magnitude and direction is

$$\mathbf{F}_{Cf} = MU_{\|}^2\frac{\mathbf{R}}{R^2}, \tag{2.6}$$

with **R** being the particle position. By the assumption of strictly **B**-parallel particle motion, there is no need to distinguish the gyro-center and particle positions here. Using (2.5) we arrive at the averaged gyro-center drift velocity

$$\overline{\mathbf{U}}_{DC} = \frac{MU_{\|}^2}{q}\frac{\mathbf{R} \times \mathbf{B}}{B^2R^2}, \tag{2.7}$$

where electrons and positive ions drift in opposite directions $\perp\mathbf{B}$.

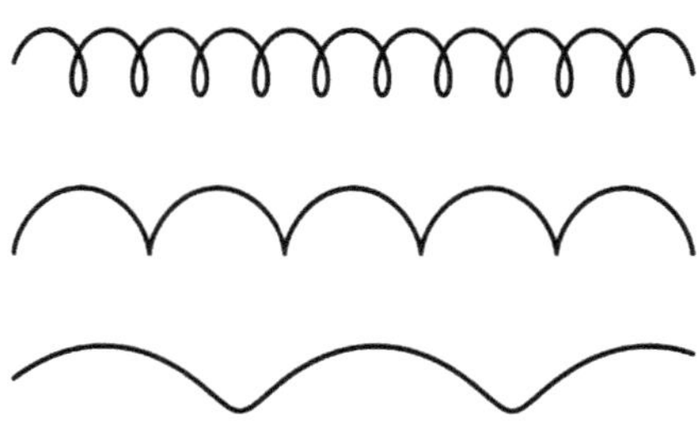

Figure 2.2. Three examples for cycloid motion for a magnetic field perpendicular to the plane of the paper and a constant force in the vertical direction. The top case is found when the initial particle velocity is in the direction opposite to $\mathbf{U}_D$, as given in (2.5) or alternatively larger in magnitude and in the same direction as $\mathbf{U}_D$, the middle trajectory results when the initial velocity is zero in the rest frame, and the bottom case when the initial non-vanishing particle velocity is smaller in magnitude than $\mathbf{U}_D$. The Larmor radius is the same in all cases, but the vertical force is varying.

If we take a thermal velocity U_{th} to be representative for $U_{\parallel}$ we find curvature drifts $MU_{th}^2/(qBR)$. If the electron and ion temperatures are equal or at least comparable, we have $MU_{th,i}^2 \approx mU_{th,e}^2$, and find comparable curvature drifts for electrons and ions. This reminds us to be a bit cautious: it is so tempting to argue that because of their small masses, the electrons are unaffected by gravitational or centrifugal forces, at least in comparison with the forces on ions. At equal temperatures, the average electron velocities are much larger than for the ions and this compensates for the small electron mass.

The result (2.7) assumes implicitly that the particle velocity vector is parallel to the magnetic field lines, i.e. that $r_L = 0$. It is tempting to assume that a finite Larmor radius is of little consequence, but it is not so (Chen 1984). If we bend magnetic field lines to have some finite curvature, we necessarily introduce a magnetic field variation with $\nabla B \perp \mathbf{B}$. To demonstrate this we express the radius of curvature R in terms of the magnetic field by the following arguments (Chen 1984): for a current free region and DC-magnetic fields we have $\nabla \times \mathbf{B} = 0$ together with the general result $\nabla \cdot \mathbf{B} = 0$. Assuming that the magnetic field lines are bent, so they are locally well approximated by a segment of a circle, we can assume $\mathbf{B}$ to have a θ-component only. This model then implies that ∇B has an r-component only. In imagination, you can take a long solenoid and bend it into a torus or doughnut. Then we have a magnetic field variation with only a θ-component of $\mathbf{B}$. For positions inside the torus, we have $B_\theta \sim 1/r$, where r is measured from the symmetry center. We can here use such a toroidal magnetic field as a local approximation for curved magnetic field lines and have

$$\nabla \times \mathbf{B}\,|_z = \frac{1}{r}\frac{\partial}{\partial r}(rB_\theta) = 0,$$

giving $B_\theta \sim 1/r$. If we locally have a curved magnetic field with $B \sim 1/R$, this will necessarily imply $\partial B_\theta/\partial r \approx B_\theta/r$, or more generally $\nabla|B|/|B| = -\mathbf{R}/R^2$, with $\mathbf{R}$ being a vector in the direction *from* the center of curvature. A curvature of a magnetic field will necessarily imply a magnetic field gradient, i.e. a decrease of the magnetic field in the direction of the radius of curvature pointing outwards. The motion of particles with finite Larmor radii in inhomogeneous magnetic fields with $\nabla B \perp \mathbf{B}$ will be discussed separately in section 2.1.7.

Note that parts of the discussions in the latter part of this section are explicitly based on current free conditions where $\nabla \times \mathbf{B} = 0$. A current distribution in the plasma implies modifications of the results and a significant part of the studies of plasma confinement deals with the case where the current density is finite and spatially varying. In terms of magnetohydrodynamics (MHD) (Chen 1984) we assumed the plasma pressure to be negligible in comparison to the magnetic field pressure, or $n\kappa T/(\frac{1}{2}B^2/\mu_0) \equiv \beta \ll 1$.

2.1.5 Polarization drifts, $\mathrm{d}E/\mathrm{d}t \neq 0$

Consider a plasma in a homogeneous stationary magnetic field, and a time-varying electric field perpendicular to $\mathbf{B}$. Generally, the electric field can vary in both intensity and direction, but for simplicity we here assume the direction to remain constant, so that we have $\mathrm{d}E/\mathrm{d}t \neq 0$. We follow a charged particle in its frame of reference moving with the instantaneous $\mathbf{E} \times \mathbf{B}/B^2$-drift. For an electric field which is slowly varying, this is not an inertial frame, and the charged particle will experience a time-varying velocity, i.e. an acceleration which by a co-moving observer can be interpreted as an effective gravitational force, $\mathbf{g}$, originating from the variation in the $\mathbf{E} \times \mathbf{B}/B^2$-velocity. This effective, or virtual, gravity gives rise to an $M\mathbf{g} \times \mathbf{B}/(qB^2)$-velocity, varying with time, see (2.5). The acceleration experienced by an observer moving with the particle is readily determined as $\mathbf{g} = -(1/B^2)\,\mathrm{d}(\mathbf{E} \times \mathbf{B})/\mathrm{d}t$. Since $\mathbf{B}$ was assumed constant we have $\mathbf{F} = M\mathbf{g} = -(M/B^2)(\mathrm{d}\mathbf{E}/\mathrm{d}t \times \mathbf{B})$, which can be directly inserted into (2.5) to give the polarization drift. The result can also be seen as an expansion to first order in a small quantity ω/Ω_c, where ω here denotes a characteristic frequency for the time variation of the fields experienced by the particle. Note that the polarization drift has opposite sign for electrons and positively charged ions. Inserting $\mathbf{F} = -(M/B^2)(\mathrm{d}\mathbf{E}/\mathrm{d}t \times \mathbf{B})$ in (2.5), we find the polarization drift

$$\overline{\mathbf{U}}_p = \frac{M}{qB^2}\frac{\mathrm{d}}{\mathrm{d}t}\mathbf{E}, \tag{2.8}$$

which appears as a first order correction to the $\mathbf{E} \times \mathbf{B}/B^2$-velocity.

For the case where the electric field is varying harmonically with time, $\mathbf{E} = \mathbf{E}_0 \exp(-\mathrm{i}\omega t)$, we find the simpler result

$$\overline{\mathbf{U}}_p = -i\,\frac{\omega}{\Omega_c}\frac{\mathbf{E}_0 \exp(-\mathrm{i}\omega t)}{B}, \tag{2.9}$$

from (2.8). The result (2.9) clarifies the expansion in ω/Ω_c mentioned before and also the meaning of 'slowly varying electric fields' by $\omega \ll \Omega_c$.

The velocity (2.8) should be added to $\overline{\mathbf{U}} = E_0 \times \mathbf{B}/B^2$. The net result is an elliptical motion of the gyro-center in the plane perpendicular to $\mathbf{B}$. The polarization drift of the electrons is negligible in comparison to that of the ions. The ellipse describing the ion motion is therefore 'fatter' than the one for the electrons.

We can discuss the present problem from a different point of view. Consider a charged particle moving in a plane perpendicular to a homogeneous magnetic field and an inhomogeneous stationary electric field, $\mathbf{E} = -\nabla\phi$. The simplest model we can imagine assumes that the gyro-center of the particle at any time moves with the local $\mathbf{E}(\mathbf{r}) \times \mathbf{B}/B^2$-velocity. We will find that this model is basically inconsistent, by violating energy conservation. The remedy of the shortcomings will be the polarization drift. The important point here is that as it moves through an inhomogeneous electric field, the particle experiences $\partial\mathbf{E}/\partial t \neq 0$, so the electric field in the moving particle frame varies in time, although in the fixed laboratory frame the field was time-stationary, but varying in space. To emphasize different aspects of inhomogeneous electric fields we ignore finite Larmor radius corrections.

Let the total energy of the particle be given as $\mathcal{W} \equiv \frac{1}{2}MU^2 + q\phi$. We decompose the velocity into the local $\overline{\mathbf{U}} = \mathbf{E}(\mathbf{R}(t)) \times B/B^2$-velocity, obtained at the particle gyro-center position $\mathbf{R}(t)$, and the velocity of the cyclotron motion $\mathbf{U}_g$ where $U_{gx} = U_g \cos(\omega_c t)$ and $U_{gy} = U_g \sin(\omega_c t)$ and have $\mathcal{W} = \frac{1}{2}M\overline{U}^2 + \frac{1}{2}MU_g^2 + q\phi$. Later on we will recognize $\frac{1}{2}MU_g^2/B$ as the magnetic moment of the particle, being a constant of motion (Chandrasekhar 1960). We will here simply assume that $\frac{1}{2}MU_g^2$= const. but note that as a special illustrative case we are free to consider a particle released with zero velocity, in which case $U_g = 0$ (and finite Larmor radius effects are certainly vanishing).

We now differentiate the particle energy to find $\mathrm{d}\mathcal{W}/\mathrm{d}t = M\overline{\mathbf{U}} \cdot \mathrm{d}\overline{\mathbf{U}}/\mathrm{d}t + q\,\mathrm{d}\phi/\mathrm{d}t$. We have $\overline{U}$ being the gyro-center velocity which depends on time after release only, while $\phi(\mathbf{R}(t))$ is the electrostatic potential at the particle gyro-center position, which by assumption varies with time. With the definition of $\overline{\mathbf{U}}$, we have $\mathrm{d}\mathcal{W}/\mathrm{d}t = (M/B^2)\mathbf{E} \cdot \mathrm{d}\mathbf{E}/\mathrm{d}t + q\,\mathrm{d}\phi/\mathrm{d}t$. With $\phi(\mathbf{r})$ being constant in time, we have $\mathrm{d}\phi/\mathrm{d}t = \partial\phi/\partial t + \mathbf{U}_p \cdot \nabla\phi = -\mathbf{U}_p \cdot \mathbf{E}$, where $\mathbf{U}_p$ is for the moment unspecified. For a conservative system the total energy (kinetic + potential) of the charged particle must be conserved. This requires $\mathrm{d}\mathcal{W}/\mathrm{d}t = (M/B^2)\mathbf{E} \cdot \mathrm{d}\mathbf{E}/\mathrm{d}t - q\mathbf{U}_p \cdot \mathbf{E} = (M/B^2)\mathbf{E} \cdot (\mathrm{d}\mathbf{E}/\mathrm{d}t - (qB^2/M)\mathbf{U}_p) = 0$, or $\mathbf{U}_p = (M/qB^2)\mathrm{d}\mathbf{E}/\mathrm{d}t$, which is the polarization drift (2.9). We thus find that in order to have energy conservation, it is necessary to introduce a drift velocity $\mathbf{U}_p \neq 0$. We have then a velocity component parallel to $\mathbf{E}$, which gives rise to a change in kinetic energy, but this is exactly compensated by the change in potential energy. *If* the gyro-center velocity is taken to be strictly the local $\mathbf{E}(\mathbf{r}) \times \mathbf{B}/B^2$-velocity, then the scalar product of this velocity with $\mathbf{E}$ is vanishing, and the force cannot do any work. Since the particle is moving in a spatially inhomogeneous potential, its potential energy *has* to change, and it must have a velocity component parallel to $\mathbf{E}$.

Finite Larmor radius (FLR) effects and polarization drifts appear as higher order corrections to the basic $\mathbf{E}(\mathbf{r}) \times \mathbf{B}/B^2$-velocity. The respective ordering of the two corrections is not always evident. In one limit, with time varying spatially constant electric fields there is no FLR-correction and only the polarization drift appears. As an alternative limit, we can have for instance $\mathbf{E} = \{0, E_y(y), 0\}$ with $\mathbf{B} = \{0, 0, B\}$. For this case we have only the FLR-corrections to a temporally constant $\mathbf{E} \times \mathbf{B}/B^2$-drift in the $\hat{\mathbf{x}}$-direction, derived at the y-position of the gyro-center.

2.1.6 Finite Larmor radius corrections

In the case of the electric field component in the direction $\perp\mathbf{B}$ is spatially varying, we have to consider the possibility that the electric field varies with time as it is sampled by the gyrating particle, i.e. the electric field variation over the Larmor orbit cannot necessarily be assumed to be constant. We assume, however, also here that the characteristic frequency for the variations in the electric field is much smaller than the cyclotron frequency. For almost all relevant problems it will be the modifications in the $\mathbf{E} \times \mathbf{B}/B^2$-drift of the *ion* component which is relevant; the electron Larmor radius will usually be negligible in comparison, and the guiding-center model sufficient for describing their motion perpendicular to a magnetic field in low frequency electric fields.

Consider a particle with charge q gyro-orbiting in a homogeneous magnetic field. The x-component of the electric field, for instance, relevant for its guiding center drift can formally be written as $E_x(\mathbf{r}) = \int E_x(\overline{\mathbf{R}})\delta(\overline{\mathbf{R}} - \mathbf{r})\mathrm{d}X\mathrm{d}Y$, with $\overline{\mathbf{R}}$ being the position of the guiding center and $\mathrm{d}X\mathrm{d}Y$ the area element in the integration over $\overline{\mathbf{R}}$. As long as the particle experiences an electric field which in the rest frame is quasi-stationary, the *effective* electric field relevant for its *finite Larmor radius corrected* orbit will be the average of the electric field experienced along this Larmor orbit. This average is indicated by an overline

$$\overline{E_x(\mathbf{r})} \equiv \frac{1}{2\pi r_L}\int E_x(\overline{\mathbf{R}})\delta\Big(|\overline{\mathbf{R}} - \mathbf{r}| - r_L\Big)\mathrm{d}X\mathrm{d}Y, \tag{2.10}$$

where r_L is the Larmor radius of the particle. We represent the electric field by its Fourier series, i.e. write $\mathbf{E}(\overline{\mathbf{R}}) = \sum_{\mathbf{k}}\mathbf{E}_{\mathbf{k}}\exp(i\mathbf{k}\cdot\overline{\mathbf{R}})$ and have

$$\overline{E_x(\mathbf{r})} = \sum_{\mathbf{k}} E_{x\mathbf{k}}\frac{1}{2\pi r_L}\int \exp\Big(i\mathbf{k}\cdot\overline{\mathbf{R}}\Big)\delta\Big(|\overline{\mathbf{R}} - \mathbf{r}| - r_L\Big)\mathrm{d}X\mathrm{d}Y, \tag{2.11}$$

and similarly for the E_y-component.

We now use (Champeney 1973) the relation for the Bessel function of the first kind, J_0, of zeroth order

$$J_0(kr_L) = \frac{1}{2\pi r_L}\int_0^{\infty}\int_0^{2\pi} s\delta(s - r_L)\exp(iks\cos\phi)\mathrm{d}s\mathrm{d}\phi$$

to obtain from (2.11) the result

$$\overline{\mathbf{E}(\mathbf{r})} = \sum_{\mathbf{k}} J_0(kr_L)\mathbf{E}_{\mathbf{k}}\exp(i\mathbf{k}\cdot\mathbf{r}). \tag{2.12}$$

From this result we can obtain the finite Larmor radius (FLR)-corrected $\overline{\mathbf{E}} \times \mathbf{B}/B^2$-drift for the particle's guiding center.

It is important to distinguish this drift velocity from a *fluid* velocity. In the latter case we take the average velocity of particles passing through a small area-element, and this is will in general give a result different from the guiding center drift.

Applications of the result (2.12) are quite feasible numerically (Knorr *et al* 1988). For analytical use, we might prefer a simpler expression. To obtain this, we restrict the analysis to electric fields varying only gently on a scale given by r_L. We then recall the one-to-one correspondence between multiplication by wavenumber $\mathbf{k}$ in the Fourier domain, and differentiation with respect to position in configuration space. We can make a series expansion of the Bessel function J_0 and then 'translate' the result into a differential operator by the replacement $i\mathbf{k} \to \nabla$. We have $J_0(\xi) = \sum_{n=0}^{\infty}(-1)^n(\xi/2)^{2n}/(n!)^2 \approx 1 - \xi^2/4$ for small ξ, giving

$$\overline{\mathbf{E}(\mathbf{r})} \approx \sum_{\mathbf{k}} \left(1 - \frac{1}{4}(kr_L)^2\right)\mathbf{E}_{\mathbf{k}} \exp(i\mathbf{k} \cdot \mathbf{r}), \tag{2.13}$$

implying that to lowest order in r_L^2, the FLR-corrected $\mathbf{E} \times \mathbf{B}/B^2$-drift of the guiding center is

$$\overline{\mathbf{U}}_{\mathbf{E}\times\mathbf{B}} \approx \frac{1}{B^2}\left(1 + \frac{1}{4}r_L^2\nabla^2\right)\mathbf{E}(\mathbf{r}) \times \mathbf{B}. \tag{2.14}$$

Physically, the reason for the FLR-corrections is clear: considering a plane electrostatic wave, for instance, with wavenumber $\mathbf{k} \perp \mathbf{B}$, we find that a particle with its gyro-center at maximum electric field will actually spend a significant part of the time in regions of weaker fields, as it is gyrating with a finite Larmor radius. The average guiding center drift is consequently smaller than what would be obtained by using the electric field value at the guiding center (Chen 1984).

For numerical use, the relation (2.14) has an unpleasant effect by enhancing very small wavelength noise components, and therefore one might prefer the more suitable result (2.12) for those applications.

In this section we only discussed the FLR-corrections to the single particle drift velocity. We might add that if the contribution of the particles to the bulk plasma density is needed, then for consistency the particles have to be 'smeared out' along their Larmor radius, i.e.

$$\overline{n(\mathbf{r})} \equiv \frac{1}{2\pi r_L} \int n\left(\overline{\mathbf{R}}\right)\delta\left(|\overline{\mathbf{R}} - \mathbf{r}| - r_L\right)\mathrm{d}X\,\mathrm{d}Y,$$

where $n(\overline{\mathbf{R}})$ represents a continuous guiding center particle density. The ensuing calculations being identical to those already carried out before (Knorr *et al* 1988), to lowest order in an expansion of the Larmor radius we find

$$\overline{n(\mathbf{r})} \approx \left(1 + \frac{1}{4}r_L^2\nabla^2\right)n(\mathbf{r}),$$

as in (2.14). When calculating the electrostatic potential we can use Poisson's equation with $\overline{n_{e,i}(\mathbf{r})}$ inserted for the electron and ion species. Usually we will find also here that FLR-corrections are important for the ion dynamics only, and $\overline{n_e(\mathbf{r})} \approx n_e(\mathbf{r})$.

- **Exercise:** Consider a particle with charge q gyrating with a Larmor radius r_L in a stationary potential variation $\phi(\mathbf{r})$, where $\mathbf{r} \perp B$. Determine the gyro-averaged electrostatic potential of the particle.

2.1.7 Particle drifts when $\nabla B \perp \mathbf{B}$

Assume now that the electric field is vanishing, $\mathbf{E} = 0$, but the magnetic field, albeit constant in time, has a gradient in intensity in the direction $\perp\mathbf{B}$. Magnetic fields cannot do work on a particle, so we know from the outset that the magnetic field can only change the direction of the velocity vector, not its magnitude. As a specific

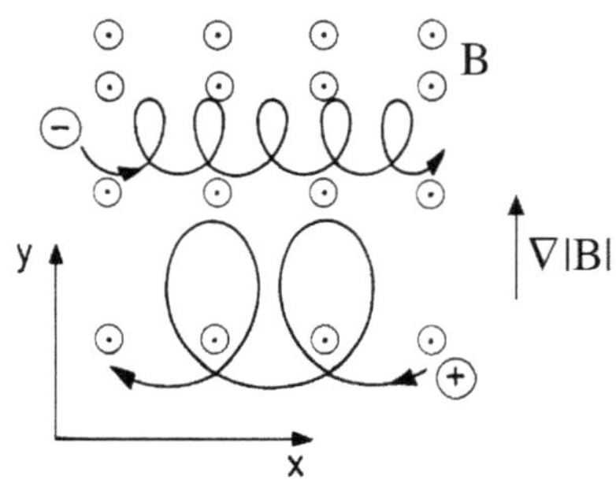

Figure 2.3. Schematic illustration of a particle moving in an inhomogeneous magnetic field, where the gradient is perpendicular to **B**.

example we consider $\mathbf{B} \equiv \{B_x, B_y, B_z\} = \{0, 0, B_0(y)\}$, giving $\nabla|B| \perp \mathbf{B}$, see figure 2.3. Note that $\nabla \cdot \mathbf{B} = 0$ for the model field, so it is physically acceptable but $\nabla \times \mathbf{B} \neq 0$, so there needs to be a spatially distributed current density to support this magnetic field. The radius of curvature for a particle trajectory is stronger in a part of the gyrating orbit, where the particle is located at the stronger magnetic field, as compared to its value half a gyro-period later. As a result, the particles will drift in a direction $\perp\mathbf{B}$ as well as $\perp\nabla B$, as illustrated in figure 2.3. To derive this drift velocity we simplify the problem by considering only gentle variations in the magnetic field, and assume that we can approximate the magnetic field locally by $\mathbf{B} \approx \mathbf{B}_0(1 + y/\ell)$ where ℓ is a characteristic scale length for the variation and $\mathbf{B}_0 \| \hat{z}$. Mathematically, we make a Taylor expansion of the magnetic field's y variation around the position of the particles gyro-center, and have $\ell = B/|\nabla B|$. By assuming that the variation of this field is small on a scale length determined by the particle's Larmor radius, r_L, we retain only the first term in this expansion. Since the magnetic field has no dependence on the coordinate z for the present case, there is no reason to be concerned with the z-component of velocities. With the Lorentz force $\mathbf{F} \equiv q\mathbf{U} \times \mathbf{B}$ we have

$$M\frac{\mathrm{d}}{\mathrm{d}t}\mathbf{U} = q\mathbf{U} \times \mathbf{B} \approx q\mathbf{U} \times \mathbf{B}_0 + q\mathbf{U} \times \mathbf{B}_0\frac{y_0(t)}{\ell}, \tag{2.15}$$

where we have to use the expression for the magnetic field entering the Lorentz force at the actual particle position $y = y_0(t)$. We assume ℓ to be large, and the second term in the expression for the Lorentz force in (2.15) is therefore small. It is convenient to rewrite (2.15) as

$$M\frac{\mathrm{d}}{\mathrm{d}t}\mathbf{U} - q\mathbf{U} \times \mathbf{B}_0 = q\mathbf{U} \times \mathbf{B}_0\frac{y_0(t)}{\ell}. \tag{2.16}$$

We expect the y-component of the particle position to oscillate (at least approximately) between $\pm r_L$, and the right-hand side of (2.16) is therefore in magnitude of order of (r_L/ℓ) compared to the second term on the left side. We express this by the notation $O(r_L/\ell)$, where O means 'of the order of'. We can always write the solution of (2.16) in the form $\mathbf{U}(t) \equiv \mathbf{U}_0(t) + \widetilde{\mathbf{U}}(t)$, where $\mathbf{U}_0$ is the solution of (2.16) with the right-hand side omitted. By integration of $\mathbf{U}$ we obtain the y-component of the particle position as $y_p(t) \equiv y_0(t) + \tilde{y}(t)$. Since the right hand side of (2.16) is small, we expect $\widetilde{\mathbf{U}}(t)$ and $\tilde{y}(t)$ to be small corrections to $\mathbf{U}_0$ and $y_0(t)$, the corrections being $O(r_L/\ell)$. We have y_0/ℓ itself being $O(r_L/\ell)$ by construction. Then we have products like $\widetilde{\mathbf{U}}(t)y_0(t)/\ell$, and similar, to be $O(r_L^2/\ell^2)$, which we assume to be so small that they can be ignored. This will be the basic assumption in the ensuing analysis. We expect that $\tilde{y}$ is $O^2(r_L/\ell)$ since it is a small correction to $y_0(t)$, so $\tilde{y}$ will not appear in the analysis.

As the first step, we ignore the right-hand side of (2.16), and solve $M\mathrm{d}\mathbf{U}_0/\mathrm{d}t - q\mathbf{U}_0 \times \mathbf{B}_0 = 0$. Take a positive charge, $q > 0$, and the magnetic field in the positive $\hat{\mathbf{z}}$-direction. To this first approximation we find for the particle positions

$$\{x_0; y_0\} = \{r_L \sin(\Omega_c t); r_L \cos(\Omega_c t)\},$$

and velocity components

$$\{U_{0x}; U_{0y}\} = \{r_L\Omega_c \cos(\Omega_c t); -r_L\Omega_c \sin(\Omega_c t)\},$$

where $U_\perp = r_L\Omega_c \equiv \sqrt{U_{0x}^2 + U_{0y}^2}$, see also (2.3). We have here $\overline{\mathbf{U}}_0 \equiv (\Omega_c/2\pi) \times \int_0^{2\pi/\Omega_c} \mathbf{U}_0\, \mathrm{d}t = 0$.

Ignoring small terms, as outlined before, we find the approximate expression

$$M\frac{\mathrm{d}}{\mathrm{d}t}\widetilde{\mathbf{U}} - q\widetilde{\mathbf{U}} \times \mathbf{B}_0 = q\mathbf{U}_0 \times \mathbf{B}_0\frac{y_0(t)}{\ell}, \tag{2.17}$$

where the right-hand side is a force term that we will analyze in more detail.

We will now include the effects of the right-hand side in (2.17) by an iteration using the first order expressions. The term on the right-hand side of (2.16) gives a correction of order r_L/ℓ to the zero order solutions (x_0, y_0) and U_{0x}, U_{0y}. In an expansion in the parameter r_L/ℓ, we find to lowest order the x-component of the additional force $\mathbf{F}$ entering the right hand side in (2.17) as

$$\begin{aligned} F_x &\approx qU_{0y}(t)B_0\frac{y_0(t)}{\ell} \\ &= -\, q\; U_\perp \sin(\Omega_c t)B_0\frac{r_L \cos(\Omega_c t)}{\ell}, \end{aligned}$$

where we here retain only the terms containing $\mathbf{U}_0$, considering those with $\widetilde{\mathbf{U}}(t)$ later. We inserted the velocity expression for a positive charge, assuming the magnetic field pointing out of the paper as in figure 2.3.

We are not interested here in the fine details of the particle orbit, only in the average. To obtain this, we average in time over one cyclotron period, $2\pi/\Omega_c$, the averaging here again indicted by an 'overline'. Upon this averaging over time, we find $\overline{F_x} \equiv (\Omega_c/2\pi) \int_0^{2\pi/\Omega_c} F_x \mathrm{d}t = 0$, so that this term is uninteresting.

We also have

$$F_y \approx -qU_{0x}(t)B_0\frac{y_0(t)}{\ell}$$
$$= -q\ U_\perp \cos(\Omega_c t)B_0\frac{r_L\cos(\Omega_c t)}{\ell}.$$

Averaging over time again, we now find

$$\overline{F_y} \equiv \frac{\Omega_c}{2\pi}\int_0^{2\pi/\Omega_c} F_y \mathrm{d}t = -\frac{1}{2}q\ U_\perp B_0\ \frac{r_L}{\ell}.$$

We consequently have an average force on the particle acting in the $\hat{\mathbf{y}}$-direction. We now have the time-averaged expression for the velocity correction $\widetilde{\mathbf{U}}(t)$ as

$$M\overline{\frac{\mathrm{d}\widetilde{\mathbf{U}}}{\mathrm{d}t}} - q\overline{\widetilde{\mathbf{U}}} \times \mathbf{B}_0 = \overline{F_y}\hat{y}.$$

According to (2.5) for $q > 0$ the force $\overline{F_y}$ gives rise to a steady particle drift in the direction of $-\hat{\mathbf{y}} \times \mathbf{B}_0$ which is also the $\mathbf{B}_0 \times \nabla B$ direction. The drift velocity has a constant magnitude $\frac{1}{2}|U_\perp r_L/\ell|$ and $\overline{\mathrm{d}\widetilde{\mathbf{U}}/\mathrm{d}t} = 0$. The analysis is readily repeated for $q < 0$. The resulting drift velocity vector is given by

$$\overline{\mathbf{U}}_{\nabla B} = \pm\frac{1}{2}U_\perp r_L\frac{\mathbf{B} \times \nabla B}{B^2}, \tag{2.18}$$

where we used the definition of $1/\ell = |\nabla B|/B$ given before, and noted the proper direction of the average force obtained, i.e. the $\mathbf{B} \times \nabla B$-direction. We need not distinguish the small difference between B_0 and B in the denominator of (2.18). Note that the drift (2.18) is opposite in direction for electrons and ions as indicated by the $\pm$-sign. We can insert the definition of $r_L = MU_\perp/(qB)$ from (2.3) and obtain a different version of (2.18) as

$$\overline{\mathbf{U}}_{\nabla B} = \frac{MU_\perp^2}{2qB}\frac{\mathbf{B} \times \nabla B}{B^2} \equiv \mu\frac{\mathbf{B} \times \nabla B}{qB^2}, \tag{2.19}$$

which need not have the $\pm$-sign in front since the particle charge now appears explicitly. Note the similarity between (2.19) and a curvature drift (2.7), but note also the difference in factor $\frac{1}{2}$. We also introduced a version of the $\mathbf{B} \times \nabla B$-velocity containing the magnetic moment μ to be discussed in section 2.1.8.

To have an order of magnitude estimate, it might be an advantage to rewrite $\overline{\mathbf{U}}_{\nabla B}$. We can take $\frac{1}{2}MU_\perp^2 \sim \kappa T$ and find

$$|\overline{\mathbf{U}}_{\nabla B}| \approx \frac{\kappa T}{qB}\left|\frac{\nabla B}{B}\right| \approx \frac{C_s^2}{\Omega_{ci}}\left|\frac{\nabla B}{B}\right| \equiv C_s\frac{a_i}{\mathcal{L}},$$

introducing the sound speed $C_s = \sqrt{\kappa T/M}$ and an effective Larmor radius $a_i \equiv C_s/\Omega_{ci}$ as distinct from r_L. A characteristic local length scale for the magnetic field variations is introduced as $\mathcal{L} \equiv |B|/|\nabla B|$. As said, this is only meant to be an order of magnitude estimate, but it can be useful. The length scale a_i is important for discussing the role of the ion polarization drift in the analysis of electrostatic drift waves. The sound speed introduced here assumes that the temperature T is representative for both ion and electron species. When $T_e \gg T_i$ as often found in laboratory and space, we will have $C_s \approx \sqrt{\kappa T_e/M}$ and the estimate for $|\overline{\mathbf{U}}_{\nabla B}|$ will apply only for the electron $\mathbf{B} \times \nabla B$-drift.

Physically, the origin of the $\overline{\mathbf{U}}_{\nabla B}$-drift is easy to explain; since a stationary magnetic field cannot change the energy of a charged particle, the absolute value of the ion or electron velocity has to remain constant along the orbit. The radius of the curvature of the particle trajectory then has to be small where the magnetic field is strong, i.e. at the top on figure 2.3, and larger at the bottom, where the magnetic field intensity is a little less. The average drift, including its direction, is then readily explained by inspection of figure 2.3. Typically, electron Larmor radii are small, and one might wonder why the electron nonetheless has an averaged drift, when it hardly samples any variation in the magnetic field? The point here is that although the magnetic field variation is indeed small over the electron Larmor radius, the electrons cyclotron frequency is usually so much larger so it receives many 'kicks', as compared to an ion at the same kinetic energy.

- **Exercise:** Note that the curvature drifts discussed in section 2.1.4 will in general add to the $\nabla B \perp \mathbf{B}$ drifts: we cannot use one of them to reduce the other by twisting the magnetic field lines in some smart way (Chen 1984). It is instructive to make a sketch of physically realistic inhomogeneous magnetic field configurations to verify that this is indeed so.

$\nabla B \perp \mathbf{B}$ *drifts in fluid models*

Anticipating the discussion of a fluid model of magnetized plasmas in section 3.4, we discuss $\nabla B \perp \mathbf{B}$ plasma drifts in a fluid model here. As mentioned, we have electrons and positive ions drifting in opposite directions due to the $\nabla B \perp \mathbf{B}$ drift. It is, however, important to emphasize that this does not necessarily imply that we have a current density everywhere in the plasma! It is important first to make clear what we mean by a current density. Imagine that we take a small vertical cut $A - A'$ on figure 2.3, and let the cut represent a small area element perpendicular to the plane of the figure. Figure 3.6 illustrates a related construction. The length of $A - A'$ is supposed to be much smaller than the local Larmor radius. The local current through the area element is then given by the net number of charges crossing this area element per second. If the plasma is uniformly distributed, i.e. the number density of particles and their temperatures are independent of position, at least in the vicinity of $A - A'$, we will find that the number of, say, negative particles per second. coming from above $A - A'$ and propagating to the right on figure 2.3 will be exactly canceled by the negative particles passing through $A - A'$ from below, because the densities are the same by assumption and their thermal velocities cannot change either with position, since a stationary magnetic field is unable to change the energy of a charged particle, as stated before. The same arguments will, of course, also hold for the ion component. Any current density in fluid models induced by the $\nabla B \perp \mathbf{B}$ drift is therefore restricted to regions of inhomogeneities in plasma density or temperature (Braginskií 1965). Note here that by assuming a constant plasma temperature and constant particle density in an inhomogeneous magnetic field we imply an inhomogeneous guiding center density since the particles are 'smeared out' along circlets with different radii, see section 2.1.6. The present discussion illustrates how careful we have to be when going from single particle models to fluid models. The question will be discussed in more detail in section 3.4.

- **Exercise:** Using a simple dipole approximation for the Earth's magnetic field, consider first the motion of electrons and ions in an altitude around 110 km (ionospheric E-region) over the magnetic pole. Assume that some external mechanism is imposing a horizontal electric field of intensity $E = 50$ mV m^{-1} in the East–West direction in the region above the magnetic pole. What is the direction and magnitude of the $\mathbf{E} \times \mathbf{B}$-drift velocity of the electrons and for hydrogen ions? Ignore possible collisions with neutrals.

  Consider now charged particles at the magnetic Equator, again at altitude 110 km. Let a particle have vanishing **B**-parallel velocity. What is the direction and magnitude of the $\nabla B \times \mathbf{B}$-drift velocity of electrons and hydrogen ions if their energies are 0.5 eV? Consider now 0.5 eV particles with directed velocities *along* the magnetic field lines. What is the direction and magnitude of the curvature-drift velocity of electrons and hydrogen ions? Ignore now magnetic field inhomogeneities, but include the effect of the Earth's gravitational field on atomic oxygen ions and on the electrons. What are the magnitudes and directions of the drift velocities for these two species?
- **Exercise:** A point particle with mass M and dipole moment $\mathbf{m}$ moves in a circular orbit with radius R around a fixed magnetic dipole with moment $\mathbf{m}_0$. The two moment vectors $\mathbf{m}$ and $\mathbf{m}_0$ are anti-parallel and perpendicular to the plane of the orbit. Calculate the velocity of the orbiting particle in terms of the given parameters. Is the orbit stable against small perturbations?

2.1.8 Particle motion when $\nabla B \parallel \mathbf{B}$

Consider now the case where $\nabla B \parallel \mathbf{B}$, this is typically the feature associated with a *magnetic mirror*. Assume for simplicity that the magnetic field is axisymmetric, with $B_\theta = 0$ and $\partial \mathbf{B}/\partial\theta = 0$. Since we have assumed the magnetic field lines to converge or diverge, there will be a radial component, B_r of the magnetic field. Writing $\nabla \cdot \mathbf{B} = 0$ in cylindrical coordinates, we have

$$\frac{1}{r}\frac{\partial}{\partial r}(rB_r) + \frac{\partial}{\partial z}B_z = 0. \tag{2.20}$$

As an approximation we find the radial magnetic field component locally near the symmetry axis as

$$B_r = -\frac{1}{2}r\frac{\partial}{\partial z}B_z. \tag{2.21}$$

The z-component of the Lorentz force on a particle with charge q and azimuthal velocity U_θ gives $F_z = \frac{1}{2} q U_\theta \, r \, \partial B_z / \partial z$. We average this force component over one period of gyration to obtain

$$\overline{F_z} = \frac{1}{2} q U_\theta r_L \frac{\partial}{\partial z} B_z = -\frac{1}{2} \frac{U_\theta^2}{B} \frac{\partial}{\partial z} B_z, \tag{2.22}$$

independent of the sign of the charge q, where we used that the direction of U_θ depends on the sign of q. Charged particles always experience a force pointing away from a region with an axially strong magnetic field. Introducing the magnetic moment defined as

$$\mu \equiv \frac{1}{2} \frac{m \, U_\perp^2}{B}, \tag{2.23}$$

we can write

$$\overline{F_z} = -\mu \frac{\partial}{\partial z} B_z. \tag{2.24}$$

The magnetic moment can be interpreted as the product of the area of the circular orbit ('circlet') traced out by the particle, πr_L^2, and the current associated with its motion, J. The current is found by taking a small ring perpendicular to the particle orbit and counting the net charge crossing. The number of charges q passing in 1 s is $(qB/m)/2\pi$ so that $J = (q^2 B/m)/2\pi$, while $\pi r_L^2 = \pi m^2 U_\perp^2/(qB)^2$, the product giving μ defined in (2.23). It is interesting that the charge q vanishes from the result.

As the particle moves along the z-axis, subject to the force F_z, it enters regions with changing magnetic fields. We find, however, that the **B**-perpendicular velocity is changing in such a way that the magnetic moment remains constant. We demonstrate this by considering the z-component of Newton's law

$$m \frac{\partial}{\partial t} U_z = \overline{F_z} = -\mu \frac{\partial}{\partial z} B_z. \tag{2.25}$$

We multiply on both sides of this equation with U_z, and use $U_z = \mathrm{d}z/\mathrm{d}t$ to get

$$\frac{1}{2} m \frac{\partial}{\partial t} U_z^2 = \overline{F_z} = -\mu \frac{\partial}{\partial t} B_z, \tag{2.26}$$

i.e. a particle moving in the $\hat{\mathbf{z}}$-direction will experience a time-varying magnetic field component B_z as indicated.

Since a stationary magnetic field cannot produce any work on a gyrating particle, the kinetic energy $\frac{1}{2}(U_z^2 + U_\perp^2) = \frac{1}{2}(U_z^2 + \mu B_z)$ must be conserved. As a consequence we find

$$\frac{\mathrm{d}}{\mathrm{d}t} \mu = 0, \tag{2.27}$$

as readily demonstrated. In the frame of reference moving with the gyro-center of the particle we find that the magnetic field appears to be changing with time. In this frame an observer will find an azimuthal electric field due to $\partial\mathbf{B}/\partial t$, and the change in the velocity component of the particle perpendicular to $\mathbf{B}$ will in this frame be increased due to this electric force. Again we note that the origin of electric and magnetic forces does not have any absolute distinction.

If we have two magnetic mirrors, i.e. two regions with strong axial magnetic field enhancements, a charged particle can be trapped between them. Some fusion device schemes rely on such configurations, and magnetic mirrors are also important for the motion of particles in geophysical and astrophysical plasmas. In particular the Earth's magnetic field is strongest at the magnetic poles, and weaker at the equator. These two magnetic mirrors are able to trap even energetic particles. These regions of the Earth's magnetosphere are called the *radiation belts*.

- **Exercise:** Consider an ion with mass M and velocity U located above the magnetic pole of the Earth. Determine the distance from the center of the Earth where the magnetic mirror force is balancing the gravitational force. Use the dipole approximation for the Earth's magnetic field. Is this position stable or unstable?

2.2 Adiabatic invariants

It is well known in classical mechanics that for a system having a periodic motion, in terms of generalized momentum and position p and q, the action integral $\oint p\,\mathrm{d}q$ taken over a period is a constant of the motion.

The generic periodic motion for magnetized plasmas is the gyration of charged particles with angular frequency Ω_c. We take here $p = MU_\perp r$, the angular momentum, and $q = \theta$ being the angle of the coordinate vector with respect to a reference line. The action integral becomes $4\pi\mu M/|q|$, where we introduced the magnetic moment $\mu \equiv MU_\perp^2/2B$. Thus μ is a constant of motion as long as $M/|q|$ remains constant. (This will not hold for instance for charged dust grains moving in a magnetic field, since their charges will in general vary with time.)

As the reader may verify, we can also interpret the magnetic moment as the magnetic flux through the Larmor orbit multiplied with $q^2/(2\pi m)$. This multiplier is a constant for the given particles. In this sense, conservation of magnetic moment is consistent with the concept of frozen-in magnetic field lines (Chen 1984, Pécseli 2012).

A second invariant can be associated with a particle trapped between two magnetic mirrors. In this case the particle bounces back between the two mirrors, and we can define a 'bounce frequency'. A constant of motion is the $\frac{1}{2}MU_\parallel \mathrm{d}s$, where $\mathrm{d}s$ is an element of the path length of the guiding center motion along a magnetic field line. We introduce a *longitudinal invariant* J as

$$J = \int_a^b U_\parallel \mathrm{d}s, \tag{2.28}$$

where the two 'turning points' are denoted a and b. Consider for instance a charged particle trapped between two approaching magnetic mirrors. Assume that we consider the problem in a frame of reference where mirror a is at rest and mirror

b is approaching with a velocity V. By reflection from this moving mirror the charged particle gains a velocity increment of $2V$. The frequency with which it encounters the moving mirror is then increased so that the rate of velocity change becomes

$$\frac{\mathrm{d}}{\mathrm{d}t}U_{\parallel} = \frac{U_{\parallel}}{2L}2V = \frac{U_{\parallel}}{L}\frac{\mathrm{d}\,L}{\mathrm{d}t} \tag{2.29}$$

at an instant where the two mirrors are separated by a distance L. We used $V = -\mathrm{d}L/\mathrm{d}t$. Using (2.29) we find $\mathrm{d}(U_{\parallel}L)/\mathrm{d}t = 0$, i.e. $U_{\parallel}L = \text{const}$.

The trapping of a charged particle between two magnetic mirrors is not perfect, however; it seems evident that for instance a particle moving parallel to, or nearly parallel to, the magnetic field lines will pass through such a mirror. In order to estimate a critical angle between the magnetic field and the velocity vector, we recall the invariance of μ, and write the appropriate value at the position of maximum magnetic field, B_0 and at the turning point at a magnetic field B_1

$$\frac{1}{2}MU_{0\perp}^2/B_0 = \frac{1}{2}MU_{1\perp}^2/B_1.$$

At any position, we can define a local 'pitch angle' for the particle as the angle between the velocity vector and the magnetic field at the gyro-center, $\theta = \arcsin(U_{\perp}/U_0)$, where $U_0 \equiv |\mathbf{U}|$, see figure 2.4. At the turning point we have by definition of a reflection that $U_{\parallel} = 0$. We assume that the particle has a **B**-parallel velocity $U_{0\parallel}$ at the position where the magnetic field is B_0. Conservation of energy implies $U_{1\perp}^2 = U_{0\parallel}^2 + U_{0\perp}^2 \equiv U_0^2$. We therefore have

$$\frac{B_0}{B_1} = \frac{U_{0\perp}^2}{U_{1\perp}^2} \equiv \sin_0^2\Theta. \tag{2.30}$$

The ratio B_0/B_1 is called the *mirror ratio*. We find that a particle with a pitch angle larger than Θ_0 will be trapped between the mirrors, while particles with small pitch angle will escape. This will in general be an irreversible process; a particle once lost is unlikely to be 'recaptured'. Since microscopic processes, collisions, fluctuations, etc,

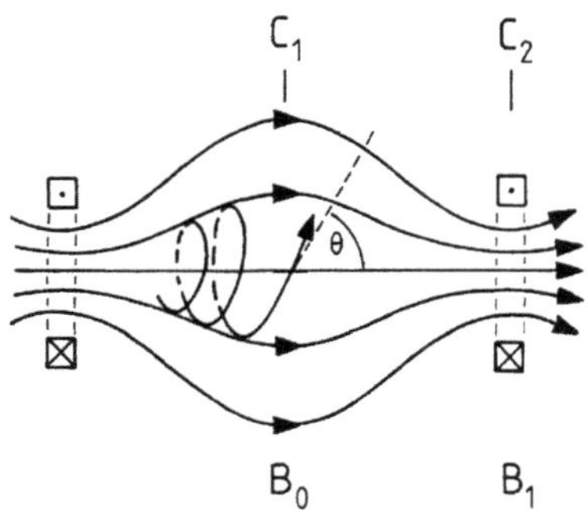

Figure 2.4. Schematic and idealized illustration of a particle moving between two magnetic mirrors. The magnetic field is minimum at C_1 with a value B_0 and maximum at C_2, with a value B_1 at one of the coils.

will give rise to scatterings of the particles, there will always be some which are scattered into the *loss cone*, and consequently any magnetic mirror will eventually be depleted, unless its particle content is maintained somehow. Since the magnetic field coils take-up some space, it might in reality be difficult to obtain mirror ratios smaller than 1/5 in a laboratory, implying that particles with pitch angles in the range {0°; 25°} will be lost right away.

Summary: As a conclusion of this section we present a short summary of particle motions in time-stationary electric and magnetic fields:

- **E**||**B**:

$$U_{\|} = \frac{q}{M}E_{\|}t + U_{0\|} \qquad \text{and} \qquad R_{\|} = \frac{q}{2M}E_{\|}t^2 + U_{0\|}t + R_{0\|}.$$

- **E**⊥**B**:

$$\overline{\mathbf{U}}_{\mathbf{E}\times\mathbf{B}} = \frac{1}{B^2}\left(1 - \frac{1}{4}r_L^2\nabla^2\right)\mathbf{E}(\mathbf{r}) \times \mathbf{B},$$

  where the lowest order FLR-corrections were included. We have $E =$ const. as a special case.

- **F**⊥**B**:

$$\mathbf{U}_D = \frac{\mathbf{F}\times\mathbf{B}}{qB^2},$$

  where **F** is an arbitrary force, gravity for instance, in which case $\mathbf{F} = M\mathbf{g}$.

- $\nabla B \perp \mathbf{B}$:

$$\overline{\mathbf{U}}_{\nabla B} = \pm\frac{1}{2}U_{\perp}r_L\frac{\mathbf{B}\times\nabla B}{B^2} = \frac{MU_{\perp}^2}{2qB}\frac{\mathbf{B}\times\nabla B}{B^2}.$$

- $\nabla B \| \mathbf{B}$:

$$\overline{F}_z = -\mu\frac{\partial}{\partial z}B_z$$

  is the *mirror force* with μ being the magnetic moment.

In all cases the overline indicates time-averaging over the cyclotron period.

2.3 Plasma instabilities explained by single particle motions

A number of plasma instabilities can be explained by considering single particle motions and some examples are given here. To simplify the problems we use sharp interfaces between vacuum and plasma or between different plasma conditions; warm and cold plasmas, for instance.

2.3.1 Gravitational or Rayleigh–Taylor instability

In figure 2.5 we give a simple, almost self-explanatory illustration of what is known as the 'gravitational instability', also called the 'Rayleigh–Taylor instability' by its analog in neutral fluids. In the case of a magnetized plasma placed in a gravitational field with acceleration $\mathbf{g} \perp \mathbf{B}$, we find that a strictly horizontal interface will be a possibility, at least formally. Charged particles are streaming with individual velocities $\mathbf{U}_g \equiv M\mathbf{g} \times \mathbf{B}/(qB^2)$ along the interface and will not cross it. The proposed flow field has $\nabla \cdot \mathbf{u}_g = 0$ for both ions and electrons. The bulk flow is incompressible so that no charge separation develops for homogeneous conditions. We again use capital letters for individual particle velocities while fluid velocities are given lower case letters.

A plane interface is a possible steady state solution for the plasma flow, but as we will see not necessarily a stable one. To demonstrate this, assume that the interface is given a slight perturbation as indicated in figure 2.5. Electrons are so light that they are hardly moved by gravity, but the ions move and cross the bended interface giving rise to a slight charge separation as shown. The ion velocity field is no longer divergence free. The charge separation gives rise to electric fields $\mathbf{E}$ as shown in figure 2.5 and these fields will in turn give a bulk $\mathbf{E} \times \mathbf{B}/B^2$ motion of both plasma species. On the bottom side the initial perturbation will be enhanced and the charge separation also increases to enhance the $\mathbf{E} \times \mathbf{B}/B^2$ velocity: the system is unstable. On the top side (see figure 2.5) the initial perturbation is diminished, and the system is stable there. The problem has its equivalent for neutral fluids where a heavy fluid supported by a lighter one becomes unstable.

A related 'Richtmyer–Meshkov instability' is found when two fluids of different density are accelerated, where the acceleration vector has a component normal to the interface between the two fluids. This can happen at the passage of a shock wave,

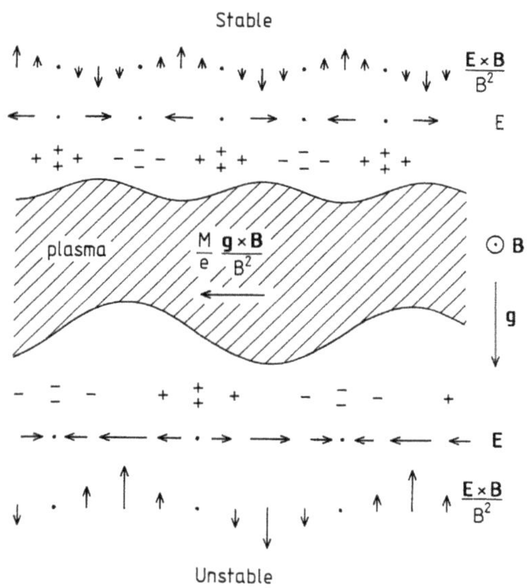

Figure 2.5. Geometry for a simple physical description of the gravitational or Rayleigh–Taylor instability. The plasma region is shaded, white regions are vacuum.

and the instability is believed to be important for some astrophysical phenomena. This instability can be considered an impulsive-acceleration limit of the Rayleigh–Taylor instability.

2.3.2 Polarization by magnetic field gradient flows

Consider a plasma in an inhomogeneous magnetic field. A toroidal field as illustrated in the top part of figure 2.6 will be representative: for the more general case a local approximation in terms of a near-toroidal field can be used by introducing the curvature radius (Chen 1984). It is readily found that the magnetic field inside the torus varies with x-position (the radial direction in figure 2.6) as $B \sim 1/x$. The resulting $\nabla|B|$- and curvature-drifts of electron and ions are illustrated in the bottom part of figure 2.6. We have here $B^{-1}\nabla B = -\mathbf{r}/r^2$ with $\mathbf{r}$ measured from the center of the torus, i.e. the origin of the major radius. The average of the combined $\nabla|B|$- and curvature-drifts can then be written as

$$\begin{aligned}\overline{\mathbf{U}}_{\nabla B+\mathbf{R}} &= \frac{M\left\langle \frac{1}{2}U_\perp^2 + U_\parallel^2 \right\rangle}{q} \frac{\mathbf{R}\times\mathbf{B}}{R^2\mathbf{B}^2} \\ &= \frac{M\left\langle \frac{1}{2}U_\perp^2 + U_\parallel^2 \right\rangle}{q} \frac{\mathbf{R}\times\mathbf{B}}{R^2\mathbf{B}^2} \\ &= \frac{2MU_{th}^2}{3q}\frac{\mathbf{R}\times\mathbf{B}}{R^2\mathbf{B}^2},\end{aligned} \tag{2.31}$$

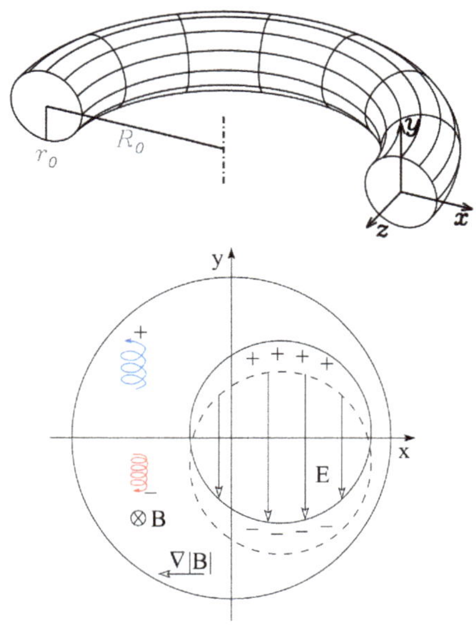

Figure 2.6. Illustration of plasma in a toroidal magnetic field (top). The major radius in the torus is R_0. A cross-section of the plasma column with ∇B-drifts of electrons and ions indicated (bottom). The average drift velocities will not be the same for electrons and ions when $T_e \neq T_i$.

where the magnetic field is obtained at the gyro-center position $\mathbf{r} = \mathbf{R}$. We used $M\langle \frac{1}{2}U_\perp^2 + U_\parallel^2 \rangle = 2\kappa T$ and $U_{th}^2 \equiv 3\kappa T/M$. Positive charges drift in one direction, negative charges in the opposite one, as indicated. Note that the electron and ion temperatures might be different so the drift velocities might differ. The resulting charge separation gives rise to an electric field $\mathbf{E}$ which displace the entire plasma column by the $\mathbf{E} \times \mathbf{B}/B^2$-velocity in the direction of the major radius $\mathbf{R}_0$ or $\hat{\mathbf{x}}$-direction in figure 2.6. By considering both polarities of the magnetic field in figure 2.6 it is easily seen that in either case the toroidal plasma column will move in the outwards direction and the plasma will ultimately be lost at the wall.

In all its simplicity, the present outline illustrates the inherent problem in toroidal geometries. Much of plasma fusion research has been concerned with means for avoiding this plasma loss. One method that has proved successful is to deform the toroidal magnetic field so that charged particles can move along magnetic field lines to short-circuit the electric fields that develop.

Strictly speaking, the summary given here does not concern an instability: it is rather a lack of equilibrium.

2.3.3 Temperature gradient instability in inhomogeneous magnetic fields

Consider conditions with inhomogeneous magnetic fields with $\nabla|B| \perp \mathbf{B}$ where there is a plasma temperature gradient $\nabla T \perp \mathbf{B}$. The temperatures as well as the gradients might be different for electrons and ions, and might also differ for different ion species, should there be more than one present. Even if the plasma density is uniform we might encounter an instability caused by a mechanism reminiscent of the one discussed for the gravitational or Rayleigh–Taylor instability.

The gradient in magnetic field gives rise to a combination of ∇B- and curvature-drifts here as well (Chen 1984). We have the result (2.31) which is here in the $\hat{\mathbf{y}}$-direction with magnitude $\pm U_{th}^2/(R\Omega_{ci}) = \pm U_{th}\langle r_L \rangle / R_0$, where we used $\mathrm{R} \approx R_0$.

To simplify the illustration we assume a sharp boundary between a cold and a warm plasma. For slowly varying phenomena we will expect that possible temperature gradients along magnetic field lines will smooth out, so we are concerned only with a plasma temperature variation in the direction perpendicular to $\mathbf{B}$ in, say, the $\hat{\mathbf{y}}$-direction. The velocity field for both ions and electrons is divergence free in this case and no internal charge separation develops. A plane interface is also a possibility for the present problem, although just as in section 2.3.1 it may not be a stable solution. Considering figure 2.7 we perturb the interface and follow charged particles orbiting in the inhomogeneous magnetic field. The figure is most easily understood if we let the cold plasma temperature vanish, so that the charged particles become immobile in the unshaded region. When warm ions entering from the middle region cross the interface they give rise to positive space charges, and warm electrons do the same on the other side of a perturbation of the interface. The velocities of electrons and ions vary along y because of the 'undulations' in the interface, the $\hat{\mathbf{y}}$-axis some places passing through cold and slow particles, other places through warm and fast particles. The flow fields at the interface are no longer divergence free. The difference in the guiding center velocity of the hot and cold particles gives rise to a charge separation just as in figure 2.5.

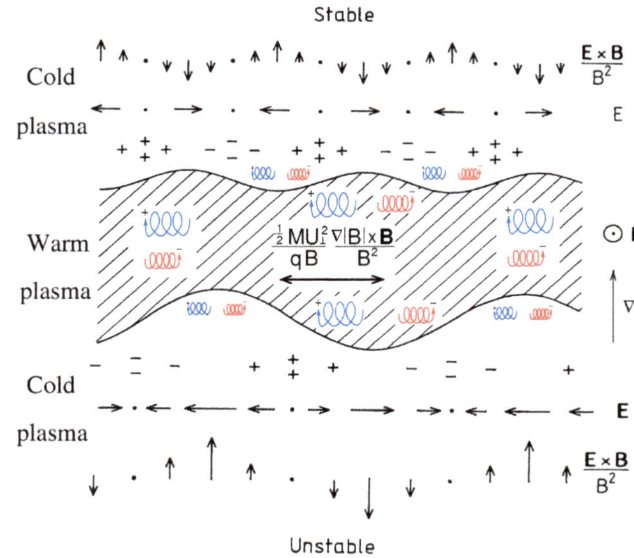

Figure 2.7. Illustration of single particle motions for cases with a gradient in plasma temperature, $\nabla T \perp \mathbf{B}$, with the resulting charge distributions and electric fields. To emphasize the similarity of this instability with the gravitational Rayleigh–Taylor instability we reused figure 2.5 with some modifications, introducing a sharp interface between warm and cold plasma. The warm plasma region is shaded, cold plasma regions are here white.

Several combinations can be envisaged: hot electrons and cold ions on one side, cold electrons and hot ions on the other side of the interface, etc. The reader may find it interesting to sort out the stability conditions for the different combinations.

The present outline has relevance for the so-called ion temperature gradient (ITG)-mode that is also important in high temperature fusion experiments (Cowley *et al* 1991, Weiland 2000).

There is reason for a word of caution: the gravitational instability of section 2.3.1 has an almost one-to-one counterpart in neutral fluids. It could be tempting to expect the temperature gradient having the same. Following this line of thought we consider figure 2.7 and note that when the drifts are due to a curvature of the magnetic field, we have an effective centrifugal force pointing down in the figure having the effect of gravity. The gradient in magnetic field can also be seen as a force pointing down, as discussed in section 2.1.7. Our experience from the neutral atmosphere is that for conditions where warm air in the bottom and cold air is on the top, we will have a buoyancy instability, with hot air rising. Be aware that for the plasma case illustrated in figure 2.7 the situation is the opposite, but also recall that the present section deals with single particle motions and not dynamic fluid models.

References

Boyd T J M and Sanderson J J 1969 *Plasmas Dynamics* (London: Thomas Nelson and Sons)

Braginskií S I 1965 Transport processes in a plasma ed M A Leontovich *Reviews of Plasma Physics* vol 1 (New York: Consultants Bureau) pp 205–311

Champeney D C 1973 *Fourier Transforms and their Physical Applications* (London: Academic)
Chandrasekhar S 1960 *Plasma Physics* (Chicago, IL: University of Chicago Press) Notes compiled by S K Trehan after a course given by S Chandrasekhar
Chen F F 1984 *Introduction to Plasma Physics and Controlled Fusion* vol 1 2nd edn (New York: Plenum)
Corum J F 1980 Relativistic covariance and rotational electrodynamics *J. Math. Phys.* **21** 2360–4
Cowley S C, Kulsrud R M and Sudan R 1991 Considerations of ion-temperature-gradient-driven turbulence *Phys. Fluids B: Plasma Phys.* **3** 2767–82
Feynman R P, Leighton R B and Sands M 1963 *The Feynman Lectures on Physics* (New York: Addison-Wesley)
Irvine W M 1964 Electrodynamics in a rotating system of reference *Physica* **30** 1160–70
Knorr G, Hansen F R, Lynov J P, Pécseli H L and Rasmussen J J 1988 Finite Larmor radius effects to arbitrary order *Phys. Scripta* **38** 829–34
Parks G K 2004 *Physics of Space Plasmas, an Introduction* 2nd edn (Cambridge, MA: Westview Press)
Pécseli H L 2012 *Waves and Oscillations in Plasmas* (London: Taylor and Francis)
Schiff L I 1939 A question in general relativity *Proc. Nat. Acad. Sci.* **25** 391–5
Trocheris M G 1949 Electrodynamics in rotating frames of reference *Phil. Mag.* **40** 1143–54
Trubnikov B A 1965 Particle interactions in fully ionized plasmas ed M A Leontovich *Reviews of Plasma Physics* vol 1 (New York: Consultants Bureau) pp 105–204
Weiland J 2000 *Collective Modes in Inhomogeneous Plasma, Kinetic and Advanced Fluid Theory* (Bristol: Institute of Physics Publishing)

IOP Publishing

Low Frequency Waves and Turbulence in Magnetized Laboratory Plasmas and in the Ionosphere

Hans Pécseli

Chapter 3

Collisions in magnetized plasmas

In an unmagnetized plasma it is usually assumed implicitly that particles propagate along straight lines of orbit between collisions, which is after all a rather trivial limit, see for instance section 2.1. In the following we give a brief discussion of the case where the charged particles are confined by a magnetic field. The basic orbit of a charged particle in a constant magnetic field is a circle, with the Larmor radius r_L, and an orbital period given by the cyclotron frequency as $2\pi/\Omega_c$. When a light charged particle collides with a heavy neutral (i.e. electrons colliding with a neutral atom or molecule), we find that its guiding center is displaced by a distance which is at most $2r_L$. If energy exchange of the charged particle and the neutral is important, the guiding center displacement can be modified somewhat, but r_L will remain as a useful order of magnitude. Collisions between charged particles are in the following given special considerations, as they are the only ones relevant for collision processes in fully ionized plasmas. Details of the collisional cross-sections, mean free paths and collision frequencies are found in the literature (Sivukhin 1965, Trubnikov and Leontovich 1965, Braginskiĭ 1965, Chen 1984, Pécseli 2015), mostly with emphasis on collisions between charged particles. A brief general introduction to collisional processes is given in appendix C.

3.1 Basic physics

We can make simplified, yet illustrative models describing the collisions of charged particles. In unmagnetized plasma, the models resemble those for collisions between neutrals, and the most important differences lie in the analysis of the cross-sections. More interesting for plasma sciences are particle collisions in magnetic fields, where we here restrict the analysis to homogeneous magnetic fields. It turns out that the presence of an inhomogeneous electric field, $\mathbf{E} \perp \mathbf{B}$, gives rise to significant modifications of the simplest models, so that case is also treated.

doi:10.1088/978-0-7503-1251-6ch3

3.1.1 Particle collisions in magnetic fields

The case where two charged magnetized particles collide is somewhat more interesting. In order to describe this process, we first write their gyro-center positions

$$\begin{aligned}\mathbf{r}_c^{(1)} &= \mathbf{R}^{(1)} + \frac{M_1}{q_1 B^2}\mathbf{U}_1 \times \mathbf{B}\\ \mathbf{r}_c^{(2)} &= \mathbf{R}^{(2)} + \frac{M_2}{q_2 B^2}\mathbf{U}_2 \times \mathbf{B},\end{aligned} \tag{3.1}$$

where $\mathbf{R}^{(j)}$ is the position of particle no. j, and $\mathbf{U}_j$ its velocity, with a corresponding notation for particle masses M_j and charges q_j, see also figure 3.1. The particles are assumed to collide at some time t_0 where $\mathbf{R}^{(1)} = \mathbf{R}^{(2)}$. Assume that the particles have different velocity components along the magnetic field, so that we need only consider one collision. The construction implied in (3.1) is easily understood by use of figure 3.1.

We assume now that the time-duration of the collision is short. This means that the particle positions slightly before and slightly after the collision are virtually the same, only the directions and magnitudes of the velocity vectors have changed, with a corresponding change in the gyro-center positions. This implies that

$$\begin{aligned}\Delta\mathbf{r}_c^{(1)} &= \frac{M_1}{q_1 B^2}\Delta\mathbf{U}^{(1)} \times \mathbf{B}\\ \Delta\mathbf{r}_c^{(2)} &= \frac{M_2}{q_2 B^2}\Delta\mathbf{U}^{(2)} \times \mathbf{B},\end{aligned} \tag{3.2}$$

where $\Delta\mathbf{U}^{(j)}$ is the change in velocity vector for particle j. Conservation of momentum implies

$$M_1\Delta\mathbf{U}^{(1)} + M_2\Delta\mathbf{U}^{(2)} = 0.$$

With (3.2) this gives

$$q_1\Delta\mathbf{r}_c^{(1)} + q_2\Delta\mathbf{r}_c^{(2)} = 0. \tag{3.3}$$

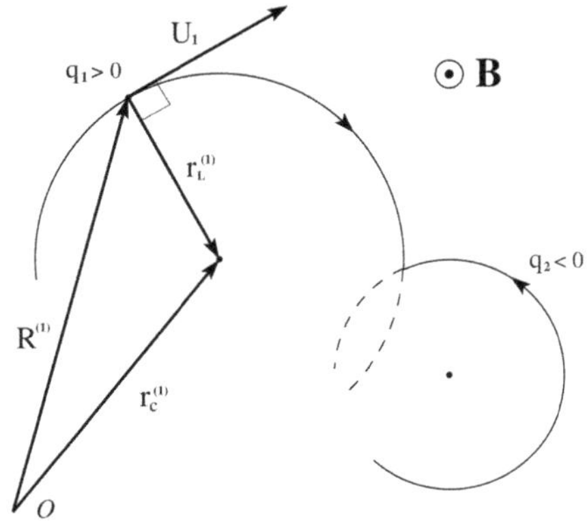

Figure 3.1. Explanation of position-vectors used in (3.1). The vector $\mathbf{r}_L$ has the length of the Larmor radius for given velocity U, and points from the particle to the gyro-center. We have $\mathbf{r}_L \perp \mathbf{U}$.

If $q_1 = q_2$ we have $\Delta\mathbf{r}_c^{(1)} + \Delta\mathbf{r}_c^{(2)} = 0$, i.e. the center of mass for two particles is not changed by collisions. If, however, $q_1 = -q_2$ we have $\Delta\mathbf{r}_c^{(1)} = \Delta\mathbf{r}_c^{(2)}$. In this latter case the gyro-center of the two particles moves in the same direction, and the magnitude of *both* 'jumps' is the same. If the two particles in question are a singly charged ion and an electron, we can again argue that the electron gyro-center displacement can be at most an electron Larmor radius, r_{Le}. This implies that the *ion* gyro-center displacement can also be at most r_{Le}. Also we note that in a coarse grained description, with a spatial resolution larger than the Larmor radii, we cannot have any charge separation neither by collision between like or by differently charged particles. This means that if we start out with a vanishing electric field before the collisions (as implicitly assumed in the analysis), the collisions will not induce any electric fields at a later time.

Elements of the arguments rely critically on the assumption that the duration of the collision is very small compared to $1/\Omega_c$. This assumption is not necessarily fulfilled for collisions between charged particles, but the conclusions remain qualitatively correct.

- **Exercise:** Discuss the case where $|q_1| \neq |q_2|$ in (3.3).

If we consider collisions between magnetized charged particles and neutrals, the physics changes somewhat. In this latter case, the 'jump-length' for the charged particles is *still* given by their appropriate Larmor radius. Since the characteristic length scale is different from ions and electrons, we now *can* have charge separations in the case of collisions taking place in a plasma density gradient.

3.1.2 Collisions in electric and magnetic fields with $E \perp B$

In section 3.1.1 we discussed collisions between magnetized charged particles, with the implicit condition that the electric field was vanishing in the direction perpendicular to **B**. We will now consider the case where we have an electric field $\perp B$. (The consequences of an electric field component *parallel* to **B** are not particularly interesting, and will not be discussed here.) We also assume here that the magnetic field is constant in direction as well as intensity. The case with a *constant* electric field is of course trivial, since we can change the frame of reference to one moving with a velocity $\mathbf{E} \times \mathbf{B}/B^2$, and in this frame the electric field vanishes. A non-trivial case arises when $\mathbf{E} = \mathbf{E}(\mathbf{r})$, ignoring time varying electric fields. We take the geometry where **B** is in the $\hat{\mathbf{z}}$-direction and **E** in the $\hat{\mathbf{y}}$-direction, with $\mathbf{E} = E(y)\hat{\mathbf{y}}$ see also figure 3.2. Charged particles will be drifting in the $\mathbf{E} \times \mathbf{B}$-directions, with modifications due to the finite Larmor radius effects, as discussed in section 2.1.6. For the electric fields entering the analysis we have $\nabla \cdot \mathbf{E} \neq 0$ which means that the field is supposed to be maintained by some unspecified charges, while $\nabla \times \mathbf{E} = 0$ for stationary conditions.

We now introduce a laboratory reference frame in such a way that the collision between the two particles takes place at the origin, and select the velocity of this frame in such a way that the electric field vanishes at $y = 0$, i.e. $\mathbf{E}(0) = 0$. The particles collide at the origin, but their gyro-center will generally be at some y-position above or below the x-axis where it is moving with the local gyro-center drift velocity $\mathbf{U}_{dr}$. This choice of reference frame is always possible. In this frame of reference, the particle orbit velocity will in general have a contribution from the drift velocity of the gyro-center $\mathbf{U}_{dr}$, which is in the $\hat{\mathbf{x}}$-direction. The velocity in the circular gyro-orbit is found by changing the frame of reference to moving with the gyro-center. The geometry of the problem is illustrated in figure 3.2.

In order to describe the collision, we first write the gyro-center positions of the two particles at the time of collisions as

$$\begin{aligned} \mathbf{r}_c^{(1)} &= \mathbf{R}^{(1)} + \frac{M_1}{q_1 B^2}\left(\mathbf{U}_{lab}^{(1)} - \mathbf{U}_{dr}^{(1)}\right) \times \mathbf{B} \\ \mathbf{r}_c^{(2)} &= \mathbf{R}^{(2)} + \frac{M_2}{q_2 B^2}\left(\mathbf{U}_{lab}^{(2)} - \mathbf{U}_{dr}^{(2)}\right) \times \mathbf{B}. \end{aligned} \tag{3.4}$$

In general $\mathbf{R}^{(j)}$ is the position of particle no. j, and $\mathbf{U}_{lab}^{(j)}$ is its velocity in the laboratory frame of reference. The corresponding particle masses are M_j and charges q_j. The particles are assumed to collide at some time t_0. With the choice of reference frame, discussed before, we have $\mathbf{R}^{(1)} = \mathbf{R}^{(2)} = 0$ here.

In order to obtain the vector directed from the particle to its gyro-center, we have to use the velocity in the local circular Larmor orbit, and therefore subtract the drift velocity $\mathbf{U}_{dr}$ from the absolute laboratory velocity, $\mathbf{U}_{lab}$. The FLR-corrected drift velocity is here given by

$$\mathbf{U}_{dr}(y) = \frac{1}{B}\left(1 + \frac{1}{4} r_L^2 \frac{\partial^2}{\partial y^2}\right) E(y)\hat{\mathbf{x}}, \tag{3.5}$$

from section 2.1.6.

Just as in section 3.1.1, we can write the change in gyro-center position from *before* to *after* the collision as

$$\begin{aligned} \Delta\mathbf{r}_c^{(1)} &= \frac{M_1}{q_1 B^2}\left(\Delta\mathbf{U}_{lab}^{(1)} - \Delta\mathbf{U}_{dr}^{(1)}\right) \times \mathbf{B} \\ \Delta\mathbf{r}_c^{(2)} &= \frac{M_2}{q_2 B^2}\left(\Delta\mathbf{U}_{lab}^{(2)} - \Delta\mathbf{U}_{dr}^{(2)}\right) \times \mathbf{B}. \end{aligned} \tag{3.6}$$

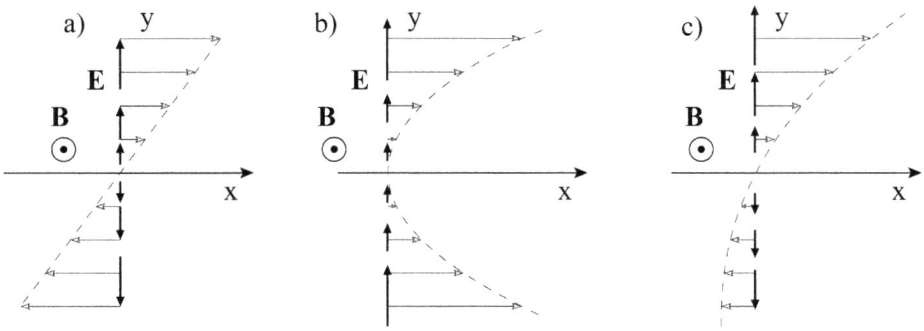

Figure 3.2. Geometry for collisions of charged particles in electric and magnetic fields, with a linear profile $E_y(y) \approx E_0 y/\ell_1$ in a), $E_y(y) \approx E_0 y^2/\ell_2^2$ in b) and $E_y(y) \approx E_0(y/\ell_1 + y^2/\ell_2^2)$ in c). The collision between the two charged particles takes place at the origin, $(x, y) = (0, 0)$. By the proper change to a moving frame of reference, it is always possible to achieve a vanishing electric field at the origin.

Conservation of momentum implies here that

$$M_1 \Delta \mathbf{U}_{lab}^{(1)} + M_2 \Delta \mathbf{U}_{lab}^{(2)} = 0,$$

giving the vector relation

$$q_1 \Delta \mathbf{r}_c^{(1)} + q_2 \Delta \mathbf{r}_c^{(2)} = -\frac{1}{B^2}\left(M_1 \Delta \mathbf{U}_{dr}^{(1)} + M_2 \Delta \mathbf{U}_{dr}^{(2)} \right) \times \mathbf{B}.$$

As it stands, the relation (3.6) is unfortunately of little use, and we have to perform some manipulations, since momentum conservation does not directly imply conditions on the drift velocities $\mathbf{U}_{dr}^{(j)}$. With the simplified assumption for the $\hat{\mathbf{y}}$-direction of $\mathbf{U}_{dr}^{(j)}$ used here, we might, however, state that $q_1 \Delta x_c^{(1)} + q_2 \Delta x_c^{(2)} = 0$ for this case. Only the discussion of changes in the y-components remains.

The approximation $E_y(y) \approx E_0 y/\ell_1$
In order to relate $\Delta \mathbf{r}_c^{(j)}$ to $\Delta \mathbf{U}_{dr}^{(j)}$ we now simplify the problem by first assuming $\mathbf{E}(y) \approx (E_0 y/\ell)\hat{\mathbf{y}}$, see figure 3.2. The FLR-correction which originates from the second derivative in (3.5) vanishes for this case. With the choice $\mathbf{R}^{(1)} = \mathbf{R}^{(2)} = 0$ in (3.4), we then have

$$y_c^{(j)} = -\frac{1}{\Omega_{c,j}}\left(U_{x,lab}^{(j)} - \frac{E_0}{B}\frac{y_c^{(j)}}{\ell} \right),$$

or

$$y_c^{(j)} = -\frac{1}{\Omega_{c,j}}\frac{U_{x,lab}^{(j)}}{1 - E_0/\left(B\ell\Omega_{c,j}\right)},$$

giving

$$q_j \Delta y_c^{(j)} = -\frac{M_j \Delta U_{x,lab}^{(j)}}{B - E_0/\left(\ell\Omega_{c,j}\right)}.$$

By considering the x-component of the basic relations, it is trivially found that

$$q_j \Delta x_c^{(j)} = -\frac{M_j}{B}\Delta U_{y,lab}^{(j)},$$

with the present assumptions. Note that the two particles change their relative kinetic energies and momenta during the collision, but the potential energy in the electrostatic electric field does not change since the particle positions are basically the same just before and just after the collision. From momentum conservation we have that $M_1 \Delta U_{x,lab}^{(1)} = -M_2 \Delta U_{x,lab}^{(2)}$, implying

$$q_1 \Delta y_c^{(1)}\left(1 - \frac{E_0}{B\,\ell\Omega_{c,1}} \right) + q_2 \Delta y_c^{(2)}\left(1 - \frac{E_0}{B\,\ell\Omega_{c,2}} \right) = 0.$$

1: If we take $q_1 = q_2 = q$ and $M_1 = M_2 = M$, then this implies $\Delta y_c^{(1)} + \Delta y_c^{(2)} = 0$. We have trivially from the previous result that $\Delta x_c^{(1)} + \Delta x_c^{(2)} = 0$. Consequently, we argue that for a simple linear electric field profile illustrated in figure 3.2(*a*), collisions between like particles do not give rise to any net particle transport, i.e. this case is similar to the one without electric field, as far as this question is concerned. To have net particle transport in this case, we must have a finite second derivative of the electric field.

2: Alternatively, we can take $q_1 = -q_2 = -q$ and $M_1 = M_2 = M$ giving $\Omega_{c,1} = -\Omega_{c,2} = \Omega_c$. We still have $\Delta x_c^{(1)} = \Delta x_c^{(2)}$ for the $\hat{\mathbf{x}}$-direction, while for the $\hat{\mathbf{y}}$-direction we find

$$\Delta y_c^{(1)}\left(1 - \frac{E_0}{B\,\ell\Omega_c}\right) = \Delta y_c^{(2)}\left(1 + \frac{E_0}{B\,\ell\Omega_c}\right).$$

The sign of the charge q enters through Ω_c. For strong electric fields we find a saturation giving $\Delta y_c^{(1)} \approx -\Delta y_c^{(2)}$, independent of E_0 and B. For this case we have no charge separation either, but the physical relevance of this limit is questionable.

For weak electric fields, we have the approximation

$$\Delta y_c^{(1)} = \Delta y_c^{(2)}\left(1 + 2\frac{E_0}{B\,\ell\Omega_c}\right).$$

Neither for collisions between oppositely charged particles do we find any charge separation on the Larmor radius scale in the $\hat{\mathbf{x}}$-direction, but *along* the direction of the electric field, i.e. in the $\hat{\mathbf{y}}$ direction, we have a charge separation due to the current induced by the electric field in a collisional plasma.

The approximation $E_y(y) \approx E_0(y/\ell_1 + y^2/\ell_2^2)$

We have found that two identical magnetized charged particles that collide in a linearly varying electric field do not change the position of their center of mass. It will now be demonstrated that a *quadratic* electric field variation $\mathbf{E}(y) \approx E_0(y/\ell_1 + y^2/\ell_2^2)\hat{\mathbf{y}}$, see figure 3.2(*c*), will in general give rise to a net particle transport by collisional interactions. Using (3.5) we now find

$$y_c^{(j)} = -\frac{1}{\Omega_{c,j}}\left(U_{x,lab}^{(j)} - \frac{E_0}{B}\left(\frac{\left(y_c^{(j)}\right)^2}{\ell_2^2} + \frac{1}{2}\frac{r_{L,j}^2}{\ell_2^2} + \frac{y_c^{(j)}}{\ell_1}\right)\right),$$

with $r_{L,j}^2\Omega_{c,j}^2 \equiv (U_{x,lab} - U_{x,dr})^2 + (U_{y,lab} - U_{y,dr})^2 = 2\mathcal{W}_{kin}^{(j)}/M_j + U_{x,dr}^2 + U_{y,dr}^2 - 2(U_{x,lab}U_{x,dr} + U_{y,lab}U_{y,dr})$. Here $U_{y,dr} = 0$ by assumption so we have $r_{L,j}^2\Omega_{c,j}^2 = 2\mathcal{W}_{kin}^{(j)}/M_j + U_{x,dr}^2 - 2U_{x,lab}U_{x,dr}$. We introduced the particle kinetic energy $\mathcal{W}_{kin}^{(j)} \equiv \frac{1}{2}M_j(U_{x,lab}^{(j)2} + U_{y,lab}^{(j)2})$.

We have here

$$\Delta y_c^{(j)} = -\frac{1}{\Omega_{c,j}}\left(\Delta U_{x,lab}^{(j)} - \frac{E_0}{B}\left(\frac{\Delta\left(y_c^{(j)}\right)^2}{\ell_2^2} + \frac{1}{2}\frac{\Delta r_{L,j}^2}{\ell_2^2} + \frac{\Delta y_c^{(j)}}{\ell_1}\right)\right)$$
$$= -\frac{1}{\Omega_{c,j}}\left(\Delta U_{x,lab}^{(j)} - \frac{E_0}{B}\left(\frac{\Delta\left(y_c^{(j)}\right)^2}{\ell_2^2} + \frac{\Delta y_c^{(j)}}{\ell_1}\right) - \frac{E_0}{B}\frac{\Delta r_{L,j}^2}{2\,\ell_2^2}\right). \tag{3.7}$$

The change in gyro-radius can be expressed as

$$\Delta r_{L,j}^2 = 2\frac{\Delta \mathcal{W}_{kin}^{(j)}}{M_j\Omega_{c,j}^2} + \frac{E_0^2}{B^2\Omega_{c,j}^2}\Delta\left(\frac{y_c^{(j)}}{\ell_1} + \frac{\left(y_c^{(j)}\right)^2}{\ell_2^2}\right)^2$$
$$- 2\frac{E_0}{B\Omega_{c,j}^2}\Delta\left(\frac{U_{x,lab}^{(j)}y_c^{(j)}}{\ell_1} + \frac{U_{x,lab}^{(j)}\left(y_c^{(j)}\right)^2}{\ell_2^2}\right).$$

To lowest order in the assumed small quantity E_0/B we have $\Delta r_{L,j}^2 \approx 2\Delta\mathcal{W}_{kin}^{(j)}/(M_j\Omega_{c,j}^2)$.

Assuming for simplicity that the two colliding particles have the same masses $M_1 = M_2 \equiv M$ giving $|\Omega_{c,1}| = |\Omega_{c,2}|$, we have from momentum conservation that $\Delta U_{x,lab}^{(1)} = -\Delta U_{x,lab}^{(2)}$, and also $\Delta\mathcal{W}_{kin}^{(1)} = -\Delta\mathcal{W}_{kin}^{(2)}$. To lowest order in a velocity parameter E_0/B, assumed small compared to the particle velocity U_{lab}, we have

$$q_1\Delta y_c^{(1)} + q_2\Delta y_c^{(2)} = \frac{ME_0}{B^2}\left(\frac{\Delta\left(y_c^{(1)}\right)^2}{\ell_2^2} + \frac{\Delta\left(y_c^{(2)}\right)^2}{\ell_2^2} + \frac{\Delta y_c^{(1)}}{\ell_1} + \frac{\Delta y_c^{(2)}}{\ell_1}\right) \tag{3.8}$$

or

$$\Delta y_c^{(1)}\left(q_1 - \frac{ME_0}{B^2\ell_1}\right) + \Delta y_c^{(2)}\left(q_2 - \frac{ME_0}{B^2\ell_1}\right)$$
$$= \frac{ME_0}{B^2\ell_2^2}\left(2y_0^{(1)}\Delta y_c^{(1)} + \left(\Delta y_c^{(1)}\right)^2 + 2y_0^{(2)}\Delta y_c^{(2)} + \left(\Delta y_c^{(2)}\right)^2\right). \tag{3.9}$$

Considering now the particularly simple case where also $q_1 = q_2 = q$, we have

$$q\Delta y_c^{(1)}\left(1 - 2y_0^{(1)}\frac{ME_0}{qB^2\ell_2^2}\frac{1}{1 - \dfrac{ME_0}{qB^2\ell_1}}\right)$$
$$+ q\Delta y_c^{(2)}\left(1 - 2y_0^{(2)}\frac{ME_0}{qB^2\ell_2^2}\frac{1}{1 - \dfrac{ME_0}{qB^2\ell_1}}\right) \tag{3.10}$$
$$= \frac{ME_0}{B^2\ell_2^2}\left(\left(\Delta y_c^{(1)}\right)^2 + \left(\Delta y_c^{(2)}\right)^2\right).$$

For consistency with the basic assumption, we have both scale lengths larger than the relevant Larmor radii, $\ell_1 \gg r_L$ and $\ell_2 \gg r_L$. It is also evident that we require $y_0 \leqslant r_L$, otherwise the two particles will not collide. Since also $E_0/B < U_{th}$, using the thermal velocity U_{th} as representative, we have

$$\frac{E_0 M}{qB^2\ell_1} = \frac{E_0}{B}\frac{1}{\Omega_c\ell_1} = \frac{E_0}{BU_{th}}\frac{U_{th}}{\Omega_c\ell_1} = \frac{E_0}{BU_{th}}\frac{r_L}{\ell_1} \ll 1,$$

and therefore

$$y_0^{(j)}\frac{ME_0}{qB^2\ell_2^2}\frac{1}{1-\dfrac{ME_0}{qB^2\ell_1}} \approx y_0^{(j)}\frac{ME_0}{qB^2\ell_2^2}\left(1+\frac{ME_0}{qB^2\ell_1}\right) \approx \frac{y_0^{(j)}}{\ell_2}\frac{E_0/B}{U_{th}}\frac{r_L}{\ell_2} \ll 1.$$

Since the right-hand side of (3.10) is positive for positive E_0, and the coefficients of $\Delta y_c^{(1)}$ and $\Delta y_c^{(2)}$ in (3.10), i.e. the two parentheses, are close to unity, we can argue that $\Delta y_c^{(1)} + \Delta y_c^{(2)}$ is positive for $q > 0$ and negative for $q < 0$. We thus find that two magnetized electrons, or like ions, colliding in an electric field with finite second derivative *will* experience a net displacement of their guiding centers, in contrast to the case where $E_0 = 0$. The net displacement is inversely proportional to ℓ_2^2.

The discussion of the present section addressed collisions between two particles only. In a plasma we have many particles colliding in a small reference volume, and the net transport is obtained by averaging the displacement resulting from all the collisions (Simon 1955b).

3.1.3 Collisions with neutrals

At first sight it might be expected that collisions between charged particles and neutrals could, at least in principle, be described in the same way as collisions between charged particles. Given the proper cross-sections the modeling indeed follows quite a similar pattern, but there is, however, one important difference: as we have seen, collisions between charged particles in magnetized plasmas do not lead to any macroscopic charge separation. When colliding with a neutral, an ion or electron can in principle move a distance given by the Larmor radius it has after a possible energy loss during the collision, the motion being allowed in any direction. This means that collisional interactions between charged particles and neutrals give rise, in general, to charge separations and corresponding electric fields. These problems will be discussed in more detail in sections 3.5 and 3.6.

Collisions between charged particles are in general relatively simple to describe, at least in the classical limit. Collisions with neutrals are in comparison much more complicated. We can have charge exchange collisions, where a neutral gives up an electron to a passing ion, which then continues as a neutral: in effect we can imagine an ion (usually fast) being instantaneously replaced by another ion, which was neutral before (and usually slow). In numerical simulations, this process is relatively easy to model, given a table for the proper cross-section and its velocity variation. Electrons and ions can have elastic collisions with neutrals, this will usually be the case when their energy is too small for excitation or ionization. If these latter processes have to be taken into account, the analysis can soon become rather messy. Accurate modeling of inelastic collisions can be quite complicated.

3.2 Auroral arcs

The standard discussions of collisional resistivity refer to homogeneous conditions. We will here discuss some extensions for the case where the media are inhomogeneous. The model will have particular reference to the case of an auroral arc, but the results will also have more general applications.

We consider here a case where the magnetic field lines can be taken vertical, as appropriate for polar regions, see figure 3.3. We assume that an electric field constant in space and time is imposed on the ionospheric plasma. As discussed before, it is unlikely that the **B**-parallel component of this electric field will be large, so we make from the outset the simplifying assumption that the electric field has a dominant horizontal component, and ignore the altitude variation, i.e. **E** has no z-dependence. Since the plasma conductivity varies significantly with altitude (Boström 1964), we anticipate that the current density has a corresponding altitude variation. The most important feature to be discussed here will however be an inhomogeneity of the plasma conductivity in the *horizontal* direction. Precipitation of fast electrons, or in some cases fast ions, along magnetic field lines give rise to *auroral arcs*, characterized by such a 'band' of high plasma conductivity. We will discuss here the currents generated by electric fields under such conditions. The altitude variation of the currents is of minor interest in this context, so we integrate Ohm's law with respect to the vertical coordinate. Writing Ohm's law in tensor form as $\mathbf{J}(z) = \boldsymbol{\sigma}(x, y, z) \cdot \mathbf{E}$, we find after integration with respect to the z-variable $\mathbf{I} = \boldsymbol{\Sigma} \cdot \mathbf{E}$, with $\mathbf{I}(x, y) \equiv \int \mathbf{J}(x, y, z)\mathrm{d}z$, and the height integrated conductivity tensor being $\boldsymbol{\Sigma}(x, y) \equiv \int \boldsymbol{\sigma}(x, y, z)\mathrm{d}z$. We note that while the dimension of σ is $[\sigma] = \mathrm{S\,m^{-1}}$, we have $[\Sigma] = \mathrm{S}$, where $\mathrm{S} \equiv \mathrm{Ohm}^{-1}$. (We have S to be the abbreviation

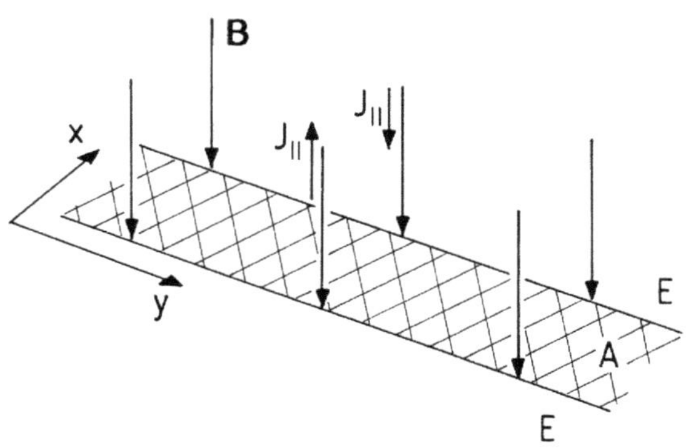

Figure 3.3. Illustration for the discussion of electric fields and currents in auroral arcs. The shaded region is the arc, labeled *A*. The surrounding poorly conducting regions are labeled *E*, not to be confused with the ionospheric E-region. Magnetic field lines are vertical.

of 'Siemens' and see sometimes the notation mho for Ohm^{-1}, i.e. Ohm spelled backwards!) Similarly, we have $[J] = \mathrm{A\,m^{-2}}$, while $[I] = \mathrm{A\,m^{-1}}$. With the geometry of figure 3.3 we have $\mathbf{I}$ and $\boldsymbol{\Sigma}$ depending on the x-coordinate only.

We now identify two basically different regions; the arc, denoted by A, where the conductivity is large, and the outer regions, which are at first set to be empty, labeled E, where the conductivity is poor, or vanishing. This E should not be confused with the ionospheric E-region! At first we set the conductivity here to be vanishing. The intricacy of the problem can be described as follows; the electric field component in the direction parallel to the arc drives a current in this direction by the Pedersen conductivity (named so after the Danish engineer P O Pedersen). This electric field component also tries to set up a current in the perpendicular direction by Hall conductivity, but since this cannot flow into the poorly conducting D and E-regions, it accumulates charges on the boundaries of the arc instead. These charges give rise to electric fields, which by their Pedersen current exactly balance the original Hall currents. These new charge induced electric fields will however now give rise to a Hall current flowing along the arc, and the current in this direction can be significantly larger than what we would have in a horizontally homogeneous plasma, without the arc.

To give an analytical formulation of the ideas outlined before, we recall first the boundary condition for electric fields, stating that the tangential component is the same at the two sides of a boundary. Using the indexes A and E, we have for the electric field components $E_y^E = E_y^A$. The y-component of the height integrated current density is then $I_y^A = I_{y,P}^A + I_{y,H}^A = \Sigma_P^A E_y^A + \Sigma_H^A E_x^A$, with the subscript P being an abbreviation of 'Pedersen', and H for 'Hall'. A term like $\Sigma_H^A E_y^A$ does not contribute to a current in the $\hat{\mathbf{y}}$-direction. For stationary conditions, we have $\nabla \cdot \mathbf{J} = 0$, from the charge continuity equation applied to quasi-neutral conditions. Since we here have no currents flowing in the vertical direction by assumption, we have for the height integrated current density that $I_x^A = I_x^E$. From $\mathbf{J}_\perp = \sigma_P \mathbf{E}_\perp + \sigma_H \hat{\mathbf{b}} \times \mathbf{E}_\perp$ we find $I_x^A = \Sigma_P^A E_x^A - \Sigma_H^A E_y^A$. Since we here assumed a very poor conductivity in the E-region, we have $I_x^E \approx 0$, and therefore also $I_x^A = 0$, implying $\Sigma_P^A E_x^A = \Sigma_H^A E_y^A = \Sigma_H^A E_y^E$, where we also used the electric field boundary condition. Consequently, we find $E_x^A \approx E_y^E \Sigma_H^A / \Sigma_P^A$. Inserting this into the expression for the height integrated current density we find

$$I_y^A = \left(\Sigma_P^A + \frac{\left(\Sigma_H^A\right)^2}{\Sigma_P^A} \right) E_y^E, \tag{3.11}$$

expressing the current density in terms of one electric field component in the poorly conducting region. In an auroral arc we often have $\Sigma_H^A / \Sigma_P^A \gg 1$, implying $I_y^A \approx E_y^E (\Sigma_H^A)^2 / \Sigma_P^A$. This 'electrojet current' can be very large, exceeding $1\ \mathrm{A\,m^{-1}}$, for typical parameters as $\Sigma_H^A \sim 30$ S, $\Sigma_P^A \sim 10$ S and $E_y^E \sim E_y^A \sim 10\ \mathrm{mV\,m^{-1}}$. Well, this may not sound so impressive, but imagine an arc of 100 or 1000 km width! This gives a net current of $\sim 10^6$ A.

For low altitudes, the electrojet current density vanishes for two reasons: the mobility of the charge carriers (electrons as well as ions) is small due to collisions with neutrals, and also because their density is small. For high altitudes we again have the plasma density decreasing, but here the collisional drag is vanishing. As a consequence, the Pedersen mobility becomes negligible and the Hall mobility is practically the same for both electrons and ions. Consequently, we have electrons and ions drifting in the same direction with nearly the same velocity, and the electrojet current density is again vanishing. The electrojet current is confined to a relatively narrow altitude region in the ionospheric E and F regions.

- **Exercise:** Repeat the analysis of this section, by assuming that the conductivity in the E-region is poor as compared to the A-region, but that Σ^E is not *entirely* negligible. Demonstrate that $I_x^A = \Sigma_P^A E_x^A - \Sigma_H^A E_y^E$, where

$$E_x^A = \frac{\Sigma_P^E}{\Sigma_P^A} E_x^E + \left(\frac{\Sigma_H^A - \Sigma_H^E}{\Sigma_P^A} \right) E_y^E. \tag{3.12}$$

  By this, we again express the electrojet current by plasma parameters Σ, and electric field components in the poorly conducting regions.

We now allow for the possibility where the charge accumulation can escape, at least partly, by flowing along magnetic field lines, as indicated in figure 3.3 by $J_\|$, which will here be interpreted as *sheath currents* $I_\|$, noting the dimensional difference. This implies that $I_x^A = I_x^E + I_\|$. We still have $I_y^A = \Sigma_p^A E_y^A + \Sigma_H^A E_x^A$, and retain the previous assumption of $I_x^E \approx 0$. The currents $J_\|$ are the *Birkeland currents.* Note that the sign of the Birkeland currents are opposite on the two sides of the arc; this is easily understood by a simple physical discussion of the origin of these currents. We readily find $\Sigma_p^A E_y^A - \Sigma_H^A E_x^A = I_\|$, giving

$$E_x^A = \frac{\Sigma_H^A}{\Sigma_P^A} E_y^E + \frac{I_\|}{\Sigma_P^A}.$$

We find that the Birkeland currents enter as corrections to E_x^A. If this turns out to be a significant correction, the arc is often called a 'Birkeland arc', otherwise 'polarization arc'. The difference can at least in principle be determined by measurements of E_x^A and E_x^A, given the plasma parameters. The Birkeland currents can only flow in the upper parts of the ionosphere, since the collisions with neutrals become much more frequent at diminishing altitudes such as the D-region, and the downward mobility of charge carriers is strongly reduced there. On the other hand, we also have to keep in mind that in order to determine the Birkeland current, we have to know the *entire* circuit; the current cannot just end up in thin air. A discussion of the complete ionospheric circuit, including the dynamo effect generating the electric fields (Kelley 1989), is outside the scope of the present treatise.

- **Exercise:** Repeat the analysis of this section, by assuming that the conductivity in the E-region is poor as compared to the A-region, but that Σ^E is not *entirely* negligible. Include the Birkeland currents in the analysis, and derive the modified version of (3.12). Give a physical explanation of the opposite directions of the Birkeland currents at the opposite sides of the arc in figure 3.3.

- **Exercise:** Consider a model ionosphere, consisting of a mixture of neutral atoms with density $n_n = 10^{18}$ m^{-3} and an ionized component (plasma) with ion density $n_i = 10^{13}$ m^{-3}. The plasma is charge neutral. We have a homogeneous magnetic field of 50×10^3 nT. All temperatures are taken to be 1000 K. How much is this in eV? Assume that the cross-section, σ, for electrons colliding with neutral is 10^{-16} cm^2 for electrons with the thermal velocity, and use this cross-section as representative also for the ion neutral collisions. What is the free mean path for electron–neutral collisions? What is a typical collision frequency?

  Assume that the effective ion mass is that of nitrogen atoms. What is the electric conductivity in the direction parallel to **B**, for electron and for ions? What is the contribution to the electric conductivity in the direction perpendicular to **B** for electrons and for ions? Discuss the limiting case where you let $\sigma \to 0$. Estimate the ratio σ_P/σ_H in this limit.

3.2.1 A simple model for an auroral arc

Previously, in section 3.2, we gave a formal outline of auroral arcs and the Birkeland currents generated by externally imposed stationary large scale electric fields. The analysis was simplified by reducing it to a two-dimensional problem by introducing surface currents. By reference to figure 3.4 we can consider a special realistic case where the detailed processes are readily understood.

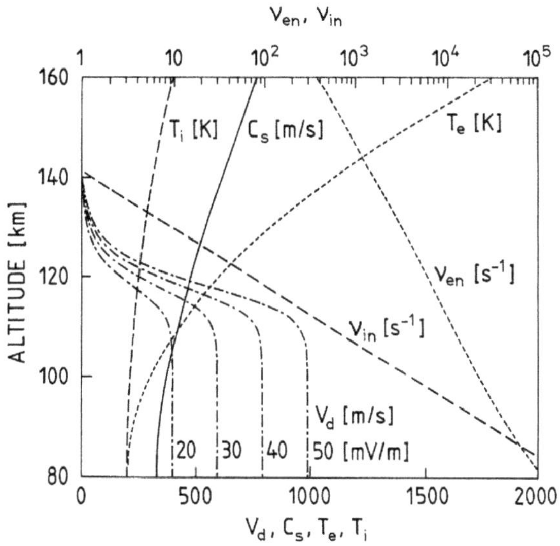

Figure 3.4. Ionospheric plasma parameters relevant for conditions over northern Scandinavia. V_d is the velocity difference between ion and electron drifts in the direction perpendicular to the magnetic field, for different externally imposed DC electric fields, given in mV m^{-1}. For the present conditions, the electrojet current is confined to altitudes below approximately 120 km. The data summarized in the figure were obtained for conditions over northern Scandinavia, in connection with the ROSE rocket campaign in 1988 and 1989, see special issue of the *Journal of Atmospheric and Terrestrial Sciences*, **54** (1992). Plasma conditions have a significant diurnal variation, and also change with latitude (Brekke 2013).

We consider here a specific current generating mechanism. The driving mechanism (or 'battery') is a steady state horizontal electric fields $\mathbf{E}_0$ imposed perpendicular to $\mathbf{B}$ in the ionospheric E-region (Garcia and Pécseli 2013). This condition occurs frequently in the polar regions (Kelley 1989) where $\mathbf{E}_0$ originates from the interaction between the Earth's magnetic field and the solar wind: we will not be concerned with the details of this interaction, and will assume $\mathbf{E}_0$ to be *a priori* given. (We introduced the notation $\mathbf{E}_0$ to distinguish this horizontal electric field from the vertical component $\overline{\mathbf{E}}$.) The electron and ion mobilities are different in the E-region since $\omega_{ce} \gg \nu_{en}$ while $\Omega_{ci} \leqslant \nu_{in}$ in terms of electron and ion collision frequencies (ν_{en} and ν_{in}) with the neutral background gas. Electron–ion collisions are negligible in comparison. The neutral collision frequencies decrease rapidly with altitude, see figure 3.4. For altitudes above some 120–130 km the electrons and ions are both drifting with essentially the same $\mathbf{E}_0 \times \mathbf{B}/B^2$-velocity. We model the E-region as a collisional horizontal 'slab' of thickness D with $\omega_{ce} \gg \nu_{en}$ while $\Omega_{ci} \leqslant \nu_{in}$. For higher altitudes above approximately 120 km, we completely ignore collisions, see figure 3.4: as what concerns the steady state electron–ion velocity difference, this is a good approximation. The neutral drag on the ions is significant in this region, while it is small or negligible on the electrons (Stubbe 1968). At lower altitudes, in the D-region for instance, the neutral drag becomes substantial on both electrons and ions.

With a finite electric field $\mathbf{E}_0$ we have a net current electron flow (the electrojet current) in the E-region due the differences in net electron and ion steady state drifts (Primdahl and Spangslev 1977). The model for the current generation can also be understood in the $\mathbf{E}_0 \times \mathbf{B}$ moving frame, where the neutral component is in motion. In this frame also the localized density enhancement (see figure 3.4) will be moving. To interpret the free energy driving the current as an electric field $\mathbf{E}_0$ or a neutral wind is merely a question of choosing the frame of reference. As an approximation we let the electrons drift with the full $\mathbf{E}_0 \times \mathbf{B}/B^2$-velocity while the ions are at rest on average with respect to the neutral background, and we find the electrojet current density to be $J_0 \approx -en_0E_0/B$ giving a net current $I_0 \approx -Den_0E_0/B$ per length unit along $\mathbf{E}_0$.

If we have a local plasma density enhancement with a steady state density gradient perpendicular to $\mathbf{B}$ with $\nabla_\perp n_0 \| \mathbf{E}_0 \times \mathbf{B}$ in the rest frame as in figure 3.5, we have a local enhancement of the net current in that region since $\mathbf{E}_0$ is imposed externally and constant giving a constant velocity and the number of charge carriers is locally enhanced. The total electron current $I_0 + I_1$ in the enhanced density region ($n_0 + \Delta n$ between the two intervals a and b in figure 3.5) is not compensated at the boundaries of the magnetic flux tube with enhanced plasma density: in regions a and b the current $\mathbf{I}_1$ therefore has to expand along the vertical magnetic field lines as illustrated in figure 3.5. The current cannot escape downward into the D-region because of the high collisionality there with a corresponding low electron mobility. Upwards, in the vertical direction along magnetic field lines into the F-region, we can have a current propagating (Primdahl and Spangslev 1977, Primdahl *et al* 1987). This current will flow along magnetic field lines perpendicular to $\nabla_\perp n_0$ and can, as we shall see, give rise to a current driven drift wave instability. For illustration we took the magnetic fields to be vertical, but an inclination with respect to the vertical does not constitute a substantial change.

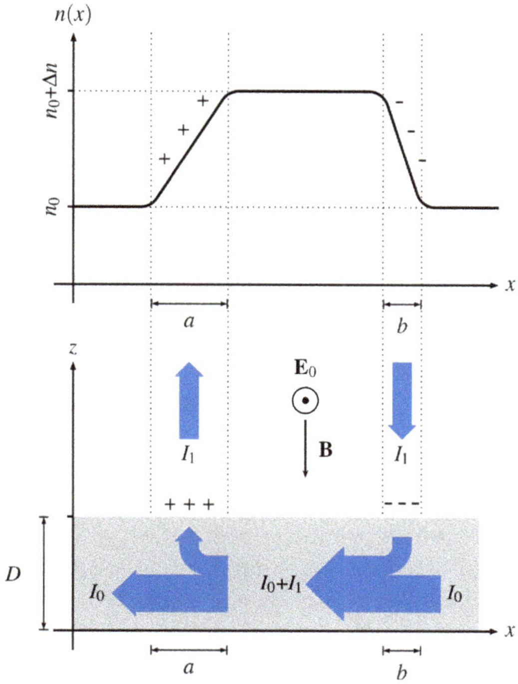

Figure 3.5. Illustration (Garcia and Pécseli 2013) of a vertical cut along a magnetic flux tube with enhanced plasma density. At the top we show details of the geometry and density variations, while the bottom of the figure illustrates the currents generated by a steady state electric field $\mathbf{E}_0 \perp \mathbf{B}$. The thickness of the E-region is $D \approx 10-20$ km.

The current I_1 in figure 3.5 is carried by the electrons, both in the E- and F-regions. With a horizontal density variation we have in general an asymmetry between the part facing the $\mathbf{E}_0 \times \mathbf{B}$-drift and the opposite side. Due to the abundance of electrons in the E-region, we expect that a quasi-steady state is achieved rapidly where electrons are flowing from the high plasma density E-region into the lower density in the F-region. On the side facing the $\mathbf{E}_0 \times \mathbf{B}$-drift in region a in figure 3.5 the electrons have to flow from a lower plasma density into the larger density in the E-region. Ion motion can be of importance here.

We note that the conservation of net current $(I_0 + I_1)$ through the cut between regions a and b in figure 3.5 can act as an amplification mechanism for current densities and thereby average electron flow velocities. In regions outside the flux tube with enhanced plasma density we have a net electrojet current $I_0 = J_0 D = en_0DE_0/B$ per unit length in the direction perpendicular to $\nabla_\perp n_0$, so by Kirchhoff's laws we find for instance at region b in figure 3.5 where J_1 is the current density, that $eD\Delta nE_0/B \approx J_1 b = be(n_0 + \Delta n)u$, giving the vertical electron drift velocity estimate $u \approx (\Delta n/(n_0 + \Delta n))(D/b)E_0/B$. In the ionospheric E-region (Kelley 1989) we frequently find $E_0 \gtrsim 20\,\mathrm{mV\,m^{-1}}$ which gives $E_0/B \approx C_s$, so we can argue that substantial magnetic field aligned electron drift velocities (i.e. field aligned currents) can be achieved by this mechanism.

3.3 Quasi-neutrality

As an approximation for charged particles with collective interactions it can at times be assumed that the density fluctuations are quasi-neutral, i.e. make what is often called 'the plasma approximation' (Chen 1984). This means that we in the analysis let $n_e = n_i \equiv n$. Now, this cannot generally be true since it renders the right-hand side of Poisson's equation identically zero. On scales comparable to or smaller than the Debye length we *can* have large deviations from charge neutrality. If we, however, restrict the analysis to scale-lengths much larger than the Debye length, it is often safe to assume that no significant differences between n_e and n_i can occur; if by chance a large plasma volume lost even a small fraction of, say, the electrons, very large electric fields would be set up, which immediately would correct the imbalance. If the length scales are large, as assumed, this implies then that the gradients of the electric fields are small, even if the fields as such are non-zero. By the plasma approximation we can argue that Poisson's equation becomes redundant, in the sense that, to the given approximation, it simply states $0 \approx 0$. By the quasi-neutrality or plasma approximation we reduce the number of variables by one, and by omitting Poisson's equation we also reduce the number of equations by one.

The arguments outlined here can be substantiated by considering Poisson's equation in normalized form. We introduce the normalized densities $\eta_e \equiv n_e/n_0$ and $\eta_i \equiv n_i/n_0$, assuming that we have a suitable normalizing reference density n_0, which is the same for electrons as well as ions. The electrostatic potential is normalized as $\psi \equiv e\phi/\kappa T_e$, whereby a local thermal equilibrium for the electrons is implicitly assumed. Implicitly, we hereby restrict the arguments to apply for low frequency phenomena; only then can we hope for the assumption of local thermal equilibrium for the electrons to be accurate. Finally, it is assumed that the perturbations have a well defined scale length, $\mathcal{L}$. Writing Poisson's equation in terms of new variables, with $\xi \equiv r/\mathcal{L}$, we readily find

$$\left(\frac{\lambda_D}{\mathcal{L}}\right)^2 \frac{\partial^2}{\partial \xi^2}\psi = \eta_i - \eta_e. \tag{3.13}$$

If we now assume that all the normalized variables ψ, η_i and η_e are of the same order of magnitude, while $\mathcal{L} \gg \lambda_D$, we argue that to lowest order accuracy we must have $\eta_i \approx \eta_e$, to have both sides of (3.13) equally small. (Note that by normalizing x by $\mathcal{L}$, we ensure that $\partial^2\psi/\partial\xi^2$ is also of order unity, when ψ is unity.) This is the analytical basis for the quasi-neutral assumption. It allows us to consider the space-time varying bulk plasma density $n(\mathbf{r}, t)$, rather than the individual electron and ion densities. This can in some cases be a great simplification.

If we take $n_e \approx n_i \equiv n$ in continuity equations for electrons and ions, $\partial n_{e,i}/\partial t + \nabla \cdot (n_{e,i}\mathbf{u}_{e,i}) \approx 0$, and take their difference, the result is $\nabla \cdot n(\mathbf{u}_e - \mathbf{u}_i) \approx 0$, or in terms of current, $\nabla \cdot \mathbf{J} \approx 0$, i.e. the currents in the plasma are divergence free in this limit. Note that the quasi-neutral assumption is not based on any direct assumption of linearization; it can also be applicable for nonlinear or large amplitude conditions.

Ignoring Poisson's equation, we note that we can no longer obtain the continuity equation for current from the generalized Ampere law. For consistency with the quasi-neutral approximation, or the 'plasma approximation', we now have to omit Maxwell's displacement current from the general Ampere's law, and have to use the original static form of Ampere's law as it also enters Magneto Hydrodynamic models. This assumption implies again that we assume $\omega \mathbf{E}_\perp / c^2$ negligible as compared to $\mu_0 \mathbf{J}$. Note that the assumption of quasi-neutrality has no association with the waves being electromagnetic or electrostatic! The assumption can be argued also without explicitly introducing the electrostatic potential as in (3.13).

When measured on the relevant slow time-scale, the cancellation of charges will be essentially instantaneous; for high frequency phenomena this assumption will no longer be applicable, and the quasi-neutral approximation cannot be used (evidently) for instance for electron plasma waves, even when their wavelength is large compared to the Debye length. For Langmuir waves, $\omega \mathbf{E}_\perp / c^2$ is *not* negligible: on the contrary, we have the plasma currents and Maxwell's displacement current balancing each other, $\omega \mathbf{E}_\perp / c^2 \approx \mu_0 \mathbf{J}$.

As a heuristic reminder of the plasma approximation we may assume that Poisson's equation becomes redundant in the limit where we, formally, let $\varepsilon_0 \to 0$. To avoid infinite electric fields in this case, the local charge density must vanish to the same approximation, i.e. $n_e \approx n_i$. The ratio of two numbers both going to zero can be anything, and we can assume that Poisson's equation in this limit simply adjusts to be consistent with the electric fields found from other equations. If, by chance, we come over a general solution to a plasma problem, derived without making the plasma approximation from the outset, we can find the simplified quasi-neutral result by letting $\varepsilon_0 \to 0$ everywhere. This means that we let λ_{De} and λ_{Di} both go to zero, while at the same time ω_{pe} and Ω_{pi} both go to infinity; the plasma approximation is valid for large scales and small frequencies. Another way of seeing this can be found by recalling that $\lambda_D \omega_p = U_{th}$. The thermal velocity U_{th} of either species is unaffected by the assumption of quasi-neutrality, so *if* we assume the Debye length to be negligible we *must* assume the corresponding plasma frequency to be very large, i.e. much larger than any frequency we want to include in the analysis. This latter argument holds for both electrons and ions. A detailed discussion of quasi-neutral conditions, including also electromagnetic modes in non-uniform magnetized plasma, is given by Shukla *et al* (2001).

- **Exercise:** Consider an infinitely wide and 10cm thick slab of immobile singly charged ions with density $n_0 = 10^{14}$ cm^{-3}. Assume for the sake of argument that the electron density is $n_e \neq n_0$. The two surfaces of the slab are at the same potential, and the potential difference between the surface and the center of the slab is $\phi = 5$ kV. Calculate n_e.

3.4 Plasma diamagnetism

It can be demonstrated that a magnetized plasma supports a diamagnetic current. The relevant analysis can be found in many textbooks, e.g. Chen (1984). In a fluid model we argue that in steady state conditions a gradient in particle pressure in the direction perpendicular to **B** has to be balanced by some other force. The only simple candidate is a $\mathbf{J} \times \mathbf{B}$-force density where **J** is a current density. In standard magnetohydrodynamics this result is found quite easily (Chen 1984, Pécseli 2012), and can be best understood in detail by considering the motions of single particles. The entire analysis will not be repeated here, but illustrated schematically in figure 3.6. In figure 3.6(*a*) the density is increasing downwards, while the temperature is taken constant. Assume that an observer counts the number of charges passing through a small area element $A - A'$. With the geometry implied in figure 3.6(*a*), it is evident that more positive particles pass through $A - A'$ from left to right than from right to left. Similarly, there are more negative particles passing from right to left than vice versa. There will therefore be a net current along the equi-density contour, as indicated. Similar arguments apply for the case where the plasma density is constant, but the temperature is spatially varying as in figure 3.6(*b*). In the early days of plasma physics this diamagnetic current was seen as paradoxical, since it is associated with particles that do not move on average. This and a few other related paradoxes are discussed in detail by, for instance, Braginskií (1965). Analytical expressions for the electron and ion contributions to the diamagnetic currents will be derived in section 3.5.

We will find that electrostatic drift wave propagate essentially in the direction of the electron diamagnetic drift. The diamagnetic currents exist as steady state only under strictly collisionless conditions. In collisional plasmas, with electron–ion collisions or with collisions between charged particles and neutrals, the currents are dissipated and it is an important problem for plasma confinement to estimate the time scales for the 'smoothing out' of the density gradients.

The magnetized plasma can be considered as composed of quasi-particles or 'circlets' with a magnetic moment $\mu = -\frac{1}{2}mU_\perp^2/B$ associated with each circlet, see also section 2.1.8. The direction of the magnetic moment is along the magnetic field. The negative sign means that the charged particle in its gyro-orbit reduces the local

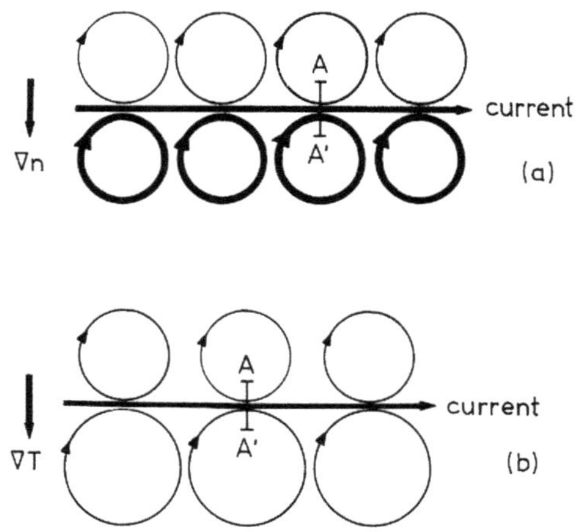

Figure 3.6. Schematic illustration of the origin of the diamagnetic current, showing the gyration orbit of positively charged particles in a magnetic field perpendicular to the plane of the paper. In (*a*) we have a gradient in density, but constant temperature, while (*b*) illustrates the case with constant plasma density, but with a temperature gradient. Even though the individual particles are not displaced *on average*, an observer will find a net current through a cut $A - A'$. This is the diamagnetic current.

magnetic field slightly. The magnetization per unit volume of a collisionless plasma is then

$$\mathbf{M} = -\sum_j \frac{n_j}{2} \frac{m_j \langle U_{\perp,j}^2 \rangle}{B^2} \mathbf{B}$$

where the summation by j is over the plasma species with guiding center densities n_j and mass m_j. Be careful here not to confuse the vector $\mathbf{M}$ with a particle mass. It is important to emphasize that n_j is the circlet density and not the density of plasma particles: if all circlets are identical the two densities will be proportional. More generally we have that the plasma particle (or particles) tracing out the circlet in their Larmor orbit can be seen as 'smearing out' their contribution to the particle density along the circular orbit. Details of the mathematical modeling is given in section 2.1.6. In the case of $r_L^2 \nabla^2 n \ll n$ the densities of guiding centers and particles will be close.

The contribution to the current density of these spatially distributed dipoles is $\nabla \times \mathbf{M}$. In terms of the magnetic moment density we find a current density $\mathbf{J} = \nabla \times \mathbf{M}$. With $\frac{1}{2} m_j \langle U_{\perp,j}^2 \rangle = \kappa T_j$ for the species j, we find for the case with constant temperature and a plasma density varying in the direction $\perp\mathbf{B}$ a current density contribution from the j-species $\mathbf{J}_j = (\kappa T_j / B) \nabla n_j$. The diamagnetic drift velocity associated with this current is $\mathbf{u}_{Dj} = \mathbf{J}_j / (q_j n_j)$. In particular, at a 'sharp' boundary or plasma–vacuum interface we have the surface current density $\mathbf{K}_j = \hat{\mathbf{n}} \Delta \times \mathbf{M}_j$ where $\hat{\mathbf{n}}$ is a unit vector perpendicular to the interface and Δ denotes the change in $\mathbf{M}$.

- **Exercise :** Discuss plasma conditions where the Biermann-battery mechanism is effective (Biermann 1950). Assume that you have an inhomogeneous plasma where the density as well as the electron temperature have a gradient, so that the direction of the two gradients are not parallel. For simplicity take $\nabla T_e \perp \nabla n$.

3.5 Diffusion in fully ionized plasmas

We now consider some model problems for illustrating fluid descriptions of classical diffusion processes. We assume collisional diffusion to be a slow process, which is characterized by time scales longer than the collision times. In that case we can write the fluid momentum equations for magnetized electrons and ions in a quasi-stationary form. Physically this means that the force terms including collisional friction have to balance each other to make the left-hand side small. The temporal variation will then appear through the continuity equations only.

First we consider the case where the collisions are mediated by electron–ion interactions, and we ignore neutrals. Ions and electrons are assumed to have the charges $\pm e$

$$-\nabla p_e - e n_e (\mathbf{E} + \mathbf{u}_e \times \mathbf{B}) + \mathbf{P} = 0$$
$$-\nabla p_i + e n_i (\mathbf{E} + \mathbf{u}_i \times \mathbf{B}) - \mathbf{P} = 0,$$

with the momentum exchange term $\mathbf{P} = \xi e^2 n_e n_i(\mathbf{u}_i - \mathbf{u}_e)$ with the standard expression in terms of the plasma resistivity ξ (van Kampen and Felderhof 1967, Pécseli 2012). The collisions are elastic, so momentum lost by an electron is found as momentum gain by an ion, and vice versa. We have $\xi \approx m\nu_{e,i}/(e^2 n) = \nu_{e,i}/(\varepsilon_0 \omega_{pe}^2)$ with $\nu_{e,i} \approx \omega_{pe}/N_p$ where $N_p \equiv \lambda_{De}^3 n$ is the plasma parameter, see also the discussion in chapter 6. More accurately we have

$$\xi \equiv \frac{1}{\sigma} = \frac{m^{1/2} Z e^2 \ln \Lambda}{64\pi\varepsilon_0^2 (\kappa T)^{3/2}}, \tag{3.14}$$

where eZ is the ion charge, assuming all ions to be in the same charged state. We denote the Spitzer logarithm by $\ln \Lambda$, with $\Lambda \approx 12\pi N_p$ (Spitzer 1956). The resistivity depends on plasma density only through the Spitzer logarithm, which is a weak dependence (Chen 1984). Variations in plasma temperature are more important, but when considering a laboratory plasma experiment, the density is important nonetheless. In order to have meaningful experimental conditions we would like that the length of the plasma column L is many collisional mean free paths. Otherwise it will be the plasma losses (mostly to end plates) that control the conditions, and these are not accounted for by the analysis. The electron–ion collisional mean free path scales approximately as $l_c \approx N_p \lambda_D \sim (\kappa T)^2 n^{-1}$, apart from a Spitzer logarithm. The requirement $L \gg l_c$ is fulfilled for large plasma densities n.

Assuming quasi-neutrality, $n_e \approx n_i = n$, see section 3.3, we can simplify the equations to

$$-\nabla p_e - en\big(\mathbf{E} + \mathbf{u}_e \times \mathbf{B}\big) + \xi e^2 n^2\big(\mathbf{u}_i - \mathbf{u}_e\big) = 0$$
$$-\nabla p_i + en\big(\mathbf{E} + \mathbf{u}_i \times \mathbf{B}\big) - \xi e^2 n^2\big(\mathbf{u}_i - \mathbf{u}_e\big) = 0.$$

The present formulation in terms of plasma resistivity has a slight advantage by having ξ independent of plasma density to logarithmic approximation. We will assume that the plasma has a density gradient in the $\hat{\mathbf{x}}$-direction, perpendicular to the magnetic field in the $\hat{\mathbf{z}}$-direction, see figure 3.7.

Referring to the discussion in section 3.1.1, we expect no charge separations to build up during diffusion processes, so the electric fields vanish. A necessary requirement is then that the electron and ion velocities along the density gradient are equal, and we therefore set $u_{ix} = u_{ex} \equiv u_x$. A detailed calculation will give the

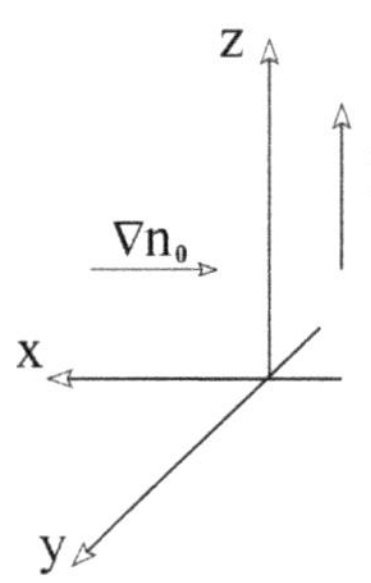

Figure 3.7. Basic geometry for the diffusion problems.

same result, but we might as well use the little *a priori* knowledge we have! With isothermal conditions for the slow diffusion processes, we use $p_{e,i} = \kappa T_{e,i} n$ and find, after a little algebra left to the reader as an exercise, that

$$u_{ey} = -\frac{\kappa T_e}{eB}\frac{1}{n}\frac{\partial n}{\partial x} \tag{3.15}$$

and

$$u_{iy} = \frac{\kappa T_i}{eB}\frac{1}{n}\frac{\partial n}{\partial x}. \tag{3.16}$$

These are the diamagnetic currents. Similarly we have for both electrons and ions

$$u_x = -\xi\frac{\kappa(T_e + T_i)}{B^2}\frac{\partial n}{\partial x}.$$

It is implicitly assumed that we are dealing with a low β plasma, so that the diffusion currents do not modify the magnetic field appreciably. Note that the analysis does not rely on any linearization of the basic equations.

Heuristically we can argue that by taking the time variation of the plasma momentum density (i.e. the left sides of the electron and ion momentum equations) to be negligible, we do not exclude the possibility of the time variation of the plasma density alone to be significant. Inserting the velocity components into the electron or the ion continuity equations we find that the y-component of the velocities do not contribute, and we end up with the diffusion equation in the form

$$\frac{\partial}{\partial t}n - \frac{\partial}{\partial x}D_\perp\frac{\partial}{\partial x}n = 0,$$

where $D_\perp \equiv n\xi\kappa(T_e + T_i)/B^2$. The electron mass enters through ξ. We note a rather favorable magnetic field dependence (that is, favorable when we want to confine a slightly collisional plasma by a magnetic field). The expression for the Spitzer resistivity (3.14) states that ξ is only weakly density dependent (via $\log\Lambda$), and scales with temperature as $T^{-3/2}$, so we have

$$D_\perp \sim \frac{n}{B^2\sqrt{\kappa T}}. \tag{3.17}$$

Generally, we expect that for a magnetically confined warm plasma, the plasma density will decrease as we approach the walls of the device, seemingly implying $D \to 0$, according to (3.17), which might lead us naively to expect good confinement. Unfortunately, at the same time we also have decreasing temperatures at the plasma edges, and this will tend to *increase* D, and the diffusion coefficient remains finite everywhere. Nevertheless, it is important to keep in mind that in fully ionized plasmas, the diffusion coefficient will depend on the spatial variables, in general.

The diffusion equation will have the form of a continuity equation if we introduce a *diffusion velocity* $-(D_\perp/n)\partial n/\partial x$. Introducing a length scale $L \equiv n/(\partial n/\partial x)$ we can write the diffusion velocity as the ratio $-D_\perp/L$. The diffusion flux is $-D_\perp\partial n/\partial x$. For steady state conditions we have a constant particle flux given by $-D_\perp \partial n/\partial x$ which we recognize as Fick's law.

3.5.1 Diffusion as a random walk

The plasma diffusion caused by electron–ion collisions can be described qualitatively as a random walk, with more details of these processes summarized in section 3.5.1. Consider for simplicity a one-dimensional model on a grid with a step of length $\pm\Delta x$ and a fixed time interval Δt between steps. Let each step be independent of all the others. After N steps the position is $X = \sum_{j=1}^{N} \Delta x_j$ where each of the steps has the same length but the direction can be in positive or negative direction with equal probability. If the process is to be repeated many times it is meaningful to look for a prediction of the average position which is $\langle X \rangle = \sum_{j=1}^{N} \langle \Delta x_j \rangle = 0$. This result could be anticipated by symmetry as well. To find a non-trivial result, we estimate the average mean square displacement $\langle X^2 \rangle = \sum_{j=1}^{N} \sum_{k=1}^{N} \langle \Delta x_j \Delta x_k \rangle$. Since Δx_j and Δx_k are independent for $j \neq k$, the average of that product is zero and we have most of the terms in the sum vanishing. Surviving terms are those for $j = k$ giving $\langle X^2 \rangle = \sum_{j=1}^{N} \langle \Delta x_j^2 \rangle = N \Delta x^2$, where the sum is simplified since all the terms are equal. The time duration of N steps is $t = N\Delta t$ so that $\langle X^2 \rangle = t \Delta x^2/\Delta t$, implying $\langle X^2 \rangle^{1/2} \sim \sqrt{t}$ which is one of the characteristics of diffusion processes. The diffusion coefficient will here be $D = \Delta x^2/\Delta t$.

By dimensional reasoning and a little common sense, we might estimate the diffusion coefficient due to electron–ion collisions. We recall that a typical step-length for an electron–ion pair per collision in a magnetized plasma is of the order of the electron Larmor radius r_{Le}, see section 3.1.1, with an average time between collisions being $1/\nu_{ei}$. We have then the estimate $D_\perp \approx r_{Le}^2 \nu_{ei}$. Now, $r_{Le}^2 = (mU_{te})^2/(eB)^2$ and $\nu_{ei} \approx \xi n e^2/m$, giving $D_\perp \approx \xi \kappa T_e n/B^2$, which is essentially the previous result. Note that it is important for the argument that we have electrons and ions to move together at a collision.

Unfortunately, many experiments indicated a scaling of the diffusion coefficient more like $D \sim 1/B$, in variance with the foregoing result (3.17). It was suggested by Bohm (1949) that the observed diffusion was not collisional but *anomalous*. He suggested an estimate in terms of fluctuating electric fields, where the characteristic velocity is $E_\perp/B$ in the direction perpendicular to $\mathbf{B}$. We here recall (again) that a diffusion coefficient can be estimated by the square of a characteristic velocity and a typical time-step. By a dimensional estimate, we then have $D \sim (E_\perp/B)^2/\nu_c$. Bohm (1949) assumed that the magnetized plasma could be considered as a random superposition of field aligned 'tubes' of radius r_L, and fluctuating potentials $\phi \sim \kappa T/e$, implying $E_\perp \sim \kappa T/(e r_L)$. With the estimate $\nu_c \sim \Omega_{ci}$ corresponding to a displacement for about every gyro-period, we finally have

$$D_B \sim \frac{\kappa T}{eB}, \tag{3.18}$$

which is the *Bohm diffusion coefficient*. The expression (3.18) does not contain the particle mass: if we take the 'tube diameter' to be the ion Larmor radius, then we take ν_c to be the ion cyclotron period, and correspondingly if we take the electrons. In either case we find the particle mass to cancel.

For confinement of a hot fusion plasma, (3.18) predicts a rather unattractive scaling with magnetic field as well as plasma temperature, at least when compared to the classical expression (3.17). The analysis giving (3.18) was very sketchy indeed, but, as already mentioned, many experimental observations gave reasons to take it seriously, at least concerning its predictions of the scaling with the magnetic field intensity. The Bohm results were later supported by theoretical studies by Taylor and McNamara (1971), demonstrating that the electric field fluctuations in thermal equilibrium would give rise to a diffusion coefficient with those characteristics.

A relation between the classical diffusion coefficient $D_\perp$ and the Bohm alternative D_B can be stated as

$$D_\perp \approx \frac{(mU_{te})^2}{(eB)^2}\nu_{ei} \approx \frac{m\kappa T}{(eB)^2}\nu_{ei} \approx D_B\frac{\nu_{ei}}{\omega_{ce}} \equiv D_B\frac{1}{K_b}, \tag{3.19}$$

introducing an equivalent for the *Kubo number* (Bakunin 2008), here written as $K_b \equiv \omega_{ce}/\nu_{ei}$. The dimensionless Kubo number is constructed usually as the ratio of two length scales, generally defined as the ratio of the distance a particle travels during an auto-correlation time over the correlation distance. We can also use a definition as the ratio of the correlation time and the time it takes a particle to travel a correlation distance. Taking the correlation time of the collisional diffusion to be ν_{ei}^{-1} and a correlation distance to be the Larmor radius (as in effect assumed by Bohm), we arrive at (3.19). The Kubo number is often used to classify transport regimes in turbulent systems.

3.6 Diffusion in partially ionized plasmas

In some respects, the diffusion process become somewhat more interesting, when we consider collisions between neutrals and charged particles. In this case we can write the quasi-stationary fluid momentum equations as

$$-\nabla p_e - en(\mathbf{E} + \mathbf{u}_e \times \mathbf{B}) - nm\nu_e\mathbf{u}_e = 0$$
$$-\nabla p_i + en(\mathbf{E} + \mathbf{u}_i \times \mathbf{B}) - nM\nu_i\mathbf{u}_i = 0.$$

where we again assumed quasi-neutrality, and introduced the electron and ion neutral collision frequencies as ν_e and ν_i, respectively. Strictly speaking we would also here have to address relative motion between plasma components and neutrals by writing relative velocities between electrons/ions and neutrals with velocity $\mathbf{u}_n$, but with little loss of generality we assume the neutral component to be at rest, $\mathbf{u}_n = 0$. The assumption is not merely a trivial choice of reference frame, since a change to a moving frame will in general induce an electric field $\mathbf{E}_0$. For the reference conditions we take both $\mathbf{E}_0 = 0$ and $\mathbf{u}_n = 0$. The quasi-stationary assumption can be illustrated here by comparing the time derivative of the velocity $\partial\mathbf{u}_{e,i}/\partial t \to -i\omega\mathbf{u}_{e,i}$ with the collision term $\nu_{e,i}\mathbf{u}_{e,i}$. The inertia term of the electron or ion momentum equations can be ignored when $\nu_{e,i} \gg \omega$. We also assume that the electric fields are *electrostatic*,

$\mathbf{E} = -\nabla\phi$, which can hardly be considered as a restriction. By the form of (3.6) it is implicitly assumed that the neutrals are at rest. This is seemingly a minor restriction, but if we should decide to change the frame of reference to change the electric field for instance, the collisions terms in (3.6) have to be modified, e.g. as $nm\nu_e(\mathbf{u}_e - \mathbf{u}_n)$ where $\mathbf{u}_n$ is the velocity of the neutral component, and similarly for the ion collisions.

Assuming again plasma conditions to vary only with the x-coordinate, we readily obtain

$$u_{iy} = -\frac{eB}{M\nu_i}u_{ix} \equiv -\frac{\Omega_{ci}}{\nu_i}u_{ix}.$$

Using the result obtained for u_{iy} we find

$$u_{ix} = -\frac{1}{nM}\frac{\kappa T_i \partial n/\partial x + ne\partial\phi/\partial x}{\nu_i\left(1 + (\Omega_{ci}/\nu_i)^2\right)}. \tag{3.20}$$

From this result it is trivial to obtain an electron velocity as

$$u_{ex} = -\frac{1}{nm}\frac{\kappa T_e \partial n/\partial x - ne\partial\phi/\partial x}{\nu_e\left(1 + (\omega_{ce}/\nu_e)^2\right)}, \tag{3.21}$$

where we might have $u_{ix} \neq u_{ex}$ as discussed in the following.

For u_{iy} we now have

$$u_{iy} = \frac{\Omega_{ci}}{nM}\frac{\kappa T_i \partial n/\partial x + ne\partial\phi/\partial x}{\left(\nu_i^2 + \Omega_{ci}^2\right)}. \tag{3.22}$$

For comparison we list the diamagnetic drift velocity $u_{Di} = (\kappa T_i/neB)\partial n/\partial x$, see (3.16). Similar expressions can be obtained for the y-component of the electron velocity, recalling that we have to keep the sign of the charge in the gyro-frequencies. We identify two components in the ion drift (and correspondingly in the electron drift), the diamagnetic flow related to the density gradient and the $\mathbf{E} \times \mathbf{B}/B^2$-flow, both with modifications due to the collisions with neutrals as given by the denominator.

3.6.1 Short-circuited electric fields

We can now encounter two basically different cases: one where the electric field is obtained self consistently from the equations, and another where we can assume that the electrostatic potential is short-circuited by some 'external' agency, i.e. $\phi = \text{const}$. We consider the latter case first, since it is the simplest. In this case we obtain a diffusion flux (i.e. *velocity* × *density*) for ions as

$$nu_{ix} = -\frac{\kappa T_i/M}{\nu_i\left(1 + (\Omega_{ci}/\nu_i)^2\right)}\frac{\partial n}{\partial x},$$

and for electrons as

$$nu_{ex} = -\frac{\kappa T_e/m}{\nu_e\left(1 + (\omega_{ce}/\nu_e)^2\right)}\frac{\partial n}{\partial x}.$$

From this we trivially identify diffusion coefficients as

$$D_i \equiv \frac{\kappa T_i/M}{\nu_i\left(1 + (\Omega_{ci}/\nu_i)^2\right)}$$

and

$$D_e \equiv \frac{\kappa T_e/m}{\nu_e\left(1 + (\omega_{ce}/\nu_e)^2\right)},$$

finding in particular for weakly collisional plasmas, where $\omega_{ce,i} \gg \nu_{e,i}$,

$$\frac{D_i}{D_e} = \frac{T_i m}{T_e M}\left(\frac{\omega_{ce}}{\Omega_{ci}}\right)^2 \frac{\nu_i}{\nu_e} \approx \left(\frac{T_i m}{T_e M}\right)^{3/2}\left(\frac{\omega_{ce}}{\Omega_{ci}}\right)^2 \frac{\ell_{ce}}{\ell_{ci}},$$

in terms of collisional mean free paths $\ell_c \approx U_{th}/\nu$ for electrons and ions, expressed in terms of their thermal velocities and collision frequencies. If $T_e \approx T_i$ and $\ell_{ci} \approx \ell_{ce}$, we find as an estimate

$$\frac{D_i}{D_e} \approx \sqrt{\frac{M}{m}} \gg 1,$$

implying that ions diffuse much faster than electrons in this case. This is not surprising since we can write as an estimate $D \sim r_L^2 \nu$ in terms of a typical distance for a 'jump' and ν is the corresponding jump-frequency. For magnetized particles we estimate the jump distance by their Larmor radius. Assuming the temperatures to be comparable we have that the Larmor radii, $r_L = U_{th}/\Omega_c$, scale with the particle mass as $r_L \sim \sqrt{M}$. The collision mean free path ℓ_c scales with the inverse scattering cross-section, which is comparable for ions and electrons, while the thermal velocities scale as $\sim 1/\sqrt{M/m}$. Since $\ell_c \sim U_{th}/\nu$ we have the collision frequency to scale as $\nu \sim 1/\sqrt{M}$. Ions collide less frequently than electrons, but when they collide they 'jump' large distances as determined by the ion Larmor radius. In the case of electrons and ions diffusing independently, we expect the ion diffusion coefficient to be larger by a factor $\sim\sqrt{M/m}$. In such a case quasi-neutrality can be maintained by an external agency that short-circuits the electric fields that would otherwise build up. The physical mechanism outlined here (involving short-circuiting end plates) was seemingly first identified by Simon (1955a) to explain unexpectedly large plasma losses in experiments with significant collisional interactions between charged particles and a neutral background gas.

3.6.2 Self-consistent electric fields

We now retain the self-consistent electric field, $\phi \neq$ const. and ignore the possibility of external short-circuiting. In order to retain a time stationary quasi-neutral plasma, we must require that $u_{ix} = u_{ex}$. This gives

$$\frac{1}{nM}\frac{\kappa T_i \partial n/\partial x + ne\partial\phi/\partial x}{\nu_i\left(1 + (\Omega_{ci}/\nu_i)^2\right)} = \frac{1}{nm}\frac{\kappa T_e \partial n/\partial x - ne\partial\phi/\partial x}{\nu_e\left(1 + (\omega_{ce}/\nu_e)^2\right)}.$$

By using the previous definitions of D_e and D_i we can write this expressions as

$$D_i\left(\frac{1}{n}\frac{\partial n}{\partial x} + \frac{e}{\kappa T_i}\frac{\partial\phi}{\partial x}\right) = D_e\left(\frac{1}{n}\frac{\partial n}{\partial x} - \frac{e}{\kappa T_e}\frac{\partial\phi}{\partial x}\right),$$

or

$$e\frac{\partial\phi}{\partial x} = \left(D_e - D_i\right)\frac{1}{n}\frac{\partial n}{\partial x} \Big/ \left(\frac{D_e}{\kappa T_e} + \frac{D_i}{\kappa T_i}\right). \tag{3.23}$$

In principle we have solved the problem, since the potential can now be eliminated, to give a diffusion type equation for the density evolution. The resulting diffusion coefficient can be made neater for the case where the plasma is isothermal, i.e. $T_e = T_i \equiv T_0$. We can make the approximation $(D_e - D_i)/(D_e + D_i) \approx (1/D_i)(D_e - D_i)(1 - D_e/D_i) \approx (2D_e - D_i)/D_i$ by ignoring terms of order D_e^2 or smaller, recalling $D_i \gg D_e$. We then have

$$\frac{e}{\kappa T_0}\frac{\partial\phi}{\partial x} = \frac{D_e - D_i}{D_e + D_i}\frac{1}{n}\frac{\partial n}{\partial x} \approx \left(2\frac{D_e}{D_i} - 1\right)\frac{1}{n}\frac{\partial n}{\partial x}, \tag{3.24}$$

where $(2D_e/D_i - 1) < 0$ so that the directions of $\partial n/\partial x$ and the electric field are the same. Ions are escaping faster to make the low density side of the gradient positive, so $\mathbf{E} = -\nabla\phi$ points into the plasma.

By insertion into (3.20) and (3.21), we find for this case

$$u_{ix} = u_{ex} = -2\frac{D_e}{D_i}\frac{\kappa T_0}{nM}\frac{\partial n/\partial x}{\nu_i\left(1 + (\Omega_{ci}/\nu_i)^2\right)} = -2D_e\frac{1}{n}\frac{\partial n}{\partial x}.$$

The ion and electron diffusion velocities are both in the opposite direction of the density gradient.

From the velocities $u_{ix} = u_{ex}$ we find the diffusion equation for the ion density (which is the same as for the electron density by the assumption of quasi-neutrality)

$$\frac{\partial}{\partial t}n = 2D_e\frac{\partial^2}{\partial x^2}n.$$

We need D_e to be a constant with constant electron temperatures and collision frequencies, so that D_e can be taken outside the differentiation. The assumption of quasi-neutrality 'locks' the ion and electron diffusion equations together, but the electric field induced by the large ion diffusion drags the electrons along to enhance the effective diffusion coefficient by a factor 2 in comparison to what would be obtained if the electrons alone were to control the process. It is in this sense interesting that it is the *ion* pressure that gives rise to the ambipolar electric field. If the electron temperature was very large, $T_e > T_i\sqrt{M/m}$, we see from (3.24) that the electric field would change sign and the electrons would drag the ions along. Such large temperature ratios are, however, unrealistic for plasmas found in nature.

We see that a basic difference between diffusion due to electron–ion collisions and collisions between charged particles and neutrals is that in the latter case we can have an ambipolar electric field of non-trivial magnitude, see (3.23).

The ion diamagnetic drift has an added component due to the electric field that gives an $\mathbf{E} \times \mathbf{B}/B^2$-drift. We can estimate this by using (3.22). Taking $T_e \approx T_i \equiv T_0$ as before we can use (3.24) and find

$$u_{iy} = 2\frac{\Omega_{ci}\kappa T_0}{nM\left(\nu_i^2 + \Omega_{ci}^2\right)}\frac{D_e}{D_i}\frac{\partial n}{\partial x}.$$

Recalling that $D_e/D_i \sim \sqrt{m/M} \ll 1$ this result implies that the $\mathbf{E} \times \mathbf{B}/B^2$-drift here almost cancels the ion diamagnetic drift.

For the electron drift we have a result similar to (3.22)

$$u_{ey} = -\frac{\omega_{ce}}{nm}\frac{\kappa T_e \partial n/\partial x - ne\partial\phi/\partial x}{\left(\nu_e^2 + \omega_{ce}^2\right)} \approx -\frac{1}{nm\omega_{ce}}\left(\kappa T_e\frac{\partial n}{\partial x} - ne\frac{\partial \phi}{\partial x}\right),$$

where the approximation is based on the assumption $\nu_e \ll \omega_{ce}$ usually met in the ionospheric E- and F-regions as well as in many laboratory experiments. For the electrons we find for $T_e \approx T_i \equiv T_0$ a simple result

$$u_{ey} = -\frac{2\kappa T_0}{nm\omega_{ce}}\left(1 - \frac{D_e}{D_i}\right)\frac{\partial n}{\partial x}.$$

In this case the correction to the electron diamagnetic drift becomes small.

The electron–ion velocity difference becomes

$$u_{iy} - u_{ey} = 2\kappa T_0\left(\frac{\Omega_{ci}}{nM\left(\nu_i^2 + \Omega_{ci}^2\right)}\frac{D_e}{D_i} + \frac{1}{nm\omega_{ce}}\left(1 - \frac{D_e}{D_i}\right)\right)\frac{\partial n}{\partial x}. \tag{3.25}$$

Ignoring terms of the order of D_e/D_i we have $u_{iy} - u_{ey} \approx 2u_{De}$, i.e. twice the electron diamagnetic drift velocity for this particular limit.

- **Exercise:** We argued that the direction of the electric field is solely determined by the difference in mobility of electron and ions, and therefore independent of the direction of the magnetic field. The directions of the electron and ion diamagnetic fields depend on the magnetic field direction. Demonstrate how the different magnetic field directions affect the directions of the velocities and velocity differences discussed in this section.

3.6.3 Unmagnetized plasmas

The present study concerns magnetized plasma, but it might be instructive to give a brief account, for reference, on how to describe diffusion in unmagnetized plasmas. The results also apply for the diffusive motion along magnetic field lines. We write the basic equations assuming quasi-neutrality here

$$qn\mathbf{E} - \nabla p - nM\nu_i\mathbf{u} = 0 \tag{3.26}$$

for the ions, and by replacing the proper symbols we obtain the expression for the electrons. We find

$$n\mathbf{u} = \frac{qn\mathbf{E}}{M\nu_i} + \frac{\kappa T}{M\nu_i}\nabla n,$$

for isothermal processes, which is appropriate for diffusion. We define a particle flux as, e.g. $\mathbf{\Gamma}_i \equiv \pm\mu n\mathbf{E} - D\nabla n$ with $\mu \equiv |q|/(M\nu_i)$ and $D_i \equiv \kappa T/(M\nu_i)$. We have the so-called "Einstein relation" $D_i = (\kappa T/|q|)\mu_i$, where μ_i is the ion mobility. We have $\mu_e \gg \mu_i$. Requiring the electron and ion fluxes to be the same, $\Gamma_i = \Gamma_e$, for maintaining quasi-neutrality as the density gradient is eroded by diffusion, we need $\mu_i n\mathbf{E} - D_i\nabla n = -\mu_e n\mathbf{E} - D_e\nabla n$ giving

$$\mathbf{E} = \frac{D_i - D_e}{\mu_i + \mu_e}\frac{\nabla n}{n}.$$

Using this relation in (3.26) we can write $\mathbf{\Gamma}_i = \mathbf{\Gamma}_e = -D_a\nabla n$ with

$$D_a \equiv \frac{\mu_i D_e + \mu_e D_i}{\mu_i + \mu_e} \approx D_i + \frac{\mu_i}{\mu_e}D_e.$$

using $\mu_e \gg \mu_i$ and $D_e \gg D_i$. Physically this result means that the electrons diffuse to build up an electric field that make the ions diffuse at a faster rate than they would due to their thermal motion alone. A magnetic field changes the physics significantly. Note that the limit $\nabla n \to \infty$ is not well posed in part because we cannot assume quasi-neutrality there. This limitation is important close to an absorbing wall, for instance, where a plasma sheath develops.

References

Bakunin O G 2008 *Turbulence and Diffusion: Scaling Versus Equations* (*Springer Series in Synergetics*) (Berlin: Springer)

Biermann L 1950 Uber den ursprung der magnetfelder auf sternen und im interstellaren raum *Z. Naturforsch Teil* A **5** 65–71

Bohm D 1949 *The Characteristics of Electrical Discharges in Magnetic Fields* (New York: McGraw Hill) p 201

Boström R 1964 A model for the Auroral electrojet *J. Geophys. Res.* **69** 4983–99

Braginskiĭ S I 1965 Transport processes in a plasma ed M A Leontovich *Reviews of Plasma Physics* vol 1 (New York: Consultants Bureau) pp 205–311

Brekke A 2013 *Physics of the Upper Polar Atmosphere* 2nd edn (Berlin: Springer)

Chen F F 1984 *Introduction to Plasma Physics and Controlled Fusion* vol 1, 2nd edn (New York: Plenum)

Garcia O E and Pécseli H L 2013 Models for electrostatic drift waves with density variations along magnetic field lines *Geophys. Res. Lett.* **40** 5565–9

Kelley M C 1989 *The Earth's Ionosphere, Plasma Physics and Electrodynamics,* Vol 43 of *International Geophysics Series* (San Diego, CA: Academic)

Pécseli H L 2012 *Waves and Oscillations in Plasmas* (London: Taylor and Francis)

Pécseli H L 2015 Spectral properties of electrostatic drift wave turbulence in the laboratory and the ionosphere *Ann. Geophysicae* **33** 875–900

Primdahl F, Marklund G and Sandahl I 1987 Rocket observation of E-B-field correlations showing up- and down-going Poynting flux during an auroral breakup event *Planet. Space Sci.* **35** 1287–95

Primdahl F and Spangslev F 1977 Cross-polar cap horizontal E region currents related to magnetic disturbances and to measured electric fields *J. Geophys. Res.* **82** 1137–43

Shukla P K, Stenflo L and Pokhotelov O A 2001 Comment on 'Role of quasineutrality in drift waves' *Phys. Plasmas* **8** 368

Shukla P K, Stenflo L and Pokhotelov O A 2001 *Phys. Plasmas* **8** 3519–21

Simon A 1955a Ambipolar diffusion in a magnetic field *Phys. Rev.* **98** 317–8

Simon A 1955b Diffusion of like particles across a magnetic field *Phys. Rev.* **100** 1557–9

Sivukhin D V 1965 Motion of charged particles in electromagnetic fields in the drift approximation *Reviews of Plasma Physics* vol 1 (New York: Consultants Bureau) pp 1–4

Spitzer L 1956 *Physics of Fully Ionized Gases,* Vol 3 of *Interscience Tracts on Physics and Astronomy* (New York: Interscience)

Stubbe P 1968 Frictional forces and collision frequencies between moving ion and neutral gases *J. Atmospheric Terrestrial Phys.* **30** 1965–85

Taylor J B and McNamara B 1971 Plasma diffusion in two dimensions *Phys. Fluids* **14** 1492–9

Trubnikov B A 1965 Particle interactions in fully ionized plasmas ed M A Leontovich *Reviews of Plasma Physics* vol 1 (New York: Consultants Bureau) pp 105–204

van Kampen N G and Felderhof B U 1967 *Theoretical Methods in Plasma Physics* (Amsterdam: North-Holland)

Part II

Electrostatic fluid models

IOP Publishing

Low Frequency Waves and Turbulence in Magnetized Laboratory Plasmas and in the Ionosphere

Hans Pécseli

Chapter 4

The basic model for drift waves

The geometry of the problem is shown in figure 4.1 giving the directions of the magnetic field and the density gradient. The magnetic field will be assumed homogeneous in all of what follows. Guided by the experimental observations we consider low frequency electrostatic waves well below the ion cyclotron frequency

$$\omega \ll \Omega_{ci}, \tag{4.1}$$

and assume that the waves are propagating in a direction almost perpendicular to the magnetic field so the **B**-parallel component of the wave vector $k_{\parallel}$ is small in such a way that

$$U_{thi} \ll \frac{\omega}{k_{\parallel}} \ll U_{the}, \tag{4.2}$$

in terms of the ion and electron thermal velocities U_{thi} and U_{the}, respectively. In the present summary we assume the Alfvén velocity V_A to be larger than the electron thermal velocity, $V_A > U_{the}$, so the drift waves do not couple to any low frequency electromagnetic modes. In this section we consider electrostatic waves.

We can illustrate an equivalent of assumption (4.2) in a different way: for the electrostatic waves considered here the electric field is parallel to the wavevector: $\mathbf{E} \parallel \mathbf{k}$. For an ion moving along the magnetic field line $\mathbf{B}$ we have $|U_{\parallel}| \approx |(e/M)E_{\parallel}/\omega|$ assuming that the electric field is harmonically varying with a frequency ω. For the ion velocity perpendicular to $\mathbf{B}$ we use $|U_{\perp}| \approx |E_{\perp}/B|$ assuming (4.1). Similar results can be obtained for the electron motion. With $\mathbf{E} = E\hat{\mathbf{k}}$ we can express the inequality for the ion velocity components $U_{\perp} \gg U_{\parallel}$ as $E_{\perp}/B \gg (e/M)E_{\parallel}/\omega$. Concerning the electron velocity we require $U_{e\perp} \ll U_{e\parallel}$ or $E_{\perp}/B \ll (e/m)E_{\parallel}/\omega$. With $\mathbf{E} = E\hat{\mathbf{k}}$ we can write these two inequalities as

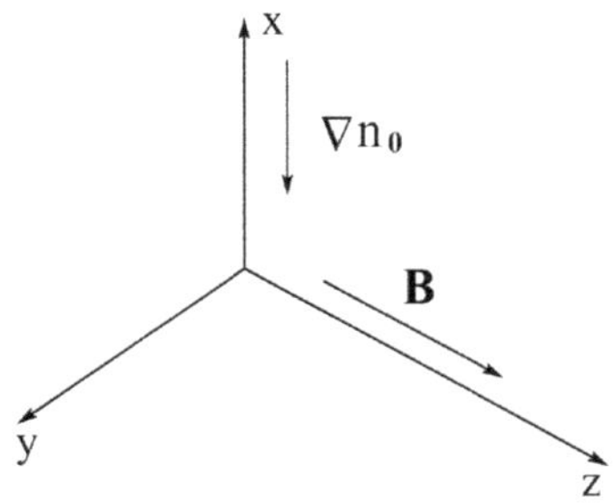

Figure 4.1. Geometry of the problem, showing the density gradient and magnetic field vector.

$$k_\perp \gg k_\parallel \frac{\Omega_{ci}}{\omega} \qquad \text{and} \qquad k_\perp \ll k_\parallel \frac{\omega_{ce}}{\omega},$$

or

$$1 \ll \frac{\Omega_{ci}}{\omega} \ll \frac{k_\perp}{k_\parallel} = \frac{1}{\tan\theta_k} \ll \frac{\omega_{ce}}{\omega},$$

where θ_k is the angle between the **k**-vector and the line normal to **B**. Consistent with (4.2) we require that $\theta_k \ll 1$, but still so that $\tan\theta_k \gg \omega/\omega_{ce}$. Since $\omega_{ce} \gg \Omega_{ci}$ it is easy to fulfill both requirements at the same time. The relation (4.2) is thus a rather mild restriction for many practical situations.

Consistently with (4.1) and (4.2) the ion fluid velocity in the direction $\perp$**B** is approximated by the $\mathbf{E}\times\mathbf{B}/B^2$-velocity, while a **B**-parallel ion velocity component is ignored (Kadomtsev 1965, Chen 1984). The ion continuity equation then becomes

$$\frac{\partial}{\partial t} n_i - \nabla \cdot \left(\frac{\nabla\phi \times \mathbf{B}}{B^2} n_i\right) = 0, \tag{4.3}$$

where the electrostatic potential was introduced, i.e. $\mathbf{E} = -\nabla\phi$. Note that ion inertia does not enter the derivation of (4.3). It is implied that an ion gyro-center (which here also specifies the ion position) assumes the $\mathbf{E}\times\mathbf{B}/B^2$-velocity instantaneously.

The **B**-parallel component of the fluid electron momentum equation gives for that velocity component

$$n_e m\left(\frac{\partial}{\partial t} u_{e\parallel} + \mathbf{u}_e \cdot \nabla u_{e\parallel}\right) = -\kappa T_e \frac{\partial}{\partial z} n_e + e n_e \frac{\partial}{\partial z}\phi, \tag{4.4}$$

where the electron motion is assumed to be isothermal in view of (4.2). The electron inertia effects, i.e. the left-hand side of (4.4) can be neglected, since it is $(\omega/k_z)/U_{the}$ times smaller than the right-hand side. We can substantiate this approximation by normalizing velocities by a characteristic velocity $\mathcal{U}$, potential by $\kappa T_e/e$, time by a characteristic time τ and displacement by a characteristic length $\mathcal{L}$, expecting $\mathcal{L}/\tau \approx \mathcal{U}$. From (4.4) we find after a little algebra the normalized nonlinear equation

$$\frac{m}{\kappa T_e}\frac{\mathcal{L}\mathcal{U}}{\tau}\left(\frac{\partial}{\partial t^{(n)}} u_{e\parallel}^{(n)} + \frac{\tau\mathcal{U}}{\mathcal{L}} \mathbf{u}_e^{(n)} \cdot \nabla^{(n)} u_{e\parallel}^{(n)}\right) = -\frac{1}{n_e}\frac{\partial}{\partial z^{(n)}} n_e + \frac{\partial}{\partial z^{(n)}}\phi^{(n)} \tag{4.5}$$

where the exponent $^{(n)}$ denotes normalized quantities. We anticipate $\mathcal{L}/\tau \approx \mathcal{U} \sim C_s = \sqrt{\kappa T_e/M}$, so the coefficient to the left side of (4.5) becomes approximately $m/M \ll 1$. In order to have the right hand side small as well we require

$$\frac{\partial}{\partial z}\ln n_e \approx \frac{e}{\kappa T_e}\frac{\partial}{\partial z}\phi, \tag{4.6}$$

which corresponds to assuming isothermally Boltzmann distributed electrons. This assumption is independent of the assumption of quasi-neutrality, although both are often made simultaneously.

Physically we can argue that the finite electron pressure attempts to 'push' electrons away from the wave crests by moving along magnetic field lines, thereby creating a slight deficit of electrons there as compared to the local ion density, while the wave troughs, on the other hand will be slightly 'overpopulated' by electrons. This process continues until electric fields have been built-up, which arrests further electron motion. These electric fields are usually called *ambipolar* electric fields. We can estimate their intensity by arguing that the electric field force $-en_eE = en_e\partial\phi/\partial z$ must balance the electron pressure $-\kappa T_e \partial n_e/\partial z$, giving $e\phi/\kappa T_e \approx \ln n_e$, i.e. precisely the relation for Boltzmann distributed electrons argued before. The ambipolar electric fields will subsequently set the ions into motion so that low frequency ion waves can propagate even though the ion temperature is negligible. The time it takes to build-up the ambipolar electric field is approximately $\lambda_{\|}/U_{the}$ which can be estimated as $(\lambda_{\|}/\lambda_{De})/\omega_{pe}$. This we assume to be a short time in the following, at least compared to $\lambda_{\|}/C_s$.

Now, strictly speaking, you cannot take the logarithm of a dimensional quantity, here the density n_e. It has to be normalized somehow, and as long as the normalization parameter is independent of z, the relation (4.6) can be made correct. We use the x-dependent unperturbed density, which by integration of (4.6), gives the isothermal Boltzmann relation

$$n_e = n_0(x)\exp(e\phi/\kappa T_e), \tag{4.7}$$

where we here take T_e to be a constant. The reason for choosing this particular normalization in (4.6) is that it gives $n_e = n_0$ when the plasma is unperturbed, $\phi = 0$. A spatially and temporally varying electrostatic potential implies that $n_e(\mathbf{r}, t) = n_0(x) + n_1(\mathbf{r}, t)$ For small perturbations, we can linearize (4.7) to give $n_1(\mathbf{r}, t)/n_0(x) \approx e\phi(\mathbf{r}, t)/\kappa T_e$.

Physically, (4.7) implies that the electrons can flow along the magnetic field lines from wave crest to wave trough in order to establish an isothermal Boltzmann equilibrium in agreement with the inequality (4.2). The process is quasi-static in the sense that the electrons can be assumed to adjust instantaneously when seen on the slow time-scale (4.1) characterizing the ion dynamics. The coefficient $n_0(x)$ in (4.7) can be understood as an imposed boundary condition at a position where $\phi = 0$. This is a natural interpretation in connection with e.g. Q-machine experiments where the plasma is produced by surface ionization on a hot plate with good electrical conductivity, but with an inhomogeneous radial temperature which implies a corresponding variation in the plasma density.

Finally, Poisson's equation becomes redundant by the assumption of quasi-neutrality, $n_e \approx n_i \equiv n$, an approximation which can be readily justified for perturbations with scale-length much larger than the electron Debye-length, $\lambda_D \equiv (\varepsilon_0 \kappa T_e / e^2 n_e)^{1/2}$. In this limit only the bulk plasma density n will enter the analysis. Implicitly it is assumed that also $\omega \ll \Omega_{pi}$, the ion plasma frequency. Noting that an exponential coefficient $\exp(e\phi/\kappa T_e)$ cancels in each term, the equations (4.3) and (4.7) are then readily combined to

$$\frac{\partial}{\partial t}\phi - \left(\frac{\kappa T_e}{eBn_0(x)}\frac{\mathrm{d}n_0(x)}{\mathrm{d}x}\right)\frac{\partial}{\partial y}\phi = 0. \tag{4.8}$$

Recalling the general vector relations $(\mathbf{a} \times \mathbf{b}) \cdot \mathbf{c} = (\mathbf{c} \times \mathbf{a}) \cdot \mathbf{b} = (\mathbf{b} \times \mathbf{c}) \cdot \mathbf{a}$ we can introduce $\nabla n_0(x) \equiv (\mathrm{d}n_0(x)/\mathrm{d}x)\hat{\mathbf{x}}$ with $\hat{\mathbf{x}}$ being a unit vector and write $(\nabla\phi \times \mathbf{B}) \cdot \hat{\mathbf{x}} = (\mathbf{B} \times \hat{\mathbf{x}}) \cdot \nabla\phi = (B\, \partial\phi/\partial y)\,\hat{\mathbf{x}}$ and find in vector notation

$$\frac{\partial}{\partial t}\phi - \frac{\kappa T_e}{e}\frac{\nabla\phi \times \mathbf{B}}{B^2} \cdot \frac{\nabla n_0(x)}{n_0(x)} = 0. \tag{4.9}$$

The ion motion is incompressible, since $\nabla \cdot (\nabla\phi \times \mathbf{B}) = 0$ with constant $\mathbf{B}$. Nevertheless, the fluctuations are associated with variations in plasma density, since the ions move in and out along a density gradient, which is here taken in the $\hat{\mathbf{x}}$-direction, see figure 4.1.

We can take a plane wave model for the potential variation as $\phi = \phi_0(x) \exp(-i\omega t + ik_y y)$ and by (4.8) obtain an x-dependent dispersion relation in the form $\omega = U_{De}(x)k_y$. Equation (4.8) thus shows that all perturbations propagate without change in shape in the $\hat{\mathbf{y}}$-direction $\perp \nabla n_0$ and $\mathbf{B}$, with the electron diamagnetic drift velocity

$$u_{De} = -\frac{\kappa T_e}{eB}\frac{1}{n_0(x)}\frac{\mathrm{d}n_0(x)}{\mathrm{d}x} \equiv -\frac{C_s^2}{\Omega_{ci}}\frac{1}{n_0(x)}\frac{\mathrm{d}n_0(x)}{\mathrm{d}x}. \tag{4.10}$$

The perturbations in one x-position do not affect the perturbations at other x-positions (Manheimer 1977) since no derivatives $\partial\phi/\partial x$ appear, even though $\phi = \phi(x, y, z, t)$, with the restriction implied in the initial assumptions where we implicitly assumed the z-variation to have a long wavelength.

Note that the result (4.8) is linear in $\phi(x)$, although no linearization assumptions were actually made! It could have been expected that the exponential in the Boltzmann distribution for electrons would give rise to a strong nonlinearity, but as it is found, all these exponentials cancel out.

In a sense we might be surprised to see the electron diamagnetic drift appears: this drift (and its associated diamagnetic current) is a phenomenon associated with plasma dynamics strictly perpendicular to $\mathbf{B}$ and we have nowhere taken this limiting case, on the contrary, see the discussion of (4.2). From basic dimensional reasoning (see for instance appendix E) we can, however, argue that if we need a characteristic velocity, this is the only combination available given the characteristic quantities we have. These parameters are the thermal energy for the electrons κT_e

with dimension $[\kappa T_e] = ML^2/T^2$ (recall that the ions were taken to be cold), the elementary charge e with dimension $[e] = AT$, the magnetic field with dimension $[B] = M/(AT^2)$, and the length scale $\mathcal{L} \equiv n_0/n_0'$ with dimension $[n_0/n_0'] = L$. The ion mass, for instance, does not appear because the $\mathbf{E} \times \mathbf{B}/B^2$ gyro-center motion ignores ion inertia. We can write the dimension matrix as

	κT_e	e	B	$\mathcal{L}$
T	-2	1	-2	0
L	2	0	0	1
M	1	0	1	0
A	0	1	-1	0

It is readily seen (here left to the reader as an exercise) that the only dimensionally correct combination giving a velocity is the electron diamagnetic drift. A numerical constant cannot be accounted for this way, but according to (4.10) it is unity. By introducing an 'effective' ion Larmor radius $a_i \equiv C_s/\Omega_{ci}$, with $C_s^2 \equiv \kappa T_e/M$, and a 'scale length' for the density gradient, $\mathcal{L} \equiv n_0/|dn_0/dx|$, we can interpret the magnitude of the electron diamagnetic drift as being the ion sound speed weighted by the ratio of a microscopic and macroscopic length scale, $|U_{De}| = C_s(a_i/\mathcal{L})$. Also the diamagnetic drift is divergence free, $\nabla \cdot \mathbf{u}_{De} = 0$, for constant and homogeneous magnetic fields.

Electrostatic drift waves appear as if propagated by an electron diamagnetic drift that is supported by particles that are not moving on average, see figure 3.6. This is in principle not different from surface waves on water, for instance. In this latter case the waves are also propagating while small parcels of fluid are performing a circular motion. This can be seen most easily by letting a small piece of cork, or similar, float on the water surface and watch it while a surface wave propagates by.

We note that it was important for the derivation of (4.9) that the ion $\mathbf{E} \times \mathbf{B}/B^2$-motion was incompressible. This requires $\mathbf{B}$ to be a constant vector, which again implies that the β-value of the plasma must be low. If $\mathbf{B} = \mathbf{B}(\mathbf{r})$, then the ion motion is no longer incompressible, $\nabla \cdot (\nabla\phi \times \mathbf{B}) \neq 0$, and the analysis requires non-trivial modifications.

Although the result (4.9) has obvious shortcomings in many respects, the basic arguments can, on the other hand, not be *entirely* wrong, and consequently we expect (4.9) to be the 'backbone' of any more detailed mathematical description of electrostatic drift waves, be it linear or nonlinear.

- **Exercise:** Consider a simplified model, where Poisson's equation is retained but the ion velocity is still identified by the $\mathbf{E} \times \mathbf{B}/B^2$-velocity, ignoring any motion parallel to $\mathbf{B}$. Derive a nonlinear dynamic equation for the electrostatic potential by use of the ion continuity equation (5.1), the assumption of isothermal Boltzmann distribution for the electrons (4.7). Discuss the resulting equation, and estimate the relative magnitude of the various terms for different scale-lengths, measured with respect to the electron Debye length.
- **Exercise:** Estimate the length scale $\mathcal{L} \equiv n/|\mathrm{d}n/\mathrm{d}x|$ required to make the electron diamagnetic drift comparable to the ion sound speed.

4.1 Physical picture

It might be advantageous to present the results in the foregoing section in the form of a physical picture, following (Chen 1965b). Thus figure 4.2 illustrates the basic elements of the derivation. A solid line gives equi-density contour of the perturbation and because of (4.7) the equi-potential lines are in phase with this. This will remain correct even without linearizing (4.7). At position ② in figure 4.2 the ion gyro-center velocity is maximum while the density perturbation, n, is vanishing here. A little later, particles from the denser region of the initial distribution have been brought in and n becomes positive at position ②. Similarly, at position ④ the density will decrease and the perturbation moves upward in figure 4.2 in the same direction as the diamagnetic drift. To find the magnitude of the phase velocity we proceed as when deriving (4.8) to find (4.10). The average value of the ion cross-field velocity in the region of density increase (shaded region) is zero, just as in the region of density depletion; the waves are marginally stable, and there is no net plasma transport along the density gradient.

One important modification from this simplified description comes from charge separation effects we did not take into account. First, since E_y is time-varying, ion inertia will cause the ions to drift in the y-direction as well as the x-direction. Second, since the ions have a much larger Larmor radius than the electrons, and since E_y varies in space, the ions will see an average E_y smaller than what the electrons see. The $\mathbf{E} \times \mathbf{B}$ drift will be smaller for ions, and there will be a charge separation because this drift is along ∇n_0. This finite-Larmor-radius (FLR) effect will be covered in a later section. These two small effects both tend to cause a charge

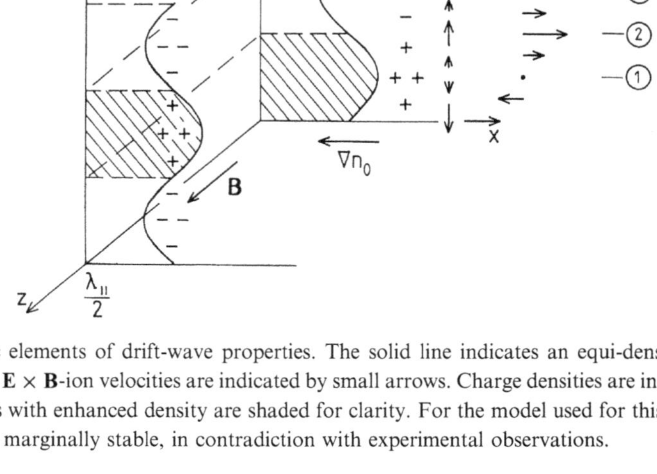

Figure 4.2. Basic elements of drift-wave properties. The solid line indicates an equi-density contour, while electric fields and $\mathbf{E} \times \mathbf{B}$-ion velocities are indicated by small arrows. Charge densities are indicated by + and − symbols. Regions with enhanced density are shaded for clarity. For the model used for this figure we find the drift-waves to be marginally stable, in contradiction with experimental observations.

build-up and hence an additional electric field E_y. The resulting equation will now also contain derivatives of ϕ with respect to x. However, if k_z is large enough, the flow of electrons along B will cancel the charges building-up so as to satisfy the Boltzmann relation (4.7), and the waves remain linearly stable.

4.2 Instability of small amplitude drift waves

The previous basic analysis accounts for some basic drift-wave properties, but not their instability. A detailed analysis is needed to account for this but some basic features can be understood from basic principles, and the consequences of a growth rate of the waves can be discussed in some detail (Manheimer 1977). Referring to the geometry of the problem, see figure 4.1, we thus obtain the x-component of the fluctuating ion density flux $n_i\mathbf{u}$, and integrate it over one wavelength along the electron diamagnetic drift direction y as

$$\Gamma_i \equiv \frac{1}{\lambda_y}\int_0^{\lambda_y} n_i u_{ix}\,\mathrm{d}y \tag{4.11}$$

$$= -\frac{1}{\lambda_y}\int_0^{\lambda_y} n_i \frac{1}{B}\frac{\partial\phi}{\partial y}\,\mathrm{d}y, \tag{4.12}$$

where $\lambda_y \equiv 2\pi/k_y$. We ignore contributions from products like $n_0\mathbf{u}$, since they trivially average to zero. Linearizing (4.3) we have

$$\frac{\partial}{\partial t}n_i - \frac{\nabla\phi \times \mathbf{B}}{B^2}\cdot \nabla n_0(\mathbf{r}) = 0,$$

or

$$\frac{\partial n_i}{\partial t} - \frac{\partial\phi}{\partial y}\frac{n_0'(x)}{B} = 0,$$

and find

$$n_i(\mathbf{r}, t) = \frac{k_y}{\omega}\frac{1}{B}\frac{\partial n_0}{\partial x}\phi_0(x)\exp\big(-\mathrm{i}(\omega t - k_y y)\big) + \text{c.c.} \tag{4.13}$$

for a plane wave with potential $\phi(\mathbf{r}, t) = \phi_0(x)\exp(-\mathrm{i}(\omega t - k_y y)) + \text{c.c.}$ Assume that we have $\omega \equiv \omega_r + i\gamma$, without here accounting for the physical origin of γ. With this assumption we reduce (4.12) to

$$\Gamma_i = -\frac{2\gamma}{\omega_r^2 + \gamma^2}k_y^2\frac{|\phi_0|^2}{B^2}\frac{\partial n_0}{\partial x}.$$

From this result we learn that the sign of Γ_i is determined by the product $\gamma\,\partial n_0/\partial x$. Any growth or damping of the waves gives rise to a net ion flux. If $\gamma > 0$ we find the ion flux to be in the direction of *decreasing* density, opposite for the case where the waves are damped. Intuitively, this might be precisely what we would expect for an unstable wave on a density gradient. A complex ω implies a phase shift between n_i

and ϕ as seen from (4.13). Consequently we might search for the origin of an instability in physical mechanisms that give rise to phase differences between density and potential. The criterion of quasi-neutrality implies that the electron and ion fluxes must be equal, and the particle fluxes together with the damping/growth-rates must adjust to this.

We have here seen that an imaginary part of ω gives rise to a net plasma transport in the direction of decreasing plasma density. A heuristic argument will provide a collisional mechanism that gives rise to the need for net plasma diffusion to be compensated by a corresponding imaginary part of ω corresponding to wave growth.

4.3 Spatially varying magnetic fields

It was important for the derivation of (4.9) that the ion $\mathbf{E} \times \mathbf{B}/B^2$-motion was incompressible. This requires $\mathbf{B}$ to be a constant vector, which again implies that the β-value of the plasma must be low. If $\mathbf{B} = \mathbf{B}(\mathbf{r})$, then the ion motion is no longer incompressible and the analysis requires non-trivial modifications.

We will briefly discuss the consequences of $\mathbf{B} = \mathbf{B}(\mathbf{r})$. In this case we find by simplifying $\nabla\phi \cdot \nabla \times \mathbf{B}/B^2$

$$\nabla \cdot \frac{\nabla\phi \times \mathbf{B}}{B^2} = -\frac{2}{B^3}(\nabla\phi \times \mathbf{B}) \cdot \nabla B \neq 0.$$

We assumed that $\beta \ll 1$ so that the magnetic field can be approximated with the value it would have in the absence of plasma, which implies $\nabla \times \mathbf{B} = 0$. In the same way we find that the diamagnetic drift is also compressible

$$\nabla \cdot \frac{\nabla p_e \times \mathbf{B}}{B^2} = -\frac{2}{B^3}\left(\nabla p_e \times \mathbf{B}\right) \cdot \nabla B \neq 0.$$

It is very tempting here to assume something like $\{B_x, B_y, B_z\} = \{0, 0, B_0(1 + x/\ell)\}$ but such a magnetic field will violate the assumption of $\nabla \times \mathbf{B}(\mathbf{r}) = 0$, so a better model is needed! Now also the model for $\nabla n_0(\mathbf{r})$ becomes more complicated, since we will in general assume that the equi-density contours follow magnetic flux surfaces. At first sight it might seem surprising that the diamagnetic flow is compressible for $\mathbf{B} = \mathbf{B}(\mathbf{r})$, but we should recall that with inhomogeneous magnetic fields, we have true particle drifts as discussed in section 2.1.7, while the standard diamagnetic drift for $\mathbf{B} =$ const. has no gyro-center motions associated.

- **Exercise:** Take the ion continuity equation $\partial n/\partial t + \nabla \cdot (\mathbf{u}\, n) = 0$, and assume Boltzmann distributed electrons as in (4.7). Assume that to a good approximation we have the ion velocity $\mathbf{u} = -\nabla\phi \times \mathbf{B}/B^2$ now with $\mathbf{B} = \mathbf{B}(\mathbf{r})$. Derive the local differential equation for ϕ. Obtain the dispersion relation for this case.

  For solving the problem, we have the ion continuity equation

  $$\partial n/\partial t + \nabla \cdot (\mathbf{u}\, n) = \partial n/\partial t + n\nabla \cdot \mathbf{u} + \mathbf{u} \cdot \nabla n = 0$$

  and Boltzmann distributed electrons $n_e = n_0(\mathbf{r}) \exp(e\phi/\kappa T_e)$. We have here

$$\nabla \cdot \frac{\nabla p_e \times \mathbf{B}}{B^2} = -\frac{2}{B^3}\left(\nabla p_e \times \mathbf{B}\right) \cdot \nabla B.$$

With the quasi-neutrality approximation $n_e \approx n_i = n$ and $\mathbf{u} = -\nabla\phi \times \mathbf{B}/B^2$ we find

$$\frac{\partial}{\partial t}\phi - \frac{\kappa T_e}{e}\frac{\nabla\phi \times \mathbf{B}}{B^2} \cdot \frac{\nabla n_0(\mathbf{r})}{n_0(\mathbf{r})} - 2\frac{\kappa T_e}{eB^3}\nabla\phi \times \mathbf{B} \cdot \nabla B(\mathbf{r}) = 0.$$

The local dispersion relation can be written in a compact form as

$$\omega + \frac{\kappa T_e}{e}\frac{\mathbf{k} \times \mathbf{B}}{B^2} \cdot \left(\frac{\nabla n_0(\mathbf{r})}{n_0(\mathbf{r})} + 2\frac{\nabla B(\mathbf{r})}{B(\mathbf{r})}\right) = 0.$$

Note that even for homogeneous plasmas, $\nabla n_0 = 0$, we have propagating oscillations due to the magnetic field inhomogeneity.

References

Chen F F 1965 Resistive overstabilities and anomalous diffusion *Phys. Fluids* **8** 912–9

Chen F F 1984 *Introduction to Plasma Physics and Controlled Fusion* vol 1 2nd edn (New York: Plenum)

Kadomtsev B B 1965 *Plasma Turbulence* (New York: Academic)

Manheimer W M 1977 *An Introduction to Trapped-Particle Instability in Tokamaks* ERDA Critical Rev. Ser., National Technical Inf. Serv. Rep. TID–27157

Chapter 5

Simplified linear wave analysis

The result of the foregoing section concerning the locality of the drift wave fluctuations seemed somewhat peculiar. This is in a certain respect a consequence of ignoring finite ion inertia as will be demonstrated in the following. Along with ion inertia we also retain a **B**-parallel ion motion. These modifications of the foregoing analysis will imply that the fluctuations become non-local, i.e. the dynamic equation contains derivatives of all spatial variables, in general. This generalization of the analysis is achieved at the expense of an explicit linearization of the basic equations. The model discussed here will refer to plane geometry. The experimentally more relevant cylindrical geometry has been studied by, for instance, Ellis *et al* (1980).

The generalization of the simple results given before assumes that the basic fluid equations are the ion continuity equation

$$\frac{\partial n}{\partial t} + \nabla \cdot (n\mathbf{u}) = 0, \tag{5.1}$$

and the momentum equation for the cold ions

$$M\left(\frac{\partial}{\partial t}\mathbf{u} + \mathbf{u} \cdot \nabla\mathbf{u}\right) = e(-\nabla\phi + \mathbf{u} \times \mathbf{B}). \tag{5.2}$$

This latter equation is rewritten in a more suitable form by taking the cross-product with **B**

$$\mathbf{u}_{\perp} = -\frac{\nabla\phi \times \mathbf{B}}{B^2} - \frac{M}{eB^2}\left[\frac{\partial}{\partial t}(\mathbf{u} \times \mathbf{B}) + \mathbf{u} \cdot \nabla(\mathbf{u} \times \mathbf{B})\right], \tag{5.3}$$

doi:10.1088/978-0-7503-1251-6ch5

where it was explicitly used that **B** is a constant vector. The term in angular brackets is the inertial drift. The **B**-parallel component of (5.2) remains

$$M\left(\frac{\partial}{\partial t}u_{\parallel} + \mathbf{u}\cdot\nabla u_{\parallel}\right) = -e\frac{\partial}{\partial z}\phi. \tag{5.4}$$

The set of equations is completed with (4.7) for isothermally Boltzmann distributed electrons. With the assumption of quasi-neutrality $n_e \approx n_i \equiv n$ there is no distinction between ion and electron density. The ion thermal motion will, for the time being, be ignored and an ion pressure term is omitted from (5.2). According to (5.2) the inertial effects are determining the **B**-parallel ion velocity component, $u_{\parallel}$. It is logically possible to assume that the **B**-perpendicular ion fluid velocity is still given by the $\mathbf{u}_{\mathbf{E}\times\mathbf{B}} = \mathbf{E}\times\mathbf{B}/B^2$-velocity. For generality we retain the effects of ion inertia also in the **B**-perpendicular direction, at least to lowest order by including the corresponding lowest order correction to $\mathbf{U}_{\mathbf{E}\times\mathbf{B}}$, namely the ion polarization drift. This is achieved by iterating once the **B**-perpendicular part of (5.3), to give

$$\mathbf{u}_{\perp} = -\frac{\nabla\phi\times\mathbf{B}}{B^2} - \frac{M}{eB^2}\left(\frac{\partial}{\partial t} - \frac{1}{B^2}\nabla\phi\times\mathbf{B}\cdot\nabla\right)\nabla_{\perp}\phi. \tag{5.5}$$

For consistency with the iteration scheme, both terms in the brackets must be small, i.e. $\nabla_{\perp}\phi$ must be slowly varying in time as well as space, with the subscript $\perp$ denoting directions perpendicular to **B**. It is implicit in the derivation that ion polarization effects are to be considered as small corrections to the $\mathbf{U}_{\mathbf{E}\times\mathbf{B}}$-drifts. With the second term in (5.5) retained, the ion motion perpendicular to the magnetic field is no longer incompressible, i.e. $\nabla_{\perp}\cdot\mathbf{u}_{\perp} \neq 0$. Inserting (4.7) with the assumption of quasi-neutrality into (5.1), the plasma density can be eliminated to give

$$\frac{\partial}{\partial t}\phi + \frac{\kappa T_e}{e}\nabla\cdot\mathbf{u} + \mathbf{u}\cdot\nabla\phi + \frac{\kappa T_e}{en_0(x)}\frac{dn_0(x)}{dx}\mathbf{u}\cdot\hat{\mathbf{x}} = 0, \tag{5.6}$$

with $\hat{\mathbf{x}}$ being a unit vector in x-direction, and $\mathbf{u} = \{\mathbf{u}_{\perp}, u_{\parallel}\}$ is the ion fluid velocity vector.

5.1 Linear dispersion relation

The set of nonlinear differential equations (5.4), (5.5) and (5.6) can be solved in particular cases only. In order to obtain a dispersion relation for drift waves, the equations are linearized, with stationary solutions being $\phi = 0$ and $\mathbf{u} = 0$. The linearized equations are

$$M\frac{\partial}{\partial t}u_{\parallel 1} = -e\frac{\partial}{\partial z}\phi_1, \tag{5.7}$$

$$\mathbf{u}_{\perp 1} = -\frac{\nabla_{\perp}\phi_1\times\mathbf{B}}{B^2} - \frac{M}{eB^2}\frac{\partial}{\partial t}\nabla_{\perp}\phi_1, \tag{5.8}$$

and

$$\frac{\partial}{\partial t}\phi_1 + \frac{\kappa T_e}{e}\left(\nabla_\perp \cdot \mathbf{u}_{\perp 1} + \frac{\partial}{\partial z}u_{\parallel 1}\right) + \frac{\kappa T_e}{e n_0(x)}\frac{\mathrm{d}n_0(x)}{\mathrm{d}x}\mathbf{u}_{\perp 1}\cdot\hat{\mathbf{x}} = 0. \tag{5.9}$$

These equations can be combined to give

$$\begin{aligned}&\frac{\partial^2}{\partial t^2}\left(\phi_1 - C_s^2\frac{1}{\Omega_{ci}^2}\nabla_\perp^2\phi_1\right) - C_s^2\frac{\partial^2}{\partial z^2}\phi_1\\ &-\frac{C_s^2}{\Omega_{ci}}\frac{n_0'(x)}{n_0(x)}\left(\frac{\partial^2}{\partial y\partial t}\phi_1 + \frac{1}{\Omega_{ci}}\frac{\partial^3}{\partial x\partial t^2}\phi_1\right) = 0,\end{aligned} \tag{5.10}$$

where we introduced the abbreviation $n_0'(x)$ for $\mathrm{d}n_0(x)/\mathrm{d}x$. A plane wave perturbation in the form

$$\phi_1(x, y, z, t) = \phi(x)\mathrm{e}^{\mathrm{i}(k_y y + k_z z - \omega t)} \tag{5.11}$$

is inserted into (5.10). The result is

$$\begin{aligned}&\frac{\omega^2 C_s^2}{\Omega_{ci}^2}\left(\frac{\mathrm{d}^2\phi}{\mathrm{d}x^2} + \frac{n_0'(x)}{n_0(x)}\frac{\mathrm{d}\phi}{\mathrm{d}x}\right)\\ &-\left(\omega^2 + \frac{\omega^2 C_s^2}{\Omega_{ci}^2}k_y^2 - C_s^2 k_z^2 + \frac{\omega C_s^2}{\Omega_{ci}}\frac{n_0'(x)}{n_0(x)}k_y\right)\phi = 0.\end{aligned} \tag{5.12}$$

Equation (5.12) is a second-order differential equation for $\phi(x)$ with variable coefficients and must be solved together with some boundary conditions, for instance taking $\phi = A_1$ and $\phi = A_2$ at the boundaries $x = R_1$ and $x = R_2$. Other terms would have appeared if we had used cylindrical rather than rectilinear coordinates. The eigenfunctions of ϕ give the variation of wave amplitude with x, and the eigenvalues ω give the frequency. The frequencies are real; in the basic equations used so far there are no mechanisms which make drift waves unstable in a linearized description.

If n_0 is some arbitrary function of x, obtained from an experiment for instance, it is clear that the wave equation, (5.12) must be solved numerically for general cases. However, there are several ways to get an approximate analytic result (Chen 1970). First, it is sometimes possible by a suitable approximation for the function $n_0(x)$ to reduce the wave equation to a standard form, so that the solution $\phi(x)$ can be written in terms of Whittaker's functions or Hermite polynomials, etc. Second, the use of a 'quadratic form' allows one to estimate $\omega(k)$ by using a trial function $\phi(x)$. Third, by a suitable transformation the wave equation can sometimes be put into the form

$$W'' + Q(x)W = 0, \tag{5.13}$$

which can be solved by the Wentzel–Kramers–Brillouin (WKB) approximation. The latter approximation is, in principle, valid only if $\partial\phi/\partial x \gg \partial\phi/\partial y$. Since, in practice, the opposite is usually true, it is a better approximation to neglect x-derivative $\partial/\partial x$

altogether (Chen 1970). This amounts to simply using the equation for the turning points of (5.13). In the same spirit one can neglect the derivatives of ϕ in (5.12). What remains gives an algebraic equation of $\omega(k)$, which can be interpreted as the 'local' dispersion relation, i.e. local at the position x. This is the simplest approximation and is almost universally employed also for the linearly unstable case considered later. We consequently here assume $\mathrm{d}n_0(x)/\mathrm{d}x/n_0(x) = \mathrm{d}\ln n_0(x)/\mathrm{d}x$ to be constant. This assumption is somewhat restrictive, of course, but it is not unphysical. It can be realized by an exponential density profile, $n_0(x) = N\exp(-x/L)$, where L is a scale-length.

We can separate the arguments concerning (5.12) into two steps:

$$\left(\omega^2 + \frac{\omega^2 C_s^2}{\Omega_{ci}^2}k_y^2 - C_s^2 k_z^2 + \frac{\omega C_s^2}{\Omega_{ci}}\frac{n_0'(x)}{n_0(x)}k_y\right)\phi = 0,$$

i.e. the local relation which leaves no constraints on $\phi(x)$, and then

$$\frac{\mathrm{d}^2\phi}{\mathrm{d}x^2} + \frac{n_0'(x)}{n_0(x)}\frac{\mathrm{d}\phi}{\mathrm{d}x} \approx 0.$$

This is a homogeneous special case of the more general Sturm–Liouville equation

$$f(x)\frac{\mathrm{d}^2\phi}{\mathrm{d}x^2} + f'(x)\frac{\mathrm{d}\phi}{\mathrm{d}x} + f_0(x)y = g(x),$$

which has been studied in detail. A remaining constraint is that $\phi(x)$ should have constant asymptotes (possibly two different ones) as $x \to \infty$ and $x \to -\infty$, implying that the electric fields vanish there. Apart from this constraint we have no information on $\phi(x)$ in the present local approximation. We shall not discuss methods for estimating the eigenfunction $\phi(x)$ in any more detail here, but it might be worth emphasizing that the problem deserves more attention than it has received in the literature.

The local dispersion relation obtained by the foregoing arguments is readily obtained from (5.12) as

$$\omega^2\left(1 + a_i^2 k_y^2\right) - \omega k_y u_{De} - C_s^2 k_z^2 = 0, \tag{5.14}$$

where $C_s = \sqrt{\kappa T_e/M}$ is the ion sound speed, u_{De} was defined in (4.10) and apart from a factor $\sqrt{2}$ the quantity $a_i = C_s/\Omega_{ci}$ can be interpreted as an effective ion Larmor radius obtained as if the ions had the temperature of the electrons. With the present exponential density profile we have $u_{De} = C_s^2/(\Omega_{ci}L)$, independent of the spatial coordinate x. The y-component of the wave vector is perpendicular to the density gradient as well as the magnetic field. The second term in the parenthesis in (5.14) originates from the ion polarization, i.e. ion inertia, and the last term from **B**-parallel ion motion, which is again inertial. (Inertial terms are here easily recognized by explicitly containing the ion mass M.) For large k_z the dispersion relation approaches the one for ion acoustic waves. One branch, propagating in the direction of the electron diamagnetic drift has $|\omega/k_z| > C_s$ and is called 'accelerated sound', the one propagating in the opposite direction with $|\omega/k_z| < C_s$ is the 'decelerated sound' branch (Kadomtsev 1965). In the limit of $k_z = 0$ and $a_i = 0$ the dispersion relation for the fast branch becomes identical to that obtained from (4.8). The slow branch propagates in the direction opposite to the electron diamagnetic drift. (Since ion temperature was ignored in the derivation, we have no ion-diamagnetic drift.) It is customary to introduce the ion drift wave frequency $\omega^* = k_y u_{De}$, and the dispersion relation can then, with $a_i = 0$, be visualized in terms of the two variables ω/ω^* and $k_z C_s/\omega^*$ as shown in figure 5.1, where we defined $\omega^* = k_y u_{De}$. Retaining ion polarization, we obtain a small dispersion of drift waves, most important in the limit of $k_z \to 0$ where $\omega = k_y u_{De}/(1 + a_i^2 k_y^2) \approx k_y u_{De}(1 - a_i^2 k_y^2)$.

It was assumed in the foregoing that the propagation vector was in the y, z-plane. It is of course possible to allow for a k_x-component of the wavevector also, having an open boundary at, say, $x \sim \infty$. This problem is however only consistent with the assumptions for the strictly exponential density profile, and will not be discussed here.

5.2 Physical picture

The results in the foregoing section can be presented in the form of a physical picture, following Chen (1965*b*). Thus figure 5.2 illustrates the basic elements of the derivation. A solid line gives the equi-density contour of the perturbation as in figure 4.2 and because of (4.7) the equi-potential lines are in phase with this here as well. As indicated, the ion velocities deviate in general from the $\mathbf{E} \times \mathbf{B}/B^2$ velocities because of the finite ion inertia, i.e. the ion polarization drifts, see section 2.1.5. If the electric fields were steady, an ion would drift with the local $\mathbf{E} \times \mathbf{B}$ velocity. Since, however, E_y is fluctuating the ions will experience an acceleration in the x-direction. The ion inertia opposes this acceleration and this can be expressed as an equivalent or effective gravitational force $M\mathbf{g} = -M\partial\mathbf{u}/\partial t$. This in turn induces a drift $\mathbf{u} = (M/eB^2)\mathbf{g} \times \mathbf{B}$ giving

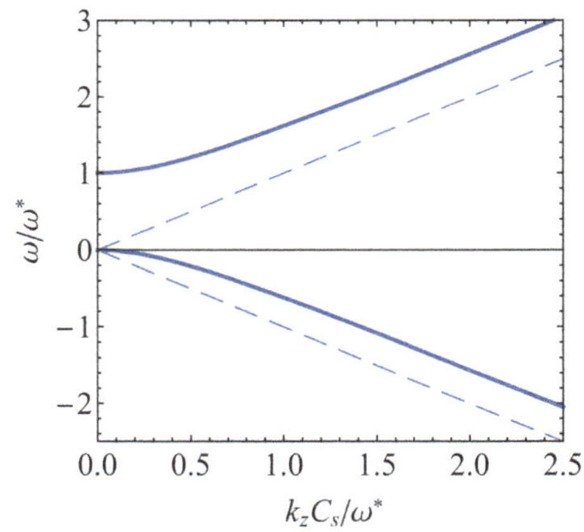

Figure 5.1. Dispersion relation for electrostatic drift waves with the assumption of cold ions with $\omega^* = k_y u_{De}$ being the drift frequency. The dashed lines give the sound wave dispersion relations. To simplify the presentation, we ignored here the ion polarization terms, i.e. assumed $a_i = 0$.

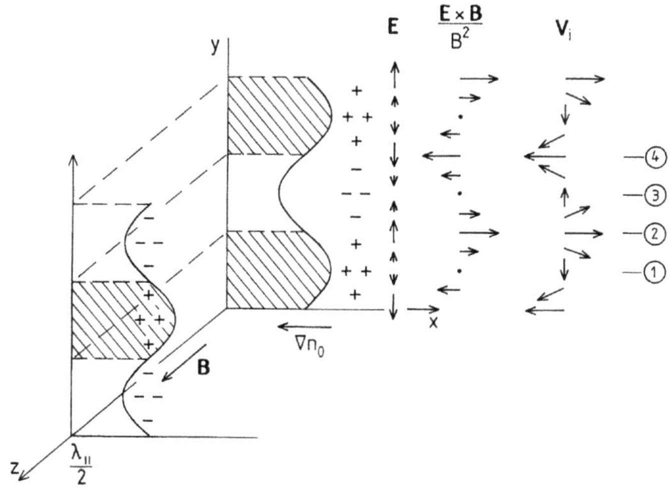

Figure 5.2. Basic elements of drift-wave properties with ion-inertia included. The solid line indicates an equi-density contour, while electric fields and ion velocities are indicated by small arrows. Charge densities are indicated by + and − symbols. Regions with enhanced density are shaded for clarity. If ion polarization drifts were to be ignored, we would find the ion velocity vectors to be the same as the $\mathbf{E} \times \mathbf{B}/B^2$ vectors.

$$u_y = \frac{M}{eB}\frac{\partial u_x}{\partial t} = \frac{M}{eB^2}\frac{\partial E_y}{\partial t} \rightarrow -i\frac{\omega}{\Omega_{ci}}\frac{E_y}{B}.$$

At position ② in figure 5.2 the electric field E_y is maximum and hence $\partial E_y/\partial t = 0$ and u_y then vanishes as well. At a position ① on the other hand u_x vanishes but not u_y and the density perturbation differs from the case where ion inertia is neglected. Note that also in this case the average value of the ion cross-field velocity in the region of density increase (shaded region) is zero, just as it is in the region of density depletion.

5.2.1 The electron velocity

The *electron* fluid velocity was nowhere considered in the discussions of the simplified linear theory. The *z*-component (4.4) of the electron moment equation gives the Boltzmann distributed electrons, and this was all the information we needed. The bulk electron fluid velocity can be determined nonetheless, and can have interest for a detailed understanding of the physical processes. To the present accuracy, the **B**-perpendicular component of the electron momentum equation is

$$-\kappa T_e \nabla_\perp n + en\nabla_\perp \phi - en\mathbf{u}_{e\perp} \times \mathbf{B} \approx 0, \tag{5.15}$$

in the limit of negligible electron inertia. To obtain an unperturbed steady state solution, we assume that there is no steady state electric field and obtain the usual gradient or diamagnetic electron drift velocity $\mathbf{u}_{e\perp 0} = (\kappa T_e/e\, n_0 B^2)\nabla n_0 \times \mathbf{B}$.

Considering fluctuating quantities, we find by insertion of Boltzmann distributed electrons in the linearized version of (5.15) that this approximation gives identically $\mathbf{u}_{e\perp 1} = 0$. Although this result may appear strange, it has a simple physical interpretation; the fluctuating electron fluid velocity in the **B**-perpendicular direction has two contributions, the electron $\mathbf{E}_1 \times \mathbf{B}/B^2$-drift and the gradient diamagnetic drift associated with the density variation of the wave. With Boltzmann distributed electrons, these two contributions happen to cancel exactly. The two velocities are however of different nature; the diamagnetic drift is, as already mentioned, not a particle drift in contrast to the $\mathbf{E}_1 \times \mathbf{B}/B^2$-drift.

The cancellation which implies $\mathbf{u}_{e\perp 1} = 0$ in the summary before is in some sense coincidental; if we assume an anisotropic electron pressure tensor with elements $p_\perp = \kappa T_{e\perp} n$ and $p_\parallel = \kappa T_{e\parallel} n$, we have the $\mathbf{B}$-parallel component of the momentum equation (4.4) giving the Boltzmann distribution $n = n_0(x)\exp(e\phi/\kappa T_{e\parallel})$ while

$$\begin{aligned}\mathbf{u}_{e\perp} &= \frac{\kappa T_{e\perp}}{enB^2}\nabla_\perp n \times \mathbf{B} - \frac{\kappa T_{e\parallel}}{eB^2}\frac{n_0(x)}{n}\nabla_\perp \frac{n}{n_0(x)} \times \mathbf{B} \\ &= \frac{\kappa T_{e\perp}}{enB^2}\nabla_\perp n \times \mathbf{B} \\ &\quad - \frac{\kappa T_{e\parallel}}{eB^2}\frac{1}{n}\left(\nabla_\perp n - \frac{n}{n_0(x)}\nabla_\perp n_0(x)\right) \times \mathbf{B},\end{aligned} \tag{5.16}$$

where $n = n_0(x) + n_1$. The unperturbed diamagnetic velocity obtained for $n_1 = 0$ is the same as before, but this time derived from $T_{e\perp}$. The perturbed velocity is now

$$\mathbf{u}_{e\perp 1} = \frac{\kappa\left(T_{e\perp} - T_{e\parallel}\right)}{eB^2}\nabla_\perp \frac{n_1}{n_0(x)} \times \mathbf{B}. \tag{5.17}$$

Note that $\mathbf{u}_{e\perp 1}$ can take either sign, depending on the temperature ratio $T_{e\perp}/T_{e\parallel}$.

For collisional plasmas we expect $T_{e\perp} = T_{e\parallel}$, and the discussion here is irrelevant, but for drift waves in collisionless plasmas, to be discussed in a subsequent chapter, we might have $T_{e\perp} \neq T_{e\parallel}$. Such an analysis needs a kinetic description.

5.2.2 Divergence-free currents

Taking the difference between the electron and ion continuity equations in the quasi-neutral limit, $n_e \approx n_i \equiv n$ we find that the fluctuating plasma current is divergence free, $\nabla \cdot \mathbf{j} = 0$. Taking first the simple case where the ion velocity along $\mathbf{B}$ is ignored and the perpendicular part is $-\nabla\phi \times \mathbf{B}/B^2$, we find the ion current contribution to the divergence of the linearized plasma current to be $-e\nabla_\perp \cdot (n_0(x)\nabla\phi \times \mathbf{B}/B^2) = -e(n_0'(x)/B)\partial\phi/\partial y$. As mentioned before when discussing equation (5.15), the electrons will not contribute to the $\mathbf{B}$-perpendicular part, since this part of their fluid velocity vanishes to the present approximation. The component of the electron velocity *along* $\mathbf{B}$ can be obtained from the linearized electron continuity equation $\partial n_1/\partial t + n_0(x)\partial u_{ez1}/\partial z = 0$ giving $u_{ez1} = (\omega/k_z)n_1/n_0(x)$. The fact that the electron velocity does not enter the electron force balance does *not* imply that this velocity vanishes. To the present approximation, thus only the electrons contribute to $\partial j_z/\partial z$. You can, if you like, consider j_z as the *return current* driven by $\mathbf{j}_\perp$. The foregoing argument gives $\partial j_z/\partial z = -i\omega e n_1$ by the electron continuity equation. With Boltzmann distributed electrons, $n_1/n_0 = e\phi/\kappa T_e$, we readily find that the criterion for $\nabla_\perp \cdot \mathbf{j}_\perp = -\partial j_z / \partial z$ becomes $\omega = \omega^*$. When the ion polarization term is included, the additional contribution to the **B**-perpendicular component of net linearized current becomes $(Mn_0(x)/B^2)\partial\nabla_\perp\phi/\partial t$. The requirement of divergence-free fluctuating plasma currents imply a shift in frequency, i.e. a slight deviation from ω^*. As also evident from figure 5.2, the ion polarization drift gives rise to accumulation of charge, inducing a (small) **B**-parallel additional plasma current contributing to j_z.

The modifications introduced by anisotropic electron temperatures giving the result (5.17) do not effect the conclusions in this section, since $\nabla \cdot \mathbf{u}_{e\perp 1} = 0$ also in that case.

5.3 Waves in rotating systems: Rossby waves

The mathematical similarity between the magnetic Lorentz force and the Coriolis force allows some interesting comparisons between the theory of electrostatic drift waves and what are called Rossby waves in rotating planetary atmospheres (Brekhovskikh and Goncharov 1985, Hasegawa 1985, Diamond *et al* 2010). We present a short summary here. The general basic equations are the continuity equation for mass density ρ

$$\frac{\partial}{\partial t}\rho + \mathbf{u} \cdot \nabla\rho = 0,$$

assuming incompressible flows, $\nabla \cdot \mathbf{u}$. The equation of motion is

$$\frac{\partial}{\partial t}\mathbf{u} + \mathbf{u} \cdot \nabla\mathbf{u} = -\frac{1}{\rho}\nabla p - g\nabla z - 2\boldsymbol{\Omega} \times \mathbf{u},$$

where the gravitational acceleration g is in the negative z-direction, as indicated. The frequency of angular rotation is Ω: in principle it can have a spatial variation, as we will see shortly. Often we find the notation where $F \equiv 2\Omega$ is the Coriolis parameter. In addition we have

$$\frac{\mathrm{d}}{\mathrm{d}t}p = C_s^2\frac{\mathrm{d}\rho}{\mathrm{d}t}, \qquad C_s^2 = \frac{\partial p}{\partial\rho}.$$

This notation implies that it is already determined how the pressure depends on mass density: for instance for adiabatic ideal gas dynamics we have $(p/p_0) = (\rho/\rho_0)^\gamma$ with $\gamma = C_p/C_V$ being the ratio of heat capacities at constant pressure and constant volume, respectively. For long wavelength sound waves in neutral gases this is a natural condition since temperatures there even out by thermal conduction (as described by a diffusion equation) which is ineffective at large scales. For other processes the mass density dependence of the pressure requires scrutiny.

With boundary conditions on the bottom and top of the fluid layer, these basic equations allow for a rich variety of wave-modes, linear as well as nonlinear. The simplest one relevant here is one where we treat Ω as a constant. Ignoring gravity, $g = 0$, we find a set of linear equations in the form

$$\frac{1}{\rho_0}\nabla_\perp^2 p = \frac{\partial^2}{\partial z \partial t}w + F\left(\frac{\partial v}{\partial x} - \frac{\partial u}{\partial y}\right),$$

where we used the momentum equation and introduced the velocity vector components as $\mathbf{u} \equiv \{u, v, w\}$ and the two-dimensional operator $\nabla_\perp \equiv \hat{\mathbf{x}}\partial/\partial x + \hat{\mathbf{y}}\partial/\partial y$. We also have

$$\frac{\partial}{\partial t}\left(\frac{\partial v}{\partial x} - \frac{\partial u}{\partial y}\right) = F\frac{\partial w}{\partial z}.$$

Note that we have not assumed the variations to be strictly two-dimensional and $\perp \boldsymbol{\Omega}$.

With a little algebra (Brekhovskikh and Goncharov 1985) we find the basic equation

$$\frac{\partial^2}{\partial t^2}\nabla^2 w + F^2\frac{\partial^2}{\partial z^2}w = 0.$$

The corresponding dispersion relation is

$$\omega^2 = F^2\frac{k_z^2}{k_x^2 + k_y^2 + k_z^2} \equiv F^2\cos^2\theta,$$

where θ is the angle between the wave vector $\mathbf{k}$ and the vertical z-direction. This is the basic equation for inertial or gyroscopic waves in a rotating fluid. When gravity is included as well, the waves become gyroscopic gravitational waves.

5.3.1 The β-plane approximation

The foregoing result assumed Ω to be a constant. We now consider conditions on a rotating planet, as shown schematically in figure 5.3. We use here the β-plane

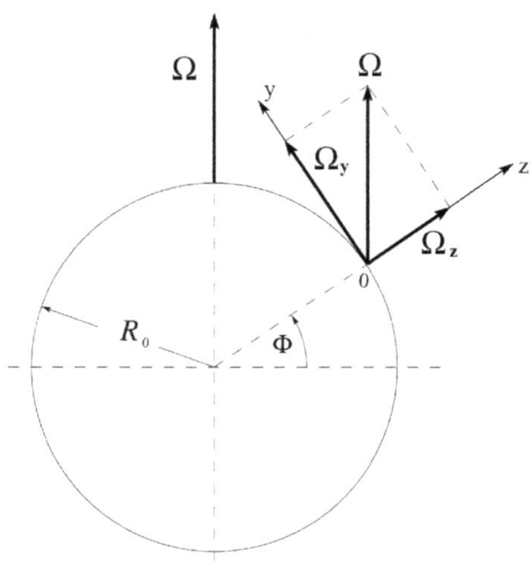

Figure 5.3. Geometry for discussing waves on the surface of a rotating Earth.

approximation where we approximate $F = 2\Omega \sin\phi \approx F_0 + \beta y$ with $F_0 = 2\Omega \sin\phi_0$ and $\beta = 2\Omega \cos\phi_0/R_0$ where the various quantities are defined in figure 5.3. We introduce the stream-line function Ψ so that the velocity components become $u = \partial\Psi/\partial y$ and $\nu = -\partial\Psi/\partial x$. We then have

$$\text{rot}\,\mathbf{u} \equiv \nabla \times \mathbf{u} = \left(\frac{\partial v}{\partial x} - \frac{\partial u}{\partial y}\right)\nabla z = -\left(\nabla_\perp^2\Psi\right)\nabla z,$$

with $\nabla z = \hat{\mathbf{z}}$. With a little algebra, where we now assume a *barotropic flow* without any z-variations, we have

$$\frac{\partial}{\partial t}\nabla_\perp^2\Psi + \beta\frac{\partial}{\partial x}\Psi = 0.$$

The dispersion relation for the linear wave-modes is

$$\omega = -\beta\frac{k_x}{k_x^2 + k_y^2}, \tag{5.18}$$

being anisotropic in the horizontal plane. For $\beta > 0$, the Rossby wave with positive frequency can propagate only in the direction opposite to the x-direction (westward in figure 5.3) with $k_x < 0$. In some sense these waves are similar to electrostatic drift waves, which can propagate only along the electron diamagnetic drift, and not in the opposite direction.

The group velocity vector for these Rossby waves is

$$\mathbf{u}_g = \left\{\frac{\partial\omega}{\partial k_x}, \frac{\partial\omega}{\partial k_y}\right\} = \frac{\beta}{k^2}\{\cos 2\alpha, -\sin 2\alpha\},$$

where $k_x = -k\cos\alpha$ and $k_y = k\sin\alpha$. The wave-fronts with $\psi=$ const. are perpendicular to $\mathbf{k}$. Fluid velocities are tangential on the stream-lines, and so are particle velocities. The Rossby waves are thus *transverse* in nature.

- **Exercise:** Demonstrate that the pressure variation associated with the linear barotropic Rossby wave is given as

$$p = -\rho_0\left(F + i\beta\frac{k_y}{k^2}\right)\Psi,$$

  and discuss the consequences for a model where a surface condition is included. (Such a surface condition was ignored in the summary here.)

5.3.2 Alternative model for the Rossby waves

Considering the importance of Rossby waves it is found relevant to include an alternative model for these waves (Tritton 1988). In the solution presented before, the fluid depth was uniform and we included the variation of the Coriolis parameter in the β-plane approximation. Here we retain a constant Coriolis parameter, but allow for a gradient in the fluid depth h. Linearizing the model equations from the outset we have

$$\frac{\partial}{\partial t}\mathbf{u} + 2\mathbf{\Omega} \times \mathbf{u} = -\frac{1}{\rho}\nabla p, \tag{5.19}$$

ignoring here the effects of gravity. We take fluid velocities to be u, v, w in the x, y, z-directions, respectively. We assume that the fluid depth varies uniformly in the y-direction so that

$$\frac{\partial h}{\partial x} = 0 \qquad \text{and} \qquad \frac{\partial h}{\partial y} = \gamma,$$

with γ being a constant. Locally this is an acceptable model.

We assume now that the waves retain the properties of the geostrophic flow

$$\frac{\partial u}{\partial z} = \frac{\partial v}{\partial z} = 0 \qquad \text{but} \qquad \frac{\partial w}{\partial z} \neq 0.$$

A posteriori we can demonstrate the consistency of this approach by showing that the pressure variation obtained by integrating the z-component of (5.19) produces negligible modifications of the x and y-components of the velocities as long as the inhomogeneity parameter γ is small. Since now u and v are taken independent of z, the condition for incompressible flows $\nabla \cdot \mathbf{u} = 0$ gives

$$\frac{\partial u}{\partial x} + \frac{\partial v}{\partial y} + \frac{\partial w}{\partial z} = 0,$$

implying that $\partial \omega / \partial z$ is constant with respect to z to the present approximation.

Referring to figure 5.4 we take $w = 0$ at $z = 0$, while $w = w_h$ at the top boundary $z = h$. A model consistent with these assumptions gives

$$\frac{\partial w}{\partial z} = \frac{w_h}{h} \qquad \text{or} \qquad w_h = v_h \tan\alpha = \gamma v_h = \gamma v,$$

the latter equality originating from $\partial v / \partial z = 0$. The incompressibility condition becomes

$$\frac{\partial u}{\partial x} + \frac{\partial v}{\partial y} + \frac{\gamma v}{h} = 0.$$

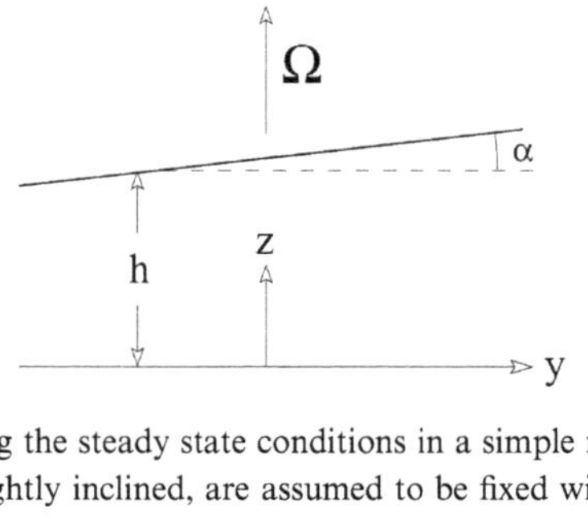

Figure 5.4. Diagram for describing the steady state conditions in a simple model for Rossby waves. The two planes, one at $z = 0$, the other slightly inclined, are assumed to be fixed with 'no-slip' flow conditions.

The x and y-components of (5.19) give

$$\begin{aligned}\frac{\partial u}{\partial t} - 2\Omega v &= -\frac{1}{\rho}\frac{\partial p}{\partial x} \\ \frac{\partial v}{\partial t} + 2\Omega u &= -\frac{1}{\rho}\frac{\partial p}{\partial y}.\end{aligned} \tag{5.20}$$

From these equations it is a simple matter to obtain

$$\frac{\partial^2 v}{\partial x \partial t} + 2\Omega \frac{\partial u}{\partial x} = 0,$$

which leads to

$$\frac{\partial^2 v}{\partial x \partial t} - \frac{2\Omega \gamma}{h} v = 0.$$

A complete analysis will have both h and v to be varying, but this will make the second term inconsistent with the linearization of the original equations used here. We therefore take $2\Omega/h$ to be locally constant and can then find a dispersion relation

$$\omega k_x - \frac{2\Omega \gamma}{h} = 0,$$

giving the phase velocity $\omega/k_x = h\omega^2/2\Omega\gamma$ and group velocity $\mathrm{d}\omega/\mathrm{d}k_x = -h\omega^2/2\Omega\gamma$. The waves are thus strongly dispersive, basically consistent with (5.18) for $k_y = 0$. The difference in sign is of no consequence, since it refers to a chosen direction of the depth-gradient in figure 5.4, i.e. the sign of the angle α and thereby γ. We find that v and $\partial u/\partial x$ are in phase (use the continuity equation with $\partial v/\partial y = 0$) so u and v are $\pi/2$ out of phase. A more detailed comparison of the two models for Rossby waves shown here is given by Tritton (1988). The moral of the two models discussed in this section 5.3 is that Rossby waves can propagate in any rotating system with a horizontal gradient of some basic parameter, the Coriolis force or the depth of the fluid, for instance.

References

Brekhovskikh L and Goncharov V 1985 *Mechanics of Continua and Wave Dynamics (Series on Wave Phenomena)* (Berlin: Springer)

Chen F F 1965 Resistive overstabilities and anomalous diffusion *Phys. Fluids* **8** 912–9

Chen F F 1970 Linear drift waves UCLA Lecture Notes unpublished

Diamond P H, Itoh S-I and Itoh K 2010 *Modern Plasma Physics* vol I Physical Kinetics of Turbulent Plasmas (Cambridge: Cambridge University Press)

Ellis R F, Marden-Marshall E and Majeski R 1980 Collisional drift instability of a weakly ionized argon plasma *Plasma Phys.* **20** 113–32

Hasegawa A 1985 Self-organization processes in continuous media *Advances Phys.* **34** 1–42

Kadomtsev B B 1965 *Plasma Turbulence* (New York: Academic)

Tritton D J 1988 *Physical Fluid Dynamics* 2nd edn (Oxford: Oxford Science Publications)

IOP Publishing

Low Frequency Waves and Turbulence in Magnetized Laboratory Plasmas and in the Ionosphere

Hans Pécseli

Chapter 6

Resistive drift waves with cold ions

The simplest model accounting for the linear instability of electrostatic drift waves assumes cold ions. This limit is considered here.

6.1 Instability of small amplitude electrostatic drift waves

Although it was easy to derive the real dispersion relation $\omega = \omega(\mathbf{k})$ in section 5.1, it is not so easy to see why the wave can become linearly unstable. An imaginary part of ω is found to originate from a phase difference between fluctuations in density and potential. As long as (4.7) obtains, ω will be real and equal to ku_{De}. We can model this effect by first postulating, for the sake of illustration, a phase change between ϕ and n as

$$\frac{n}{n_0} = (1 - i\Delta)\frac{e\phi}{\kappa T_e}. \tag{6.1}$$

From the simplest description of the ion dynamics we have the relation $n/n_0 = (\omega^*/\omega)e\phi/\kappa T_e$, which ultimately implies $\omega \approx \omega^*(1 + i\Delta)$. The plasma will be unstable for $\Delta > 0$ and stable for $\Delta < 0$. The purpose of a detailed study will therefore be to account for the physical mechanisms which give rise to the phase difference between density and electrostatic potential (Manheimer 1977).

In order to have an instability, k_z must be so small that electrons cannot flow unimpeded from wave crest to trough along $\mathbf{B}$. Then (4.7) will be violated, and it will be possible to have E_y and E_x different from what is rigorously prescribed by (4.7). There are several mechanisms which can affect electron parallel mobility:

a) electron–neutral collisions
b) electron–ion collisions
c) kinetic plasma effects, electron Landau damping in particular
d) inductance
e) electron inertia.

doi:10.1088/978-0-7503-1251-6ch6

These mechanisms can be considered as 'triggering' the instability by allowing (4.7) to be violated, so that net charges can build-up. Early analytical studies of unstable drift waves (Timofeev 1963) allowed for varying electron temperatures, introducing a dynamic equation for the variation of T_e. The present analysis assumes the electron heat conductivity to be high and take $T_e = \text{const}$.

Effect a) is dominant in weakly ionized gases, such as in laboratory-experimental conditions in a positive column, in the ionospheric E-region or the lower parts of the F-region.

Effect b) is the finite resistivity of a fully ionized plasma, and is the mechanism we shall consider in detail in the following.

Effect c) has to do with a distortion of the electron distribution function due to the Landau interaction with the wave and electrons with $u_z \approx \omega/k_z$. This is the mechanism responsible for drift instabilities in collisionless plasmas and its proper description needs kinetic modeling.

Effect d) is the impedance to the electron flow provided by the magnetic field the flow creates. This effect is important only at high $\beta \equiv n\kappa T/(B^2/2\mu_0)$, when $\omega/k_z \approx V_A$, the Alfvén speed.

Effect e) might be expected to be the dominant mechanism in the absence of collisions (Chen 1970), but it turns out that the effect has the wrong phase and does not cause an instability except in higher order.

Since b) is the dominant mechanism in for instance Q-machine experiments (Hendel *et al* 1968), we shall derive the dispersion relation for that case. As it turns out, the results will apply qualitatively for the case a) as well with some subtle restrictions. This limit will be relevant also for some ionospheric conditions in the E and F-regions, respectively.

The simplest description of unstable drift-waves can be found from the fluid equations when the resistivity ξ is finite and T_i is vanishing. In this case, there are no finite ion Larmor radius effects since $r_L = 0$, and finite resistivity together with ion inertia (i.e. the ion polarization drifts) causes the instability. The special case where also the polarization drift is ignored is a limiting case. The geometry of the problem is the one shown in figure 4.1.

6.2 Basic equations

As already mentioned, the analysis will rely on a fluid model. This is an approximate description, valid only for collisional plasmas, although some of the results will remain qualitatively correct also for collision *less* plasmas. A number of simplifying assumptions are listed in the following.

a) $T_i = 0$, $T_e > 0$
b) $\xi \neq 0$ (finite resistivity caused by electron–ion collisions with frequency ν_{ei})
c) $\nabla \times \mathbf{E} = 0$, $\mathbf{B} = \text{const.}$ (low-β, electrostatic assumption)
d) $m/M \simeq 0$ (electron inertia is ignored)
e) $n_i = n_e \equiv n$ (quasi-neutrality, or 'the plasma approximation' valid for $k\lambda_D \ll 1$ and $\omega \ll \Omega_{pi}$, the ion plasma frequency)
f) $U_{thi} \ll \omega/k_z \ll U_{the}$ in terms of ion and electron thermal velocities
g) $\nabla p_e = \kappa T_e \nabla n$ (isothermal electrons)

First we summarize some basic elements of collisional interaction in plasmas, distinguishing partially and fully ionized cases. For partially ionized plasmas (as found in for instance the ionospheric E-region or the lower parts of the F-region) collisions with neutrals can dominate electron–ion collisions. A simple and often used model introduces a term of the form $-\nu_{in} n M(\mathbf{u}_i - \mathbf{u}_n)$ in the ion momentum equation with a similar term in the electron momentum equation. We have ν_{in} and ν_{en} representing the frequency of collisional momentum losses from the ions and electrons, respectively, due to collisions with neutrals moving at a bulk velocity $\mathbf{u}_n$. The momentum losses of electrons and ions are unrelated in the sense that an electron can collide with a neutral irrespective of ion collisions. The times ν_{in}^{-1} and ν_{en}^{-1} thus in general are not the average times between encounters between the neutrals and the corresponding charged plasma particles, although for the electrons it will be nearly so: an electron will on average lose almost all of its momentum when colliding with a neutral. For the ions the mass ratio of the particles entering the process needs to be considered.

For electron–ion collisions in fully ionized plasmas the conditions are different: here the momentum lost by electrons has to enter as a gain for the ions and vice versa. This is usually made clear by including a collision term $\mathbf{P}_{ie}$ in the ion momentum equation and a term $\mathbf{P}_{ei}$ in the electron momentum equation with $\mathbf{P}_{ie} = -\mathbf{P}_{ei} \equiv \mathbf{P}$. Ion–ion and electron–electron collisions evidently do not count here since the momentum exchanged remains in the respective species. It is generally assumed (van Kampen and Felderhof 1967, Pécseli 2012) that we can write the momentum exchange term in the form

$$\mathbf{P} = \vartheta n_e n_i (\mathbf{u}_i - \mathbf{u}_e). \tag{6.2}$$

The constant of proportionality ϑ is positive, but otherwise unspecified for the moment. We proposed the expression (6.2) without demonstrating any detailed insight into the collisional interactions mediating the momentum exchange. The justification of (6.2) is that it is the simplest expression obeying the requirement that $\mathbf{P}$ must be a vector in the direction of the relative velocity, that it vanishes where the two velocities are equal, and vanishes also when either of the particle densities vanish. The momentum exchange is mediated via electron–ion collisions with an average frequency $\nu_{e,i}$. This is not a contact frequency since electrons and ions are never actually in contact, calling it an encounter frequency may be more correct. The average vector momentum transfer per encounter is of the order of $m(\mathbf{u}_i - \mathbf{u}_e)$. For realistic conditions we can assume quasi-neutrality and let $n_e \approx n_i \equiv n$, see section 3.3.

As an illustration we can take $\mathbf{u}_i \approx 0$ and argue that a colliding electron loses almost all of its momentum (which is a vector quantity) while losing very little of its kinetic energy (which is a scalar quantity), very much like a tennis ball bouncing off a wall. The average time between encounters and the corresponding time for momentum loss is thus approximately the same for electrons colliding with ions. The number of electron collisions per sec per m^{-3} is $\nu_{e,i} n$, with $\nu_{e,i}$ being the number of electron–ion collisions per section Hence $\mathbf{P} \sim \nu_{e,i} n m(\mathbf{u}_i - \mathbf{u}_e)$ giving $\vartheta \sim \nu_{e,i} m/n$. Note that ϑ has a dimension *mass* × *length*^{3} × *time*$^{-1}$. We recall here that the plasma conductivity $\sigma \approx e^2 n/(m\nu_{e,i})$ implying $\sigma \sim e^2/\vartheta$. In terms of the plasma resistivity $\xi \equiv 1/\sigma$ we have the momentum exchange term in the form $\mathbf{P} = \xi e^2 n^2(\mathbf{u}_i - \mathbf{u}_e)$ where ξ is given by the Spitzer expression. We have $\xi = m\nu_{ei}/(ne^2)$ where

$$\nu_{ei} = \frac{ne^4}{64\pi m^2 \varepsilon_0^2}\left(\frac{m}{\kappa T_e}\right)^{3/2} \ln(\Lambda), \tag{6.3}$$

with $\ln(\Lambda)$ being the *Spitzer logarithm*, where $\Lambda = 12\pi n\lambda_D^3$. The expression (6.3) assumes singly charged ions. This is a trivial restriction.

We could ask the same question as before but now seen for ions colliding with electrons. The encounter rates will be the same for electrons colliding with ions as for ions colliding with electrons, but the momentum losses are different. An ion loses on average a fraction of its momentum when colliding with an electron, the fractional loss being of the order of m/M. A statistical analysis of the transition to equilibrium is summarized by Pécseli (2012). The time between encounters and time between momentum loss is significantly different when seen by following the ions. If we by $\nu_{i,e}$ denote the average frequency for ion momentum losses, we write $-\nu_{i,e}Mn(\mathbf{u}_i - \mathbf{u}_e)$ for the collisional momentum loss in the ion momentum equation. It is of course important here to distinguish $\nu_{i,e}$ and $\nu_{e,i}$. The foregoing arguments give $\nu_{i,e} \approx (m/M)\nu_{e,i}$ so the net result is that the average collisional losses have the same magnitude but opposite signs in the momentum equations for ions and electrons. For collisions between plasma species and neutrals no such simple relation is given *a priori*.

The basic equations for the present analysis can now be listed as follows

$$Mn\left(\frac{\partial \mathbf{u}_i}{\partial t} + \mathbf{u}_i \cdot \nabla \mathbf{u}_i\right) = en(-\nabla\phi + \mathbf{u}_i \times \mathbf{B}) - n^2 e^2 \xi(\mathbf{u}_i - \mathbf{u}_e) \tag{6.4}$$

and

$$0 = -en(-\nabla\phi + \mathbf{u}_e \times \mathbf{B}) - \kappa T_e \nabla n - n^2 e^2 \xi(\mathbf{u}_e - \mathbf{u}_i), \tag{6.5}$$

which are the momentum equations for ions and electrons with the simplifications listed before. In retaining the electron collision term on the right-hand side of (6.5) we implicitly assume $\omega \ll \nu_{ei}$. In addition we have the continuity equations for the ions and electrons

$$\frac{\partial n}{\partial t} + \nabla \cdot (n\mathbf{u}_i) = 0 \tag{6.6}$$

$$\frac{\partial n}{\partial t} + \nabla \cdot (n\mathbf{u}_e) = 0. \tag{6.7}$$

With the quasi-neutrality assumption there is no distinction between ion and electron densities in these equations, so their difference becomes $\nabla \cdot (n(\mathbf{u}_e - \mathbf{u}_i)) = 0$.

In a linearized model the assumption of isothermal electron fluctuations is not crucial for the analysis. Since Boltzmann's constant occurs only in the ∇p_e term, an adiabatic equation of state will give the same result as these equations if κ is replaced by $\frac{5}{3}\kappa$ in the final answer.

The full set of nonlinear equations (6.4)–(6.7) cannot be solved analytically. The dispersion relation for the fluctuations can however be obtained by a standard linearization procedure and the stability of the waves subsequently analyzed.

6.3 Equilibrium

We assume $u_{0z} = 0$ and $n_0 = n_0(x)$ in the unperturbed state. Since electron–ion collisions do not give rise to any charge separation, i.e. no ambipolar electric fields, the assumption of $E_0 = 0$ is justified. Setting $\partial/\partial t = 0$ for steady state conditions we obtain

$$Mn_0(\mathbf{u}_{i0} \cdot \nabla \mathbf{u}_{i0}) = en_0\mathbf{u}_{i0} \times \mathbf{B} - n_0^2 e^2 \xi(\mathbf{u}_{i0} - \mathbf{u}_{e0}) \tag{6.8}$$

$$0 = -en_0\mathbf{u}_{e0} \times \mathbf{B} - \kappa T_e \nabla n_0 - n_0^2 e^2 \xi(\mathbf{u}_{e0} - \mathbf{u}_{i0}). \tag{6.9}$$

Neglecting $\mathbf{u}_{i0} \cdot \nabla \mathbf{u}_{i0}$ for the time being, adding (6.9) to (6.8), and taking the cross-product with $\mathbf{B}$, we obtain

$$\mathbf{j}_0 = \kappa T_e \frac{\mathbf{B} \times \nabla n_0}{B^2}, \tag{6.10}$$

where $\mathbf{j}_0 \equiv en_0(\mathbf{u}_{i0} - \mathbf{u}_{e0})$ is the diamagnetic current. The y components of (6.8) and (6.9) yield $(\mathbf{u}_{i0} \times \mathbf{B})_y = (\mathbf{u}_{e0} \times \mathbf{B})_y = \xi j_0$. Using (6.10), we find

$$u_{i0x} = u_{e0x} = -\frac{\xi \kappa T_e}{B^2} n_0'(x). \tag{6.11}$$

This is the classical diffusion velocity due to electron–ion Coulomb collisions (being in the radial direction for cylindrical geometry). There is no current associated with this motion since electrons and ions move with the same velocity in the same direction, see the discussion in section 3.5. Since this radial diffusion is assumed to be a slow process compared to the wave frequencies we are considering, we shall neglect u_{0x}. In effect we are assuming that $\partial n_0/\partial t \ll \omega^* n_0$, where ω^* is a characteristic drift wave frequency, and consider n_0 constant in time. In the following we are thus linearizing the equations around something which is not really an *exact* equilibrium solution for the initial equations (this is a trick often met in similar investigations). The procedure will be justified *a posteriori* by demonstrating that the drift waves can have characteristic time scales which are much shorter than the time it takes the density gradient to decay by classical diffusion alone.

The complexity induced by the electron–ion collisional drag as represented by ξ in (6.9) is readily noted. Even though we assume the ions to be cold, and thus not contributing to a diamagnetic drift we find that if we were to retain all terms, the electron diamagnetic drift will induce an ion motion through the electron–ion-momentum exchange, etc. It is an advantage to simplify the problem by ignoring some of the steady state velocities.

The term $\mathbf{u}_{i0} \cdot \nabla \mathbf{u}_{i0}$ automatically vanishes to the desired accuracy when we set $u_{i0x} = 0$, since u_{i0y} can depend on the coordinate x only, because of the assumed homogeneity in the y-direction, see figure 4.1. The initial assumption for obtaining (6.10) is fulfilled to the desired accuracy. The x-components of (6.8) and (6.9) give

$$u_{i0y} = 0, \qquad u_{e0y} = -\frac{\kappa T}{eB}\frac{n_0'(x)}{n_0} \equiv u_{De}. \tag{6.12}$$

With the given approximation for the equilibrium, there is only an electron diamagnetic drift of the electrons in the 'azimuthal' direction. It should be emphasized here that the analytical expression results from our assumption of uniform electron temperatures which applies for many relevant physical conditions. More generally it should be recalled that the driving mechanisms for drift-wave instabilities are gradients in *pressure*, $\nabla(n_0(x)T_0(x))$, so the steady-state equations will generally contain also terms with $T_0'(x)$, see also figure 3.6. The generalization to the case where n_0 is constant and $T_e = T_e(x)$ requires some attention though.

6.4 Perturbations

We neglect classical diffusion by keeping ξ only in the $\hat{\mathbf{z}}$-component of the electron equation of motion and introduce the plane wave assumption as follows

$$\begin{aligned}
\phi_1(x, y, z, t) &= \phi(x)\mathrm{e}^{\mathrm{i}(k_y y + k_z z - \omega t)} \\
\mathbf{u}_1(x, y, z, t) &= \mathbf{u}(x)\mathrm{e}^{\mathrm{i}(k_y y + k_z z - \omega t)} \\
n_1(x, y, z, t) &= \eta(x) n_0(x)\mathrm{e}^{\mathrm{i}(k_y y + k_z z - \omega t)},
\end{aligned} \tag{6.13}$$

so that the relative density variation is $\eta = (n_1/n_0)\mathrm{e}^{-\mathrm{i}(k_y y + k_z z - \omega t)}$ at any position x. This combination may not appear entirely obvious at first, but it turns out to be convenient.

6.4.1 Electron dynamics

By linearizing the $\hat{\mathbf{x}}$-component of the electron momentum equation (6.5), we have

$$0 = en_0(-\phi' + u_{ey}B) + \kappa T_e(n_0\eta' + \eta n_0') + en_0\eta u_{e0}B. \tag{6.14}$$

The last two terms cancel by virtue of (6.12). Defining

$$\chi \equiv \frac{e\phi}{\kappa T_e},$$

we then have

$$u_{ey} = \frac{\kappa T_e}{eB}\frac{\partial(\chi - \eta)}{\partial x}. \tag{6.15}$$

Similarly, the y and z components of (6.5) give

$$iu_{ex} = k_y \frac{\kappa T_e}{eB}(\chi - \eta) \tag{6.16}$$

$$u_{ez} = ik_z \frac{\kappa T_e}{eB} \omega_{ce}\tau_{ei}(\chi - \eta), \tag{6.17}$$

where

$$\omega_{ce}\tau_{ei} = \frac{eB}{m\nu_{ei}} = \frac{B}{n_0 e \xi}$$

is the important dimensionless quantity characterizing the resistivity. Since we retain the effect of resistivity only in the **B**-parallel dynamics, we have no τ_{ei} entering u_{ex} and u_{ey}. Note that the velocity components u_{ex} and u_{ey} vanish for Boltzmann distributed electrons, $\chi = \eta$, as expected. At first sight it might be surprising to see the electron cyclotron frequency ω_{ce} appear: this quantity explicitly contains the electron mass m, and we have in the beginning assumed the electrons to be essentially massless particles! Note, however, that m appears explicitly in the plasma conductivity ξ, and ω_{ce} is introduced by 'juggling around' with ξ.

The z-parallel electron dynamics is retained not only because of the large electron thermal velocity for finite electron temperatures; what is important here is that even a minute **B**-parallel electric field component will give the light electrons a large fluid velocity. The ions on the other hand will hardly move at all, due to their large mass. The **B**-perpendicular dynamics is quite different; here the ions move in the lowest approximation with the **E** × **B**-velocity, independent of mass.

Linearizing the electron continuity equation (6.7), we have

$$\left(\omega - k_y u_{De}\right) n_1 + i n_0 \left(u'_{ex} + i k_y u_{ey} + i k_z u_{ez}\right) + i u_{ex} n'_0 = 0. \tag{6.18}$$

Substituting for u_{ex}, u_{ey}, and u_{ez} from (6.15)–(6.17), we find that the x-derivatives cancel out (i.e. the ones referring to the coordinate axis parallel to ∇n_0). This is a fortunate feature of the fluid equations for massless particles, although it rests on the requirement of isotropic electron temperatures, i.e. $T_{e\perp} = T_{e\parallel}$. Finally we have an equation which does not depend on the 'shape' of the perturbation in the x-direction:

$$\left(\omega - k_y u_{De}\right)\eta - \left(k_y u_{De} + i k_z^2 \frac{\kappa T_e}{eB} \omega_{ce}\tau_{ei}\right)(\chi - \eta) = 0. \tag{6.19}$$

To simplify the notation we now introduce the following normalized quantities in agreement with most of the literature on drift waves

$$\begin{aligned}
\omega^* &= k_y u_{De} = -k_y \frac{\kappa T_e}{eB}\frac{n'_0}{n_0} \equiv -k_y \frac{\kappa T_e}{m\omega_{ce}}\frac{n'_0}{n_0} \\
\sigma_\parallel &= \frac{k_z^2}{k_y^2}(\omega_{ce}\tau_{ei})\Omega_{ci}, \qquad a_i^2 = \frac{\kappa T_e}{M\Omega_{ci}^2} \\
b &= k_y^2 a_i^2, \qquad b\sigma_\parallel = k_z^2 \frac{\kappa T_e}{eB}(\omega_{ce}\tau_{ei}),
\end{aligned} \tag{6.20}$$

see also appendix A summarizing the normalizations used. The quantity $a_i\sqrt{2}$ is, as before, the ion Larmor radius obtained using the electron temperature, and b is a common parameter in small ion Larmor radius expansions. The expression for the conductivity $\sigma_{\|}$ depends on the direction of the wavevector $\mathbf{k}$, but not its magnitude. With this notation, (6.19) becomes

$$\eta = \chi \frac{\omega^* + ib\sigma_{\|}}{\omega + ib\sigma_{\|}}. \tag{6.21}$$

This relation between the density and potential perturbations is a particularly useful one because it is derived solely on the basis of the electron equations and it must be obeyed regardless of the complications for the ion dynamics to be introduced in the following sections. (One exception, which is immaterial in the present context, is the case where there are no density gradient, $\omega^* = 0$, and infinite resistivity, $b\sigma_{\|} = 0$. In this case the relation between η and χ is undetermined because of the idealizations in the basic assumptions.) It is important to note that the 'de-coupling' of electron and ion modeling obtained by (6.21) is only valid for small k_z because we implicitly violate the strict collisional momentum exchange balance in the basic equations (6.4)–(6.5). The ion motion along magnetic field lines, i.e. $\widetilde{u}_z$, is ignored by assuming that only the $\widetilde{\mathbf{E}} \times \mathbf{B}/B^2$ ion velocity together with its correction due to ion polarization drifts is important. The two sound wave branches in figure 5.1 are no longer accounted for, only the part where $\omega \approx \omega^*$.

At first sight it seems wrong that the ion mass enters (6.21), even though it is derived solely from electron equation, but this is due to creative multiplication and division by the ion mass M. An implicit wavenumber dependence enters (6.21) through the dispersion relation $\omega = \omega(\mathbf{k})$ to be derived in the following. In the limit $\sigma_{\|} \to \infty$ the relation (6.21) becomes $\eta = \chi$, which is simply equivalent to the linearized version of the Boltzmann relation (4.7). Finite $b\sigma_{\|}$ introduces the phase shift between η and χ which is necessary for instability.

6.4.2 Ion dynamics

To obtain a dispersion relation, we need another relation between η and χ to be derived from the ion dynamics. In order to simplify this part we note the electron–ion momentum exchange has two consequences: i) to inhibit the free flow of electrons along magnetic field lines, and ii) to mediate a collisional diffusion flux along ∇n_0 as discussed in section 3.5. The first effect is adequately accounted for by (6.21). Since we here will take u_{ix} small, we will not gain any accuracy of significance by retaining the full momentum exchange for the $\mathbf{B}$-parallel ion dynamics, so we ignore this. We will use a perturbation analysis with respect to a postulated steady state density gradient as discussed in section 6.3. Consequently we also ignore the ion-electron momentum contribution in describing the $\mathbf{B}$-perpendicular ion dynamics since it is these terms that account for the ∇n_0-parallel ion diffusion flux. A convenient consequence of this approximation is that the electron velocity $\mathbf{u}_e$ no longer appears in the ion momentum equations: the coupling between the electron and ion dynamics is solely due to the collective electric fields.

Linearizing (6.4) we set $\mathbf{u}_{i0} = 0$ and have for the ions

$$-i\omega\mathbf{u}_i = \frac{e}{M}(-\nabla\phi + \mathbf{u}_i \times \mathbf{B}). \tag{6.22}$$

Since we took $T_i \approx 0$, we have no zero order ion diamagnetic drifts. The momentum exchange between electrons and ions in the z-direction is omitted in (6.22), since it will be immaterial due to the assumptions given in the following. As a consequence the collision term is retained only in the electron dynamics which implies that the momentum exchange balance is not exactly fulfilled. This simplification introduces an error, or inaccuracy, of the same order of magnitude as the one originating from ignoring the collisional diffusion velocity u_{i0x}. This assumption would seemingly also imply that the analysis is similar to the one we would obtain by starting out with electron–neutral collisions and retaining a collisionless description for the ions. For most of the analysis it is indeed so, but one non-trivial difference is associated with the different diffusion rate for electrons and ions when collisions with neutrals is dominating; as a result, a DC electric field builds up, in such a way that the net flux of electrons and ions becomes the same, see also section 3.6. Such a DC electric field is omitted in the present analysis.

For frequencies $\omega \ll \Omega_{ci} \equiv eB/M$, the iterated solution of (6.22) is

$$\begin{aligned} iu_{ix} &= \frac{k_y\phi}{B} - \frac{\omega}{\Omega_{ci}}\frac{\phi'}{B}, \\ u_{iy} &= \frac{\phi'}{B} - \frac{\omega}{\Omega_{ci}}\frac{k_y\phi}{B}, \\ u_{iz} &= \frac{ek_z}{M\omega}\phi \approx 0, \end{aligned} \tag{6.23}$$

where a prime $'$ denotes $\partial/\partial x$ here and in the following. In the last equation of (6.23) we have indicated that we shall consider k_z so small that u_{iz} can be neglected; this amounts to neglecting the transition from drift waves to ion acoustic waves at large k_z/k_y, see figure 5.1. The first term in u_{ix} and u_{iy} is simply the $\mathbf{E} \times \mathbf{B}/B^2$ drift, and the second term is the ion inertia effect mentioned earlier, where only the lowest order correction, the ion-polarization drift, is retained just as in section 2.1.5.

The last equation needed here is the linearized form of the ion continuity equation (6.6), here rewritten in the form

$$-i\omega n_1 + n_0\left(u_{ix}' + ik_y u_{iy}\right) + u_{ix}n_0' = 0. \tag{6.24}$$

Substituting for u_i from (6.23) and for n_1 from (6.21), we obtain the differential equation

$$\phi''(x) + \frac{n_0'(x)}{n_0(x)}\phi'(x) - \left(k_y^2 + \frac{\Omega_{ci}}{\omega}\frac{n_0'(x)}{n_0(x)}k_y - \frac{\omega^* + ib\sigma_\parallel}{\omega + ib\sigma_\parallel}\frac{\Omega_{ci}^2}{C_s^2}\right)\phi(x) = 0. \tag{6.25}$$

The equation (6.25) reproduces (5.12) in the limit $\sigma_\parallel \to \infty$ as it should.

As a final test we can estimate the relative magnitude between the ion fluid velocity *along* and *perpendicular* to the magnetic field for the low frequency case relevant here, for uniform plasma densities and cold ions. We take as an estimate $\mathbf{u}_\perp \approx \mathbf{E} \times \mathbf{B}/B^2$, while $u_z \approx i(e/M)E_z/\omega$. We find

$$\frac{|\mathbf{u}_\perp|}{u_z} \approx \frac{E_\perp}{E_z}\frac{\omega}{\Omega_{ci}},$$

recalling that we implicitly assumed $\omega \ll \Omega_{ci}$ by using $\mathbf{u}_\perp \approx \mathbf{E} \times \mathbf{B}/B^2$. It is evidently justified to ignore the **B**-parallel ion motion, provided $E_z/E_\perp = k_z/k_\perp \ll \omega/\Omega_{ci} \approx \omega^*/\Omega_{ci}$, i.e. for electrostatic waves propagating almost perpendicular to **B**. We note that in the foregoing estimate we cannot simply replace M by m to get the corresponding result for the electrons, since for that case the electron pressure is essential for the **B**-parallel motion.

6.5 Dispersion relation

Equation (6.25) is a complex second-order differential equation for $\phi(x)$ with variable coefficients and must be solved together with some boundary conditions–typically, $\phi = 0$ at $x = 0$ and $x = R$. Other terms would have appeared had we used cylindrical rather than rectilinear coordinates. The solution of such an equation is the principal problem in drift-wave theory. The eigenfunctions of ϕ give the variation of wave amplitude with x, and the complex eigenvalues of ω give the frequency and growth rate.

If n_0 is some *a priori* known function of x, obtained from an experiment for instance, the equation (6.25) can in general only be solved numerically. There are, however, several ways to get an approximate analytic dispersion relation just as in the case of (5.12). Different methods were outlined before in section 5.1 after (5.12) and (5.13). These discussions apply here as well.

If we neglect ϕ'' and ϕ' in (6.25) and multiply through by $a_i^2\omega(\omega + ib\sigma_\parallel)$, we obtain the desired dispersion relation for $T_i = 0$ as

$$\omega^2 + i\sigma_\parallel(\omega(1 + b) - \omega^*) = 0. \tag{6.26}$$

Note that the waves are unstable for all combinations of k_z and k_y. The only natural frequency for the problem is $\omega^* = u_{De}k_y$, which contains a wavenumber k_y. We have an unstable solution

$$\omega = \frac{1}{2}\left(-i\sigma_\parallel(1 + b) + \sqrt{4i\sigma_\parallel\omega^* - \sigma_\parallel^2(1 + b)^2}\right)$$

and a damped solution

$$\omega = \frac{1}{2}\left(-i\sigma_\parallel(1 + b) - \sqrt{4i\sigma_\parallel\omega^* - \sigma_\parallel^2(1 + b)^2}\right).$$

For $\sigma_\parallel/\omega^* \to 0$ we find for the unstable solution $\omega \approx (1 + i)\sqrt{\omega^*\sigma_\parallel/2} - i(1 + b)\sigma_\parallel/2$. For small $\sigma_\parallel$ we thus have $\Re\{\omega\} \approx \Im\{\omega\} \approx \sqrt{\omega^*\sigma_\parallel/2}$ independent of b.

In the limit of large $\sigma_{\|}/\omega^*$ and small b it is on the other hand clear that $\omega \approx \omega^*$. Solving for $\omega - \omega^*$, we obtain an expression for the growth rate in this limit in the form

$$\omega - \omega^* = -b\,\omega^* + i\frac{\omega^{*2}}{\sigma_{\|}}, \tag{6.27}$$

where we have substituted ω^* for ω on the right-hand side. We have $\Re\{\omega\} \approx \omega^*(1 - b)$ and $\Im\{\omega\} \approx \omega^{*2}/\sigma_{\|}$. The ion polarization drift included here through b lowers the drift wave frequency slightly from the reference value ω^*. Note that the growth rate, $\Im\{\omega\}$, is proportional to the resistivity. The other root of (6.26) is heavily damped in this limit and is unimportant. In the older literature, drift waves would be called 'overstable', since both the real and imaginary parts of the frequency are non-zero. The term 'instability' was then reserved for exponentially and aperiodically growing modes. It seems that this distinction has disappeared from the modern literature on plasma physics.

The wavenumber variation of the unstable root of the full expression (6.26) is shown for $b = 0$ in figure 6.1. Due to the fortunate combination of parameters, it is here possible to present the dispersion relation by a single curve, although the frequency ω is a function of two variables, k_z and k_y. In reality k_y is often determined by cylindrical geometry selecting some mode-numbers, and typically we have $\lambda_y \sim 2\pi R$, implying $k_y \sim 1/R$, with R being the characteristic radius of the plasma column. We can understand k_y as a fixed wavenumber, selected by the geometry of the experimental set-up, and assume k_z to be the variable wavenumber component. For the present simplified model, we find that the growth rate can increase indefinitely with increasing k_z. The range of k_z-variations is, however, in general limited by the finite length of a plasma column. Here the boundary conditions at the ends, sheath conditions in particular, are important. Inclusion of the u_{zi} terms would have made Re (ω) approach the line C_s (acoustic velocity) at large k_z/k_y. The results are valid in the limit where $\omega/k_z \ll V_A$, and $\omega/k_z \ll U_{the}$ in terms of the Alfvén

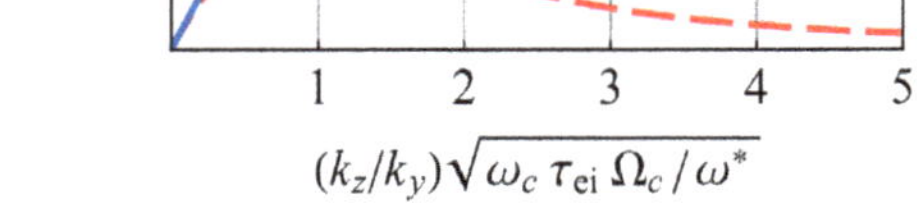

Figure 6.1. Dispersion relation for resistive electrostatic drift waves with the assumption of cold ions and ignoring ion polarization drifts by taking $b = 0$, with $\omega^* = k_y u_{De}$ being the drift frequency and $\sigma_{\|} = (k_z/k_y)^2 \omega_{ce}\tau_{ei}\Omega_{ci}$. The full line gives the real part, dashed line the imaginary part of ω/ω^*. For small b the linear growth rate is maximum for $k_y/k_z \approx \sqrt{\omega_{ce}\tau_{ei}\Omega_{ci}/\omega^*}$ with magnitude $\Im\{\omega^{(m)}\} \approx \frac{1}{4}\omega^*$. The drift frequency ω^* contains a wavenumber k_y, but in a cylindrical geometry the azimuthal wavenumber will usually be fixed as, for instance, $k_y \approx 1/R$ where R is the radius of the plasma column.

velocity, V_A, and the electron thermal velocity U_{the}. For the low-β plasmas considered here, the Alfvén velocity is much larger than the acoustic velocity as well as the electron thermal velocity.

As k_z is reduced we find an increase in growth rate but also a reduction in oscillation frequency. This feature can be explained by noting that the electron mass enters only in the combination $m\nu_{e,i}$. This means that an increase in collisionality and thereby resitivity is equivalent to an increase in electron mass. Although this term does not enter as an inertia term (this was ignored in the model equations for the electron dynamics) it corresponds to an effective increase in mass loading of the drift waves as the electrons have to collide many times when moving along **B** from wave crest to wave trough. The decrease in oscillation frequency is a consequence of this (Hendel *et al* 1968).

In the limit of vanishing resistivity, $\sigma_{\|} \to \infty$ in (6.26), we obtain the real frequency $\omega = \omega^*/(1 + b)$, i.e. the drift-waves are marginally stable as in chapter 4. In other words; if the electron flow along magnetic field lines is unimpeded so the electrons can maintain an isothermal Boltzmann equilibrium, then the waves remain stable irrespective of the inclusion of ion inertia in the equations for the ion dynamics, in agreement with the results of section 5.1.

6.6 Amplitude and phase relations

By inserting ω from (6.27) into (6.21) we find that $|\chi| < |\eta|$ in general, i.e. the potential drop between wave-top and wave-crest is decreased, for a given density perturbation, when the electron flow along magnetic field lines is inhibited by collisions. This was, after all, to be expected. In figure 6.2 we illustrate the variation of the ratio $|\chi/\eta|$ and in figure 6.3 the corresponding phase angle Ψ between η and χ as a function of $\sqrt{\sigma_{\|}} = (k_z/k_y)\sqrt{\omega_{ce}\Omega_{ci}\tau_{ei}/\omega^*}$ and b fixed. From (6.21) we have

$$\Psi \equiv \mathrm{ArcTan}\left(\frac{\Im\left[\dfrac{\omega^* + ib\sigma_{\|}}{\omega + ib\sigma_{\|}}\right]}{\Re\left[\dfrac{\omega^* + ib\sigma_{\|}}{\omega + ib\sigma_{\|}}\right]}\right).$$

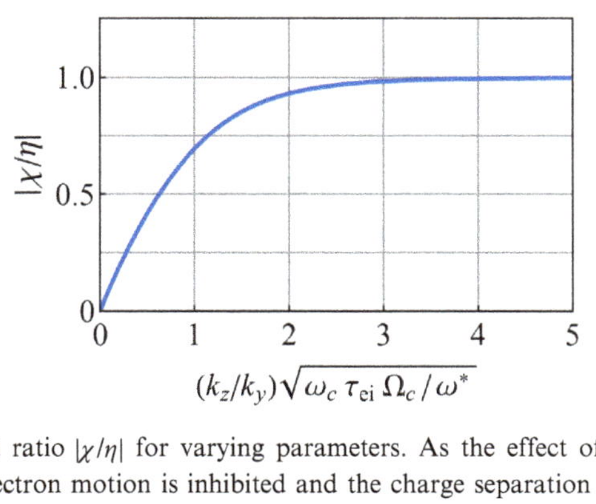

Figure 6.2. The normalized ratio $|\chi/\eta|$ for varying parameters. As the effect of resistivity increases (small conductivities σ) the free electron motion is inhibited and the charge separation reduced for a given density perturbation, resulting in a reduced electrostatic potential. We took $b = 0$ for simplicity.

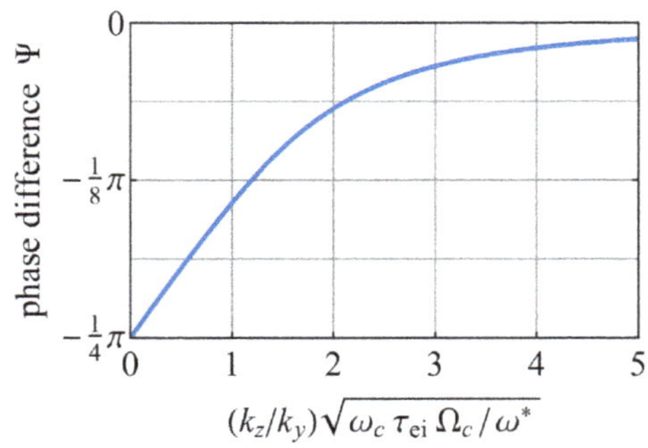

Figure 6.3. The phase angle, Ψ, between η and χ for varying parameters. We took $b = 0$ for simplicity.

For short **B**-parallel wavelengths we have $\omega \approx \omega^*$ giving $|\eta| \sim |\chi|$ and $\Psi \sim 0$, and the electrons are nearly in local Boltzmann equilibrium. As $\lambda_\parallel$ is increased, the phase angle increases and $|\eta|$ exceeds $|\chi|$; when the electron motion is impeded by collisions it takes a larger perturbation in density, as compared to the collisionless case, to build-up a potential variation as already mentioned. For small $\sigma_\parallel/\omega^*$ we have

$$\Psi \approx \mathrm{ArcTan}\left(\frac{\mathfrak{I}\left\{\dfrac{1}{(1+i)\sqrt{\frac{1}{2}\sigma_\parallel/\omega^*}}\right\}}{\mathfrak{R}\left\{\dfrac{1}{(1+i)\sqrt{\frac{1}{2}\sigma_\parallel/\omega^*}}\right\}}\right) = -\frac{\pi}{4},$$

independent of b.

6.7 Physical picture

The results from the foregoing section can be explained physically following Chen (1965b). Thus figure 6.4 illustrates the basic elements of the derivation. A solid line gives an equi-density contour of the perturbation as in figure 4.2. The ion velocities deviate as in figure 5.2 also here from the $\mathbf{E} \times \mathbf{B}$ velocities because of the finite ion inertia. At position ② in figure 6.4 the electric field E_y is maximum and hence $\partial E_y/\partial t = 0$ and u_y then vanishes also. At a position ① on the other hand u_x vanishes but u_y is maximum. In the vicinity of position ② ions are brought in from the equi-density contour of the perturbation as in figure 4.2. The ion velocities deviate as in figure 5.2 also here from the $\mathbf{E} \times \mathbf{B}$ velocities because of the finite ion inertia. At position ② in figure 6.4 the electric field E_y is maximum and hence $\partial E_y/\partial t = 0$ and u_y then vanishes also. At a position ①, on the other hand, u_x vanishes but u_y is maximum. In the vicinity of position ② ions are brought in from the main distribution by u_x while u_y, which points away from ②, acts to deplete ions. Conversely in the neighborhood of position ④ ions are depleted by being brought

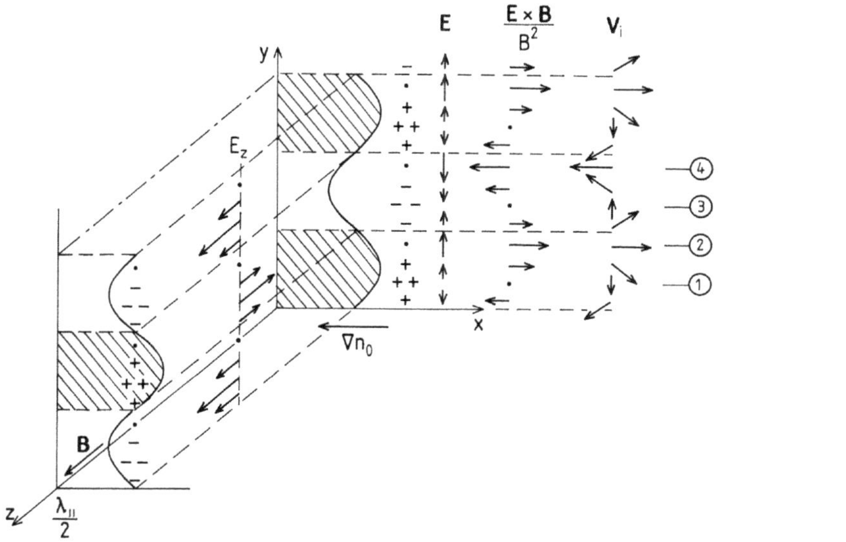

Figure 6.4. Basic elements of drift-wave properties for unstable conditions with ion inertia and finite resistivity included. The solid line indicates an equi-density contour, while electric fields and ion velocities are indicated by small arrows. Charge densities are indicated by + and – symbols. Regions with enhanced density are shaded for clarity. The ions have an elliptical orbit with major axis along the density gradient. When the polarization term due to ion inertia is ignored, this ellipse collapses to a straight line parallel to the density gradient.

back into the main distribution by a velocity u_x while the converging velocity component u_y acts to increase the local ion density.

Because of finite conductivity, the assumption of local Boltzmann equilibrium (4.7) is no longer accurate and the equi-potential lines are no longer in phase with the equi-density contours. The electrons can only short-circuit the ion charges by flowing along magnetic field lines, but because of the collisional drag this requires an excess electric field, over the normally existing field which balances the electron pressure gradient along **B**. As the wave propagates in the direction perpendicular to **B** as well as ∇n_0, a perturbation in ion density is propagating also *along* **B**. Because of their reduced mobility the electrons cannot follow this motion and lag slightly behind, giving rise to a charge separation. The net result is a downward shift in the potential distribution, as indicated in figure 6.4. In the (x, y)-plane, the potential curves are shifted in the negative y-direction. This shift causes a component of u_x to appear in phase with the density perturbation n. Thus the average u_x is positive in the region where n is positive and vice versa. A positive u_x causes an increase in n where it is large already (shaded regions) due to the presence of the gradient ∇n_0, while a negative u_x causes a further decrease in density at positions where n is negative; consequently the perturbation will grow in time. This example also illustrates the importance of drift waves for anomalous transport of plasma across magnetic field lines. A full study of that problem under stationary conditions (in a statistical sense) will require an analysis of the saturated turbulent stage of the linear instability (Hu *et al* 1997).

As the **B**-parallel wavelength is increased, it takes a longer time for the electrons to transport excess charge away from the (x, y)-plane. Hence the frequency of the oscillations decreases. Finally, at $k_z = 0$ the wave ceases to exist. This limit corresponds to a static perturbation, i.e. an arbitrary two-dimensional charge distribution can be generated by charging-up flux tubes. Since classical diffusion was ignored, these charges do not annihilate, and the perturbations remain indefinitely in the present description. This $\omega = 0$ limit for $k_z = 0$ is in this sense not physically correct; the missing small imaginary part of ω is associated with the basic assumptions concerning the zero order, or unperturbed state. The singularity in figure 6.2 which is caused by $\omega \to 0$ for $k_z \to 0$ is therefore also unphysical.

In one respect the results obtained in this section are unphysical; they predict linearly unstable waves for arbitrarily large wavenumbers k_y, i.e. arbitrarily small (non-zero) **B**-perpendicular wavelengths. Even with the assumption of vanishing ion temperature, the basic equations break down when the scales become comparable to the electron Larmor radius. In reality, however, it is more likely to be the finite *ion*-Larmor radius which sets the limit for applicability of the foregoing analysis. In the following sections, these finite ion Larmor radius effects will be studied in detail.

6.7.1 Model consistency

As already mentioned, we are making a stability analysis of a steady state plasma gradient, which is not *really* a steady state, since the collisional diffusion is eroding inhomogeneities in the plasma due to classical diffusion mediated by the electron-ion collisions. These very same collisions give rise to the finite plasma conductivity, which is responsible for the resistive drift instability. For the model to be consistent we require that the characteristic time for the drift wave dynamics is much shorter than the time it takes for the density gradient to be 'flattened out'. We consider first the latter process. We have the diffusion equation for the **B**-perpendicular dynamics to be $\partial n/\partial t = \nabla_\perp \cdot D\nabla_\perp n$, with D being the diffusion coefficient. In analogy with the continuity equation we can define a diffusion flux as $\mathbf{u}_D n = -D\nabla_\perp n$, implying a diffusion velocity $\mathbf{u}_D$ which we have obtained already (6.11). Since ions and electrons diffuse 'together' by electron-ion collisions, there are no charge separations associated with this process, and the diffusion velocity is the same for both species. Introducing a characteristic length scale as $\mathcal{L} \equiv n/|\nabla_\perp n|$, we obtain a characteristic time for the variation of the density gradient as $\tau_0 \equiv \mathcal{L}/|\mathbf{u}_D|$. The density scale length is a constant only for an exponential profile, but we here take a representative value, just as in the foregoing analysis. Using (6.11) we have $|u_D| = \kappa T_e \nu_{ei}/(\omega_{ce}^2\, m\mathcal{L})$, giving $\tau_0 = \omega_{ce}^2\, m\mathcal{L}^2/(\kappa T_e \nu_{ei})$. If we take the characteristic time for the drift waves to be $1/\omega^*$, we require $\tau_0 \gg 1/\omega^*$ for the quasi-stationary density gradient model to be applicable. By simple manipulations we obtain the estimate

$$\frac{\omega_{ce}}{\nu_{ei}} \gg \frac{1}{k_y\mathcal{L}}. \tag{6.28}$$

One might, however, have argued that the characteristic time scale is *not* the drift wave frequency as such, but the linear growth rate: after all, we will be able to actually observe a drift wave only after a time $>1/\Im\{\omega\}$. In (6.27) we found $\Im\{\omega\} \approx \omega^{*2}/\sigma_{\|}$, and it is readily shown that $\tau_0 \gg 1/\Im\{\omega\}$ implies

$$\frac{1}{k_y a_i}\left(\frac{k_z}{k_y}\right)^2 \ll 1, \tag{6.29}$$

independent of ν and $\mathcal{L}$. We note that we generally have $k_z \ll k_y$ for drift waves, but on the other hand, we also have $k_y a_i \ll 1$. The inequality (6.29) can be fulfilled for $k_z^2 \ll k_y^3 a_i$ which can easily be achieved.

- **Exercise:** Carry out the calculations giving the inequalities (6.28) and (6.29).

Implicit in the foregoing discussion we assumed that the appropriate growth rate for the drift instability was a fairly small one, obtained for $(k_z/k_y)\sqrt{\omega_{ce}\Omega_{ci}/\omega^*\nu_{ei}} \gg 1$, where $\Re\{\omega\} \approx \omega^*$, see figure 6.1. By some simple manipulations, we can write the normalized variable on the horizontal axes of figure 6.1, 6.2 and 6.3 as

$$\frac{k_z}{k_y}\sqrt{\frac{\omega_{ce}\Omega_{ci}}{\omega^*\nu_{ei}}} = \frac{k_z\ell_c}{k_y a_i}\sqrt{\frac{\nu_{ei}}{\omega^*}}, \tag{6.30}$$

where we introduced the collisional mean free path $\ell_c \equiv \sqrt{\kappa T_e/m}/\nu_{ei}$ for the electrons colliding with ions. We require in general $k_y a_i \ll 1$ as well as $k_z\ell_c \ll 1$ for the stability analysis of section 6.5 to be appropriate. It is then justified to assume $k_z\ell_c/k_y a_i \sim 1$. For relevant plasmas, we expect $\nu_{ei} \gg \omega^*$, and (6.30) is then a large number, in general. The approximations giving (6.28) and (6.29) are therefore justified.

One *might* argue that we can always obtain a growth rate $\Im\{\omega\} \approx \omega^*/4$, see figure 6.1, by taking k_z small enough. In reality, this will for most relevant experimental conditions correspond to wavelengths much larger than the length of a device, and this might be difficult to argue, even when the sheath conditions at the ends are taken into account (Chen 1965a, 1979). The analysis becomes significantly complicated if the end conditions are taken into account: when a particle is lost to the end-plate, its momentum is lost from the plasma as well. When a new particle is emitted to replace it from the surface (i.e. a hot plate in a Q-machine (Motley 1975)) this particle has to be 'synchronized' with the wave, a process which again costs energy. The complexity of these processes are often jointly called *end-plate damping*. It *is* of course feasible to maintain a steady state plasma gradient (after all, we find one in the ionosphere, for instance), but this has to be done by sources and sinks, and these will effectively act in the same way as the end-plate damping. The important conclusion here is that a truly steady state plasma gradient can only be maintained by some mechanisms which inevitable influence the growth rate of the instability, and they should therefore be included in a detailed analysis. This problem is discussed in little more detail in chapter 11, for ionospheric conditions where there are no obvious constraints on the wavelength parallel to the magnetic field.

- **Exercise:** Consider a Q-machine plasma (Motley 1975) with density $n_0 \approx 10^{15}$ m^{-3}, temperature 0.2 eV, and cesium ions. Assume that the plasma column of length 1 m and diameter 4 cm is terminated at both ends by a hot plate, i.e. double-ended conditions, which can be considered as conducting in the sense that the waves have nodes there. Calculate the collisional electron mean free path, the electron–ion collision frequency and estimate the maximum growth rate of the resistive drift wave instabilities in this plasma. Note that for Q-machines, the electron temperatures are generally only slightly larger than the ion temperatures, so that the stability analysis of section 6.5 can only be taken as approximative here.

References

Chen F F 1965a Excitation of drift instabilities in thermionic plasmas *Plasma Phys. (J. Nucl. Energy Part C)* **7** 399–417

Chen F F 1965b Resistive overstabilities and anomalous diffusion *Phys. Fluids* **8** 912–9

Chen F F 1970 Linear drift waves UCLA Lecture Notes unpublished

Chen F F 1979 Axial eigenmodes for long-$\lambda_{||}$ waves in plasmas bounded by sheaths *Phys. Fluids* **22** 2346–58

Hendel H W, Chu T K and Politzer P A 1968 Collisional drift waves – identification, stabilization, and enhanced plasma transport *Phys. Fluids* **11** 2426–39

Hu G, Krommes J A and Bowman J C 1997 Statistical theory of resistive drift-wave turbulence and transport *Phys. Plasmas* **4** 2116–33

Manheimer W M 1977 *An Introduction to Trapped–Particle Instability in Tokamaks* ERDA Critical Rev. Ser., National Technical Inf. Serv. Rep. TID–27157

Motley R W 1975 *Q Machines* (New York: Academic)

Pécseli H L 2012 *Waves and Oscillations in Plasmas* (London: Taylor and Francis)

Timofeev A V 1963 Instability of a weakly ionized nonuniformly heated plasma in a uniform magnetic field *Sov. Phys. Techn. Phys.* **7** 959–62

Timofeev A V 1962 *Russian original Zh. Tech. Fiz.* **32** 1297–301

van Kampen N G and Felderhof B U 1967 *Theoretical Methods in Plasma Physics* (Amsterdam: North-Holland)

IOP Publishing

Hans Pécseli

Chapter 7

Resistive drift waves with warm ions

7.1 Basic equations

The modifications of the results of the foregoing sections due to finite ion temperatures are relatively trivial as long as the electron–ion temperature ratio is large. In this case we can simply include a pressure term $-\kappa T_i \nabla n_i$ in the expression for the force density on the ion component. When the two temperatures are comparable, the analysis becomes more complicated due to finite ion Larmor radius (FLR) effects, and the results of the foregoing sections are modified, the drift wave stability conditions in particular. We have reason to expect that this will be so because of the effective ion Larmor radius a_i (the Larmor radius derived by the ion mass and electron temperature) that we saw entering in chapter 6. It is reasonable to expect that also the *true* ion Larmor radius (as distinct from a_i) has importance, on levels where a_i is important. For $T_i > 0$, we shall here replace (6.4) with

$$Mn\left(\frac{\partial \mathbf{u}}{\partial t} + \mathbf{u} \cdot \nabla \mathbf{u}\right) = en(-\nabla \phi + \mathbf{u} \times \mathbf{B}) - \nabla p - \nabla \cdot \Pi + Mn\mathbf{g}, \tag{7.1}$$

where we again assumed quasi-neutrality. The subscript i has been suppressed on the variables $\mathbf{u}$ and n since there will be no possibilities for misunderstandings here. We neglected classical diffusion and split the divergence of the ion stress tensor with T_i constant into an isotropic part $\nabla p = \kappa T_i \nabla n \equiv \kappa T_i \nabla \cdot \mathbf{1} n$ and an anisotropic part $\nabla \cdot \mathbf{\Pi}$. (In the following analysis we could write a slightly more general expression by assuming adiabatic ion dynamics. In a linearized model the change is small, however. Taking $p/p_0 = (n/n_0)^\gamma$ with $\gamma = C_p/C_V$ we linearize this expression with $p = p_0 + \widetilde{p}$ with $\widetilde{p} \ll p_0$ to have $1 + \widetilde{p}/p_0 = (1 + \widetilde{n}/n_0)^\gamma \approx 1 + \gamma\widetilde{n}/n_0$. Omitting ~ for simplicity we find in the linearized limit $\nabla p = \gamma_i \kappa T_i \nabla n$ and can allow for adiabatic ion dynamics. In the following we take $\gamma_i = 1$ for simplicity since the change in the linear analysis is close to trivial: at any stage we can replace T_i by γT_i.) The tensor $\mathbf{\Pi}$

doi:10.1088/978-0-7503-1251-6ch7

is called the magnetic viscosity tensor (Kaufman 1960, Roberts and Taylor 1962, Braginskií 1965, Chen 1970), and consists of a part connected with ion-ion collisions and a part which remains in the limit $\nu_{ii} \to 0$. This collisionless viscosity is simply the FLR effect mentioned earlier: the ions have a modified $\mathbf{E} \times \mathbf{B}$ drift in a non-uniform electric field. In addition, FLR-effects act to 'smear-out' the ion charge distribution, i.e. the ions are to be considered as being distributed along their gyro-circle rather than associating their position with the gyro-center. It should be emphasized that the viscosity tensor in the form given here is correct only to lowest order in an expansion of $(kr_L)^2$ where k is a wavenumber characterizing the perturbation and r_L is the ion Larmor radius. A study retaining FLR-effects to all orders in $(kr_L)^2$ is based on plasma kinetic theory, where gyro-kinetic equations have been extensively used in studies of low frequency drift waves in collisionless plasmas.

In (7.1) we have also included a term $Mn\mathbf{g}$ accounting for a constant gravitational acceleration $\mathbf{g}$, mostly because it does not complicate the analysis to do so. The gravitational force can also represent the centrifugal force of a rotating plasma column, but in that case g is no longer a constant, but varies with radius.

Postponing collisional viscosity to the next section, we may write the relevant components of $\mathbf{\Pi}$ in the collisionless case as follows, correct to first order in $k^2 r_L^2$ as already mentioned

$$\begin{aligned} \Pi_{yy} = -\Pi_{xx} &= \frac{1}{2}\frac{n\kappa T_i}{\Omega_{ci}}\left(\frac{\partial u_y}{\partial x} + \frac{\partial u_x}{\partial y}\right) \\ \Pi_{xy} = \Pi_{yx} &= \frac{1}{2}\frac{n\kappa T_i}{\Omega_{ci}}\left(\frac{\partial u_x}{\partial x} - \frac{\partial u_y}{\partial y}\right). \end{aligned} \tag{7.2}$$

We shall neglect u_{zi} as in the foregoing sections. Since $\omega_{ce} \gg \Omega_{ci}$, we see right away that viscosity is unimportant for the electron dynamics in the present case, where all relevant spatial scales are much larger than the electron Larmor radius.

After some simple algebra, the contribution in (7.1) from the collisionless viscosity tensor with constant ion temperature, T_i, can be given a more compact formulation as

$$\begin{aligned} \nabla \cdot \mathbf{\Pi} = \frac{n\kappa T_i}{\Omega_{ci}}\Bigg[&\hat{\mathbf{z}} \times \nabla^2 \mathbf{u} - \left(\nabla u_y - \hat{\mathbf{z}} \times \nabla u_x\right)\frac{\partial \ln n}{\partial x} \\ &- \left(\nabla u_x + \hat{\mathbf{z}} \times \nabla u_y\right)\frac{\partial \ln n}{\partial y}\Bigg], \end{aligned} \tag{7.3}$$

with $\hat{\mathbf{z}}$ being a unit vector in the z-direction. When we linearize the equations, we will assume $n = n_0(x)$, and the last term in (7.3) vanishes.

The physics behind collisionless viscosity can be explained by a simple example: assume we have an initially unperturbed plasma density with a simple ion fluid velocity field in the form $\mathbf{u} = \{u_x(x, t), 0, 0\}$, where we have z along the magnetic field lines. This flow is compressible, $\nabla \cdot \mathbf{u} \neq 0$. We here ignore the term $u_x \partial n/\partial x$ in the ion continuity equation since it vanishes at $t = 0$ and is presumed to remain small for a transient time interval. For such a short initial time interval the ion continuity equation then gives $\partial \ln n/\partial t = -\partial u_x/\partial x$ or $\partial^2 \ln n/\partial t\partial x = -\partial^2 u_x/\partial x^2$ implying that a density gradient builds up with ∇n in the $\hat{\mathbf{x}}$-direction. This density variation has an associated electron diamagnetic drift by (4.10), which can be re-written to give $\partial u_{De}/\partial t = -(\kappa T_e/eB)\partial^2 \ln n/\partial t\partial x = (\kappa T_e/eB)\partial^2 u_x/\partial x^2$ with $\mathbf{u}_{De}$ perpendicular to the $\hat{\mathbf{x}}$-direction. The contribution of this term to $Mn\partial\mathbf{u}/\partial t$ in (7.1) becomes $(n\kappa T_e/\Omega_{ci})\partial^2 u_x/\partial x^2$, which is precisely the first term in (7.3) for the given geometry. This is a finite ion temperature effect. The last two terms in (7.3) contribute when the gradients in the plasma density increase. We thus find that due to the collisionless ion viscosity an inhomogeneous velocity field in, say, the $\hat{\mathbf{x}}$-direction induces an ion drift velocity in the perpendicular direction. In some respect this is reminiscent of momentum transfer due to viscous effects (Pécseli 2012), but that effect is mediated by irreversible collisional processes, so it may be somewhat misleading to generally introduce the word 'viscosity' in relation to the part of the $\mathbf{\Pi}$-tensor discussed here.

- **Example:** When velocities are expressed as $\mathbf{E} \times \mathbf{B}/B^2$-velocities, we can write

$$\begin{aligned} \Pi_{yy} = -\,\Pi_{xx} &= \frac{1}{2}\frac{n\kappa T_i}{B\Omega_{ci}}\left(\frac{\partial^2\phi}{\partial x^2} - \frac{\partial^2\phi}{\partial y^2}\right) \\ \Pi_{xy} = \Pi_{yx} &= \frac{1}{2}\frac{n\kappa T_i}{B\Omega_{ci}}\left(-2\frac{\partial\phi}{\partial x\partial y}\right), \end{aligned} \tag{7.4}$$

or in terms of a new tensor $\mathbf{Q}$

$$\mathbf{\Pi} \equiv \frac{1}{2}\frac{n\kappa T_i}{B\Omega_{ci}}\mathbf{Q},$$

where in particular

$$\nabla \cdot \mathbf{Q} = \begin{pmatrix} -\dfrac{\partial}{\partial x}\nabla_\perp^2\phi \\ -\dfrac{\partial}{\partial y}\nabla_\perp^2\phi \end{pmatrix}.$$

These relations illustrate some basic mathematical properties of the collisionless ion viscosity. The analysis in the present chapter uses slightly more general expressions.

7.2 Equilibrium

As a first step we find a steady state solution. In addition to u_{e0} found before, there is now also an ion fluid drift velocity u_{i0} given by

$$\mathbf{u}_{i0} = (u_{Di} - u_G)\hat{\mathbf{y}}, \tag{7.5}$$

where the two contributions are an ion diamagnetic drift and a gravitational drift. The first is not a true particle drift, while the second one is, with

$$u_{Di} = \frac{\kappa T_i}{eB}\frac{n_0'}{n_0} \qquad \text{and} \qquad u_G \equiv \frac{g}{\Omega_{ci}}.$$

We have assumed n_0'/n_0 = constant and $\mathbf{g} = g\hat{\mathbf{x}}$ = constant, so that u_{i0} = constant and $\mathbf{u}_0 \cdot \nabla \mathbf{u}_0 = 0$, together with $\mathbf{\Pi}_0 = 0$. Equation (7.5) is the only place where the effect of gravity will appear. With the small electron mass, the effect of gravity on the electron dynamics can be ignored and (6.21) remains unchanged.

- **Comment on centrifugal forces:** It is tempting to let the gravitational acceleration g be due to a curvature in B, and set it equal to $\kappa T/MR_c$, where R_c is the radius of curvature and $\sqrt{\kappa T/M}$ is a characteristic velocity. This is however only formally correct. The bending of the field lines can only be achieved by introducing a gradient in magnetic field and the ∇B-drift must then be taken into account as well, giving modifications of the equilibrium and the subsequent analysis (Chen 1984). To substantiate this argument we assume that we succeeded in bending the magnetic field lines to have a curvature R in some region of space. Since we deal with low β-plasmas we require $\nabla \times \mathbf{B} = 0$ as in vacuum, i.e. the presence of the plasma does not influence the magnetic field in any appreciable way. In cylindrical coordinates we then find that $\nabla \times \mathbf{B}$ only has a $\hat{\mathbf{z}}$-component, since the assumed magnetic field, $\mathbf{B}$, only has a $\hat{\boldsymbol{\theta}}$-component. We then have $(\nabla \times \mathbf{B})_z = (1/r)\partial(rB_\theta)/\partial r = 0$ which implies $B_\theta \sim 1/r$; a curvature of the magnetic field is accompanied by a variation in intensity of the same field along the radius of curvature. Consequently, the curvature drifts will necessarily be accompanied by ∇B-drifts (Chen 1984).

  The curvature and gradient-B drifts can in principle be included in the analysis as well, but are not considered here. An important consequence of the curved magnetic field lines is that the diamagnetic drift associated with the density gradient is no longer divergence-free. This will have importance for the so-called ballooning modes which also involve magnetic field perturbations. Similarly, we note that the electrostatic $\mathbf{E} \times B/B^2$-drift is no longer incompressible in the case where the magnetic field varies in space.

7.3 Perturbation

The linear part of (7.1) is

$$M\left(\frac{\partial \mathbf{u}}{\partial t} + \mathbf{u}_0 \cdot \nabla \mathbf{u}\right) = e(-\nabla\phi + \mathbf{u} \times \mathbf{B}) - \kappa T_i \nabla \eta - \frac{1}{n_0}\nabla \cdot \mathbf{\Pi}_1, \tag{7.6}$$

where we used $\nabla n/n = \nabla \ln n = \nabla \ln(n_0 + n_1) \approx \nabla n_0/n_0 + \nabla\eta$. We shall make the local approximation from the outset and neglect x-derivatives of $\mathbf{u}$, η and ϕ. In evaluating $\mathbf{\Pi}_1$, there will be terms $n_0\nabla\mathbf{u}$ and $n_1\nabla u_0$; the latter vanishes because we have taken u_0 = constant. Only waves propagating along $\hat{\mathbf{y}}$, in the direction perpendicular to the magnetic field as well as the density gradient, will be studied, i.e. $k_z \approx 0$ and $k_x = 0$ also. With these assumptions, (7.2) yields

$$\Pi_{yx} = -\frac{1}{2}ik_y\frac{n_0\kappa T_i}{\Omega_{ci}}u_y, \qquad \Pi'_{yx} = -\frac{1}{2}\,ik_y\,\frac{\kappa T_i}{\Omega_{ci}}n'_0u_y$$
$$\Pi_{xx} = -\Pi_{yy} = -\frac{1}{2}ik_y\frac{n_0\kappa T_i}{\Omega_{ci}}u_x, \qquad \Pi'_{xx} = -\frac{1}{2}\,ik_y\,\frac{\kappa T_i}{\Omega_{ci}}n'_0u_x, \tag{7.7}$$

with a prime $'$ again denoting differentiation with respect to x. To simplify the notation we define

$$\omega_i \equiv k_y u_{Di} = k_y\,\frac{\kappa T_i}{M\Omega_{ci}}\frac{n'_0}{n_0}$$
$$r_L^2 \equiv \frac{2\kappa T_i}{M\Omega_{ci}^2}, \qquad \theta \equiv \frac{T_i}{T_e}, \tag{7.8}$$

so that

$$\omega_i = -\theta\omega^* \qquad \text{and} \qquad b = k_y^2\frac{r_L^2}{2\theta}.$$

We recognize u_{Di} as the ion diamagnetic drift, and ω_i is in this sense the ion equivalent of ω^* although with opposite sign. It is possible to eliminate ω_i everywhere: most places it will simply imply a multiplier $(1 + \theta)$ on ω^*. It is however an advantage to keep ω_i explicitly as a reminder for the ion diamagnetic drift.

In (6.20) we had a definition that is seemingly different from the parameter b given here, but at closer inspection, the two expressions are found to be the same. See also the summary of notations in appendix A.

The x and y components of (7.6) give

$$\left(\omega - k_y u_0 + \frac{1}{2}\omega_i\right)u_x = i\Omega_{ci}\left(1 - \frac{1}{2}\theta b\right)u_y$$
$$\left(\omega - k_y u_0 + \frac{1}{2}\omega_i\right)u_y = k_y\,\frac{\kappa T_e}{M}(\chi + \theta\eta) - i\Omega_{ci}\left(1 - \frac{1}{2}\theta b\right)u_x. \tag{7.9}$$

The waves are electrostatic and $\mathbf{E}$ is along $k\|\hat{\mathbf{y}}$, but the ions have an oscillating velocity component also along $\hat{\mathbf{x}}$. The equations (7.9) refer to the *ion* dynamics, and the electron temperature appears solely because of the normalizations used for χ and θ. The equations for u_x and u_y may be solved writing $\alpha = \omega - k_y u_0 + \frac{1}{2}\omega_i$, $\beta = 1 - \frac{1}{2}\theta b$ and eliminating u_y and u_x, respectively. We then have

$$\left(\alpha^2 - \beta^2\Omega_{ci}^2\right)u_x = i\beta\Omega_{ci}k_y\frac{\kappa T_e}{M}(\chi + \theta\eta)$$
$$\left(\alpha^2 - \beta^2\Omega_{ci}^2\right)u_y = \alpha k_y\frac{\kappa T_e}{M}(\chi + \theta\eta). \tag{7.10}$$

When the effect of gravity is small, the term α^2 can be dropped for most relevant cases when $T_i \leqslant T_e$, because it is of relative order b^2. This can be seen by writing

$$\left|\frac{\alpha}{\beta\Omega_{ci}}\right| \approx \frac{k_y u_0}{\Omega_{ci}} = k_y \frac{\kappa T_i}{M\Omega_{ci}^2}\frac{1}{L} = \frac{k_y^2 r_L^2}{2k_y L} \lesssim b \quad \text{for} \quad k_y L > 1, \quad \theta \lesssim 1,$$

with L being a typical length scale for the density gradient. With this simplification, (7.10) becomes

$$u_x = -i\frac{k_y(\chi + \theta\eta)}{\Omega_{ci}\left(1 - \frac{1}{2}\theta b\right)}\frac{\kappa T_e}{M}, \tag{7.11}$$

$$u_y = -\frac{\left(\omega - k_y u_0 + \frac{1}{2}\omega_i\right)k_y(\chi + \theta\eta)}{\Omega_{ci}^2\left(1 - \frac{1}{2}\theta b\right)^2}\frac{\kappa T_e}{M}. \tag{7.12}$$

The coefficients $1/(1 - \frac{1}{2}\theta b)$ and $1/(1 - \frac{1}{2}\theta b)^2$ enhance the u_x and u_y velocity components as compared to their values for $T_i = 0$.

The linearized ion equation of continuity is expressed as

$$\left(\omega - k_y u_0\right)\eta - k_y u_y + i u_x \frac{n_0'}{n_0} = 0. \tag{7.13}$$

Inserting (7.11) and (7.12) into (7.13), we find a fortunate cancellation of terms which is due to fluid drifts which are not real particle drifts (Chen 1970), and end up with

$$\left(\omega - k_y u_0 - \theta\omega^*\right)\eta + \left(b\left(\omega - k_y u_0\right) - \omega^*\right)\chi = 0, \tag{7.14}$$

by omitting second order terms in b.

A dispersion relation is obtained by eliminating η or χ, using (6.21). It is relatively easy to retain the α^2-term in (7.10), but the resulting dispersion relation will be considerably more complicated, which makes the effort hardly worthwhile.

- **Comment:** The cancellation of terms referred to before can be argued in more detail, to make the physics more evident. It is readily shown that the viscosity term gives a contribution $\mathbf{u}_\pi \equiv (1/n_0 eB)\hat{\mathbf{z}} \times \nabla \cdot \mathbf{\Pi}$ to the ion fluid velocity. This term is however canceled by the polarization drift when inserted into the $\nabla \cdot (n\mathbf{u})$-term in the continuity equation (Weiland 1992). To substantiate this statement, we first calculate $\nabla \cdot (n_0(x)\mathbf{u}_\pi)$ and find using (7.3), including only terms which are linear in n_0'/n_0

$$\begin{aligned}\nabla \cdot (n_0(x)\mathbf{u}_\pi) &= \mathbf{u}_\pi \cdot \nabla n_0 + n_0 \nabla \cdot \mathbf{u}_\pi \\ &= -\frac{1}{4}r_L^2\left(\nabla n_0 \cdot \nabla_\perp^2 \mathbf{u} + n_0 \nabla_\perp^2 \nabla \cdot \mathbf{u} + n_0\left(n_0'/n_0\right)\nabla^2 u_x\right).\end{aligned} \tag{7.15}$$

Assuming $\nabla \cdot \mathbf{u} \approx 0$, we find $\nabla \cdot (n_0(x)\mathbf{u}_\pi) \approx -\frac{1}{2}r_L^2 \nabla n_0 \cdot \nabla^2 \mathbf{u}$.

We then consider the polarization drift $\mathbf{u}_p \equiv (\hat{\mathbf{z}} \times \partial\mathbf{u}/\partial t + \mathbf{u} \cdot \nabla(\hat{\mathbf{z}} \times \mathbf{u}))/\Omega_{ci}$ which contains two terms. By linearization we calculate the contribution from the last term in $\mathbf{u}_p$ to $\nabla \cdot (n\mathbf{u})$, obtaining

$$\begin{aligned} &\frac{n_0}{\Omega_{ci}}(\mathbf{u}_{Di} \cdot \nabla)\nabla \cdot (\hat{\mathbf{z}} \times u) \\ &= -\frac{1}{2}r_L^2(\hat{\mathbf{z}} \times \nabla n_0) \cdot \nabla\left(\frac{\partial u_x}{\partial y} - \frac{\partial u_y}{\partial x}\right), \end{aligned} \tag{7.16}$$

where the zero order velocity $\mathbf{u}_{Di} = \frac{1}{2}r_L^2\Omega_{ci}(\hat{\mathbf{z}} \times \nabla n_0)/n_0$ was again assumed constant. Combining the contributions to $\nabla \cdot (n\mathbf{u})$ from $\mathbf{u}_\pi$ and $\mathbf{u}_p$ we find after simple manipulations the result $\frac{1}{2}r_L^2\nabla n_0 \cdot \nabla(\nabla \cdot \mathbf{u}) \approx 0$ again with the assumption of incompressibility; the two contributions cancel to the desired accuracy. Physically this result is a consequence of diamagnetic drifts being non-convective, since the gyrating ions do not move on average. Note however that the cancellation is not *exact* if we include terms of higher order than those considered here, and it relies in particular on the assumption that the fluctuations in plasma density are incompressible to lowest order, $\nabla \cdot \mathbf{u} \approx 0$ when we approximate ion velocities by the $\mathbf{E} \times \mathbf{B}/B^2$-drifts with constant $\mathbf{B}$.

Consequently, to the accuracy argued here the only contribution we find is from the time derivative part of the polarization drift, i.e.

$$\nabla \cdot \big(n(\mathbf{u}_\pi + \mathbf{u}_p)\big) \approx \nabla \cdot \left[\frac{n_0}{\Omega_{ci}}\frac{\partial}{\partial t}(\hat{\mathbf{z}} \times \mathbf{u})\right].$$

This expression contains two parts

$$-\frac{1}{2}n_0r_L^2\frac{\partial}{\partial t}\nabla^2\frac{e\phi}{\kappa T_i},$$

from the $\mathbf{E} \times \mathbf{B}$-drift and

$$-\frac{1}{2}n_0r_L^2\frac{\partial}{\partial t}\nabla^2\frac{n}{n_0},$$

from perturbations in the ion diamagnetic drift. We have again assumed that ∇n_0 is small, and ignored higher order terms.

7.4 Dispersion relation

The electron equations are unchanged, so we use (6.21) to eliminate η. We also use (7.5) for u_0. After a little algebra that includes division by b and using $\theta\omega^* = -\omega_i$, the expression (7.14) yields

$$\omega\big(\omega - k_yu_0\big) - \frac{\omega^*}{b}\big(k_yu_0 - \omega_i\big) + i\sigma_{\|}\big(\omega^*(1 + \theta) + (1 + b)(\omega - ku_0)\big) = 0.$$

Using $u_0 = u_{Di} - u_G$ and $\omega_i = k_y u_{Di}$, this result can be expressed as the dispersion relation

$$\begin{aligned} &\omega(\omega - \omega_i) + k_y u_G\left(\frac{\omega^*}{b} + \omega\right) \\ &+ i\sigma_{\|}\left[\omega - \omega^* + k_y u_G + b\left(\omega - \omega_i + k_y u_G\right)\right] = 0. \end{aligned} \tag{7.17}$$

The last term $bk_y u_G$ in (7.17) can be ignored for wavelengths much longer than the ion Larmor radius. The dispersion relation (7.17) can readily be analyzed in various limits as shown in the following.

7.4.1 Pure drift wave

For $u_G = 0$, we have the following dispersion relation for resistive drift waves

$$\omega(\omega - \omega_i) + i\sigma_{\|}[\omega - \omega^* + b(\omega - \omega_i)] = 0, \tag{7.18}$$

where the last b-containing term is usually small, as mentioned. In the limit of $b \to 0$ and $\theta \to 0$ the previous result for resistive drift waves with cold ions is recovered. In the limit of $\sigma_{\|}/\omega^* \gg 1$, we have

$$\omega - \omega^* \approx i\frac{\omega^*(\omega^* - \omega_i)}{\sigma_{\|}} = i\frac{\omega^{*2}(1 + \theta)}{\sigma_{\|}}. \tag{7.19}$$

Compared to (6.27), the growth rate is increased by the factor $1 + \theta$, representing the increase in total plasma pressure. This is reasonable from an energetic viewpoint. Kinematically, the growth rate is increased through the FLR effect in the ion drifts. The ion diamagnetic drift is opposite in direction to the electron diamagnetic drift, so the net result of a finite ion temperature is an increase in the diamagnetic current which acts as the energy source for the instability.

With the approximation of neglecting the term containing b in (7.18), the result is a simple modification of the dispersion relation shown in figure 6.1. In particular the combination of independent variables used in that figure is retained.

In the limit of infinite conductivity along magnetic field lines, $\sigma_{\|} \to \infty$, we find

$$\omega = \frac{\omega^*(1 - \theta b)}{1 + b}. \tag{7.20}$$

In this limit where the electrons are isothermally Boltzmann distributed, the FLR effects thus give rise to a dispersion of the waves given by $(1 - \theta b)/(1 + b) \approx (1 - \theta b)(1 - b) \approx 1 - b(1 + \theta)$.

7.4.2 Resistive-*g* mode

If we retain the $k_y u_G$ terms in (7.17) and take at the same time the limit $\sigma_{\|}/\omega^* \gg 1$, noting that $k_y \omega^*/b = -\Omega_{ci} n_0'/n_0$, we find

$$\omega \approx \omega^* - k_y u_G + i\frac{\omega^*(\omega^* - \omega_i) - gn_0'/n_0}{\sigma_\|}. \tag{7.21}$$

Depending on your point of view, (7.21) then shows either the increase in drift wave growth rate when $gn_0' < 0$, or the decrease in gravitational instability growth rate (see next subsection) due to finite k_z entering through $\sigma_\|$. Note that g has to be inserted with its sign included. This *gravitational instability* is often called the *Rayleigh–Taylor instability*, and it was first discussed for neutral fluids with a vertical density gradient in a constant gravitational field.

The pure gravitational mode with $k_z = 0$, as implied in figure 2.5, has a simple physical explanation, see section 2.3.1 and also the discussion by Rosenbluth and Longmire (1957), Lehnert (1962) or Chen (1984).

- **Example:** We discuss here the standard analysis of the gravitational instability following Rosenbluth and Longmire (1957). Consider the case where a uniform cold collisionless plasma is supported against gravity by a horizontal magnetic field. Assume that there is a sharp boundary between plasma and vacuum. In the case where this boundary is unperturbed, it is readily demonstrated that the proposed conditions form an equilibrium. In particular, we have a current driven by gravity, $\mathbf{J}_0 = ne(\mathbf{u}_i - \mathbf{u}_e)$ with $\mathbf{u}_i = (Mg/eB)\hat{\mathbf{y}}$ and $\mathbf{u}_e = -(mg/eB)\hat{\mathbf{y}} \approx 0$, with $-e$ being the electron charge. The equilibrium is however not stable as demonstrated in the following. Let the boundary be slightly perturbed by a spatially harmonic displacement as

$$\Delta x = a \sin(ky), \qquad \text{with} \qquad a \ll \lambda = 2\pi/k. \tag{7.22}$$

The perturbation (7.22) is assumed initially. For later times we assume that the coefficient can be time varying, i.e. $a = a(t)$. The currents are divergence-free and do not cause any change in the charge densities within the plasma.

First we calculate the surface charge density σ resulting from the perturbation. At a sharp boundary we have from the continuity equation the surface charge density $\partial\sigma/\partial t = -\hat{\mathbf{n}} \cdot \mathbf{J}$ where $\hat{\mathbf{n}}$ is the unit vector being normal to the surface. Here the current $\mathbf{J}$ is the *entire* current. However, since we use a linearized description it suffices to retain only $\mathbf{J}_0$, and ignore the products of small terms Consequently, $\hat{\mathbf{n}} \cdot \mathbf{J} \approx J_0 \cos\alpha_1$, where α_1 is the angle between the y-axis and $\hat{\mathbf{n}}$, see figure 7.1. We have $\cot\alpha_1 = \tan\alpha_2 = \mathrm{d}\Delta x/\mathrm{d}y = ka\cos(ky)$, where Δx is again the displacement of the boundary, noting the sign convention for α_1 and α_2. We have

$$\cos\alpha_1 = \frac{\cot\alpha_1}{\sqrt{1 + \cot^2\alpha_1}} = \frac{ak\cos(ky)}{\sqrt{1 + a^2k^2\cos^2(ky)}}.$$

With $ak \ll 1$ we have $\cos\alpha_1 \approx ak\cos(ky)$, giving

$$\frac{\partial\sigma}{\partial t} \approx \frac{n_0 Mg}{B} a(t)k\cos(ky), \tag{7.23}$$

using $M \gg m$ by ignoring the electron gravitational drift.

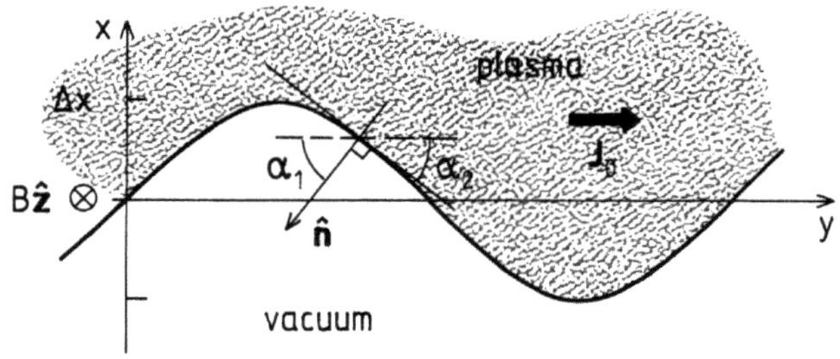

Figure 7.1. The geometry of the gravitational or Rayleigh–Taylor instability in magnetized plasmas. The surface displacement of the plasma from $x = 0$ is Δx, and the peak amplitude of the displacement is a. For small perturbation amplitudes, we have α_1 close to $\pi/2$ while α_2 is small. The magnetic field is in the positive $\hat{z}$-direction and gravity in the negative $\hat{x}$-direction.

Inside the plasma we have no charge separations because of the assumed homogeneous density, so Poisson's equation gives $\nabla^2\phi = 0$ there. Assuming no z-dependence of the electrostatic potential this implies $\partial^2\phi/\partial x^2 + \partial^2\phi/\partial y^2 = 0$. Assuming separation of variables as $\phi(x, y, t) = \phi_x(x, t)\phi_y(y, t)$ we have

$$\frac{1}{\phi_x}\frac{\partial^2\phi_x}{\partial x^2} = -\frac{1}{\phi_y}\frac{\partial^2\phi_y}{\partial y^2} = A^2,$$

where A^2 is a positive constant, to be determined. We expect ϕ to be an oscillating function of y, so the constant cannot be taken negative. Consequently, we have inside the plasma $\phi_x = \phi_{0x}\exp(-Ax)$ and $\phi_y = \phi_{0y}\cos(Ay) + \phi_{1y}\sin(Ay)$. From this we find that the electrostatic field components are $E_x = -\partial\phi/\partial x \equiv -\phi_y\partial\phi_x/\partial x = A\phi_y\phi_{0x}\exp(-Ax)$ and $E_y = -\partial\phi/\partial y \equiv -\phi_x\partial\phi_y/\partial x = A\phi_x(\phi_{0y}\sin(Ay) - \phi_{1y}\cos(Ay))$.

Following Rosenbluth and Longmire (1957) we have to deal with a somewhat tricky question in order to determine the electric field from the surface charge density. Classical electromagnetism distinguishes two types of charges; free and bound charges. When we assume that we place a charged metal sphere in the vicinity of a dielectric and thereby polarizing it, then the charges on the sphere are free charges while the electrons and ions constituting the dielectric are bound charges. The charges on the sphere create an electric field which polarizes the dielectric, i.e. redistributes slightly the charges in it, thereby creating polarization fields. The importance in the distinction arises by recalling that electrostatic polarization fields **D** begin and end on 'true' or free charges, while the electric field **E** has bound as well as free charges as source. Intuitively, one might expect that the surface charges σ should be considered as 'bound' charges, since they are constituents of the dielectric, in the present case the plasma. However, these charges do not originate from *electric* polarizations but from gravitational forces, and are therefore not recognized as polarization charges in electromagnetism. As far as Maxwell's equations are concerned, σ is constituted by free charges, implying by $\hat{\mathbf{n}}\Delta \cdot \mathbf{D} = \sigma$ that

$$D_{0x}^{\text{plasma}} - D_{0x}^{\text{vacuum}} \approx \sigma, \tag{7.24}$$

since $\hat{\mathbf{n}}$ is nearly parallel to $\hat{\mathbf{x}}$ when the amplitude Δx is small, see figure 7.1. The interpretation of the superscripts *plasma* and *vacuum* is self-evident. On the other hand at the interface, i.e. at the position $x = 0$, we also have, using the polarization description, that

$$\varepsilon_0\left(E_{0x}^{\text{plasma}} - E_{0x}^{\text{vacuum}}\right) = \sigma + \sigma', \tag{7.25}$$

where σ' denotes the polarization charge density induced at the surface by the electric fields.

For symmetry reasons, the electric fields in the plasma and in vacuum are of equal magnitude and pointing in opposite directions, either towards the charges (where they are negative), or away from them (where they are positive), i.e. $E_{0x}^{\text{plasma}} = -E_{0x}^{\text{vacuum}}$.

From (7.24) we have

$$\varepsilon_0\varepsilon_r E_{0x}^{\text{plasma}} - \varepsilon_0 E_{0x}^{\text{vacuum}} = \sigma, \tag{7.26}$$

giving $E_{0x}^{\text{plasma}} = (\sigma/\varepsilon_0)/(1 + \varepsilon_r)$ and $\sigma' = \sigma(1 - \varepsilon_r)/(1 + \varepsilon_r)$. If we remove the plasma but keep the surface charge we have $\sigma' = 0$ and $\varepsilon_r = 1$ to give $E_{0x}^{\text{plasma}} = \frac{1}{2}\sigma/\varepsilon_0$ in this case.

The relative dielectric constant for a magnetized plasma is at these low frequency magnetic field aligned perturbations

$$\varepsilon_r = 1 + \frac{n_0 M c^2}{B^2/\mu_0} \equiv 1 + \frac{n_0 M}{\varepsilon_0 B^2} = 1 + \frac{a_i}{\lambda_D}.$$

For relevant problems we have $\varepsilon_r \gg 1$, implying $\varepsilon_r \approx \mu_0 M c^2 n_0 / B^2$.

We now first postulate the solution for the surface charge in the form $\sigma = \sigma_0 \cos(ky)$ and then determine the resulting electric fields. With the expressions for the electric fields obtained before we find the electric field in the plasma to have the components

$$E_x = \frac{\sigma_0}{(1 + \varepsilon_r)\varepsilon_0} \exp(-kx)\cos(ky) \approx \frac{B^2}{n_0 M}\sigma_0 \exp(-kx)\cos(ky)$$

and

$$E_y = \frac{\sigma_0}{(1 + \varepsilon_r)\varepsilon_0} \exp(-kx)\sin(ky) \approx \frac{B^2}{n_0 M}\sigma_0 \exp(-kx)\sin(ky),$$

where we inserted $c^2 = 1/\mu_0\varepsilon_0$, see also figure 7.2. This electric field gives rise to an $\mathbf{E} \times \mathbf{B}/B^2$-velocity with components

$$u_x = \frac{E_y}{B} \approx \frac{B}{n_0 M}\sigma_0 \exp(-kx)\sin(ky),$$

and

$$u_y = -\frac{E_x}{B} \approx -\frac{B}{n_0 M}\sigma_0 \exp(-kx)\cos(ky).$$

In particular, we find that the velocity component u_x is in phase with the original surface perturbation Δx. It causes the amplitude $a(t)$ to increase,

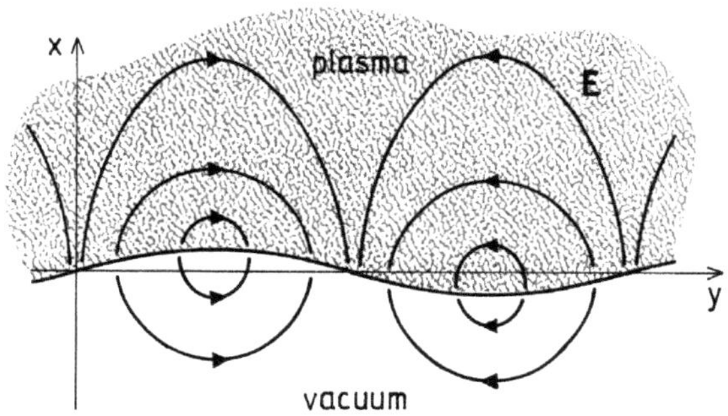

Figure 7.2. The spatial variation of the electric fields in the Rayleigh–Taylor instability, shown schematically. The surface displacement is given as $\Delta x(t) = a(t)\sin(ky)$.

hinting already by this that the condition is unstable. Consistently with the present linearized analysis we discard the effect of u_y on σ. We obtain an expression for the maximum value of u_{0x} obtained at $x = 0$ and where $\sin(ky) = 1$ by combining $u_{0x} \equiv \mathrm{d}a(t)/\mathrm{d}t$ with (7.23) to give

$$u_{0x}^{\max} = \frac{\mathrm{d}a(t)}{\mathrm{d}t} = \frac{\sigma_0(t)B}{n_0 M}, \tag{7.27}$$

and by differentiation

$$\frac{\mathrm{d}^2 a}{\mathrm{d}t^2} = \frac{B}{n_0 M}\frac{\mathrm{d}\sigma_0(t)}{\mathrm{d}t} = \frac{B}{n_0 M}\frac{n_0 Mgk\, a(t)}{B} = gk\, a(t). \tag{7.28}$$

Solving this differential equation we find

$$a(t) = a_0 \exp\left(\pm t\sqrt{gk}\,\right), \tag{7.29}$$

where a_0 is determined by the initial condition. The + sign corresponds to an aperiodic instability. The growth rate $\sqrt{gk}$ is independent of B. If we change the direction of gravity, i.e. the sign of g, we find stable oscillations with period $\sqrt{|g|k}$, see also figure 2.5. For this latter case, the waves appear as surface waves, very similar to waves on a water surface.

Another view: Given the importance of the problem (also historically) it might be worthwhile to take a different approach, making the introduction of the dielectric constant more explicit. Here we treat (at least in principle) all charged particles as free charges, and write an expression for their motion in the plane perpendicular to the homogeneous magnetic field. With the model (the 'polarization model') we have in this case $\mathbf{D} = \mathbf{E}/\varepsilon_0$.

Let the velocity of a plasma ion be given by the sum of the gravitational drift, the $\mathbf{E} \times \mathbf{B}/B^2$-drift plus the ion polarization drift

$$\mathbf{u}_i = \frac{Mg}{eB}\hat{\mathbf{y}} + \frac{\mathbf{E}\times\mathbf{B}}{B^2} + \frac{1}{\Omega_{ci}}\frac{\partial}{\partial t}\frac{\mathbf{E}}{B},$$

while the drift velocity of an electron is $\mathbf{u}_e = \mathbf{E} \times \mathbf{B}/B^2$, with the given approximations. These expressions are true for all plasma particles, in particular also for those at the surface. The current density is obtained as

$$\mathbf{J} = en(\mathbf{u}_i - \mathbf{u}_e) = \frac{Mng}{B}\hat{\mathbf{y}} + \frac{en}{\Omega_{ci}}\frac{\partial}{\partial t}\frac{\mathbf{E}}{B},$$

where $n = n_0$ inside the plasma and is vanishing outside. From the surface charge continuity equation we determine

$$\frac{\mathrm{d}\sigma}{\mathrm{d}t} = -\hat{\mathbf{n}} \cdot \mathbf{J} = -\frac{Mng}{B}ak\cos(ky) - \frac{Mn}{B^2}\frac{\mathrm{d}}{\mathrm{d}t}\hat{\mathbf{n}} \cdot \mathbf{E}. \tag{7.30}$$

Implicitly, we made an approximation by placing the time-varying direction $\hat{\mathbf{n}}$ under the differential operator, but this is justified by the linear approximation, where $\hat{\mathbf{n}}$ is nearly parallel to $\hat{\mathbf{x}}$.

The surface charge condition for electric fields states that $\hat{\mathbf{n}}\Delta \cdot \mathbf{E} = \sigma/\varepsilon_0$ in standard notation. The electric field component parallel to the charged surface is the same on both sides, while the perpendicular component changes by σ/ε_0. Since in the present case we have only one source of electric field, i.e. the surface charge, we find that the normal component of the electric field changes sign when passing through the surface. For symmetry reasons we have in this model $E = \frac{1}{2}\sigma/\varepsilon_0$. Using this result in (7.30) we find

$$\frac{\mathrm{d}\sigma}{\mathrm{d}t} = -\left(\frac{2Mn_0g}{B}\right)\frac{ka(t)\cos(ky)}{2 + \dfrac{Mn_0}{\varepsilon_0 B^2}}, \tag{7.31}$$

We recognize $1 + \varepsilon_r$ in the denominator.

To lowest approximation, the bulk plasma velocity (electrons and ions together) is $\mathbf{u} = \mathbf{E} \times \mathbf{B}/B^2$, with components $u_x = E_y/B$ and $u_y = -E_x/B$ where $E_x = (\zeta_0/\varepsilon_0)\exp(-kx)\cos(ky)$ and $E_y = (\zeta_0/\varepsilon_0)\exp(-kx)\sin(ky)$, where we introduced $\sigma(t, y) = \zeta_0(t)\cos(ky)$ recalling that σ_0 was used for a different quantity before. We will be interested in the motion in the x-direction, perpendicular to the plasma surface. The maximum plasma displacement is proportional to the amplitude $a(t)$ of the perturbation in surface position, so we have $\mathrm{d}a/\mathrm{d}t \sim u_x = E_y/B$.

Using (7.31) we find

$$\frac{\mathrm{d}^2 a(t)}{\mathrm{d}t^2} = \frac{Mn_0 g}{\varepsilon_0 B^2}\frac{ka(t)}{2 + \dfrac{Mn_0}{\varepsilon_0 B^2}}, \tag{7.32}$$

which contains our previous result (7.29) in the limit of large plasma densities. The present analysis includes also the limit of small densities, where we find a growth rate $\sqrt{\frac{1}{2}Mn_0gk/\varepsilon_0 B^2} = \sqrt{\frac{1}{2}gk(\varepsilon_r - 1)}$, which is small for small plasma densities: a magnetic field supports a dilute plasma better than a dense one against gravity.

7.4.3 Finite Larmor radius stabilization of flute modes

The *flute mode* at $k_z = 0$ in the dispersion relation in figure 6.1 can be made unstable by gravity as shown in the previous section. When $k_z = 0$ we have $\sigma_{\|} = 0$ and (7.17) becomes

$$\omega^2 + \left(k_y u_G - \omega_i\right)\omega - g n_0'/n_0 = 0. \tag{7.33}$$

In the absence of the middle term, one obtains the usual growth rate $\gamma = (-g n_0'/n_0)^{1/2}$ for what is known as the interchange, Rayleigh–Taylor, or gravitational flute instability resulting when $g n_0'/n_0 < 0$. In the limit of vanishing ion temperature, the instability is simple to describe phenomenologically, see for instance figure 2.5. Electrons with their small mass are hardly affected by gravity, while the ions drift with a $\mathbf{u}_G = M\mathbf{g} \times \mathbf{B}/eB^2$-velocity to set-up polarization electric fields in an initial perturbation. These in turn enhance (instability) or diminish (stability) the initial perturbation, depending on the sign of $\mathbf{g} \cdot \nabla n_0$.

The $k_y u_G$ term in (7.33) arises from finite ion inertia, and the ω_i term from finite ion Larmor radius. Together, these terms lower the growth rate and reduce it to zero for sufficiently large k_y. This is known as FLR stabilization. The stabilization originates from the 'smearing out' of the ion charge distribution due to the finite Larmor radii, which counteracts the charge separation caused by the ion gravity drift, see figure 2.5. The full dispersion relation is readily obtained as

$$\omega = \frac{1}{2}\left(\omega_i - k_y u_G\right) \pm \frac{1}{2}\sqrt{\left(\omega_i - k_y u_G\right)^2 + 4g\frac{n_0'}{n_0}},$$

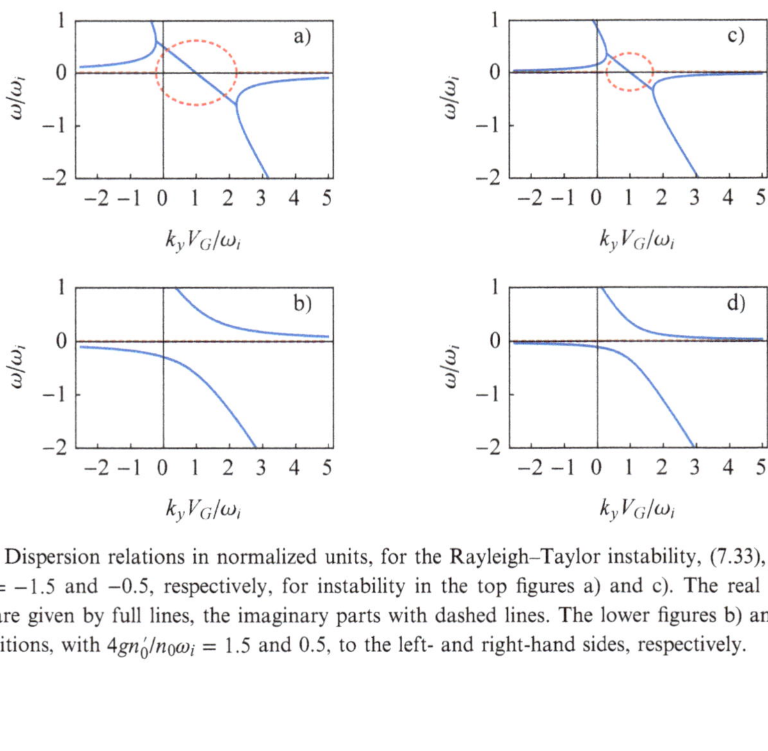

Figure 7.3. Dispersion relations in normalized units, for the Rayleigh–Taylor instability, (7.33), shown with $4g n_0'/n_0\omega_i = -1.5$ and -0.5, respectively, for instability in the top figures a) and c). The real parts of the frequency are given by full lines, the imaginary parts with dashed lines. The lower figures b) and d) are for stable conditions, with $4g n_0'/n_0\omega_i = 1.5$ and 0.5, to the left- and right-hand sides, respectively.

with the condition for instability being $gn_0'/n_0 < -\frac{1}{4}(\omega_i - k_y u_G)^2$. The dispersion relation is illustrated in figure 7.3 for $4gn_0'/n_0\omega_i = -1.5$ and -0.5, as well as 0.5 and 1.5. The most unstable perturbation has $k_y = \omega_i/u_G$, and has vanishing phase velocity. The FLR-stabilization of the gravitational instability is sometimes seen as a counterpart of the stabilizing surface tension for the fluid version of the original Rayleigh–Taylor instability. It is not obvious that this analogy is of any help.

We can again consider the relation (6.21) between normalized density η and normalized electrostatic potential χ. In the limit $\sigma_\parallel \to 0$ we find now a finite ratio $|\eta/\chi|$ and the relative phase angle is also changed from the zero ion temperature case, see figure 6.3. The finite ion temperature modifies ω entering (6.21) even in the limit of vanishing gravitational drifts.

References

Braginskiĭ S I 1965 Transport processes in a plasma ed M A Leontovich *Reviews of Plasma Physics* vol 1 (New York: Consultants Bureau) pp 205–311

Chen F F 1970 Linear drift waves UCLA Lecture Notes unpublished

Chen F F 1984 *Introduction to Plasma Physics and Controlled Fusion* vol 1 2nd edn (New York: Plenum)

Kaufman A N 1960 Plasma viscosity in a magnetic field *Phys. Fluids* **3** 610–6

Lehnert B 1962 Stability of an inhomogeneous plasma in a magnetic field *Nucl. Fusion Supplement* **1** 135–42

Pécseli H L 2012 *Waves and Oscillations in Plasmas* (London: Taylor and Francis)

Roberts K V and Taylor J B 1962 Magnetohydrodynamic equations for finite Larmor radius *Phys. Rev. Lett.* **8** 197–8

Rosenbluth M N and Longmire C L 1957 Stability of plasmas confined by magnetic fields *Annals Phys.* **1** 120–40

Weiland J 1992 Low frequency modes associated with drift motions in inhomogeneous plasmas Technical Report CTH–IEFT/PP–1992–17, Chalmers Technical University, Göteborg, Sweden

Chapter 8

Electrostatic drift waves with viscosity due to ion-ion collisions

Equation (7.18) predicts that drift waves are always unstable regardless of k_y and k_z, as long as they are both finite. Experimentally, it was found that drift waves could be stabilized by lowering B, and hence increasing $k_y^2 r_L^2$. To explain this, we must include the collision terms in the ion viscosity tensor $\mathbf{\Pi}$. We take the full tensor and take the limit $\Omega_{ci}^2 \tau_{ii}^2 \gg 1$ with τ_{ii} being the ion–ion collision time (Braginskií 1965). In this limit we have for motions perpendicular to $\mathbf{B}$ the collisional part of $\nabla_\perp \cdot \mathbf{\Pi}$ in the form

$$\nabla_\perp \cdot \mathbf{\Pi}_c = -\mu_\perp \nabla_\perp^2 \mathbf{u}, \qquad \mu_\perp \equiv \frac{n\kappa T_i}{4\Omega_{ci}^2 \tau_{ii}},$$
$$\tau_{ii} = \frac{32\pi M^2 \varepsilon_0^2}{ne^4 \ln(\Lambda)} \left(\frac{\kappa T_i}{M} \right)^{3/2}, \tag{8.1}$$

where $\ln(\Lambda)$ is the usual Coulomb logarithm. To account for ion–ion collisions we will here merely add a term $\mu_{\perp 0} \nabla_\perp^2 \mathbf{u}$ to the right-hand side of (7.6). The full equation then becomes

$$Mn\left(\frac{\partial \mathbf{u}}{\partial t} + \mathbf{u} \cdot \nabla \mathbf{u} \right) = en(-\nabla\phi + \mathbf{u} \times \mathbf{B}) - \nabla p - \nabla \cdot \Pi + \mu_\perp \nabla_\perp^2 \mathbf{u},$$

with gravity ignored.

Dividing (7.6) by Mn_0 we have the collisional viscosity term in the form

$$-k_y^2 \mu_\perp / Mn_0 = -\frac{1}{4} k_y^2 \theta a_i^2 / \tau_{ii} = -\frac{1}{4} \theta b \nu_{ii} \equiv -\sigma_\perp, \tag{8.2}$$

where $\nu_{ii} = 1/\tau_{ii}$ is the ion–ion collision frequency. The solution of (7.6) which replaces (7.11) is now

$$u_x \frac{n_0'}{n_0} = i\omega * \frac{(\chi + \theta\eta)}{1 - \frac{1}{2}\theta b} \quad \text{(no change)}, \tag{8.3}$$

and for (7.12)

$$u_y = -k_y a_i^2 \left(\omega - k_y u_0 + \frac{1}{2}\omega_i + i\sigma_\perp \right) \frac{(\chi + \theta\eta)}{\left(1 - \frac{1}{2}\theta b\right)^2}. \tag{8.4}$$

We assumed $|\omega + i\sigma_\perp|^2 \ll \Omega_{ci}^2$. The $\sigma_\perp$ term represents an ion flux in the y-direction which can short-circuit the fluctuating electric field in that direction if b is large enough; when this happens, the wave is damped out (Chen 1970).

Some qualitative features of the stabilization of drift waves by ion viscosity can be anticipated from the outset. Thus inspection of the linear dispersion relation shown in figure 6.1 indicates that the growth rate is small for very large and very small values of the parameter $\sigma_\parallel = (k_z/k_y)^2 \omega_{ce} \tau_{ei} \Omega_{ci}$. With ion viscosity being a small effect, we expect stabilization to appear primarily where the growth rate is small, i.e. at very large or very small B-parallel wavelengths.

We omit the analysis of the Rayleigh–Taylor instability in this section so that $u_G = 0$ here.

8.1 Dispersion relation

The expressions (8.3) and (8.4) for the ion fluid velocity are now inserted into the ion equation of continuity (7.13), and η is eliminated by the electron equation (6.21). Some lengthy but otherwise straightforward algebra (Chen 1970) gives the dispersion relation

$$\left(\omega + i\sigma_\perp\right)(\omega - \omega_i) + i\sigma_\parallel\left[\omega - \omega^* + b(\omega - \omega_i) + ib\sigma_\perp(1 + \theta)\right] = 0. \tag{8.5}$$

A simplification of the calculations is achieved by noting that the contributions from the terms $Mn_0\mathbf{u}_0 \cdot \nabla\mathbf{u}$ and the collisionless part of $\nabla \cdot \mathbf{\Pi}_1$ in (7.6) happen to cancel to the order desired (Hendel *et al* 1968). While this is not exactly true, since the term $b(\omega - \omega_i)$ in (8.5) would be replaced by $b(\omega + \theta\omega^*)$, there is no practical difference for $\omega \approx \omega^*$.

To illustrate the variation of (8.5) with wavenumber components k_z and k_y we rewrite (8.5) by introducing ω/ω^* with $\omega_i = -\theta\omega^*$ to find

$$\left(\frac{\omega}{\omega^*} + i\frac{\sigma_\perp}{\omega^*}\right)\left(\frac{\omega}{\omega^*} + \theta\right) + i\frac{\sigma_\parallel}{\omega^*}\left[\frac{\omega}{\omega^*} - 1 + b\left(\frac{\omega}{\omega^*} + \theta\right) + ib\frac{\sigma_\perp}{\omega^*}(1 + \theta)\right] = 0.$$

We have $\omega^* \sim k_y$ and $\sigma_\perp \sim k_y^2$ giving $\sigma_\perp/\omega^* \sim k_y$, and $b \sim k_y^2$ giving $b\sigma_\perp/\omega^* \sim k_y^3$, see appendix A. On the other hand, $\sigma_\parallel \sim (k_z/k_y)^2$ depends on the direction of the

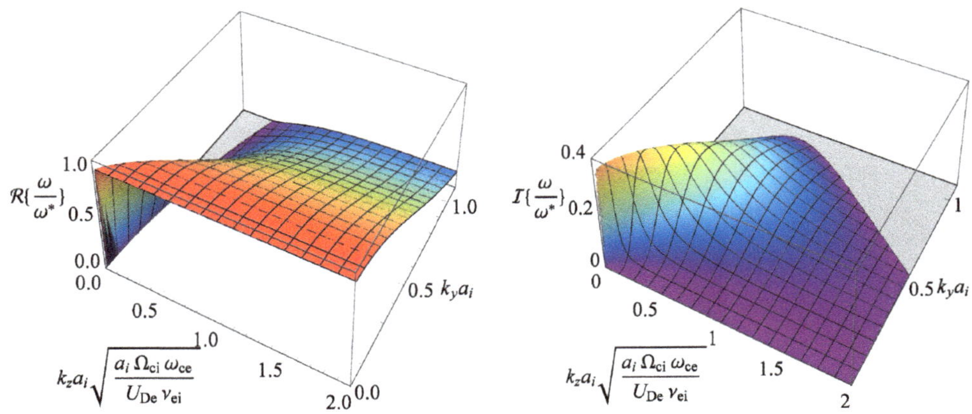

Figure 8.1. Normalized dispersion relation $\Re\{\omega/\omega^*\}$ in a) and growth rate $\Im\{\omega/\omega^*\}$ in b) for electrostatic drift waves obtained by (8.5) shown as functions of k_z and k_y. The model contains two parameters with the given normalizations. We used here $\theta = 1/2$ and $\nu_{ii}a_i/u_{De} = 1$.

wavevector and not its magnitude. We have $\sigma_\parallel/\omega^* \sim k_z^2/k_y$. The limit $(k_z, k_y) \to (0, 0)$ is a singular point. Note the change in sign with k_y in some of the terms The dispersion relation and growth rate in particular as obtained by (8.5) are shown in figure 8.1 as functions of normalized k_z and k_y. We note stabilization of the drift waves both in the long and short **B**-parallel wavelength limits. When interpreting the numerical values in figure 8.1 we keep in mind that ω^* also contains k_y, but at least in laboratory experiments this wavenumber is usually fixed by some boundary conditions while k_z has a wider range of variability. A solution corresponding to waves moving in the direction of the ion diamagnetic drift is heavily damped and is not shown. The threshold $\Im\{\omega\} = 0$ is found from figure 8.1. To obtain the numerical values for $\Im\{\omega\}$ for all **k** we multiply the results in figure 8.1 by $\omega^* = u_{De}k_y$. The growth rate will have a local maximum (Hendel *et al* 1968) for some wavevector $\mathbf{k}^{(m)}$. For modeling a diffusion coefficient in the weak turbulence limit we take the $k_\perp^{(m)}$-component of this wavenumber together with the corresponding maximum linear growth rate $\Im\{\omega\}$ and form (Kadomtsev 1965) the diffusion coefficient for transport across magnetic field lines in the form $D_\perp \sim \Im\{\omega(\mathbf{k}^{(m)})\}/k_\perp^2$.

An alternative presentation of the stability conditions was used by, for instance Hendel *et al* (1968), see figure 8.2. The stabilization of the drift waves in the long and short **B**-parallel wavelength limits is seen better in this presentation. The growth rate has a maximum value for some (λ_z, λ_y).

Equation (8.5) can be simplified in the long-$\lambda_\parallel$ and short-$\lambda_\parallel$ limits as discussed in the following (Chen 1970).

8.1.1 Long-$\lambda_\parallel$ limit

Consider first $\omega \leqslant \mathcal{O}(\omega^*)$ and $\sigma_\perp \leqslant \mathcal{O}(\omega^*)$, and assume $b\sigma_\parallel \ll \omega^*$. The two terms containing b in (8.5) will be neglected compared to the first term in the equation. The dispersion relation then becomes

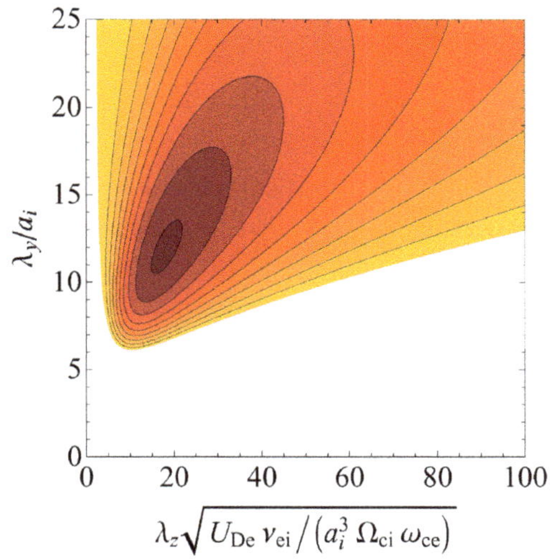

Figure 8.2. Normalized growth rate $\Im\{\omega/\omega^*\}$ for electrostatic drift waves obtained by (8.5) shown here as functions of normalized λ_z and λ_y, see also figure 8.1. The contours give $\Im\{\omega\}$, where the normalization can be found in figure 8.1(b).

$$(\omega + i\sigma_\perp)(\omega - \omega_i) + i\sigma_\|(\omega - \omega^*) = 0, \tag{8.6}$$

which recovers the previous result (7.18) in the limit $\nu_{ii} \to 0$.

We can obtain an explicit expression for the growth rate provided b can satisfy the inequality $\sigma_\| \gg \omega^* \gg b\sigma_\|$. Solving (8.6) for $\omega - \omega^*$ and letting as before $\omega \approx \omega^*$ on the right-hand side, we obtain

$$\omega - \omega^* = -\frac{\sigma_\perp}{\sigma_\|}(\omega^* - \omega_i) + \frac{i\omega^*(\omega^* - \omega_i)}{\sigma_\|}. \tag{8.7}$$

In this limit, the growth rate is not affected by $\sigma_\perp$, but the real part of $\omega - \omega^*$ shows that the assumption $\omega \approx \omega^*$ that we made is valid only for $\sigma_\perp \ll \sigma_\|$, which need not always be satisfied. For more general solutions we consider individual cases.

To obtain the condition for viscous stabilization, equation (8.6) is rewritten as follows

$$\omega^2 - (\omega_i - i\sigma)\omega - i\omega^*\left(\sigma_\| - \theta\sigma_\perp\right) = 0, \tag{8.8}$$

where $\sigma \equiv \sigma_\| + \sigma_\perp$. If the quantity $\Delta\sigma \equiv \sigma_\| - \theta\sigma_\perp$ vanishes, we have a real solution $\omega = 0$ and a damped solution $\omega = \omega_i - i\sigma$. To examine the stability threshold, we should therefore consider $\omega \approx 0$. Assuming $\omega \ll \omega^*$ and neglecting the ω^2 term, we obtain

$$\omega = \frac{i\omega^*\Delta\sigma}{i\sigma - \omega_i}. \tag{8.9}$$

For $\sigma \ll |\omega_i|$, this simplifies to

$$\omega = \frac{\omega^*}{\omega_i^2}\sigma\Delta\sigma + \frac{i\Delta\sigma}{\theta}. \tag{8.10}$$

For $\sigma \gg |\omega_i|$, we obtain instead

$$\omega = \omega^* \frac{\Delta\sigma}{\sigma} + \frac{i\theta\omega^{*2}}{\sigma}\frac{\Delta\sigma}{\sigma}. \tag{8.11}$$

In either case, $\sigma_{\|} = \theta\sigma_{\perp}$ is a necessary and sufficient condition for marginal stability. If $\sigma_{\|}$ is held constant by keeping the ratio $(k_z/k_y)^2$ constant and $\sigma_{\perp}$ decreased from this value by increasing k_y, both $\Im(\omega)$ and $\Re(\omega)$ increase from 0. Physically, this means that the increased ion conductivity across magnetic field lines short-circuits wave-crests and wave-troughs. When, on the other hand, $\sigma_{\perp}$ is *in*creased from the threshold value $\Delta\sigma = 0$, we find a slightly damped solution which propagates in the direction *opposite* to that of the electron diamagnetic drift, i.e. the *ion* diamagnetic drift direction. Alternatively, we can keep $\sigma_{\perp}$ with a constant k_y and increase $\sigma_{\|}$ by turning the wave vector by increasing $(k_z/k_y)^2$, and find that the waves become unstable. This part can be understood by considering the dispersion relation in figure 6.1 in the long $\lambda_{\|}$-limit, where it is seen that an increase in $\sigma_{\|}$ gives a corresponding increase in growth rate, which in the present case compensates the damping effect of ion viscosity. Similarly, we can explain also the stabilization by *de*crease in $\sigma_{\|}$ for fixed $\sigma_{\perp}$.

- **Exercise:** Prove that the general solution of (8.6) for $\sigma_{\|} \gg \omega^* \gg b\sigma_{\|}$, found by straightforward expansion, is

$$\omega = \omega^* \frac{\Delta\sigma}{\sigma} + \frac{i(1+\theta)\omega^{*2}}{\sigma}\frac{\sigma_{\|}\Delta\sigma}{\theta\sigma^2}, \tag{8.12}$$

which reduces to (8.7), (8.10), and (8.11) in the proper limits.

8.1.2 Amplitude and phase relations

Since $\omega = 0$ at threshold, the phase shift relation (6.21) reduces to

$$\eta = \chi \frac{\omega^* + ib\sigma_{\|}}{ib\sigma_{\|}} \approx \frac{\omega^*}{ib\sigma_{\|}}\chi. \tag{8.13}$$

In the long-$\lambda_{\|}$ limit, we find n_1 and ϕ_1 to be 90° out of phase at threshold, and $|\eta| \gg |\chi|$. In the limit where viscosity vanishes, we found n_1 and ϕ_1 to be in phase when $\Im(\omega) \to 0$, which occurred for $\sigma_{\|} \to \infty$, see figure 6.3.

More generally, we can consider the relation (6.21) between normalized density η and normalized electrostatic potential χ. The finite ion viscosity modifies ω entering (6.21). In the limit $\sigma_{\|} \to 0$, often called the 'flute-mode' limit, we find a finite ratio $|\eta/\chi|$ and the relative phase angle is also changed from the zero ion temperature case, see figure 6.3. Solving (8.5) for the case where $\sigma_{\|} = 0$ we find the solutions $\omega = \omega_i$ and $\omega = -i\sigma_{\perp}$. Using (6.21) we find $\eta/\chi = \omega^*/\omega_i = -T_e/T_i \equiv -1/\theta$, and $\eta/\chi = i\omega^*/\sigma_{\perp}$, respectively. The latter result explicitly contains k_y through $\sigma_{\perp}$ and ω^*. For large k_y, i.e. wavelengths shorter than the scale-length for the density gradient, we find $|\chi| \gg |\eta|$, which is a property often found to characterize flute-modes.

8.1.3 Short-$\lambda_{\|}$ limit

If we keep the terms with b in (8.5), we obtain another stabilization threshold, this time for short $\lambda_{\|}$. For these short $\lambda_{\|}$, the plasma density and the electrostatic potential are nearly in phase, and the electrons are close to be in locally isothermal Boltzmann distribution. For large $\sigma_{\|}/\omega^*$, we would expect $\omega \approx \omega^*$, see for instance the dispersion relation in figure 6.1. Consequently we let $\omega = (1+\epsilon)\,\omega^*$ and expand in the small parameter ϵ to obtain

$$-\epsilon\,\omega^* = (1+\theta)\frac{\omega^{*2} - b\sigma_{\|}\sigma_{\perp} + i(\sigma_{\perp} + b\sigma_{\|})\omega^*}{i\sigma_{\|}(1+b) + (2+\theta)\omega^* + i\sigma_{\perp}}. \tag{8.14}$$

For $\sigma_{\|} \gg \omega^*$, $\sigma_{\|} \gg \sigma_{\perp}$, this reduces after some algebra to

$$\omega - \omega^* = \frac{1+\theta}{\sigma_{\|}}\left[-(\sigma_{\perp} + b\sigma_{\|})\omega^* + i(\omega^{*2} - b\sigma_{\|}\sigma_{\perp})\right]. \tag{8.15}$$

There is a stability threshold at $b\sigma_{\|}\sigma_{\perp} = \omega^{*2}$ where the last imaginary term vanishes in (8.15). In un-normalized terms we have the criterion, here written out term-by-term

$$\frac{k_y^2 r_L^2}{2\theta} \times \left(\frac{k_z}{k_y}\right)^2 \omega_{ce}\tau_{ei}\Omega_{ci} \times \frac{1}{4}\theta\frac{k_y^2 r_L^2}{2\theta}\nu_{ii} = k_y^2\left(\frac{\kappa T_e}{eB}\right)^2,$$

where several elements cancel. We have $r_L^2 \equiv 2\kappa T_i/M\Omega_{ci}^2$.

As mentioned, we have both long-$\lambda_{\|}$ and short-$\lambda_{\|}$ stabilization points. With the definitions (6.20) and (8.2), the long-$\lambda_{\|}$ stabilization criterion, $\theta\sigma > \sigma_{\|}$, works out to be

$$\frac{1}{4}\frac{T_i}{T_e}\frac{k_y^4}{k_z^2}\frac{\kappa T_i}{M\Omega_{ci}^2}\frac{m}{M\Omega_{ci}^2}\frac{\nu_{ii}}{\nu_{ei}} > 1, \tag{8.16}$$

or

$$\frac{B}{k_y} < \sqrt{\lambda_{\|}} \times \text{const.}, \tag{8.17}$$

for given plasma parameters. For given $\lambda_{\|}$, the wave is stabilized at sufficiently small B/k_y. The collision frequencies ν_{ii} and ν_{ei} are proportional to the plasma density, see (6.3) and (8.2), and the right-hand side of (8.17) is therefore proportional to $\sqrt{n}$.

The short-$\lambda_{\|}$ stabilization criterion, $b\sigma_{\|}\sigma_{\perp} > \omega^{*2}$, becomes

$$\frac{1}{4}b^2\frac{T_i^2}{T_e^2}\frac{k_z^2}{k_y^2}\omega_{ce}\Omega_{ci}\frac{\nu_{ii}}{\nu_{ei}} > \omega^{*2}. \tag{8.18}$$

The ratio ν_{ii}/ν_{ei} depends on the mass- and temperature-ratios. Taking $\nu_{ii}/\nu_{ei} \approx 4(m/M)^{1/2}$ and $T_i = T_e$, we obtain

$$k_z^2 > \left(\frac{m}{M}\right)^2\left(\frac{n_0'}{n_0}\right)^2. \tag{8.19}$$

The short-$\lambda_{\|}$ stabilization criterion found here is independent of B and k_y. This result has, however, never been observed (Chen 1970).

The effects of $\sigma_{\|}$ and $\sigma_{\perp}$ can qualitatively be explained by considering the transport of ions and electrons (Hendel *et al* 1968) in the direction $\perp\mathbf{B}$ and $\|\mathbf{B}$. The characteristic time for electrons to move a distance k_z^{-1} by random walk (see section 3.5.1) is estimated by the ratio of the square of a characteristic length and the appropriate diffusion coefficient. For the present case this gives $t_{\|} \sim (\nu_{ei} k_z^2 \ell_c^2)^{-1}$. By similar arguments, the diffusive motion of ions over a distance k_y^{-1} has a characteristic time scale $t_{\perp} \sim (k_y^2 a_i^2 \nu_{ii})^{-1}$. The latter **B**-transverse ion motion has a stabilizing effect by 'short circuiting' wave crest and wave trough. The electron motion along **B** has a dual effect: on one hand it provides the phase difference between plasma density and potential necessary for instability, and on the other hand it slows down the electron motion along **B** which is a stabilizing effect (Hendel *et al* 1968), see also the discussion of figure 6.1.

8.1.4 An apparent paradox

We found in the previous analysis that ion–ion collisions can stabilize drift waves. This stabilization can be understood as a short circuiting of the electric field from wave-crest to wave-trough. This observation is however apparently conflicting with the well known result that ion–ion (or electron–electron) collisions do not give rise to any plasma conductivity or plasma transport across magnetic field lines, see section 3.1.1. The basic result was that for two particles of the same kind, the center of-mass is the same *after* the collision as it was before. The derivation of this standard result is, however, based on the assumption of vanishing or spatially constant electric fields. In the present case the electric field of the wave is harmonically varying in space, and ion–ion collisions *will* in this case give rise to a net current across magnetic fields in response to this inhomogeneous electric field. The physical details of collisions for these conditions were discussed in section 3.1.2.

References

Braginskiĭ S I 1965 Transport processes in a plasma ed M A Leontovich *Reviews of Plasma Physics* vol 1 (New York: Consultants Bureau) pp 205–311

Chen F F 1970 Linear drift waves *UCLA Lecture Notes* unpublished

Hendel H W, Chu T K and Politzer P A 1968 Collisional drift waves–identification, stabilization, and enhanced plasma transport *Phys. Fluids* **11** 2426–39

Kadomtsev B B 1965 *Plasma Turbulence* (New York: Academic)

Chapter 9

Experimental results

Drift waves have been studied experimentally in great detail. One of the most versatile experimental set-ups have been the *Q-machine* (Motley 1975), being a device for producing an alkali plasma of relatively low temperature. The plasma can be studied by means of *Langmuir probes* (Huddlestone and Leonard 1965, Fahleson 1967, Hutchinson 1987, Allen 1992, Piel 2010, Pécseli 2012). Historically, it was one of the first devices to produce a steady state plasma with a modest noise level, and to celebrate this, is was given the name Q-machine, where the Q is an abbreviation for *Quiet*.

9.1 The Q-machine

The Q-machine, in its standard version, see figure 9.1, consists of a vacuum vessel at low background pressure, $p < 10^{-4}$ Pa, where the plasma is produced by surface or contact ionization on a hot metal plate. The plasma is confined by an axial magnetic field, produced by coils outside the vessel. The cathode, is heated to temperatures in excess of 2000 °C by electron bombardment from a hot filament (not shown) placed on the back side of the hot plate. Typical electron energies for the heating of the plate are 2000 keV, with 0.5 A currents. The hot plate is fairly thick, 5 mm or so, while the supporting cylinder is thin, in order to reduce the heat conduction away from the hot plate. In order to avoid this support being exposed to the damaging impact from the energetic electrons inside the cylinder, a certain minimum magnetic field is usually necessary for the standard operation of most Q-machines. The figure shows a Q-machine in single ended operation. They can be in *double ended* operation as well, with a hot cathode in both ends. The length of the plasma column can be several meters, but many devices are shorter than this.

Contact ionization was discovered by Langmuir and Kingdon (1925), who found that an electron could be 'stripped' from an atom upon contact with a metallic

doi:10.1088/978-0-7503-1251-6ch9

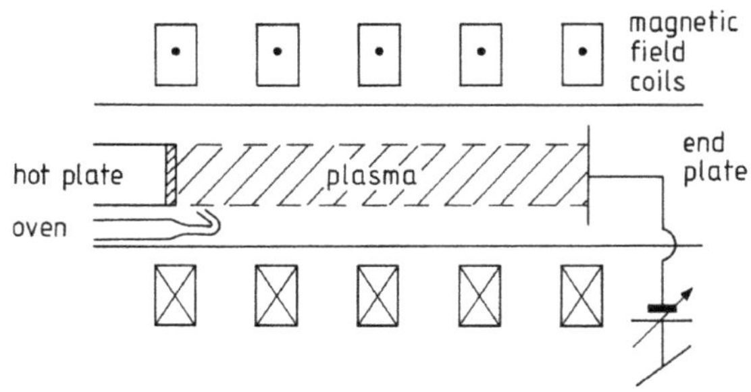

Figure 9.1. Schematic representation of a Q-machine. The 'oven' indicates the source of the neutral gas (usually alkali metals) to be ionized upon contact with the hot plate, or cathode.

surface, and that the efficiency approached 100% when the work function of the metal exceeded the ionization potential of the atom. The physical explanation of this process need not concern us here. Work functions of metals are usually of the order of a few electron volts (eV), while the ionization energy of for instance hydrogen is 13.6 eV. This does not look promising, so it is evident that we have to look for materials with particularly low ionization energies. Alkali metals are such candidates; the energies are 5.14 eV for Na, 4.34 eV for K, and 3.89 eV for Cs, for instance (Kaye and Laby 1995), so, if we are prepared to live with the inconveniences of alkali metals, it is no problem to achieve an ionization of the appropriate metal vapor. In order to have a macroscopically charge neutral plasma, the metal plate must also be able to supply the electrons. This can be achieved by heating it to temperatures large enough to give an appreciable Richardson emission of electrons. As we shall see, this requires temperatures in excess of 1000 °C, which in effect restricts the choice of metals to tungsten or tantalum, with melting points 3380 °C and 3000 °C, respectively. We will in the first part of the following summary assume that the confining axial magnetic field is sufficiently strong to justify a description in one spatial dimension.

The electrons are assumed to be emitted by the Richardson effect, expressed as

$$I_R = AT^2 \mathrm{e}^{-eW/\kappa T}, \tag{9.1}$$

where T is the temperature of the cathode, $A = 120$ A cm^{-2} K^{-2} is a constant, and W the work function of the cathode material. Assuming the velocity distribution of the emitted electrons just outside the cathode surface to be a half Maxwellian (the other half is the one inside the cathode),

$$f(u) = \sqrt{\frac{m}{2\pi\kappa T}}\,\mathrm{e}^{-\frac{1}{2}mu^2/\kappa T},$$

we have

$$I_R = en_0 \int_0^\infty uf(u)\mathrm{d}u = en_0\sqrt{\frac{\kappa T}{2\pi m}},$$

where n_0 is the electron density at the hot plate surface, and e, m being the electron charge and mass, respectively. Negative electrons, with charge $-e$ flow away from the cathode, giving a current flowing in the direction *into* the cathode.

We can now anticipate two basically different scenarios: one where the hot plate is very efficient in emitting electrons, and one where the ions are emitted in abundance. The two cases are usually denoted *electron rich* and *ion rich* conditions, respectively. Q-machines are usually operated in the former condition, so we here assume that there is a surplus of electrons being emitted. This implies that there will be a negative space charge in front of the hot plate, which will 'push' the surplus of electrons back into the plate, and the macroscopically charge neutral plasma column will assume a negative potential, $-\Delta\Phi_p$, with respect to the hot plate, assumed to be at ground potential. The region with a net charge density will have a spatial extent of the order of the Debye length. Only electrons with a velocity large enough to surmount this potential barrier will reach into the plasma. Assume that this minimum electron velocity is u_{eM}, i.e. $\frac{1}{2}mu_{eM}^2 = e\Delta\Phi_p$. The current associated with the electrons entering the plasma column is then

$$I_1 \equiv en_0 \int_{u_{eM}}^{\infty} uf(u)\mathrm{d}u = I_R \exp\bigl(-e\Delta\Phi_p/\kappa T\bigr), \tag{9.2}$$

as readily shown. One might argue simply that $W + e\Delta\Phi_p$ is an 'effective' ionization energy for the cathode plus the surplus charge region. Physically, the interpretation of $e\Delta\Phi_p$ is that of the potential energy (with respect to ground) of a plasma electron with zero velocity.

We assume now that the metal plate terminating the plasma column, see figure 9.1, is sufficiently negative to reflect *all* electrons. Literally speaking, this would of course require an infinitely large negative potential, but we assume that the error by using a finite potential is negligible. The current contribution from the reflected electrons is then

$$I_2 \equiv en_p \int_0^{\infty} uf(u)\mathrm{d}u = n_p e\sqrt{\frac{\kappa T}{2\pi m}}, \tag{9.3}$$

where now n_p is the electron density in the plasma column, $n_p \neq n_0$. The temperature of the electrons at the hot plate surface and in the plasma are the same, so there is no need to distinguish $f(u)$ from the electron velocity distribution in the plasma column. Since we assumed *all* electrons to be reflected, we cannot have any net electron current through the system, and require $I_1 = I_2$, which determines the plasma potential as

$$\Delta\Phi_p = \frac{\kappa T}{e}\ln\left(\frac{AT^2}{en_p\sqrt{\kappa T/(2\pi m)}}\right) - W. \tag{9.4}$$

Since the plasma in the main part of the column outside the surface charge area must be macroscopically charge neutral, we have n_p being the equal to the electron as well as the ion plasma density there. The plasma potential will depend on the plasma density, which is reasonable after all: if we have many ions available, they can compensate more of the electrons, and only a small $\Delta\Phi_p$ is necessary to reflect the surplus electrons.

Note that for small variations in the plasma density, $n_p = n_{p0} + \widetilde{n}_p$, with corresponding variations in plasma potential $\Delta\Phi_p = \Delta\Phi_{p0} + \widetilde{\Phi}_p$, the relation (9.4) implies

$e\widetilde{\Phi}_p/\kappa T \approx -\widetilde{n}_p/n_{p0}$. Note the minus sign! A *decrease* in plasma density, $\widetilde{n}_p < 0$, implies that the plasma potential becomes more negative, $\widetilde{\Phi}_p > 0$ with the present sign-convention, as argued before. Physically this can be explained by noting that with a given temperature the hot plate is emitting electrons at a constant rate: if the ion density deceases, the plasma potential must become more negative to reflect the surplus electrons.

- **Example:** For relevant parameters as $T = 2500$ K, $n_p = 10^{16}\ \mathrm{m}^{-3}$, and $W = 4.1$ eV for a tantalum cathode, we find $\Delta\Phi_p = 1.6$ V, negative with respect to the cathode ground potential.

Ions, on the other hand, are *accelerated* through the space charge region, and are *lost* at the negatively biased end plate. In order to obtain an expression for the ion distribution function in the plasma we assume that for a given neutral beam illumination of the hot plate, see figure 9.1, we have a current density associated with ions in the velocity interval $(v, v + \mathrm{d}v)$ just at the cathode surface to be

$$I_{i1} = n_{i0} e v \sqrt{\frac{M}{2\pi\kappa T}}\, \mathrm{e}^{-\frac{1}{2}Mv^2/\kappa T} \mathrm{d}v.$$

Since ions must be emitted in thermal equilibrium with the cathode material, we here take the temperature to be that of the cathode. The local ion density is n_{i0}. After acceleration through the space charge, an ion has an energy $\frac{1}{2}Mu^2 = \frac{1}{2}Mv^2 + e\Delta\Phi_p$, for $u > 0$. *In* the plasma, these very same ions give a current contribution

$$I_{i2} = n_{i0} e u \sqrt{\frac{M}{2\pi\kappa T}}\, \mathrm{e}^{-\left(\frac{1}{2}Mu^2 - e\Delta\Phi_p\right)/\kappa T} \mathrm{d}u,$$

since $u\mathrm{d}u = v\mathrm{d}v$. Assume that the ion velocity distribution *in* the plasma is $g(u)$, normalized to unity, $\int_{-\infty}^{\infty} g(u)\mathrm{d}u = 1$, with the ion density being $n_i = n_p$, as already mentioned. We can then write $I_{i2} = n_p e u g(u)\mathrm{d}u$, and find

$$n_{i0} u \sqrt{\frac{M}{2\pi\kappa T}}\, \mathrm{e}^{-\left(\frac{1}{2}Mu^2 - e\Delta\Phi_p\right)/\kappa T} = n_p g(u).$$

It would be tempting here to argue that $n_p = n_{i0} \exp(e\Delta\Phi_p/\kappa T)$, but this would be an error! The problem is that because of the net acceleration through the sheath, only ions with velocities $u > e\Delta\Phi_p \equiv u_{iM}$ are present in the plasma column, outside the space charge region. We have required $g(u)$ to be normalized, but have

$$\sqrt{\frac{M}{2\pi\kappa T}} \int_{u_{iM}}^{\infty} \exp\left(-\frac{1}{2}Mu^2/\kappa T\right)\mathrm{d}u \neq 1,$$

so the normalization of $g(u)$ must be ensured in a more subtle manner. To have the correct normalization, we require

$$n_p = n_{i0} e u \sqrt{\frac{M}{2\pi\kappa T}}\, \mathrm{e}^{\frac{e\Delta\Phi_p}{\kappa T}} \int_{u_{iM}}^{\infty} \mathrm{e}^{-\frac{1}{2}Mu^2/\kappa T} \mathrm{d}u,$$

giving

$$g(u) = \frac{1}{1 - \operatorname{erf}\sqrt{e\Delta\Phi_p/\kappa T}}\sqrt{\frac{2M}{\pi\kappa T}}\,\mathrm{e}^{-\frac{1}{2}Mu^2/\kappa T}, \quad \text{for} \quad u \geqslant \sqrt{\frac{2\pi\kappa T}{M}} \tag{9.5}$$

and $g(u) = 0$ for $u < \sqrt{2\pi\kappa T/M}$. We introduced the *error function* as

$$\operatorname{erf}\left(x/\sqrt{2}\right) \equiv \sqrt{2/\pi}\int_0^x \exp\left(-\frac{1}{2}y^2\right)\mathrm{d}y.$$

In figure 9.2 we show the variation of the ion distribution function for varying normalized plasma potentials, $e\Delta\Phi_p/\kappa T = 0$, 0.5, 1.5 and 2.5.

We note that the average velocity $\langle u\rangle = \int ug(u)\mathrm{d}u$ seems to increase with $e\Delta\Phi_p/\kappa T$. We can readily calculate

$$\langle u\rangle = \frac{\exp\left(-e\Delta\Phi_p/\kappa T\right)}{1 - \operatorname{erf}\left(\sqrt{e\Delta\Phi_p/\kappa T}\right)}\sqrt{\frac{2}{\pi}},$$

as shown in figure 9.3. We note an effective drift velocity already for $e\Delta\Phi_p/\kappa T = 0$, but this is just a consequence of the ion distribution being a drifting 'half Maxwellian', even in the case where the plasma potential happens to be ground potential. The net current through the Q-machine is relatively small, because the ion velocities are rather modest, in spite of the acceleration, and plasma densities in Q-machines seldom exceed 10^{16} m^{-3}.

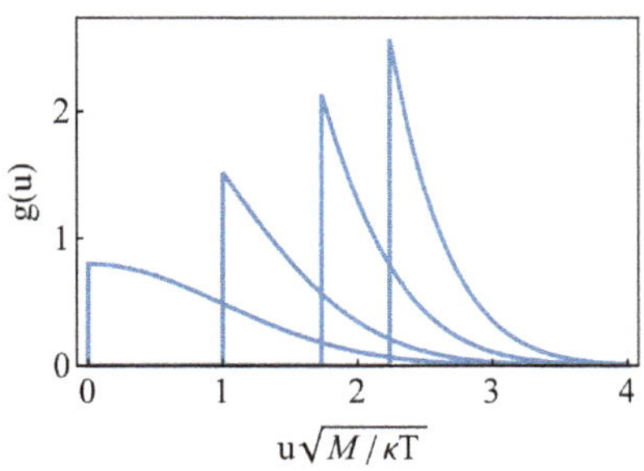

Figure 9.2. Variation of the ion distribution function with normalized plasma potential.

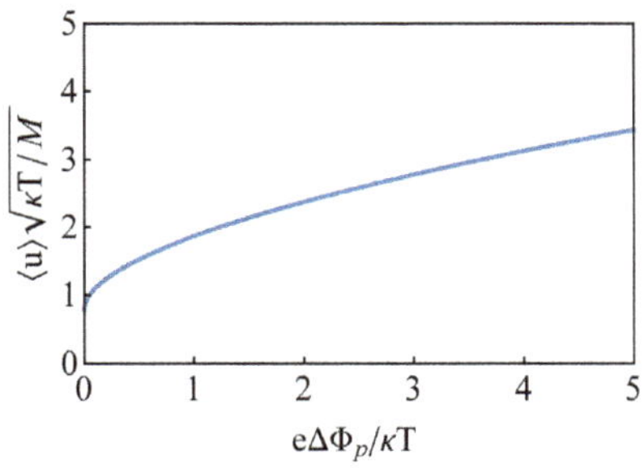

Figure 9.3. Variation of the normalized drift velocity with normalized plasma potential.

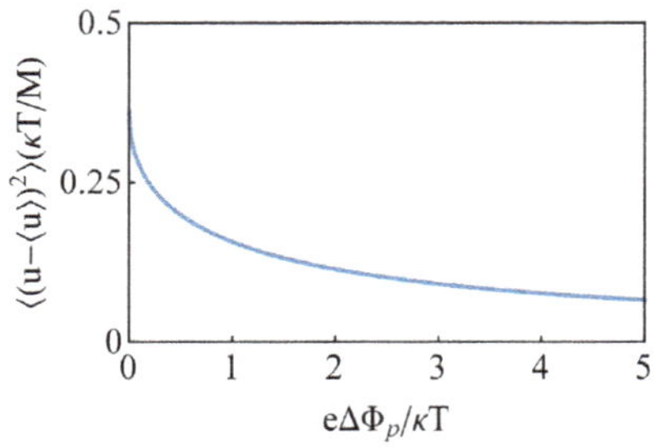

Figure 9.4. Variation of the normalized effective temperature of the ion distribution with normalized plasma potential.

Finally, we can obtain an 'effective temperature' for the ion distribution as $T_{\rm eff} \equiv (M/\kappa)\langle(u - \langle u\rangle)^2\rangle = \int(u - \langle u\rangle)^2 g(u)\mathrm{d}u$. Also this result has an analytical expression, but it is rather lengthy, and need not be reproduced here. The result is shown graphically in figure 9.4. The effective 'cooling' due to the bulk ion acceleration over the potential drop $\Delta\Phi_p$ is sometimes called 'adiabatic cooling', and is generic to similar phenomena also in a wider context. It is important to note that the cooling only affects the **B**-parallel part of the velocity distribution function. The **B**-perpendicular part has the same Maxwellian as at the hot plate surface, and the ion velocity distribution is therefore strongly anisotropic. The *electron* component can be assumed to be in thermal equilibrium with the hot plate, and therefore temperature, but not ions, are in equilibrium.

The analysis outlined before can be repeated for the *ion rich* conditions, usually obtained by operating the hot plate at low temperatures. *Some* heating will always be necessary to maintain a clean cathode surface, but the contact ionization is effective also at temperatures below ~1000°C, where the Richardson emission is rather ineffective. In this case the plasma potential will become positive, in order to reflect surplus ions back to the cathode. This mode of operation is found to be more noisy that the electron rich operation, and is seldom used.

If the temperature of the hot plate was ideally constant, and the illumination by the neutral was ideally uniform, the plasma density would be expected to be constant within the plasma column terminated by the hot plate. (Actually, even this is a bit optimistic, since the surface ionization depends on the crystal surface orientation, and after long time operation at high temperatures, the cathode material will often be crystallized to some degree. This can be observed visually, quite easily.) In reality, of course, all these idealized conditions are far from realized, and the plasma column will have radial density gradients. These should not be confused with the strong gradient at the edge of the column, i.e. at the *scrape off layer*. The plasma conditions in this region are quite complicated, in particular because of the strong radial electric fields there. Fluctuations in this region should not be confused with the drift waves in the form described in the foregoing sections.

9.2 Langmuir probes

Fluctuations in plasma density and floating potential can be detected by a Langmuir probe. By measuring potential differences between two probes it is possible to obtain an estimate for electric field components as well (Fahleson 1967), although the practical use of the method is restricted to cases where scale lengths (or wave lengths) of the potential fluctuations are much larger than the probe separation. Finite scale lengths gives uncertainties in the amplitudes as well as directions of the measured electric fields (Sato *et al* 2012). Basically, a Langmuir probe consists of a metal wire with a small exposed tip immersed in the plasma. The probe can be biased by a variable voltage source to collect ions or electrons from the plasma, or its potential can be left floating.

9.2.1 Plane probes

A simple one-dimensional model serves to illustrate the basic principles of the probe. We can for instance consider a strongly magnetized plasma, and let the probe be a disc with a diameter much larger than the ion Larmor radius, which is again much larger than the *electron* Larmor radius, for normal conditions. The plane of the probe is perpendicular to the magnetic field. Assume that the plasma potential is Φ_p, and that the probe is biased to a certain potential $\Phi \leqslant \Phi_p$. The probe potential will reflect all electrons with velocities *below* $u_{eM} \equiv \sqrt{2|e(\Phi - \Phi_p)|/m}$. Electrons with larger velocities will reach the probe, and be absorbed. Even when $\Phi = \Phi_p$, only particles (ions or electrons) with a velocity directed towards the probe will be absorbed, i.e. only one half of the distribution. For large positive probe potentials, Ψ, all electrons are absorbed, and all ion reflected, and vice versa for very negative potentials. The physical meaning of the plasma potential is, as already mentioned, that a charged particle with zero velocity has the potential energy $q\Phi_p$ with respect to a suitably defined ground.

The electron current to the probe will be

$$I_{ep} = -en\mathcal{A}\int_{u_{eM}}^{\infty} u\, f_e(u)\mathrm{d}u$$

in terms of the electron velocity distribution function, $f_e(u)$, while $\mathcal{A}$ is the probe collector area. In addition to I_{ep} we have an ion current, which is $I_{ip} = -en\mathcal{A}\int_0^{\infty} u\, f_i(u)\mathrm{d}u$, which is a constant as long as $\Phi \leqslant \Phi_p$. Obviously, we have $I_{ip} \ll I_{ep}$, unless $\Phi \ll \Phi_p$. It is only for very large negative probe potentials we are able to detect the ion current, and I_{ip} is often ignored in discussions of Langmuir probe performance. The outline given here for the ion current ignores the Bohm condition (Chen 1984) originating from ion acceleration across a pre-sheath. This condition will be of importance for large electron–ion temperature ratios. In this limit we need ions to arrive at the sheath edge at sonic or supersonic speeds for the sheath to be stable.

Remembering the $u_{eM} \sim \sqrt{|\Phi - \Phi_p|}$ dependence, it is now a simple matter to differentiate I_{ep} with respect to Φ, and we find

$$\frac{\mathrm{d}}{\mathrm{d}\Phi}I_{ep} = \frac{\mathrm{d}\,I_{ep}}{\mathrm{d}\sqrt{2\left|e(\Phi - \Phi_p)\right|/m}}\frac{\mathrm{d}}{\mathrm{d}\Phi}\sqrt{2\left|e(\Phi - \Phi_p)\right|/m}$$
$$= -\frac{1}{2}en\mathcal{A}\frac{2e}{m}f_e\left(\sqrt{2\left|e(\Phi - \Phi_p)\right|/m}\right) \tag{9.6}$$
$$\propto f_e\left(\sqrt{2\left|e(\Phi - \Phi_p)\right|/m}\right).$$

It is thus possible to obtain the electron energy distribution function by a relatively simple differentiation of the current to a Langmuir probe, at least in the present idealized one-dimensional operation.

- **Exercise:** Assume that the electron energy distribution is a Maxwellian. Demonstrate that you can determine the electron temperature by plotting the logarithm of the probe current as a function of the probe potential, Ψ. Demonstrate that the electron probe current in the present simple model can be expressed as $I_{ep} = \frac{1}{4}\langle|u|_e\rangle ne\mathcal{A}$, where the symbol $\langle\rangle$ represents averaging over the entire three-dimensional Maxwellian, giving $\langle|u|_e\rangle = \sqrt{8\kappa T_e/\pi m}$. The factor 1/4 in I_{ep} is obtained by the requirement that the particles collected must have their velocity vector directed towards the collecting surface.

Assuming the velocity distributions of the electrons and ions to be Maxwellians (although not necessarily at the same temperatures), we can give analytical expressions for the probe current contributions, I_{ep} and I_{ip}. We find (see also the foregoing exercise) the two probe current contributions as a function of probe potential Φ measured with respect to the plasma potential Φ_p

$$I_{ep}(\psi) = \frac{1}{4}\langle|u|_e\rangle ne\mathcal{A}e^{-e|\Phi-\Phi_p|/\kappa T_e} \quad \text{for} \quad \Phi < \Phi_p,$$

and

$$I_{ip}(\psi) = -\frac{1}{4}\langle|u|_i\rangle ne\mathcal{A}e^{-e|\Phi-\Phi_p|/\kappa T_i} \quad \text{for} \quad \Phi > \Phi_p.$$

The complete probe current $I_p = I_{ep} + I_{ip}$ in this model is shown in figure 9.5. The plasma potential Φ_p is taken to be negative in this case. We note another characteristic potential, the 'floating potential', Ψ_{fl}, where the net probe current vanishes, $I_p = 0$. Note that the probe is hit by an equal number of electrons and ions per second when $\Phi = \Phi_p$. To have a vanishing net current we must have somewhat less electrons than ions arriving per second, since electrons, with their large thermal velocities, give a large contribution per electron to the probe current as compared to the ions. It is only for quite unrealistic electron/ion temperature ratios that the electron and ion current contributions are equal. Now letting Φ_{fl} be the probe potential with respect to the plasma potential, we find for a Maxwellian velocity distribution

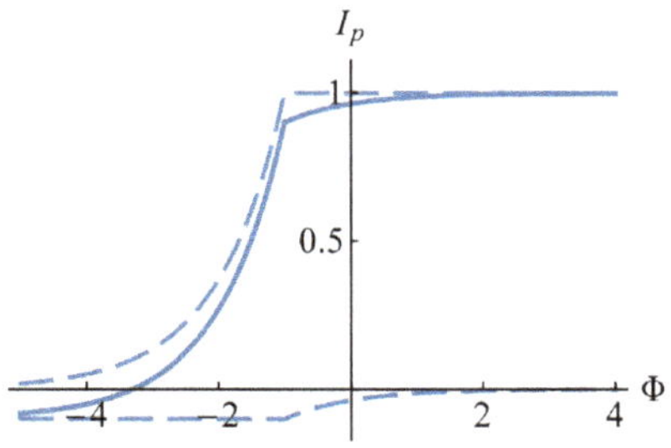

Figure 9.5. Illustration of an idealized probe characteristic for a one-dimensional model. The full line shows the (idealized) probe current, while the dashed lines show the electron and ion current contributions. For the sake of illustration we took an ion to electron mass ratio of $M/m = 100$. Otherwise, the ion current would hardly be noticeable. We choose here to show the electron current positive: there are no established conventions for this.

$$\frac{1}{4}\langle|u|_e\rangle ne\mathcal{A}e^{-|e\Psi_{fl}|/\kappa T_e} = \frac{1}{4}\langle|u|_i\rangle ne\mathcal{A},$$

i.e.

$$\Psi_{fl} = \ln\left(\frac{\langle|u|_e\rangle}{\langle|u|_i\rangle}\right)\frac{\kappa T_e}{e}. \tag{9.7}$$

For Maxwellian plasmas we readily find

$$\Psi_{fl} = \ln\left(\sqrt{\frac{T_e}{T_i}\frac{M}{m}}\right)\frac{\kappa T_e}{e}.$$

Obviously we find $\Psi_{fl} \to -\infty$ for $M/m \to \infty$, since the ion saturation current vanishes in this limit, and cannot compensate any finite electron probe current.

If a probe, or any other object for that matter, is left floating in the plasma, it assumes the floating potential, and not, as one might naively have expected, the plasma potential. Thus, the plasma potential as such is not quite as easy to measure as one might hope. For low frequency phenomena in Maxwellian plasmas, we can assume that the difference between Ψ_{fl} and Φ_p are constant, and the floating potential is, in a sense, just as good as the plasma potential. In plasmas with electron beams, for instance, the floating potential can be very sensitive to details in the distribution functions.

When the probe is much smaller than the ion Larmor radius, the ion trajectory through the sheath can be quite complicated, and requires a careful study. For most investigations, we require small probes in order to avoid disturbances of the plasma, and these problems can be relevant. As long as only the electron current to the probe is used for the diagnostics, the previous results can be used for strongly magnetized plasmas as long as the probe diameter is much larger than a typical electron Larmor radius: this is usually not a particularly restrictive requirement.

- **Exercise:** A Langmuir probe at its standard operation will draw a relatively large electron current from the plasma when biased to the plasma potential.

In some cases this can cause unacceptable disturbances. These can be reduced by using a double probe, where the potential is applied not to one probe with respect to ground, but as a difference potential between two probes. Make a sketch of the arrangement, and discuss the probe performance, assuming Maxwellian distributions for the plasma species.

Because of the absorbed ions, the probe surface will eventually be coated (although pump oil, etc will usually pose a greater problem in this respect), and the work function of a probe is often critically depending on the surface conditions. Usually, probes are heated by some means to retain a clean surface.

Concerning more detailed discussion on Langmuir probe performance and interpretation of results, we refer to the literature (Huddlestone and Leonard 1965, Fahleson 1967, Hutchinson 1987, Piel 2010). In particular, two limiting cases need attention, one where the shielding distance is thin compared to the probe sizes and the opposite limit with very thin probes. In the former limit, it is often allowed to assume the local geometry to be plane. Sheath limited and orbit limited probe geometries are distinguished for the two cases mentioned (Allen 1992). Probes used for flowing plasmas, in particular, have been discussed by Oksuz and Hershkowitz (2004). For these cases, the interpretation of the Debye shielding has to be reanalyzed, in particular when the flow velocity of one of the plasma species becomes comparable to or even larger than the respective thermal velocities.

9.2.2 Time varying conditions

Most of the discussion in section 9.2.1 dealt with information of basic plasma properties such as density and temperature. These data will have interest for specifying time stationary conditions. As illustrated in the following, care should be taken when modeling probe performance for dynamic, time varying conditions; for instance, by numerical simulations. Specifically the current collected by a probe, or any object in contact with hot plasmas, needs attention for time varying conditions.

Many numerical as well as analytical studies of Langmuir probe performance account for currents collected by the probes to be given as the number of particles crossing the probe surface per time unit. Such models ignore the effects of charges (Stratton 1941) induced in metallic surfaces by the moving charged plasma particles that happen to be in the vicinity of the probe surface. For steady state conditions methods counting charged particles crossing the surface happen to give the correct results, but in principle they fail for detecting dynamic plasma phenomena. The error resulting from ignoring the induced charges is estimated in the present note. The problem can become pertinent when relating analytical or numerical results to the actual dynamic variations of plasma parameters, the density in particular. The discussion applies for sheath limited as well as orbit limited probe geometries (Allen 1992).

We discuss first a simple realistic physical model for an electric circuit that is sufficiently simple to allow a transparent demonstration of moving charges inducing currents into an electric circuit. A large capacitor C is here made up of two plates,

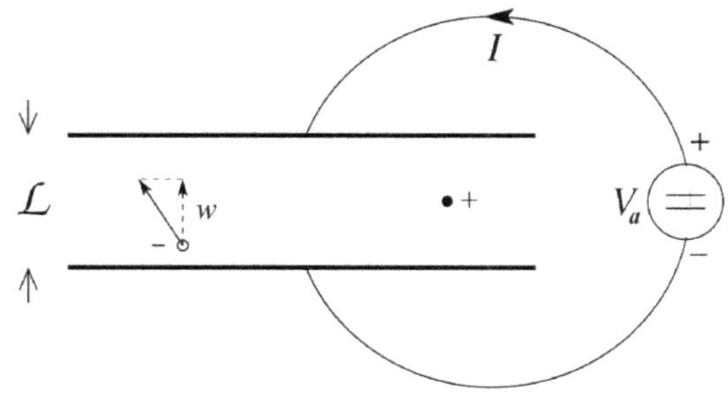

Figure 9.6. A simple physical model consisting of a pair of capacitor plates with area $\mathcal{A}$, separated by a distance $\mathcal{L}$. The capacitor contains one ion at rest and one moving electron.

assumed to be ideally conducting, with area $\mathcal{A}$ with separation $\mathcal{L}$, see figure 9.6. When $\mathcal{L}^2 \ll \mathcal{A}$ we have $C \approx \varepsilon_0 \mathcal{A}/\mathcal{L}$. The z-direction is here chosen to be perpendicular to the plates. A charged particle placed between the two plates will be accelerated in the positive or negative z-direction, depending on charge and potential polarity.

If we on the other hand assume the two terminals of the capacitor to be short circuited, there will not be any charge accumulations on the plate and the electron and the ion will propagate along straight line orbits, at least until they hit one of the plates. During such time intervals any of the charged particles will induce a current in the circuit by inducing charges q on the capacitor plates. This current will be recorded by an external detector. The electron–ion pair here represents the plasma and the current the diagnostic.

Applying a voltage difference V_a between the capacitor plates by some external battery we have an electric field $E_a = -V_a/\mathcal{L}$ between the plates. At any position between the plates we have a potential $V(z) = V_a z/\mathcal{L}$. A particle with charge $-e$ released at $t = 0$ with velocity w_0 will at later times have $w = w_0 + teV_a/(m\mathcal{L})$. The charge $-e$ is inducing charges on the condenser plates. The potential distribution for the problem with one ideally conducting plate and one fixed charge can be found analytically by the 'method of images' (Stratton 1941). A moving negative charge will repel electrons from the capacitor plates into the connecting wires, leaving the positively charged ions behind. The induced charges q are here the net charge on a plate and should not be confused with the image charge used for the 'method of images'.

The charge q induced on a plate by the moving particle can be determined by the following argument (Davenport and Root 1958, Pécseli 2000). The amount of work, U, done to induce a charge q on a plate, when the applied potential across the capacitor is V_a, must equal the energy W gained (or lost) by a charged particle moving through a certain potential difference V. The 'image charge current' is necessary: without it there would be no current through the battery and the moving charge could gain or lose energy without affecting the energy source, i.e. the battery.

Let the charged particle start at some position z. While $U = qV_a$ we have $W = -\Delta V\, e = -V_a e\, \Delta z/\mathcal{L}$, i.e. $q = -e\Delta z/\mathcal{L}$ with Δz being the spatial displacement of the moving charged particle. The current is obtained by $i = \mathrm{d}q/\mathrm{d}t$, giving the intuitively reasonable result

$$i = -\frac{ew}{\mathcal{L}}, \tag{9.8}$$

with $w \equiv \mathrm{d}\Delta z/\mathrm{d}t$ being the particle velocity component perpendicular to the plates, taken to be the $\hat{z}$-direction. Velocities $\perp\hat{\mathbf{z}}$ do not contribute.

The current pulse through the external circuit, i.e. the battery and connecting wires, induced by one moving electron will be

$$i = -\frac{ew}{\mathcal{L}} = -\frac{ew_0}{\mathcal{L}} + t\frac{e^2 V_a}{m\mathcal{L}^2}, \tag{9.9}$$

as long as $t < t_a$, with t_a being the arrival time of the electron at one of the capacitor plates. For $t > t_a$ we have $i = 0$. Note that V_a can take both signs. The moving charge does not contribute particularly much to the current at the instant where it hits a capacitor plate; at that time it only neutralizes the induced charge. The contribution of ions to (9.8) is ignored here, since they in general have low velocities. The role of ions is solely to maintain overall charge neutrality. The pulse-form found here enters as an acceptable model for 'shot noise' in diodes, as discussed in more detail in section 20.3.

We can disconnect the wire between the two plates. For this *open* circuit the current density J is exactly canceled by Maxwell's displacement current, i.e. the $\varepsilon_0\partial\mathbf{E}/\partial t$ term in Maxwell's equations. In this case one plate will assume a time varying potential V_c with respect to the other one. Since there is no external connection to the plate, the potential will generally be denoted 'floating potential'. Usually the floating potential is assumed to be the one resulting from a balance between electron and ion currents arriving at the surface, but as we have seen in the model there will be a long time interval where there are no charged particles arriving at the surface, but we have a stationary current nonetheless. This is not a paradox, but a consequence of accounting for induced charges.

If we short circuit the two condenser plates without a battery setting $V_a = 0$ we will have w constant. Assume that we have N electrons-ion pairs present, and let again the ions be at rest and all electrons having constant velocity component w in the $\hat{z}$-direction. The net current is then $I = Ni = -eNw/\mathcal{L} = -ewn\mathcal{A}$, where we introduced the constant electron density $n \equiv N/(\mathcal{L}\mathcal{A})$, with $\mathcal{A}$ being the area of the capacitor plates. The steady state electron current density is $J = -ewn$.

The example considered so far assumed one electron velocity. Averaging over a stationary Maxwellian velocity distribution is at least formally trivial.

General stationary conditions

In the simple model discussed in the previous section we can assume one of the plates to serve as ground while the other one represents a large Langmuir probe detecting the stationary current induced by the moving charged particles. The model gives a stationary current until the moving particle arrives at one of the capacitor plates.

To obtain truly stationary conditions we can make a small opening in one of the plates and inject a steady state flow of moving charged particles as illustrated case i) in figure 9.7. Each particle contributes with the current (9.8), giving the net current $ewN/\mathcal{L}$. The number of particles arriving at the capacitor plate per time unit is however also $ewN/\mathcal{L}$, implying that a simple counting of charges arriving at the plate will for this time stationary example give the correct current.

If we now generalize the model to have a statistical distribution of particles in thermal equilibrium with vanishing net drift, we will at any time with equal probability find a particle with velocity $+\mathbf{w}$ and $-\mathbf{w}$ and the net current vanishes. All there will be left is a fluctuating shot-noise due to the discreteness of the particles.

To generalize the foregoing results to account for more familiar probe geometry we assume a spherical probe with radius r_0 and a shielding distance R as shown in figure 9.8. The previous results can be applied as long as $R - r_0 \ll r_0$. In this limit we can estimate the current induced by any one moving electron as $i \approx -e\mathbf{w} \cdot \hat{\mathbf{r}}/(R - r_0)$, where $\hat{\mathbf{r}}$ is the unit vector pointing from the origin to the appropriate particle. Under stationary conditions the current contribution of the particles will be obtained correctly by counting the number of charged particles collected on the probe surface at $r = r_0$. Numerical or analytical models for determining steady state parameters like plasma density or temperature are thus

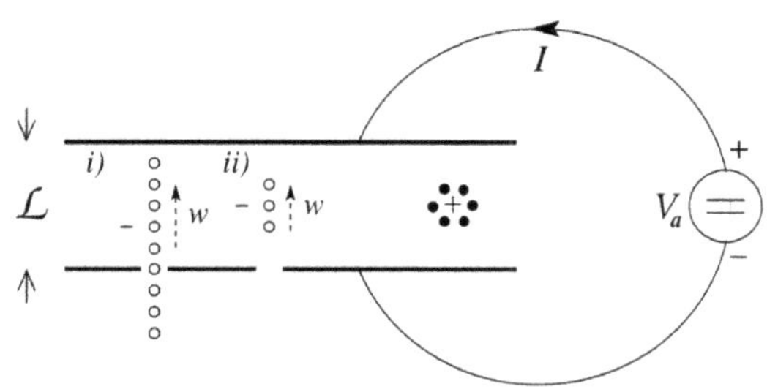

Figure 9.7. A generalization of the model from figure 9.6. The capacitor contains N ions at rest and N moving electrons, all with the same velocity w. For this example, the two plates are short circuited, so the velocities w remain constant. By case i) we illustrate a long, essentially time stationary, flow of particles, while case ii) shows a small burst of particles injected through an opening in the bottom plate.

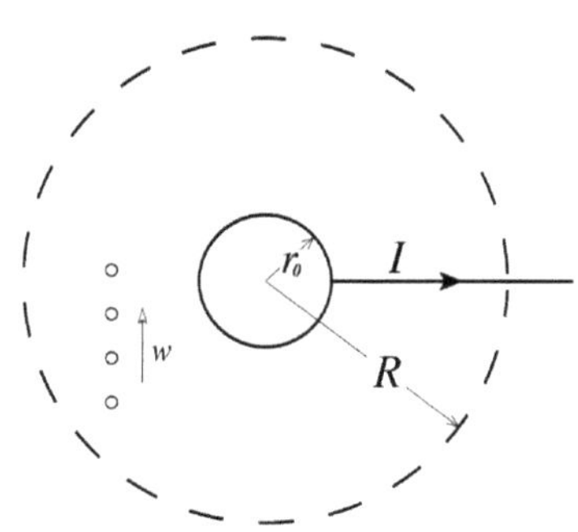

Figure 9.8. Generalization of the conditions by ii) in figure 9.7, for non-stationary conditions where a short burst of particles passes in the vicinity of a spherical probe with radius r_0.

not affected by induced charges. Classical probe theory applies for these stationary conditions.

Non-stationary conditions
An example of non-stationary conditions is illustrated as case ii) in figure 9.7 where the spatial length of the burst is ℓ. A short burst of such particles passing by the probe without reaching the probe surface will induce a current in the connecting probe wire. Analytical or numerical models based on counting particles crossing the probe surface will miss this contribution to the probe current.

To give an estimate of the error in ignoring the contribution from a burst like the one shown as case ii) in figure 9.7 we can consider a low frequency problem where the electrons can be assumed to be in a local isothermal Boltzmann equilibrium, and Langmuir probes are used to detect ion acoustic wave phenomena (Chen 1984). Considering the electron component under these conditions we can at any time find two electrons compensating the induced current, and we need only be concerned with the ions. Assuming the spatial extent of an ion burst to be $\ell < \mathcal{L}$ and its velocity to be w we find a duration of the induced burst to be $\tau \sim \ell / w$ as an order of magnitude. Taking $\ell \sim \lambda_D$, the Debye length, and $w \sim \sqrt{T_i/M}$, the ion thermal speed, we find $\tau \sim \Omega_{pi}^{-1}$, the inverse ion plasma period. This estimate indicates that analytical or numerical models studying the dispersive limit of ion sound waves by counting particles crossing the probe surface will give misleading results. For very low frequencies $\omega \ll \Omega_{pi}$, on the other hand, the error will be negligible since the wavelengths will be much longer than ℓ and the induced charge signal will be small. The current response in the connecting wire to a propagating particle burst is proportional to its velocity. Deducing a plasma temperature from a non-stationary probe current also has to take into account the variation in velocity response for different particle velocities.

9.2.3 Emissive probes

As seen from (9.7), we have the floating potential being identical to the plasma potential in case $\langle |u|_e \rangle = \langle |u|_i \rangle$, as would be the case for instance if the electron and ion masses as well as temperatures were the same. Except for some 'esoteric' plasmas, we will not encounter such cases, but note on the other hand that an electron *emitted* from the probe gives a current corresponding to an ion *entering* the probe. By constructing heated, electron emitting probes, one can by careful choice of parameters obtain conditions where the floating potential of the probe is indeed very close to the plasma potential. It should be emphasized that it is *not* a sufficient condition to have a copious amount of electrons being emitted from the probe: in case the electron emission is too strong, the floating potential becomes larger than the plasma potential. Ideally, the conditions should be chosen such that the electron emission current + the ion collection current for a large negative probe bias equals the electron collection current for a large positive bias.

9.3 Experimental observations of drift waves

Electrostatic drift waves have been extensively studied in Q-machine plasmas (Politzer 1971, Motley 1975), with one of the most detailed, now classic, investigations carried out by Hendel *et al* (1968). Q-machines offer good access to the plasma and the conditions allow a wide range of parameters to be analyzed. Because of their importance for anomalous transport, drift waves have been studied in fusion related experiments as well.

The Q-machine investigations of Hendel *et al* (1968) were carried out at relatively large plasma densities, 10^{17}–10^{18} m^{-3}, so that ion viscosity damped most of the large azimuthal mode-numbers. Some of their results are summarized in figure 9.9, as obtained by detecting density fluctuations by Langmuir probes. In particular, we note that the $m = 2, 3, 4$ modes appear one at a time with only a slight overlap, as the magnetic field is increased. It was found that the natural frequency of oscillations was close to $\frac{1}{2}\omega^*$, which is what we expect to be the linearly most unstable wave

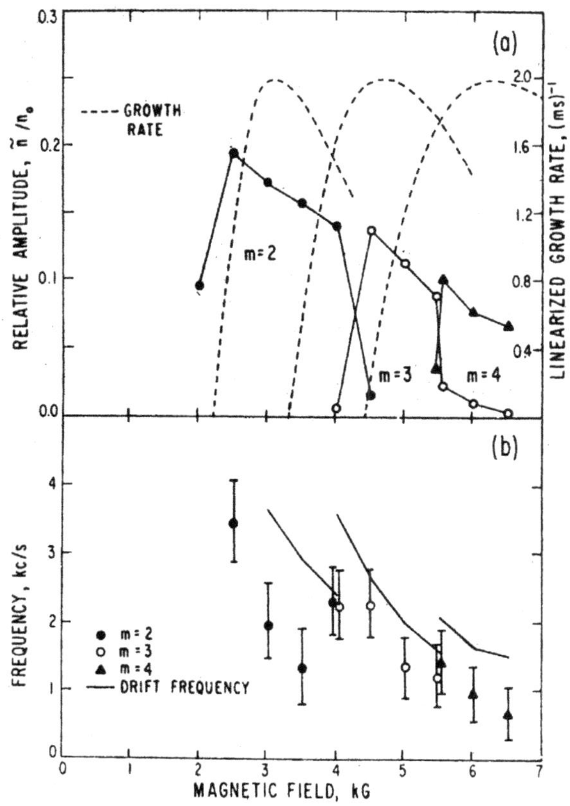

Figure 9.9. Experimental results of drift wave investigations, carried out in a Q-machine plasma at large densities. Figure (a) shows relative mode amplitudes for varying magnetic field, with dashed lines giving the linearized growth rate of the instability for reference. Figure (b) shows the variation of the frequency for mode numbers $m = 2, 3, 4$ for varying B. The full line gives the drift wave frequency ω^* for reference. The figure is reproduced from Hendel *et al* (1968) with permission. Copyright 1968, AIP Publishing LLC.

frequency, see figure 6.1. To obtain this result it was necessary to compensate for the azimuthal rotation of the entire plasma column, which is induced by a radial electric field component, inherent in the set-up. The amplitude of the relative density fluctuations, $\tilde{n}/n_0$, was found to maximize at the position with largest density gradient, as expected. Considering the uncertainties involved in the measurements, the agreement with analytical results can be considered as good. We bear in mind that the rotation of the plasma column gives rise to a slight increase in the azimuthal current by the centrifugal force acting predominantly on the ions. These small corrections were unaccounted for in the theoretical results. A perfect agreement should after all not be expected since the theoretical results we have derived are based on a local theory, where gradients ϕ' and ϕ'' were ignored in deriving the dispersion relation.

The theoretical expression for marginal stability of drift waves (8.17), indicates a $\sqrt{n}$ scaling of B/k_y. Experimental results by Hendel *et al* (1968) indicate that this predicted variation is in excellent agreement with observations.

By introducing both Cs and K atoms in the system, Hendel *et al* (1968) were able to change the average mass, $\overline{M}$ by a factor of more than 3 (from 39 to 132 in atomic mass units), and found results indicating a scaling $\overline{M}^{3/8}$ of $B/k_\perp$, again in good agreement with (8.16)–(8.17). The parameter scaling is in near excellent agreement with theory, but a numerical agreement concerning the threshold is obtained only by scaling down the magnetic field by a factor 1.5, or, if you like, the threshold value of ν_{ii} is $(1.5)^4 \approx 5$ times what it should be for a given magnetic field. Considering the approximations made in the theoretical analysis, such a disagreement might be considered as minor.

Other experiments were carried out by Schlitt and Hendel (1972) in a modified double ended Q-machine where the hot plates were movable along the axis of the device. The separation L between the two hot plates (i.e. the length of the plasma column) could be varied, and thereby also the **B**-parallel wavelength. The range of variation available was $L = 15$–135 cm. The hot plates act as ideal conductors, ensuring the presence of waves-nodes at these positions, thus determining the maximum parallel wavelengths (Chen 1979). The curves for marginal stability in a plane spanned by the parameters $B/k_\perp$ and $\lambda_\parallel$ could be studied, as illustrated in figure 9.10 for two parameter sets. The figure demonstrates three of the stability characteristics of collisional drift waves (Schlitt and Hendel 1972):

1. Low field stability—the plasma is stable for all parallel wavelengths for $B/k_\perp < 0.35$.
2. Short wavelength stability—as $\lambda_\parallel$ is decreased, the critical value of $B/k_\perp$ increases until a critical parallel wavelength (here $\lambda_\parallel = 60$ cm) is reached. For shorter $\lambda_\parallel$ the plasma is stable for all $B/k_\perp$.
3. High field stability—the four observations of critical values for $B/k_\perp > 0.8$ are points where the $m = 1$ mode is stabilized by **B**-parallel ion motion for large $B/k_\perp$.

Schlitt and Hendel (1972) also investigated the growth rate of the linearly unstable drift waves. It was found that the waves could be at least partially

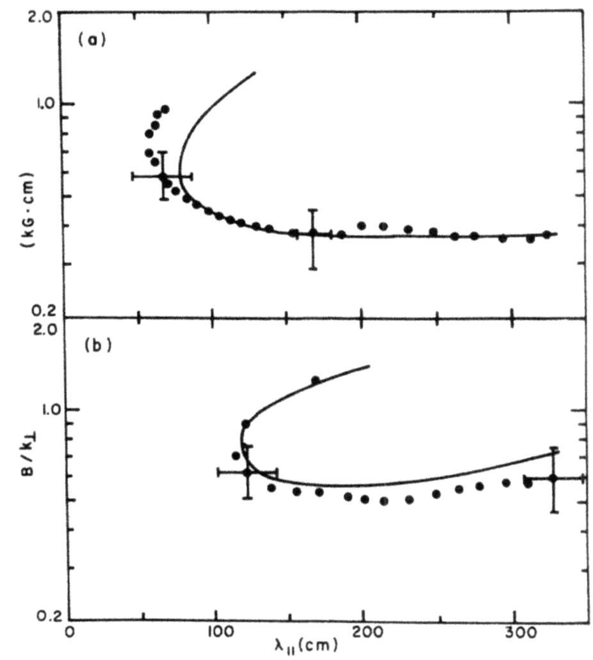

Figure 9.10. Experimental results of drift wave investigations carried out in a Q-machine plasma at large densities. Solid lines give theoretical curves with magnetic field scaled by a factor 1.5. In (a) we have $n = 2.5 \times 10^{10} \text{cm}^{-3}$, in (b) $n = 2 \times 10^{11}\ \text{cm}^{-3}$. The figure is reproduced from Schlitt and Hendel (1972) with permission. Copyright 1972, AIP Publishing LLC.

suppressed by applying a short pulse (duration 20 μs) with amplitude −25 V to one of the end plates (the precise mechanism for the observed stabilization seems not to be understood, nor studied in any detail). After the 'switch-off' of the pulse, the drift waves were found to grow from the noise level, and their growth rate could be determined. In the *stable* regime, waves could be *excited* by application of a −5 V pulse, and their subsequent damping investigated. Also these results were in good agreement with theory. Also Rogers and Chen (1970) investigated the linear growth rate of the drift instabilities, using a feed-back stabilization technique which could be switched on and off. When the stability condition was changed from stable to unstable, the growth of the instability, as well as the formation of the stationary saturated fluctuation level, could be directly investigated. These results were also in good agreement with analytical results.

All in all, it can be concluded that the linear theory of resistive drift waves seems to be on firm ground when compared with experimental results. It is therefore reasonable to continue the investigations by also studying the nonlinearly saturated state of the instability. It may be worth noting, that in none of the experiments summarized here was a turbulent state observed (at least it was not described). The presence of predominantly one mode only, has been attributed to a nonlinear 'mode competition' (Lashinsky 1969), which is a phenomenon also observed in optical systems, for instance. This observation is consistent with our expectations of drift waves being generally only weakly unstable, and that additional sources of free energy are required for driving the system into a strongly turbulent state. A rotation of the plasma column, or a sheared velocity field are likely candidates. Indeed, the experiments where strongly turbulent low frequency fluctuations were observed by Pécseli *et al* (1983) were characterized by a strong nearly uniform or 'solid body' rotation, giving a radial centrifugal force. The electron motion is almost unaffected by this force, but it gives rise to a reduction in the ion rotation frequency and a corresponding net enhancement of the azimuthal plasma current, as compared to the 'natural' diamagnetic current. This enhancement of the free energy available for the instability gave rise to strongly turbulent conditions in that experiment.

References

Allen J E 1992 Probe theory—the orbital motion approach *Phys. Scripta* **45** 497–503

Chen F F 1979 Axial eigenmodes for long-$\lambda_{||}$ waves in plasmas bounded by sheaths *Phys. Fluids* **22** 2346–58

Chen F F 1984 *Introduction to Plasma Physics and Controlled Fusion* vol 1 2nd edn (New York: Plenum)

Davenport W B and Root W L 1958 *An Introduction to the Theory of Random Signals and Noise* (New York: McGraw-Hill)

Fahleson U 1967 Theory of electric field measurements conducted in magnetosphere with electric probes *Space Sci. Rev.* **7** 238–62

Hendel H W, Chu T K and Politzer P A 1968 Collisional drift waves–identification, stabilization, and enhanced plasma transport *Phys. Fluids* **11** 2426–39

Huddlestone R H and Leonard S L (ed) 1965 *Plasma Diagnostic Techniques* (New York: Academic)

Hutchinson I H 1987 *Principles of Plasma Diagnostics* (Cambridge: Cambridge University Press)

Kaye G W C and Laby T H 1995 *Tables of Physical and Chemical Constants* 16th edn (London: Longmans)

Langmuir I and Kingdon K H 1925 Thermionic effects caused by vapors of alkali metals *Proc. Roy. Phys. Soc. Ser.* A **107** 61–79

Lashinsky H 1969 *Turbulence in Fluids and Plasmas* vol 18 *of Microwave Research Institute Symposia Series* (New York: Polytechnic Press,) chapter Periodic pulling and the transition to turbulence in a system with discrete modes pp 29–46

Motley R W 1975 *Q Machines* (New York: Academic)

Oksuz L and Hershkowitz N 2004 Understanding Mach probes and Langmuir probes in a drifting, unmagnetized, non-uniform plasma *Plasma Sources Sci. Techn.* **13** 263–71

Pécseli H L 2000 *Fluctuations in Physical Systems* (Cambridge: Cambridge University Press)

Pécseli H L 2012 *Waves and Oscillations in Plasmas* (London: Taylor and Francis)

Pécseli H L, Mikkelsen T and Larsen S E 1983 Drift wave turbulence in a low-β plasma *Plasma Phys.* **25** 1173–97

Piel A 2010 *Plasma Physics: An Introduction to Laboratory, Space, and Fusion Plasmas* (Heidelberg: Springer)

Politzer P A 1971 Drift instability in collisionless alkali metal plasmas *Phys. Fluids* **14** 2410–25

Rogers K C and Chen F F 1970 Direct measurements of drift wave growth rates *Phys. Fluids* **13** 513–6

Sato H, Pécseli H L and Trulsen J 2012 Fluctuations in the direction of propagation of intermittent low frequency ionospheric waves *J. Geophys. Res.* **117** A03329
Schlitt L G and Hendel H W 1972 Effects of parallel wavelength on the collisional drift instability *Phys. Fluids* **15** 1578–89
Stratton J A 1941 *Electromagnetic Theory* (New York: McGraw-Hill)

IOP Publishing

Low Frequency Waves and Turbulence in Magnetized Laboratory Plasmas and in the Ionosphere

Hans Pécseli

Chapter 10

Velocity shear driven instabilities

Gradient drift instabilities can be seen as surface waves, associated with a density gradient in the direction perpendicular to a magnetic field. A uniform plasma flow of both species along or across magnetic field lines will not have any effect in an otherwise uniform plasma, but if there are gradients in this flow, they can also give rise to instabilities, which can have significant growth rates. Since it is not easy to avoid electric fields perpendicular to magnetic field lines, it is very common to have gradients in the $\mathbf{E}_0 \times \mathbf{B}/B^2$-drifts and these velocity gradient instabilities are quite common and might be confused with drift instabilities since they occur in the same frequency range. Current driven instabilities are found when there are relative electron–ion drifts, and these will be discussed later. Velocity shear instabilities have a close counterpart in neutral flows where the problem has been studied intensively (Miles 1961).

We here consider two cases: one where the velocity gradient is derived from velocities $\mathbf{u}_i \perp \mathbf{B}$ and one where we have $\mathbf{u}_i \parallel \mathbf{B}$. The nature of the bulk plasma flow is different in the two cases. In the first case the flow is the $\mathbf{E}_0(\mathbf{r}_\perp) \times \mathbf{B}/B^2$-velocity to lowest order, while the second case has no *a priori* relation between a steady state electric field and plasma velocity. In particular we can have different electron and ion flows in the $\parallel\mathbf{B}$ case, but this case is not considered here.

10.1 Shear in ion velocities with $u_i \perp B$

We consider first the ion velocity gradient illustrated in figure 10.1. For this problem we can distinguish two different models for the electron dynamics, i.e. one with and one without Debye shielding.

doi:10.1088/978-0-7503-1251-6ch10

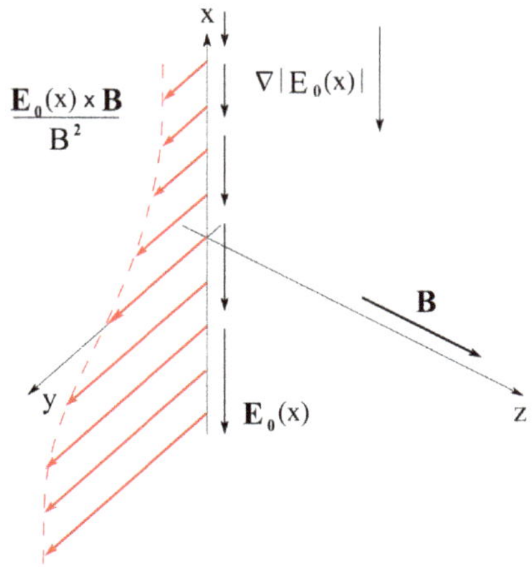

Figure 10.1. Illustration of a velocity shear with $\mathbf{u}_0(x) \perp \mathbf{B}$.

10.1.1 Velocity shear instabilities without electron shielding

Here we first consider electrostatic structures in a homogeneous magnetic field, with the assumption that the scale lengths along **B** are infinite. The physical arguments for assuming Boltzmann distributed electrons are then not valid. Instead we assume a continuity equation for the electrons similar to (4.3) as used for the ions. The dynamics of the two species are coupled through Poisson's equation

$$\nabla^2 \phi = \frac{e}{\varepsilon_0}(n_e - n_i). \tag{10.1}$$

The velocities $\perp \mathbf{B}$ are assumed to be given as the $\mathbf{E}_0 \times \mathbf{B}/B^2$-velocities for both species. Subtraction of the two continuity equations and use of Poisson's equation gives the nonlinear equation

$$\left(\frac{\partial}{\partial t} - \frac{1}{B^2}\nabla_\perp \phi \times \mathbf{B} \cdot \nabla_\perp\right)\nabla^2 \phi = 0, \tag{10.2}$$

which uniquely determines the evolution of the electrostatic plasma potential when the initial condition is given. The incompressible bulk plasma velocity is $\mathbf{u} = -\nabla_\perp \phi \times \mathbf{B}/B^2$, and the vorticity deduced from it is $|\nabla_\perp \times \mathbf{u}| = |\nabla_\perp^2 \phi|/|B|$, which by Poisson's equation (10.1) shows that the charge density uniquely determines the vorticity in the present simplified two-dimensional model. The equation (10.2) is inherently non-linear; upon linearization the result is trivially $\partial \phi / \partial t = 0$.

Consider here a basic flute-type model (10.2), and assume that for some reason, we have an inhomogeneous background potential $\phi_0(x)$ in the system. This implies that a consistent electric field is of the form $\mathbf{E}_0 = -\nabla \phi_0(x) = \{E_x(x), 0\}$, which in turn imposes a sheared plasma velocity $\mathbf{u}_0 = \{0, E_x(x)/B\}$ in the direction transverse to the magnetic field lines. We note that in this form $\phi_0(x)$ is an *exact* steady state solution of (10.2). We now make a perturbation analysis, and introduce $\phi = \phi_0(x) + \phi_1$. Ignoring second order terms, we find by trivial manipulations the equation

$$\frac{\partial}{\partial t}\nabla^2\phi_1 + \frac{E_x(x)}{B}\hat{\mathbf{y}} \cdot \nabla\nabla^2\phi_1 + \hat{\mathbf{z}} \times \nabla\phi_1 \cdot \left(\frac{\partial^2}{\partial x^2}\frac{E_x(x)}{B}\hat{\mathbf{x}}\right) = 0 \tag{10.3}$$

or

$$\frac{\partial}{\partial t}\nabla^2\phi_1 + u_0\frac{\partial}{\partial y}\nabla^2\phi_1 - u_0''\frac{\partial}{\partial y}\phi_1 = 0, \tag{10.4}$$

where $u_0'' \equiv \mathrm{d}^2u_0(x)/\mathrm{d}x^2$ is the gradient of the zero order vorticity. The equation (10.4) is inhomogeneous only with respect to the x-variable, so we are free to Fourier transform it with respect to y as well as time t, i.e. write $\phi_1 = \phi(x)\exp(-i(\omega t - k_y y))$. By insertion, we find the equation

$$\frac{\mathrm{d}^2}{\mathrm{d}x^2}\phi(x) - k_y^2\phi(x) - \frac{k_y u_0''(x)}{k_y u_0(x) - \omega}\phi(x) = 0. \tag{10.5}$$

For general profiles, $u_0''(x)$, the equation (10.5) can be solved only by numerical methods (Horton *et al* 1987). We consider a special case with $u_0(x) = V\tanh(x/a)$, and find here a simplified solution by analyzing the inner and outer regions, respectively.

The outer solution is obtained by taking large values of $|x|$, where $u_0''(x) \approx 0$. Physically acceptable solutions for the electrostatic potential must be finite for $|x| \to \infty$, recalling that the potential ϕ_1 is the deviation from the unperturbed potential $\phi_0(x)$, associated with the sheared $u_0(x)$ velocity. In this case it is easy to obtain $\phi(x) = C\exp(k_y x)$ for $x < 0$ and $\phi(x) = D\exp(-k_y x)$ for $x > 0$. For the inner solution, we can take $u_0(x) \approx Vx/a$, and have $u_0''(x) \approx 0$. Consequently, also in this case it is easy to solve (10.5). The general solution is $\phi(x) = A\exp(k_y x) + B\exp(-k_y x)$. The full solution is then obtained by fitting the inner and outer solutions at $x = a$ and $x = -a$. In effect we hereby approximate the true variation $u_0(x)$ with $-u_0$ for $x < -a$, u_0 for $x > a$, and $u_0 x/a$ for $-a < x < a$. We argue on physical grounds that the potential ϕ must be a continuous function at $x = \pm a$, otherwise we would experience infinite electric fields. At the discontinuities we *can* have discontinuous electric fields, but this we have to live with. We then have

$$\begin{aligned} x = a &: D\mathrm{e}^{-k_y a} = A\mathrm{e}^{k_y a} + B\mathrm{e}^{-k_y a} \\ x = -a &: C\mathrm{e}^{-k_y a} = A\mathrm{e}^{-k_y a} + B\mathrm{e}^{k_y a}. \end{aligned} \tag{10.6}$$

From (10.5) we can obtain two boundary conditions at $x = a$ and $x = -a$. We thus integrate (10.5) from $x = -a - \epsilon$ to $x = -a + \epsilon$, and then let $\epsilon \to 0$. In this limit only the derivative terms contribute, and we find the *jump conditions* (Chandrasekhar 1961, Baumjohann and Treuman 1997, Marshall 2001) to be

$$\frac{\mathrm{d}}{\mathrm{d}x}\phi\Bigg[_{-a^-}^{-a^+} - \frac{k_y u_0'(x)}{k_y u_0(x) - \omega}\phi\Bigg[_{-a^-}^{-a^+} = 0, \tag{10.7}$$

where $x = -a \pm \epsilon \to -a^{\pm}$ for $\epsilon \to 0$. In obtaining (10.7) we used that $u_0''(x)$ approximates a $\delta(x + a)$-function, which contributes no matter how small ϵ is, while all other terms, containing $u_0'(x)$ for instance, contribute with terms becoming negligible when $\epsilon \to 0$.

The condition (10.7) gives that $\mathrm{d}\phi/\mathrm{d}x$ is discontinuous at $x = -a$ and

$$k_y\left(A\mathrm{e}^{-k_y a} - B\mathrm{e}^{k_y a}\right) - k_y C\mathrm{e}^{-k_y a} - \frac{k_y u_0/a}{-k_y u_0 - \omega}\left(A\mathrm{e}^{-k_y a} + B\mathrm{e}^{k_y a}\right) = 0.$$

Using (10.6), we find

$$B\left(2 - \frac{k_y u_0/a}{k_y u_0 + \omega}\right) = \frac{k_y u_0/a}{k_y u_0 + \omega}\mathrm{e}^{-2k_y a}A.$$

By an entirely similar procedure at $x = a$ we find

$$A\left(2 - \frac{k_y u_0/a}{k_y u_0 - \omega}\right) = \frac{k_y u_0/a}{k_y u_0 - \omega}\mathrm{e}^{-2k_y a}B.$$

Simple manipulations finally give the dispersion relation (Horton *et al* 1987)

$$\omega^2 = \left(\frac{u_0}{2a}\right)^2\left[\left(1 - 2k_y a\right)^2 - \mathrm{e}^{-4k_y a}\right]. \tag{10.8}$$

We have two purely imaginary solutions for $k_y a < k_c a \approx 0.628$, where one corresponds to an aperiodic instability. For $k_y > k_c$ we find two propagating waves with dispersion relation

$$\omega = \pm\left(\frac{u_0}{2a}\right)\sqrt{\left(1 - 2k_y a\right)^2 - \mathrm{e}^{-4k_y a}}. \tag{10.9}$$

see also figure 10.2.

The maximum growth rate for the instability is obtained for $k_y a = \frac{1}{2}$, with the value $\omega = i\frac{1}{2}(u_0/a)/\exp(1)$, where we write $\exp(1)$ to avoid a possible confusion with the electron charge e on a bad day. For this particular case we find $B = iA$, so the corresponding eigenfunction is

$$\phi_m = \cosh(x/2a) + i\sinh(x/2a) \quad \text{for} \quad -a < x < a,$$

while $\phi_m = 1 + i\exp(1)$ for $x < -a$, and $\phi_m = \exp(1) + i$ for $x > a$. The potential variation associated with this simple anti-symmetric eigensolution is shown in figure 10.3. The mode amplitude is here $[2\cosh^2(x/2a) - 1]^{1/2}$, while the phase is $\theta(x) = \arctan[\tanh(x/2a)]$. Note that if the width of the shear layer $a \to \infty$ in such a

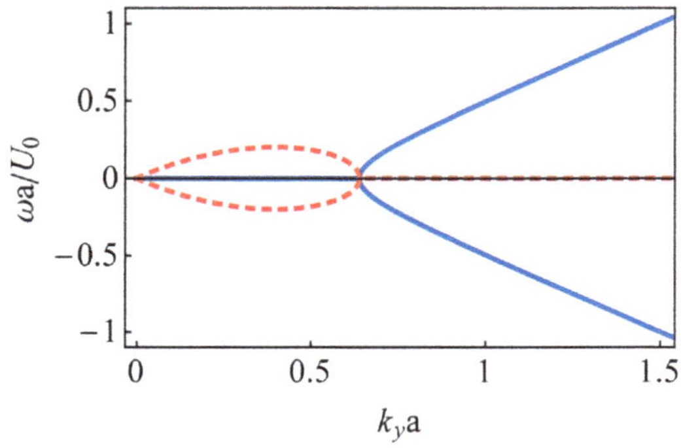

Figure 10.2. Dispersion relation for a Kelvin–Helmholtz, or shear flow, instability with velocities $\perp\mathbf{B}$. Full lines give the real parts, dashed lines the imaginary parts of ω.

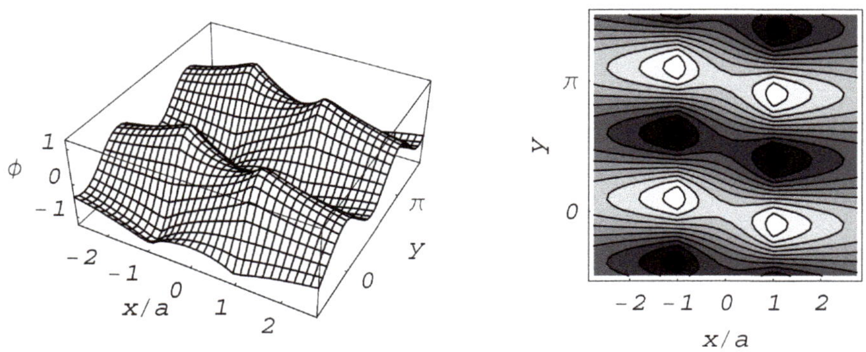

Figure 10.3. Potential variation associated with the most unstable eigenfunction for the shear flow. The electrostatic potential is shown as a surface to the left, as contours to the right. The velocity shear is confined to the interval $-1 < x/a < 1$.

way that u_0/a = const., we have a constant value for the maximum growth rate, but the most unstable wavenumber $k_y \to 0$, and also $k_c \to 0$. In this sense we can argue that for an infinitely wide shear, there is no finite wavelength instability, indicating that a necessary condition for instability is the *Rayleigh condition*, that the gradient of vorticity $u_0'' = 0$ for some x-value in the profile. The theorem is closely related to the one known as *Fjørtoft's theorem* (Marshall 2001). If we let the transition region shrink in order to let the velocity shear become a step function, i.e. let $a \to 0$ while keeping u_0 constant, we find that the growth rate of the most unstable perturbation increases indefinitely.

- **Exercise:** Determine the eigenfunction associated with the marginally stable mode, where $k_y = k_c$, see also figure 10.2.

- **Exercise:** Determine, in terms of a, the wavenumber k_c associated with the marginally stable mode, $\omega = 0$, for the case $u_0(x) = V\tanh(x/a)$, without making use of the approximative velocity profile used for obtaining (10.8).

The potential variation in figure 10.3 is interesting by resembling two parallel alternating vortex chains. Since local shears will inevitably appear in any realizations of two-dimensional flows with large amplitude perturbations, this observation may give a hint for the seemingly universal presence of coupled vortex-like perturbations in two-dimensional turbulence. Numerical simulations for this problem, performed by Horton *et al* (1987), demonstrate that this conjecture is to some extent justified.

- **Example:** Mathematically, the relation (10.5) is a special case of the Taylor–Goldstein equation (Taylor 1931, Goldstein 1931), derived for neutral fluids in a constant gravitational field

$$\frac{\mathrm{d}^2}{\mathrm{d}x^2}w(x) - k_y^2 w(x) - \frac{u_0''(x)}{u_0(x) - \omega/k_y}w(x) + \frac{N^2}{\left(u_0(x) - \omega/k_y\right)^2}w(x) = 0, \quad (10.10)$$

  where w is the complex amplitude of the vertical velocity. Equation (10.10) also allows a change in fluid density through the buoyancy frequency $N \equiv \sqrt{-(g/\rho)\mathrm{d}\rho/\mathrm{d}x}$, with g being the (constant) gravitational acceleration. The standard Rayleigh–Taylor stability problem is obtained for vanishing flow velocity, $u_0 = 0$, with $N \neq 0$. Instabilities of neutral fluids as described by (10.10) have been studied in great detail, theoretically as well as experimentally by e.g. Lawrence *et al* (1991).

10.1.2 Velocity shear instabilities with electron shielding

In chapter 17 we discussed shear instabilities in the 'flute limit', where the perturbations are assumed to be strictly magnetic field aligned, see for instance (17.6). If we, as before, allow for a long **B**-parallel wavelength so that electrons can flow along magnetic field lines to be in an electrostatic Boltzmann equilibrium we modify (10.5) to become

$$\frac{\partial}{\partial t}\left(\nabla^2\phi_1 - \frac{1}{\lambda_D^2}\phi_1\right) + \frac{E_x(x)}{B}\hat{\mathbf{y}} \cdot \nabla\nabla^2\phi_1 + \hat{\mathbf{z}} \times \nabla\phi_1 \cdot \left(\frac{\partial^2}{\partial x^2}\frac{E_x(x)}{B}\hat{\mathbf{x}}\right) = 0 \quad (10.11)$$

or

$$\frac{\partial}{\partial t}\left(\nabla^2\phi_1 - \frac{1}{\lambda_D^2}\phi_1\right) + u_0\frac{\partial}{\partial y}\nabla^2\phi_1 - u_0''\frac{\partial}{\partial y}\phi_1 = 0. \quad (10.12)$$

We find, after a Fourier transform with respect to t and y, the equation

$$\frac{\mathrm{d}^2}{\mathrm{d}x^2}\phi(x) - k_y^2\phi(x) - \frac{k_y u_0''(x) - \omega/\lambda_D^2}{k_y u_0(x) - \omega}\phi(x) = 0. \quad (10.13)$$

Unfortunately, this is not at all so easy to solve as (10.5), although the differences between the two basic equations appear to be modest, at first sight. The problem arises mainly in the intermediate region where $u_0'' = 0$. If $\lambda_D \to \infty$, the last term in

(10.5) and the equation is relatively simple to deal with, as we have seen. The outer solutions of (10.13), where $u = V =$ const. are easily obtained just as before as

$$\phi(x) = C\cos\left(\frac{x\sqrt{\omega - k_y^3V\lambda_D^2 + k_y^2\lambda_D^2\omega}}{\lambda_D\sqrt{k_yV - \omega}}\right) + D\sin\left(\frac{x\sqrt{\omega - k_y^3V\lambda_D^2 + k_y^2\lambda_D^2\omega}}{\lambda_D\sqrt{k_yV - \omega}}\right).$$

The result can also be expressed in terms of exponential functions. With λ_D finite, we have a problematic explicit x-dependence remaining through $u_0(x) = Vx/a$ in the denominator of (10.13). The inner solution is for this case

$$\begin{aligned}\phi(x) &= A\exp\left(-k_yx + a\omega/V\right)U\left(-\frac{a\omega}{2k_y^2V\lambda_D^2}, 0, 2k_yx - 2a\omega/V\right)\\ &+ B\exp\left(-k_yx + a\omega/V\right)L\left(\frac{a\omega}{2k_y^2V\lambda_D^2}, -1, 2k_yx - 2a\omega/V\right),\end{aligned}$$

in terms of hypergeometric functions $U(a, b, c) \equiv (1/\Gamma(a))\int_0^\infty \exp(-zt)t^{(a-1)}$ $(1 + t)^{(b-a-1)}\mathrm{d}t$, and Laguerre polynomia $L(n, a, x)$ that satisfy the differential equation $xy'' + (a + 1 - x)y' + ny = 0$. The result can also be expressed in terms of AiryAi and AiryBi functions. The 'jump-condition' remains here the same as (10.7), since $\phi(x)$ is continuous also here.

10.2 Shear in ion velocities with $u_i \| B$

The most relevant formulation of the problem uses a cylindrical symmetry as relevant for instance for Q-machine experiments (D'Angelo 1965, D'Angelo and von Goeler 1966). To simplify the notation we use here a Cartesian coordinate system. The electron dynamics is simplified by assuming the electrons can flow along magnetic field lines to assume a local isothermal Boltzmann distribution at all times. This allows elimination of the electric field term in the dynamic ion momentum equation as we have seen before. The ion temperature is assumed to be vanishing, implicitly assuming $T_e \gg T_i$. The ion momentum equation is then reduced to

$$n\left(\frac{\partial \mathbf{u}}{\partial t} + \mathbf{u}\cdot\nabla\mathbf{u}\right) = -en\nabla\phi + en\mathbf{u}\times\mathbf{B}.$$

We assume steady state conditions where $n_0(\mathbf{r}) \sim \exp(-x/\mathcal{L})$ so that $\nabla n_0 = -n_0(x)\hat{\mathbf{x}}/\mathcal{L}$. This density gradient gives rise to a diamagnetic drift of electrons but not for the ions because of the assumption $T_i \approx 0$. The steady state ion velocity assumed here is $\mathbf{u}_{i0}(\mathbf{r}) = u_{z0}(x)\hat{\mathbf{z}}$, with $\mathbf{B} = B\hat{\mathbf{z}}$, see illustration in figure 10.4. We take the zero order density gradient ∇n_0 to be parallel to the velocity gradient: this is not logically necessary, but on the other hand it seems to be the most reasonable assumption for a cylindrical geometry so we retain it here as well. The steady state ion continuity equation is exactly satisfied by the proposed equilibrium.

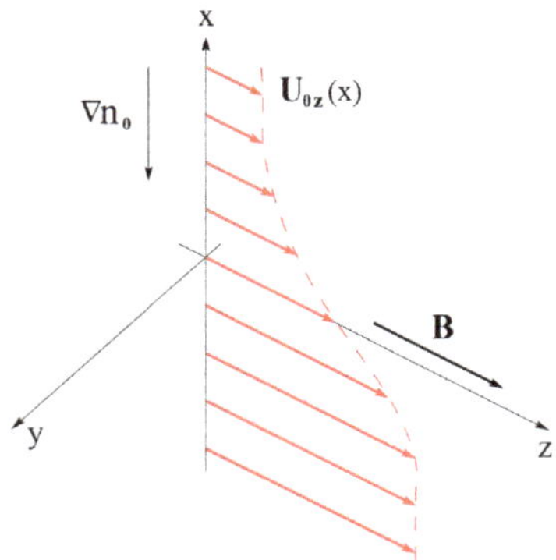

Figure 10.4. Illustration of a velocity shear with $\mathbf{u}_0(x)||\mathbf{B}$.

The linearized dynamic ion continuity equation is

$$\frac{\partial n_i}{\partial t} + n_0\nabla \cdot \mathbf{u}_i + n_i\overbrace{\nabla \cdot \mathbf{u}_{i0}}^{=0} + \mathbf{u}_i \cdot \nabla n_0 + \mathbf{u}_{i0} \cdot \nabla n_i = 0, \qquad (10.14)$$

where $\nabla \cdot \mathbf{u}_0 = 0$ for the given conditions. We introduce now the expression $e\phi = \kappa T_e n_e/n_0$ for the perturbations in potential and electron density, and use the quasi-neutrality condition $n_e \approx n_i$. Then the linearized ion momentum equation becomes

$$n_0\frac{\partial \mathbf{u}_i}{\partial t} + n_0(\mathbf{u}_i \cdot \nabla \mathbf{u}_{i0} + \mathbf{u}_{i0} \cdot \nabla \mathbf{u}_i) + C_s^2\nabla n_i - \Omega_{ci}\left(n_0\mathbf{u}_i \times \hat{\mathbf{z}} + n_i\overbrace{\mathbf{u}_{i0} \times \hat{\mathbf{z}}}^{=0}\right) = 0,$$

with $C_s \equiv \sqrt{\kappa T_e/M}$ being the ion sound velocity. Using

$$\frac{1}{n_0}\nabla n_i = \nabla\frac{n_i}{n_0} + \frac{n_i}{n_0}\frac{\nabla n_0}{n_0} = \nabla\frac{n_i}{n_0} - \frac{n_i}{n_0}\frac{\hat{\mathbf{x}}}{\mathcal{L}},$$

we can write normalized versions of these two equations with $\eta \equiv n_i/n_0$ in the form

$$\frac{\partial \eta}{\partial t} + \mathbf{u}_{i0} \cdot \nabla\eta + \nabla \cdot \mathbf{u}_i + \mathbf{u}_i \cdot \frac{\nabla n_0}{n_0} + \eta\overbrace{\mathbf{u}_{i0} \cdot \frac{\nabla n_0}{n_0}}^{=0} = 0,$$

and

$$\frac{\partial \mathbf{u}_i}{\partial t} + \mathbf{u}_{i0} \cdot \nabla\mathbf{u}_i + \mathbf{u}_i \cdot \nabla\mathbf{u}_{i0} + C_s^2\nabla\eta - \Omega_{ci}\mathbf{u}_i \times \hat{\mathbf{z}} = 0,$$

where we use $\nabla n_0 = -n_0(x)\hat{\mathbf{x}}/\mathcal{L}$. With the given conditions we have $\mathbf{u}_{i0} \cdot \nabla n_0 = 0$ and also $\mathbf{u}_{i0} \times \hat{\mathbf{z}} = 0$.

Since spatial inhomogeneities are here in the **x**-direction, we can make Fourier transforms in time and in the spatial **y**- and **z**-directions to give a set of equations

$$
\begin{aligned}
-i\left(\frac{\partial}{\partial x} - \frac{1}{\mathcal{L}}\right)u_x + k_y u_y + k_z u_z - \Omega\eta &= 0, \\
\Omega u_x - i\Omega_{ci} u_y + iC_s^2 \frac{\partial \eta}{\partial x} &= 0, \\
i\Omega_{ci} u_x + \Omega u_y - C_s^2 k_y \eta &= 0, \\
i u_x \frac{\partial u_{z0}}{\partial x} + \Omega u_z - C_s^2 k_z \eta &= 0,
\end{aligned}
\tag{10.15}
$$

introducing the Doppler shifted frequency $\Omega \equiv \omega - k_z u_{z0}$ where generally $\Omega = \Omega(x)$.

The important term here is $u_x \mathrm{d}u_{z0}/\mathrm{d}x$ in (10.15), which is accounting for the velocity shear (D'Angelo 1965). A detailed analysis requires the set of equations (10.15) to be solved for eigenvalues and eigenfunctions varying with x. The present analysis will be restricted to a local approximation, very much like the one argued in section 5.1, so we also here take $\partial\eta/\partial x \approx 0$ and $\partial u_x/\partial x \approx 0$. By this local assumption we also here ignore propagation in the x-direction. Requiring that the set of equations (10.15) has a non-trivial solution within the present approximations we obtain a local dispersion relation in the form

$$
\begin{aligned}
&\Omega^4 - \left(C_s^2\left(k_y^2 + k_z^2\right) + \Omega_{ci}^2\right)\Omega^2 - \Omega\Omega_{ci}\frac{k_y C_s^2}{\mathcal{L}} \\
&+ C_s^2 k_z \Omega_{ci}\left(k_z \Omega_{ci} - k_y \frac{\mathrm{d}u_{z0}}{\mathrm{d}x}\right) = 0.
\end{aligned}
\tag{10.16}
$$

This is a fourth order equation in Ω which is not too difficult to solve. In particular, for $\mathcal{L} \to \infty$ where we have no density gradient, the equation becomes of second order in Ω^2. For a general analysis we normalize (10.16) by $\Omega' \equiv \Omega/\Omega_{ci}$, $K_{y,z} \equiv k_{y,z}C_s/\Omega_{ci}$, and introduce dimensionless parameters $\alpha_n \equiv C_s/(\mathcal{L}\Omega_{ci})$ and $\beta_u \equiv \mathrm{d}u_{z0}/\mathrm{d}x/\Omega_{ci}$, noting that the two parameters can assume both signs. We then have

$$
\Omega'^4 - \left(1 + K_y^2 + K_z^2\right)\Omega'^2 - \Omega' K_y \alpha_n + \left(K_z^2 - K_z K_y \beta_u\right) = 0. \tag{10.17}
$$

Two of the solutions are real, corresponding to stable wave propagation, while two other roots form a complex conjugate pair, one of them corresponding to instability. Since we have assumed quasi-neutrality, there is no small length scale in the problem (i.e. the Debye length) and there is no length scale to arrest the wave growth as $K \to \infty$. The remedy is to retain Poisson's equation, but we will leave that as an exercise to the reader.

As a test we can take the case where $\alpha_n = \beta_u = 0$. Then we have $\Omega'^4 - (1 + K_y^2 + K_z^2)\Omega'^2 + K_z^2 = 0$ giving the stable solution $\Omega'^2 = \frac{1}{2}(1 + K_y^2 + K_z^2) \pm \sqrt{\frac{1}{4}(1 + K_y^2 + K_z^2)^2 - K_z^2} > 0$. For $K_z = 0$ we find $\Omega' = 0$ and $\Omega' = \pm\sqrt{1 + K_y^2}$, the latter solution being the normalized ion cyclotron waves, the former a degenerated ion sound wave that does not propagate $\perp\mathbf{B}$. For $K_y = 0$ we have $\Omega'^2 = \frac{1}{2}(1 + K_z^2) \pm \frac{1}{2}(1 - K_z^2)$ giving $\Omega' = 1$ which is the degenerated ion cyclotron wave, and $\Omega' = \pm K_z$ which is the sound wave solution. The gradients in density and velocity couple these wavemodes.

Some simplifications can be justified. We can safely assume that the ion cyclotron frequency is the largest characteristic frequency in the system and assume $\Omega^2 \ll \Omega_{ci}^2$, $C_s^2/\mathcal{L}^2 \ll \Omega_{ci}^2$ as well as $k_{y,z}^2 C_s^2 \ll \Omega_{ci}^2$. This allows a simpler second order dispersion relation (D'Angelo 1965) in the form

$$\Omega^2 - \Omega\frac{k_y C_s^2}{\Omega_{ci}\mathcal{L}} - C_s^2 k_z^2\left(1 - \frac{k_y}{k_z}\frac{1}{\Omega_{ci}}\frac{du_{z0}(x)}{dx}\right) = 0. \tag{10.18}$$

Numerical solutions of the normalized equation are shown in figure 10.5. It is comforting that for $u_{z0}(x) = 0$ this equation reproduces the simple drift-wave dispersion relation (5.14) without the ion polarization term. Although Ω_{ci} is large, the last term in the right hand parenthesis of (10.18) can become significant for small k_z. The condition for (10.18) having unstable solutions is readily found to be given as

$$\frac{du_{z0}}{dx} > \frac{k_z}{k_y}\Omega_{ci} + \frac{C_s^2}{\mathcal{L}^2}\frac{k_y}{4k_z\Omega_{ci}}.$$

By varying $k_z\Omega_{ci}/k_y$ we find that the minimum threshold for the velocity shear is for $k_z\Omega_{ci}/k_y = \frac{1}{2}|C_s/\mathcal{L}|$, and we determine the lowest threshold value for the instability as

$$\frac{du_{z0}}{dx} > \left|\frac{C_s}{\mathcal{L}}\right|.$$

Steep density gradients (i.e. small $\mathcal{L}$) are stabilizing the velocity shear instability. Velocity shears have been suggested as the mechanism for exciting fluctuations in the auroral regions of the ionosphere (Basu and Coppi 1989), but collisions need to be included in a full analysis of this problem.

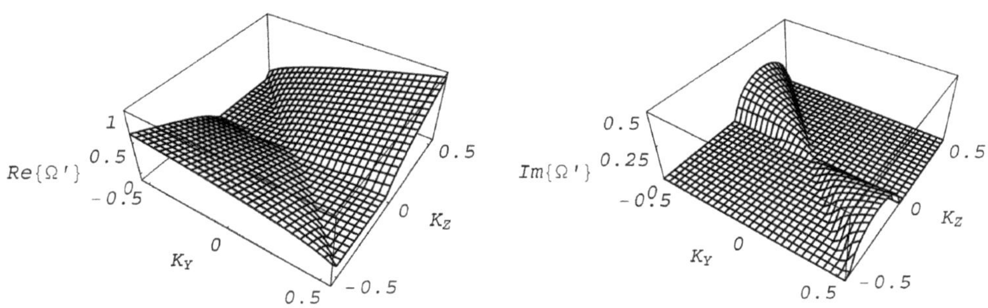

Figure 10.5. Illustration of the real and imaginary parts of Ω' in (10.18), here shown for a selected set of parameters $\alpha_n = .5$ and $\beta_u = 1$. Note that waves grow in a direction at an angle to the magnetic field, which is here taken in the z-direction. Recall the definition $\Omega \equiv \omega - k_z u_{z0}$.

Other instabilities, the inhomogeneous energy-density driven (IEDD) instability for instance, closely related to velocity shear instabilities are described in the literature (Koepke *et al* 1994).

References

Basu B and Coppi B 1989 Velocity shear and fluctuations in the auroral regions of the ionosphere *J. Geophys. Res. Space Phys.* **94** 5316–26

Baumjohann W and Treuman R A 1997 *Basic Space Plasma Physics* (London: Imperial College Press)

Chandrasekhar S 1961 *Hydrodynamic and Hydromagnetic Stability* (London: Oxford University Press)

D'Angelo N 1965 Kelvin–Helmholtz instability in a fully ionized plasma in a magnetic field *Phys. Fluids* **8** 1748–50

D'Angelo N and von Goeler S 1966 Investigation of the Kelvin–Helmholtz instability in a cesium plasma *Phys. Fluids* **9** 309–13

Goldstein S 1931 On the stability of superimposed streams of fluids of different densities *Proc. Roy. Soc. London Ser.* A **132** 524–48

Horton W, Tajima T and Kamimura T 1987 Kelvin–Helmholtz instability and vortices in magnetized plasma *Phys. Fluids* **30** 3485–95

Koepke M E, Amatucci W E, Carroll J J and Sheridan T E 1994 Experimental verification of the inhomogeneous energy-density driven instability *Phys. Rev. Lett.* **72** 3355–8

Lawrence G A, Browand F K and Redekopp L G 1991 The stability of a sheared density interface *Phys. Fluids* A **3** 2360–70

Marshall J S 2001 *Inviscid Incompressible Flow* (New York: Wiley)

Miles J W 1961 On the stability of heterogeneous shear flows *J. Fluid Mech.* **10** 496–508

Taylor G I 1931 Effect of variation in density on the stability of superposed streams of fluid *Proc. Roy. Soc. London Ser.* A **132** 499–523

IOP Publishing

Low Frequency Waves and Turbulence in Magnetized Laboratory Plasmas and in the Ionosphere

Hans Pécseli

Chapter 11

Ionospheric conditions

As already mentioned, the analysis of linearly unstable drift waves as it is presented in the foregoing sections, will apply qualitatively also for the case where the resistivity is caused by electron–neutral collisions. One exception is however cases where there are DC-electric fields, $\mathbf{E}_0$, imposed on the plasma; assuming that the field is homogeneous, it can be transformed out of the problem by a change of frame of reference to one moving with the $\mathbf{E}_0 \times \mathbf{B}/B^2$-velocity, provided we only have collisions between electrons and ions. In the other case, where a neutral component is important, there will appear a neutral wind in this new frame, i.e. no simplification is gained by a change of reference. As these particular conditions are often encountered in the ionosphere as well as in many laboratory experiments, it is worthwhile to give them special consideration (Sudan *et al* 1973, 1997, Kelley 1989). Ion–electron collisions will be ignored here. This is appropriate for large neutral densities as in the ionospheric E-region.

11.1 Collisions with neutrals

First, we comment on collisional regimes. The analysis will be restricted to the case where the electron cyclotron frequency is much larger than the electron–neutral collision frequency, $\omega_c \gg \nu_e$, while the ion cyclotron frequency is assumed to be smaller than the ion–neutral collision frequency, $\Omega_{ci} \ll \nu_i$. Such a parameter regime exists, since for a given magnetic field the cyclotron frequencies scale with the inverse mass, while for comparable ion and electron temperatures, the collision frequencies, on the other hand, scale approximately as the square-root of the inverse mass. This can be argued by noting that the cross-sections, σ, for ion–neutral and electron–neutral collisions are approximately the same, so the mean free paths for collisions, $\ell_c = 1/(n\sigma)$ are also approximately the same for both species. Taking the cross-section

doi:10.1088/978-0-7503-1251-6ch11

for collisions to be constant (as for 'billiard-ball' collisions) the collision frequency is then $\nu \approx U_{th}/\ell_c$ with $U_{th} = \sqrt{\kappa T/m}$ being the appropriate thermal velocity. Consequently, $\Omega_{ci}/\nu_i \approx (eB/M)\ell_c\sqrt{M/\kappa T_i}$ and $\omega_{ce}/\nu_e \approx (eB/m)\ell_c\sqrt{m/\kappa T_e}$. Assuming that $\Omega_{ci} \approx \nu_i$, we find $\omega_c/\nu_e \approx \sqrt{M/m}\sqrt{T_i/T_e} \gg 1$ for $T_e \sim T_i$, as stated.

Alternatively, we can take $(\nu_i/\Omega_{ci})^2 \approx \sqrt{M/m}$, and obtain in this particular limit the approximation $(\omega_{ce}/\nu_e)^2 \approx (T_i/T_e)\sqrt{M/m} \gg 1$, for $T_e \sim T_i$. The foregoing discussion ignored the variations of cross-sections with velocity. These effects will in general strengthen the arguments, since cross-sections usually decrease at large velocities. Consequently, it is possible to have $\omega_{ce}/\nu_e \gg 1$ and $\Omega_{ci}/\nu_i \ll 1$ also in cases where $T_e \gg T_i$.

As a numerical example we have in the ionospheric E-region at an altitude of 110 km over northern Scandinavia: $\nu_i \approx 600\ \mathrm{s}^{-1}$ and $\nu_e \approx 1.5 \cdot 10^4\ \mathrm{s}^{-1}$. The temperatures are $T_i \approx 200$ K and $T_e \approx 400$ K. The magnetic fields are $B \approx 50\ \mu$T giving $\Omega_{ci} \approx 180\ \mathrm{rad\ s}^{-1}$, corresponding to an average ion mass of 31 AMU. For the same conditions $\omega_{ce} = 8.8 \cdot 10^6\ \mathrm{rad\ s}^{-1}$, giving $\Omega_{ci}/\nu_i \approx 0.3 < 1$ and $\omega_{ce}/\nu_e \approx 600 \gg 1$. These values are highly variable with altitude, where figure 3.4 serves as an illustration. Plasma conditions have a significant diurnal variation, and change also with latitude (Brekke 2013).

As we increase the altitude, the neutral density decreases and the electron–neutral collision frequency decreases with it. The electron–ion collision frequency decreases with plasma density as given by (6.3). There seems to be a consensus concerning a transition altitude of approximately 200 km: below this we have neutral dominated electron dynamics, above it we find electron–ion or resistivity dominated electron dynamics. In particular in the E-region we can assume the electron dynamics to be dominated by neutral collisions. The conditions change with day/night-time conditions and to some extent also solar activity.

11.2 Basic equations

The basic equation here is the ion momentum equation

$$M\left(\frac{\partial \mathbf{u}_i}{\partial t} + \mathbf{u}_i \cdot \nabla \mathbf{u}_i\right) = e(\mathbf{E}_0 - \nabla\phi) - M\nu_i \mathbf{u}_i, \tag{11.1}$$

where only collisions with neutrals are retained, ion–electron collisions are assumed to be immaterial (Sudan *et al* 1973). Since we consider cold ions, $T_i \approx 0$, we have no ion pressure term and can divide all terms by the ion density, so that n does not appear in (11.1). According to the foregoing arguments the effect of the magnetic field on the ions can be ignored since an ion will, on average, never complete a full gyro-orbit when $\Omega_{ci} \ll \nu_i$. The DC-electric field is given a special symbol $\mathbf{E}_0$ and ϕ refers to a fluctuating quantity. For simplicity we take $\mathbf{E}_0 \perp \mathbf{B}$. Ion viscosity is ignored with the assumption $T_e \gg T_i$, i.e. the perturbations are driven essentially by the DC-electric field and the electron pressure. To be consistent we omitted also a $\kappa T_i \nabla n$-term in (11.1) as already stated. Due to the frequent collisions with the neutrals, the ions are isothermal and have a temperature which cannot deviate much from the temperature of the cold neutral gas, in general. We assumed quasi-neutrality form the outset and take $n_e \approx n_i \equiv n$.

In the electron momentum equation, electron inertia is ignored as before, giving

$$0 = -en(\mathbf{E}_0 - \nabla\phi + \mathbf{u}_e \times \mathbf{B}) - \kappa T_e \nabla n - mn\nu_e \mathbf{u}_e. \qquad (11.2)$$

It is a great advantage here that T_e is a constant. We can write (11.2) as

$$0 = -e(\mathbf{E}_0 - \nabla\phi + \mathbf{u}_e \times \mathbf{B}) - \kappa T_e \nabla \ln n - m\nu_e \mathbf{u}_e. \qquad (11.3)$$

This equation can be rewritten in the more convenient form for the **B**-parallel component

$$u_{e\parallel} = \frac{e}{\nu_e m}\frac{\partial}{\partial z}\phi - \frac{\kappa T_e}{\nu_e m}\frac{\partial}{\partial z}\ln n, \qquad (11.4)$$

while the **B**-perpendicular component is found to be

$$\mathbf{u}_{e\perp} = \frac{\omega_c^2}{\omega_c^2 + \nu_e^2}\left(\frac{\hat{\mathbf{b}} \times \nabla_\perp\phi}{B} + \frac{\nu_e}{\omega_c}\frac{\nabla_\perp\phi}{B} - \frac{\kappa T_e}{m\,\omega_c}\hat{\mathbf{b}} \times \nabla_\perp \ln n \right.$$
$$\left. - \frac{\kappa T_e\,\nu_e}{m\,\omega_c^2}\nabla_\perp \ln n - \frac{\nu_e}{\omega_c}\frac{\mathbf{E}_0}{B} + \frac{\mathbf{E}_0 \times \hat{\mathbf{b}}}{B}\right). \qquad (11.5)$$

In the electron and ion continuity equations an ionization production term q, and a recombination term $\alpha n_e n_i$, are included with α being the recombination coefficient. These terms will be important in particular for ionospheric applications. For the ions we have

$$\frac{\partial n}{\partial t} + \nabla \cdot (n\mathbf{u}_i) = q - \alpha n^2. \qquad (11.6)$$

With the assumption of quasi-neutrality we have $\alpha n_e n_i \approx \alpha n^2$, which together with the electron continuity equation implies $\nabla \cdot (n(\mathbf{u}_e - \mathbf{u}_i)) = 0$. In the present analysis we consider q and α as constants, both being implicitly determined by the neutral gas density, which is assumed constant. For large-scale phenomena in the ionosphere for instance, an altitude variation of the neutral density implies a corresponding variation of q as well as α and of course also of the collision frequencies. The momentum equations (11.1) and (11.2) do not contain terms originating from ionization or recombination; when pairs of electrons and ions recombine they carry their entire momentum with them, and ions and electrons are injected into the plasma by ionization without carrying any *a priori* momentum.

We can write (11.6) as

$$\frac{\partial \ln n}{\partial t} + \mathbf{u}_i \cdot \nabla \ln n + \nabla \cdot \mathbf{u}_i = \frac{1}{n}(q - \alpha n^2), \qquad (11.7)$$

which can be useful later. When we linearize (11.7) with respect to small fluctuations n_1 around an average density n_0 we have $\ln n \rightarrow \ln n_0 + n_1/n_0 \equiv \ln n_0 + \eta$ which will be true for any spatial variation of n_0 consistent with (11.7), where the productions and losses can vary with spatial position, as in the ionosphere for instance. For steady state conditions without significant ion drifts we have $q \approx \alpha n_0^2$. The problem of the normalization of $\ln n_0$ is usually of little concern.

11.3 Equilibrium

We assume here that the homogeneous DC-electric and magnetic fields are imposed externally. Due to the collisions with neutrals it is strictly speaking not possible to maintain a gradient in plasma density within the framework given by (11.1)–(11.6), as any such gradient present initially will be eroded by diffusion, just as in the foregoing case with electron–ion collisions. If we however, as before, assume this to be slow process, it is possible to make a stability analysis of a quasi-stationary equilibrium.

With the parameters assumed for the present study (high ion–neutral collisionality), we ignore the effect of the magnetic field on the ion component but not on the electrons. Assuming a steady state spatially varying density $n_0(x)$, the direction of $\mathbf{E}_0$ is taken to be parallel or anti-parallel with ∇n_0, i.e. $\mathbf{E}_0 = E_0\hat{\mathbf{x}}$. See figure 11.1 for definition of the coordinate system. From (11.1), the equilibrium ion velocity with $T_i \approx 0$ is determined by the ion mobility as

$$U_{i0x} \approx \frac{\Omega_{ci}}{\nu_i}\frac{E_0}{B}, \tag{11.8}$$

assuming $\Omega_{ci} \leqslant \nu_i$. We also find

$$u_{i0y} \approx \left(\frac{\Omega_{ci}}{\nu_i}\right)^2 \frac{|\mathbf{E}_0 \times \mathbf{B}|}{B^2} \ll U_{i0x}.$$

The electron equation (11.3) or (11.5) gives the equilibrium velocity components

$$u_{e0x} = -\left(\frac{\kappa T_e}{m}\frac{n_0'}{n_0} + \omega_{ce}\frac{E_0}{B}\right)\frac{\nu_e}{\nu_e^2 + \omega_{ce}^2}, \tag{11.9}$$

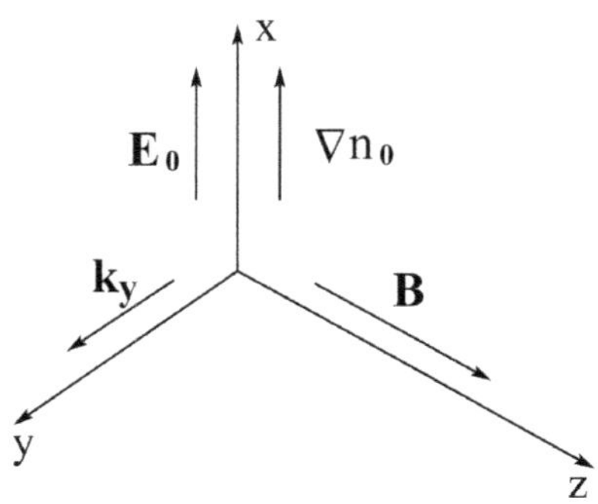

Figure 11.1. Reference coordinate system for the analysis of the Farley–Buneman and gradient drift instabilities. The external electric field $\mathbf{E}_0$ is here chosen to be parallel to the density gradient ∇n_0.

which is a combination of the diffusion drift along the density gradient and a so called Pedersen drift along the DC-electric field. In addition we have

$$u_{e0y} = -\left(\frac{\kappa T_e}{m}\frac{n_0'}{n_0} + \omega_{ce}\frac{E_0}{B}\right)\frac{\omega_{ce}}{\nu_e^2 + \omega_{ce}^2}, \tag{11.10}$$

which is combined Hall and gradient drifts, impeded by the collisions with neutrals. The drifts u_{i0x} and u_{e0x} erode the density gradient, but this is a slow process. The difference in electron and ion mobilities give rise to an ambipolar electric field (see section 3.6), which we assume to be included in E_0. In the following we take the density gradient to be quasi-stationary and ignore u_{i0x} which is small due to the high collision frequency ν_i and ignore also u_{e0x} which is small because $\nu_e \ll \omega_{ce}$. Note that an error of corresponding magnitude arises in the continuity equations, since now the assumed density gradient will not necessarily be compatible with the given model for production and recombination.

11.4 Perturbations

The analysis is simplified by again assuming $n_0'/n_0 \approx 1/L$ = constant (Sudan *et al* 1973). This can be a good local approximation, while globally it is only correct for an exponential density variation that is unphysical since it includes arbitrarily large plasma densities. Note that the sign of L is important.

Taking $\mathbf{u}_{i0} \approx 0$ and linearizing the basic equations we find the continuity equation for the ions

$$\frac{\partial \eta}{\partial t} + \mathbf{u}_i \cdot \frac{\nabla n_0}{n_0} + \nabla \cdot \mathbf{u}_i = -2\alpha n_0 \eta, \tag{11.11}$$

after division by n_0. The production term q vanishes with αn_0^2.

From the electron continuity equation we have

$$\frac{\partial \eta}{\partial t} + \mathbf{u}_{0e} \cdot \nabla \eta + \mathbf{u}_e \cdot \frac{\nabla n_0}{n_0} + \nabla \cdot \mathbf{u}_e = -2\alpha n_0 \eta. \tag{11.12}$$

By the basic assumptions, we ignored a zero order ion drift, having $\mathbf{u}_{0i} \approx 0$, while similarly, we implicitly assumed $\mathbf{u}_{0e} \approx \{0, u_0, 0\}$.

The ion momentum equation becomes

$$M\frac{\partial \mathbf{u}_i}{\partial t} = -e\nabla\phi - M\nu_i\mathbf{u}_i. \tag{11.13}$$

It is important here to note that we have two basically different reasons for loss of ion momentum. At recombination, it is thus assumed that all of the ion momentum is lost. If this was all, the ion would propagate from ionization to recombination without collisional or frictional momentum loss, and in that case we were to use (11.13) with $\nu_i = 0$. The collisional losses included in (11.13) account for a steady 'frictional' momentum loss due to many collisions with neutrals. If we include electron–ion collisions (as for resistive drift waves) we had to include a similar term but with opposite sign in the electron momentum equation to retain overall momentum conservation. For ion–neutral collisions we assume the neutrals to carry the momentum and simply consider it lost.

Using $\ln(n_0 + n_1) \approx \ln(n_0) + \eta$ in the linearized electron momentum equation we find approximately

$$0 = e(\nabla\phi - \mathbf{u}_e \times \mathbf{B}) - \kappa T_e \nabla\eta - m\nu_e \mathbf{u}_e. \tag{11.14}$$

The discussion of the electron momentum equation follows the one from chapter 6, although it is again emphasized that details in the electron–ion and electron–neutral collisions are different. The comments concerning ion momentum losses apply to electrons as well. For the analysis in the following section 11.4.1 we will only need the components of (11.13) and (11.14) in the direction $\perp\mathbf{B}$.

11.4.1 Wave propagation perpendicular to both ∇n_0 and B

The analysis can become rather lengthy, so first we consider the simple case with waves propagating strictly in the direction along the $\mathbf{E}_0 \times \mathbf{B}$-direction which is perpendicular to both ∇n_0 and $\mathbf{B}$. The geometry is shown in figure 11.1.

Fourier transforming equations (11.13)–(11.14), the following relations are obtained, where only waves propagating in the y-direction are considered here, i.e. $\mathbf{k} = \{0, k_y, 0\}$

$$(-i\omega - \nu_i)u_{iy} - ik_y \frac{e}{M}\phi = 0, \tag{11.15}$$

while we have $u_{ix} = 0$, to the present approximation. The electron momentum equation gives for the y-component

$$-\omega_{ce}u_{ex} + \nu_e u_{ey} + ik_y\left(\frac{\kappa T_e}{m}\eta - \frac{e}{m}\phi\right) = 0, \tag{11.16}$$

and for the x-component

$$-\omega_{ce}u_{ey} + \nu_e u_{ex} = 0, \tag{11.17}$$

where a term drops out because $k_x = 0$ was assumed.

The continuity equation (11.11) gives for the ions

$$\left(-i\omega + 2\alpha n_0\right)\eta + ik_y u_{iy} = 0. \tag{11.18}$$

The spatial variation of the recombination term αn_0 is of little physical significance, so it is treated as a constant here and in the following.

For the electrons we find by the continuity equation (11.12)

$$\left(-i\omega + ik_y u_0 + 2\alpha n_0\right)\eta + ik_y u_{ey} + u_{ex}\frac{1}{L} = 0, \tag{11.19}$$

where u_0 is the zero-order electron drift u_{e0y}, as given by (11.10). A term containing the density gradient remains in the electron equation (11.19) since the electron fluid velocity retains an x-component even when $\mathbf{k} = \{0, k_y, 0\}$, in contrast to the ions.

The five equations (11.15)–(11.19) for the five unknowns ϕ, η, u_{ex}, u_{ey}, u_{iy} can be solved by standard methods. The criterion for non-trivial solutions gives the dispersion relation

$$\begin{aligned} 2\alpha n_0\Bigg[& k_y\left(m\nu_e^2 + M\nu_i\nu_e - m\omega_{ce}^2\right) + \frac{M\omega_{ce}\omega}{L} - iM\left(\frac{\nu_i\omega_{ce}}{L} - k_y\nu_e\omega\right)\Bigg] \\ & - ik_y\left(m\nu_e^2\omega + M\nu_e\nu_i\omega - m\omega_{ce}^2\omega\right) \\ & - ik_y^2\left[\frac{m\omega_{ce}U_{the}^2}{L} + mu_0\left(\omega_{ce}^2 - \nu_e^2\right)\right] \\ & - \frac{M\omega_{ce}\omega(\nu_i + i\omega)}{L} + k_yM\nu_e\omega^2 + k_y^3m\nu_eU_{the}^2 = 0, \end{aligned} \tag{11.20}$$

where $\omega = \omega(k_y) \equiv \Re\{\omega\}(k_y) + i\Im\{\omega\}(k_y)$ is in general complex. We introduced the thermal velocity $U_{the} = \sqrt{\kappa T_e/m}$ for electrons. With a little algebra, where we ignore terms of relative magnitude $(\nu_e/\omega_{ce})^2$, the relation (11.20) can be expressed as

$$\begin{aligned} \omega - k_yu_0 = & \frac{\psi}{\nu_i}\left[(\omega + i2\alpha n_0)(i\omega - \nu_i) - ik_y^2C_s^2\right] \\ & \times\left(1 - i\frac{\Omega_{ci}}{\nu_ek_yL}\right) - i2\alpha n_0, \end{aligned} \tag{11.21}$$

with $\psi \equiv \nu_e\nu_i/(\Omega_{ci}\omega_{ce})$. We assumed the ions to be unmagnetized from the outset, but the ion cyclotron frequency Ω_{ci} appears due to a multiplication of ω_{ce} by m and a division by M. Inserting $\omega = \Re\{\omega\} + i\Im\{\omega\}$ and separating real and imaginary parts, we find for instance for the real part

$$\begin{aligned} \Re\{\omega\} - k_yu_0 = & \frac{\psi}{\nu_i}\Bigg[- \nu_i\Re\{\omega\} - 2\Re\{\omega\}\Im\{\omega\} - 2\alpha n_0\Re\{\omega\} \\ & + \frac{\Omega_{ci}}{\nu_ek_yL}\Big(\Re\{\omega\}^2 - k_y^2C_s^2 - \Im\{\omega\}^2 - \Im\{\omega\}\nu_i \\ & - 2\alpha n_0(\Im\{\omega\} + \nu_i)\Big)\Bigg]. \end{aligned} \tag{11.22}$$

To obtain the dispersion relation it is first assumed that the imaginary part $\Im\{\omega\}$ is small which implicitly assumes αn_0 to be small as well. Ignoring small terms we find as a first approximation

$$\Re\{\omega\} \approx \frac{k_yu_0}{1 + \psi}. \tag{11.23}$$

To the same approximation, the imaginary part of ω is obtained as

$$\Im\{\omega\} \approx \frac{\psi}{1 + \psi}\left[\frac{\Re\{\omega\}^2 - k_y^2C_s^2}{\nu_i} + \frac{\Re\{\omega\}\omega_{ce}}{\nu_ek_yL}\right] - 2\alpha n_0. \tag{11.24}$$

The imaginary part contains essentially three terms:

1. A recombination term which always gives damping. This damping originates from the loss of wave momentum by the recombination. When new ions and electrons are created by ionization to replace those being lost, these new particles have to receive momentum in order to participate in the organized wave-motion, and the corresponding energy has to be taken from the wave itself.
2. A purely gradient depending term, i.e. the second one in the angular brackets, which depending on the sign of L contributes to an instability or a damping.
3. The first term in the angular brackets of (11.24) gives instability if $\Re\{\omega\}^2 > k_y^2 C_s^2$. This instability can be present also for vanishing density gradients, $L \to \infty$. In this limit the plasma becomes unstable essentially when the electron drift across the magnetic field lines exceeds the ion acoustic sound velocity. This instability is often called the Farley–Buneman instability after the two scientists who discovered it simultaneously (Farley 1963, Buneman 1963), or the type I irregularity after the classification used in describing radar back-scattering from the ionosphere. The Farley–Buneman instability is due to ion inertia, while the gradient instability (causing what is usually called the type II irregularity) is also recovered when ion inertial effects are ignored and the ion motion described by a simple mobility term. The gradient instability is closely related to the one described by Simon (1963) and Hoh (1963) and was for a time named after these authors.

Often one finds the name 'two-stream instability' for the Farley–Buneman instability, referring to the free energy being associated with the electrons streaming through the ions. This is, however, a misnomer. Note that the growth rate $\Im\{\omega\}$ is proportional to ν_e; we can find, formally, a set of equations, which retain the electron stream, but where the electron collisions are ignored. The two-stream situation remains, but the contribution to the growth rate in (11.24) disappears. The Farley–Buneman instability is *resistive*; the electron collisions cause the phase lag between fluctuations in density and electric field, which gives rise to the instability.

It is important that the foregoing arguments rely on the assumption of small $\Im\{\omega\}$, i.e. the results are valid close to threshold. For general conditions the dispersion relation (11.21) should be solved numerically. Also the approximation $n_0'/n_0 \approx 1/L = \text{const.}$ deserves some attention here. For large wavenumbers, i.e. short wavelengths this local approximation will often be adequate. However, the dispersion relation (11.21) is particularly sensitive to the long wavelength-limit, and here an approximation of a locally exponential density gradient will usually be difficult to justify. A complete understanding of this long wavelength-limit requires a solution of the full eigenvalue problem as in the discussion of (6.25). Such an analysis is, unfortunately, rather complicated in general, and it is carried out only seldomly.

Extensions of the analysis of ionospheric instabilities to large amplitude coherent modes has been discussed by Nekrasov *et al* (1992).

11.4.2 Propagation in arbitrary directions

The foregoing analysis was restrictive by considering only wave propagation in the direction perpendicular to both ∇n_0 and $\mathbf{B}$ and even with this simplification, the analysis became rather cumbersome. For wave propagation in arbitrary directions, the analysis becomes even worse, so we here merely quote the result (Fejer *et al* 1975, Kelley 1989) valid in the same simplified limit as (11.23) and (11.24), i.e. $\Re\{\omega\} \leqslant \nu_i$ and $\Im\{\omega\} \ll \Re\{\omega\}$. We have

$$\Re\{\omega\} \approx \frac{\mathbf{k} \cdot (\mathbf{u}_{e0} + \Psi \mathbf{u}_{i0})}{1 + \Psi}. \tag{11.25}$$

To the same approximation, the imaginary part of ω is obtained as

$$\begin{aligned} \Im\{\omega\} \approx \frac{1}{1+\Psi}\Bigg\{ & \frac{\Psi}{\nu_i}\left(\left(\Re\{\omega\} - \mathbf{k}\cdot\mathbf{u}_{i0}\right)^2 - k^2 C_s^2\right) \\ & + \frac{1}{k^2 L}\Bigg[\left(\Re\{\omega\} - \mathbf{k}\cdot\mathbf{u}_{i0}\right)\left(\frac{\nu_i}{\Omega_{ci}} k_y + \Psi_0 k_z\right) \\ & - \left(\Re\{\omega\} - \mathbf{k}\cdot\mathbf{u}_{e0}\right)\left(\frac{\nu_i \Omega_{ci}}{\nu_i^2 + (\Re\{\omega\} - \mathbf{k}\cdot\mathbf{u}_{i0})^2} k_y - k_z\right)\Bigg]\Bigg\} - 2\alpha n_0, \end{aligned} \tag{11.26}$$

with

$$\Psi \equiv \frac{\nu_i \nu_e}{\Omega_{ci}\omega_{ce}}\left(\frac{k_\perp^2}{k^2} + \frac{\omega_{ce}^2}{\nu_e^2}\frac{k_z^2}{k^2}\right) \equiv \Psi_0\left(\frac{k_\perp^2}{k^2} + \frac{\omega_{ce}^2}{\nu_e^2}\frac{k_z^2}{k^2}\right),$$

where k_z is the wavelength component along $\mathbf{B}$, and x is along the density gradient. The results summarized here relax the simplifying assumption from before, where we assumed the average ion velocity to vanish. Even with $\Omega_{ci} \ll \nu_i$ we will have some ion mobility, and in general $\mathbf{u}_{i0} \neq 0$. Note the large numerical factor ω_{ce}/ν_e on k_z/k in Ψ: this is a consequence of the large electron mobility along magnetic field lines. Even a small electric field component along this direction can accelerate electrons to significant velocities.

As an illustration we show in figure 11.2 the real and imaginary part (only the positive part) of the approximate dispersion relation for the Farley–Buneman instability for the wavevector $\mathbf{k}$ in the y–z-plane. In the real part, $\Re\{\omega\} = \Re\{\omega\}(\mathbf{k})$, we ignored the contribution from u_{i0}. Note that the axes are not dimensionless: we have no reference length scale when we assume quasi-neutrality (or the 'plasma approximation') since the Debye length then no longer enters the problem. All axes have now dimension of length and we can use any length scale for normalization on the three axes: the resulting ratio of the real part of the phase velocity and sound speed will come out correctly. The imaginary part has to be corrected by $-2\alpha n_0$ to account for ionization and recombination, see (11.26).

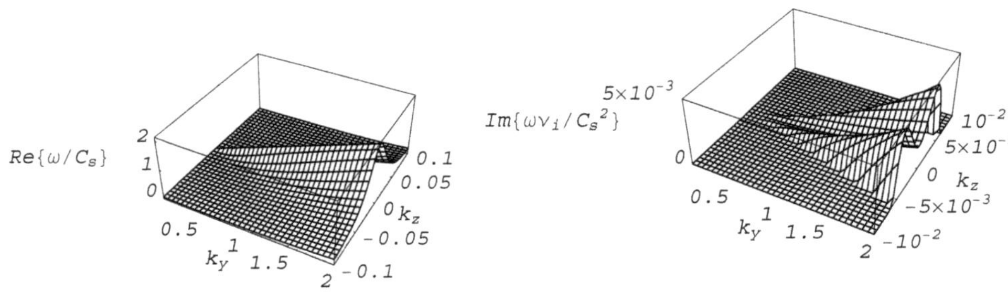

Figure 11.2. Real and imaginary part of the dispersion relation for the Farley–Buneman instability for **k** in the $y - z$-plane. Note that the axes are not dimensionless and that the scales on the two k_z-axes are different. We took $\nu_i = 5\Omega_{ci}$ and $\nu_e = \omega_{ce} \times 10^{-3}$ which gives $\Psi_0 = 0.005$ and $u_{e0} = 1.05\ C_s$. The gradient driven instability is not included.

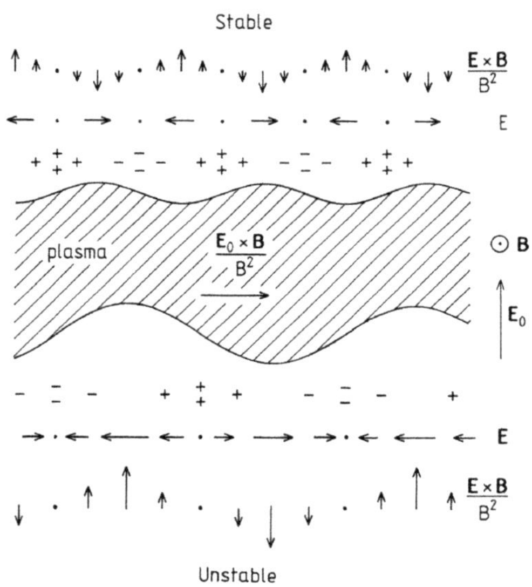

Figure 11.3. Geometry for a simple description of the gradient instability of a plasma with neutral collisions and an externally imposed DC-electric field. $\mathbf{E}_0$ is the macroscopic 'external' electric field, which is here taken to be constant in space, while **E** is a space-time varying perturbation. To emphasize the similarity of this collisional drift instability with the gravitational Rayleigh–Taylor instability we re-used figure 2.5 with some modifications.

The physical mechanism for the Farley–Buneman instability has been discussed, for instance, by Kelley (1989).

The results summarized here have one flaw by indicating that the plasma becomes stable when $\mathbf{E}_0 = 0$ in which case $\mathbf{u}_{e0} = 0$, $\mathbf{u}_{i0} = 0$ giving $\Re\{\omega\} = 0$ and $\Im\{\omega\} < 0$. Since resistivity due to neutral collisions will always give rise to deviations from a simple isothermal Boltzmann distribution of electrons, we would expect the plasma to be unstable with a small growth rate also in this limit.

We found in section 3.6.2 that collisions with neutrals in magnetized plasmas with a density gradient $\nabla n \perp \mathbf{B}$ always give rise to an electric field given as

$$e\frac{\partial \phi}{\partial x} = \frac{D_e - D_i}{\dfrac{D_e}{\kappa T_e} + \dfrac{D_i}{\kappa T_i}} \frac{1}{n} \frac{\partial n}{\partial x}. \tag{11.27}$$

Because of the differences in neutral-collisional drags there will therefore always be some relative velocity between the electrons and ions in the Hall-direction, $\mathbf{u}_{e0} - \mathbf{u}_{i0} \neq 0$, see (3.25). The wave dynamics, the stability conditions in particular, will be influenced by the electric field (11.27). It is, however, difficult to imagine realistic conditions where this ambipolar electric field causes the ion–electron velocity difference to exceed the ion sound velocity so this electric field will mostly give rise to a damping. As an illustration we again use $T_i \approx T_e \equiv T$ and $D_i \gg D_e$ to give a quasi-stationary electric field by $e\partial\phi/\partial x \approx -\kappa T \partial \ln n/\partial x$. This electric field gives an E_0/B-drift of the same order of magnitude as the diamagnetic drift. For this velocity to be of the order of the sound speed, we need the density length scale to be comparable to the effective ion Larmor radius, $a_i \approx \mathcal{L}$ which would be a rather extreme condition.

To allow for a detailed comparison between the analytical results and observations it is important to have good models for the composition of the ionosphere (see appendix B) and for the collisional cross-sections. Some relevant, although selected, references are Pack and Phelps (1961), Frost and Phelps (1964), Takayanagi and Itikawa (1970a, 1970b), Itikawa (1973), see also appendix C. The electron momentum and energy losses in the ionosphere are mostly due to elastic collisions, and of rotational and vibrational excitations of atoms and molecules. In modeling the ion–neutral collisions we note the different nature of elastic and inelastic collisions as compared to charge exchange collisions (Ott and Farley 1975, St-Maurice and Schunk 1979, St-Maurice 1978). The differences are illustrated by numerical simulations by Miloch *et al* (2013). Charge exchange collisions can induce ion velocity distributions that are similar to loss-cone distributions: these are unstable in kinetic models (Rosenbluth and Post 1965).

11.5 Gradient instability

Conceptually, the gradient instability is very similar to the Rayleigh–Taylor instability discussed previously (Sudan *et al* 1997). The similarity can be made evident by considering figure 11.3 as compared to figure 2.5. Due to their small mobility, caused by large collision frequencies $\nu_i \gg \Omega_{ci}$, the ions are almost immobile. The electrons on the other hand are assumed to drift with essentially the $\mathbf{E}_0 \times \mathbf{B}$-velocity, apart from small corrections of the order of ν_e/ω_{ce}. A small initial perturbation gives rise to polarization electric fields which enhance the perturbation (instability) if $\mathbf{E}_0 \cdot \nabla n_0 > 0$. Depending on point of view, it can be argued that on the other hand a gradient in plasma density has a stabilizing effect on the Farley–Buneman instability when $\mathbf{E}_0 \cdot \nabla n_0 < 0$ since the wave-growth due to the Farley–Buneman instability is diminished by the damping due to the gradient in this case.

In the long wavelength limit it is possible to give a dispersion relation for the gradient instability without the restriction of small growth rates. To this end ion inertia effects are ignored and the ion motion is described by a simple mobility as $\mathbf{u}_i = -(e/M\nu_i)\nabla\phi$ which can be obtained from (11.13). The main consequence of this modification is that the parenthesis $(i\omega - \nu_i)$ in (11.21) is reduced to $-\nu_i$ and the resulting dispersion relation becomes of first order in ω. With some small terms being ignored, the expression for the complex frequency ω becomes

$$\omega = \left(\frac{\Omega_{ci}}{\omega_{ce}}\right)\frac{k_y^2 U_{the}^2\left[1 - 1/(k_y L)^2\right] + ik_y u_0 \omega_{ce}^2}{(\nu_e\omega_{ce}/k_y L) + i(\nu_e\nu_i + \omega_{ce}\Omega_{ci})} - i2\alpha n_0. \tag{11.28}$$

The term containing the electron thermal velocity, U_{the}, is important for

$$\frac{L\nu_e}{u_0}\left(\frac{r_{Le}}{L}\right)^2 \gg k_y L \gg \frac{u_0}{L\nu_e}\left(\frac{L}{r_{Le}}\right)^2,$$

where r_{Le} is the electron Larmor radius. For intermediate wavenumbers $k_y \approx 1/L$ we can use the approximation given by e.g. Kudeki *et al* (1982) as

$$\Re\{\omega\} \approx \left(\frac{k_y u_0}{1 + \psi}\right)\left(\frac{1}{1 + (k_0/k_y)^2}\right), \tag{11.29}$$

for the real part of the frequency and

$$\Im\{\omega\} \approx \left(\frac{k_0 u_0}{1 + \psi}\right)\left(\frac{1}{1 + (k_0/k_y)^2}\right) - 2\alpha n_0, \tag{11.30}$$

for the growth rate. We introduced $\psi = \nu_e\nu_i/(\omega_{ce}\Omega_{ci})$ and $k_0 = (\nu_i/\Omega_{ci}L)/(1 + \psi)$. The waves are now dispersive, with the longest wavelengths propagating with the smallest phase velocity. Note that the sound velocity no longer appears explicitly; ion sound waves exist only in models retaining ion inertia. Incidentally, *some* evidence was found by analyzing data from instrumented rockets (Iranpour *et al* 1997) for such a wave dispersion, but the results had a large uncertainty.

If we ignore the damping due to recombination losses, and let $L \to \infty$, implying $k_0 \to 0$, we can recover (11.23) for the real part of the frequency from (11.28) in the limit where the term with ω_{ce}^2 dominates in the numerator.

Modifications of the results for the gradient instability due to finite ion inertia were discussed by for instance Ossakow *et al* (1978), considering the limit of short wavelengths, $k \gg 1/L$. They found a constant growth rate, see also a review by Tsunoda (1988). The same limit can also be recovered in the inertialess model of (11.28) giving a growth rate varying approximately with k_y^2.

We note that it is possible to envisage conditions where the perturbation electric fields associated with, for instance, a gradient instability become so strong that they can drive secondary instabilities in the direction *along* the phase fronts. This will be a second order effect, not to be discussed here, but the phenomenon is important nonetheless for a detailed understanding of the saturated instability.

The nonlinear saturated stage of the instabilities discussed here can modify the plasma properties of the low altitude ionosphere. Theories of heating by electrojet waves has been developed in several papers (St-Maurice *et al* 1981, 1986, St-Maurice and Schlegel 1983, St-Maurice and Laher 1985).

11.5.1 Model equations

For completeness, we give a set of equations containing the basic physics of the 'inertia-free' gradient instability. We have from the electron and ion continuity and momentum equations

$$\frac{\partial}{\partial t}n + \hat{\mathbf{z}} \times \nabla_\perp \psi \cdot \nabla_\perp n = D_a \nabla_\perp^2 n, \tag{11.31}$$

$$\nabla_\perp \cdot \left(\beta_1 n(\hat{\mathbf{z}} \times \nabla_\perp \psi) + n\mathbf{u}_0 + n\nabla_\perp \psi\right) = \beta_2 D_a \nabla_\perp^2 n, \tag{11.32}$$

with

$$D_a = \frac{\mu_\perp^i D_\perp^e + \mu_\perp^e D_\perp^i}{\mu_\perp^i + \mu_\perp^e},$$

$$u_0 = \frac{\mu_\perp^i \mu_H^e + \mu_\perp^e \mu_H^i}{\left(\mu_\perp^i + \mu_\perp^e\right)^2}(u_{0e} - u_{0i}),$$

$$D' = \frac{\mu_\perp^i \mu_H^e + \mu_\perp^e \mu_H^i}{\left(\mu_\perp^i + \mu_\perp^e\right)^2}\left(D_\perp^e - D_\perp^i\right),$$

$$\psi = \frac{\mu_\perp^i \mu_H^e + \mu_\perp^e \mu_H^i}{\mu_\perp^i + \mu_\perp^e}\phi,$$

$$\beta_1 = \frac{\mu_H^e - \mu_H^i}{\mu_\perp^i + \mu_\perp^e}, \qquad \beta_2 = \frac{D'}{D_a},$$

as discussed by, for instance, Ott and Farley (1974), in terms of electron and ion cross-field diffusion coefficients, $D_\perp^e$ and $D_\perp^i$ for electrons and ions respectively, the corresponding Pedersen mobilities, $\mu_\perp^e$, $\mu_\perp^i$ and the Hall mobilities μ_H^e, μ_H^i, while $\mathbf{u}_{0e}$, $\mathbf{u}_{0i}$ are the zero order ambient plasma drifts across the magnetic field, which is assumed to be in the z-direction. We denote ϕ the fluctuating electrostatic potential, distinguished from the zero order potential variation implied in the steady state electric field, while n is the density. The $\hat{\mathbf{z}} \times \nabla_{\perp}\psi$-term originates from the $\mathbf{E} \times \mathbf{B}/B^2$-drift of the electrons, where their cyclotron frequency is much larger that the electron–neutral collision frequency as argued before. Gradients in mobilities have been neglected. The model equations implicitly assume spatial scales to be much larger than the ion Larmor radii as well as collisional mean free paths.

In writing (11.31) we transformed the reference frame to one moving with

$$\mathbf{u}_{\phi} = \frac{\mathbf{u}_{0e} + \left(\mu_{\perp}^{e}/\mu_{\perp}^{i}\right)\mathbf{u}_{0i}}{1 + \mu_{\perp}^{e}/\mu_{\perp}^{i}},$$

assumed to be uniform. The second term in (11.31) represents a nonlinear convection, with $\hat{\mathbf{z}} \times \nabla_{\perp}\psi$ being a divergence free velocity.

11.6 Sound waves for stable conditions

Propagating acoustic-like disturbances generated by sudden impulses (such as nuclear explosions) have been observed to split into two components in the lower ionosphere. The resulting two disturbances propagate vertically into the upper ionosphere at different speeds, the slower at ordinary sound speed in a neutral gas and the faster at almost twice the speed of sound.

The faster disturbance travels at the speed of an ion-acoustic wave in the plasma. Strictly speaking, such an identification is prohibited theoretically by the large damping, or coupling, of the ionic mode to the neutral particle medium through ion–neutral elastic collisions; however, if inelastic collisions are included in the theory, the principal exothermic charge-exchange and charge-transfer reactions in the ionosphere provide enough additional coherent momentum in the charged particles to offset the losses through elastic collisions. It can be shown quantitatively that in some regions of the ionosphere ion-acoustic wave propagation can occur almost without losses (Wickersham 1970).

For the phase velocity of magneto-acoustic waves in loss-free plasmas we have the general expression

$$\frac{\omega}{k} = c\sqrt{\frac{C_s^2 + V_A^2}{c^2 + V_A^2}},$$

with V_A being the Alfvén velocity, c the speed of light in vacuum and C_s is the ion acoustic speed. The full analysis also takes into account the collisions with neutrals, electron–ion collisions being of minor importance in the lower parts of the ionosphere. For small temperatures, i.e. small sound speeds, the wave becomes the modified Alfvén wave. The limit of small magnetic fields give small V_A and we then have the ordinary sound wave.

Table 11.1. Comparison of observed and theoretical wave propagation speeds (Wickersham 1970), all in units of km s^{-1}. The theoretical expressions include damping by collisions and finite conductivity.

Height in km	Speed of a slow disturbance	Speed of a fast disturbance	Speed of sound for revised U.S. standard neutral atmosphere	Theoretical speed of a modified Alfvén wave	Theoretical speed of an ion-acoustic wave
100	0.356	0.356	0.293	0.051	0.494
150	0.431	0.659	0.625	0.850	0.598
200	0.526	0.918	0.750	1.84	0.876
250	0.684	1.28	0.825	3.44	1.37
300	0.812	1.59	0.885	6.04	1.55
350	0.961	1.78	0.945	9.95	1.79
400	1.110	2.02	0.995	15.5	1.98

The standard expression for the Alfvén wave dispersion relation was generalized by, for instance, Pécseli and Engvold (2000) to include collisions with neutrals and finite conductivity. The linear dispersion relation for the waves was found in the form

$$\omega^2 = k^2 V_A^2 + k^2 \frac{\nu_{in}}{\mu_0 \sigma}\left(1 - i\frac{\omega}{\nu_{in}} - \frac{\nu_{ni}}{\nu_{ni} - i\omega}\right) - i\omega\nu_{in} + i\omega\frac{\nu_{in}\nu_{ni}}{\nu_{ni} - i\omega}, \tag{11.33}$$

with $V_A = \sqrt{B_0^2/\rho_0\mu_0}$, being the Alfvén velocity. Introducing the normalized frequency $\tau \equiv \omega/\nu_{in}$ and the skin-depth $\delta \equiv \sqrt{2/(\omega\mu_0\sigma)}$ with σ being the conductivity we rewrite (11.33) as

$$\tau^2 = (k\delta)^2 V_A^2/(\delta\nu_{in})^2 + \frac{1}{2}\tau(k\delta)^2 - i\tau\left(1 + \frac{1}{2}\tau(k\delta)^2\right) - \frac{1}{2}\tau(i + (k\delta)^2)\frac{\eta}{\eta - i\tau}, \tag{11.34}$$

in terms of the ratio between collision frequencies, $\eta \equiv \nu_{ni}/\nu_{in} = \rho_0/\rho_n$, with ρ_n being while $\nu_{ni} \neq \nu_{in}$ are the neutral–ion and ion–neutral collision frequencies, recalling here that ν_{ni} and ν_{in} account for momentum losses and are merely encounter rates.

When collisions with neutrals become abundant, the phase velocity of Alfvén waves will be decreased even with negligible damping by finite conductivity, because through collisions, the ions also have to set the neutrals in motion, thus increasing the effective density to be used for determining the Alfvén velocity.

References

Brekke A 2013 *Physics of the Upper Polar Atmosphere* 2nd edn (Berlin: Springer)

Buneman O 1963 Excitation of field aligned sound waves by electron streams *Phys. Rev. Lett.* **10** 285–7

Farley D T 1963 Two-stream plasma instability as a source of irregularities in the ionosphere *Phys. Rev. Lett.* **10** 279–82

Fejer B, Farley D, Balsley B and Woodman R 1975 Vertical structure of the VHF backscattering region in the equatorial electrojet and the gradient drift instability *J. Geophys. Res.* **80** 1313–24

Frost L S and Phelps A V 1964 Momentum-transfer cross sections for slow electrons in He, Ar, Kr, and Xe from transport coeff-icients *Phys. Rev.* **136** A1538–45

Hoh F C 1963 Instability of Penning-type discharges *Phys. Fluids* **6** 1184–91

Iranpour K, Pécseli H L, Trulsen J, Bahnsen A, Primdahl F and Rinnert K 1997 Propagation and dispersion of electrostatic waves in the ionospheric E region *Ann. Geophys.* **15** 878

Itikawa Y 1973 Effective collision frequency of electrons in gases *Phys. Fluids* **16** 831–5

Kelley M C 1989 *The Earth's Ionosphere, Plasma Physics and Electrodynamics, of International Geophysics Series* vol 43 (San Diego, CA: Academic)

Kudeki E, Farley D T and Fejer B 1982 Long wavelength irregularities in the equatorial electrojet *Geophys. Res. Lett.* **9** 684–7

Miloch W J, Pécseli H L and Trulsen J K 2013 Unstable ring-shaped ion distribution functions induced by charge-exchange collisions *Plasma Phys. Control. Fusion* **55** 124006

Nekrasov A K, Kurchashov Y P, Ivanov V N, Pokhotelov O A, Bharuthram R, Shukla P K and Stenflo L 1992 Dynamics of density irregularities in the E-region of the ionosphere *Phys. Scripta* **46** 369–74

Ossakow S L, Chaturvedi P K and Workman J B 1978 High-altitude limit of the gradient drift instability *J. Geophys. Res.: Space Phys.* **83** 2691–3

Ott E and Farley D T 1974 The *k* spectrum of ionospheric irregularities *J. Geophys. Res.* **79** 2469–72

Ott E and Farley D T 1975 Microinstabilities and production of short-wavelength irregularities in auroral-F region *J. Geophys. Res.* **80** 4599–602

Pack J L and Phelps A V 1961 Drift velocities of slow electrons in helium, neon, argon, hydrogen, and nitrogen *Phys. Rev.* **121** 798–806

Pécseli H L and Engvold O 2000 Modeling of prominence threads in magnetic fields: levitation by incompressible MHD waves *Solar Phys.* **194** 73–86

Rosenbluth M N and Post R F 1965 High-frequency electrostatic plasma instability inherent to loss-cone particle distributions *Phys. Fluids* **8** 547–50

Simon A 1963 Instability of a partially ionized plasma in crossed electric and magnetic fields *Phys. Fluids* **6** 382–8

St-Maurice J-P 1978 On a mechanism for the formation of VLF electrostatic emissions in the high latitude F region *Planet. Space Sci.* **26** 801–16

St-Maurice J-P, Hanuise C and Kudeki E 1986 On the dependence of the phase velocity of equatorial irregularities on the polarization electric field and theoretical implications *J. Geophys. Res.: Space Physics* **91** 13493–505

St-Maurice J-P and Laher R 1985 Are observed broadband plasma wave amplitudes large enough to explain the enhanced electron temperatures of the high-latitude E region? *J. Geophys. Res.: Space Physics* **90** 2843–50

St-Maurice J-P and Schlegel K 1983 A theory of coherent radar spectra in the auroral E-region *J. Geophys. Res.* **88** 4087–95

St-Maurice J-P, Schlegel K and Banks P M 1981 Anomalous heating of the polar E region by unstable plasma waves 2. theory *J. Geophys. Res.: Space Physics* **86** 1453–62

St-Maurice J-P and Schunk R W 1979 Ion velocity distributions in the high latitude ionosphere *Rev. Geophys.* **17** 99–134

Sudan R N, Akinrimisi J and Farley D T 1973 Generation of small-scale irregularities in the equatorial electrojet *J. Geophys. Res.* **78** 240–8

Sudan R N, Gruzinov A V, Horton W and Kukharkin N 1997 Convective turbulence in weakly ionized plasma *Phys. Rep.* **283** 95–119

Takayanagi K and Itikawa Y 1970a Elementary processes involving electrons in the ionosphere *Space Science Rev.* **11** 380–450

Takayanagi K and Itikawa Y 1970b The rotational excitation of molecules by slow electrons *Advances in Atomic and Molecular Physics* vol. 6 ed D Bates and I Esterrnan (New York: Academic) pp 105–53

Tsunoda R T 1988 High-latitude F region instabilities: a review and synthesis *Rev. Geophys.* **26** 719–60

Wickersham A F J 1970 Detection and analysis of ion-acoustic waves in the ionosphere *Pure Appl. Geophys.* **80** 271–87

IOP Publishing

Low Frequency Waves and Turbulence in Magnetized Laboratory Plasmas and in the Ionosphere

Hans Pécseli

Chapter 12

Inhomogeneous plasma temperatures

The standard models for electrostatic drift waves summarized in the foregoing chapters assume constant background plasma temperatures, with only the plasma density being inhomogeneous. In the following we discuss some of the consequences of temperature inhomogeneities. The most interesting case is found for $\Omega_{ci} < \Omega_{pi}$.

12.1 Gradients in electron temperature

For low frequency electrostatic waves the gradients in electron temperature require special attention because the ion sound velocity, $C_s = \sqrt{T_e/M}$, depends on the local variation in $T_e(\mathbf{r})$ while it is independent of the plasma density. In the terminology of geometrical optics we can argue that the index of refraction for these waves is spatially varying. It is perfectly feasible to make a ray-tracing analysis for this case as well.

The study is here carried out for a slab geometry with a homogeneous magnetic field along the z-axis and $T_e = T_e(x)$ as discussed by e.g. Nishida and Hirose (1977), Guio *et al* (2001), Guio and Pécseli (2010) or Guio and Pécseli (2016). We are primarily interested in the case where the wave frequency is well below the ion cyclotron frequency. However, this problem contrasts in an interesting way with its high frequency counterpart. Both cases are therefore discussed. In the linearized analysis used in the present study, a density gradient will be immaterial.

It might be appropriate first to add a few general comments concerning the absence of the $\mathbf{B}_0$-perpendicular electron dynamics in the analysis. Ignoring electron inertia, as usual for low frequency ion acoustic waves, we have

$$en\nabla\phi - \nabla(n\kappa T_e) + en\mathbf{u}_e \times \mathbf{B}_0 = 0. \qquad (12.1)$$

As an illustration we first take the case where T_e = constant, but assume that the unperturbed plasma has a density gradient in the direction perpendicular to $\mathbf{B}_0 = B_0\hat{\mathbf{z}}$, i.e. $n_0 = n_0(x)$. The $\mathbf{B}_0$-parallel component of (12.1) gives the Boltzmann distributed electrons $n = n_0(x)\exp(e\phi/\kappa T_e)$. By insertion of this relation into (12.1) we find the component perpendicular to $\mathbf{B}_0$ as $en_0\mathbf{u}\times\mathbf{B}_0 = \kappa T_e\nabla_\perp n_0$, giving $u_y = (\kappa T_e/eB_0)d\ln n_0(x)/\mathrm{d}x$, which is nothing but the electron diamagnetic drift associated with the unperturbed plasma density gradient. The equations have not been linearized to obtain this result. The electron density can be fluctuating as well as the plasma potential, but there will not be any electron fluid velocity component in the direction perpendicular to $\mathbf{B}_0$. We have an exact cancellation of the fluctuations in the electron diamagnetic drifts and the $-\nabla\phi\times\mathbf{B}_0/B_0^2$-drifts associated with the fluctuations in plasma density, see also section 5.2.1. We now take n_0 = const. and $T_e = T_e(x)$. The $\mathbf{B}_0$-parallel component (12.1) is unaffected, giving again the Boltzmann distribution of electrons now with a varying electron temperature. Using this in (12.1), we find again without any linearization that $u_y = (1 - e\phi/\kappa T_e(x))(\kappa T_e(x)/eB_0)d\ln\kappa T_e(x)/\mathrm{d}x \equiv (1 - \ln n/n_0)(\kappa T_e(x)/eB_0)d\ln\kappa T_e(x)/\mathrm{d}x$. In this case there is a fluctuating component of the bulk electron velocity, but its direction is here along the phase fronts of the low frequency waves, and therefore it does not contribute to any charge separations in the present limit.

12.2 The low frequency case, $\omega < \Omega_{ci}$

In this section we consider a simple analytical model which assumes a vanishing ion temperature $T_i = 0$ in the quasi-neutral limit where $n_e \approx n_i \equiv n$. The basic equations are the ion continuity and momentum equations as well as the z-component of the electron momentum equation where the electron inertia has been neglected. Since the electron fluid velocity will not appear explicitly in the following, we need not specify the ion velocity, so a subscript $_i$ is omitted. The relevant equations are, respectively, in their linearized form

$$\frac{\partial n}{\partial t} + n_0\nabla\cdot\mathbf{u} = 0 \tag{12.2}$$

$$M\frac{\partial\mathbf{u}}{\partial t} - e(-\nabla\phi + \mathbf{u}\times\mathbf{B}_0) = 0 \tag{12.3}$$

$$-en_0\frac{\partial\phi}{\partial z} + \frac{\partial}{\partial z}\left[n\kappa T_e(x)\right] = 0, \tag{12.4}$$

where n and $\mathbf{u}$ are the density and velocity perturbations, T_e is the electron temperature, ϕ the electrostatic potential ($\mathbf{E} = -\nabla\phi$), $\mathbf{B}_0$ the magnetic field, e and M the ion charge and mass, respectively. Assuming that the relevant frequencies are well below the ion-cyclotron frequency Ω_{ci}, we iterate (12.3) in order to obtain the standard approximation for the ion velocity component perpendicular to the magnetic field

$$\mathbf{u}_\perp = -\frac{\nabla_\perp \phi \times \mathbf{B}_0}{B_0^2} - \frac{1}{B_0 \Omega_{ci}} \frac{\partial \nabla_\perp \phi}{\partial t}. \tag{12.5}$$

The first term is the $\mathbf{E} \times \mathbf{B}$ drift velocity and the second term the ion polarization drift. With a little standard algebra we obtain a differential equation for the electrostatic potential in the relevant two spatial dimensions (x, z)

$$\frac{e}{\kappa T_e(x)} \frac{\partial^2 \phi}{\partial t^2} - \frac{1}{B_0 \Omega_{ci}} \frac{\partial^2}{\partial t^2} \frac{\partial^2 \phi}{\partial x^2} - \frac{e}{M} \frac{\partial^2 \phi}{\partial z^2} = 0. \tag{12.6}$$

The $\mathbf{B}_0$-perpendicular component of the electron dynamics does not enter the analysis with the assumption of Boltzmann distributed electrons. As mentioned, we here assume that the ion temperature is vanishing for simplicity. The results for $T_i \neq 0$ are not significantly different as long as we assume isothermal conditions.

For a homogeneous plasma with constant electron temperature T_e, we find the standard dispersion relation

$$\omega^2 = \frac{k_z^2 C_s^2}{1 + k_x^2 C_s^2 / \Omega_{ci}^2}, \tag{12.7}$$

for $\omega^2 \ll \Omega_{ci}^2$, where $C_s = \sqrt{\kappa T_e / M}$ is the ion sound acoustic velocity and C_s/Ω_{ci} takes the role of an effective Larmor radius. The dispersion relation is shown in figure 12.1 (upper panel), together with a vectorial presentation of the group velocity $\mathbf{u}_g$, deduced from (12.7) (lower panel). We note that $\mathbf{k} \cdot \mathbf{u}_g = k_z C_s / (1 + k_x^2 C_s^2 / \Omega_{ci}^2)^{3/2}$, and the directions of $\mathbf{k}$ and $\mathbf{u}_g$ are different. In figure 12.2 we show the variation of the angle between the wave vector and the group velocity for varying wave vector components, k_x and k_z.

Retaining the spatially varying electron temperature, we can still take the Fourier transform in time and in space, along the z-axis, to obtain

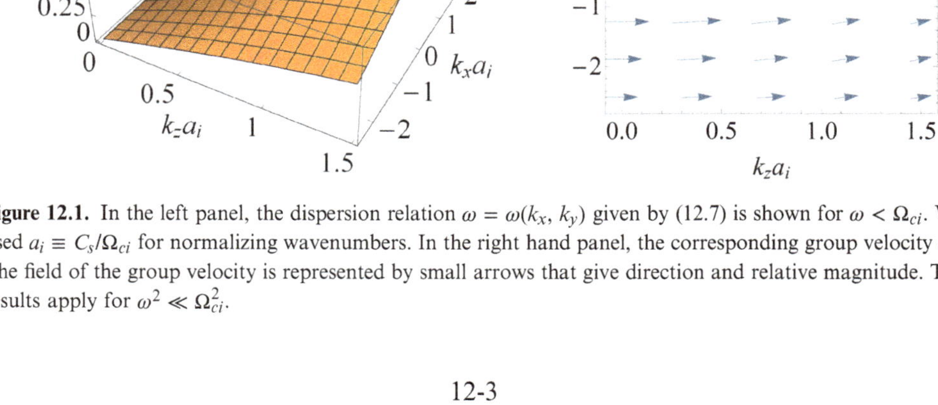

Figure 12.1. In the left panel, the dispersion relation $\omega = \omega(k_x, k_y)$ given by (12.7) is shown for $\omega < \Omega_{ci}$. We used $a_i \equiv C_s/\Omega_{ci}$ for normalizing wavenumbers. In the right hand panel, the corresponding group velocity $\mathbf{u}_g$. The field of the group velocity is represented by small arrows that give direction and relative magnitude. The results apply for $\omega^2 \ll \Omega_{ci}^2$.

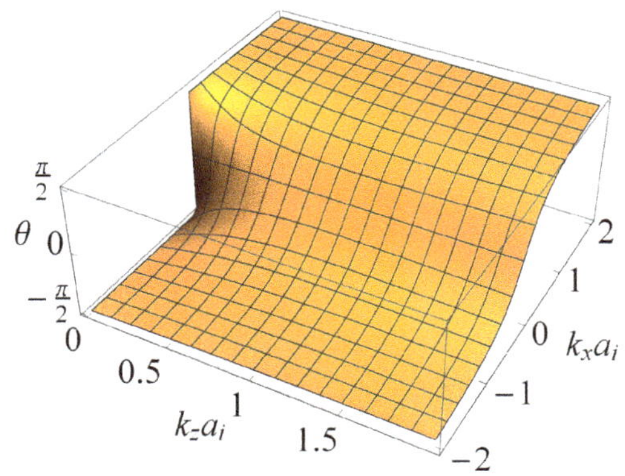

Figure 12.2. The variation of the angle between the wave vector and the group velocity for varying wave vector components, k_x and k_z. The point $(k_x, k_z) = (0, 0)$ is singular. The results apply for $\omega^2 \ll \Omega_{ci}^2$.

$$\frac{\mathrm{d}^2\phi}{\mathrm{d}x^2} = \left(\frac{M\omega^2}{\kappa T_e(x)} - k_z^2\right)\frac{\Omega_{ci}^2}{\omega^2}\phi. \tag{12.8}$$

In the limit where we ignore the ion polarization drift, the left side of (12.8) vanishes, and the result is

$$\frac{\omega^2}{k_z^2} = \frac{\kappa T_e(x)}{M}. \tag{12.9}$$

Physically, (12.9) implies that in this limit each magnetic field line supports oscillations with the local sound velocity, irrespective of what happens at another field line. The system has a local, but no global dispersion relation. If excited externally, the system will support a continuum of phase velocities in the range determined by the minimum and maximum values of the electron temperature. In many ways this low frequency limit of magnetized plasmas has similarities with Langmuir oscillations with cold electrons in spatially varying plasma densities (Barston 1964), and also with Alfvén waves propagating in inhomogeneous magnetic fields (Uberoi 1972).

For a more general analysis of (12.8) it is advantageous to introduce the normalized variables $\gamma^2 \equiv M\omega^2/(\kappa T_0 k_z^2)$ and $\xi \equiv x\Omega_{ci}/C_s$, in terms of a reference temperature T_0. The reference temperature can be defined as $T_0 \equiv \frac{1}{2}(T_1 + T_2)$, where T_2 and T_1 are the maximum and minimum values the electron temperature attains. We find

$$\frac{\mathrm{d}^2\phi}{\mathrm{d}\xi^2} = \left(\frac{T_0}{T_e(\xi)} - \frac{1}{\gamma^2}\right)\phi. \tag{12.10}$$

For later reference we define $k_{T_j}^2 \equiv 1/\gamma^2 - T_0/T_j$ in the case where $\gamma^2 > T_j/T_0$ and $1/\lambda_{T_j}^2 \equiv T_0/T_j - 1/\gamma^2$ otherwise, with $j = 1,2$. The relation (12.10) can be interpreted as an operator equation with ϕ being the eigenfunction and $1/\gamma^2$ the corresponding eigenvalue.

12.2.1 Monotonic electron temperature variation

Let us assume first that the variation in the electron temperature is monotonic, decreasing from T_2 at $\xi \to -\infty$ to T_1 at $\xi \to \infty$. For a given value of (ω, k_z), we require that (12.10) has physically acceptable solutions which are bounded at infinity. For the case where $\gamma^2 > T_2/T_0$ this is not possible, since the solution for ϕ diverges either at $\xi \to \infty$ or $\xi \to -\infty$. In the case where $\gamma^2 < T_1/T_0 < T_2/T_0$ we have solutions of (12.10) with $\phi \to \exp(ik_{T_1}\xi)$ for $\xi \to \infty$, and $\phi \to \exp(ik_{T_2}\xi)$ for $\xi \to -\infty$. In the case where $\gamma^2 = T_e(\xi)/T_0$ for some value of ξ, we find solutions for ϕ which decay as $\exp(-\xi/\lambda_{T_1})$ when $\xi \to \infty$, and oscillate for $\xi \to -\infty$. Physically this result implies that the pre-selected value of the phase velocity ω/k_z must resonate with the natural sound velocity at a certain point to have such an acceptable solution. As long as such an internal resonance point exists, i.e. that $T_1/T_0 < \gamma^2 < T_2/T_0$, it is possible to obtain physically acceptable solutions for any combination of ω and k_z. For an arbitrary temperature profile $T_e(\xi)$, the solution for ϕ can only be obtained numerically for the general case. In figure 12.3 we show examples for numerical solutions, assuming an electron temperature described by $T_0/T_e(x) = 1 + \arctan(x)/\pi$ and in figure 12.4, an electron temperature described by $T_e(x)/T_0 = 1 - \frac{1}{2}\Delta + \Delta\exp(-x^2/W^2)$. Some local solutions can however be found analytically. Assume for instance $T_0/T_e(\xi) \approx \xi$ in a certain interval, then we readily find $\phi = \mathrm{Ai}(\xi - 1/\gamma^2)$ locally, in terms of the Airy function, $\mathrm{Ai}(x)$.

The physical reason for the Airy function-like solution is easy to explain. Taking $\gamma^2 = 1$, we have a resonance with the local sound velocity at $\xi = 0$ with $k_\xi = 0$. For

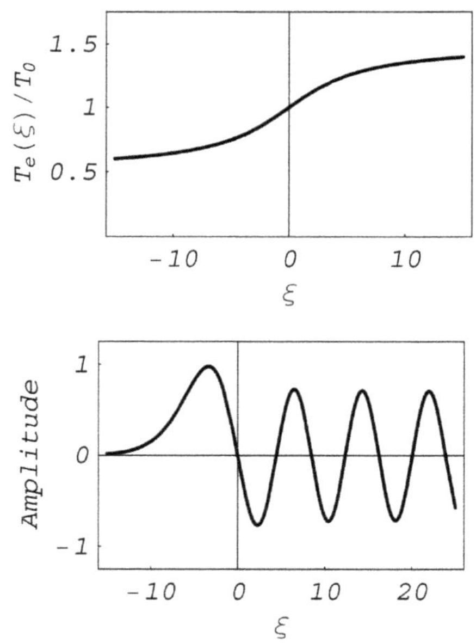

Figure 12.3. Numerical solution of (12.10), assuming a monotonically increasing electron temperature variation $T_e(\xi)/T_0 = 1 + \frac{1}{\pi}\arctan(\xi/5)$). The first panel shows the variation of $T_e(\xi)$, while the second panel presents a solution for the electrostatic potential ϕ for $\gamma^2 = 0.723299$.

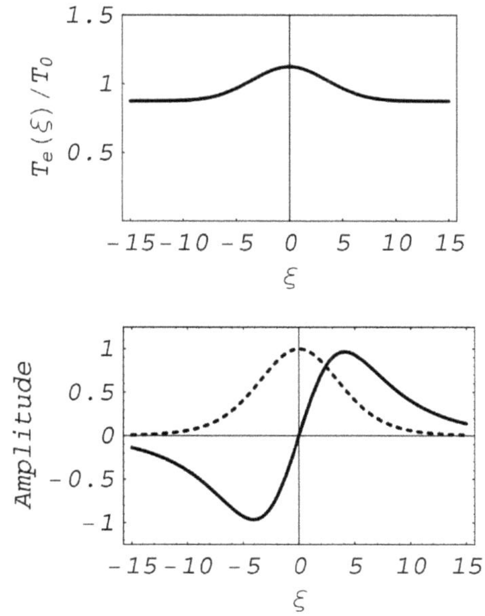

Figure 12.4. Numerical solution of (12.10), assuming an electron temperature profile with a local enhancement given by $T_e(\xi)/T_0 = 1 - \Delta/2 + \Delta \exp(-\xi^2/W^2)$, with $W = 5$ and $\Delta = 1/4$. The waveguide mode solutions for the potential are plotted for $\gamma^2 = 1.03327$ and $\gamma^2 = 0.91094$.

the local phase fronts at $\xi < 0$ to follow those at $\xi = 0$, we require that the local component of the phase velocity parallel to the magnetic field should be equal at all ξ. Since here C_s is increasing for $\xi < 0$, we must require a local wavenumber $k_\xi \neq 0$ here, according to (12.7) to compensate the increase in T_e. On the other hand, for $\xi > 0$ there is no possibility to have local phase velocities matching that at $\xi = 0$, and the solution for ϕ must be evanescent.

12.2.2 Compact electron temperature variations

Let us next assume that the electron temperature variation is compact, first with a local enhancement with a maximum value T_2 at $\xi = 0$, and $T_e(\xi) \to T_1$ for $|\xi| \to \infty$. For $\gamma^2 > T_2/T_0$ there are no physically acceptable solutions. In the case $\gamma^2 < T_1/T_0$, we have solutions of (12.10) with $\phi \to \exp(ik_{T_1}\xi)$ for $|\xi| \to \infty$, having the form of 'radiating' solutions. If on the other hand, $T_1/T_0 < \gamma^2 < T_2/T_0$, it is possible to obtain waveguide modes for selected values of γ. Examples are shown in figure 12.4. The basic features are the evanescent solutions at large $|\xi|$, and the possibility for oscillations around the maximum temperature region at $\xi \approx 0$. Note that a local analytical solution can be obtained by assuming spatial variations of the temperature as $T_0/T_e(\xi) \approx \xi^2$.

If the variation of $T_e(\xi)$ can be characterized by a single length scale $\mathcal{L}$, the continuum limit is recovered for $\mathcal{L} \to \infty$. Here, the density of the eigenvalues becomes larger with $\mathcal{L}$ increasing and form a continuum in the limiting case, irrespective of the actual functional form of $T_e(\xi)$.

Using (12.10) we find, by a simple manipulation the relation

$$\frac{\mathrm{d}}{\mathrm{d}\xi}\left(\frac{\mathrm{d}\phi}{\mathrm{d}\xi}\right)^2 = \left(\frac{T_0}{T_e(\xi)} - \frac{1}{\gamma^2}\right)\frac{\mathrm{d}\phi^2}{\mathrm{d}\xi}. \tag{12.11}$$

From (12.11) we find, for instance, that at a zero crossing of ϕ (i.e. at a local minimum of ϕ^2) we have a local extremum of $(\mathrm{d}\phi/\mathrm{d}\xi)^2$, in general, irrespective of $T_e(\xi)$.

Parts of the foregoing discussion assumed a compact local temperature enhancement. The discussion for a local depletion is entirely similar, with the result that 'radiating' solutions at $|\xi| \to \infty$ will here exist also for the case where $\gamma^2 = T_e(\xi)/T_0$ for some value of ξ. An example is shown in figure 12.5.

The discussion here focused on the localized modes with a discrete set of eigenvalues γ. In addition to these we have solutions extending to $|\xi| \to \infty$. These are modes that are locally modified by the presence of the local temperature variation, and the reduce to plane waves when $T_e(\xi)$ reduces to a constant. The discrete modes will disappear in this limit, and are thus modes that are characteristics of the temperature variation.

12.3 The high frequency case, $\omega > \Omega_{ci}$

Within the simple fluid model discussed here, we might also consider the case of waves with frequency well above the ion cyclotron frequency. The plasma is still magnetized and can support the spatially varying electron temperature, but due to

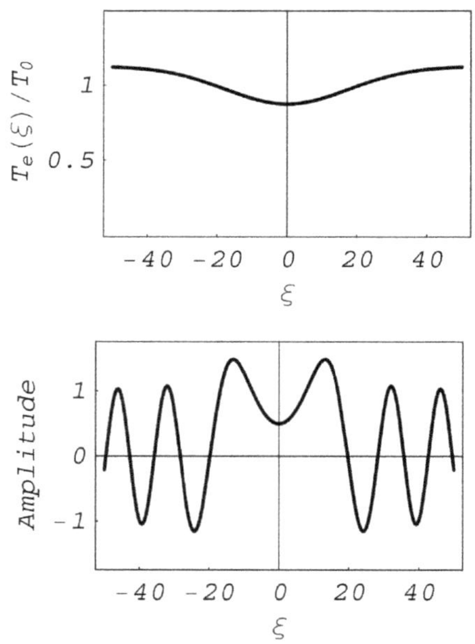

Figure 12.5. Numerical solution of (12.10) assuming a localized depletion in electron temperature given by $T_e(\xi)/T_0 = 1 + \Delta/2 - \Delta\exp(-\xi^2/W^2)$ with the parameters $W = 5$ and $\Delta = 1/4$. A solution has been calculated for $\gamma^2 = 0.9025$.

the assumed high frequency we can consider the ions as unmagnetized. The basic equations are trivially obtained from (12.2)–(12.4) but by omitting the magnetic force $\mathbf{U} \times \mathbf{B}$. The resulting differential equation for the potential becomes

$$\frac{\mathrm{d}^2\phi}{\mathrm{d}x^2} = \left(k_z^2 - \frac{M\omega^2}{\kappa T_e(x)}\right)\phi, \tag{12.12}$$

which is basically the same as (12.8), apart from a difference in a numerical coefficient and an important change in sign. The physical arguments are readily reformulated to account for this case as well. The resulting radiation pattern is similar to the one anticipated by Huld *et al* (1990) from the coupling between a beam mode and the background acoustic mode in a plasma, where the injected ion beam has a finite cross-section.

12.4 Generalizations for $T_i \neq 0$

We considered the modifications of the results due to finite ion temperatures, and also included for generality a collision frequency ν for ions colliding with a neutral component. Rather than (12.3) we now have

$$nM\frac{\partial \mathbf{u}}{\partial t} = -en(\nabla\phi - \mathbf{u} \times \mathbf{B}_0) - \nabla(n\kappa T_i) - nM\nu\mathbf{u}. \tag{12.13}$$

After some simple algebra, we find the modifications in the expressions for the ion velocity due to the finite ion pressure, and we end up with a new differential equation for the electrostatic potential

$$\begin{aligned}&\left(\frac{k_z^2\kappa}{M(\omega + i\nu)}\big(T_e(x) + T_i(x)\big) - \omega\right)\phi \\ &\quad + \frac{\kappa T_e(x)(\omega + i\nu)}{M\big(\Omega_{ci}^2 - (\omega + i\nu)^2\big)}\frac{\mathrm{d}^2}{\mathrm{d}x^2}\left(\phi\left[1 + \frac{T_i(x)}{T_e(x)}\right]\right) = 0.\end{aligned} \tag{12.14}$$

Basically, the finite ion pressure adds to the electron pressure, but in addition $T_i \neq 0$ has some other consequences by changing the character of the differential equation for the linearized potential. In order to apply a fluid model, as done here, we implicitly assume $T_i/T_e \ll 1$, implying that the new terms appearing in (12.14) are small. With $\nu \neq 0$ we have now complex eigenvalues for the equation, but this is merely describing the wave damping due to the collisions.

With a little work, the simple results for $T_i = 0$, (12.10) and (12.12) can be generalized somewhat. Assuming as before that the electrostatic potential varies only in the z- and x-directions, we can obtain the full ion velocity from the linearized equations, and ultimately find the relation

$$\frac{\mathrm{d}^2\phi}{\mathrm{d}x^2} = \left(\omega^2 - \Omega_{ci}^2\right)\left(\frac{k_z^2}{\omega^2} - \frac{M}{\kappa T_e(x)}\right)\phi, \tag{12.15}$$

or in dimensionless units

$$\frac{d^2\phi}{d\xi^2} = \left(\Omega^2 - 1\right)\left(\frac{1}{\gamma^2} - \frac{T_0}{T_e(\xi)}\right)\phi, \tag{12.16}$$

with two new dimensionless parameters $\Omega \equiv \omega/\Omega_{ci}$ and $1/\gamma^2 \equiv (T_0/M)(k_z/\omega)^2$. For the simple case with a simple temperature proportionality, $T_e(x) = \alpha T_i(x)$, with $\alpha =$ const. and $\nu = 0$, the result (12.14) is an almost trivial modification of (12.15).

These equations contain the foregoing results of (12.10) and (12.12) as special limits, but the present derivation may not be as transparent. The results (12.15) and (12.16) have one interesting implication, however. If we thus assume that γ is constant, (12.16) can still have physically acceptable solutions with finite ϕ for all values of Ω, except for only one singular point at $\Omega = 1$. These solutions disappear in the case of $T_1 = T_2 = T_0$, implying $T_e(\xi) = T_0$ and consequently $\gamma = 1$, where then only a potential ϕ independent of ξ is acceptable. For either sign, the factor $(\Omega^2 - 1)$ will formally act to re-scale the ξ variation of a solution of (12.10) or (12.12).

12.5 Linear drift waves with inhomogeneous electron temperatures

The basic equations for a simplified analysis for the case where $T_e = T_e(x)$ are again the linearized ion continuity equation

$$\frac{\partial}{\partial t} n_1 + \nabla \cdot (n_0(x)\mathbf{u}_1) = 0 \tag{12.17}$$

and the ion momentum equation, assuming cold ions for simplicity,

$$\frac{\partial}{\partial t}\mathbf{u}_1 = -\frac{e}{M}\left(\nabla\phi_1 - \mathbf{u}_1 \times \mathbf{B}_0\right). \tag{12.18}$$

Also for the present case we assume the electrons to flow along magnetic field lines in the $\hat{\mathbf{z}}$-direction to maintain a local isothermal Boltzmann equilibrium as in (12.4). With n_0 and T_e varying only with the x-variable we have

$$n_1 = n_0(x)\frac{e\phi_1}{\kappa T_e(x)}. \tag{12.19}$$

It is also here implied that the x-coordinate is in the direction perpendicular to the homogeneous magnetic field (Nycander 1989).

By simple manipulations of the $\mathbf{B}_0$-perpendicular component of the ion momentum equation we find

$$\frac{\partial^2}{\partial t^2}\mathbf{u}_{\perp 1} = -\frac{e}{M}\frac{\partial}{\partial t}\nabla_\perp\phi_1 - \frac{e^2}{M^2}\nabla_\perp\phi_1 \times \mathbf{B}_0 - \Omega_c^2\mathbf{u}_{\perp 1}. \tag{12.20}$$

Ignoring $\partial^2\mathbf{u}_{\perp 1}/\partial t^2$ as compared to $\Omega_c^2\mathbf{u}_1$, we find the usual result

$$\mathbf{u}_{\perp 1} = -\frac{1}{B_0\Omega_c}\frac{\partial}{\partial t}\nabla_\perp\phi_1 - \frac{1}{B_0^2}\nabla_\perp\phi_1 \times \mathbf{B}_0. \tag{12.21}$$

From the $\mathbf{B}_0$-parallel component of the ion momentum equation we find

$$\frac{\partial}{\partial t}u_{\parallel 1} = -\frac{e}{M}\frac{\partial}{\partial z}\phi_1. \tag{12.22}$$

After insertion into the ion continuity equation we find

$$\frac{\partial^2}{\partial t^2}\left(\phi_1 - \frac{C_s^2(x)}{\Omega_c^2}\nabla_\perp^2\phi_1\right) - C_s^2(x)\frac{\partial^2}{\partial z^2}\phi_1 \tag{12.23}$$

$$-C_s^2(x)\frac{M}{e}\left(\frac{1}{B_0^2}\nabla_\perp\frac{\partial}{\partial t}\phi_1 \times \mathbf{B}_0 + \frac{1}{B_0\Omega_c}\frac{\partial^2}{\partial t^2}\nabla_\perp\phi_1\right)\cdot\frac{\nabla_\perp n_0(x)}{n_0(x)} = 0, \tag{12.24}$$

where we defined $C_s^2(x) \equiv \kappa T_e(x)/M$, and used again the abbreviation $n_0'(x)$ for $\mathrm{d}n_0(x)/\mathrm{d}x$. Note that due to the cold plasma assumption, $T_e(x)$ enter only at the time differential in the ion continuity equation, while $n_0(x)$ has a spatial differential as in the standard linear drift wave analysis.

Equation (12.24) is solved with

$$\phi_1(x, y, z, t) = \phi(x)e^{i(k_y y + k_z z - \omega t)}. \tag{12.25}$$

The result is

$$\begin{aligned}&\frac{\omega^2 C_s^2(x)}{\Omega_c^2}\left(\frac{\mathrm{d}^2\phi}{\mathrm{d}x^2} + \frac{n_0'(x)}{n_0(x)}\frac{\mathrm{d}\phi}{\mathrm{d}x}\right)\\ &-\left(\omega^2 + \frac{\omega^2 C_s^2(x)}{\Omega_c^2}k_y^2 - C_s^2(x)k_z^2 + \frac{\omega C_s^2(x)}{\Omega_c}\frac{n_0'(x)}{n_0(x)}k_y\right)\phi = 0.\end{aligned} \tag{12.26}$$

equation (12.26) is a second-order differential equation for $\phi(x)$ with variable coefficients and must be solved together with some boundary conditions, just as (5.12). Little is known about the solution in the case of arbitrary variation of $n_0(x)$ and $C_s^2(x)$. In order to obtain a simpler equation, we make the assumption of a locally linear variation of density as well as electron temperature, i.e. let $n_0(x) = n_0(0)(1 + \beta_n x + \cdots)$ and $T_e(x) = T_e(0)(1 + \beta_T x + \cdots)$. Considering β_n as well as β_T as small quantities, with $(1 + \beta_T x)^{-1} \approx 1 - \beta_T x$ we find

$$\begin{aligned}&\frac{\omega^2 C_{s0}^2}{\Omega_c^2}\left(\frac{\mathrm{d}^2\phi}{\mathrm{d}x^2} + \beta_n\frac{\mathrm{d}\phi}{\mathrm{d}x}\right)\\ &-\left(\omega^2(1 - x\beta_T) + \frac{\omega^2 C_{s0}^2}{\Omega_c^2}k_y^2 - C_{s0}^2 k_z^2 + \beta_n\frac{\omega C_{s0}^2}{\Omega_c}k_y\right)\phi = 0,\end{aligned} \tag{12.27}$$

where $C_{s0}^2 \equiv \kappa T_e(0)/M$. From (12.27) we recover (5.12) by taking $\beta_T = 0$. In the present case we will not find a local model by ignoring the first parenthesis in (12.27) since the x-variable enters explicitly in the second parenthesis.

The equation (12.26) or its simplified version (12.27) can be seen as eigenvalue equation determining ω and the corresponding eigenfunction $\phi(x)$ for given temperature and density variations. We note however that the case with $\beta_T = 0$ and $\beta_T \neq 0$ are significantly different. In the former case, all coefficients in (12.27) are constant, while an explicit x-dependence remains in the latter case, even in the present very simplified formulation. Since we are only interested in solutions of (12.27) in the vicinity of the reference point for the expansion of $n_0(x)$ and $C_s^2(x)$, here taken to be the origin, the relevant region of the spatial variation is restricted to small x. We therefore anticipate that for temperature gradients to be of importance we have to require $\beta_T \gg \beta_n$.

The assumption of Boltzmann distributed electrons is restrictive. For low frequency phenomena it is easily justified but it might fail for higher frequencies where a dynamic electron model is needed. Instability of drift waves in electron temperature gradients have been investigated experimentally (Moon *et al* 2013).

12.6 Ion temperature gradient modes

We can have shears in thermal velocities as well as in bulk plasma flow velocities. One such case will be discussed here: the ion temperature gradient mode or the 'ITG-mode' that is also found important in high temperature fusion experiments (Cowley *et al* 1991, Weiland 2000). We assume here homogeneous magnetic fields and allow for ion motion along magnetic field lines. The basic physics will be apparent also for homogeneous plasma densities, so we make this simplifying assumption here. For the low frequency, slow phase velocities relevant here we can safely assume the perturbations to be electrostatic, $\mathbf{E} = -\nabla\phi$. The ion velocity in the direction perpendicular to $\mathbf{B}$ is approximated as $\mathbf{u}_\perp \approx -\nabla\phi \times \mathbf{B}/B^2$ just as for the simple limit of drift wave modeling in chapter 4. Taking $\nabla \equiv \nabla_\perp + \hat{\mathbf{z}}\partial/\partial z$ we write the ion continuity equation as

$$\left(\frac{\partial}{\partial t} + \mathbf{u}_\perp \cdot \nabla_\perp + u_\parallel \frac{\partial}{\partial z}\right) n = -n \frac{\partial u_\parallel}{\partial z}. \tag{12.28}$$

Variations in the ion density are solely due to compression along magnetic field lines since $\nabla \cdot \mathbf{u}_\perp = 0$. The component of the ion momentum equation along $\mathbf{B}$ gives an equation

$$nM\left(\frac{\partial}{\partial t} + \mathbf{u}_\perp \cdot \nabla_\perp + u_\parallel \frac{\partial}{\partial z}\right) u_\parallel = -\frac{\partial p}{\partial z} - ne\frac{\partial \phi}{\partial z}, \tag{12.29}$$

while the component $\perp\mathbf{B}$ becomes

$$nM\left(\frac{\partial}{\partial t} + \mathbf{u}_\perp \cdot \nabla_\perp + u_\parallel \frac{\partial}{\partial z}\right) \mathbf{u}_\perp = -\nabla_\perp p + ne(\mathbf{u}_\perp \times \mathbf{B} - \nabla_\perp \phi), \tag{12.30}$$

since $\mathbf{u} \cdot \nabla \equiv \mathbf{u}_\perp \cdot \nabla_\perp + u_\parallel \partial/\partial z$.

Assuming quasi-neutrality we do not distinguish electron and ion densities. We anticipate (as for drift waves) phase velocity components along the magnetic field lines to be larger than ion thermal velocities and take the adiabatic equation of state for the ions, using $n = n_0(p/p_0)^{1/\gamma}$ where $p = \kappa T_i n$. This inserted into the equation of continuity gives after some simple manipulations

$$\left(\frac{\partial}{\partial t} + \mathbf{u}_\perp \cdot \nabla_\perp + u_\parallel \frac{\partial}{\partial z}\right) p = -\gamma p \frac{\partial}{\partial z} u_\parallel, \tag{12.31}$$

where the consequences of compressibility appear explicitly by the right-hand side. The spatial derivative-terms on the left side account for the convection of pressure perturbations.

The equation for pressure (and not density) is most relevant here, since this is what we need in the force term on the right hand side of (12.29). We might, however, just as well obtain a dynamic equation for the ion temperature by inserting $n = n_0(T_i/T_{oi})^{1/(\gamma-1)}$ into the ion continuity equation to find

$$\left(\frac{\partial}{\partial t} + \mathbf{u}_\perp \cdot \nabla_\perp + u_\parallel \frac{\partial}{\partial z}\right) T_i = -(\gamma - 1) T_i \frac{\partial}{\partial z} u_\parallel$$

for the present low frequency dynamics in magnetized plasmas. More generally we would have $(\partial/\partial t + \mathbf{u} \cdot \nabla) T_i = -(\gamma - 1) T_i \nabla \cdot \mathbf{u}$. At first sight it might seem that we have a choice for γ but since we have explicitly used the equation of state for an ideal gas, the only acceptable value is $\gamma = 5/3$ for adiabatic ideal monatomic gas dynamics.

Allowing for $\nabla_\perp T_{0i} \neq 0$, we have a steady state solution as one where $n = n_0$, $u_\parallel = 0$, $\nabla \phi = 0$ with an ion diamagnetic drift velocity $\mathbf{u}_{0i\perp} = -\nabla_\perp p_0 \times \mathbf{B}/(n_0 e B^2)$ with $\nabla_\perp p_0 = n_0 \nabla_\perp \kappa T_{0i}$. The electron dynamics is also here simplified by assuming the electrons to be isothermally Boltzmann distributed at all times, $n = n_0 \exp(e\phi/\kappa T_e)$.

The nonlinear development is interesting, but we here restrict the analysis to a stability analysis of small amplitude perturbations. We specify the geometry so that $\mathbf{B} \| \hat{\mathbf{z}}$ and $\nabla_\perp T_{0i} \| \hat{\mathbf{x}}$ giving $\mathbf{u}_{0i\perp} = \kappa T'_{0i} \hat{\mathbf{y}}/(eB)$ with $T'_{0i} \equiv \mathrm{d}T_{i0}/\mathrm{d}x$.

We linearize the basic equation with $n = n_0 + \tilde{n}$, etc. and make the plane wave assumption $\tilde{n} = n_1(x) \exp(\mathrm{i}k_z z + \mathrm{i}k_y y - \mathrm{i}\omega t)$, etc, without including k_x in the exponent. For the Boltzmann distributed electrons we have $n_1 = n_0 e\phi_1/\kappa T_e$, while for instance (12.31) gives

$$(\omega - k_y u_{0i\perp}) p_1 - i \mathbf{u}_{1\perp} \cdot \nabla_\perp p_0 = \gamma k_z \kappa T_{0i} n_0 u_{1\parallel},$$

with $p_1 = \gamma \kappa T_{0i} n_1$ and $\mathbf{u}_{1\perp} = -\nabla_\perp \phi_1 \times \mathbf{B}/B^2$. We have $\mathbf{u}_{1\perp} \cdot \nabla_\perp p_0 = -ik_y(\phi_1/B)\mathrm{d}p_0/\mathrm{d}x$. This term accounts for the mixing of cold and warm ions being responsible for the ultimate wave growth. The source of free energy is the ion temperature gradient, or, if you like, the ion drift current.

By (12.29) we find

$$n_0 M(\omega - k_y u_{0i\perp}) u_{1\parallel} = k_z p_1 + k_z n_0 e \phi_1.$$

The ion continuity equation gives

$$(\omega - k_y u_{0i\perp}) n_1 = n_0 k_z u_{1\parallel}.$$

By eliminating n_1, $u_{1\|}$ and introducing $\Omega \equiv \omega - k_y u_{0i\perp}$ to simplify the notation, we find a condition for non-trivial solutions for ϕ_1 in the form

$$\left(\Omega^2 - k_z^2 C_s^2\right) = k_z^2 C_s^2 \frac{T_{0i}'}{T_{0i}} \frac{k_y U_{thi}^2}{\Omega \Omega_{ci}\left(1 + \frac{\gamma T_{0i}}{T_e}\right)}, \tag{12.32}$$

where $C_s \equiv \sqrt{\kappa(T_e + \gamma T_i)/M}$ is the sound speed for warm adiabatic ions and $U_{thi}^2 \equiv \kappa T_{0i}/M$. We took the inverse length scale $T_{0i}'/T_{0i} \equiv 1/L_T$ to be constant: as a local approximation this can be justified to the same accuracy that we took n_0'/n_0 to be a constant in the studies of electrostatic drift waves. By the right hand term in the dispersion relation (12.32) we can identify a characteristic frequency

$$\omega_T^* \equiv \frac{k_y U_{thi}^2}{\Omega_{ci}\left(1 + \frac{\gamma T_{0i}}{T_e}\right)} \frac{T_{0i}'}{T_{0i}},$$

reminiscent of the drift frequency with a sign that here depends on the sign of T_{0i}'. We have $k_y u_{0i\perp}/\omega_T^* = 1 + \gamma T_{0i}/T_e$.

Normalizing frequencies by ω_T^* we find the dispersion relation in the form

$$\frac{\Omega}{\omega_T^*}\left(\frac{\Omega^2}{\omega_T^{*2}} - \frac{k_z^2 C_s^2}{\omega_T^{*2}}\right) = \frac{k_z^2 C_s^2}{\omega_T^{*2}}. \tag{12.33}$$

We have two types of natural modes of oscillation $\Omega = 0$ and $\Omega = \pm C_s k_z$ that are coupled by the ion temperature gradient. The two type of modes are the sound waves and ion temperature striations that are stationary in the chosen frame of reference. The solutions of (12.33) are shown in figure 12.6.

By construction, the direction of wave propagation is restricted to the (y, z)-plane. Fourier transforming along the x-direction gives problems due to the ion temperature inhomogeneity, but the question of a possible instability is settled already by the present analysis.

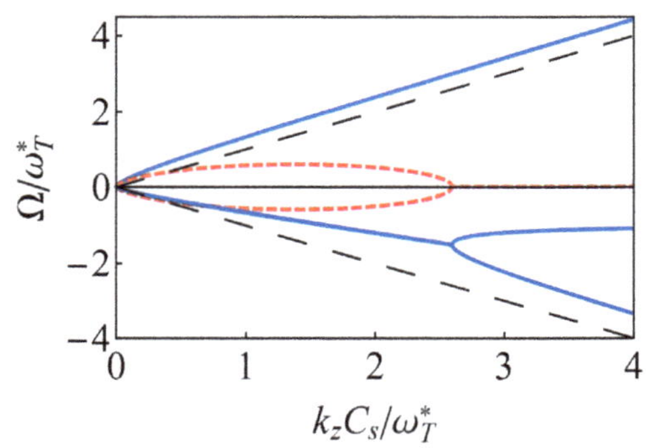

Figure 12.6. Real (full lines) and imaginary parts (dashed lines) of $\Omega' \equiv \Omega/\omega_T^*$ in (12.33) shown as a function of $k_z C_s/\omega_T^*$. The fixed frame frequency is $\omega = \Omega + k_y u_{0i}$. Thin dashed lines give the ion sound velocity for reference.

The physical mechanism for the ITG-instability was discussed by, for instance, Cowley *et al* (1991). The basic mechanism is the mixing of cold and warm ions by the $-\nabla\phi \times \mathbf{B}/B^2$-drifts along the temperature gradient. The instability requires $k_{\|} \neq 0$ so the ion temperature perturbations can relax along $\mathbf{B}$. The resulting density imbalance creates potential variations due to the electron motion. These potential variations give rise to changes in the electric fields, that modify the $-\nabla\phi \times \mathbf{B}/B^2$-drifts. At the same time, the perturbations propagate as sound waves in magnetized plasmas. The detailed dynamics can be understood best by considering the linear relation between plasma density and ion pressure

$$p_1 = \left(1 + \frac{k_y}{\Omega_{ci}\Omega}\frac{\kappa T_e}{M}\frac{T_{0i}'}{\gamma T_{0i}}\right)\gamma n_1 \kappa T_{0i}.$$

When $T_{0i}' = 0$, the relation reduces to the linearized adiabatic ion equation of state. For $T_{0i}' \neq 0$ the second term enters as a correction which may change sign for some parameter values. If the parenthesis becomes negative, the ions appear as having negative compressibility: an increase in plasma density reduces the ion pressure. These conditions are unstable.

The ITG-instability is interesting also by being effective even for Boltzmann distributed electrons: in studies of isothermal drift waves supported by density gradients we learned that a mechanism for disrupting the Boltzmann equilibrium was needed to trigger an instability. It might have been expected that it is the diamagnetic drift (i.e. $\mathbf{u}_{Di} = -\nabla p_i \times \mathbf{B}/(neB^2)$ and similarly for the electrons with $e \rightarrow -e$) that supports unstable drift waves, irrespective of how the pressure gradient is established, i.e. by gradients in density or temperature, or both, so that isothermally Boltzmann distributed electrons would give stability in all cases. This is not so: a density gradient enters the linearized continuity equation, a temperature gradient does not, so the mathematical modeling of the two cases becomes different.

The ion temperature gradient case studied in this section has an equivalent for electron temperature gradients. In that case we need a dynamic model for the electrons. The waves are usually restricted to high frequencies.

References

Barston E M 1964 Electrostatic oscillations in inhomogeneous cold plasmas *Ann. Phys. (NY)* **29** 282–303

Cowley S C, Kulsrud R M and Sudan R 1991 Considerations of ion-temperature-gradient-driven turbulence *Phys. Fluids B: Plasma Phys.* **3** 2767–82

Guio P, Børve S, Pécseli H L and Trulsen J 2001 Low frequency waves in plasmas with spatially varying electron temperature *Ann. Geophysicae* **18** 1613–22

Guio P and Pécseli H L 2010 Collisionless plasma shocks in striated electron temperatures *Phys. Rev. Lett.* **104** 085002

Guio P and Pécseli H L 2016 Weakly nonlinear ion waves in striated electron temperatures *Phys. Rev.* E **93** 043204

Huld T, Pécseli H L and Rasmussen J J 1990 Ion-acoustic wave propagation in plasma with ion beams having a finite cross section *IEEE Trans. Plasma Sci.* **18** 149–58

Moon C, Kaneko T and Hatakeyama R 2013 Dynamics of nonlinear coupling between electron-temperature-gradient mode and drift-wave mode in linear magnetized plasmas *Phys. Rev. Lett.* **111** 115001

Nishida Y and Hirose A 1977 Observation of refraction and convergence of ion acoustic waves in a plasma with a temperature gradient *Plasma Phys.* **19** 447–53

Nycander J 1989 The effect of the electron temperature gradient on nonlinear drift waves in plasmas *Phys. Scripta* **39** 758–63

Uberoi C 1972 Alfvén waves in inhomogeneous magnetic fields *Phys. Fluids* **15** 1673–5

Weiland J 2000 *Collective Modes in Inhomogeneous Plasma, Kinetic and Advanced Fluid Theory* (Bristol: Institute of Physics Publishing)

Chapter 13

Waves in a gravitational plasma ionosphere

Drift waves represent dynamical features associated with density gradients in magnetized plasmas, and can be seen as surface waves on a diffuse boundary. It is an essential feature of the foregoing analysis that these density gradients are perpendicular to magnetic field lines. We can have inhomogeneities with other geometries, one important example being vertical density gradients in gravitational fields. This case offers a simple model for density gradients having a component also along the magnetic field lines (Pécseli 2016). Also other models with similar properties can be found (Doucet *et al* 1974, D'Angelo *et al* 1976, Pécseli 2016). The analysis of neutral and plasma atmospheres is very similar and the wavetypes found have some similarities as well.

13.1 Stable and unstable stratifications

The basic and simplest steady state model for a gravitational atmosphere is based on the equilibrium model without flow

$$0 = -\nabla p - nM\mathbf{g} - ne\nabla\phi, \tag{13.1}$$

where we assume a constant gravitational field appropriate for altitudes much smaller than the Earth radius R_e. For stationary conditions we can take the electric field to be electrostatic, $\mathbf{E} = -\nabla\phi$. For modeling a neutral atmosphere we set $e = 0$. For equilibrium conditions the most natural equation of state is isothermal $p = n\kappa T$ where T is the constant ion temperature or the neutral gas temperature, depending on the problem. This model gives

$$0 = -\kappa T\nabla \ln\left(\frac{n}{n_0}\right) - Mg\nabla z - e\nabla\phi, \tag{13.2}$$

with the constant gravitational acceleration being $\mathbf{g} \parallel \hat{\mathbf{z}}$.

doi:10.1088/978-0-7503-1251-6ch13

For the plasma problem we assume an adequate electron model to be an isothermal Boltzmann distribution $n_e(x) = n_0 \exp(e\phi/\kappa T_e)$ as used several times before, with the reference density n_0 being defined as $n_e(x)$ at a position where $\phi = 0$. With quasi-neutrality $n_e \approx n_i \equiv n$ this model gives

$$e\phi = -\frac{Mgz}{1 + T/T_e}, \tag{13.3}$$

which is correct also for $T = T(z)$ and $T_e = T_e(z)$. If T and T_e are constant we find a constant electric field $\mathbf{E} = Mg\hat{\mathbf{z}}/(1 + T/T_e)$ which results from the force balance on the ions of pressure, gravity and the electric field set up by the electron pressure. The result (13.3) becomes particularly simple for cold ions, $T = 0$. The expression for the plasma density becomes

$$n = n_0 \exp\left(-\frac{Mgz}{\kappa(T + T_e)}\right). \tag{13.4}$$

For an isothermal neutral gas atmosphere we find the result by setting $T_e = 0$. The result (13.4) can be seen as a limiting case for the classical expressions of Pannekoek (1922) and Rosseland (1924). We note that the assumption of quasi-neutrality should be considered with some care: according to (13.4) we have the Debye length increasing for increasing z and it is therefore always possible to find an altitude where the local value for λ_D exceeds any prescribed value.

The isothermal model outlined here is not realistic, although it serves as a good reference. In the neutral atmosphere as well as in the ionosphere we have the temperatures and also the gas composition to vary, sometimes significantly, with altitude. There are no universal models for such cases.

We consider now a stratified atmosphere/ionosphere different from the equilibrium solution given by (13.4). At some reference altitude z_0 we have $p(z_0)$ and $n(z_0)$. We now take a small volume of the local gas and in imagination move it to a slightly higher altitude $z = z_0 + \Delta z$. As the volume element moves up into a lower density it expands, and its internal pressure adjusts accordingly. We denote the density and pressure of this volume element as $n'(z)$ and $p'(z)$, noting that $n'(z_0) = n(z_0)$ and $p'(z_0) = p(z_0)$. For the plasma case we assume the electrons to readjust to their locally isothermal condition instantaneously during the displacement of the volume element.

The force density on the displaced volume element is

$$Mn'\frac{\mathrm{d}^2\Delta z}{\mathrm{d}t^2} = -\frac{\mathrm{d}p}{\mathrm{d}z} - en'\frac{\mathrm{d}\phi}{\mathrm{d}z} - Mn'g. \tag{13.5}$$

We have $\mathrm{d}p/\mathrm{d}z = -Mng + enE = -Mng - en\mathrm{d}\phi/\mathrm{d}z$ by (13.1). We can then reduce (13.5) to

$$\frac{\mathrm{d}^2\Delta z}{\mathrm{d}t^2} = g\frac{n - n'}{n'} + \frac{e}{M}\frac{n - n'}{n'}\frac{\mathrm{d}\phi}{\mathrm{d}z}. \tag{13.6}$$

The first term accounts for the updrift or buoyant force by the surrounding medium (Archimedes' law). We now need to find a relation between n and n'.

Assume that the motion of the volume element is sufficiently slow to allow the local pressure equilibrium to be established, giving $p'(z) = p(z)$, but on the other hand the motion is so fast that there is no time for the temperatures to be equalized so that in general $n'(z) \neq n(z)$. Taking the expansion to be adiabatic we have $p/p_0 = (n/n_0)^\gamma$ with $\gamma = C_p/C_V$ being the ratio of specific heats at constant pressure and constant volume, respectively. This will be the same for the plasma and the neutral atmosphere. The relation between a gradient in pressure and gradient in density is

$$\frac{\mathrm{d}n}{\mathrm{d}z} = \frac{n}{\gamma p}\frac{\mathrm{d}p}{\mathrm{d}z} = -\frac{n^2}{\gamma p}(Mg + eE) = -\frac{n}{\gamma \kappa T}(Mg - eE), \tag{13.7}$$

with $E = -\mathrm{d}\phi/\mathrm{d}z$. Again we set $e = 0$ when dealing with a neutral gas. The expression (13.7) is correct also for the case where we have a gradient in temperature, $T = T(z)$, in which case we have that E is no longer constant, but can be determined using for instance (13.3). A simple exponential density variation with altitude is found only when T and T_e are constant.

For small displacements Δz we can write

$$n'(z) \approx n(z_0) + \left.\frac{\mathrm{d}n'}{\mathrm{d}z}\right|_{z_0} \Delta z = n(z_0) - \frac{n(z_0)}{\gamma \kappa T(z_0)}(Mg - eE(z_0))\Delta z. \tag{13.8}$$

For the background density we have similarly

$$n \approx n(z_0) + \left.\frac{\mathrm{d}n}{\mathrm{d}z}\right|_{z_0} \Delta z.$$

From (13.6) we find

$$\frac{\mathrm{d}^2\Delta z}{\mathrm{d}t^2} \approx \left(g - \frac{e}{M}E(z_0)\right)\frac{1}{n(z_0)}\left(\left.\frac{\mathrm{d}n}{\mathrm{d}z}\right|_{z_0} + \frac{n(z_0)}{\gamma \kappa T(z_0)}(Mg - eE(z_0))\right)\Delta z \equiv -\omega_{BV}^2 \Delta z,$$

which has the form of an equation for a harmonic oscillator with

$$\omega_{BV}^2 \equiv -\left(g - \frac{e}{M}E(z_0)\right)\frac{1}{n(z_0)}\left(\left.\frac{\mathrm{d}n}{\mathrm{d}z}\right|_{z_0} + \frac{n(z_0)}{\gamma \kappa T(z_0)}(Mg - eE(z_0))\right), \tag{13.9}$$

which for a neutral atmosphere is usually denoted the Brunt–Väisälä frequency. For stable conditions it is a positive quantity, for unstable conditions negative. For a neutral gas in equilibrium we have a definition $\omega_{BV} = \sqrt{-(g/\rho)\mathrm{d}\rho(z)/\mathrm{d}z}$ with ρ being the mass density. This result differs from its plasma equivalent basically because we in the latter case deal with a mixture of an isothermal electron gas and a (possibly) adiabatic ion gas.

The local scale length for the unperturbed atmosphere is $L \equiv n(z_0)/|\mathrm{d}n/\mathrm{d}z|_{z_0}$. Since $\mathrm{d}n/\mathrm{d}z < 0$ usually, we can write (13.9) as

$$\omega_{BV}^2 = -\left(\frac{(Mg - eE(z_0))}{L} - \frac{1}{\gamma\kappa T(z_0)}(Mg - eE(z_0))^2\right)$$

$$= -\left(\frac{(Mg - eE(z_0))}{L} - \frac{M}{\gamma\kappa T(z_0)}\left(g + \frac{\kappa T_e}{Mn}\frac{dn}{dz}\bigg|_{z_0}\right)^2\right)$$

$$= -\left(\frac{g - \kappa T_e/ML}{L} - \frac{M}{\gamma\kappa T(z_0)}\left(g - \frac{\kappa T_e}{ML}\right)^2\right),$$

assuming again an isothermal Boltzmann distribution for the electrons.

For a neutral atmosphere we set $T_e = 0$ and find $\omega_{BV}^2 < 0$ in the case of $L < C^2/g$, using the local sound speed in neutral gas to be $C \equiv \sqrt{\gamma\kappa T/M}$. The atmosphere is unstable in this case and characterized by large scale temporal variation. For the equilibrium neutral atmosphere with density variation $n = n_0 \exp(-Mgz/\kappa T)$ we have $L = \kappa T/Mg$, resulting in stable conditions with $\omega_{BV} = (g/C)\sqrt{\gamma - 1}$, consistent with standard results (Yeh and Liu 1972, Brekhovskikh and Goncharov 1985).

For the plasma case with $T_e \neq 0$ in equilibrium we have (13.4), now with T being the ion temperature, giving $L = \kappa(T + T_e)/Mg$. Some simple algebra gives

$$\omega_{BV}^2 = \frac{(\gamma - 1)}{L^2}\frac{\kappa T}{M} = (\gamma - 1)\frac{g^2 M\kappa T}{\kappa^2(T_e + T)^2} = (\gamma - 1)\frac{g^2}{C_s^2}\frac{T}{T_e + T},$$

in terms of an isothermal ion sound speed $C_s \equiv \sqrt{\kappa(T_e + T)/M}$. The analysis outlined here indicates that once perturbed, a stable ionosphere or atmosphere will oscillate for all later times. Viscosity and collisions will eventually damp the oscillations.

The limiting case with $T = 0$ is singular in the sense that a small volume element of a displaced component (e.g. ions in an ionosphere with warm electrons) will not expand so that $n'(z) = n(z_0)$, and its partial pressure vanishes at all altitudes.

13.2 Wave propagation

Section 13.1 described large scale or bulk oscillations in an atmosphere in a gravitational field. To describe the propagation of waves with finite wavelengths we here use a simplified model using cold ions. In this limit $\omega_{BV} \to 0$ and the only pressure term left in the model originates from the electrons. A more complete model needs also a dynamic equation for the ion temperature, where a fluctuating ion fluid velocity having a component along the density gradient induces oscillations in the ion pressure. Most relevant models take adiabatic ion temperature variations, and the additional complications of the analysis should not be underestimated.

Assuming a vertical density variation in a gravitationally confined plasma with a magnetic field also in the vertical direction we write the linearized ion continuity equation as

$$\frac{\partial\eta}{\partial t} + \mathbf{u}\cdot\nabla\ln\bar{n} + \nabla\cdot\mathbf{u} = \frac{\partial\eta}{\partial t} + u_z\frac{\mathrm{d}\ln\bar{n}}{\mathrm{d}z} + \frac{\partial u_z}{\partial z} + \nabla_\perp\cdot\mathbf{u}_\perp = 0,$$

where we introduced $\eta \equiv n/\bar{n}(z)$. The linearized ion momentum equation becomes

$$M\frac{\partial}{\partial t}\mathbf{u} = -\,e\nabla\phi + e\mathbf{u}\times\mathbf{B},$$

where ϕ and $\mathbf{u}$ are fluctuating quantities. We have for the velocity component along $\mathbf{B}$ that

$$M\frac{\partial}{\partial t}u_z = -\,e\frac{\partial\phi}{\partial z}.$$

The equilibrium solution with $\mathbf{u} = 0$ is found in section 13.1.

The electron component is also here assumed to be a Boltzmann distribution at all times with constant temperature T_e, i.e. $n = n_0\exp(e\phi/\kappa T_e)$. We linearize this expression as

$$n \equiv \bar{n} + \tilde{n} = n_0\exp\left(\frac{e\bar{\phi} + e\tilde{\phi}}{\kappa T_e}\right) \approx n_0\exp\left(\frac{e\bar{\phi}}{\kappa T_e}\right)\left(1 + \frac{e\tilde{\phi}}{\kappa T_e}\right).$$

The linearized part of this condition gives $e\phi/\kappa T_e = \eta$ with $\eta \equiv \tilde{n}/\bar{n}$. Assuming quasi neutrality we use this relation to eliminate the ion density and find from the ion continuity equation

$$\frac{eM}{\kappa T_e}\frac{\partial^2\phi}{\partial t^2} - e\frac{\partial\phi}{\partial z}\frac{\mathrm{d}\ln\bar{n}}{\mathrm{d}z} - e\frac{\partial^2\phi}{\partial z^2} + M\frac{\partial}{\partial t}\nabla_\perp\cdot\mathbf{u}_\perp = 0. \tag{13.10}$$

We also used the z-component of the ion momentum equation. Taking the $\times\mathbf{B}$-product of the ion momentum equation we find

$$M\frac{\partial}{\partial t}\mathbf{u}\times\mathbf{B} = -\,e\nabla_\perp\phi\times\mathbf{B} - e\mathbf{u}_\perp B^2.$$

Inserting $\mathbf{u}\times\mathbf{B}$ from the original ion momentum equation we find

$$M\frac{\partial}{\partial t}\nabla_\perp\phi + \frac{M^2}{e}\frac{\partial^2}{\partial t^2}\mathbf{u}_\perp = -\,e\nabla_\perp\phi\times\mathbf{B} - e\mathbf{u}_\perp B^2.$$

We will need the divergence $\nabla_\perp\cdot\mathbf{u}_\perp$ and have the simpler expression

$$M\frac{\partial}{\partial t}\nabla_\perp^2\phi + \frac{M^2}{e}\frac{\partial^2}{\partial t^2}\nabla_\perp\cdot\mathbf{u}_\perp = -\,eB^2\nabla_\perp\cdot\mathbf{u}_\perp. \tag{13.11}$$

By these operations we have a set of two coupled differential equations (1) and (2). A dispersion relation is found by the plane wave test model where all quantities vary like $\exp(-\,\mathrm{i}\omega t + \mathrm{i}\mathbf{k}\cdot\mathbf{r})$ and eliminating $\nabla_\perp\cdot\mathbf{u}_\perp \to i\mathbf{k}_\perp\cdot\mathbf{u}_\perp$. We find

$$\omega^2 = C_s^2k_z^2 - i\frac{\kappa T_e}{M}k_z\frac{\mathrm{d}\ln\bar{n}}{\mathrm{d}z} + \frac{\omega^2k_\perp^2C_s^2}{\omega^2 - \Omega_{ci}^2}, \tag{13.12}$$

where we again used $C_s^2 \equiv \kappa T_e/M$. The wavetype discussed here has direct similarities with gravitational waves (Hines 1960) in neutral atmospheres.

As a test we can let $\mathrm{dln}\,\bar{n}/\mathrm{d}z = 0$ and find

$$\omega^2 = C_s^2\left(k_z^2 + \frac{\omega^2 k_\perp^2}{\omega^2 - \Omega_{ci}^2}\right)$$

which gives purely the classical ion sound waves along the z-direction and ion cyclotron waves perpendicular to $\hat{\mathbf{z}} \parallel \mathbf{B}$, while $\Omega_{ci} \to 0$ gives $\omega^2 = C_s^2(k_z^2 + k_\perp^2) \equiv C_s^2 k^2$.

For vertical propagation with $k_\perp = 0$ we find the dispersion relation

$$\omega^2 = C_s^2 k_z^2 + i\frac{\kappa T_e}{M}\frac{k_z}{L}. \tag{13.13}$$

The Brunt–Väisälä frequency does not appear since we assumed vanishing ion temperatures. For $T_i \neq 0$ it will enter as a correction to ω for small k_z.

Considering an initial value problem, we find temporally increasing waves for $k_z > 0$ propagating up, and damped waves for $k_z < 0$ propagating down. For a boundary value problem with a real frequency imposed at the boundary $z = 0$, the solution is more complicated, see figures 13.1 and 13.2. For the boundary value problem with a real imposed frequencies, the most natural case is having the exciter at $z = 0$, and the waves will propagate upwards with $\Re\{k_z\} > 0$. Some preliminary results were given by Garcia and Pécseli (2013).

The analysis assumes quasi-neutrality. This restriction can be relaxed by noting that with the assumption of isothermally Boltzmann distributed electrons, T_e originates solely from the expression $e\phi/\kappa T_e = n/n_0$. If we retain the assumption of Boltzmann distributed electrons but relax the assumption of quasi-neutrality we find by the Poisson equation that $e\phi/\kappa T_e = \eta/(1 + (k\lambda_{De})^2)$. The result, also valid for wavenumbers close to the electron Debye length, can be found from the previous

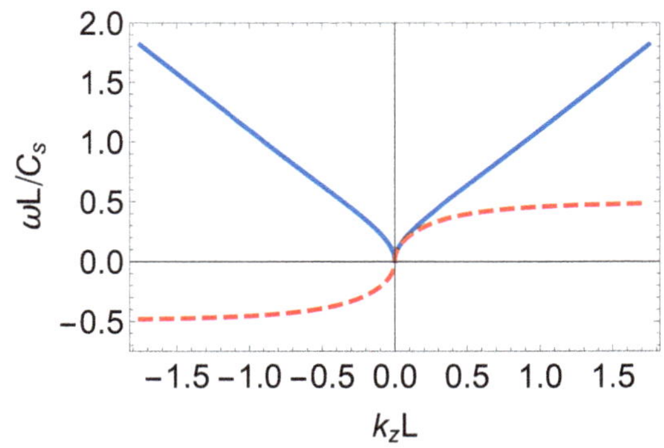

Figure 13.1. Dispersion relation for normalized real wavenumbers $k_z L$ and complex frequencies, with the full line giving the real part, dashed line the imaginary part. We have the parameter $\theta \equiv \gamma T_i/(T_e + \gamma T_i) = 0.25$. The normalized cut-off frequency $\omega_S = |L|^{-1}\sqrt{\gamma\kappa T_i/M}$ is here $\omega_S L/C_s = \sqrt{\theta}$.

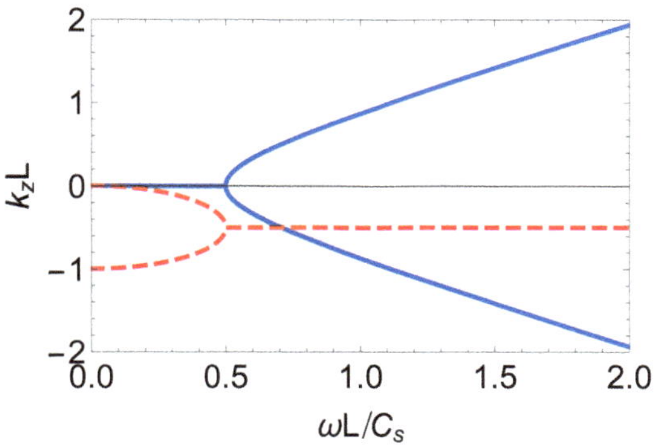

Figure 13.2. Dispersion relation for normalized real frequencies $\omega L/C_s$ and complex wavenumbers corresponding to spatial damping or growth, with the full line giving the real part, dashed line the imaginary part. See also figure 13.1. Recall that $\Im\{k_z\} < 0$ corresponds to growth for the reference waveform $\exp(-i\omega t + ik_z z)$.

dispersion relation (13.15) by replacing T_e by $T_e/(1 + (k\lambda_{De})^2)$ everywhere, in particular also in C_s. The result becomes rather lengthy and is not reproduced here.

For $k_z > 0$, the long wavelengths (small k_z) have the largest relative growth rates, $\Im\{\omega\}/\Re\{\omega\}$. The plasma here has no source of free energy, so the observed wave-growth cannot properly be called an instability (Dysthe *et al* 1975, Pécseli 2016). When the wavelength becomes much smaller than the vertical length scale L, the relative growth rate $\Im\{\omega\}/\Re\{\omega\}$ becomes small, and the waves can be assumed to propagate with the local dispersion relation, corresponding to the appropriate altitude z. For the initial value problem the physical interpretation of the wave growth is not simple, since a plane wave initial condition has an associated inhomogeneous energy density along z. The arguments are more transparent for a boundary value problem. Physically, the apparent wave growth in that case is due to the vertical energy flux being constant in the absence of any dissipation mechanism. The real part of the propagation velocity is a constant, C_s, which is independent of density. For instance, the local kinetic energy density is to the given approximation $\frac{1}{2}Mn_0\widetilde{u}^2$. Since $n_0(z)$ is decreasing with altitude, we must have $\widetilde{u}$ increasing to keep the energy flux constant. This altitude increase in amplitude of the fluctuating velocity can then appear as an instability.

The present analysis was based on a fluid model where the analysis of sound waves in neutral gases and in plasmas appear similar and significant differences appear when a kinetic analysis is applied, possibly including collisional effects as well. A kinetic plasma analysis was given by Parkinson and Schindler (1969) and Liu (1970) where the 'fake instability' (there seems to be no established terminology: 'geometric growth' has also been used) can be overcome by Landau damping for short wavelengths. For long wavelengths it was found that the wave growth can remain, albeit with a modified growth rate.

The dispersion relation for general wavevector direction is obtained from (13.15) best by normalizing frequencies by Ω_{ci} and lengths by C_s/Ω_{ci}, an effective gyro-radius. The result in these normalized units is

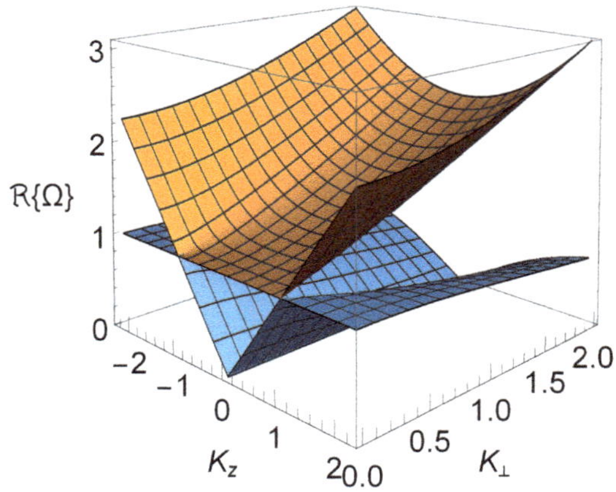

Figure 13.3. Real part of the full dispersion relation with normalized real wavenumbers and complex Ω, with $K_\perp \neq 0$. We have $L = 15$ in normalized units. Note that $\omega = \Omega_{ci}$, i.e. $\Omega = 1$, is a resonance for all K_z when $K_\perp = 0$.

$$\Omega^2 = K_z^2 + iK_z + \frac{\Omega^2 K_\perp^2}{\Omega^2 - 1}.$$

The case with $\Omega_{ci} = 0$ gives a trivial generalization of (13.17). In the limit of $K_z = 0$, i.e. for horizontal propagation, the waves no longer grow, but are still influenced by the vertical density gradient. Results for the real part of the dispersion relation are shown in figure 13.3. The growth rate $\Im\{\Omega\}$ (not shown) follows the results shown in figure 13.1 with small modifications as $K_\perp$ increases: the largest relative temporal increase $\Im\{\omega\}/\Re\{\omega\}$ (for real $\mathbf{K}$) is found for long wavelengths and low frequencies.

The analysis outlined in the present section does not address imposed boundary conditions. For waves in a neutral atmosphere it is relatively easy to formulate relevant boundary conditions, such as 'no-slip' or similar. For the plasma ionosphere the boundary conditions are no longer obvious. Note though, that a restrictive class of boundary conditions at some $z = z_0$ can be realized by superimposing waves propagating up and down (positive and negative k_z) from z_0, the waves being either *in* phase or in *counter* phase.

Qualitatively, the results for a 'fake instability' apply more generally to waves propagating along density gradients generated by some external means (Doucet *et al* 1974, D'Angelo *et al* 1976) and also to pre-sheaths (Chen 1984, Pécseli 2016). Some of these results have been confirmed experimentally.

The analysis outlined in the present chapter is based on a fluid model. The consequences of modification introduced by a kinetic analysis can be argued heuristically for the case where waves are propagating along a density gradient parallel to the gravitational acceleration. For the long wavelength quasi-neutral limit we have approximately that the linear ion Landau damping varies as $\Im\{\omega\} \approx -\alpha k$ where α is some constant depending on the electron to ion temperature ratio. When the curve for $\Im\{\omega\}$ obtained by a kinetic analysis crosses the curve obtained by (13.17) for some k_0, we can argue that if such a crossing exists, the

Landau damping dominates for $k > k_0$ while the 'geometric growth' or 'fake instability' will dominate for $k < k_0$.

- **Exercise:** Take the solution of (13.17) for the boundary value problem with real ω and complex k_z to demonstrate that the vertical energy flux associated with the waves is constant for sufficiently large frequencies. Also, explain the result for low frequencies.

References

Brekhovskikh L and Goncharov V 1985 *Mechanics of Continua and Wave Dynamics (Series on Wave Phenomena)* (Berlin: Springer)

Chen F F 1984 *Introduction to Plasma Physics and Controlled Fusion* vol 1 2 edn (New York: Plenum)

D'Angelo N, Michelsen P and Pécseli H L 1976 Ion-acoustic-waves in a density gradient *Z. Naturforsch.* A **31** 578–82

Doucet H J, Jones W D and Alexeff I 1974 Linear ion acoustic waves in a density gradient *Phys. Fluids* **17** 1738–43

Dysthe K B, Misra K D and Trulsen J K 1975 On the linear cross-field instability problem *J. Plasma Phys.* **13** 249–57

Garcia O E and Pécseli H L 2013 Models for electrostatic drift waves with density variations along magnetic field lines *Geophys. Res. Lett.* **40** 5565–9

Hines C O 1960 Internal atmospheric gravity waves at ionospheric heights *Canadian J. Phys.* **38** 1441–81

Liu C H 1970 Long wavelength ion-acoustic waves in a magneto-plasma in a gravitational field *J. Plasma Phys.* **4** 617–27

Pannekoek A 1922 Ionisation in stellar atmospheres *Bull. Astronomical Inst. Netherlands* **1** 107–22

Parkinson D and Schindler K 1969 Landau damping of long wavelength ion acoustic waves in a collision-free plasma with a gravity field *J. Plasma Phys.* **3** 13–20

Pécseli H L 2016 Low frequency electrostatic waves propagating in plasmas with parameters varying along magnetic field lines *Plasma Sources Sci. Technology* **25** 015010

Rosseland S 1924 Electrical state of a star *Mon. Not. R. Astron. Soc.* **84** 720–9

Yeh K C and Liu C H 1972 *Theory of Ionospheric Waves* Vol 17 *of International Geophysics Series* (New York: Academic)

Part III

Linear kinetic models

IOP Publishing

Low Frequency Waves and Turbulence in Magnetized Laboratory Plasmas and in the Ionosphere

Hans Pécseli

Chapter 14

Kinetic models for electrostatic drift waves

The foregoing analysis was based on a fluid model, which is appropriate for most of the resistive drift waves observed in most laboratory experiments. In particular for ionospheric and magnetospheric conditions the collision frequency is negligibly small, and the fluid model will predict drift waves to be stable. In reality, the waves remain weakly unstable due to kinetic effects. Landau damping of the low velocity part of the electron population which has a **B**-parallel velocity component in resonance with the corresponding component of the drift wave phase velocity will act to inhibit the exact isothermal Boltzmann distribution resulting from the fluid analysis. As a consequence, the drift waves will remain weakly unstable also in this collisionless limit, but the physical origin of the instability is changed.

14.1 Drift kinetic equation

Since the relevant phenomena have characteristic frequencies much below the electron gyro-frequency we use a pseudo one-dimensional description for the electron dynamics, where the **B**-perpendicular velocity is identified with the $\mathbf{E} \times \mathbf{B}/B^2$-velocity. With a constant homogeneous magnetic field, the electron gyro-center Vlasov equation, appropriate for electrostatic drift waves in homogeneous magnetic fields, takes the form (Rudakov and Sagdeev 1960, 1961, Kadomtsev 1965)

$$\frac{\partial}{\partial t} f + \nabla_\perp \cdot (\mathbf{u}_{\mathbf{E}\times\mathbf{B}} f) + u_\parallel \frac{\partial}{\partial z} f + \frac{e}{m} \frac{\partial}{\partial z} \phi \frac{\partial}{\partial u_\parallel} f = 0, \tag{14.1}$$

with $\mathbf{u}_{\mathbf{E}\times\mathbf{B}} = \mathbf{E} \times \mathbf{B}/B^2 = -\nabla\phi \times \mathbf{B}/B^2$, and $f = f(z, \mathbf{r}_\perp, u_\parallel, \mu_m, t)$ is here the electron gyro-center distribution function, in terms of the magnetic moment $\mu_m \equiv \frac{1}{2} m u_\perp^2 / B$. Since μ_m is constant, within the relevant approximations, the magnetic moment does not appear explicitly in (14.1). In the present electrostatic approximation we have

doi:10.1088/978-0-7503-1251-6ch14

$\mathbf{B}$ = const. and there is no $\mu_m \nabla B$-force on the particles. (If the magnetic field is perturbed by Alfvén waves, for instance, then we have a different situation (Mjølhus and Wyller 1988).) With $\phi = 0$ we have any function $f_0 = f_0(x, u_\parallel, \mu_m)$ being a steady-state equilibrium solution to (14.1). Note that in the present collisionless model, we have no problems in defining a 'true' steady state; compare with chapter 6. Linearizing (14.1) with respect to f_0, and Fourier transforming with respect to time and the spatial z- and y-coordinates, we obtain a relation between electron density $n(x, k_y, k_z, \omega)$ and potential $\phi(x, k_y, k_z, \omega)$

$$\begin{aligned} n &\equiv \int_{-\infty}^{\infty} f(x, k_y, k_z, u_\parallel, \mu_m, \omega)\, \mathrm{d}u_\parallel\, \mathrm{d}\mu_m \\ &= e\phi \fint \left(\frac{k_z}{m} \frac{\partial}{\partial u_\parallel} f_0 - \frac{k_y}{m\omega_{ce}} \frac{\partial}{\partial x} f_0 \right) \frac{\mathrm{d}u_\parallel\, \mathrm{d}\mu_m}{\omega - k_z u_\parallel}, \end{aligned} \tag{14.2}$$

where ω_{ce} is introduced as the electron cyclotron frequency, while $\fint$ denotes the usual Landau integration contour around the singularity at $u_\parallel = \omega/k_z$. With the assumption of quasi-neutrality, equation (14.2) gives a relation between the plasma density and the electrostatic potential to replace (6.21). Note that when linearizing (14.1), there will appear a term containing f_0 originating from the second term; this is usually not so for the unmagnetized case, where the second and third terms read $\mathbf{u} \cdot \nabla f(\mathbf{r}, \mathbf{u}, t)$.

To obtain an explicit result we assume the undisturbed distribution function to be a product, $f_0 = G(\mu_m)F(x, u_\parallel)$, where F is taken to be a displaced Maxwellian with an x-dependent density

$$F(x, u_\parallel) = n_0(x) \sqrt{\frac{m}{2\pi\kappa T_e}}\, \mathrm{e}^{-\frac{1}{2}m(u_\parallel - U)^2/\kappa T_e},$$

corresponding to a plasma with an electron current density $J_e = -en_0 U$ along the magnetic field lines. With this particular choice we have $\partial F/\partial u_\parallel = -(m/\kappa T_e)(u_\parallel - U)F$ and $\partial F/\partial x = -\beta F$, again with $\beta \equiv -d \ln n_0(x)/\mathrm{d}x$. Properly normalized, the coefficient containing the distribution of magnetic moments $G(\mu_m)$ no longer appears after integration with respect to μ_m. A mechanism for generating magnetic field aligned currents was proposed in section 3.2.1.

It is now straightforward to express the integral in (14.2) in terms of the complex plasma dispersion function (Fried and Conte 1961)

$$Z(\zeta) = \frac{1}{\sqrt{\pi}} \fint_{-\infty}^{\infty} \frac{e^{-\gamma^2}}{\gamma - \zeta}\, \mathrm{d}\gamma \tag{14.3}$$

and its derivative

$$Z'(\zeta) = -\frac{2}{\sqrt{\pi}} \fint_{-\infty}^{\infty} \frac{\gamma e^{-\gamma^2}}{\gamma - \zeta}\, \mathrm{d}\gamma$$

where $Z'(\zeta) = -2(1 + \zeta Z(\zeta))$. We have the useful limit $Z(\zeta) \approx -2\zeta(1 - 2\zeta^2/3 + 4\zeta^4/15 - \cdots) + i\sqrt{\pi}\exp(-\zeta^2)$ for $|\zeta| \ll 1$, while the asymptotic series for $|\zeta| \gg 1$ is $Z(\zeta) \approx -\zeta^{-1}(1 + 1/(2\zeta^2) + 3/(4\zeta^4) + \cdots) + i\sigma\sqrt{\pi}\exp(-\zeta^2)$ with $\sigma = 0$ for $x > |y|^{-1}$, $\sigma = 1$ for $|x| < |y|^{-1}$, and $\sigma = 2$ for $x < -|y|^{-1}$, where $\zeta \equiv x + iy$.

We find the relation between the electron density and the electrostatic potential to be

$$\frac{n_e}{n_0} = \frac{e\phi}{\kappa T_e}\left[\frac{\omega - Uk_z - \omega^*}{k_z u_{T_e}} Z\left(\frac{\omega - Uk_z}{k_z u_{T_e}}\right) + 1\right], \tag{14.4}$$

where $u_{T_e} = \sqrt{2\kappa T_e/m}$ is the electron thermal velocity. The result (14.4) can also be used in models where we have no assumption of quasi-neutrality, so for the moment we retain the subscript $_e$ on the density.

By introducing the plasma dispersion function Z we have implicitly assumed that the electron velocity distribution function is a Maxwellian. For a collisional plasma, this choice is easy to argue, but for a collisionless case we might in principle have *any* velocity distribution. In the ionosphere, for instance, we often find a distribution that can be modeled by mixture of warm and cold Maxwellian distributions (Hasegawa 1975). The result (14.4) can be easily modified to include these cases as well by using a sum of weighted Z-functions. The density ratio of the two (or more) components is an important parameter for the plasma stability conditions.

14.2 Dispersion relations

From the foregoing analysis we found a relation between n_e/n_0 and ϕ based on the linearized electron dynamic equations. In order to obtain a dispersion relation for drift waves in kinetic descriptions, we now consider various models for the ion dynamics. The set equations will here be closed by assuming quasi-neutrality, but we could use Poisson's equation to also allow for inclusion of short wavelengths on the Debye length scale.

14.2.1 Simple model with $T_i = 0$

In the simplest model we assume $T_i = 0$ and let the **B**-perpendicular velocity be identified with the $\mathbf{E} \times \mathbf{B}/B^2$-velocity, hereby ignoring ion inertia. In this case $n/n_0 = (\omega^*/\omega)e\phi/\kappa T_e$, as shown before. The dispersion relation is obtained by eliminating n/n_0 by use of (14.4). With $\omega - Uk_z \ll k_z u_{T_e}$ we expand the plasma dispersion function for small arguments, where $Z(u) \approx i\sqrt{\pi}$ for $u \ll 1$. We find

$$1 - \frac{\omega^*}{\omega} + i\sqrt{\pi}\frac{\omega - Uk_z - \omega^*}{k_z u_{T_e}} = 0, \tag{14.5}$$

with the approximate solution

$$\omega = \omega^*\left(1 + i\sqrt{\pi}\frac{U}{u_{T_e}}\right). \tag{14.6}$$

The dispersion relation indicates that a current along the magnetic field lines is necessary to excite drift waves for collisionless plasmas. If we change sign of U, i.e. the direction of the current, it will be waves with negative values of k_z which become unstable. Although the present model is to some extent oversimplified, it demonstrates that in the long wavelength limit, kinetic drift waves are likely to be only weakly unstable in currentless thermal collisionless plasmas, i.e. $\mathfrak{I}\{\omega\} \ll \mathfrak{R}\{\omega\}$.

The present model can be generalized somewhat by retaining $\mathbf{E} \times \mathbf{B}/B^2$ for the $\mathbf{B}$-perpendicular ion velocity but using a kinetic ion equation similar to (14.1). A relation quite similar to (14.4) is then obtained for the ions, and a slightly modified dispersion relation results.

14.2.2 Finite ion inertia

The simple model in section 14.2.1 gave a vanishing growth rate in (14.6) for the drift waves in the case where $U = 0$, due to a cancellation of terms in the case where $\omega = \omega^*$. The model for the ion dynamics in section 14.2.1 was however simplified by ignoring for instance the ion polarization drift, which as already shown implies that the drift wave frequency will deviate slightly from ω^*. A somewhat more general result is based on the linearized ion continuity equation and the ion velocity $\mathbf{u}_{i\perp} = -\nabla_\perp \phi \times \mathbf{B}/B^2 - (M/eB^2)\partial \nabla_\perp \phi/\partial t$ and $\partial u_{iz}/\partial t = -(e/M)\partial\phi/\partial z$, see also chapter 5. Considering only waves propagating in the (y, z)-plane, we find the relation between perturbations in density and potential to be

$$\frac{n}{n_0} = \frac{\omega^*}{\omega}\frac{e\phi}{\kappa T_e} - C_s^2 \frac{e\phi}{\kappa T_e}\left(\frac{k_y^2}{\Omega_{ci}^2} - \frac{k_z^2}{\omega^2}\right). \tag{14.7}$$

The relation (14.7) was obtained solely by use of the equations governing the *ion* dynamics. The appearance of T_e is here purely formal due to the introduction of C_s and ω^*. By use of the electron relation (14.4), the resulting dispersion relation becomes

$$\frac{\omega^*}{\omega} - C_s^2\left(\frac{k_y^2}{\Omega_{ci}^2} - \frac{k_z^2}{\omega^2}\right) = 1 + i\sqrt{\pi}\frac{\omega - Uk_z - \omega^*}{k_z u_{T_e}}, \tag{14.8}$$

with the approximate solution

$$\omega = \frac{\omega^*}{1 + b}\left(1 + i\sqrt{\pi}\frac{\omega^* b + Uk_z(1 + b)}{(1 + b)^2 k_z u_{T_e}}\right), \tag{14.9}$$

where $b \equiv k_y^2 C_s^2/\Omega_{ci}^2 \equiv (k_y a_i)^2$ as before in (6.20), see also appendix A. When describing the ion dynamics we took the limit of $k_z \approx 0$ as most relevant for drift waves. With finite ion inertia included in the analysis, electrostatic drift waves retain a small growth rate induced by electron Landau damping even in currentless plasmas, $U = 0$.

In figure 14.1 we show an illustrative result for the growth rate of the drift instability including finite ion inertia, for a selected normalized parameter set. The result is obtained by numerical solution of (14.8). The real part of the frequency is only weakly affected by the instability, and the fast wave solution from (5.14) shown in figure 5.1 remains a good approximation.

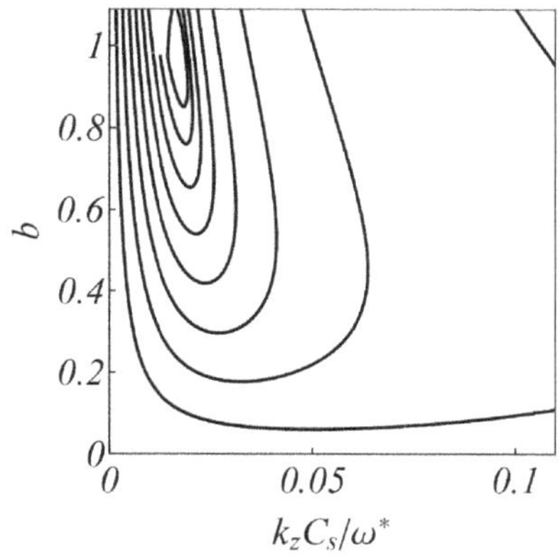

Figure 14.1. Normalized growth rate $\Im\{\omega/\omega^*\}$ for a current carrying plasma, with $u_{Te}/C_s = 0.02$ and $U/C_s = 1$. The contour separation is 0.04.

14.2.3 Warm ions with $(k_y r_L)^2 \ll 1$

The lowest order kinetic effects can be included in the ion description by taking into account the collisionless part of the ion viscosity tensor as in section 7.1. Also here we include the effect of gravity since it does not complicate the analysis to do so. Modifications in the electron dynamics by gravity are ignored with reference to the smallness of the electron mass. From (7.14) we have

$$\frac{n}{n_0} = -\left(\frac{e\phi}{\kappa T_e}\right)\frac{b(\omega - k_y U_0) - \omega^*}{\omega - k_y U_0 - \theta\omega^*}, \tag{14.10}$$

where $U_0 = U_{Di} - U_G$ as before. The dispersion relation becomes

$$-\frac{b(\omega - k_y U_0) - \omega^*}{\omega - k_y U_0 - \theta\omega^*} = 1 + i\sqrt{\pi}\frac{\omega - Uk_z - \omega^*}{k_z u_{T_e}}, \tag{14.11}$$

with the approximate solution

$$\begin{aligned}\omega = \omega^*\frac{1+\theta}{1+b} &+ k_y U_0 \\ + i\sqrt{\pi}\frac{\omega^*(1-\theta b)}{k_z u_{T_e}(1+b)^3}&\left\{\omega^*(b-\theta) + (1+b)\left(Uk_z - k_y U_0\right)\right\}.\end{aligned} \tag{14.12}$$

Re-introducing $U_0 = U_{Di} - U_G$ in the imaginary part, we find

$$\begin{aligned}\omega = \omega^*\frac{1+\theta}{1+b} &+ k_y U_0 \\ + i\sqrt{\pi}\frac{\omega^*(1-\theta b)}{k_z u_{T_e}(1+b)^3}&\left\{\omega^* b(1+\theta) + (1+b)\left(Uk_z + k_y g/\Omega_{ci}\right)\right\}.\end{aligned} \tag{14.13}$$

Depending on the direction of the gravitational force, it can for given k_z, k_y enhance or diminish an instability in current carrying plasmas with $U \neq 0$. When $U = 0$ and $g = 0$, we find that a finite ion temperature enhances the growth rate by a factor of approximately $(1 + \theta)$ as compared to (14.9), since $b\theta$ is usually small. Note that we cannot let $k_z \to \infty$, with the assumptions used in deriving (14.12).

Anomalous diffusion caused by current generated drift waves was studied by Matsumoto (1967) considering cylindrical symmetry. The result was consistent with the Bohm-scaling, albeit with a reduced numerical coefficient.

14.2.4 Kinetic ion model for $(k_y r_L)^2 \leqslant 1$

The limit with arbitrary ion Larmor radii requires a kinetic description of the ion dynamics. The basic equation is here the full ion Vlasov equation

$$\frac{\partial f}{\partial t} + \mathbf{u}_\perp \cdot \nabla_\perp f + u_\| \frac{\partial f}{\partial z} - \frac{e}{M}(\nabla_\perp \phi - \mathbf{u}_\perp \times \mathbf{B}) \cdot \nabla_{u_\perp} f - \frac{e}{M}\frac{\partial \phi}{\partial z}\frac{\partial f}{\partial u_\|} = 0. \qquad (14.14)$$

With $\phi = 0$, it is readily seen that any function $f_0 = f_0(\frac{1}{2}Mu^2, u_\|, x - u_y/\Omega_{ci})$ is an equilibrium solution of (14.14). An explicit $u_\|$ dependence will be ignored here. For the linearized problem, (14.14) is in general solved by integration along unperturbed orbits (Ichimaru 1973, Schmidt 1979).

For the present problem we simplify the analysis by assuming from the outset that only low frequency fluctuations are important, in agreement also with previous discussions. After some calculations we obtain a relation between perturbations in bulk plasma density and electrostatic potential as

$$\frac{n}{n_0} = -\frac{e\phi}{\kappa T_i}\left[\frac{\omega + \omega^*}{k_z u_{T_i}} I_0\left(\frac{1}{2}k_\perp^2 r_L^2\right)\exp\left(-\frac{1}{2}k_\perp^2 r_L^2\right) Z\left(\frac{\omega}{k_z u_{T_i}}\right) + 1\right], \qquad (14.15)$$

where Z is defined in (14.3), while I_0 is a Bessel function, and the ion Larmor radius is defined as $r_L \equiv \sqrt{2\kappa T_i/M\Omega_{ci}^2}$. (Usually the definition $r_L \equiv \sqrt{\kappa T_i/M\Omega_{ci}^2}$ is found in the literature.) Assuming quasi-neutrality, we equate the electron and ion densities to obtain the full kinetic dispersion relation as

$$\begin{aligned}\frac{\omega + \omega^*}{k_z u_{T_i}} I_0\left(\frac{1}{2}k_\perp^2 r_L^2\right)\exp\left(-\frac{1}{2}k_\perp^2 r_L^2\right) Z\left(\frac{\omega}{k_z u_{T_i}}\right)&\\ + \frac{\omega - Uk_z - \omega^*}{k_z u_{T_e}} Z\left(\frac{\omega - Uk_z}{k_z u_{T_e}}\right) &= 2,\end{aligned} \qquad (14.16)$$

which can only be solved numerically, in a general case. Here we assume again $U_{T_i} \ll \omega/k_z \ll U_{T_e}$, and use $Z(\zeta) \sim 1/\zeta$ for $|\zeta| \to \infty$ in the ion contribution to obtain

$$\left(1 + \frac{\omega^*}{\omega}\right) I_0\left(\frac{1}{2}k_\perp^2 r_L^2\right)\exp\left(-\frac{1}{2}k_\perp^2 r_L^2\right) + i\sqrt{\pi}\frac{\omega - Uk_z - \omega^*}{k_z u_{T_e}} = 2. \qquad (14.17)$$

Introducing the notation $\Lambda \equiv I_0(\frac{1}{2}k_\perp^2 r_L^2)\exp(-\frac{1}{2}k_\perp^2 r_L^2)$, in (14.17) we find the dispersion relation for $U = 0$ as

$$\omega = \frac{\omega^* \Lambda}{2 - \Lambda}\left(1 + i2\sqrt{\pi}\,\frac{\omega^*}{k_z u_{T_e}}\frac{\Lambda(1 - \Lambda)}{2 - \Lambda}\right). \tag{14.18}$$

The real and imaginary parts of this relation are shown in figure 14.2 as functions of $k_y r_L$, taking the y-component of $r_\perp$. Within the present approximations, the maximum growth rate of the kinetic drift instability is approximately $0.5\omega^{*2}/k_z U_{T_e}$, obtained for $k_y/\Omega_{ci} \approx 0.7u_{T_i}$, see figure 14.2.

For $k_\perp^2 r_L^2 \ll 1$ we can use $\Lambda \approx 1 - k_\perp^2 r_L^2$, and obtain in particular also the results from section 14.2.1. The results in this section emphasize the effects of finite ion Larmor radii for the instability growth rate, while ion polarization effects are ignored. These can be included as well, giving rise to non-trivial complications of the final results.

With $\Re\{\omega\} \approx \omega^*$ and the assumption $\omega/k_z \ll U_{Te}$, we have $\Im\{\omega\} \ll \Re\{\omega\}$, i.e. kinetic drift instabilities in currentless plasmas are weak. Drift waves in collisionless plasma have been reported (Little and Middleton 1969), albeit with externally imposed radial electric fields that enhance the instability. Laboratory experiments demonstrating drift instabilities with currents along magnetic field lines (Hatakeyama *et al* 1980, 2011) achieved substantial fluctuation levels but used large current densities.

By introducing the Z-function, the present analysis also assumes a Maxwellian velocity distribution, this time for the ions. Cases where several ion species are present can be included by a weighted sum of Z-functions, each with its own thermal velocity U_{T_i}. More complicated velocity distribution, a loss-cone distribution for instance, can be modeled by addition and subtraction of Z-functions with different arguments and weight factors, ensuring that overall charge neutrality is maintained.

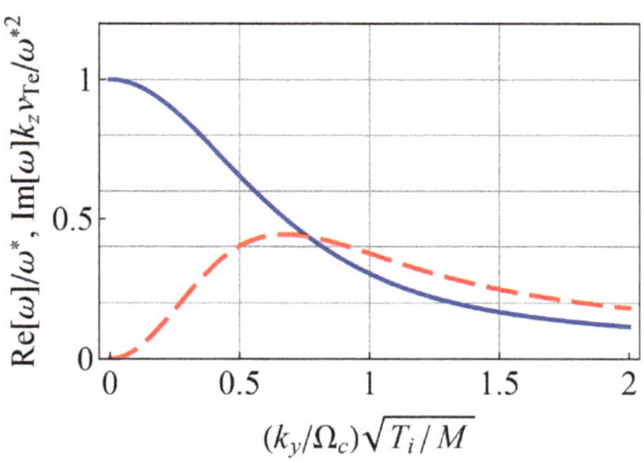

Figure 14.2. Dispersion relation for electrostatic drift waves with finite ion Larmor radii, r_L, for normalized variables with $\omega^* = k_y U_{De}$ being the drift frequency. Note that different normalizations are used for the real and imaginary parts of the frequency. With the present approximation k_z enters only through the normalization of $\Im\{\omega\}$.

- **Exercise:** Derive the maximum growth rate for kinetic drift instabilities in current carrying plasmas, where the electron velocity distribution function is a displaced Maxwellian.

14.2.5 High frequencies, $\omega > \Omega_{ci}$

Drift waves can persist at high frequencies, $\omega > \Omega_{ci}$, where the ions can no longer be assumed to be fully magnetized. Heuristically, we can argue that the instability requires that the phase velocity of the drift wave is close to the phase velocity of the ion cyclotron waves with dispersion relation $\omega^2 \approx \Omega_{ci}^2 + C_s^2 k^2$. The two dispersion relations are thus crossing when $\omega^*/k_y \sim \Omega_{ci}/k_y > C_s$. Since the ions are unmagnetized in this model, we can have ion sound waves propagating in the direction almost perpendicular to the magnetic field. The requirement given before can be expressed as $\Omega_{ci} \sim \omega^* = -k_y a_i^2 \Omega_{ci} n_0'/n_0$ or $k_y a_i^2 n_0'/n_0 \sim 1$ which means that the scale-length for the density gradient has to be comparable to the ion Larmor radius. This is a rather strong requirement of steep density gradients, but the conditions can be realized in for instance a laboratory experiment or numerical simulations.

A detailed theoretical analysis has to be based on kinetic models for ions and electrons taking into account the final ion Larmor radius effects (FLR) as found in the literature (Mikhailovskii and Timofeev 1963). Assuming here a simplified model with cold ions where we ignore the ion magnetization altogether, we have from the ion momentum and continuity equations

$$\frac{n}{n_0} = \frac{ek^2}{M\omega^2}\phi. \tag{14.19}$$

With the assumption of quasi-neutrality we find by use of the kinetic electron equations

$$\frac{\kappa T_e k^2}{M\omega^2} + \frac{\omega - \omega^*}{u_{Te} k_z} Z\left(\frac{\omega}{k_z u_{Te}}\right) = 1. \tag{14.20}$$

For oscillations with phase velocity $\omega/k_z \ll u_{T_e}$ we have a simpler expression

$$\frac{C_s^2 k^2}{\omega^2} - i\sqrt{\pi}\frac{\omega - \omega^*}{u_{Te} k_z} = 1. \tag{14.21}$$

For oscillations with $\omega \approx kC_s$ we find a growth rate

$$\gamma \approx \sqrt{\pi}(\omega^* - \omega)\frac{k}{k_z}\sqrt{\frac{m}{M}}. \tag{14.22}$$

In order to have an unstable condition, we require $\omega < \omega^*$, implying that the electron diamagnetic drift velocity is higher than the speed of sound. Note that again we have an instability which disappears by the assumption of isothermally Boltzmann distributed electrons.

14.2.6 Ion cyclotron drift instability

In the foregoing section 14.2.5 we assumed $\omega > \Omega_{ci}$. This requirement does not exclude the possibility that the wave frequency is close to one of the harmonics of the ion cyclotron frequency, and in a kinetic theory with warm ions (but not in a standard *fluid* theory) we expect strong modification of the wave dispersion relation when $\omega \approx N\Omega_{ci}$ with $N = 2, 3, \ldots$. These resonances occur at branches of electrostatic ion Bernstein modes (Hasegawa 1975). Also, we might want to consider the case where $\omega \approx \Omega_{ci}$, in which the ions are magnetized, but their velocity component in the direction $\perp\mathbf{B}$ can no longer be described by a simple $\mathbf{E} \times \mathbf{B}/B^2$-drift, not even approximately. The results of section 14.2.5 will, however, remain applicable as an approximation provided $\omega \neq N\Omega_{ci}$ for any integer value of N.

An ion cyclotron drift instability has been identified in a current carrying Q-machine plasma (Hendel and Yamada 1974). The current is enhancing an otherwise small linear growth rate.

References

Fried B D and Conte S D 1961 *The Plasma Dispersion Function* (New York: Academic)

Hasegawa A 1975 *Plasma Instabilities and Nonlinear Effects* Vol 8 *of Physics and Chemistry in Space* (Berlin: Springer)

Hatakeyama R, Moon C, Tamura S and Kaneko T 2011 Collisionless drift waves ranging from current-driven, shear-modified, and electron-temperature-gradient modes *Contrib. Plasma Phys.* **51** 537–45

Hatakeyama R, Oertl M, Märk E and Schrittwieser R 1980 Collisionless drift instability and ion heating in a current-carrying inhomogeneous plasma *Phys. Fluids* **23** 1774–81

Hendel H W and Yamada M 1974 Identification of ion-cyclotron drift instability with discrete and continuous spectra *Phys. Rev. Lett.* **33** 1076–9

Ichimaru S 1973 *Basic Principles of Plasma Physics* (Boulder, CA: Westview Press)

Kadomtsev B B 1965 *Plasma Turbulence* (New York: Academic)

Little P F and Middleton C R 1969 Noise in collision-free alkali metal plasmas *Nucl. Fusion* **9** 67–75

Matsumoto M 1967 Anomalous diffusion arising from drift instability in a cylindrical plasma *J. Phys. Soc. Japan* **22** 292–300

Mikhailovskii A B and Timofeev A V 1963 Theory of cyclotron instability in a non-uniform plasma *Sov. Phys. JETP* **17** 626–7

Mikhailovskii A B and Timofeev A V 1963 *Russian original J. Exptl. Theoret. Phys.* **44** 919–21

Mjølhus E and Wyller J 1988 Nonlinear Alfvén waves in a finite beta plasma *J. Plasma Phys.* **40** 299–318

Rudakov L I and Sagdeev R Z 1960 Oscillations of an inhomogeneous plasma in a magnetic field *Sov. Phys. JETP* **10** 952–4

Rudakov L I and Sagdeev R Z 1959 *Russian original J. Exptl. Phys. (USSR)* **37** 1337–41

Rudakov L I and Sagdeev R Z 1961 On the instability of a nonuniform rarefied plasma in a strong magnetic field *Sov. Phys. Doklady* **6** 415–7

Rudakov L I and Sagdeev R Z 1961 *Russian original Dokl. Akadem. Nauk. SSR* **138** 581–3

Schmidt G 1979 *Physics of High Temperature Plasmas* (New York: Academic)

Part IV

Linear drift Alfvén waves

Chapter 15

Electromagnetic modes

The discussions in the previous chapters considered electrostatic wave modes, drift waves in particular. For the foregoing analysis, the basic inequality for the ion thermal velocity U_{thi}, the **B**-parallel drift wave phase velocity and the electron thermal velocity U_{thi} was

$$U_{thi} \ll \frac{\omega}{k_{\parallel}} \ll U_{the} \ll V_A.$$

When the β-value of the plasma exceeds the electron to ion mass ratio, m/M, we find that we can have $\omega/k_{\parallel} \approx V_A$ and the drift waves can couple to Alfvénic modes. In this limit we can no longer ignore fluctuations in the magnetic field (Kadomtsev 1965, Hasegawa 1975, Weiland 1992, 2000). To see this we rewrite the plasma β-value as

$$\beta \equiv \frac{n\kappa T}{B^2/2\,\mu_0} = \frac{\rho\kappa T}{MB^2/2\,\mu_0} \approx \frac{C_s^2}{V_A^2} \approx \frac{U_{the}^2}{V_A^2}\frac{m}{M},$$

or $U_{the}/V_A = \sqrt{\beta M/m}$, being cavalier with factors of 2. To have $U_{the} \geqslant V_A$ we require $\beta \geqslant m/M$. The generalizations of the analysis taking these effects into consideration will be approached in this section. One basic problem, deserves, however, to be addressed from the outset; since the Alfvén wave velocity depends on density as well as magnetic field intensity, it will in an inhomogeneous plasma depend on the spatial coordinate. This is in contrast to the ion sound speed which is constant in the quasi-neutral limit, even when the distribution of plasma density is inhomogeneous, provided the electron and ion temperatures are constant. For the following analysis, we assume that the dominant effect comes from the density variation and consider the variation in the average magnetic field induced for instance by the diamagnetic drifts to be negligible, i.e. $\mathbf{B}_0 = \text{const}$. The basic linearized magnetohydrodynamic (MHD) equations for illustrating the effects mentioned here are taken to be

doi:10.1088/978-0-7503-1251-6ch15

$$\nabla \times \mathbf{E}(\mathbf{r}, t) = -\frac{\partial}{\partial t}\mathbf{B}(\mathbf{r}, t) \tag{15.1}$$

$$\mathbf{E}(\mathbf{r}, t) = -\mathbf{u}(\mathbf{r}, t) \times \mathbf{B}_0 \tag{15.2}$$

$$\rho_0(\mathbf{r})\frac{\partial}{\partial t}\mathbf{u}(\mathbf{r}, t) = \frac{1}{\mu_0}\Big(\nabla \times \mathbf{B}(\mathbf{r}, t)\Big) \times \mathbf{B}_0, \tag{15.3}$$

where we assumed the plasma to be an ideal conductor in (15.2), and neglected the pressure term in the MHD-force equation (15.3). We find

$$\frac{\partial^2}{\partial t^2}\mathbf{B}(\mathbf{r}, t) = \frac{1}{\mu_0}\nabla \times \left[\frac{1}{\rho_0(\mathbf{r})}\Big([\nabla \times \mathbf{B}(\mathbf{r}, t)] \times \mathbf{B}_0\Big) \times \mathbf{B}_0\right]. \tag{15.4}$$

Assuming a harmonic time variation with frequency ω, and $\rho_0 = \rho_0(x)$ with $\mathbf{B}_0$ being in the z-direction, we have

$$\omega^2\mathbf{B}(\mathbf{r}, \omega) = -\frac{1}{\mu_0}\nabla \times \left[\frac{1}{\rho_0(x)}\Big([\nabla \times \mathbf{B}(\mathbf{r}, \omega)] \times \mathbf{B}_0\Big) \times \mathbf{B}_0\right]. \tag{15.5}$$

It is a simple matter to calculate the vector operations involved in (15.5)

$$\begin{aligned}&[(\nabla \times \mathbf{B}) \times \mathbf{B}_0] \times \mathbf{B}_0 \\ &= -B_0^2\left(\frac{\partial}{\partial y}B_z - \frac{\partial}{\partial z}B_y\right)\hat{x} - B_0^2\left(\frac{\partial}{\partial z}B_x - \frac{\partial}{\partial x}B_z\right)\hat{y}.\end{aligned} \tag{15.6}$$

Using this result we find

$$\begin{aligned}&\nabla \times \left[\frac{1}{\rho_0(x)}[(\nabla \times \mathbf{B}) \times B_0] \times \mathbf{B}_0\right] \\ &= \frac{B_0^2}{\rho_0(x)}\left(\frac{\partial}{\partial z}\left(\frac{\partial}{\partial z}B_x\frac{\partial}{\partial x} - B_z\right)\hat{x} - \frac{\partial}{\partial z}\left(\frac{\partial}{\partial y}B_z - \frac{\partial}{\partial z}B_y\right)\hat{y}\right) \\ &- B_0^2\left(\frac{\partial}{\partial x}\left[\frac{1}{\rho_0(x)}\left(\frac{\partial}{\partial z}B_x - \frac{\partial}{\partial x}B_z\right)\right] - \frac{1}{\rho_0(x)}\frac{\partial}{\partial y}\left(\frac{\partial}{\partial y}B_z - \frac{\partial}{\partial z}B_y\right)\right)\hat{z}.\end{aligned} \tag{15.7}$$

In addition we now assume that the magnetic field fluctuations (i.e. the part superimposed on the steady state value $\mathbf{B}_0$) can be written as a linearly polarized wave.

We have two options; let the magnetic field perturbation be along equi-density contours, or alternatively in the perpendicular direction. Consider first the former case where

$$\mathbf{B} \equiv \left\{0,\ B_y(x)\exp(\mathrm{i}k_z z),\ B_z(x)\mathrm{e}^{\mathrm{i}k_z z}\right\}.$$

We omit, for simplicity, the index k_z on $B_y(x)$ and $B_z(x)$. Inserting this in (15.7) we find from the x-component that $B_z = 0$. With this result, the z-component of the equations is also satisfied. The y-component gives

$$\mu_0 \omega^2 B_y = \frac{B_0^2}{\rho_0(x)} k_z^2 B_y,$$

which has non-trivial solutions only if $\omega = \omega(x)$; each field line carries oscillations with its own frequency, determined by the local plasma mass density and B_0, i.e. $\omega^2(x) = V_A^2(x) k_z^2$ given by the local Alfvén velocity.

As the second alternative for the linearized polarization we have

$$\mathbf{B} \equiv \{B_x(x)\mathrm{e}^{\mathrm{i}k_z z}, 0, B_z(x)\mathrm{e}^{\mathrm{i}k_z z}\}.$$

We omit, for simplicity, also here the index k_z on $B_x(x)$ and $B_z(x)$. With a little work, using (15.5) and (15.7), we can obtain

$$\frac{\mathrm{d}}{\mathrm{d}x} B_x(x) + ik_z B_z(x) = 0 \tag{15.8}$$

and

$$\frac{\mathrm{d}^2}{\mathrm{d}x^2} B_x(x) + \frac{\rho_0'(x)}{\rho_0(x)} \frac{\mathrm{d}}{\mathrm{d}x} B_x(x) - B_x(x)\left(k_z^2 - \omega^2 \rho_0(x) \frac{\mu_0}{B_0^2}\right) = 0, \tag{15.9}$$

with $\rho_0' \equiv \mathrm{d}\rho_0(x)/\mathrm{d}x$. From (15.9) we determine $B_x(x)$ and subsequently find $B_z(x)$ from (15.8).

We can write what turns out to be a generic form determining $B_x(x)$, valid also for the case $B_0 = B_0(x)$

$$\frac{\mathrm{d}}{\mathrm{d}x} F(x) \frac{\mathrm{d}}{\mathrm{d}x} B_{k_z,\omega}(x) - k_z^2 F(x) B_{k_z,\omega}(x) = 0, \tag{15.10}$$

where for the present case $F(x) = \rho_0(x)\omega^2 - B_0^2 k_z^2/\mu_0$. If we had taken $\rho_0 = \text{const.}$ and $B_0 = B_0(x)$ we could still use (15.10), now with $F(x) = \omega^2 - \Omega_A^2(x)$ where $\Omega_A^2(x) \equiv k_z^2 V_A^2(x)$. It is interesting that we find an expression quite similar to (15.10) also in the analysis of Langmuir oscillations in inhomogeneous unmagnetized plasmas (Barston 1964). It can be demonstrated (the proof is left to the reader as an exercise in literature study) that (15.10) has a physically acceptable solution only if $F(x = x_0) = 0$ for some position, x_0, corresponding to a 'resonance', $\omega = \Omega_A$. It can furthermore be demonstrated (Barston 1964, Uberoi 1972, Heyvaerts and Priest 1983) that in this case the function $F(x)$ can be written in polynomial form

$$F(x) = \sum_{n=K}^{\infty} (x - x_0)^n f_n, \tag{15.11}$$

with x_0 being the resonance point where ω equals the local Alfvén frequency for given k_z, we can write the solution as

$$
\begin{aligned}
B_{k,\omega}(x) &= A_1 W(x) + B_1\Big[W(x)\ln|x - x_0| + (x - x_0)^\alpha V(x)\Big] \\
&\quad \text{for } x > x_0, \\
B_{k,\omega}(x) &= A_2 W(x) + B_2\Big[W(x)\ln|x - x_0| + (x - x_0)^\alpha V(x)\Big] \\
&\quad \text{for } x < x_0,
\end{aligned}
\tag{15.12}
$$

where A_1, A_2, B_1 and B_2 are constants, $W(x)$ and $V(x)$ are analytical in the neighborhood of x_0 and, in particular, $W(x_0) = 1$. The solution diverges logarithmically at $x = x_0$. For $K = 1$ we have $\alpha = 1$ in (15.12), while for $K > 1$ it is $\alpha = 1 - K$. It is important to note that the singularity is *integrable*. By a linear superposition of such solutions we can have

$$B_k(x, t) = \int Q_k(\omega) B_{k,\omega}(x) \exp(-\mathrm{i}\omega t)\mathrm{d}\omega, \tag{15.13}$$

where $Q_k(\omega)$ is a weight function for the individual solutions (15.12). The solution (15.13) is regular, but it also contains a continuous spectrum of phase velocities ω/k_z. The solution will in general be damped with time by a sort of phase mixing, which is reminiscent of Landau damping. Typically (Barston 1964), we find an asymptotic damping $\sim 1/t$.

It is important to emphasize that the existence of the Alfvén continuum is not by itself anything extraordinary; also a simple waveguide supports an infinity of phase velocities. For that case the allowed frequencies form a discrete set, but this is only because of the idealized boundary conditions usually applied there. The important thing is that here a singular mode corresponds to a definite phase velocity. In order to obtain a 'well behaved' magnetic field variation we have to address a *band* of phase velocities. It can be a narrow band, but it will necessarily have a finite width, which then gives rise to a damping caused by phase mixing as already mentioned, at least within the ideal MHD description used here. There is a non-trivial analogy to the description of the time evolution of plasma waves by a superposition of so called *van Kampen modes* in kinetic plasma theory (van Kampen and Felderhof 1967).

As an *approximation*, we can assume a narrow distribution for $Q_k(\omega)$ in (15.13) and associate the resulting wavefield with one well defined frequency, and ignore the temporal damping. The waves will be *local*, and propagate with the Alfvén speed corresponding to the density and magnetic fields at the position selected by x_0 in (15.12). The results will be valid in a finite (but possibly very long) time interval. Ultimately, the requirement of finite wave amplitudes necessarily impose a finite width of $Q_k(\omega)$ in (15.13), and the asymptotic evolution will be dominated by a slow damping.

It has at times been quoted as a sort of paradox that no dispersion relation exists for this problem. In reality there is (of course) nothing paradoxical in this situation, first of all because ω/k_z is a phase velocity relating only to one particular component of the wave vector. It is only seldom we would obtain a unique dispersion relation by such a procedure. More important is that a dispersion relation is, in general, obtained for homogeneous media, and we are not dealing with these here. The most simple inhomogeneous medium we can imagine is presumably a wave guide; it is vacuum inside and an ideal conductor outside. Such a waveguide does not have any single general dispersion relation either, since an imposed frequency can find many wavenumber-responses; we find a unique dispersion relation only if a certain mode structure is selected. Several similar cases can be found in plasma physics (Trivelpiece and Gould 1959, Guio *et al* 2001).

The difference in the results for the two polarizations of the magnetic field evidently stems from the fact that in the case where $B \perp \nabla n_0(x)$ there are no fluctuations in plasma density associated with the Alfvén waves; within the MHD description the plasma motion is along equi-density contours. The case is exactly opposite for the other (linear) polarization. Here the wave gives rise to a 'sloshing' of plasma in the direction along $\nabla n_0(x)$.

The phenomena discussed in this section are particularly relevant for solar conditions (Roberts 2000), where we have an inhomogeneous high β-plasma in the coronal loops, where large amplitude oscillations are often observed.

- **Exercise:** In the foregoing discussion we implicitly assumed that spatial variations in the background density and magnetic field could be discussed independently. If the finite plasma temperature is taken into account this is no longer correct since the magnetic and plasma pressures are related. Generalize the basic equation (15.9) with this taken into account.

15.1 Simplified linear theory

In order to account for the coupling of drift waves to Alfvénic waves, we relax the electrostatic wave approximation. In the general case an electric field can be expressed as a sum of two terms, one being the gradient of a scalar potential and the second the time derivative of the vector potential. We postulate here a two-potential approximation Kadomtsev 1965, where the electric field is expressed as $\mathbf{E} = -\nabla_\perp \phi - \hat{\mathbf{z}}\partial\psi/\partial z$. To understand the physical implication of this approximation we use the Maxwell equation $\nabla \times \mathbf{E} = -\partial \mathbf{B}/\partial t$ for a plane wave $\exp(-i\omega t + i\mathbf{k}\cdot\mathbf{r})$ to find

$$\mathbf{B} \equiv \{B_x, B_y, B_z\} = \{(i/\omega)k_y k_z(\phi - \psi), (i/\omega)k_x k_z(\psi - \phi), B_0\}, \tag{15.14}$$

with the unperturbed magnetic field being in the z-direction also here. This result shows that the two-potential approximation accounts for the *turning* of magnetic field lines, but does it in a way which keeps their z-components constant. Since deviations from the magnetic field intensity will be of second order in perturbed quantities, we can argue that the approximation allows bending of magnetic field lines without accounting for changes in field strength due to compression of the field, to the given accuracy. Magnetosonic waves are thus lost in the two-potential approximation, and the analysis retains only the shear Alfvén waves.

With $V_A \equiv B_0/\sqrt{nM\mu_0} \equiv c\Omega_c/\omega_{pi}$ being the local Alfvén velocity in the sense discussed before, the basic inequality for the drift-Alfvén waves is

$$U_{thi} \ll \omega/k_z \leqslant V_A \leqslant U_{the}, \tag{15.15}$$

replacing the electrostatic equivalent (4.2). In describing the electron dynamics we again assume that the electrons flow along magnetic field lines to maintain a local Boltzmann equilibrium, taking into account that now the direction of the magnetic field is fluctuating.

We now choose the polarization for the electric field used in the electrostatic analysis, i.e. assume that the dominant electric field is in the $\hat{\mathbf{y}}$-direction. In this case the perturbations of the magnetic field are in the $\hat{\mathbf{x}}$-direction, according to (15.14) and the fluctuating plasma velocity associated with the wave motion is along the bulk density gradient. The electron pressure variation at a given position has then also a contribution from the large scale density gradient which now has a small component along the *perturbed* magnetic field, even when $\nabla n_0 \perp \mathbf{B}_0$. For a linearly polarized plane wave perturbation, the result is

$$\kappa T_e\left(ik_z n + \frac{B_x}{B_0}\frac{\mathrm{d}n_0}{\mathrm{d}x}\right) = ik_z e n_0 \psi. \tag{15.16}$$

Physically, the second term in (15.16) accounts for the electrons reaching equilibrium by flowing along the perturbed magnetic field lines; mathematically, its origin is the z-component of the $\mathbf{u}_{e0} \times \mathbf{B}_1$-term in the linearized electron momentum equation with $\mathbf{u}_{e0} = -\hat{\mathbf{y}}(\kappa T_e/eB_0)n_0'/n_0$ being the electron gradient drift.

Using the x-component of (15.14) in (15.16), we find

$$\frac{n}{n_0} = \frac{e\phi}{\kappa T_e} + \frac{e}{\kappa T_e}(\psi - \phi)\left(1 - \frac{\omega^*}{\omega}\right), \tag{15.17}$$

where the drift frequency ω^* was again introduced. For the electrostatic case we have $\psi = \phi$, and then the standard result for Boltzmann distributed electrons in an electrostatic potential is trivially obtained from (15.17).

In order to obtain a relation between ϕ and ψ we first use the relation $\nabla \times \nabla \times \mathbf{E} = -\mu_0 \partial \mathbf{j}/\partial t$, where the contribution $\mu_0 \varepsilon_0 \partial^2 \mathbf{E}/\partial t^2$ from Maxwell's displacement current is ignored, in agreement with the standard description of low frequency Alfvén waves. The result is

$$j_z = \frac{1}{\mu_0 \omega} k_z k_\perp^2 (\phi - \psi). \tag{15.18}$$

To determine j_z we note that to the present quasi-neutral approximation we have $\nabla \cdot \mathbf{j} = 0$, implying $j_z = i(e/k_z)\nabla_\perp \cdot [n(\mathbf{u}_{i\perp} - \mathbf{u}_{e\perp})]$, in a self-evident notation for ion and electron velocity components. Boltzmann distributed electrons will not contribute to the B_0-perpendicular component of the linearized plasma current. The ions contribute here with the ion polarization current $(Mn_0(x)/B_0^2)\partial \nabla_\perp \phi/\partial t$ and with $-en_0(x)\nabla\phi \times B_0/B_0^2$. Calculating the divergence of the B_0-perpendicular ion current and using (15.18) we find

$$j_z = -\frac{n_0 M}{B^2}\frac{k_\perp^2}{k_z}(\omega + \omega^*)\phi, \tag{15.19}$$

where again the drift frequency was introduced. Using (15.18), the following relation between ψ and ϕ is found

$$\psi - \phi = -\phi\frac{\omega(\omega + \omega^*)}{k_z^2 V_A^2}. \tag{15.20}$$

We note that $V_A \to \infty$ implies $\psi = \phi$, i.e. the electrostatic limit is recovered as expected. Using (15.20) together with (15.17), we obtain a relation between the perturbations in density and the electrostatic potential

$$\frac{n}{n_0} = \left(1 - \frac{\omega^2 - \omega^{*2}}{k_z^2 V_A^2}\right)\frac{e\phi}{\kappa T_e}. \tag{15.21}$$

A simple illustrative result is obtained by now approximating the $\mathbf{B}_0$-perpendicular ion velocity with $-\nabla\phi \times \mathbf{B}_0/B_0^2$. With this approximation inserted into the linearized ion continuity equation we have

$$\frac{n}{n_0} = \frac{\omega^*}{\omega}\frac{e\phi}{\kappa T_e}, \tag{15.22}$$

where the electron temperature appears explicitly only because ω^* was introduced. Using again the assumption of quasi-neutral fluctuations, there is no need to distinguish the electron and the ion density, and we can combine the equations (15.21) and (15.22) to find the simplified dispersion relation

$$(\omega - \omega^*)\left(\omega^2 + \omega\omega^* - V_A^2 k_z^2\right) = 0, \tag{15.23}$$

which splits-up into two parts, the electrostatic drift waves with $\omega = \omega^*$ we had before, and a relation

$$\omega^2 + \omega\omega^* - V_A^2 k_z^2 = 0 \tag{15.24}$$

for Alfvén waves in the presence of a density gradient. This relation has two solutions; an accelerated Alfvén wave with $|\omega/k_z| > V_A$, and a decelerated Alfvén wave where $|\omega/k_z| < V_A$, the two waves propagating in opposite directions. Compare with the results of the simplified theory for the electrostatic drift waves, where accelerated and decelerated ion sound waves were found.

With the assumption of $\mathbf{E} \times \mathbf{B}$-drifting ions in (15.22), we are within the limits of ideal MHD, and expect that the electric field associated with the Alfvén waves only has a $\mathbf{B}$-perpendicular component. Using (15.14) we find $\mathbf{E} \cdot \mathbf{B} = -ik_z\psi B_0$. However, the dispersion relation (15.24) implies $\psi = 0$ when inserted into (15.20), giving the consistent result $\mathbf{E} \cdot \mathbf{B} = 0$.

- **Exercise:** Discuss the case where the perturbed electric field polarization is in the $\hat{\mathbf{x}}$-direction. Consider the case where ion inertia (i.e. the ion polarization drift) is ignored, as well as the case where it is retained in the analysis. Do you find any essential differences?

- **Exercise:** Demonstrate that alternative choices of potentials can give the same results as those obtained in this section. Choose $\mathbf{E} = -\nabla\phi - (\partial\psi/\partial t)\hat{\mathbf{z}}$, which gives $\mathbf{B}_\perp = \hat{z} \times \nabla\psi$ and also $j_\parallel = \frac{1}{\mu_0}\nabla^2\psi$ by $\nabla \times \mathbf{B} = \mu_0\mathbf{j}$.

The simplification of the result (15.23) was obtained at the expense of a slight inconsistency in the sense that the ion polarization drift was retained only in deriving (15.19), but not in (15.22). A somewhat more general result is based on the linearized ion continuity equation and the ion velocity $\mathbf{u}_{i\perp} = -\nabla_\perp\phi \times \mathbf{B}/B^2 - (M/eB^2)\partial\nabla_\perp\phi/\partial t$ and $\partial u_{iz}/\partial t = -(e/M)\partial\psi/\partial z$. With ion inertia included the results will also have validity outside ideal MHD, and we can expect a **B**-parallel component of the electric field, $\mathbf{E} \cdot \mathbf{B} \neq 0$, in general. In this limit a non-trivial particle acceleration along **B** might be associated with Alfvén waves Hasegawa 1975.

Considering only waves propagating in the (y, z)-plane, we find the relation between perturbations in density and potential to be

$$\begin{aligned}\frac{n}{n_0} &= \frac{\omega^*}{\omega}\frac{e\phi}{\kappa T_e} - \frac{k_y^2}{\Omega_c^2}C_s^2\frac{e\phi}{\kappa T_e}\\ &+ C_s^2\frac{k_z^2}{\omega^2}\frac{e\phi}{\kappa T_e}\left(1 - \frac{\omega(\omega + \omega^*)}{k_z^2V_A^2}\right).\end{aligned} \tag{15.25}$$

The relation (15.25) was obtained solely by use of the equations governing the *ion* dynamics. The appearance of T_e is again purely formal due to the introduction of C_s and ω^*. By use of the electron relation (15.21), the resulting dispersion relation becomes

$$\begin{aligned}(\omega - \omega^*)\left(\omega^2 + \omega\omega^* - V_A^2k_z^2\right) &= \frac{\omega}{\Omega_c^2}k_z^2k_y^2C_s^2V_A^2\\ &- \frac{1}{\omega}C_s^2k_z^2\left(k_z^2V_A^2 - \omega(\omega + \omega^*)\right).\end{aligned} \tag{15.26}$$

The non-vanishing right-hand side of (15.26) couples the dispersion relations on the left side of (15.23).

The **B**-parallel ion dynamics is usually ignored, and we obtain a simplified dispersion relation, where the last term in (15.26) is omitted. Hereby we effectively ignore ion sound waves along **B**, i.e. set $V_A \gg C_s$. For the same reason we will have no backward traveling (or 'slow') sound waves in this simplification. In the present limit we have only one parameter in the normalized dispersion relation, i.e. $k_y^2C_s^2/\Omega_c^2$.

For completeness we give here the full solution of (15.26) as shown in figure 15.1. This full problem contains two parameters $k_y^2C_s^2/\Omega_c^2$ and C_s^2/V_a^2. For intermediate k_zV_A/ω^*, we find that the intermediate branch in figure 15.1 transforms into the electrostatic drift wave. For large k_zV_A/ω^* we find accelerated and decelerated sound waves as in figure 5.1.

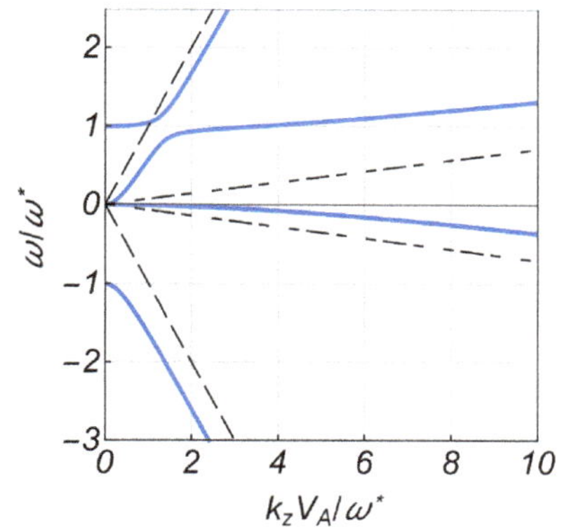

Figure 15.1. The full dispersion relation (15.26) using normalized variables. The dashed lines give the dispersion relations $\omega = \pm V_A k_z$ and dash-dotted lines $\omega = \pm C_s k_z$ for reference. The figure is obtained with the parameters $k_y^2 C_s^2/\Omega_c^2 = 5 \times 10^{-2}$ and $C_s^2/V_A^2 = 5 \times 10^{-3}$ as an illustrative example.

The dispersion relation (15.26) reduces to (5.14) in the limit where $V_A \to \infty$, as expected. If the plasma is homogeneous, i.e. $\omega^* \to 0$, we find the approximate dispersion relation

$$\omega^2 \approx k_z^2 V_A^2 \left(1 + \frac{k_y^2 C_s^2}{\Omega_c^2}\right) = k_z^2 V_A^2 \left(1 + k_y^2 a_i^2\right), \tag{15.27}$$

where we used that to the lowest approximation $\omega^2 \approx k_z^2 V_A^2$, and the last term in (15.26) will be vanishing with $\omega^* = 0$. The dispersion relation (15.27) is the well known result (Morales and Maggs 1997, Pécseli 2012) for kinetic Alfvén waves in the limit of $\beta > m/M$.

As already mentioned, the basic two-potential model for the electric field implies that magnetosonic, or compressional, modes are absent in the analysis. Intuitively, we do not expect this to be any major loss; the free energy available for the drift wave instability inherent in the diamagnetic drift is unlikely to be large, and the shear Alfvén waves do not require the same large amount of energy that would be needed to compress the magnetic field, with the result that they are most easily excited by relatively weak sources of free energy.

As the β-value of the plasma is further increased to, say $\beta \approx \sqrt{m/M}$, we can no longer ignore compressional modes. We can have the drift wave frequency ω^* resonating with a compressional Alfvén wave propagating in the direction $\perp \mathbf{B}_0 \| \hat{\mathbf{y}}$, where shear Alfvén waves need to have a propagation component $\| \mathbf{B}_0$. The propagation velocity of this wave is $\sqrt{V_A^2 + C_s^2} \approx C_s\sqrt{1 + 1/\beta} \approx C_s/\sqrt{\beta}$. The resonance condition becomes $k_y u_{De} \equiv \omega^* \approx k_y C_s/\sqrt{\beta}$. It is more convenient to express this condition in terms of the plasma density gradient generating the diamagnetic drift. We find

$$\left(\frac{\mathrm{d}\ln n_0(x)}{\mathrm{d}x}\right)^{-1} \leqslant \frac{C_s}{\Omega_{ci}}\sqrt{\beta} \equiv a_i\sqrt{\beta},$$

introducing an effective ion Larmor radius a_i as before. To find the coupling to magnetosonic waves in the sense described here, we need very steep gradients in plasma density since $\beta < 1$ for most relevant plasma conditions.

When β is further increased to approach unity as in the solar wind, the magnetic field becomes highly dynamic and the direction of the background magnetic field might become difficult to identify uniquely.

References

Barston E M 1964 Electrostatic oscillations in inhomogeneous cold plasmas *Ann. Phys. (NY)* **29** 282–303

Guio P, Børve S, Pécseli H L and Trulsen J 2001 Low frequency waves in plasmas with spatially varying electron temperature *Ann. Geophysicae* **18** 1613–22

Hasegawa A 1975 *Plasma Instabilities and Nonlinear Effects* Vol 8 of *Physics and Chemistry in Space* (Berlin: Springer)

Heyvaerts J and Priest E R 1983 Coronal heating by phase-mixed shear Alfvén waves *Astron. Astrophys.* **117** 220–34

Kadomtsev B B 1965 *Plasma Turbulence* (New York: Academic)

Morales G J and Maggs J E 1997 Structure of kinetic Alfvén waves with small transverse scale length *Phys. Plasmas* **4** 4118–25

Pécseli H L 2012 *Waves and Oscillations in Plasmas* (London: Taylor and Francis)

Roberts B 2000 Waves and oscillations in the corona *Solar Phys.* **193** 139–52

Trivelpiece A W and Gould R W 1959 Space charge waves in cylindrical plasma columns *J. Appl. Phys.* **30** 1784–93

Uberoi C 1972 Alfvén waves in inhomogeneous magnetic fields *Phys. Fluids* **15** 1673–5

van Kampen N G and Felderhof B U 1967 *Theoretical Methods in Plasma Physics* (Amsterdam: North-Holland)

Weiland J 1992 Low frequency modes associated with drift motions in inhomogeneous plasmas Technical Report CTH–IEFT/PP–1992–17 (Götebor, Sweden, Chalmers Technical University)

Weiland J 2000 *Collective Modes in Inhomogeneous Plasma, Kinetic and Advanced Fluid Theory* (Bristol: Institute of Physics Publishing)

Part V

Weakly nonlinear waves

IOP Publishing

Low Frequency Waves and Turbulence in Magnetized Laboratory Plasmas and in the Ionosphere

Hans Pécseli

Chapter 16

Classifications of turbulence conditions

For all we know, the Navier–Stokes equation forms the best available model for describing incompressible flows and in particular incompressible turbulence

$$\frac{\partial}{\partial t}\mathbf{u}(\mathbf{r},\, t) + \mathbf{u}(\mathbf{r},\, t)\cdot\nabla\mathbf{u}(\mathbf{r},\, t) = -\frac{1}{\rho}\nabla p + \nu\nabla^2\mathbf{u}(\mathbf{r},\, t), \tag{16.1}$$

with $\nabla\cdot\mathbf{u}(\mathbf{r},\, t) = 0$ for incompressible flows. The pressure is p and the (constant) fluid mass density is ρ, and ν is the kinematic viscosity. For discussing the differences and similarities between turbulence in fluids and in plasmas we might begin writing the Navier–Stokes equation in Fourier transformed form. We have

$$\frac{\partial}{\partial t}\mathbf{u}(\mathbf{k},\, t) + \sum_{\mathbf{k}'} i(\mathbf{u}(\mathbf{k}',\, t)\cdot\mathbf{k})\,\mathbf{u}(\mathbf{k}-\mathbf{k}',\, t) = -\frac{i\mathbf{k}}{\rho}p(\mathbf{k},\, t) - \nu k^2\mathbf{u}(\mathbf{k},\, t). \tag{16.2}$$

Incompressibility $\mathbf{k}\cdot\mathbf{u}(\mathbf{k},\, t) = 0$ enters as an effective equation of state (Pécseli 2012) to give

$$p(\mathbf{k},\, t) = -\frac{\rho}{k^2}\mathbf{k}\cdot\sum_{\mathbf{k}'}(\mathbf{u}(\mathbf{k}',\, t)\cdot\mathbf{k})\mathbf{u}(\mathbf{k}-\mathbf{k}',\, t).$$

Consequently we can express the spatially Fourier transformed Navier–Stokes equation in the form

$$\frac{\partial}{\partial t}u_i(\mathbf{k},\, t) + \nu k^2 u_i(\mathbf{k},\, t) = \sum_{j,\ell;\mathbf{p}} M_{(i,j,\ell)}u_j(\mathbf{p},\, t)u_\ell(\mathbf{k}-\mathbf{p},\, t), \tag{16.3}$$

where $u_i(\mathbf{k},\, t)$ is a component of the velocity vector, $M_{(i,j,\ell)}$ is the coupling coefficient or element of the coupling matrix. We recall that a differential equation is not complete before initial and/or boundary conditions are stated, but for the moment

doi:10.1088/978-0-7503-1251-6ch16

we leave out the discussion of these. If we linearize (16.3), we obtain the rather simple result that an initial condition for $u_i(\mathbf{k}, t)$ decays like $u_i(\mathbf{k}, 0)\exp(-k^2\nu t)$. In the limit of small k which corresponds to the inertial range most relevant for turbulence studies, we therefore find a relatively trivial result. The dynamics of turbulence in neutral incompressible flows is controlled by nonlinear effects in the inertial subrange. The magnitude of the nonlinear coupling coefficient is measured by the Reynolds number, which we can write as the ratio of the order of magnitude of the nonlinear term and the linear term as

$$Re = \frac{|M_{(i,j,\ell)}u^2|}{k^2\nu u} = \frac{|M_{(i,j,\ell)}u|}{k^2\nu} = \frac{u}{k\nu} = \frac{Lu}{\nu}.$$

Recalling that the dimension $[M_{(i,j,\ell)}] = \mathit{length}^{-1}$, we assumed that $M_{(i,j,\ell)}$ is of order k, which can be demonstrated by a detailed calculation of the coupling coefficients, but as an estimate it can be found simply by noting that the origin of the right-hand side of (16.2) is the $\mathbf{u} \cdot \nabla\mathbf{u}$ term in the Navier–Stokes equation. This terms scales as ku^2. We can let the smallest wave-number relevant be $k \sim 1/L$, where L is a characteristic system size (Dupree 1969). The Reynolds number can also be seen as the ratio between two time-scales, the damping time $k^2\nu$ and a nonlinear coupling time Mu. It is important to note that the Reynolds number obtained by these arguments depends on the wavenumber $\mathbf{k}$ to be considered: in the original formulation by Reynolds L was taken to be the diameter of a tube, for instance, i.e. a fixed length scale. (In all fairness it might be added that a characteristic number $Re \equiv UL/\nu$ was seemingly introduced by Stokes around 1850, but its full importance was realized somewhat later by Reynolds around 1880). Strong turbulence in fluids is found when $Re \gg 1$, implying that the nonlinear term on the right-hand side of (16.2) completely dominates the second linear term on the left side.

Classical turbulence can be understood as a state with energy input, usually mediated by an instability where the nonlinearity accounts for the interaction between many degrees of freedom, while dissipation removes the injected energy. The analysis can be restricted somewhat by requesting that there should be a steady state energy input with a corresponding steady state dissipation.

We can now try to write an expression in the form of (16.3) for the plasma dynamics in a fluid model

$$\frac{\partial}{\partial t}E_i(\mathbf{k}, t) + i\omega(\mathbf{k})E_i(\mathbf{k}, t) = \sum_{j,\ell;\mathbf{p}} C_{(i,j,\ell)}E_j(\mathbf{p}, t)E_\ell(\mathbf{k} - \mathbf{p}, t), \tag{16.4}$$

where we introduce an electric field component E_i as a reminder that we study electric fields, but we could of course have retained a velocity as well. We also changed the symbol for the coupling coefficient, merely to avoid confusion. The important change is in the left side, where we have an otherwise unspecified frequency $\omega(\mathbf{k})$ appearing. In one relevant example we would have $\omega(\mathbf{k}) \sim \omega_{pe}$, the electron plasma frequency, but more generally we let $\omega(\mathbf{k})$ account for the linear dynamics of the plasma. Already here we see a problem: a quite general model that accounts for *all* relevant phenomena in magnetized plasmas is hopelessly complicated. We have to make a selection and analyze a restricted problem, which we specify by the proper choice of $\omega(\mathbf{k})$. In some cases we might find similarities with the analysis of the Navier–Stokes equation, in other cases not. Analytical models for weak turbulence that also keep the constraints imposed by the linear dispersion relation in equations like (16.4) have been developed (Kadomtsev 1965, Krommes 2015) and in several cases found to be in good agreement with experimental observations or numerical simulations.

It is also here possible to estimate an equivalent of Reynolds number as the ratio of a nonlinear and a linear term (Dupree 1969), and we find

$$Rp = \frac{CE^2}{|\omega(\mathbf{k})|E} = \frac{CE}{|\omega(\mathbf{k})|},$$

anticipating that $\omega(\mathbf{k})$ will generally be complex, so the absolute value is taken for the order of magnitude. It is still possible to consider Rp as a ratio between two time scales, and the problem may be easier to approach by asking for the time necessary for nonlinear interaction, assuming that the relevant linear time scale is already known by some set of equations. If we consider the Korteweg–deVries equation, for instance, we can define a nonlinear time-scale as the time it takes two solitons to pass through each other (Pécseli 2012). Nothing more can be gained here without considering specific examples. The consequence is that universal laws for strong plasma turbulence are unlikely to exist, the only possibility for universality is found by restricting the problem, by studying for instance low frequency electrostatic waves in low-β-plasmas. One more reason for the lack of universal features may be that plasma turbulence is often not strong, i.e. the linear times scale (i.e. $\omega(\mathbf{k})^{-1}$ in (16.4)) need not be small, and we often find $Rp \approx 1$, so the mode interactions are restricted in some subtle way.

In many studies of seemingly turbulent plasmas it is possible, by suitable chosen diagnostic methods (Beall *et al* 1982), experimentally to determine a relation between frequency and wavevector. An example of such results is shown in figure 16.1. The results shown here use two probes and give one component of the direction of propagation. Although these experimental results are not dispersion relations in a traditional sense, their presence indicates that the $i\omega(\mathbf{k})E_i(\mathbf{k}, t)$ -term in (16.4) is not negligible in comparison to the right-hand side, and consequently the plasma conditions will not properly be called 'strongly turbulent'. The case on the right in figure 16.1, on the other hand, has features that are consistent with strongly turbulent plasma conditions.

The distinction of weak and strong turbulence has consequences for the modeling of the transport properties, or diffusion, caused by the random fluctuations. In terms of the maximum linear growth rate for electrostatic drift wave instability $\Im\{\omega^{(m)}(\mathbf{k}^{(m)})\}$ and the corresponding most unstable wavenumber $k_\perp^{(m)}$ Kadomtsev (1965) argued for an effective diffusion coefficient $D_\perp \sim \Im\{\omega^{(m)}\}/k_\perp^{(m)2}$ accounting for the turbulent plasma transport across magnetic field lines. The ideas were generalized by, for instance, Hatayama *et al* (1981). For a strong turbulence model the

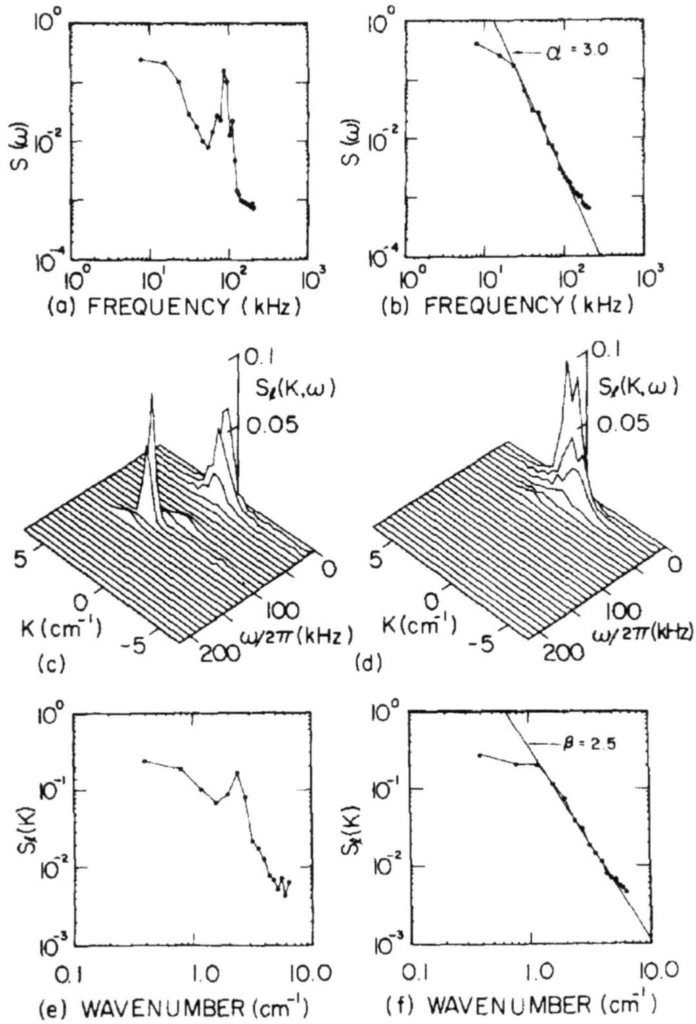

Figure 16.1. Summary figure from experimental studies of plasma density fluctuations in radio frequency (RF) discharges, for obtaining dispersion relations in a noisy background (Iwama *et al* 1979, Beall *et al* 1982). For the conditions to the left it is feasible to obtain evidence for a dispersion relation. For the fully developed turbulent conditions on the right-hand side all that remains is a 'blur' for small ω and k. Reproduced with permission from Beall *et al* (1982). Copyright 1982, AIP Publishing LLC.

information from the linear dispersion relation has little relevance and the transport properties of the turbulence has to be found by other means.

Depending on the level of ambition, turbulent flows are challenging also numerically (Frisch 1995). If we decide to resolve the Kolmogorov micro-scale $\eta \equiv (\nu^3/\epsilon)^{1/4}$ we face a really challenging problem for a strongly turbulent case. To illustrate this we write an alternative Reynolds number as a ratio of length scales by $Re = \mathrm{L}u/\nu \approx (\mathrm{L}/\eta)^{4/3}$, found by taking a characteristic velocity to be $u \approx (L\epsilon)^{1/3}$. A strongly turbulent state can have $Re \sim 10^4$ or larger as found in some tidal flows (Grant *et al* 1962). To resolve the scale L to the desired accuracy by say a spectral code in three spatial dimensions we will need in excess of $\sim 10^3 \times 10^3 \times 10^3 = 10^9$ wavenumbers, or degrees of freedom where at least in principle all of them are interacting at any time. It is safe to assume that a numerical solution with this many degrees of freedom for a time duration of, say, an eddy turn over time is (with present day computers) out of range and is likely to remain so for all foreseeable future. In reality, the situation is even worse: the estimate outlined here assumes that scales smaller than the micro-scale need not be resolved, but this is not so. The energy cascade to small scales, as mediated by the vortex-stretching (Tennekes and Lumley 1972), will pile up energy at wavenumbers $k \sim \eta^{-1}$ if the viscous subrange is unresolved. This 'bottle-neck' effect is found in many numerical simulations of turbulent flows (Lohse and Müller-Groeling 1996, Dobler *et al* 2003). Physically, the bottleneck effect arises because of an insufficient resolution of the sub-Kolmogorov scales. This gives a deficit of small scale vortices, which makes the energy cascade less effective around the Kolmogorov scale, as compared to ideal, physical conditions. State of the art numerical solutions address Reynolds numbers of the order of 10^3. 'Large eddy simulations' reduce the ambition by considering only the dynamics of the largest scales of the turbulence followed by suitable methods for a sub-grid modeling. The number of degrees of freedom in the system is thereby reduced. These methods have proved useful for design. The problem outlined here is studied at length for application in fluid dynamics, but for plasma turbulence, weak turbulence models in particular, the problem seems to have received less attention than it deserves.

When turbulence is not fully developed we may anticipate that coherent structures are not destroyed by the shearing and stretching turbulent motions, and that nonlinearly saturated plasma instabilities can be characterized by the presence of coherent structures. At first sight this could appear as a significant difference from incompressible neutral flows, but this is in reality only so when we make restricted references to homogeneous isotropic conditions, decoupled from any boundary. When turbulent boundary layers in incompressible flows are considered, large coherent structures are also found here (Schlichting and Gersten 2000, Sagaut and Cambon 2008), such as 'hair-pin vortices'. Just as a number of nonlinear equations support some basic nonlinear structure (solitons for the Korteweg–deVries and nonlinear Schrödinger equations, for instance), the Navier–Stokes equation supports a vortex which may be linear or circular ('smoke ring'). Homogeneous isotropic turbulence in neutral flows can be interpreted in terms of shearing and stretching of these basic structures due to the turbulent motions: the vicinity of a solid surface can locally reduce (although not completely cancel) the deformation of the vortices. Since many of the relevant equations describing nonlinear plasma wave phenomena have solutions in the form of coherent structures, it is plausible that they will dominate, or at least contribute to the dynamics close to boundaries in those cases where the plasma turbulence is not strong, i.e. Rp is small or close to unity.

With the diversity of phenomena possible in plasma turbulence, the diagnostic problem also becomes complicated. It is not necessarily so that we can be content with the methods from neutral turbulence, although they may represent a good starting point. Novel methods have to be devised, not only on the hardware side but also for data acquisition and handling. If coherent structures are believed to be a relevant constituent of turbulence, it is necessary to devise methods for their identification in a turbulent background.

References

Beall J M, Kim Y C and Powers E J 1982 Estimation of wavenumber and frequency spectra using fixed probe pairs *J. Applied Phys.* **53** 3933–40

Dobler W, Haugen N E L, Yousef T A and Brandenburg A 2003 Bottleneck effect in three-dimensional turbulence simulations *Phys. Rev.* E **68** 026304

Dupree T H 1969 Introduction to basic phenomena of turbulence in plasmas *Turbulence in Fluids and Plasmas* (Brooklyn, NY: Polytechnic Press) ch 1

Frisch U 1995 *Turbulence: The Legacy of A. N. Kolmogorov* (Cambridge: Cambridge University Press)

Grant H L, Stewart R W and Moilliet A 1962 Turbulence spectra from a tidal channel *J. Fluid Mech.* **12** 241–68

Hatayama A, Koshi Y and Ogasawara M 1981 Mixing length hypothesis in the dissipative trapped ion instability *J. Phys. Soc. Japan* **50** 976–9

Iwama N, Ohba Y and Tsukishima T 1979 Estimation of wave-number spectrum parameters from fixed probe-pair data *J. Applied Phys.* **50** 3197–206

Kadomtsev B B 1965 *Plasma Turbulence* (New York: Academic)

Krommes J A 2015 A tutorial introduction to the statistical theory of turbulent plasmas, a half-century after Kadomtsev's Plasma Turbulence and the resonance-broadening theory of Dupree and Weinstock *J. Plasma Phys.* **81** 205810601 (80 pages)

Lohse D and Müller-Groeling A 1996 Anisotropy and scaling corrections in turbulence *Phys. Rev.* E **54** 395–405

Pécseli H L 2012 *Waves and Oscillations in Plasmas* (London: Taylor and Francis)

Sagaut P and Cambon C 2008 *Homogeneous Turbulence Dynamics* (Cambridge: Cambridge University Press)

Schlichting H and Gersten K 2000 *Boundary Layer Theory* 8th edn (Berlin: Springer)

Tennekes H and Lumley J L 1972 *A First Course in Turbulence* (Cambridge, MA: The MIT Press)

IOP Publishing

Low Frequency Waves and Turbulence in Magnetized Laboratory Plasmas and in the Ionosphere

Hans Pécseli

Chapter 17

Weakly nonlinear waves in homogeneously magnetized plasmas

Before discussing the properties of nonlinear drift waves, it may be instructive to discuss low frequency electrostatic waves in homogeneous magnetized plasmas. In the absence of density or temperature inhomogeneities, we can discuss several simplified limits. In particular, we have the possibility of constructing stationary solutions of the basic equations. Some of these cases will be discussed in the following.

17.1 Flute modes

We consider here the limit where the **B**-parallel wavelength is infinite so that the plasma charge perturbations are strictly aligned with the magnetic field. The physical arguments for assuming Boltzmann distributed electrons are then not valid. Instead we assume a continuity equation for the electrons similar to (4.3) applied for the ions, the two species being again coupled through Poisson's equation $\nabla^2\phi = \frac{e}{\varepsilon_0}(n_e - n_i)$. Subtraction of the two continuity equations and use of Poisson's equation gives

$$\left(\frac{\partial}{\partial t} - \frac{1}{B^2}\nabla_\perp\phi \times \mathbf{B} \cdot \nabla_\perp\right)\nabla^2\phi = 0, \tag{17.1}$$

which uniquely determines the evolution of the electrostatic plasma potential when the initial condition is given. The bulk plasma velocity is $\mathbf{u} = -\nabla_\perp\phi \times \mathbf{B}/B^2$, and the vorticity deduced from it is $|\nabla_\perp \times \mathbf{u}| = \nabla_\perp^2\phi$, which by Poisson's equation shows that the charge density uniquely determines the vorticity in the present simplified two-dimensional model. The equation (17.1) is inherently nonlinear; upon linearization the result is trivially $\partial\phi/\partial t = 0$.

An equation for the evolution of the bulk plasma density can be obtained by *adding* the two continuity equations to give

$$\left(\frac{\partial}{\partial t} - \frac{1}{B^2}\nabla_\perp\phi \times \mathbf{B} \cdot \nabla_\perp\right)n = 0, \tag{17.2}$$

with $n \equiv \frac{1}{2}(n_e + n_i)$, where ϕ in (17.2) is assumed given from the solution of (17.1). It is interesting to note that there is here no distinction of the average or unperturbed part of the density $n_0(x)$ as we had before. The equation (17.2) determines the evolution of the *entire* plasma density once the potential has been determined from (17.1). The variation in bulk plasma density has no *a priori* relation to the variation in plasma potential in the present description, but is solely determined by the initial conditions. The bulk plasma density n and the electrostatic potential ϕ thus evolve by the same operator $\partial/\partial t - (1/B^2)\nabla_\perp\phi \times \mathbf{B} \cdot \nabla_\perp$, but differ in general by the initial conditions.

The plasma temperature does not appear explicitly in (17.2) and (17.1). Indirectly, we have however assumed that the ion and electron Larmor radii are small compared to the scale length of the perturbations, so FLR-effects have been ignored.

The inherently nonlinear equations (17.2) and (17.1) have been extensively studied, maybe not so much because of the physics they describe, but rather for their almost simplistic beauty which nevertheless turns out to contain a wealth of complexity.

17.1.1 Discrete vortex modes

Since the linearized form of the equations (17.2) and (17.1) are relatively trivial, we will here discuss possibilities of exact nonlinear solutions for these equations. The simplest example of such a solution is obtained by considering a 'line-charge', i.e. the case where $e(n_i - n_e) = Q\delta(r)$ in Poisson's equation, formally corresponding to a 'charging-up' of one magnetic field line. For this case, the potential variation is obtained trivially as

$$\phi(r) = \frac{Q}{2\pi\varepsilon_0}\ln r, \tag{17.3}$$

and the components of the $\mathbf{E} \times \mathbf{B}/B^2$-velocity in cylindrical coordinates being

$$\begin{Bmatrix} U_r \\ U_\theta \\ U_z \end{Bmatrix} = (Q/2\pi\varepsilon_0)\begin{Bmatrix} 1/r \\ 0 \\ 0 \end{Bmatrix},$$

the result being a non-uniform rotation of the entire plasma around the position of the charge distribution at $r = 0$. If we now take the case with *two* line vortices with opposite charges $+Q$ and $-Q$ separated by a distance, say, ℓ along the y-axis, we can argue that any one of them is simply convecting the other in such a way that they propagate as a pair in the x-direction with a velocity $U = (Q/2\pi\varepsilon_0)/\ell$. This classical solution (Lin 1943, Lamb 1945, Milne-Thomson 1968) is visualized in figure 17.1 by

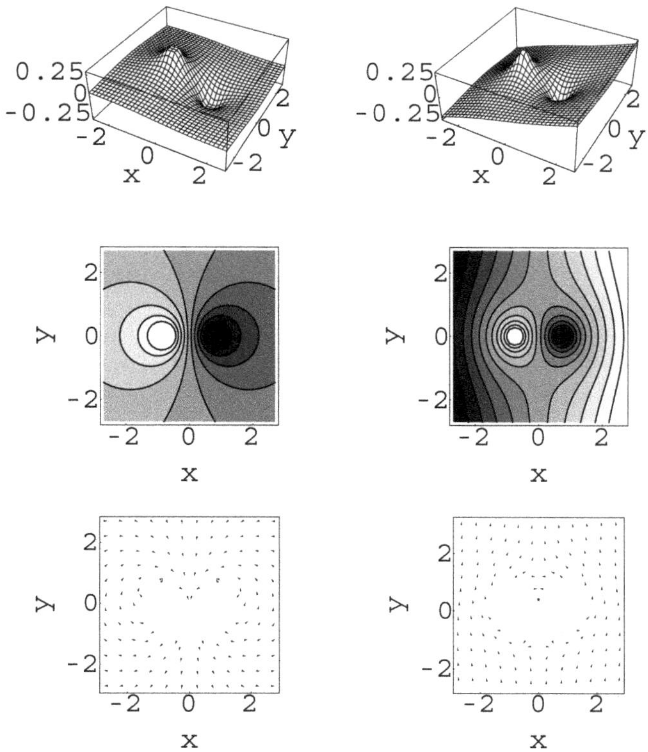

Figure 17.1. Schematic representation of dipole-solutions obtained by two line charges of opposite sign. The left column shows a 'snapshot' of equi-potentials and velocity vectors in the rest frame, the right column shows the same quantities in the co-moving frame. The longest velocity vectors have been omitted for clarity. We have $Q = 1$ and $\ell = 0.75$.

showing the electrostatic potential in the absolute as well as the co-moving frames. If we follow the vortex pair moving with velocity U, there will in this new frame be induced an electric field $\mathbf{E}_0 = -\mathbf{U} \times \mathbf{B}$, corresponding to an electrostatic potential $\phi_0 = (U/B)y$, which has to be added to the vortex potentials. The rest frame figure represents a 'snap-shot' only, while the figure is stationary in the co-moving frame.

On the other hand, if we have a pair consisting of two *equal* charges they will convect each other in opposite directions, and the result will be a rotation of the pair, without any net displacement. Finally, we might have the situation where the two charges are of unequal magnitude. In this case we still have a net circular motion, but in this case around a point which is not the average position, but closer to the strongest vortex.

- **Example:** There is a well known, and often used, analogy between electric and magnetic dipoles and their fluid counterparts. For instance, in two spatial dimensions, the equivalent of a magnetic dipole is generated by two parallel, infinitesimally close current carrying wires. The resulting magnetic field can be obtained by first noting that a single wire placed at the origin gives rise to an azimuthal magnetic field with intensity $\mu_0 I/(2\pi r)$, where I is the current.

This magnetic field can be derived from an effective potential $A = \ln(r)\mu_0 I/(2\pi)$ by $\mathbf{B} \equiv \{B_r, B_\theta\} = \{(1/r)\partial A/\partial\theta, -\partial A/\partial r\}$, being the two-dimensional equivalent of $\mathbf{B} = \nabla \times \mathbf{A}$. The magnetic field from a slightly displaced wire with current in the opposite direction is then derived from a potential $-A - dA$, where dA is the increment in A produced by displacing the wire a distance ℓ along x, where $x = r\cos\theta$. The resultant magnetic potential is thus simply

$$-\mathrm{d}A = -\ell\frac{\partial}{\partial x}\ln(r)\mu_0 I\frac{1}{2\pi} = -\ell\mu_0 I\frac{1}{2\pi r}\frac{\partial}{\partial x}r = -\ell I\mu_0\frac{1}{2\pi r}\cos\theta,$$

using $r = \sqrt{x^2 + y^2}$, giving $\partial r/\partial x = x/r = \cos\theta$. Consequently we find the magnetic field components $B_r = (\ell I\mu_0/2\pi r^2)\cos\theta$ and $B_\theta = (\ell I\mu_0/2\pi r^2)\sin\theta$. We have a simple one-to-one correspondence between the dipolar velocity field discussed before and the magnetic field discussed here, where ℓI takes the role of a dipole moment.

It is interesting to note that $|B| = \ell I\mu_0/(2\pi r^2)$ is here independent of θ. We recall that, in comparison, a magnetic dipole in three spatial dimensions, as generated by a tiny current loop, has a magnetic field intensity of $|B| = \mu_0\mathcal{M}\sqrt{(1 + 3\cos^2\theta)}/(4\pi r^3)$, where $\mathcal{M}$ is the magnetic moment.

The trajectories of the velocity field $\mathbf{u} = \mathbf{u}(x, y, t)$ are denoted *stream-lines*, and are the solutions of the two simultaneous equations $\mathrm{d}\mathbf{r}/\mathrm{d}s = \mathbf{u}$, where s is the parameter along the streamline (two equations because $\mathbf{u}$ is here a two-dimensional vector). This parameter should not be confused with time, because t is held fixed while the equations are integrated, and the resulting curves are stream-lines at the instant t. These may vary from instant to instant, and will in general not coincide with the particle paths. If the function $\mathbf{u} = \mathbf{u}(x, y)$ does not depend on time, the stream-lines and particle paths will coincide. In the frame following the double vortex, the equi potential lines are stream-lines, and also particle paths. We have also the so called *streak-lines* which are defined as the curve traced out by a plume of smoke or dye which is continuously injected from a point source at a fixed spatial position, with the dye passively following the flow (Aris 1962). The definitions are evidently not restricted to two-dimensional flows.

- **Example:** The notion of stream lines, particle paths and streak-lines can be illustrated (Aris 1962) by the simple plane, two-dimensional, flow here with $\nabla \cdot \mathbf{u} \neq 0$

$$\mathbf{u} = \left\{\frac{x}{1+t}, y\right\}.$$

For this case the stream-lines at a time t are the solutions of

$$\frac{\mathrm{d}}{\mathrm{d}s}x = \frac{x}{1+t}, \qquad \frac{\mathrm{d}}{\mathrm{d}s}y = y.$$

Keeping t constant, the stream-lines through the point $\mathbf{a}$ are

$$x = a_x e^{s/(1+t)}, \qquad y = a_y e^s,$$

which corresponds to a curve in the plane as given by

$$\frac{y}{a_y} = \left(\frac{x}{a_x}\right)^{1+t}.$$

The particle paths, on the other hand, are the solutions of

$$\frac{d}{dt}x = \frac{x}{1+t}, \qquad \frac{d}{dt}y = y,$$

giving

$$x = a_x(1+t), \qquad y = a_y e^t,$$

being curves

$$y = a_y e^{(x-a_x)/a_x}$$

in the plane. For the inverse relations defining the particle at position $\mathbf{r}_1 \equiv \{x_1; y_1\}$ at time τ we have $a_x = x_1/(1+\tau)$ and $a_y = y_1 e^{-\tau}$. Hence the streak-lines are given through

$$x = x_1\frac{1+t}{1+\tau}, \qquad y = y_1 e^{t-\tau}.$$

At $\tau = t$ we have (x_1, y_1), being the position of release, while $\tau = 0$ corresponds to the position reached by the particles which were released at the time where their source was 'switched on'.

17.1.2 Electron shielding

The previous subsection assumed the line charges to be ideally field aligned, implying that electrons as well as ions were strictly bound to their respective field lines. As a modified model, in line with the drift wave analysis used before, we can retain the field aligned description of the ions, but allow the electrons to maintain an isothermal Boltzmann distribution, $n_e(r, t) = n_0 \exp(e\phi(r, t)/\kappa T_e)$ in the electrostatic potential. Physically, this can be achieved by assuming that the perturbations are not *strictly* aligned with the magnetic field, but have a small angle with respect to the **B**-field lines, so that electrons can flow along magnetic field lines of force to maintain the stated equilibrium. The two-dimensional ion description is consistent provided the $\mathbf{E}_\perp \times \mathbf{B}/B^2$ ion velocity in the **B**-transverse direction is much larger than the ion velocity, $v_\| \sim eE_\|/(\omega M)$, *along* **B**.

To account for the electron shielding we use Poisson's equation in the form

$$\nabla_\perp^2 \phi = \frac{e}{\varepsilon_0}\left(\frac{en_0\phi}{\kappa T_e} - \tilde{n}_i\right) \qquad (17.4)$$

for electrostatic fluctuations, where we have linearized the exponential originating from the Boltzmann distribution, and have the ion density as $n_i = n_0 + \tilde{n}_i$. This relation, together with the ion continuity equation with $-\nabla\phi \times \mathbf{B}/B^2$ ion velocities, finally gives

$$\left[\frac{\partial}{\partial t} - \frac{1}{B^2}\nabla_\perp\phi \times \mathbf{B} \cdot \nabla_\perp\right]\left(\nabla_\perp^2\phi - \frac{1}{\lambda_D^2}\phi\right) = 0, \tag{17.5}$$

which is identically equal to

$$\frac{\partial}{\partial t}\left(\nabla^2\phi - \frac{1}{\lambda_D^2}\phi\right) - \frac{1}{B^2}\nabla_\perp\phi \times \mathbf{B} \cdot \nabla_\perp\nabla_\perp^2\phi = 0, \tag{17.6}$$

where the electron Debye length enters as a shielding distance. (For the corresponding, and formally quite equivalent problem of geostrophic large scale motion on a rotating planet, the Rossby radius takes the role of the Debye length.)

To make the shielding explicit, we have the line vortex solution $\phi(r) = aK_0(r)$ to the equation $\nabla_\perp^2\phi - (1/\lambda_D^2)\phi = a2\pi\delta(r)$, obtained from (17.4), where the δ-function represents a line distribution of ion excess (or deficit) charge. We let e/ε_0 be accommodated in the constant a. The approximation for the Bessel function $K_0(r) \approx \ln(\frac{1}{2}r/\lambda_D)I_0(r/\lambda_D)$ for small $r \ll \lambda_D$, corresponds to the near field limit with negligible shielding, while for large $r \gg \lambda_D$ we have exponential Debye shielding by $K_0(r) \approx \exp(-r/\lambda_D)\sqrt{\pi\lambda_D/2r}$. In figure 17.2 we show the potential associated with shielded as well as the unshielded line vortices.

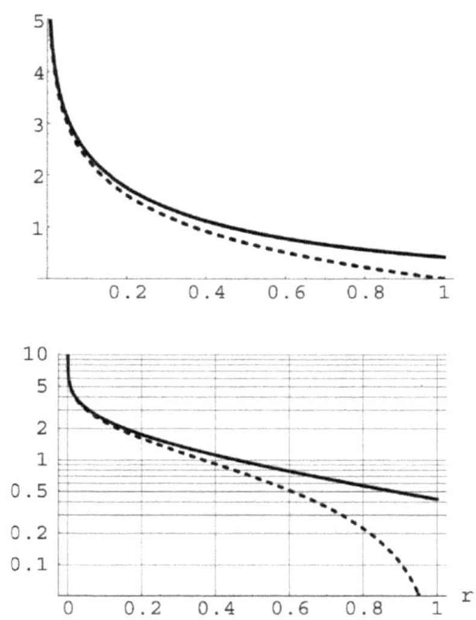

Figure 17.2. Illustration of the shielded K_0 vortex (full line) and its unshielded logarithmic counterpart (dashed line).

17.1.3 Models with many vortices

The foregoing discussion referred to superpositions of two vortices only. It seems intuitively clear that stationary vortex patterns consisting of many such vortices can be constructed (Morikawa 1960, Morikawa and Swenson 1971). In the following, we consider the simple model equation

$$\frac{\partial}{\partial t}\left(\kappa^2\phi - \nabla^2\phi\right) - \left\{\phi;\, \nabla^2\phi\right\} = 0, \tag{17.7}$$

which includes both the unshielded and shielded models for $\kappa = 0$ and $\kappa = 1$, respectively, where κ denotes a screening effect. We introduced the Poisson brackets as $\{;\}$, with

$$\{a(x, y);\, b(x, y)\} \equiv \frac{\partial a}{\partial x}\frac{\partial b}{\partial y} - \frac{\partial b}{\partial x}\frac{\partial a}{\partial y} \equiv \hat{\mathbf{z}} \cdot \nabla a \times \nabla b.$$

Point vortices (or rather line vortices) can be introduced through (17.4), here written as

$$\left(\kappa^2 - \nabla^2\right)\phi(\mathbf{r}, t) = \sum_\alpha \gamma_\alpha \delta(\mathbf{r} - \mathbf{r}_\alpha), \tag{17.8}$$

which gives the potential as

$$\phi(\mathbf{r}, t) = \sum_\alpha \frac{\gamma_\alpha}{2\pi} K_0\left(\kappa|\mathbf{r} - r_\alpha(t)|\right), \tag{17.9}$$

where K_0 is a modified Bessel function of the second kind. In the limit of $\kappa \to 0$, (17.9) reduces to

$$\phi(\mathbf{r}, t) = \sum_\alpha \frac{\gamma_\alpha}{2\pi} \ln\left(|\mathbf{r} - \mathbf{r}_\alpha(t)|\right), \tag{17.10}$$

which is a solution of (17.8) for $\kappa = 0$.

The equations for the motion of vortices are given by the standard form

$$\begin{aligned}\frac{\mathrm{d}}{\mathrm{d}t}\mathbf{r}_\alpha(t) &= \hat{\mathbf{z}} \times \nabla\phi(\mathbf{r}_\alpha, t)\\ &= \frac{\kappa}{2\pi}\sum_\beta \gamma_\beta \frac{\hat{\mathbf{z}} \times \left(\mathbf{r}_\alpha - \mathbf{r}_\beta\right)}{\left|\mathbf{r}_\alpha - \mathbf{r}_\beta\right|} K_1\left(\kappa\left|\mathbf{r}_\alpha - \mathbf{r}_\beta\right|\right).\end{aligned} \tag{17.11}$$

Physically, this expression is based on the assumption that any individual vortex, say the one at position $\mathbf{r}_\alpha$ is convected by the superimposed velocity field from all the other vortices. the superposition is *linear* since the point vortices cannot overlap. The K_1-function arises by the differentiation of K_0 in (17.9).

When the distance between the vortices is short enough for the shielding to be neglected, the dynamical behavior of the vortices is described by the following standard equation

$$\frac{\mathrm{d}}{\mathrm{d}t}\mathbf{r}_\alpha = \frac{1}{2\pi}\sum_\beta \gamma_\beta \frac{\hat{\mathbf{z}} \times (\mathbf{r}_\alpha - r_\beta)}{|\mathbf{r}_\alpha - \mathbf{r}_\beta|^2}, \tag{17.12}$$

which is obtained from (17.11) in the limit of $\kappa \to 0$, or from (17.10) directly.

Equilibrium configurations of point vortices have been discussed in (Morikawa and Swenson 1971) for the case that N vortices are equally distributed on a circle with or without a center vortex. They found, for instance, that the configuration is stable for $2 \leqslant N \leqslant 6$ in the case of $\kappa \leqslant 1.289$ without a center vortex and for $4 \leqslant N \leqslant 8$ in the case of $\kappa \leqslant 1$ without a center vortex.

Starting from the vortex equation (17.11), a more general analysis (Kono *et al* 1998b) can be made by introducing the distribution function

$$f(\mathbf{r}, t) = \sum_\alpha \gamma_\alpha \delta\big(\mathbf{r} - r_\alpha(t)\big). \tag{17.13}$$

Then (17.11) is rewritten as

$$\frac{\partial}{\partial t}f(\mathbf{r}, t) = -\frac{\kappa}{2\pi}\int \mathrm{d}\mathbf{r}' \frac{K_1\big(\kappa|\mathbf{r} - \mathbf{r}'|\big)}{|\mathbf{r} - \mathbf{r}'|}\hat{\mathbf{z}} \times (\mathbf{r} - r') \cdot \nabla f(\mathbf{r}', t) f(\mathbf{r}, t). \tag{17.14}$$

Changing from Cartesian variables (x, y) to cylindrical coordinates (r, θ) and noting that

$$\hat{\mathbf{z}} \times (\mathbf{r} - \mathbf{r}') \cdot \nabla = -r' \sin(\theta - \theta')\frac{\partial}{\partial r} + \left[1 - \frac{r'}{r}\cos(\theta - \theta')\right]\frac{\partial}{\partial \theta},$$

we have

$$\frac{\partial}{\partial t}f(r, \theta, t) + \Omega(r, \theta, t)\frac{\partial}{\partial \theta}f(r, \theta, t) + V(r, \theta, t)\frac{\partial}{\partial r}f(r, \theta, t) = 0, \tag{17.15}$$

where

$$\Omega(r, \theta, t) = \frac{\kappa}{2\pi}\int \mathrm{d}\mathbf{r}' \frac{K_1\big(\kappa|\mathbf{r} - \mathbf{r}'|\big)}{|\mathbf{r} - \mathbf{r}'|}\left[1 - \frac{r'}{r}\cos(\theta - \theta')\right] f(r', \theta', t), \tag{17.16}$$

$$V(r, \theta, t) = -\frac{\kappa}{2\pi}\int \mathrm{d}\mathbf{r}' r' \frac{K_1\big(\kappa|\mathbf{r} - \mathbf{r}'|\big)}{|\mathbf{r} - r'|}\sin(\theta - \theta') f(r', \theta', t). \tag{17.17}$$

From the foregoing equations, equilibrium configurations can be obtained by imposing the conditions that $V(r, \theta, t)$ are identically zero and $\Omega(r, \theta, t) = \Omega_0 =$ const , for which the vortex system is rigidly rotating around the origin without any displacement in the radial direction. These equilibrium configurations consist of the fixed points of the dynamical system given by (17.11). In figure 17.3 we show examples of stationary solutions consisting of many vortices.

In the following we look for the concentric structure of M rings on each of which $N_m (m = 1, \cdots, M)$ vortices are equally distributed, assuming that all the vortices are the same, i.e. $\gamma_\alpha = \gamma$, with or without a center vortex γ_0, corresponding to $\gamma_0 = \gamma$ or $\gamma_0 = 0$

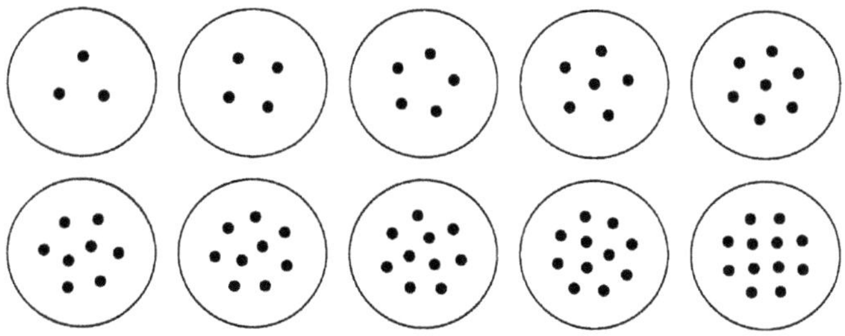

Figure 17.3. Illustration of stationary line vortex distributions (Kono and Škorić, 2010). The circles drawn around each figure are inserted merely to separate different solutions.

$$f(r,\theta,t)=\sum_{m,n}\frac{\gamma}{a_m}\delta(r-a_m)\delta(\theta-\theta_{m,n})+\frac{\gamma_0}{r}\delta(r)\delta(\theta), \tag{17.18}$$

for which following two relations hold in the equilibrium

$$V(m_0,n_0)=-\frac{\kappa^2}{2\pi}\gamma\sum_{m,n}\frac{a_m}{\sigma_{m,n}}K_1(\sigma_{m,n})\sin(\theta_{m_0,n_0}-\theta_{m,n})=0, \tag{17.19}$$

and

$$\begin{aligned}\Omega(m_0,n_0)=&\frac{\kappa^2}{2\pi}\gamma\sum_{m,n}\frac{1}{\sigma_{m,n}}K_1(\sigma_{m,n})\left[1-\frac{a_m}{a_{m_0}}\cos(\theta_{m_0,n_0}-\theta_{m,n})\right]\\&+\frac{\kappa}{2\pi}\gamma_0\frac{K_1(\kappa a_{m_0})}{a_{m_0}}\\&=\text{const},\end{aligned} \tag{17.20}$$

where

$$\sigma_{m,n}=\kappa\sqrt{a_{m_0}^2+a_m^2-2a_{m_0}a_m\cos\left(\theta_{m_0,n_0}-\theta_{m,n}\right)},$$

and (m,n) refers to the n-th vortex on the m-th circle. The coordinates (m_0, n_0) refer to an arbitrarily chosen reference vortex.

For the case where vortices on each circle are distributed so that the separation distance between neighboring vortices is the same, that is, $\theta_{m,n}=2\pi n/N_m$ where N_m is the number of vortices on m the circle, then $V(m_0, n_0)$ is always equal to zero and $\Omega(m_0, n_0)$ does not depend on n_0 because of the symmetry. There are many combinations of the number of rings and the number of vortices on each ring to make $\Omega(M_0, n_0)$ constant.

17.1.4 Hamiltonian property of vortex systems

It is here a trivial observation that the collection of vortices evolve with time so that the number of vortices is conserved. Other more interesting conserved quantities are found by noting that the equation of motion for point vortices can be written in the form of Hamilton's canonical equations (Onsager 1949, Aref 1979). For instance by insertion we can prove that

$$\gamma_i \frac{\mathrm{d}}{\mathrm{d}t} x_i = \frac{\partial}{\partial y_i} H \tag{17.21}$$

$$\gamma_i \frac{\mathrm{d}}{\mathrm{d}t} y_i = -\frac{\partial}{\partial x_i} H, \tag{17.22}$$

where

$$H = -\sum_{i,j}{}' \frac{\gamma_i \gamma_j}{4\pi} F\left(\left|\mathbf{r}_i - \mathbf{r}_j\right|\right), \tag{17.23}$$

where γ_i is the strength of the vortex number i, and F is a function accounting for the vortex interaction. The prime′ on $\sum$ indicates that the summation excludes the term where $i = j$. We have 4π rather than 2π in (17.23) since each term is counted twice with the definition used. For unshielded line vortices we have $F(|\mathbf{r}|) = \ln(|\mathbf{r}|)$, for the shielded vortices discussed in section 17.1.2 we find $F(|\mathbf{r}|) = K_0(\kappa|\mathbf{r}|)$. The Hamiltonian property of many interacting vortices is however quite general, allowing for a large class of interactions F, with one other case discussed for instance by Lynov *et al* (1991). Not all the interaction potentials F need to be physically realizable. The Hamiltonian (17.23) accounts for an effective potential energy in the interacting vortex system.

The system does not possess any kinetic energy in the usual sense. Recall here that the system has no inertia since, by assumption, a structure released in the flow *immediately* assumes the local flow velocity, as already discussed. We note the rather peculiar property of the relations (17.22); x_i and y_i are generalized coordinate and momentum, implying that as far as the Hamiltonian is concerned, one spatial coordinate of the system is a 'coordinate', and the other spatial coordinate is the generalized momentum. The expression (17.23) applies for unbounded systems. The presence of periodic boundary conditions requires that a modified potential for F is introduced (Ewald 1921).

The Hamiltonian H does not depend explicitly on time, and is therefore an integral of motion. Furthermore, H is invariant to a translation and rotation of the coordinates. These symmetries imply three conservation laws

$$\sum_i \gamma_i x_i = \text{const.} \tag{17.24}$$

$$\sum_i \gamma_i y_i = \text{const.} \tag{17.25}$$

$$\sum_i \gamma_i \left(x_i^2 + y_i^2\right) = \text{const.} \tag{17.26}$$

When the sum of all vortex strengths is non-zero, $\sum_i \gamma_i \neq 0$, the center of vorticity

$$(X,\ Y) \equiv \frac{1}{\sum_i \gamma_i}\left(\sum_i \gamma_i x_i,\ \sum_i \gamma_i y_i\right) \tag{17.27}$$

presents a fixed point for the flow. Implicitly, we already made use of this property in section 17.1.3, see also figure 17.3. Introducing a vortex separation as $\ell_{i,j} \equiv |r_i - \mathbf{r}_j|$, it can be demonstrated that

$$\begin{aligned} \frac{1}{2}\sum_i \gamma_i \gamma_j \ell_{i,j}^2 &= \left(\sum_i \gamma_i\right)\sum_j \gamma_j\left(x_j^2 + y_j^2\right) \\ &\quad - \left(\sum_i \gamma_i x_i\right)^2 - \left(\sum_i \gamma_i y_i\right)^2 \end{aligned} \tag{17.28}$$

is a constant of motion, independent of choice of the origin of the reference system. From (17.22) we can derive an equation for the square of the vortex separation $\ell_{i,j}^2$, without reference to the position of the center-of-mass of the vortices, or the orientation of the axes of the coordinate system with respect to the vortices. Considering unshielded vortices, we have

$$\frac{\mathrm{d}}{\mathrm{d}t}\ell_{i,j}^2 = \frac{2}{\pi}\sum_k{}'' \gamma_k \sigma_{i,j,k} A_{i,j,k}\left(\frac{1}{\ell_{i,k}^2} - \frac{1}{\ell_{i,k}^2}\right), \tag{17.29}$$

where the sum over k excludes $k = i$ and $k = j$. The quantity $\sigma_{i,j,k}$ defines the orientation of the triangle spanned by the three vortices i, j, k in the sense that $\sigma_{i,j,k} = +1$ if i, j, k appear in counter clockwise order, and $\sigma_{i,j,k} = -1$ for the other case. $A_{i,j,k}$ is the area spanned by the triangle formed by the three vortices. If the three vortices are on a line, we have that $\sigma_{i,j,k}$ is undefined, but that does not matter since $A_{i,j,k} = 0$ there.

Finally we note that in the present formulation the number of vortices N is trivially conserved.

- **Exercise:** Prove the relations (17.27)–(17.29).

The relation (17.29) with (17.23) and (17.28) define the problem of relative motion of N vortices. Except for the orientations as defined by $\sigma_{i,j,k}$ in (17.29), the problem is closed in the $N(N-1)/2$ variables $\ell_{i,j}$, of which $2N - 3$ are independent. The structure of relation (17.29) suggests that the interaction of just three vortices is fundamental for understanding the $N > 3$ vortex problem. For two vortices, the right-hand side of (17.29) is absent, so their relative distance is a constant of motion. The three vortex problem is Hamiltonian with six degrees of freedom and four constants of motion, where the individual vortex strengths appear as constant parameters. The three vortex problem is integrable as shown by Poincaré (1893). It turns out that some of the solutions are so unexpected (at least first time you see them) that a summary of these special results is justified.

17.1.5 Collapse of unshielded vortices

As we have seen, it is a relatively simple matter to describe the dynamics of two vortices. Also problems including *three* vortices can be treated analytically, in principle, to obtain relations for the individual vortex positions as functions of time, with the initial positions as parameters (Onsager 1949, Novikov 1976). The problem also has historical interest, in particular because of the theoretical studies of Gröbli (1877), which have been summarized by Aref *et al* (1992) and also Kono and Škorić (2010). Here we follow the disposition of Aref (1979), and introduce the parameter $\mathcal{C}$ by the definition

$$\gamma_1\gamma_2\ell_{1,2}^2 + \gamma_2\gamma_3\ell_{2,3}^2 + \gamma_3\gamma_1\ell_{3,1}^2 = 3\gamma_1\gamma_2\gamma_3\mathcal{C}, \tag{17.30}$$

where we can have $\mathcal{C} = 0$, in particular. Two of the vortices must necessarily be of the same sign, and we choose the sign convention $\gamma_1 \geqslant \gamma_2 > 0$ and γ_3 arbitrary. If we are to consider situations where two vortices have *negative* sign, that situation can be obtained trivially by a reflection. Note that the requirement of overall charge neutrality of the plasma does not restrict us to consider only cases where $\gamma_1 + \gamma_2 + \gamma_3 = 0$: there can be other line charges (i.e. line vortices) present somewhere else making up for the excess charge. As long as they are at a large distance from the three selected vortices, we need not be concerned with their effects.

For $\mathcal{C} \neq 0$, it is now an advantage to introduce dimensionless variables as

$$b_1 \equiv \frac{\ell_{2,3}^2}{\gamma_1\mathcal{C}}, \qquad b_2 \equiv \frac{\ell_{3,1}^2}{\gamma_2\mathcal{C}}, \qquad \text{and} \qquad b_3 \equiv \frac{\ell_{1,2}^2}{\gamma_3\mathcal{C}},$$

such that $b_1 + b_2 + b_3 = 3$. The variables b_j, with $j = 1, 2, 3$, can be interpreted as coordinates in a plane: the phase space for the problem, see figure 17.4. For any point, P in this diagram, the coordinates b_1, b_2, and b_3 are the distances from the three sides of the triangle, with signs indicated by arrows on figure 17.4. With the present sign convention, a phase point (b_1, b_2, b_3) must lie in any of the regions marked I, II, or III. The region I applies when all three vortices have positive sign, and $\mathcal{C} > 0$, for instance. Only parts of the three regions are accessible, however, because the lengths $\ell_{i,j}$ must satisfy the conditions for forming the sides of a triangle,

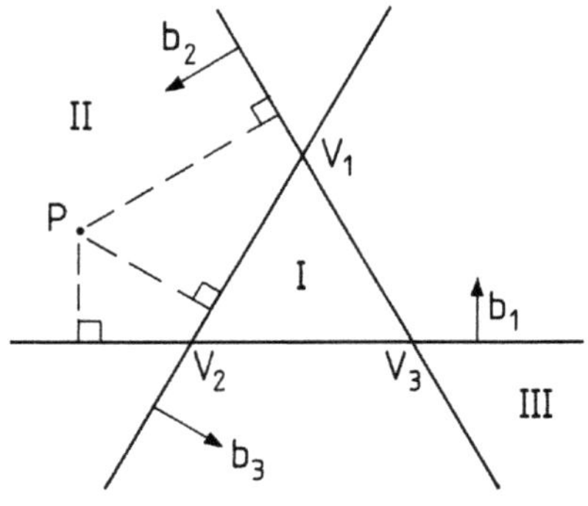

Figure 17.4. Phase space for the unshielded three vortex system. The triangle V_1 V_2 V_3 is equilateral, with height 3.

i.e. the conditions $\ell_{1,2} \leqslant \ell_{2,3}\ell_{3,1}$, $\ell_{2,3} \leqslant \ell_{3,1}\ell_{1,2}$, and $\ell_{3,1} \leqslant \ell_{1,2}\ell_{2,3}$. It is readily shown that these conditions are equivalent to the requirement

$$(\gamma_1 b_1)^2 + (\gamma_2 b_2)^2 + (\gamma_3 b_3)^2 \leqslant 2(\gamma_1\gamma_2 b_1 b_2 + \gamma_2\gamma_3 b_2 b_3 + \gamma_3\gamma_1 b_3 b_1). \tag{17.31}$$

This is a quadratic equation and thus defines a conic section, defining the *physical regions*. These can be

1. The physical regions are ellipses if $\gamma_3 > 0$, giving $\gamma_1 + \gamma_2 + \gamma_3 > 0$ and also if $\gamma_3 < 0$ with $\gamma_1 + \gamma_2 + \gamma_3 < 0$.
2. The physical region is a parabola if $\gamma_1 + \gamma_2 + \gamma_3 = 0$.
3. The physical region is a hyperbola if $\gamma_3 < 0$, while $\gamma_1 + \gamma_2 + \gamma_3 > 0$.

Within the physical regions, the phase point must follow a trajectory constrained by the constant Hamiltonian. Thus, from (17.23) we find for the special case of three unshielded vortices that

$$\ell_{1,2}^{1/\gamma_3}\ell_{2,3}^{1/\gamma_1}\ell_{3,1}^{1/\gamma_2} = \mathrm{e}^{-4\pi H/(\gamma_1\gamma_2\gamma_3)}.$$

Introducing the definitions of b_j, we rewrite this expression as

$$|b_1|^{1/\gamma_1}|b_2|^{1/\gamma_2}|b_3|^{1/\gamma_3} = \frac{\exp\left(-4\pi H/g^3\right)}{|\gamma_1|^{1/\gamma_1}|\gamma_2|^{1/\gamma_2}|\gamma_3|^{1/\gamma_3}|C|^{3h}}, \tag{17.32}$$

with $h \equiv \frac{1}{3}(1/\gamma_1 + 1/\gamma_2 + 1/\gamma_3)$ and $g \equiv (\gamma_1\gamma_2\gamma_3)^{1/3}$. To be on the safe side, we made the expressions symmetric, by putting absolute value signs everywhere. The right-hand side of (17.32) is determined solely by the initial conditions. The phase trajectories are thus level curves of constant $|b_1|^{1/\gamma_1}|b_2|^{1/\gamma_2}|b_3|^{1/\gamma_3}$, subject to the constraint that the curves must be inside the physical regions.

The proofs and discussions of the previous summary are left to the reader as an exercise, with the advice of consulting the work by, e.g., Aref (1979). Here we consider the special case where $C = 0$, which necessarily implies for instance $\gamma_3 < 0$. Evidently, we have to redefine the coordinates b_j, and take

$$b_1 \equiv \frac{\ell_{2,3}^2}{\gamma_1}, \qquad b_2 \equiv \frac{\ell_{3,1}^2}{\gamma_2}, \qquad \text{and} \qquad b_3 \equiv \frac{\ell_{1,2}^2}{\gamma_3}.$$

We have $b_1 + b_2 + b_3 = 0$. Introducing $b_3 = -b_1 - b_2$ in (17.31) we obtain an equation for the boundary of the physical region for this particular case. This equation now becomes homogeneous in b_1 and b_2, so the conic it represents is degenerate. Different possibilities are classified by the arithmetic mean $a \equiv \frac{1}{3}(\gamma_1 + \gamma_2 + \gamma_3)$. The case $a > 0$ is incompatible with $C = 0$. For $a > 0$, the physical region is a wedge in the (b_1, b_2)-plane, with its apex at the origin. For $a = 0$, the wedge collapses to a line through the origin, and the three vortices are co-linear.

The requirement of constant H in (17.23) implies for the present case that the phase trajectories are constant levels for the function

$$g(b_1, b_2) \equiv |b_1|^{1/\gamma_1}|b_2|^{1/\gamma_2}|b_1 + b_2|^{1/\gamma_3}.$$

For $h > 0$, these curves are hyperbola-like, while they are closed curves for $h < 0$, although we have to recall the restrictions implied by the physical regions in the (b_1, b_2)-plane. For $h = 0$, the situation is somewhat special. Inserting this in, for instance, the previous expression we find $g(b_1, b_2)(b_2/b_1)^{1/\gamma_2}(1 + b_2/b_1)^{-1/\gamma_1 - 1/\gamma_2}$, implying that *any* relation $b_2 = b_1\lambda$ is a phase trajectory (where then $b_3 = -(1 + \lambda)b_1)$). It follows that $\ell_{1,2}$, $\ell_{2,3}$ and $\ell_{3,1}$ are all proportional. The physically accessible region can be expressed in terms of λ, and we find the inequality

$$\gamma_1^4(1 + 1/\lambda) + \gamma_2^4(1 + \lambda) \leqslant (\gamma_1 + \gamma_2)^4.$$

We then consider the equations of motion (17.29), which for the present case become

$$\frac{\mathrm{d}}{\mathrm{d}t}\ell_{1,2}^2(t) = \frac{2}{\pi}\gamma_3\sigma A(t)\left(\frac{1}{\ell_{2,3}^2(t)} - \frac{1}{\ell_{3,1}^2(t)}\right) \tag{17.33}$$

$$\frac{\mathrm{d}}{\mathrm{d}t}\ell_{2,3}^2(t) = \frac{2}{\pi}\gamma_3\sigma A(t)\left(\frac{1}{\ell_{3,1}^2(t)} - \frac{1}{\ell_{1,2}^2(t)}\right) \tag{17.34}$$

$$\frac{\mathrm{d}}{\mathrm{d}t}\ell_{3,1}^2(t) = \frac{2}{\pi}\gamma_3\sigma A(t)\left(\frac{1}{\ell_{1,2}^2(t)} - \frac{1}{\ell_{2,3}^2(t)}\right). \tag{17.35}$$

The time evolution is self-similar in the sense that the shape of the triangle formed by the three vortices retains its shape, i.e. the angles, during the evolution, so $\sigma = \text{const}$. The *area* spanned by the vortices is proportional to either of $\ell_{i,j}^2$. For the present case, all the right-hand sides become constants! For $\ell_{3,1} > \ell_{1,2} > \ell_{2,3}$ and $\sigma = 1$, the constants are all negative, and the triangle formed by the three vortices contracts to a point within a finite time. The point is of course located at the center of vorticity. We have $\ell_{i,j}(t) = \ell_{i,j}(0)\sqrt{1 - t/\tau_c}$, where τ_c depends on the initial conditions. Depending on the initial conditions, we can have $\tau_c >$, corresponding to the collapse condition mentioned before, or we can have $\tau_c < 0$, which corresponds to an 'explosion', i.e. a divergence of the vortices as $t \to \infty$.

For a discussion of the absolute motion of the three vortices, it is convenient to introduce polar coordinates, r and ϕ, with respect to the center of vorticity. There will be simple relations between $\ell_{3,1}$, $\ell_{1,2}$, $\ell_{2,3}$ and the coordinates r_1, r_2, r_3 of the vortices. We have

$$r_1 = \ell_{2,3}\sqrt{\frac{\gamma_2 + \gamma_3}{\gamma_1 + \gamma_2 + \gamma_3}},$$

etc, implying

$$r_j(t) = r_j(0)\sqrt{1 - t/\tau_c}.$$

Since the triangle formed by the three vortices retains its shape, we have

$$\frac{\mathrm{d}}{\mathrm{d}t}\phi_1(t) = \frac{\mathrm{d}}{\mathrm{d}t}\phi_2(t) = \frac{\mathrm{d}}{\mathrm{d}t}\phi_3(t) \equiv \omega(t),$$

where

$$\omega(t) = \frac{(\gamma_1 + \gamma_2)^3 + \gamma_1^3(1 + 1/\lambda) + \gamma_2^3(1 + \lambda)}{4\pi(\gamma_1 + \gamma_2)^2 \ell_{1,2}^2(0)(1 - t/\tau_c)}.$$

The angular velocity of the collapsing vortex triplet is thus diverging as $1/\sqrt{1 - t/\tau_c}$. For the 'exploding' triplet, we find that this velocity falls off as $\sim 1/t$ for large times. It is a trivial observation that the trajectories of the vortices form three logarithmic spirals radiating out from the center of vorticity (Aref 1979, Novikov and Sedov 1979, Kimura 1987).

The analysis summarized in this section refers to a system of idealized point-vortices, but it was found (Vosbeek *et al* 1997) that the analysis also captures essential parts of the dynamics of finite size vortices.

Complex notation

An elegant formulation of the vortex motion can be obtained (Kimura 1987) by considering the (x,y)-plane as a *complex* plane, and introduce the coordinates of any vortex as $z_j \equiv x_j + iy_j$. It is then trivially demonstrated that the equation of motion can be written as

$$\frac{\mathrm{d}}{\mathrm{d}t} z_j = -\frac{1}{2\pi i} \sum_{m=1}^{N}{}' \frac{\gamma_m}{\bar{z}_j - \bar{z}_m} \qquad j = 1, 2, \ldots, N, \tag{17.36}$$

where again the terms $j = m$ are excluded in the sum by the prime. Here, the overline denotes complex conjugate.

The formulation (17.36) makes the solution for instance for collapsing vortices particularly elegant. We can introduce an assumption of self-similar motion by $z_m = k_m f(t)$, and obtain

$$k_j \bar{f} \frac{\mathrm{d}}{\mathrm{d}t} f = i \sum_{m=1}^{N}{}' \frac{\Gamma_m}{\bar{k}_j - \bar{k}_m} \qquad j = 1, 2, \ldots, N, \tag{17.37}$$

with $\Gamma_m \equiv \gamma_m/2\pi$. As a postulate, we then introduce $\bar{f}\, \mathrm{d}f/\mathrm{d}t = C \equiv A + iB$, with A and B being real constants. The problem is then reduced to solving the algebraic equations for the k-s and we obtain

$$k_j C = i \sum_{m=1}^{N}{}' \frac{\Gamma_m}{\bar{k}_j - \bar{k}_m} \qquad j = 1, 2, \ldots, N. \tag{17.38}$$

We anticipate that solutions can exist for $N > 3$ (Novikov and Sedov 1979, Chefranov 1987), but only the case $N = 3$ can be solved generally, it seems.

- **Example:** We consider the case where $\Gamma_1 = 2$, $\Gamma_2 = 2$, and $\Gamma_3 = -1$. These three vortices may collapse for properly chosen initial positions, as easily demonstrated from the foregoing analysis. In (17.37) we take

$$
\begin{aligned}
k_1 &= \frac{\Gamma_2}{\Gamma_1 + \Gamma_2}\ell = \frac{1}{2}\ell \\
k_2 &= -\frac{\Gamma_1}{\Gamma_1 + \Gamma_2}\ell = -\frac{1}{2}\ell \\
k_3 &= \sqrt{\frac{\Gamma_1 + \Gamma_2 + \Gamma_3}{\Gamma_1 + \Gamma_2}}\,\ell \cos\theta + i\sqrt{\frac{\Gamma_1 + \Gamma_2 + \Gamma_3}{\Gamma_1 + \Gamma_2}}\,\ell \sin\theta \\
&= \frac{\sqrt{3}}{2}\ell \cos\theta + i\frac{\sqrt{3}}{2}\ell \sin\theta,
\end{aligned}
$$

identifying positions as in figure 17.5.

The angle to vortex 3 from the origin is denoted θ. The center of vorticity is, for this case,

$$
\frac{\Gamma_1 k_1 + \Gamma_2 k_2 + \Gamma_3 k_3}{\Gamma_1 + \Gamma_2 + \Gamma_3} = -\frac{\sqrt{3}}{6}\ell \cos\theta - i\frac{\sqrt{3}}{6}\ell \sin\theta.
$$

We would like to have the center of vorticity at the origin, so we subtract this value from all k-s, and find

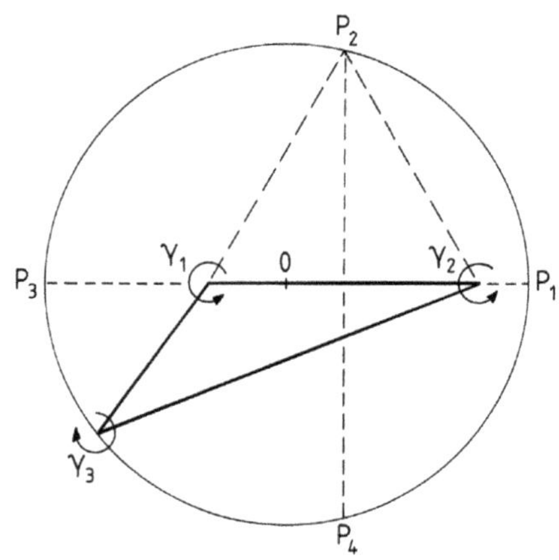

Figure 17.5. Summary of initial conditions for a vortex triplet, where position of γ_3 determines a collapsing or explosive time evolution.

$$k_1 = \frac{3 + \sqrt{3}\cos\theta}{6}\ell + i\frac{\sqrt{3}\sin\theta}{6}\ell$$
$$k_2 = \frac{-3 + \sqrt{3}\cos\theta}{6}\ell + i\frac{\sqrt{3}\sin\theta}{6}\ell$$
$$k_3 = \frac{2\sqrt{3}\cos\theta}{3}\ell + i\frac{2\sqrt{3}\sin\theta}{3}\ell.$$

This would correspond to moving the vortex triangle on figure 17.5 slightly *down*, i.e. the center of vorticity is *outside* the triangle spanned by the three vortices. We can then calculate

$$C = \frac{i}{\bar{k}_3}\left(\frac{\Gamma_1}{\bar{k}_3 - \bar{k}_1} + \frac{\Gamma_2}{\bar{k}_3 - \bar{k}_2}\right) = -\frac{6\sin 2\theta}{(5 - 3\cos 2\theta)\ell^2} + i\frac{18 - 6\cos 2\theta}{(5 - 3\cos 2\theta)\ell^2}$$
$$A = -\frac{6\sin 2\theta}{(5 - 3\cos 2\theta)\ell^2}$$
$$B = \frac{18 - 6\cos 2\theta}{(5 - 3\cos 2\theta)\ell^2}.$$

The collapse time can be calculated as

$$\tau_c = \frac{5 - 3\cos 2\theta}{12\sin 2\theta}\ell^2.$$

The smallest value of τ_c, for a given ℓ, is obtained for $\theta = \pm\frac{1}{2}\arccos 3/5$. The variations of A and B with θ are shown in figure 17.6, assuming $\ell = 1$.

The vortex collapse is algebraically unstable in the sense that a small deviation from the condition no longer leads to the collapse. In those cases three vortices first converge to some extent and then turn to diverge. Once they start to diverge, they keep diverging as long as (17.12) is applicable. This is, however, not the case for (17.11). When the vortices diverge in this case two of them eventually form a pair to travel together and the third one is left behind, because of the short range nature of

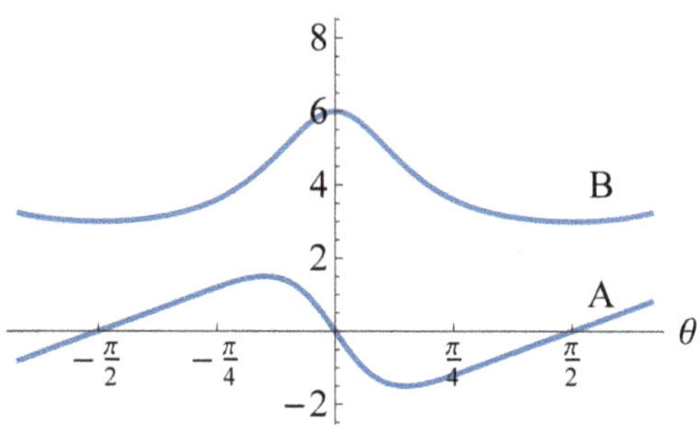

Figure 17.6. Variation of A and B with θ. The interval where $A < 0$ corresponds to collapsing solutions for the three given vortices, while $A > 0$ corresponds to the vortex 'explosion'. The three points where $A = 0$ correspond to a rigid motion of the vortex triangle.

the interaction. However since the vorticities of two vortices traveling together are opposite in sign and have to be different in magnitude, the two vortices return to the place where the third one remains. At the time they come to the position where the two sides of the triangle become equal, the pair is renewed by changing the partner. The new pair travels in a similar way as before until it comes back to the place and again, exchanging its partner (Kono *et al* 1998a).

The vortex collapse is thought to provide a key to understanding fundamental processes of strong turbulence. As discussed for instance by Novikov (1976) and Novikov and Sedov 1979), the collapse accounts for energy being systematically transferred into large scales in the spectrum, and for vortex dispersion, or 'explosion', energy cascades into the small scales. The boomerang effect described here, an effect which owes its existence to the shielding effect ($\kappa \neq 0$), consequently corresponds to a 'sloshing' of energy between large and small scales.

17.2 Steady state solutions with distributed vorticity

As already noticed, any cylindrically symmetric charge distribution is a steady state solution to the basic equations, for instance (17.1). It is, however, possible to construct more general and also more interesting solutions, such as dipoles or higher order multipoles. Physically, we might imagine such solutions obtained by a 'close packing' of line vortices so that they can be seen as a continuum in coarse grained description. This idea forms the basis of the so-called vortex-in-cell simulations of two-dimensional magnetized plasmas.

We consider here the basic equation in the unshielded form (17.1) with shielded structures to be discussed in section 18.1.4. We note first that any cylindrically symmetric continuous potential variation $\phi(\mathbf{r}) = \phi(|\mathbf{r}|)$ is a steady state solution, although a rather trivial one. More interesting stationary solutions can be found by moving to a frame of reference moving with a (for the time being unspecified) velocity $\mathbf{U}$, implying that the time variation found by a stationary observer is giving $\partial\phi/\partial t \to -\mathbf{U} \cdot \nabla\phi$. Since we have, for the time being, no bulk plasma density gradients, there is no preferred direction either in the plane perpendicular to the magnetic field, and any direction for $\mathbf{U}$ is equally acceptable. We assume $\mathbf{U}$ to be in the y-direction, i.e. $\mathbf{U} = U\hat{\mathbf{y}}$. Using normalized units, it is readily found by insertion in (17.1) that any potential variation satisfying $\nabla^2\phi = G(\phi + Ux)$, with G being an arbitrary function with constant $\mathbf{U}$ is a solution to the problem, giving $\partial\phi/\partial t \to -U\partial\phi/\partial y$. It need not, however, be a physically acceptable solution. In order to demonstrate that the problem *has* physically acceptable solutions we consider a particularly simple case where $G(\phi + Ux) = (\phi + Ux)C_1$, with C_1 being a constant of proportionality. We use normalized units so that $[\phi] = [Ux]$. In polar coordinates we write $\phi = F(r)\sin\theta$ giving $G(\phi + Ux) = C_1(F\sin\theta + Ut\sin\theta)$, with F being a function to be determined. We then have in cylindrical coordinates

$$\nabla^2\phi \to \sin\theta\left(\frac{\mathrm{d}^2}{\mathrm{d}r^2}F(r) + \frac{1}{r}\frac{\mathrm{d}}{\mathrm{d}r}F(r) - \frac{F(r)}{r^2}\right),$$

implying

$$r^2\frac{d^2}{dr^2}F(r) + r\frac{d}{dr}F(r) - F(r) = \left(r^2F(r) + Ur^3\right)C_1. \tag{17.39}$$

This equation does not, as it stands, have any physically acceptable bounded solutions. These can, however, be constructed by dividing the plane into an inner and an outer region, assigning two different values of the constant C_1 to each region. We let the boundary between the inner and outer region be a circle with radius R. With $C_1 = 0$ in the outer region and $C_1 = -1$ in the inner, we find a physically acceptable solution for the potential in the form

$$\phi(r, \theta) = UR\begin{cases}\left(\dfrac{2}{\gamma_{1,1}J_2(\gamma_{1,1})}J_1\left(\gamma_{1,1}r/R\right) + \dfrac{r}{R}\right)\sin\theta & \text{for } r < R\\ \dfrac{R}{r}\sin\theta & \text{for } r > R,\end{cases} \tag{17.40}$$

where we have $\gamma_{1,1}$ being the first zero for the Bessel function, i.e. $J_1(\gamma_{1,1}) = 0$. The solution (17.40) is made continuous at $r = R$, with a continuous derivative. We used explicitly $J_2(\gamma_{1,1}) = -J_0(\gamma_{1,1})$. The important observation is now that the solution (17.40), often called a *modon*, is completely determined by two parameters, R and U. The velocity U, in particular, enters in the peak amplitude of the potential. The solution (17.40) is shown in figure 17.7. By choosing the first zero, $\gamma_{1,1}$ of the Bessel

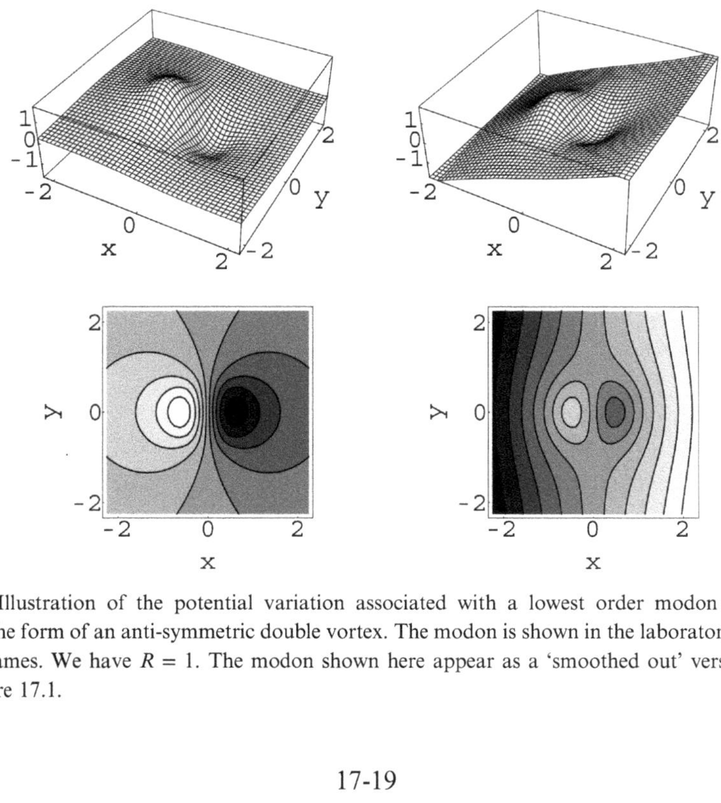

Figure 17.7. Illustration of the potential variation associated with a lowest order modon solution. This solution has the form of an anti-symmetric double vortex. The modon is shown in the laboratory as well as the co-moving frames. We have $R = 1$. The modon shown here appear as a 'smoothed out' version of the line dipole in figure 17.1.

function we selected the *lowest order* modon solution. It is evident that higher order solutions can be constructed, similarly. This is left as an exercise for the reader. Note that the polarity of the modon is arbitrary.

17.2.1 Rotating modons

The modon solution shown in for instance figure 17.7 is anti-symmetric, and propagates with constant velocity along a straight line. It is possible to construct asymmetric modons, which propagate in a circular orbit. These have been discussed in detail by for instance (Flierl *et al* 1983). Such solutions are found by

$$\nabla^2\phi = F\left(\phi + Ur\sin\theta + \frac{1}{2}\epsilon Ur^2/a\right),$$

where F is arbitrary. Here, u is the velocity along the circular track with radius R, while a is the radius of the modon, while $\epsilon \equiv a/R$. The angular velocity of the modon is then $\Omega = c\epsilon/a$. Also in this case we choose the simple linear relationship for F in the interior part, and $F = 0$ in the outer part. After some straight forward algebra, which is given in detail by (Flierl *et al* 1983), the final result is obtained as

$$\phi(r,\theta) = Ua\begin{cases} \left(\dfrac{2}{\gamma_{1,1}J_2(\gamma_{1,1})}J_1(\gamma_{1,1}r/R) - \dfrac{r}{a}\right)\sin\theta & \\ +\dfrac{1}{2}\epsilon\left(1 - \dfrac{r^2}{a^2}\right) + \dfrac{2}{\gamma_{1,1}^2}\epsilon\left(1 - \dfrac{J_0(\gamma_{1,1}r/a)}{J_0(\gamma_{1,1})}\right) & \text{for } r < a \\ -\dfrac{a}{r}\sin\theta - \epsilon\ln\dfrac{r}{a} & \text{for } r > a \end{cases} \tag{17.41}$$

In the moving modon frame the solution can be written as $\phi + Ur\sin\theta + \frac{1}{2}\epsilon\, r^2/a^2$. These solutions are shown in figure 17.8 for three values of ϵ. Note the intensification of one vortex and weakening of the other as the curvature of the path becomes greater. For $\epsilon = 0$ we recover the simple modon (17.40) which propagates with a straight line orbit. The rotating modon is not charge neutral, and the excess charge has to be compensated somewhere else in the plane, to have the plasma overall charge neutral.

It is interesting to note that the potential variation looks quite different in the absolute and in the modon frames of reference. One should bear this in mind when interpreting for instance experimental results as those of Pécseli *et al* (1985).

17.2.2 Modons versus solitons

The modons as discussed in sections 17.2 and 17.2.1 are exact solutions to nonlinear equations, and can be considered as *solitary* waves. These are, however, in general distinct from *solitons* (Drazin and Johnson 1989). For a soliton solution we

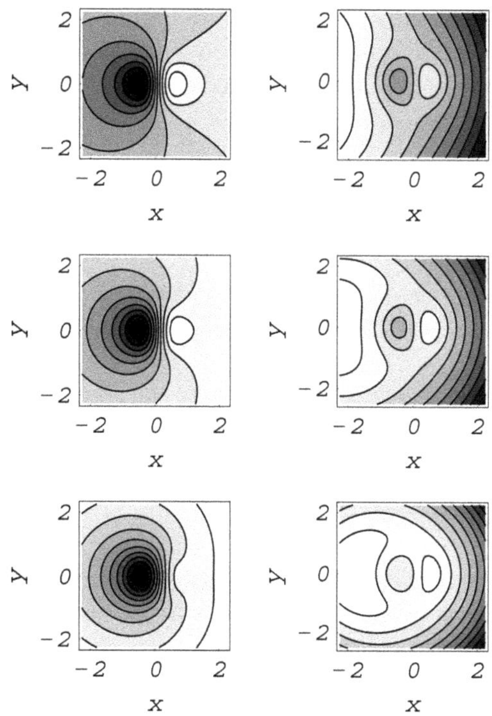

Figure 17.8. Illustration of the potential variation associated with a lowest order rotating modon solution. This solution has the form of an asymmetric double vortex. The modon is shown in the laboratory as well as the co-moving frames. The left column is the absolute frame, right column the modon frame. The upper row is for $\epsilon = 0.25$, the middle for $\epsilon = 0.5$, and the lower one for $\epsilon = 1$, in all cases we have $u = 1$ and $a = 1$.

require that two such solutions 'survive' a collision, and emerge after such an interaction without change of shape, though possibly with a slight phase shift. This feature seems to be intimately connected with the existence of an inverse scattering solution (Drazin and Johnson 1989) of the basic equations as we know it from the Korteweg–de Vries equation, or the nonlinear Scrödinger equation, to mention two. There is no such inverse scattering solution for (17.7), and no reason to expect that one will ever be found. It is true, however that two modons will emerge essentially unchanged from a head-on collision, but allowing for a non-zero 'impact parameter' of the collision, the modons become strongly deformed by such a collisional interaction. Such solutions were presented by, for instance, Larichev and Reznik (1983) for a related problem.

An inverse scattering method will usually imply that the final, asymptotic, solution of any initial value problem with finite spatial support can be described as a sum of solitons, with some additional low level fluctuation which can adequately be described by a linearized version of the basic equation. In the absence of such a solution for equations like (17.1) it might actually be argued that the modon solutions we have found with such an effort might be rather uninteresting if they require a very carefully chosen initial condition to be excited. It turns out, however, that from a random initial condition (in particular in numerical solutions of two-dimensional flows) one often finds structures emerging which have a surprising similarity with the modons. In this sense our efforts have not been wasted!

References

Aref H 1979 Motion of three vortices *Phys. Fluids* **22** 393–400

Aref H, Rott N and Thomann H 1992 Gröbli's solution of the three-vortex problem *Ann. Rev. Fluid Mech.* **24** 1–21

Aris R 1962 *Vectors, Tensors, and the Basic equations of Fluid Mechanics* (London: Prentice-Hall)

Chefranov S G 1987 Dynamics of point vortex dipoles and spontaneous singularities in three dimensional turbulent flows *Sov. Phys. JETP* **66** 85–9

Chefranov S G 1987 *russian original Zh. Eksp. Teor. Fiz.* **93** 151–8

Drazin P G and Johnson R S 1989 *Solitons: an Introduction* (Cambridge: Cambridge University Press)

Ewald P P 1921 Die berechnungen optischer und elektronischer gitterpotentiale *Ann. Phys.* **64** 253

Flierl G R, Stern M E and Whitehead J A Jr 1983 The physical significance of modons: laboratory experiments and general integral constraints *Dynam. Athmos. Oceans* **7** 233–63

Gröbli W 1877 Specielle probleme über die bewegung geradliniger paralleler wirbelfälden *Vierteljahrsschrift der Naturforschenden Gesselschaft in Zürich* **22** 37–81, 129–165

Kimura Y 1987 Similarity solution of two-dimensional point vortices *J. Phys. Soc. Jpn* **56** 2024–30

Kono M, Krane B, Pécseli H L and Trulsen J 1998a Localization and vortices in strongly magnetized plasmas *Phys. Scripta* **T75** 290–4

Kono M, Krane B, Pécseli H L and Trulsen J 1998b Vortex dynamics in magnetized plasmas *Phys. Scripta* **58** 238–45

Kono M and Škorić M M 2010 *Nonlinear Physics of Plasmas,* number 62 (*Springer Series on Atomic, Optical, and Plasma Physics*) (Heidelberg: Springer)

Lamb H 1945 *Hydrodynamics* 6th edn (New York: Dover)

Larichev V D and Reznik G M 1983 On collisions between two-dimensional solitary Rossby waves *Oceanology* **23** 545–52

Lin C C 1943 *On the Motion of Vortices in Two Dimensions,* vol 5 *of Applied mathematics series* (Toronto: The University of Toronto Press)

Lynov J P, Nielsen A H, Pécseli H L and Juul Rasmussen J 1991 Studies of the Eulerian-Lagrangian transformation in two-dimensional random flows *J. Fluid Mech.* **224** 485–505

Milne-Thomson L M 1968 *Theoretical Hydrodynamics* 5 edn (London: Macmillan)

Morikawa G K 1960 Geostrophic vortex motion *J. Meteorol* **17** 148–58

Morikawa G K and Swenson E V 1971 Interacting motion of rectilinear geostrophic vortices *Phys. Fluids* **14** 1058–73

Novikov E A 1976 Dynamics and statistics of a system of vortices *Sov. Phys. JETP* **41** 937

Novikov E A and Sedov Y B 1979 Vortex collapse *Sov. Phys. JETP* **50** 297–301

Novikov E A and Sedov Y 1979 *Russian original Zh. Eksp. Teor. Fiz.* **77** 588–97

Onsager L 1949 Statistical hydrodynamics *Il Nuovo Cimento Series 9* **6** 279–87

Pécseli H L, Rasmussen J J and Thomsen K 1985 Nonlinear interaction of convective cells in plasmas *Plasma Phys. Contr. Fusion* **27** 837–46
Poincaré H 1893 *Théorie des Tourbillons* (Paris: Deslis Fréres) Ch. IV.
Vosbeek P W C, van Geffen J H G M, Meleshko V V and van Heijst G J F 1997 Collapse interactions of finite-sized two-dimensional vortices *Phys. Fluids* **9** 3315–22

Hans Pécseli

Chapter 18

Weakly nonlinear electrostatic drift waves

The analysis in the foregoing sections was based on a linearized analysis applicable for small amplitude waves. It was an interesting feature, though, that the basic equation (4.8) of chapter 4 turned out to become linear all by itself so to speak, without requiring additional linearizations. Two essential approximations were made to derive this equation; the assumption of quasi-neutrality (the 'plasma approximation') and omission of ion polarization drifts, or in more general terms, ion inertia was ignored. In the following we outline the consequences of relaxing the latter of these two approximations.

- **Exercise:** Consider a simplified cold ion model, where the quasi-neutrality assumption is relaxed, i.e. Poisson's equation is retained, but the ion velocity is still identified by the $\mathbf{E} \times \mathbf{B}/B^2$-velocity, ignoring any motion parallel to $\mathbf{B}$. Derive a nonlinear dynamic equation for the electrostatic potential by use of the ion continuity equation (5.1), the assumption of isothermal Boltzmann distribution for the electrons (4.7). Discuss the resulting equation, and estimate the relative magnitude of the various terms for different scale-lengths, measured in terms of the electron Debye length.

18.1 The Hasegawa–Mima equation

The relaxation of the quasi-neutrality condition implies some formal modifications of the basic equations, but these are unlikely to be of physical importance in comparison to the ion polarization drift as long as the wavelengths considered are much larger than the Debye length. The consequences of including ion polarization drifts is studied in the following, but the quasi-neutral approximation is retained. First we consider the limiting case where the electrons are assumed to be in isothermal Boltzmann equilibrium (4.7). This limit was demonstrated to be linearly stable.

We first formulate the basic equations for electrostatic drift waves, using a cold ion model which has often been used (Kadomtsev 1965). Introducing the operator notation $D/Dt \equiv \partial/\partial t + \mathbf{u} \cdot \nabla = \partial/\partial t + \mathbf{u}_\perp \cdot \nabla_\perp + u_\| \partial/\partial z$ we have the ion continuity equation in the form

$$\frac{D}{Dt} \ln n = -\nabla \cdot \mathbf{u}, \tag{18.1}$$

where $n_e \approx n_i \equiv n$ by the assumption of quasi-neutrality and $\mathbf{u} \equiv \mathbf{u}_\perp + u_\| \hat{\mathbf{z}}$ is the ion velocity.

The ion momentum equation can be separated into two components, where the part perpendicular to the magnetic field can be written in the approximate form

$$\begin{aligned} \mathbf{u}_\perp &= -\frac{\nabla_\perp \phi \times \mathbf{B}}{B^2} - \frac{M}{eB^2}\frac{D}{Dt}\mathbf{u}_\perp \times \mathbf{B} \\ &\approx -\frac{\nabla_\perp \phi \times \mathbf{B}}{B^2} + \frac{M}{eB^2}\frac{D}{Dt}\nabla_\perp \phi, \end{aligned} \tag{18.2}$$

where it was explicitly used that $\mathbf{B}$ can be taken to be a constant vector. The first term on the right-hand side of (18.2) is the $\mathbf{E} \times \mathbf{B}/B^2$-drift, the second is the ion polarization drift. To the same accuracy, we rewrite the D/Dt-operator as

$$\frac{D}{Dt} \approx \frac{\partial}{\partial t} - \frac{\nabla_\perp \phi \times \mathbf{B}}{B^2} \cdot \nabla_\perp + u_\| \frac{\partial}{\partial z}. \tag{18.3}$$

The $\mathbf{B}$-parallel ion velocity component is

$$M\frac{D}{Dt}u_\| = -e\frac{\partial}{\partial z}\phi. \tag{18.4}$$

The set of equations is closed by assuming again that electrons are locally Boltzmann distributed at all times with constant temperature T_e

$$n = n_0(\mathbf{r}_\perp)\exp\left(\frac{e\phi}{\kappa T_e}\right), \qquad \text{or} \qquad \ln n = \ln n_0(\mathbf{r}_\perp) + \frac{e\phi}{\kappa T_e}, \tag{18.5}$$

where $n_0(\mathbf{r}_\perp)$ is the plasma density in the case where the electrostatic potential vanishes, $\phi = 0$. The normalization of density is not specified since it is immaterial for the applications of (18.5).

Recalling the basic model for the electrostatic drift waves in chapter 4 we ignore here the ion dynamics in the direction along magnetic field lines and take $u_\| \approx 0$. For this case (18.4) becomes redundant and the D/Dt operator becomes simpler. We obtain a closed expression by combining the nonlinear equations (18.1), (18.2) and (18.5). Using $\nabla_\perp \cdot \nabla_\perp \phi \times \mathbf{B} = 0$ and also $\nabla_\perp \cdot (\nabla_\perp \phi \times \mathbf{B} \cdot \nabla_\perp)\nabla_\perp \phi = (\nabla_\perp \phi \times \mathbf{B} \cdot \nabla_\perp)\nabla_\perp^2 \phi$ we obtain the equation

$$\begin{aligned} &\frac{\partial}{\partial t}\phi - \frac{\kappa T_e}{e}\frac{\nabla_\perp \phi \times \mathbf{B}}{B^2} \cdot \nabla_\perp \ln n_0 \\ &- \frac{C_s^2}{\Omega_c^2}\left(\frac{\partial}{\partial t} - \frac{1}{B^2}\nabla_\perp \phi \times \mathbf{B} \cdot \nabla_\perp\right)\nabla_\perp^2 \phi = 0. \end{aligned} \tag{18.6}$$

The first term is readily recognized as the simple result (4.8) obtained in chapter 4. The second term in (18.6) is the correction from the polarization drift. The ordering implied in the derivation of this equation can be made explicit as

$$\frac{1}{\Omega_c}\frac{\partial}{\partial t} \sim \frac{\Omega}{\Omega_c} \sim a_i|\nabla_\perp \ln n_0| \sim \frac{1}{k_\| U_{the}}\frac{\partial}{\partial t} \sim \epsilon,$$

where $\Omega \equiv \nabla \times \mathbf{u}_\perp$ is here the ion vorticity, a_i is the ion Larmor radius derived from the electron temperature and ϵ is a small quantity. The restriction on $k_\| U_{the}$, in terms of the electron thermal velocity U_{the} is equivalent to the condition $\omega/k_\| \ll U_{the}$. The two-dimensional description of the ion dynamics assumes that $\mathbf{u}_{i\perp} \gg u_{i\|}\hat{\mathbf{z}}$ and is not violating the constraint regarding the electron dynamics.

We see the operator $\partial/\partial t - B^{-2}\nabla_\perp\phi \times \mathbf{B} \cdot \nabla_\perp$ pop up so often that it seems natural to assign it a particular importance for describing two-dimensional turbulence.

Equation (18.6) is often called the Hasegawa–Mima equation after the two scientist who derived it in their analysis of weakly nonlinear drift waves (Hasegawa and Mima 1978). A similar equation, first derived by Charney (1948), applies for Rossby waves in the atmosphere and in the oceans, see also section 5.3. Introducing dimensionless variables by normalizing length with a_i, time with $1/\Omega_c$ and potential with $\kappa T_e/e$ we have

$$\frac{\partial}{\partial t}\left(\nabla_\perp^2\phi - \phi\right) + \left(\hat{\mathbf{z}} \times \nabla_\perp\phi\right) \cdot \nabla_\perp\left(\nabla_\perp^2\phi\right) - \beta\frac{\partial}{\partial y}\phi = 0, \tag{18.7}$$

where a constant $\beta = -\mathrm{d}\ln n_0/\mathrm{d}x$ was introduced as a measure of the electron diamagnetic drift. It is assumed that β can be considered as constant, at least locally. The notation is eased by introducing the Poisson brackets,

$$\{f;\, g\} \equiv \frac{\partial f}{\partial x}\frac{\partial g}{\partial y} - \frac{\partial f}{\partial y}\frac{\partial g}{\partial x} \equiv \hat{\mathbf{z}} \cdot (\nabla_\perp f \times \nabla_\perp g).$$

Note that $\{f; f\} = 0$, $\{f;\, g\} = -\{g; f\}$, $\{f + h;\, g\} = \{f;\, g\} + \{h;\, g\}$ and $\{f \cdot h;\, g\} = h\{f;\, g\} + f\{h;\, g\}$. These relations will be useful later on. A compact formulation of (18.7) can be obtained as

$$\frac{\partial}{\partial t}\left(\nabla_\perp^2\phi - \phi\right) - \beta\frac{\partial}{\partial y}\phi + \left\{\phi;\, \nabla_\perp^2\phi\right\} = 0, \tag{18.8}$$

or

$$\frac{\partial}{\partial t}q + \{\phi;\, q\} = 0, \tag{18.9}$$

with $q = \phi - \nabla_\perp^2\phi - \beta x$, where the $\nabla_\perp^2\phi$ term originates from the polarization drift. As an alternative form of (18.9) we can write

$$\left(\frac{\partial}{\partial t} + \mathbf{u}_\perp \cdot \nabla_\perp\right)q = 0, \tag{18.10}$$

with $\mathbf{u}_\perp \equiv -\nabla_\perp\phi \times \hat{\mathbf{z}}$. Equation (18.10) has the form of a continuity equation for q in an incompressible flow.

The equations (18.6) or (18.7) have a number of interesting properties which will be discussed in the following.

In the limit where $C_s/\Omega_c \to \infty$ and the density gradient vanishes, $\nabla_\perp n_0 = 0$, the Hasegawa–Mima equation formally reduces to the two-dimensional Euler equation

$$\left(\frac{\partial}{\partial t} + \mathbf{u}_\perp \cdot \nabla_\perp\right)\rho = 0, \tag{18.11}$$

with $\mathbf{u} \equiv -\nabla_\perp\phi \times \hat{\mathbf{z}}$, where $\rho \equiv \nabla_\perp^2\phi = \hat{\mathbf{z}} \cdot \nabla_\perp \times \mathbf{u}$ takes the role of vorticity in this two-dimensional model. Equation (18.11) has the form of a continuity equation for ρ, to be compared with (18.10). Seen from this point of view, it is expected that properties of the Hasegawa–Mima equation have many similarities with the equations characterizing two-dimensional turbulence in incompressible fluids. Being a generalization of the fluid vorticity, $q = \phi - \nabla_\perp^2\phi - \beta x$ is often called *potential vorticity*.

- **Exercise**: Prove $\nabla_\perp \cdot (\nabla_\perp\phi \times \mathbf{B} \cdot \nabla_\perp)\nabla_\perp\phi = (\nabla_\perp\phi \times \mathbf{B} \cdot \nabla_\perp)\nabla_\perp^2\phi$.

Comments on terminology:

The mathematical models for electrostatic drift waves have many features in common with the modeling of weakly nonlinear waves on rotating planets. The similarity is made conspicuous by the Charney-equation (Charney 1948) for Rossby waves. For the large-scale horizontal flow on a rotating spherical Earth, the equations can be simplified to the so-called nonlinear Charney balance equation (Pedlosky 1987). The equation enables us to determine the geo-potential field on the sphere from a given stream-function field or vice versa. In geophysical fluid dynamics, an approximation whereby the Coriolis parameter, f, is set to vary linearly in space is called a β-plane approximation (Rhines 1975). For the geophysical problem term $\mathbf{u} \times \hat{\mathbf{f}}$ originates from the Coriolis force: a similar term for the drift wave problem $q\mathbf{u} \times \hat{\mathbf{b}}$ has its origin in the Lorentz force.

On a rotating sphere such as the Earth, f varies with the sine of the latitude; in the so-called f-plane approximation this variation is ignored, and a value of f appropriate for a particular latitude is used throughout the domain. This approximation can be visualized as a tangent plane touching the surface of the sphere at the relevant latitude. A somewhat more accurate model is obtained by a linear approximation to this variability about a given latitude (in the sense of a Taylor series expansion). In analogy with the f-plane, this approximation is termed the β-plane, even though it no longer describes dynamics on a hypothetical tangent plane. The advantage of the β-plane approximation over more accurate formulations is that it does not contribute nonlinear terms to the dynamical equations.

The beta plane approximation is useful for the theoretical analysis of many phenomena in geophysical fluid dynamics since it makes the equations much more tractable, yet retains the important information that the Coriolis parameter varies in space. In particular, Rossby waves, the most important type of waves if one considers large-scale atmospheric and oceanic dynamics, depend on the variation of f as a restoring force; they do not occur if the Coriolis parameter is approximated by a constant.

18.1.1 Linearized Hasegawa–Mima equation

In its linearized version, the Hasegawa–Mima equation becomes

$$\frac{\partial}{\partial t}\left(\nabla_\perp^2\phi - \phi\right) - \beta\frac{\partial}{\partial y}\phi = 0. \tag{18.12}$$

The linear dispersion relation is easily obtained as

$$\omega = \frac{\beta k_y}{1 + k_\perp^2} \equiv \frac{\omega^*}{1 + k_\perp^2}, \tag{18.13}$$

with $k_\perp^2 \equiv k_x^2 + k_y^2$. The phase velocity is evidently restricted to a finite range with $\omega/k_y \in [0, \beta]$ for $k_\perp^2 \in [0, \infty]$. Note that we have a range of backward wave propagation, i.e. a region of wavenumbers where the wavevector and the group velocity have opposite direction. In particular we find that while the phase velocity for $k_x > 0$ has a component in the direction of decreasing density, the group velocity will be in the opposite direction. To some extent, it may be misleading to include also waves having a propagation component along the density gradient: when obtaining the basic expressions for the drift wave dynamics in section 5.1 we omitted a term like this in order to obtain a local dispersion relation.

18.1.2 Conservation laws

Considering (18.10), it is readily concluded that q is constant along the characteristics of this equation. The same conclusion will hold for any function $F(q)$. Making explicit use of the incompressibility $\nabla \cdot \mathbf{u} = \nabla \cdot \nabla\phi \times \mathbf{B}/B^2 = 0$ with $\mathbf{B}$ constant, it is easily shown that any integral (Casimir functional) $\iint F(q)\mathrm{d}x\mathrm{d}y$ will be a constant of motion, see (18.10). Alternatively $\iint q^n\mathrm{d}x\mathrm{d}y$ is a constant of motion for all n. The notation in terms of Poisson brackets is particularly useful for obtaining conservation laws when the relations $\iint f\{h; g\}\mathrm{d}x\mathrm{d}y = \iint g\{f; h\}\mathrm{d}x\mathrm{d}y = \iint h\{g; f\}\mathrm{d}x\mathrm{d}y$ are used, with the proper assumptions on the properties of the functions at $|\mathbf{r}| \to \infty$.

- **Exercise:** Using the incompressibility condition, $\nabla \cdot \mathbf{u} = 0$ together with $\partial q(\mathbf{r}, t)/\partial t + \mathbf{u} \cdot \nabla q(\mathbf{r}, t) = 0$, prove that $\mathrm{d}\iint F(q)\,\mathrm{d}x\mathrm{d}y/\mathrm{d}t = 0$ for any differentiable $F(q(\mathbf{r}, t))$.

In particular we have a quadratic constant of motion, the energy integral, $E \equiv \frac{1}{2}\iint[\phi^2 + (\nabla\phi)^2]\mathrm{d}x\mathrm{d}y$. This can be shown most simply by use of the form (18.9), which after multiplication with ϕ and integration gives

$$\iint\left[\phi\frac{\partial}{\partial t}q + \phi\{\phi; q\}\right]\mathrm{d}x\mathrm{d}y = \iint\phi\frac{\partial}{\partial t}\left(\nabla^2\phi - \phi\right)\mathrm{d}x\mathrm{d}y = 0, \tag{18.14}$$

where it was used that $\iint f\{f;\, g\}\mathrm{d}x\mathrm{d}y = 0$ for all g with the proper asymptotic relations. We note the identity $\phi\partial\nabla^2\phi/\partial t = \nabla\cdot(\phi\partial\nabla\phi/\partial t) - \frac{1}{2}\partial(\nabla\phi)^2/\partial t$. By simple manipulations of (18.14), using Green's theorem giving $\int\nabla\cdot(\phi\partial\nabla\phi/\partial t)\mathrm{d}x\mathrm{d}y = \oint\phi\partial\nabla\phi/\partial t\cdot \mathrm{d}\mathbf{n} = 0$ with $\mathbf{n}$ being a unit normal vector to the curve in the x, y-plane bounding the entire system, it is readily demonstrated that

$$\frac{\mathrm{d}}{\mathrm{d}t}\iint\frac{1}{2}\left[\phi^2 + (\nabla\phi)^2\right]\mathrm{d}x\mathrm{d}y = 0$$

stating the conservation of E. The two terms can be interpreted as an electrostatic potential energy ($n\phi \sim \phi^2$) and a kinetic energy ($n_0u^2 \sim (\nabla\phi)^2$).

In a Fourier representation this conservation law takes the form

$$\frac{\mathrm{d}}{\mathrm{d}t}\sum_{\mathbf{k}}(1 + k^2)\left|\phi_{\mathbf{k}}\right|^2 = 0, \tag{18.15}$$

by use of Parseval's theorem $\iint|f|^2\mathrm{d}x\mathrm{d}y = \sum_{\mathbf{k}}|f_{\mathbf{k}}|^2$.

Another quadratic constant of motion is the generalized enstrophy, defined as $W = \frac{1}{2}\iint[(\nabla\phi)^2 + (\nabla^2\phi)^2]\mathrm{d}x\mathrm{d}y$, which can be interpreted as being similar to the enstrophy defined for two-dimensional incompressible flows. The conservation of W is demonstrated by using (18.8), which after multiplication with $\nabla^2\phi$ and integration gives

$$\iint\nabla^2\phi\left(\frac{\partial}{\partial t}(\nabla^2\phi - \phi) - \beta\frac{\partial}{\partial y}\phi\right)\mathrm{d}x\mathrm{d}y = 0.$$

Using

$$\begin{aligned}\left(\nabla^2\phi\right)\frac{\partial}{\partial t}\phi &= \nabla\cdot\left(\nabla\phi\frac{\partial}{\partial t}\phi\right) - \nabla\phi\cdot\frac{\partial}{\partial t}\nabla\phi\\ &= \nabla\cdot\left(\nabla\phi\frac{\partial}{\partial t}\phi\right) - \frac{1}{2}\frac{\partial}{\partial t}(\nabla\phi)^2,\end{aligned}$$

and

$$\begin{aligned}\left(\nabla^2\phi\right)\frac{\partial}{\partial y}\phi &= \nabla\cdot[\nabla\phi(\nabla\phi\cdot\hat{\mathbf{y}})] - \frac{1}{2}\frac{\partial}{\partial y}(\nabla\phi)^2\\ &= \nabla\cdot\left(\nabla\phi(\nabla\phi\cdot\hat{\mathbf{y}}) - \frac{1}{2}(\nabla\phi)^2\hat{\mathbf{y}}\right),\end{aligned}$$

we obtain by use of Green's theorem and some mild restrictions on ϕ at $|\mathbf{r}| \to \infty$

$$\frac{\mathrm{d}}{\mathrm{d}t}\iint\frac{1}{2}\left[(\nabla\phi)^2 + \left(\nabla^2\phi\right)^2\right]\mathrm{d}x\mathrm{d}y = 0,$$

demonstrating the conservation of generalized enstrophy, W, as defined before. For two-dimensional homogeneous and incompressible flows the enstrophy, i.e. the space-integral of the squared vorticity $\iint(\nabla^2\phi)^2\mathrm{d}x\mathrm{d}y$ is conserved. In addition to such a term, the generalized enstrophy contains a term $(\nabla\phi)^2$, which earlier was identified with the square of the $\nabla\phi \times \mathbf{B}$-velocity. In Fourier representation the conservation law for generalized enstrophy takes the form

$$\frac{\mathrm{d}}{\mathrm{d}t}\sum_{\mathbf{k}}\left(1 + k^2\right)k^2|\phi_{\mathbf{k}}|^2 = 0, \tag{18.16}$$

obtained again by use of Parseval's theorem. Combination of the conservation laws of E and W gives trivially that also $\iint[\phi^2 - (\nabla^2\phi)^2]\mathrm{d}x\mathrm{d}y$ is conserved by the Hasegawa–Mima equation.

It might be expected that the multitude of conserved quantities implies strong restrictions of the evolution of an initial perturbation as it is described by the Hasegawa–Mima equation. This is however not necessarily so, since for the continuum system there are infinitely many degrees of freedom. In for instance numerical solutions of the equation the number of degrees of freedom is constrained when a spectral code with a finite number of Fourier modes is used (Seyler *et al* 1975, Rose and Sulem 1978). It will be found that usually only a few of the conservation laws survive this truncation in wavenumbers, conservation of E and W being two such examples. Such conservation laws are at times called 'robust'.

Some consequences for the correlation functions or structure functions of the enstrophy conservation in two-dimensional random flows were discussed by Kliatskin (1969) and also by Seyler *et al* (1975). As mentioned elsewhere, in the absence of a spectral cascade, these conditions do not correspond to truly turbulent conditions.

It should be noted that in special cases, in particular in simulations truncated at only a few Fourier modes, the number of conserved qualities can be larger (Hald 1976). Energy and enstrophy, as defined before, will remain for all these systems, irrespective of the number of Fourier components they may contain.

18.1.3 Coherent three wave interactions

In this section we carry out a somewhat restrictive, yet illustrative, analysis of three interacting drift waves with the restrictions imposed on the frequencies of the waves. Assume that initially a large amplitude wave $A_0\exp[-\mathrm{i}(\omega_0 t - \mathbf{k}_0\cdot\mathbf{r})]$ is present, where ω_0 an $\mathbf{k}_0$ satisfy the dispersion relation. This wave is a solution to the model equations (23.9) provided the two other wave amplitudes in the nonlinear term on the right-hand side vanish. We now assume that this solution to the equations is perturbed by two other small amplitude waves $A_1(t)\exp[-\mathrm{i}(\omega_1 t - \mathbf{k}_1\cdot\mathbf{r})]$ and $A_2(t)\exp[-\mathrm{i}(\omega_2 t - \mathbf{k}_2\cdot\mathbf{r})]$. Assuming $A_1 \ll A_0$ and $A_2 \ll A_0$ we linearize the equations to obtain

$$\frac{\mathrm{d}}{\mathrm{d}t}A_1(t) = \Lambda^{\mathbf{k}_1}_{\mathbf{k}_2,-\mathbf{k}_0}\left(k_1^2 - k_0^2\right)A_2^*(t)A_0(t)\mathrm{e}^{\mathrm{i}\theta t}, \tag{18.17}$$

$$\frac{\mathrm{d}}{\mathrm{d}t}A_2(t) = \Lambda^{\mathbf{k}_2}_{-\mathbf{k}_0,\mathbf{k}_1}\left(k_0^2 - k_2^2\right)A_0(t)A_1^*(t)\mathrm{e}^{\mathrm{i}\theta t}, \tag{18.18}$$

where we for the time being allow for a non-vanishing frequency mismatch $\theta = \omega_0 - \omega_1 - \omega_2$, but assume, as before, exact wavenumber matching $\mathbf{k}_1 + \mathbf{k}_2 = \mathbf{k}_0$. From (18.17)–(18.18) we can obtain an equation for, say, A_1 as

$$\frac{\mathrm{d}^2}{\mathrm{d}t^2}A_1(t) - i\theta\frac{\mathrm{d}^2}{\mathrm{d}t^2}A_1(t) = \Lambda^{\mathbf{k}_1}_{\mathbf{k}_2,-\mathbf{k}_0}\left(k_1^2 - k_0^2\right)\Lambda^{\mathbf{k}_2}_{-\mathbf{k}_0,\mathbf{k}_1}\left(k_0^2 - k_2^2\right)A_0^2A_1. \tag{18.19}$$

Assuming $A_1 \sim \exp(-i\Omega t)$, we find

$$\Omega^2 + \theta\Omega = -\Lambda^{\mathbf{k}_1}_{\mathbf{k}_2,-\mathbf{k}_0}\left(k_1^2 - k_0^2\right)\Lambda^{\mathbf{k}_2}_{-\mathbf{k}_0,\mathbf{k}_1}\left(k_0^2 - k_2^2\right)A_0^2 \tag{18.20}$$

or

$$\Omega = -\frac{1}{2}\theta \pm \frac{1}{2}\left(\theta^2 - 4\Lambda^{\mathbf{k}_1}_{\mathbf{k}_2,-\mathbf{k}_0}\left(k_1^2 - k_0^2\right)\Lambda^{\mathbf{k}_2}_{-\mathbf{k}_0,\mathbf{k}_1}\left(k_0^2 - k_2^2\right)A_0^2\right)^{1/2}. \tag{18.21}$$

The perturbation will be unstable ($\Im\{\Omega\} > 0$) if

$$4\Lambda^{\mathbf{k}_1}_{\mathbf{k}_2,-\mathbf{k}_0}\left(k_1^2 - k_0^2\right)\Lambda^{\mathbf{k}_2}_{-\mathbf{k}_0,\mathbf{k}_1}\left(k_0^2 - k_2^2\right)A_0^2 > \theta^2. \tag{18.22}$$

Evidently, the strongest instability (largest growth rate) occurs for exact frequency matching where $\theta = 0$. This requirement imposes non-trivial restrictions on the wavenumbers of the perturbations, since the assumed perfect wavenumber matching has to be retained together with the dispersion relation $\omega_j = \omega(\mathbf{k}_j)$, for all three wavenumber components $\mathbf{k}_j$, $j = 1, 2, 3$.

The stability condition (18.22) can be discussed in more detail by considering the coupling coefficients appearing in the expression. We first note that the absolute values of all the vector products entering the $\Lambda^{-\mathbf{k}}_{\mathbf{k}',\mathbf{k}''}$-s are equal, i.e. $\mathbf{k}_1 \times \mathbf{k}_0 = \mathbf{k}_0 \times \mathbf{k}_2$, due to the assumed wavenumber matching, irrespective of the magnitude of the individual wavenumbers. The sign of the coupling coefficients are then determined by the product $(k_1^2 - k_0^2)(k_0^2 - k_2^2)$. In particular $k_1 < k_0 < k_2$ implies instability for sufficiently large A_0^2, confirming the previous result that under certain conditions, energy can cascade from intermediate wavenumbers towards larger and smaller wavenumbers (Hasegawa 1985). On the other hand, the large amplitude wave in a wave-triplet cannot act as a 'donor' of wave energy for two waves *both* being associated with larger (or smaller) scales.

18.1.4 Stationary solutions

It was demonstrated in section 17.2 that stationary traveling solutions could be constructed for the unshielded flute model. Here we consider the related problem of shielded structures for the Hasegawa–Mima equation. Assume thus that such a solution exists in some frame of reference moving with the velocity U along the y-axis (Makino *et al* 1981, Kono and Škorić 2010). In this case $\phi = \phi(y - Ut, x)$

and we can write $\partial\phi/\partial t = -U\partial\phi/\partial y'$ with the coordinates x and $y' = y - Ut$ and obtain from (18.9)

$$-U\frac{\partial}{\partial y'}q + \{\phi; q\} = 0, \tag{18.23}$$

or

$$\{\phi - Ux; q\} = 0, \tag{18.24}$$

with $q = \phi - \nabla_\perp^2\phi - \beta x$ as before, and $\{-Ux; q\} = -U\partial q/\partial y'$. Since q contains ϕ explicitly, both elements of the Poisson bracket in (18.24) contains ϕ. By isolating ϕ, the relation can be rewritten in a more convenient way as

$$\left\{\phi + Ux; \nabla^2\phi + (U - \beta)x\right\} = 0. \tag{18.25}$$

A class of solutions to (18.25) can be found as $\nabla^2\phi + (U - \beta)x = G(\phi + Ux)$ where G is an arbitrary function. In particular a simple linear relationship can be assumed, giving

$$\nabla_\perp^2\phi + (U - \beta)x = C\,(\phi + Ux), \tag{18.26}$$

where C is a multiplicative constant. The solutions of this equation in terms of Bessel functions are in principle simple, but these solutions are generally not spatially localized, as would be required from physically acceptable pulse-like solutions. One way to get around this problem was suggested by e.g. Larichev and Reznik (1976), who considered two spatial regions separated by a circle with radius R, i.e. an inner and an outer part of the x, y'-plane, where the two solutions were matched on the boundary, see also a detailed discussion by Flierl *et al* (1980) of the related problem for perturbations of a thin fluid layer on a rotating planet. To demonstrate the procedure, new coordinates r and θ are introduced, with $x = r\sin\theta$, $y = r\cos\theta$ and $\phi = F(r)\sin\theta$. Choosing $C = -\kappa^2$ for the inner region, $r < R$, the relation (18.26) is then rewritten as

$$r^2\frac{\mathrm{d}^2}{\mathrm{d}r^2}F + r\frac{\mathrm{d}}{\mathrm{d}r}F + \left[(\kappa r)^2 - 1\right]F = \left[\beta - \left(1 + \kappa^2\right)U\right]r^3. \tag{18.27}$$

By choosing $C = \rho^2$ for the outer region, $r > R$, we similarly find

$$r^2\frac{\mathrm{d}^2}{\mathrm{d}r^2}F + r\frac{\mathrm{d}}{\mathrm{d}r}F - \left[(\rho r)^2 + 1\right]F = \left[\beta - \left(1 - \rho^2\right)U\right]r^3, \tag{18.28}$$

with R being a constant radius.

We now recall the relations

$$\left[r^2\frac{\mathrm{d}^2}{\mathrm{d}r^2} + r\frac{\mathrm{d}}{\mathrm{d}r} + \left(r^2 - 1\right)\right]\begin{Bmatrix} J_1(r) \\ Y_1(r) \end{Bmatrix} = 0, \tag{18.29}$$

and

$$\left[r^2\frac{\mathrm{d}^2}{\mathrm{d}r^2} + r\frac{\mathrm{d}}{\mathrm{d}r} - \left(r^2 + 1\right)\right]\begin{Bmatrix} I_1(r) \\ K_1(r) \end{Bmatrix} = 0. \tag{18.30}$$

As is well known, the solutions have the asymptotic properties

$$J_1(r \to \infty) \to \sqrt{\frac{2}{\pi r}} \cos\left(r - \frac{3\pi}{4}\right), \qquad \text{and} \qquad Y_1(r \to \infty) \to \sqrt{\frac{2}{\pi r}} \sin\left(r - \frac{3\pi}{4}\right),$$

while

$$I_1(r \to \infty) \to \frac{1}{\sqrt{2\pi r}} e^r, \qquad \text{and} \qquad K_1(r \to \infty) \to \sqrt{\frac{\pi}{2r}} e^{-r}.$$

For the *outer* solution we discard J_1 and Y_1 because they decay too slowly for compact, finite energy pulses, and I_1 because it increases indefinitely. The only acceptable outer solution of the ones listed here is therefore K_1. This solution (modified Bessel function of second kind) can be achieved by choosing

$$\rho^2 = (U - \beta)/U, \tag{18.31}$$

to remove the inhomogeneous term containing r^3 in (18.28). The results contain homogeneous two-dimensional plasmas as a special case if we let $\beta = 0$, giving $\rho = 1$. For the *outer* solution (indicated by the subscript o) we thus have

$$\phi_o(r, \theta) = -\frac{U - \beta}{\rho^2} R \frac{K_1(\rho r)}{K_1(\rho R)} \sin\theta, \tag{18.32}$$

for $r > R$.

On the other hand, for $r \to 0$ we have

$$J_1(r \to 0) \to \frac{r}{2}, \qquad \text{and} \qquad Y_1(r \to 0) \to -\frac{2}{\pi r},$$

while

$$I_1(r \to 0) \to \frac{r}{2}, \qquad \text{and} \qquad K_1(r \to 0) \to \frac{1}{r}.$$

The requirement of finite *inner* solutions discards Y_1 and K_1. Since I_1 increases monotonically for all r, it is not possible to match it smoothly with a continuous derivative to the outer K_1 solution, and it is therefore discarded as well. The acceptable J_1-solution (Bessel function of first kind) is obtained by choosing κ real in (18.27) (this solution was anticipated already by taking the negative sign for the appropriate constant).

The potential ϕ and its second derivative $d^2\phi/dr^2$ are required to be continuous on $r = R$. This condition implies for the inner solution of ϕ_i (indicated by the subscript i)

$$\phi_i(r, \theta) = \frac{U - \beta}{\kappa^2}\left[R\frac{J_1(\kappa r)}{J_1(\kappa R)} - r\left(\frac{\kappa^2}{\rho^2} + 1\right)\right] \sin\theta, \tag{18.33}$$

for $r < R$, where the term proportional to r originates from the inhomogeneous solution of (18.27), using $u = \beta/(1 - \rho^2)$.

The value of κ is determined by requiring also $d\phi/dr$ to be continuous at $r = R$. Here the $\sin\theta$ function in $\phi_i(r, \theta)$ and $\phi_o(r, \theta)$ is immaterial and we find

$$\frac{R}{\kappa}\frac{J_1'(\kappa R)}{J_1(\kappa R)} - \frac{1}{\kappa^2}\left(\frac{\kappa^2}{\rho^2} + 1\right) = -\frac{R}{\rho}\frac{K_1'(\rho R)}{K_1(\rho R)}, \tag{18.34}$$

or

$$\frac{1}{R\kappa}\frac{J_1'(\kappa R)}{J_1(\kappa R)} - \frac{1}{(R\kappa)^2} = \frac{1}{R\rho}\left(\frac{1}{R\rho} - \frac{K_1'(\rho R)}{K_1(\rho R)}\right), \tag{18.35}$$

which relates the parameters R, κ and ρ.

We can use the relation

$$\frac{sJ_2(s)}{J_1(s)} = 2s^2 \sum_{n=1}^{\infty}{}' \frac{1}{\gamma_{1,n}^2 - s^2}, \tag{18.36}$$

where $\gamma_{1,n}^2$ is an increasing sequence of zeroes of the function $J_1(r)$, i.e. $J_1(\gamma_{1,n}) = 0$, excluding though the zero at $r = 0$ as indicated by the prime, and obtain (Larichev and Reznik 1976, Petviashvili and Pokhotelov 1986)

$$\sum_{n=1}^{\infty}{}' \frac{1}{(\kappa R)^2 - \gamma_{1,n}^2} = \frac{1}{2\rho R}\left(\frac{1}{\rho R} - \frac{K_1'(\rho R)}{K_1(\rho R)}\right). \tag{18.37}$$

In figure 18.1 we show the variation of the left and right-hand sides of (18.37) with full and dashed lines, demonstrating that the relation (18.34) *has* solutions. We have solutions of (18.37) where the two lines cross. The lowest value is found numerically as $\kappa R \simeq 4.38\,\rho R$, relating κ and ρ.

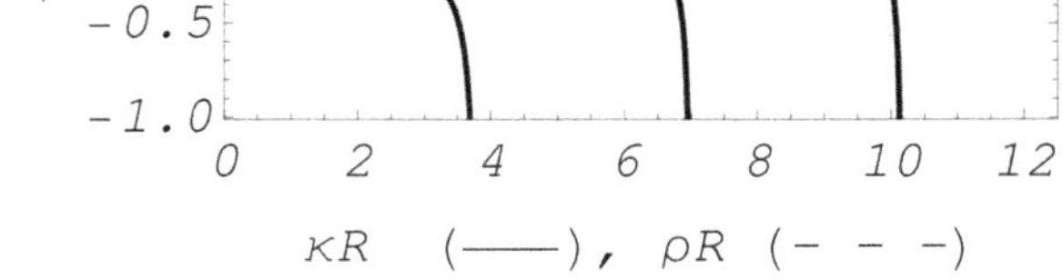

Figure 18.1. Illustration of solutions of (18.37). Full line shows the left-hand side, where the horizontal axis is in units of κR and the dashed line gives the right-hand side, with axis in units of ρR. Note the singularities at the zeroes of the $J_1(r)$-function.

Derivatives of order higher than two will in general be discontinuous. It can be argued that since the basic equation is accurate only to the approximation where first and second order derivatives are retained, there is no justification in requiring higher order derivatives to be continuous.

We can make some simplifications at small values of κR and ρR, where the solutions are close to the singularities. As an approximation, valid in the vicinity of a selected pole $\kappa R \approx \gamma_{1,m}$, we have

$$2\kappa R \sum_{n=1}^{\infty}{}' \frac{1}{(\kappa R)^2 - \gamma_{1,n}^2} \approx \frac{1}{\kappa R - \gamma_{1,m}}.$$

The properties of the solutions (18.32)–(18.33) are determined by two parameters, its radius R and velocity u. The amplitude, for instance, is determined when these two parameters are given. Alternatively, the amplitude can be prescribed to impose conditions on R and u. Evidently, there is large freedom in selecting, for instance, the number of oscillations in the inner J_1 solution before it is matched to the outer K_1 solution, and a variety of modon solutions can be constructed. When the smallest value of the solutions for κ in (18.37) is chosen, the extremum point for the modon will correspond to the first extremum of $J_1(r)$ at $r \approx 1.8$.

An important observation concerns the *velocity* of the modon. According to (18.31) we require either $u > \beta$ or $u < 0$. Implicitly, these requirements imply that the modon velocity have to be *outside* the range of the phase velocity $\{0, \beta\}$ determined by the *linearized* equations.

In the present discussion only the possibility of constructing an exact stationary solution to the basic equation was discussed, without reference to their stability and to how such solutions actually develop. Numerical solutions of the Hasegawa–Mima equation have demonstrated that isolated modons are seemingly stable and also that structures strongly reminiscent of them develop spontaneously from rather general initial perturbations. Modon solutions can be found also under inhomogeneous conditions (Horihata *et al* 1990).

- **Exercise:** You can demonstrate another version of (18.37) having the form

$$\frac{J_1'(s)}{J_1(s)} = \frac{1}{s} + 2s \sum_{n=1}^{\infty}{}' \frac{1}{s^2 - \gamma_{1,n}^2}.$$

You will need the relations $J_1'(s) = \frac{1}{2}(J_0(s) - J_2(s))$ and $J_1(s) = \frac{s}{2}(J_0(s) + J_2(s))$.

18.1.5 Drift waves with electron temperature gradients

Previously, in chapter 12, we made the observation that gradients in plasma density and electron temperature cannot be described in the same way, although one might naively think that it was only their product, i.e. the electron pressure that matters. The effect of the electron temperature gradient remains of course also in the nonlinear equation. The generalization of the Hasegawa–Mima equation to include $T_e = T_e(x)$ becomes by straightforward generalizations of the results in chapter 12

$$\frac{\partial}{\partial t}(\phi - x\beta_T\phi - \nabla^2\phi) - \beta_n\frac{\partial}{\partial y}\phi - \hat{\mathbf{z}} \times \nabla\phi \cdot \nabla\nabla^2\phi + \beta_T\phi\frac{\partial}{\partial y}\phi = 0 \qquad (18.38)$$

in normalized units. The derivation of the nonlinear equation (18.38) was also discussed in detail by Nycander (1989).

18.2 The Hasegawa–Wakatani equations

A self-evident restriction of the Hasegawa–Mima equation is that it describes linearly *stable* waves or fluctuations, i.e. any perturbation has to be imposed initially. This feature is violating the basic property of drift waves as described in the foregoing section where it was demonstrated that drift waves are linearly *un*stable. The Hasegawa–Mima equations can however readily be generalized to include also this feature.

18.2.1 Electron dynamics

Following Kadomtsev (1965), we first consider the electron dynamics and write the electron continuity equation as $\partial n/\partial t + \nabla_\perp \cdot (n\mathbf{u}_{e\perp}) + \partial_z(nu_{e\|}) = 0$. Retaining, as in the linear analysis of chapter 6, collisional resistivity only in the $\mathbf{B}$-parallel electron dynamics, we obtain first the $\mathbf{B}$-perpendicular fluctuating electron velocity as $\mathbf{u}_{e\perp} = -\nabla_\perp\phi \times \mathbf{B}/B^2 + (T_e/e)\nabla_\perp \ln n \times \mathbf{B}/B^2$ giving $\nabla_\perp \cdot \mathbf{u}_{e\perp} = 0$, see also section 5.2.1. The last term in $\mathbf{u}_{e\perp}$ does not contribute when inserted into the electron continuity equation. Consequently, we find here

$$\frac{\partial}{\partial t}\ln n + \frac{\hat{\mathbf{z}} \times \nabla_\perp\phi}{B} \cdot \nabla_\perp \ln n = \frac{1}{en}\frac{\partial}{\partial z}j_z, \tag{18.39}$$

where j_z is z-component of the electron current, $\mathbf{j} = -en\mathbf{u}_e$. Since the ion dynamics in the z-direction are assumed to be immaterial, j_z is also the *total* $\mathbf{B}$-aligned current. An expression for j_z is obtained from the z-component of the electron momentum equation (6.5), giving

$$en\frac{\partial}{\partial z}\phi - \kappa T_e\frac{\partial}{\partial z}n + \xi enj_z = 0, \tag{18.40}$$

with the plasma resistivity included as $\xi = m\nu_{ei}/(ne^2)$. From (18.40) we obtain

$$j_z = \frac{\kappa T_e}{e\xi}\frac{\partial}{\partial z}\left(\ln n - \frac{e\phi}{\kappa T_e}\right)$$

and insert it into (18.39) to obtain an equation relating density and potential fluctuations as obtained from the electron dynamics. The electron current j_z vanishes for Boltzmann distributed electrons.

18.2.2 Ion dynamics

In describing ion dynamics it is customary to retain an ion viscosity term in the momentum equation. We have the dynamic ion viscosity coefficient being $\mu_\perp$ as in chapter 8. The kinematic viscosity will here be $\mu_\perp/Mn$. At the same time, the terms containing the ion temperature are ignored, i.e. omitting ion pressure and collisionless viscosity terms; this is strictly speaking not consistent, and we will return to this question later, but the point here is that some sort of damping mechanism is needed at small scale lengths, and viscosity works fine here. The equation which contains the ion polarization current to lowest order then becomes

$$\mathbf{u}_{i\perp} = -\frac{\nabla_\perp \phi \times \mathbf{B}}{B^2} - \frac{M}{eB^2}\left(\frac{\partial}{\partial t} - \frac{1}{B^2}\nabla_\perp \phi \times \mathbf{B} \cdot \nabla_\perp\right)\nabla_\perp \phi + \frac{\mu_\perp}{neB^2}\nabla_\perp^2 \mathbf{u}_{i\perp} \times \mathbf{B}, \tag{18.41}$$

where it was implicitly assumed that the time derivative of $\nabla_\perp \phi$ is of the same order as $(\mathbf{u}_{i\perp} \cdot \nabla_\perp)\nabla_\perp \phi$, i.e. there are no fine-scale spatial variations in potential. Inserting (18.41) into the ion continuity equation in the form $\partial \ln n/\partial t + \nabla \cdot \mathbf{u}_i + \mathbf{u}_i \cdot \nabla \ln n = 0$ we find with the same approximations as in deriving (18.6)

$$\frac{\partial}{\partial t}\left(\ln n - \frac{1}{B\Omega_c}\nabla_\perp^2 \phi\right) + \frac{\hat{\mathbf{z}} \times \nabla_\perp \phi}{B} \cdot \nabla_\perp\left(\ln n - \frac{1}{B\Omega_c}\nabla_\perp^2 \phi\right) = -\frac{\mu_\perp}{neB\Omega_c}\nabla_\perp^4 \phi. \tag{18.42}$$

When evaluating $(\mu_\perp/B^2)\nabla_\perp \cdot \nabla_\perp^2 \mathbf{u}_{i\perp} \times \mathbf{B}$ we used $\nabla_\perp \times \mathbf{u}_i \approx \nabla_\perp^2 \phi/B$ in an approximation for small $\mu_\perp$, i.e. the higher order contributions to $\mathbf{u}_{i\perp}$ will be multiplied with a small quantity $\mu_\perp$, and can therefore be ignored. Quasi-neutrality was also implicitly assumed here. Approximating $\ln n \approx \ln n_0 + n_1/n_0$ as before, we obtain an equation based solely on the ion dynamics

$$\left(\frac{\partial}{\partial t} + \frac{\hat{\mathbf{z}} \times \nabla_\perp \phi}{B} \cdot \nabla_\perp\right)\left(\frac{n_1}{n_0} - \frac{1}{B\Omega_c}\nabla_\perp^2 \phi\right) + \frac{n_0'}{n_0}\frac{\partial}{\partial y}\phi = -\frac{\mu_\perp}{neB\Omega_c}\nabla_\perp^4 \phi, \tag{18.43}$$

where it was explicitly used that the density gradient is 'gentle', i.e. the length scale $n_0/|n_0'|$ is large.

18.2.3 Hasegawa–Wakatani equations

Using the normalizations $\eta \equiv n/n_0$ for density, $\chi \equiv e\phi/\kappa T_e$, $t\Omega_c$ for the electrostatic potential, r/a_i and $\mu/(a_i^2\Omega_c)$ (where we from now on omit the subscript $_\perp$ on μ), z/ℓ_c with $a_i^2 = \kappa T_e/M\Omega_c^2$ and ℓ_c being the electron–ion collisional mean free path, we finally obtain the set of Hasegawa–Wakatani equations (Hasegawa and Wakatani 1983)

$$\left(\frac{\partial}{\partial t} + \hat{\mathbf{z}} \times \nabla_\perp \chi \cdot \nabla_\perp\right)\eta + \beta\frac{\partial}{\partial y}\chi = -C\frac{\partial^2}{\partial z^2}(\chi - \eta), \tag{18.44}$$

$$\left(\frac{\partial}{\partial t} + \hat{\mathbf{z}} \times \nabla_\perp \chi \cdot \nabla_\perp\right)\nabla_\perp^2 \chi = -C\frac{\partial^2}{\partial z^2}(\chi - \eta) + \mu' \nabla_\perp^4 \chi, \tag{18.45}$$

with $\mathcal{C} \equiv \kappa T_e/(n_0 e^2 \Omega_c \xi \ell_c^2) = \kappa T_e/(\Omega_c \nu_{ei} m \ell_c^2)$ where we introduced the resistivity $\xi = m\nu_{ei}/(ne^2)$ as given before, and used $\hat{\mathbf{z}} \times \nabla_\perp \chi \cdot \nabla_\perp \ln n_0(x) = -(n_0'/n_0)\partial\chi/\partial y \equiv \beta\partial\chi/\partial y$, with $\beta = $ const. The coefficient $\mathcal{C}$ corresponds to the 'adiabaticity parameter' introduced by Camargo *et al* (1995), although it is multiplied by $k_\parallel^2$ that selects a characteristic wavenumber along **B**. We see that it is natural to use different length scalings in the directions parallel and perpendicular to **B**. In equation (18.45) originating from the description of the ion dynamics we used (18.39) and then (18.40) to eliminate the terms containing $\ln n$ in (18.42). The derivation of equation (18.44) is based solely on the assumed electron dynamics, irrespective of the assumptions concerning the ions.

If the ion viscosity vanishes and the conductivity is infinite, $\mathcal{C} \to \infty$, implying $\eta \to \chi$, the Hasegawa–Mima equation is recovered by subtracting (18.45) from (18.44). Note that while $\mathcal{C} \to \infty$ gives $\eta \to \chi$, we cannot say anything about $\mathcal{C}\partial^2(\chi - \eta)/\partial z^2$ in this limit, so the subtraction mentioned *is* necessary! In the opposite limit where $\mathcal{C} \to 0$, the equation (18.45) reduces to the convection–dissipation equation for $\nabla_\perp^2\chi$. If we also let $\mu' \to 0$ we find the two-dimensional Euler equation and $\nabla_\perp^2\chi$ is convected by the $\hat{\mathbf{z}} \times \nabla_\perp\chi$-flow. We can then, at least formally, imagine the variation of χ obtained from (18.45) to be inserted into (18.44), thus acting as a 'drive' for η. The coefficient $\mathcal{C}$ controls the phase change between density and electrostatic potential and thereby the linear growth rate of the waves.

Frequently, the equation (18.44) is completed by a loss term in the form $\mu_\eta \nabla_\perp^2 \eta$, which makes it appear similar to the $\mu'\nabla_\perp^2(\nabla_\perp^2\chi)$-term in (18.45). The introduction of such a term is however purely *ad hoc*, and actually in variance with the basic assumptions, as discussed in section 5.3. For computational purposes (Camargo *et al* 1995) with a modest number of spectral modes, it is often advantageous to have an exponent larger than 4 in the viscosity term, in order to have a well defined spectral subrange where the nonlinearity is important but where the influence of viscosity is negligible. For such cases a *hyper*-viscosity term of e.g. the form $(-1)^{p+1}\mu'\nabla_\perp^{2p}\nabla_\perp^2\chi$ can be introduced, with an even exponent $2p$ for integer p.

18.2.4 Linearized Hasegawa–Wakatani equations

The set of nonlinear coupled Hasegawa–Wakatani equations can be studied in the linearized limit. In the linear form, the equations become

$$\frac{\partial}{\partial t}\eta + \beta\frac{\partial}{\partial y}\chi = -\mathcal{C}\frac{\partial^2}{\partial z^2}(\chi - \eta), \tag{18.46}$$

$$\frac{\partial}{\partial t}\nabla_\perp^2\chi = -\mathcal{C}\frac{\partial^2}{\partial z^2}(\chi - \eta) + \mu'\nabla_\perp^4\chi. \tag{18.47}$$

By Fourier transform, it is readily shown that (18.46) reproduces (6.21). Assuming plane wave solutions, we obtain the dispersion relation

$$\omega^2 + i\omega\left[Ck_z^2\left(1 + \frac{1}{k_\perp^2}\right) + \mu' k_\perp^2\right] - Ck_z^2\left(i\frac{\omega^*}{k_\perp^2} + \mu' k_\perp^2\right) = 0, \tag{18.48}$$

where the drift frequency $\omega^* = \beta k_y$ was again introduced. In the limit of infinite conductivity, $C \to \infty$, the solution is simply

$$\omega = \frac{\omega^*}{1 + k_\perp^2} - i\frac{\mu' k_\perp^4}{1 + k_\perp^2}, \tag{18.49}$$

reproducing the standard result for drift waves, i.e. (5.14) with $k_z = 0$, where in addition a damping due to the ion viscosity is recovered. Ignoring ion viscosity, on the other hand, we find

$$\omega^2 + i\omega Ck_z^2\left(1 + \frac{1}{k_\perp^2}\right) - iCk_z^2\frac{\omega^*}{k_\perp^2} = 0, \tag{18.50}$$

which is simply the result (6.26) for the collisional or resistive drift wave instability. The plasma becomes stable for $C = 0$ as well as for $\omega^* = 0$, as expected. If we imagine (ignoring the laws of physics!) setting the plasma resistivity $\xi < 0$ in C, we find that electrostatic drift waves become *stable* in the present model, while other previously stable modes now become unstable. With $\xi < 0$, electrons will be diffusing *opposite* the gradient in the direction on *increasing* plasma density, see also discussion in section 6.1, and the drift waves become stabilized.

The Hasegawa–Wakatani equations thus form a closed set including driving in terms of the linear instability as described here, and dissipation in terms of a simple model for ion viscosity. For the time being, the set of equations presents one of the most appealing models for studies of nonlinear drift waves and drift wave turbulence, and has also been analyzed numerically in great detail (Camargo *et al* 1995, Korsholm *et al* 1999).

18.2.5 Conservation laws for the Hasegawa–Wakatani equations

As demonstrated before, the equations (18.45) and (18.44) contain the Hasegawa–Mima equation in certain limits. This equation had the basic conserved integrals, energy $E \equiv \frac{1}{2}\iint[\phi^2 + (\nabla\phi)^2]\mathrm{d}x\mathrm{d}y$ and enstrophy $W \equiv \frac{1}{2}\iint[(\nabla\phi)^2 + (\nabla^2\phi)^2]\mathrm{d}x\mathrm{d}y$. or the Hasegawa–Wakatani equation these two quantities can be generalized as $E \equiv \frac{1}{2}\iint[\eta^2 + (\nabla_\perp\chi)^2]\mathrm{d}x\mathrm{d}y\mathrm{d}z$ and $W \equiv \frac{1}{2}\iint(\eta - \nabla_\perp^2\chi)^2\mathrm{d}x\mathrm{d}y\mathrm{d}z$. It is customary to interpret the term η^2 in E as a potential energy density, and $\nabla_\perp^2\chi$ as a kinetic energy density to the given accuracy. This latter interpretation is indeed tempting, but should be considered with care; as we have seen, to lowest order the $\mathbf{E} \times \mathbf{B}$-motion is inertialess, and in that case the notion of kinetic energy does not really make sense.

Since the Hasegawa–Wakatani equations describe a linearly unstable situation with the possibility of growing waves, the generalized energy and enstrophy is, however, no longer conserved by (18.45) and (18.44), and it is interesting to study their evolution in time. After some calculations, given by e.g. Camargo *et al* (1995), the following results can be obtained

$$\frac{\mathrm{d}}{\mathrm{d}t}E = \Gamma_\eta - \Gamma_\chi - \mathcal{D}_E, \tag{18.51}$$

$$\frac{\mathrm{d}}{\mathrm{d}t}W = \Gamma_\eta - \mathcal{D}_W, \tag{18.52}$$

with

$$\Gamma_\eta \equiv -\beta \int \eta \frac{\partial}{\partial y}\chi \,\mathrm{d}x\mathrm{d}y\mathrm{d}z, \tag{18.53}$$

$$\Gamma_\chi \equiv \mathcal{C} \int \left(\frac{\partial}{\partial z}(\eta - \chi)\right)^2 \mathrm{d}x\mathrm{d}y\mathrm{d}z, \tag{18.54}$$

$$\mathcal{D}_E \equiv -\int \left(\eta\mathcal{D}_\eta - \chi\mathcal{D}_\chi\right) \mathrm{d}x\mathrm{d}y\mathrm{d}z, \tag{18.55}$$

$$\mathcal{D}_W \equiv -\int \left(\eta - \nabla_\perp^2\chi\right)\left(\mathcal{D}_\eta - \mathcal{D}_\chi\right)\mathrm{d}x\mathrm{d}y\mathrm{d}z. \tag{18.56}$$

The damping terms were abbreviated as $\mathcal{D}_\eta = \mu'\nabla_\perp^4\chi$ and $\mathcal{D}_\chi = \mu_\eta\nabla_\perp^2\eta$, but other (even) damping exponents can be inserted as models, e.g. $\mathcal{D}_\chi = \mu_\eta\nabla_\perp^{2p}\chi$, depending on the sort of problem being considered, as discussed later.

To prove for instance (18.51) we write

$$\frac{\mathrm{d}}{\mathrm{d}t}E = \frac{1}{2}\int \frac{\partial}{\partial t}\left(\eta^2 + (\nabla_\perp\chi)^2\right) \mathrm{d}x\mathrm{d}y\mathrm{d}z,$$

and integrate by parts to obtain

$$\frac{\mathrm{d}}{\mathrm{d}t}E = \int\left(\eta\frac{\partial}{\partial t}\eta - \chi\frac{\partial}{\partial t}\nabla_\perp^2\chi\right) \mathrm{d}x\mathrm{d}y\mathrm{d}z.$$

The two partial time derivatives are now inserted from the Hasegawa–Wakatani equations, to give

$$\begin{aligned}\frac{\mathrm{d}}{\mathrm{d}t}E = \int -\eta&\left((\hat{\mathbf{z}} \times \nabla_\perp\chi \cdot \nabla_\perp)\eta + \beta\frac{\partial}{\partial y}\chi + \mathcal{C}\frac{\partial^2}{\partial z^2}(\chi - \eta) - \mu_\eta\nabla_\perp^2\eta\right)\\ -\chi&\left((\hat{\mathbf{z}} \times \nabla_\perp\chi \cdot \nabla_\perp)\nabla_\perp^2\chi - \mathcal{C}\frac{\partial^2}{\partial z^2}(\chi - \eta) + \mu'\nabla_\perp^4\chi\right)\mathrm{d}x\mathrm{d}y\mathrm{d}z.\end{aligned} \tag{18.57}$$

The terms containing $\hat{\mathbf{z}} \times \nabla_\perp\chi \cdot \nabla_\perp$ do not contribute to the integral. The two linear coupling terms, those containing $\mathcal{C}$, are combined to give

$$\frac{\mathrm{d}}{\mathrm{d}t}E = \int\left(-\beta\eta\frac{\partial}{\partial y}\chi + C(\chi - \eta)\frac{\partial^2}{\partial z^2}(\chi - \eta) - \mu_\eta\eta\nabla_\perp^2\eta + \mu'\chi\nabla_\perp^4\chi\right)\mathrm{d}x\mathrm{d}y\mathrm{d}z. \quad (18.58)$$

By partial integration of the second term we readily find

$$\frac{\mathrm{d}}{\mathrm{d}t}E = \int\left(-\beta\eta\frac{\partial}{\partial y}\chi - C\left(\frac{\partial}{\partial z}(\chi - \eta)\right)^2 - \mu_\eta\eta\nabla_\perp^2\eta + \mu'\chi\nabla_\perp^4\chi\right)\mathrm{d}x\mathrm{d}y\mathrm{d}z, \quad (18.59)$$

which is the desired result (18.51). The relation (18.52) can be proven in a similar way.

The relations (18.55) and (18.56) can be simplified somewhat in the case where $\mu' = \mu_\eta$, and the same powers, n, are used for both $\nabla_\perp^n$-operators, but there is no *a priori* reason for this. As an illustration we consider $\mathcal{D}_W$ in (18.56) for the general case which is almost always used, with spatial operators $\nabla_\perp^{2p}$ and $\mu' = \mu_\eta = \nu$ giving

$$\begin{aligned}\mathcal{D}_W &= -\nu\int\left(\eta - \nabla_\perp^2\chi\right)\left(\mathcal{D}_\eta - \mathcal{D}_\chi\right)\mathrm{d}x\mathrm{d}y\mathrm{d}z \\ &\equiv -\nu\int(-1)^{p+1}\left(\eta - \nabla_\perp^2\chi\right)\nabla_\perp^{2p}\left(\eta - \nabla_\perp^2\chi\right)\mathrm{d}x\mathrm{d}y\mathrm{d}z \\ &= \nu\int\left(\nabla_\perp^p\left(\eta - \nabla_\perp^2\chi\right)\right)^2\mathrm{d}x\mathrm{d}y\mathrm{d}z.\end{aligned} \quad (18.60)$$

Similar calculations for $\mathcal{D}_E$ give

$$\mathcal{D}_E = \nu\int\left(\left(\nabla_\perp^p\eta\right)^2 + \left(\nabla_\perp^{p+1}\chi\right)^2\right)\mathrm{d}x\mathrm{d}y\mathrm{d}z. \quad (18.61)$$

The quantities (18.53)–(18.56) constitute sources and sinks of wave energy for the nonlinear evolution of initial conditions. Only Γ_η can act as a source, the other quantities are sinks. The density gradient β, enters as a coefficient to Γ_η, which represents the rate at which free energy is extracted from the gradient, see also the discussion of plasma flux in figure 6.4.

Writing the x-component of the normalized velocity as $u_x = -\partial\chi/\partial y$, we can interpret $\Gamma_\eta = \beta\int nu_x\mathrm{d}x\mathrm{d}y$ as the net plasma flux along the density gradient in the x-direction. The term Γ_χ represents the rate with which energy is being dissipated. These two quantities account for the collisional drift wave kinetics in the non-viscous limit. It is important that they are active only to the extent that the electron dynamics deviates from isothermal, i.e. if $\eta = \chi$, then both Γ_χ and Γ_η vanishes, in agreement with the foregoing results based on the linearized equations. The two sinks, $\mathcal{D}_\eta$ and $\mathcal{D}_\chi$, constitute viscous dissipation. Physically, these arise through collisional diffusion of electrons across magnetic field lines and due to ion viscosity, respectively.

Note that Γ_χ vanishes for flute modes, where there is no variation in density or potential along the magnetic field lines. Numerical simulations by e.g. Cheng and Okuda (1977) indicate that, in particular, the anomalous transport is strongly influenced by these modes, which can be excited by nonlinear coupling from drift modes with $k_z \neq 0$ to flute- or convective cell-modes with $k_z = 0$.

The most obvious damping mechanism, smoothing out of the density gradients to reduce the linear growth rate, seems to be missing in the Hasegawa–Wakatani equations by the assumption of $n_0 = n_0(x)$, independent of t. However, the term $\hat{\mathbf{z}} \times \nabla_\perp \chi \cdot \nabla_\perp \eta$ can contain a modification to the β in the gradient term $\beta \partial\chi/\partial y$, provided the plasma density η develops a time stationary component with a non-trivial variation along x, but being essentially constant along the y-direction. This observation is, however, of little avail as far as stabilization of the waves is concerned; if we introduce $\eta \equiv \bar{\eta}(x, t) + \tilde{\eta}(x, y, t)$ in (18.53) we find $\Gamma_\eta = -\beta \int \tilde{\eta}\, (\partial\chi\partial y)\, \mathrm{d}x\mathrm{d}y\mathrm{d}z$, independent of any $\bar{\eta}(x, t)$.

18.2.6 Comments on the Hasegawa–Wakatani equations

In deriving the Hasegawa–Wakatani equations only the ion viscosity term contains remnants of a finite ion temperature. This is not consistent, as for instance the collisionless part of the ion viscosity tensor is absent from the analysis although it enters to the same order as collisional viscosity. With ion pressure and ion viscosity included, the expression for the ion velocity replacing (18.41) becomes

$$\begin{aligned}\mathbf{u}_{i\perp} = &-\frac{\nabla_\perp \phi \times \mathbf{B}}{B^2} - \frac{M}{eB^2}\left(\frac{\partial}{\partial t} + \mathbf{u}_i \cdot \nabla_\perp\right)\mathbf{u}_{i\perp} \times \mathbf{B} \\ &+ \frac{\mu_\perp}{neB^2}\nabla_\perp^2 \mathbf{u}_{i\perp} \times \mathbf{B} \\ &- \frac{\kappa T_i}{neB^2}\nabla_\perp n \times \mathbf{B} - \frac{1}{neB^2}(\nabla_\perp \cdot \mathbf{\Pi}) \times \mathbf{B},\end{aligned} \tag{18.62}$$

which is inserted into the ion continuity equation. It was demonstrated previously in section 7.4 that here the second term in the polarization drift cancels the contribution from the collisionless part of the viscosity tensor. The only remaining new term was shown to be $\partial\nabla^2\eta/\partial t$, while $\partial\nabla^2\chi/\partial t$ was included already. Consequently we find that with the present approximations, equation (18.45) is modified to

$$\left(\frac{\partial}{\partial t} + \hat{\mathbf{z}} \times \nabla_\perp \chi \cdot \nabla_\perp\right)\nabla_\perp^2 \chi + \theta \frac{\partial}{\partial t}\nabla_\perp^2 \eta = -C\frac{\partial^2}{\partial z^2}(\chi - \eta) + \mu' \nabla_\perp^4 \chi, \tag{18.63}$$

with $\theta = T_i/T_e$, while (18.44) remains unchanged. Strictly speaking, also the term containing the collisional viscosity is modified, but since this term is often modeled *ad hoc* there is little point in elaborating it in further detail. Also, it was implicitly assumed that θ is small, i.e. terms containing products of θ, χ and η are ignored. These can easily be retained, but the resulting equations become so complicated that analytical approaches are prohibitive.

Strictly speaking the modification of the Hasegawa–Wakatani equations advocated here is not complete. In terms of the tensor $\mathbf{Q}$ introduced in (7.4) we have more generally a term

$$\frac{\kappa T_i}{2n_0 M B \Omega_{ci}^2}\Big(\nabla n_0 \times (\nabla \cdot \mathbf{Q}) + \nabla n_1 \times (\nabla \cdot \mathbf{Q})\Big) \cdot \hat{\mathbf{z}},$$

where ∇n_0 and ∇n_1 are of same order with the ordering of terms used for the Hasegawa–Wakatani equations, where n_0 is large and varies of a large length scale, while $n_1 \ll n_0$ is small and varies on a length scale short in comparison to $n_0/|\nabla n_0|$. In (18.63) we retained only what corresponds to the first one of these two terms The consequences of the second term for the conservation laws is not clarified.

In the limit of Boltzmann distributed electrons, the linearized dispersion relation based on (18.44) and (18.63) does not reproduce (7.20) due to differences in the basic assumptions. The dispersion relation with finite electron resistivity, where $\chi \neq \eta$, is slightly more complicated than (18.50) or (18.48)

$$\omega^2 + i\omega\left[Ck_z^2\left(1 + \theta + \frac{1}{k_\perp^2}\right) + \mu' k_\perp^2 - i\theta\omega^*\right] - Ck_z^2\left(\frac{i\omega^*}{k_\perp^2} + \mu' k_\perp^2\right) = 0. \tag{18.64}$$

The modification introduced by θ in the parenthesis with C as a coefficient is of minor importance, but the growth rate of the linear instability will be non-trivially modified by the changes in the other term.

The modified result (18.63) can be re-expressed in a more natural form by noting that the new term on the left side, accounting for the effects of a small but non-zero ion temperature, enters as a small correction. To lowest order we have $\partial\eta/\partial t \approx -\beta\partial\chi/\partial y$ from (18.46). By the ordering implied in the Hasegawa–Wakatani as well as the Hasegawa–Mima equations, we have here $\beta \sim \nabla_\perp\eta$, see section 18.1, and we have one more term contributing from the ion viscosity tensor. We can rewrite (18.63) to the desired accuracy as

$$\begin{aligned}\left(\frac{\partial}{\partial t} + \hat{\mathbf{z}} \times \nabla_\perp\chi \cdot \nabla_\perp\right)\nabla_\perp^2\chi - \theta\left(\nabla_\perp\eta \times \nabla_\perp\nabla_\perp^2\chi \cdot \hat{\mathbf{z}} + \beta\frac{\partial}{\partial y}\nabla_\perp^2\chi\right)& \\ = -C\frac{\partial^2}{\partial z^2}(\chi - \eta) + \mu'\nabla_\perp^4\chi.&\end{aligned} \tag{18.65}$$

In this form the equations (18.44) and (18.65) incidentally reproduce the dispersion relation (7.20) if the electrons are assumed to be Boltzmann distributed. In the limit of finite electron resistivity $\chi \neq \eta$, with the modified model (18.65), the linear dispersion relation becomes

$$\begin{aligned}\omega^2 + i\omega\left[Ck_z^2\left(1 + \frac{1}{k_\perp^2}\right) + \mu' k_\perp^2 - i\theta\omega^*\right]& \\ - Ck_z^2\left(i\omega^*\left(\frac{1}{k_\perp^2} - \theta\right) + \mu' k_\perp^2\right) = 0.&\end{aligned} \tag{18.66}$$

In the flute limit, $k_z \to 0$, we have $\omega = 0$ but also backwards propagating weakly damped waves, $\omega = -\theta\omega^* - i\mu' k_\perp^2$. These latter wave-modes depend critically on the ion temperature. The modifications introduced in (18.63) are expected to be significant for describing the coupling between drift- and flute-type modes. The splitting of the double root at $\omega = 0$ induced by $\theta \neq 0$ will be conspicuous for instance in numerical simulations for large times $t > 1/(\theta\omega^*)$, manifested physically by the 'beating' between the two wave-modes.

In the limit of $\mathcal{C} \to \infty$ we still have $\eta = \chi$ for Boltzmann distributed electrons. By subtracting (18.44) from (18.65) in this limit, a modified Hasegawa–Mima equation is obtained in the form

$$\left(\frac{\partial}{\partial t} + (\hat{\mathbf{z}} \times \nabla_\perp \phi) \cdot \nabla_\perp\right)\left((1 + \theta)\nabla_\perp^2 \phi - \phi\right) - \beta \frac{\partial}{\partial y}\left(\theta \nabla_\perp^2 \phi + \phi\right) = 0,$$

expressed in terms of normalized potential ϕ, with dissipation by ion viscosity ignored. The modifications due to the finite (but small) ion temperature will be important when $\theta\nabla_\perp^2\phi \geqslant \phi$. It is easily demonstrated that also the modified Hasegawa–Mima equation has the basic conserved integrals, energy $E \equiv \frac{1}{2}\iint[\phi^2 + (1 + \theta)(\nabla\phi)^2]\mathrm{d}x\mathrm{d}y$ and enstrophy $W \equiv \frac{1}{2}\iint[(1 + \theta)(\nabla\phi)^2 + (\nabla^2\phi)^2]\mathrm{d}x\mathrm{d}y$.

The Hasegawa–Wakatani equations describe the linear instability of resistive drift waves in a model that adequately contains many of the basic features of the analysis in chapter II. They also contain nonlinear couplings that account for couplings between the various modes being excited by the instability. The weakly nonlinear set of equations discussed in this chapter thus form the basis for studies of weak drift wave turbulence, and in particular its relevance and importance for describing anomalous transport due to resistive drift-wave turbulence (Hu *et al* 1997).

An important discovery in turbulent two-dimensional flows, seemingly a consequence of the inverse cascade, concerns formation of zonal flows. In reality these flows have been observed for a long time, and the 'belts' on the surface of Jupiter may represent such formations (Hasegawa 1985). Zonal flows are important for neutral flows on rotating planets (Pedlosky 1987, Connaughton *et al* 2010) and also in magnetized plasmas (Tynan *et al* 2009, Kono and Škorić 2010), fusion experiments in particular (Fujisawa 2009, Itoh *et al* 2006). This field of research is still in rapid progress and will not be discussed further in the present work. Interested readers should consult the literature, noting that several review papers are already available, as listed before.

References

Camargo S J, Biskamp D and Scott B 1995 Resistive drift-wave turbulence *Phys. Plasmas* **2** 48–62

Charney J G 1948 On the scale of atmospheric motions *Geophys. Publ. Norwegian Acad. of Sciences (Geofysiske Publikasjoner)* **17** 251–65

Cheng C Z and Okuda H 1977 Formation of convective cells, anomalous diffusion, and strong plasma turbulence due to drift instabilities *Phys. Rev. Lett.* **38** 708–11

Connaughton C P, Nadiga B T, Nazarenko S V and Quinn B E 2010 Modulational instability of rossby and drift waves and generation of zonal jets *J. Fluid Mech.* **654** 207–31

Flierl G R, Larichev V D, McWilliams J C and Reznik G M 1980 The dynamics of baroclinic and barotropic solitary eddies *Dynam. Atmos. Oceans* **5** 1–41

Fujisawa A 2009 A review of zonal flow experiments *Nucl. Fusion* **49** 013001

Hald O H 1976 Constants of motion in models of two-dimensional turbulence *Phys. Fluids* **19** 914–5

Hasegawa A 1985 Self-organization processes in continuous media *Advances Phys.* **34** 1–42
Hasegawa A and Mima K 1978 Pseudo-three-dimensional turbulence in magnetized nonuniform plasma *Phys. Fluids* **21** 87–92
Hasegawa A and Wakatani M 1983 Plasma edge turbulence *Phys. Rev. Lett.* **50** 682–6
Horihata S, Irie H and Sato M 1990 Stationary drift-rossby vortices in shear flows *J. Phys. Soc. Jpn* **59** 1242–51
Hu G, Krommes J A and Bowman J C 1997 Statistical theory of resistive drift-wave turbulence and transport *Phys. Plasmas* **4** 2116–33
Itoh K, Itoh S-I, Diamond P H, Hahm T S, Fujisawa A, Tynan G R, Yagi M and Nagashima Y 2006 Physics of zonal flows *Phys. Plasmas* **13** 055502
Kadomtsev B B 1965 *Plasma Turbulence* (New York: Academic)
Kliatskin V I 1969 On static theory of 2-dimensional turbulence *J. Appl. Math. Mech.-USSR* **33** 864–6
Kono M and Škorić M M 2010 *Nonlinear Physics of Plasmas* number 62 (*Springer Series on Atomic, Optical, and Plasma Physics*) (Heidelberg: Springer)
Korsholm S B, Michelsen P K and Naulin V 1999 Resistive drift wave turbulence in a three-dimensional geometry *Phys. Plasmas* **6** 2401–8
Larichev V D and Reznik G M 1976 Two-dimensional solitary rossby waves *Doklady Akad. Nauk. SSSR* **231** 1077–9
Makino M, Kamimura T and Taniuti T 1981 Dynamics of two-dimensional solitary vortices in a low-β plasma with convective motion *J. Phys. Soc. Japan* **50** 980–9
Nycander J 1989 The effect of the electron temperature gradient on nonlinear drift waves in plasmas *Phys. Scripta* **39** 758–63
Pedlosky J 1987 *Geophysical Fluid Dynamics* 2nd edn (New York: Springer)
Petviashvili V I and Pokhotelov O A 1986 Solitary vortices in plasmas *Sov. J. Plasma Phys.* **12** 651 Petviashvili V I and Pokhotelov O 1986 *Russian original Fiz. Plazmy* **12** 1127–44
Rhines P B 1975 Waves and turbulence on a β-plane *J. Fluid Mech.* **69** 417–43
Rose H A and Sulem P L 1978 Fully developed turbulence and statistical-mechanics *J. Physique* **39** 441–84
Seyler C E, Salu Y, Montgomery D and Knorr G 1975 Two-dimensional turbulence in inviscid fluids or guiding center plasmas *Phys. Fluids* **18** 803–13
Tynan G R, Fujisawa A and McKee G 2009 A review of experimental drift turbulence studies *Plasma Phys. Control. Fusion* **51** 113001

Part VI

Randomly varying fields and turbulence

IOP Publishing

Low Frequency Waves and Turbulence in Magnetized Laboratory Plasmas and in the Ionosphere

Hans Pécseli

Chapter 19

Elements of statistical analysis

Assume a stochastic variable defined by a range of values, each associated with a certain probability. The variable is denoted X and the values it can assume x. The variable may assume discrete values only, e.g. it can be the number of particles in a volume element, each value, x_j, associated with a probability $P(x_j)$. Heuristically, we might interpret the probability for an event as the ratio between the number of 'desired' events, divided by the total number of relevant events. For simple cases like honest dies or card games, this interpretation is adequate, but it is insufficient as a general definition.

The sample space for relevant events may be discrete and finite, or it can be discrete and *in*finite. The variable can be continuous, for instance the voltage output of a noisy amplifier, with the probability of X having a value in a narrow range x, $x + \mathrm{d}x$ being $P(x)\mathrm{d}x$. Finally combinations of the two may occur, i.e. a mixture of discrete and continuous states as often encountered in atomic physics. The set of values, or *states*, that X can assume may be multi-dimensional, in which case it can conveniently be written as a vector $\mathbf{X}$. An example is the velocity components of a randomly moving particle or the wind velocity vector in the atmosphere.

By statistical averaging we understand the process based on the assumption that, ideally, infinitely many realizations are available. The average value of, say, X is then obtained by *ensemble averaging*

$$\langle X \rangle = \lim_{x\to\infty} \frac{{}^1x + {}^2x + {}^3x + \cdots + {}^Nx}{N} \tag{19.1}$$

where the indexes refer to the label of individual realizations in the ensemble. If done in practice, only a finite number, N, of realizations are available, and only an *estimate*, i.e. an *approximation* to the actual average can be achieved this way.

doi:10.1088/978-0-7503-1251-6ch19

In terms of the normalized probability density $P(x)$, the mathematical process of averaging is expressed as

$$\langle X \rangle = \int_{-\infty}^{\infty} x P(x) \mathrm{d}x, \tag{19.2}$$

for a continuous variable or, alternatively, in terms of a sum for discrete variables. It is assumed that the probability density, $P(x)$, is *a priori* known. More generally, the average, or mean, of any quantity $f(X)$ is defined as $\langle f(X) \rangle = \int_{-\infty}^{\infty} f(x) P(x) \mathrm{d}x$. Often we find the term *expectation value* $E[f(X)]$ for $\langle f(X) \rangle$. The term expectation value might be somewhat misleading; assume an honest die and calculate the average or expectation of the number $\mathcal{N}$ of dots. With the numbers $\mathcal{N} = 1, 2, 3, 4, 5,$ and 6 being equally probable, we easily obtain $\langle \mathcal{N} \rangle = 3.5$. It is however unwise ever to actually *expect* the number 3.5 to come up, as only integer numbers can occur.

The average value of a randomly varying quantity is often very easy to estimate and is used as one of the most basic quantities for characterizing a random variable. The *median a* of a distribution of a continuous random variable is the value where $\int_{-\infty}^{a} P(x) \mathrm{d}x = \frac{1}{2}$ and $\int_{a}^{\infty} P(x) \mathrm{d}x = \frac{1}{2}$. For a symmetric distribution, the median equals the average. For discrete variables we have to be a little more careful. Often also the 'mode', being the variable where the probability density assumes its maximum value, is important. A practical problem is that many important distributions have several maxima, where some symmetries give two of them equal.

It may be important to note that even though an event has zero probability, this does not necessarily mean that the event is physically impossible. The probability for an honest die to show 6 at *every* trial is zero, but there are no physical laws prohibiting this from actually happening.

Averages may not exist, i.e. $\int_{-\infty}^{\infty} f(x) P(x) \mathrm{d}x$ can diverge for particular choices of $f(x)$. It may be in variance with intuitive expectations, but it might even be so that the average value $\langle X \rangle$ found by the choice $f(x) = x$ does not exist. All we can require is after all that $P(x)$ can be normalized, and nothing more. The probability density $P_n(x) \equiv n \exp(-x) I_n(x)/x$ for $x > 0$ and $P_n(x) = 0$ otherwise, defined here in terms of the modified Bessel function $I_n(x)$ can be normalized, but it has no average.

Probabilities can be expressed in terms of a *distribution function* or cumulative distribution function (Hogg and Craig 1970), $Pr(x)$, with the relation

$$Pr(x) = \int_{-\infty}^{x} P(x') \mathrm{d}x'$$

for the one variable case. While $P(x)\mathrm{d}x$ denotes the probability for finding x in the small interval $\{x, x + \mathrm{d}x\}$, the distribution function $Pr(x)$ gives the probability for finding x in the interval $\{-\infty, x\}$. For discrete variables the probability distribution is sometimes preferred, in order to avoid the δ-functions which have to be introduced in the probability density for this case. The formulation in the following will only be using probability densities.

- **Exercise:** Consider a random variable x and assume that you have found an estimate $\frac{1}{N}\sum_{j=1}^{N} x_j$ for the average $\langle x\rangle$. Determine the mean-square average error on this estimate.

  Solution: the error e is given as the root mean square (RMS) of the average difference between what you have and what you should have, i.e.

$$e^2 \equiv \left\langle \left(\langle x\rangle - \frac{1}{N}\sum_{j=1}^{N} x_j \right)^2 \right\rangle.$$

  We find

$$e = \langle x\rangle^2 + \frac{1}{N^2}\left\langle \left(\sum_{j=1}^{N} x_j \right)^2 \right\rangle - 2\left\langle \langle x\rangle \frac{1}{N}\sum_{j=1}^{N} x_j \right\rangle.$$

  The last sum is easy: each term in the sum becomes $\langle x\rangle$ and we have N of these, so we find all in all $-2\langle x\rangle^2$ for this term. The second term is written as $\frac{1}{N^2}\sum_{k=1}^{N}\sum_{j=1}^{N}\langle x_j x_k\rangle$. The double sum contains two sorts of terms: N terms where $j = k$ and $N(N-1)$ terms where $j \neq k$. In the former case we have $\langle x_j^2\rangle = \langle x^2\rangle$ for each term, in the latter case we find $\langle x_j x_k\rangle = \langle x_j\rangle\langle x_k\rangle = \langle x\rangle^2$. Consequently we have $e = \langle x\rangle^2 + \frac{1}{N^2}(N\langle x^2\rangle + N(N-1)\langle x\rangle^2) - 2\langle x\rangle^2$ giving $e = \frac{1}{N}(\langle x^2\rangle - \langle x\rangle^2)$. We have $e \to 0$ for $N \to \infty$ as expected, in such a way that the RMS error decreases as $1/\sqrt{N}$.

  The discussion is easily generalized to higher order moments, but the expressions can become rather cumbersome.

19.1 One variable probabilities

First we discuss the case where we deal with probabilities of one variable, say $P(x)$. Often we are not really interested in all the information contained in $P(x)$, but are content with averages as $\langle X^n\rangle$, for some values of n. This information is more readily obtained for instance from moment generating functions.

19.1.1 Generating functions

The average of $\exp(\alpha x)$, called the *moment generating function*, is defined as (Bendat 1958, Hogg and Craig 1970)

$$M(\alpha) \equiv \langle \exp(\alpha X)\rangle = \int_{-\infty}^{\infty} e^{\alpha x} P(x)\mathrm{d}x. \tag{19.3}$$

In particular we find by (19.3) the derivatives

$$\frac{d^n M(\alpha)}{d\alpha^n} = \int_{-\infty}^{\infty} x^n e^{\alpha x} P(x) dx.$$

The moment generating function is evidently a function of the variable α and serves to generate the averages, or *moments*, of X, i.e. $\langle X \rangle$, $\langle X^2 \rangle$, etc obtained by evaluating $dM(\alpha)/d\alpha$, $d^2M(\alpha)/d\alpha^2$, etc at $\alpha = 0$. The individual terms in a Taylor expansion of $M(\alpha)$ will consequently contain the averages $\langle X^n \rangle$ in increasing order.

19.1.2 Characteristic functions

Not every probability density has an associated moment generating function, and it is often advantageous to consider the *characteristic function* defined as the Fourier transform of the probability density

$$CH(\alpha) \equiv \langle \exp(i\alpha X) \rangle = \int_{-\infty}^{\infty} e^{i\alpha x} P(x) dx. \tag{19.4}$$

With rather mild restrictions on the variation of $P(x)$ at large $|x|$, the moments of X are then given by

$$\langle X^n \rangle = (-i)^n \frac{d^n}{d\alpha^n} CH(\alpha) \bigg|_{\alpha = 0}, \tag{19.5}$$

the subscript indicating that the derivative is to be taken at $\alpha = 0$.

For a discrete variable, the characteristic functions

$$CH(\alpha) \equiv \sum_{j}^{\infty} P_j e^{i\alpha x_j} \tag{19.6}$$

can be introduced, where P_j is the probability of the j-th event. Here, the characteristic function is a sum of exponentials.

For the case where the variable j takes on only integer values, a definition in terms of the z-transform (Oppenheim and Schafer 1975) gives the generating function

$$\Gamma(z) \equiv \langle Z^n \rangle = \sum_{n=-\infty}^{\infty} z^n P_n. \tag{19.7}$$

Here n is a *lattice type* random variable. We recognize $\Gamma(1/z)$ as the z-transform of P_n with n taking on only integer values (Papoulis 1991). (The Γ-function introduced here should not be confused with the Gamma-function which interpolates $n!$) Differentiating (19.7) k times we obtain $d^k\Gamma(z)/dz^k = \langle n(n-1)\cdots(n-k+1)z^{n-k} \rangle$, which for $z = 1$ becomes $\langle n(n-1)\cdots(n-k+1) \rangle$.

- **Exercise:** Derive the moment generating function for a Poisson distribution.
- **Exercise:** Obtain the moment generating function and the characteristic function for the binomial distribution.

- **Exercise:** Derive the moment generating function for a Gaussian distribution

$$P(x) = (2\pi A)^{-1/2} \exp\left(-\frac{1}{2}x^2/A\right),$$

and demonstrate that $\langle X^4 \rangle = 3\langle X^2 \rangle^2$.
- **Exercise:** Demonstrate that the characteristic function for a variable Z which is the sum of two independent random variables, $Z = X + Y$, is the product of the individual characteristic functions. Generalize this result to a sum of arbitrary many independent variables.

19.1.3 Change of variable

An important question concerns the change in variables; often we know *a priori* the probability density for a certain event, and want to determine the probability density for something else which depends in a deterministic way on the outcome of this event. As a simple example, we may consider X to be a temperature and Y to be the length of a metal rod which is varying due to thermal expansion in some possibly nonlinear way. Assume that the probability density, $P_X(x)$, of the event X is known, and that another event Y is a deterministic consequence of X, i.e. $Y = f(X)$. It is then at least in principle straightforward to determine the probability density $P_Y(y)$. The probability that Y has a value in the range $\{y;\, y + \Delta y\}$ is

$$P_Y(y) = \int \delta[f(x) - y] P_X(x) \mathrm{d}x. \tag{19.8}$$

In the case where there is a one-to-one correspondence between X and Y, the transformation is readily expressed as

$$P_Y(y)\mathrm{d}y = P_X(x)\mathrm{d}x. \tag{19.9}$$

In cases where there are problems with the sign (probabilities better be positive numbers!), this expression is to be interpreted as $P_Y(y) = P_X(x)|J|$, with J being the Jacobian determinant (for generalizations to the multi-variable problem). The expression (19.9) can be most useful.

- **Exercise:** Assume that a circle with radius R is placed randomly with its center on the x-axis, with positions uniformly distributed in the interval $\{0;\, \mathcal{L}\}$. Determine the probability density of cord-lengths along the y-axis, i.e. the distribution of the lengths of segments of the y-axis inside the circle. Let the ratio $R/\mathcal{L}$ be arbitrary. Repeat the problem, now with a *sphere* with radius R placed randomly in the x, z-plane, with its center uniformly distributed in a square $\{0;\, \mathcal{L}\}$, $\{0;\, \mathcal{L}\}$.
- **Exercise:** Assume that a variable X has a Gaussian distribution, $P(x) = (2\pi\sigma^2)^{-1/2} \exp(-\frac{1}{2}x^2/\sigma^2)$. Consider the variable $Y = X^2$ and demonstrate that its probability density is given by the *chi-square* probability density

$$P(y) = \frac{1}{\sigma\sqrt{2\pi y}} \exp\left(-\frac{y}{2\sigma^2}\right), \tag{19.10}$$

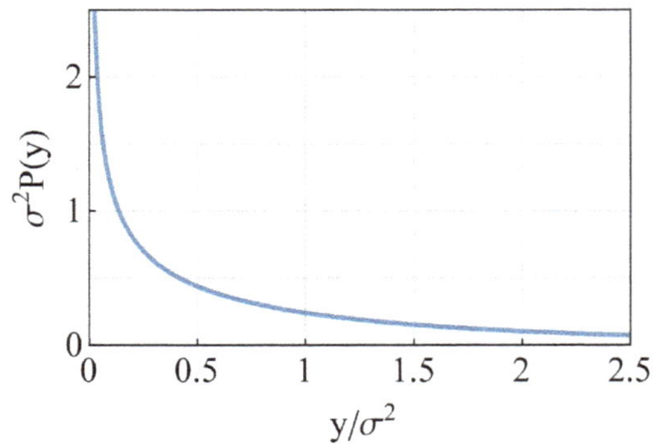

Figure 19.1. The normalized chi-square distribution as given by (19.10).

for $y \geqslant 0$ and $P(y) = 0$ otherwise, see figure 19.1. Demonstrate that $\langle Y \rangle = \sigma^2$ and $\langle Y^2 \rangle = 3\sigma^4$. Prove explicitly that $P(y)$ is indeed normalized, $\int_0^\infty P(y)\mathrm{d}y = 1$.

More generally we have that chi-squared distribution (or χ^2-distribution) with k degrees of freedom is the distribution of a sum of the squares of k independent standard normal random variables. The chi-squared distribution is used in so called 'chi-squared tests' for goodness of fit of an observed distribution to a theoretical one, and in confidence interval estimation for a population standard deviation of a normal distribution from a sample standard deviation.

- **Exercise:** Assume that a variable X has a Gaussian distribution, $P(x) = (2\pi\sigma^2)^{-1/2}\exp(-\frac{1}{2}x^2/\sigma^2)$. Consider another variable $Y = (2/\pi)^{-1/2} \int_0^X \exp(-\frac{1}{2}\gamma^2)\mathrm{d}\gamma$ for $X \geqslant 0$, while $Y = 0$ for $X < 0$. Derive the probability density for Y.
- **Example:** We can derive the cumulative probability density for a two parameter Gamma distribution by the following pragmatic arguments. We know that the Gamma function $\Gamma(k)$ with $k > 0$ can be expressed by the integral (Hogg and Craig 1970) as

$$\Gamma(k) = \int_0^\infty x^{k-1}\mathrm{e}^{-x}\mathrm{d}x.$$

We can introduce a new variable $y \equiv x\theta$ with $\theta > 0$ and find

$$\Gamma(k) = \int_0^\infty \left(\frac{y}{\theta}\right)^{k-1} \mathrm{e}^{-y/\theta}\frac{1}{\theta}\mathrm{d}y.$$

From this expression we can conclude that the function

$$P(y;\, k,\, \theta) = \frac{1}{\theta^k}\frac{1}{\Gamma(k)}y^{k-1}\mathrm{e}^{-y/\theta} > 0$$

is normalized to unity for $y > 0$ as well as all k, $\theta > 0$, see also figure 19.2. We let $P(y;\, k,\, \theta) = 0$ for $y < 0$. In reality the two requirements (it is positive and

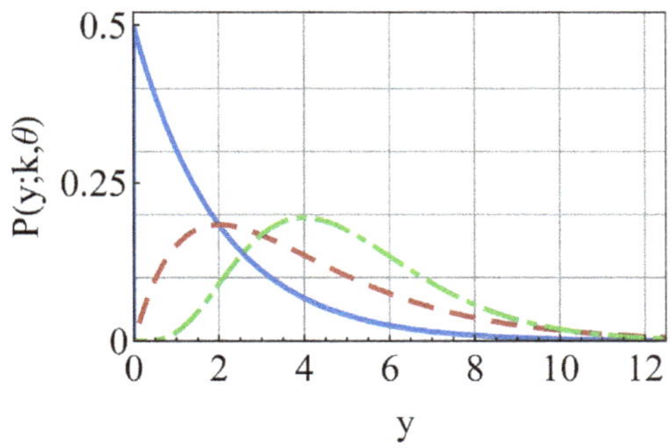

Figure 19.2. The Gamma distribution $P(y;\, k,\, \theta)$ as a function of y for three parameter sets: $(k, \theta) = (1, 2), (2, 2)$ and $(5, 1)$ with full, dashed and dot-dashed lines, respectively.

normalized) are all we can require from a probability density (Pécseli 2000), so we can introduce $P(y;\, k,\, \theta)$ as the two parameter Gamma distribution. The Gamma distribution has been used to model, for instance, the size of insurance claims and rainfalls.

19.2 Multi-variable probabilities

The random variable can be multi-dimensional, as already mentioned, i.e. the corresponding probabilities can depend on many different variables. Averages are defined by a simple generalization of (19.2) as

$$\langle f(X_1,\, X_2 \ldots X_N)\rangle = \int_{-\infty}^{\infty} f(x_1,\, x_2 \ldots x_N) P(x_1,\, x_2 \ldots x_N) \mathrm{d}x_1 \mathrm{d}x_2 \ldots \mathrm{d}x_N, \tag{19.11}$$

where $P(x_1,\, x_2 \ldots x_N)$ is the joint probability density for the variables $x_1,\, x_2 \ldots x_N$.

Moment generating functions and characteristic functions can be defined for multi-variate distributions as well. For instance for a bivariate probability density, $P(x_1,\, x_2)$, we have

$$CH(\alpha,\, \xi) \equiv \langle \exp(i\alpha X_1 + i\xi X_2)\rangle = \int_{-\infty}^{\infty} \mathrm{e}^{i\alpha x_1 + i\xi x_2} P(x_1,\, x_2) \mathrm{d}x_1 \mathrm{d}x_2. \tag{19.12}$$

Averages or moments are then given by

$$\langle X_1^n X_2^m \rangle = (-i)^{n+m} \frac{\mathrm{d}^{n+m}}{\mathrm{d}\alpha^n \mathrm{d}\xi^m} CH(\alpha,\, \xi)\bigg|_{\alpha=\xi=0}. \tag{19.13}$$

A probability can be *conditional*, with $P(x|y)$ giving the probability for the event x, given with certainty the event y. By Bayes' rule we have

$$P(x|y) = \frac{P(x,\, y)}{P(y)}, \tag{19.14}$$

where $P(x,\, y)$ is the joint probability for x and y. Often (19.14) is taken as the definition of a conditional probability. Bayes' rule has self-evident generalizations to multi-variable probabilities. It is often easier to predict or argue expressions for *conditional* probabilities than for their *un*conditional counterparts. The inverted version of (19.14) can then sometimes be used to obtain the full joint probability. We have $P(x, y) = P(y)P(x|y) = P(x)P(y|x)$.

Quite generally we have $P(y) = \int_{-\infty}^{\infty} P(x, y)\mathrm{d}x$ for a continuous variable, or a corresponding sum for discrete variables. Then we can write

$$\begin{aligned}\int_{-\infty}^{\infty} P(x|y)\mathrm{d}x &= \int_{-\infty}^{\infty} \frac{P(x, y)}{P(y)}\mathrm{d}x \\ &= \frac{1}{P(y)}\int_{-\infty}^{\infty} P(x, y)\mathrm{d}x \\ &= \frac{1}{P(y)}P(y) = 1.\end{aligned}$$

For continuous variables, we can thus argue that $P(x|y)$ has the properties of a probability density (PDF) of one continuous random variable (Hogg and Craig 1970).

For a function $f(x)$ we have the conditional expectation value

$$\langle f(x)|y\rangle \equiv \int_{-\infty}^{\infty} f(x)P(x|y)\mathrm{d}x.$$

Similarly, we have a conditional variance

$$\left\langle (x - \langle x|y\rangle)^2 \middle| y\right\rangle = \left\langle x^2|y\right\rangle - \langle x|y\rangle^2.$$

- **Examples:** A harmonic oscillation $\tau_1 \cos(t + \tau_2)$ with a random phase τ_2 and a random amplitude τ_1 represents a sample function with two parameters.

  A function which alternates between +1 and −1 at N random times $t = \tau_n$ with $n = 1, 2, \ldots, N$ represents a sample function with N parameters.

19.2.1 Correlation

The *covariance* of two random scalar variables X and Y can be defined as

$$\begin{aligned}Cov &\equiv \langle (X - \langle X\rangle)(Y - \langle Y\rangle)\rangle \\ &\equiv \langle XY\rangle - \langle X\rangle\langle Y\rangle.\end{aligned} \tag{19.15}$$

The *correlation coefficient* between two random variables X and Y can be defined as

$$C_{\{X,Y\}} \equiv \frac{\langle (X - \langle X\rangle)(Y - \langle Y\rangle)\rangle}{\sqrt{\langle (X - \langle X\rangle)^2\rangle\langle (Y - \langle Y\rangle)^2\rangle}}. \tag{19.16}$$

For the case $X = Y$ we have $C = 1$, identically. If on the other hand X and Y are independent variables, we readily find $C = 0$. For complex variables, we define $Cov \equiv \langle XY^*\rangle - \langle X\rangle\langle Y^*\rangle$, with the asterix denoting complex conjugate.

19.2.2 Stochastic processes

Consider a measurable quantity $Y(t)$ function of some variable t. Let $Y_x(t)$ denote an ensemble of such functions of t, labeled by a random parameter x with given distribution $P(x)$. When t stands for time it is customary to call $Y(t)$ a stochastic process. A *sample function* or *a realization* of the process is obtained for one particular value of x. An ensemble of such sample functions thus constitutes a stochastic process. The averaging is understood as an ensemble average in the sense indicated in (19.1).

- **Examples:** A particularly simple example for a stochastic process is one where each sample function is for all times a constant x which varies over the ensemble.

  A harmonic oscillation $\cos(t + x)$ with a random phase x represents a sample function with one parameter.

Let $P_1(y_1, t_1)\mathrm{d}y_1$ be the probability that $Y(t)$ at time $t = t_1$ lies in the interval $\{y_1; y_1 + \mathrm{d}y_1\}$. Then

$$P_1(y_1, t_1) = \int_{-\infty}^{\infty} \delta\left[Y_x(t_1) - y_1\right]P(x)\mathrm{d}x. \tag{19.17}$$

The ensemble average of $Y(t_1)$ is then obtained by

$$\langle Y(t_1)\rangle = \int_{-\infty}^{\infty} y_1 P_1(y_1, t_1)\mathrm{d}y_1,$$

and similarly for other averages.

Generalizing the foregoing expression we let $P_2(y_1, t_1; y_2, t_2)\mathrm{d}y_1\mathrm{d}y_2$ be the joint probability that $Y(t_1)$ lies is the interval $\{y_1; y_1 + \mathrm{d}y_1\}$ at time t_1 and $Y(t_2)$ in the interval $\{y_2; y_2 + \mathrm{d}y_2\}$ at time t_2. Quite generally we have

$$P_N(y_1, t_1; y_2, t_2; \ldots ;y_N, t_N) = \int_{-\infty}^{\infty} \delta\left[Y_x(t_1) - y_1\right] \cdots \delta\left[Y_x(t_N) - y_N\right]P(x)\mathrm{d}x. \tag{19.18}$$

It is natural to assume a time ordering $t_n \geqslant t_{n-1} \geqslant \cdots \geqslant t_2 \geqslant t_1$, noting that some of the times can actually be equal. One can thus define a hierarchy of probabilities P_1, P_2, P_3, $\cdots$ which serve to describe the stochastic process.

Again it can be argued that the probability densities contain too much information. For practical purposes, a reduced information contained in the first few moments of the distributions will often suffice. Such moments are for a continuous one point probability density defined as

$$\langle Y^n\rangle \equiv \int_{-\infty}^{\infty} y_1{}^n P_1(y_1)\mathrm{d}y_1, \tag{19.19}$$

for $n = 1, 2, \ldots$. The average value, for $n = 1$, is often known *a priori* by symmetry arguments. The standard deviation, or root-mean-square, $\sigma \equiv \langle (Y - \langle Y \rangle)^2 \rangle^{1/2}$ measures the level of fluctuations. The *skewness*, $S \equiv \langle (Y - \langle Y \rangle)^3 \rangle / \sigma^3$, measures the lack of symmetry of the signal, while the *kurtosis*, $K \equiv \langle (Y - \langle Y \rangle)^4 \rangle / \sigma^4$ is a measure for the excursion of the signal to values in excess of the standard deviation (Cramér 1946). For a Gaussian random signal we have $S = 0$ and $K = 3$. Using these figures as reference, the kurtosis is sometimes defined as $K \equiv \langle (Y - \langle Y \rangle)^4 \rangle / \sigma^4 - 3$ (Abramowitz and Stegun 1972). With this definition we would have $K = 0$ for a Gaussian random process. Signals characterized by a large kurtosis are often 'bursty' in nature. See figure 19.3 for illustrative examples. Such signals are often called 'intermittent'. The literature often uses a Gaussian process as a reference and denotes an actual signal as 'leptokurtic' if its kurtosis is larger than for a Gaussian. You sometimes find the term 'super Gaussian'. A platykurtic distribution will have smaller kurtosis than a Gaussian. You sometimes find the term 'sub Gaussian' for these. The probability density describing the outcome of a coin-toss game is one example for an extreme platykurtic distribution. A distribution having the same kurtosis as a Gaussian but without actually being one, is called mesokurtic. Evidently, the moments in (19.19) can be defined for stationary as well as non-stationary random processes.

Inequality for skewness and kurtosis

For a process where the standard deviation exists, we can obtain a general inequality for the skewness and the kurtosis (Pearson 1929, Wilkins 1944). In particular for meteorological applications in studies of concentration fluctuations, these results have been met with some interest.

For some variable x we have the trivial relations $\langle (x\langle x^4 \rangle - x^2 \langle x^3 \rangle)^2 \rangle \geqslant 0$ or $\langle (x\langle x^3 \rangle - x^2 \langle x^2 \rangle)^2 \rangle \geqslant 0$, giving $\langle x^4 \rangle \geqslant \langle x^3 \rangle^2 / \langle x^2 \rangle$. The equality sign applies for $\langle x^2 \rangle = 0$, where we have the 'null identity'. More generally we have $\mathcal{K} \geqslant \mathcal{S}^2 + 1$ when $\sigma \neq 0$. For completeness we give the simplest proof here by taking the dimensionless vector $\mathbf{X} \equiv \{1, x, x^2\}$ and then obtaining the expectation value of the tensor $\mathbf{XX}^\dagger$, where † denotes 'transpose'. With $\langle x \rangle = 0$ we then have

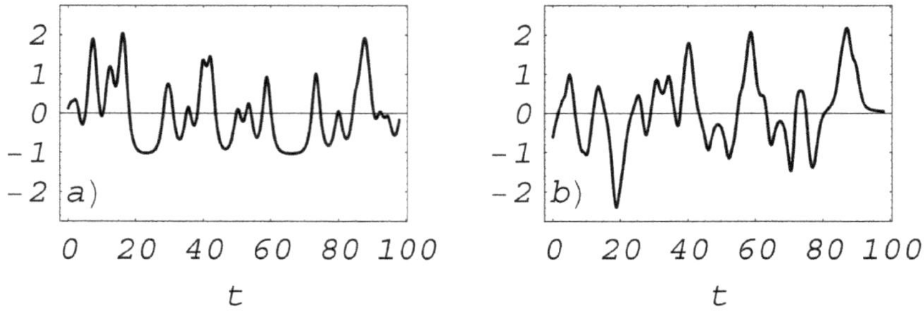

Figure 19.3. Illustration of a signal having positive skewness in (a). A signal with vanishing skewness but large kurtosis is illustrated in (b). Both signals have zero average.

$$\langle \mathbf{X}\mathbf{X}^{\dagger}\rangle = \begin{pmatrix} 1 & 0 & \langle x^2\rangle \\ 0 & \langle x^2\rangle & \langle x^3\rangle \\ \langle x^2\rangle & \langle x^3\rangle & \langle x^4\rangle \end{pmatrix}.$$

Using a vector $\mathbf{Y} \equiv \{\langle x^2\rangle^2, \langle x^3\rangle, -\langle x^2\rangle\}$, we construct the scalar $\mathbf{Y}^{\dagger} \cdot \langle \mathbf{X}\mathbf{X}^{\dagger}\rangle \cdot \mathbf{Y}$, which is identical to $\langle(\mathbf{Y}^{\dagger} \cdot \mathbf{X})(\mathbf{X}^{\dagger} \cdot \mathbf{Y})\rangle \equiv \langle(\mathbf{Y}^{\dagger} \cdot \mathbf{X})(\mathbf{Y}^{\dagger} \cdot \mathbf{X})^{\dagger}\rangle \geqslant 0$. Consequently we have $\langle x^4\rangle\langle x^2\rangle - \langle x^3\rangle^2 - \langle x^2\rangle^6 \geqslant 0$. If $\langle x^2\rangle = 0$, we have trivially that $\langle x^3\rangle = 0$. When $\langle x^2\rangle > 0$ this gives the desired result $\mathcal{K} \geqslant \mathcal{S}^2 + 1$, which remains correct also when $\langle x\rangle \neq 0$. A more general, but also more lengthy, analysis containing the foregoing results is given by Mole and Clarke (1995).

The results summarized here have general validity. They have been quoted elsewhere (Labit *et al* 2007) with particular reference to general Beta distributions.

Schwarz inequality

The *Schwarz inequality* (sometimes known as the *Cauchy–Schwarz inequality*) is valid for arbitrary integration limits and under general conditions

$$\int [f(t)]^2 \mathrm{d}t \int [g(t)]^2 \mathrm{d}t \geqslant \left[\int f(t)g(t)\mathrm{d}t\right]^2, \tag{19.20}$$

with equality occurring when $f(t)/g(t)$ equals a constant (Bendat 1958). All integrals have the same support, but the integration limits are otherwise unspecified, apart from the obvious requirement that the integrals must exist.

An alternative form, in terms of ensemble averages, can be obtained by introducing two new functions $f \equiv \phi/\sqrt{\langle\phi^2\rangle}$ and $g \equiv \psi/\sqrt{\langle\psi^2\rangle}$, derived from some arbitrary ϕ and ψ. We have the correlation coefficient $C_{f,g} \equiv \langle fg\rangle$. We can have $\langle f\rangle \neq 0$ and $\langle g\rangle \neq 0$, but choose here the compact definition of $C_{f,g}$. To determine the bounds of $C_{\{f,g\}}$ we use the relation

$$\langle(f + pg)^2\rangle = 1 + p^2 + 2pC_{\{f,g\}} \geqslant 0, \tag{19.21}$$

for an arbitrary number p. From (19.21) we obtain $-1 \leqslant p^* C_{\{f,g\}}$ with $p^* \equiv 2p/(1 + p^2)$. We can have p^* positive or negative. Considering both cases we obtain

$$-\frac{1}{|p^*|} \leqslant C_{\{f,g\}} \leqslant \frac{1}{|p^*|}.$$

It is easily demonstrated that $|p^*| \leqslant 1$, so that $|C_{\{f,g\}}| \leqslant 1$. Alternatively, we can express this result as $\langle\phi^2\rangle\langle\psi^2\rangle \geqslant \langle\phi\psi\rangle^2$, which is the equivalent of (19.20) with ensemble averages replacing time averages.

- **Exercise:** Prove the Schwarz inequality by observing that the integral $\int [af(t) + bg(t)]^2 \mathrm{d}t > 0$ for all real a and b, unless $af(t) + bg(t) = 0$.

A generalization of the theorem is called *Hölder's inequality*, see for instance Titchmarsh (1950).

- **Exercise:** A simple example of Schwarz inequality can be made to state that for two vectors **A** and **B** in three-dimensional space we have $|\mathbf{A}|\,|\mathbf{B}| \geqslant |\mathbf{A}\cdot\mathbf{B}|$. Demonstrate this result, and give it a geometrical interpretation. Show that the equality sign applies when $\mathbf{A} = c\mathbf{B}$ where c is an arbitrary constant of either sign. Does the dimensionality of space matter? If so, how?

Jensen's inequality

Let a continuous function $f = f(t)$ be defined on an interval I, and $\{\xi, f(\xi)\}$ be a point on its graph. A tangent line with slope λ is passing through $\{\xi, f(\xi)$ so that the curve of f is entirely above the tangent, i.e.

$$f(t) \leqslant f(\xi) + (t - \xi)\lambda \tag{19.22}$$

for all t in I. By a convex function f we imply the relation (19.22) to be fulfilled for all ξ in I. In 1905 Jensen showed (Feller 1968, Hansen and Pedersen 1982) that a (mid-point) convex continuous function satisfies

$$f\left(\sum_i \lambda_i t_i\right) \leqslant \sum_i \lambda_i f(t_i) \tag{19.23}$$

for any convex combination $\{\lambda_i\}$ of points $\{t_i\}$ in I. Taking

$$a = \begin{pmatrix} \sqrt{\lambda_1} & 0 & \dots & 0 \\ \vdots & \vdots & \vdots & \vdots \\ \sqrt{\lambda_n} & 0 & \dots & 0 \end{pmatrix}, \qquad x = \begin{pmatrix} t_1 & \dots & 0 \\ \vdots & \vdots & \vdots \\ 0 & \dots & t_n \end{pmatrix}, \tag{19.24}$$

and considering f as a function on self-adjoint matrices (with $f(0) = 0$), the inequality (19.23) reads

$$f(a^* x a) \leqslant a^* f(x) a. \tag{19.25}$$

We shall say that a continuous, real function f on I satisfies Jensen's operator inequality if (19.25) holds for any self-adjoint matrix x with spectrum in I and every matrix a with $||a|| \leqslant 1$.

The most often used form of Jensen's inequality is found as

$$\langle f(t)\rangle \leqslant f(\langle t\rangle), \tag{19.26}$$

stating that the expectation value of a function at random variable t is larger than or equal to the functional value of the expectation value of that variable, given the basic assumption. In a different notation with the identical information we have $E[f(t)] \leqslant f(E[t])$, with $E[\cdot]$ again denoting the expectation value. The relation (19.26) follows directly from (19.22) if we introduce the particular case where $\xi = \langle t\rangle \equiv E[t]$ and then take the expectation value of (19.22).

Wirtinger's inequality
The first and simplest version of Wirtinger's inequality for a periodic function $f(x)$ with continuous derivatives f' where $\int_0^{2\pi} f(x)\mathrm{d}x = 0$ can be expressed as

$$\int_0^{2\pi} f'^2(x)\mathrm{d}x \geqslant \int_0^{2\pi} f^2(x)\mathrm{d}x,$$

where the equality sign applies if and only if $f(x) = s\sin(x) + b\cos(x) \equiv c\sin(x+d)$ for some constants a, b, c, d. The present result can be seen as a one-dimensional case of Poincaré's inequality.

A related inequality is expressed as

$$a^2 \int_0^a |f'(x)|^2\mathrm{d}x \geqslant \pi^2 \int_0^a |f|^2(x)\mathrm{d}x, \qquad (19.27)$$

where $f(0) = f(a) = 0$.

The proof of the first version follows by use of the expansion

$$f(x) = \frac{1}{2}a_0 + \sum_{n\geqslant 1}\left(\frac{a_n}{\sqrt{\pi}}\sin(nx) + \frac{b_n}{\sqrt{\pi}}\cos(nx)\right),$$

and then using Parseval's identity. Note that here $a_0 = 0$ by assumption.

19.2.3 Correlation functions

With an ensemble of time records available for the fluctuating quantity Y, we can construct averages as $\langle Y^n(t_1)Y^m(t_2)\rangle$ referring to different times t_1 and t_2 in the same record. Particularly important is the correlation function defined as

$$R(t_1, t_2) \equiv \langle Y(t_1)Y(t_2)\rangle = \iint_{-\infty}^{\infty} y_1(t_1)y_2(t_2)P_2(y_1, t_1; y_2, t_2)\mathrm{d}y_1\mathrm{d}y_2, \qquad (19.28)$$

where $R(t_1, t_1) \neq R(t_2, t_2)$ in general, see also (19.15) and (19.16). As mentioned, the function $R(t_1, t_2)$ refers to events taken at two different times but in the same record. This is sometimes emphasized by adding subscripts as $R_{YY}(t_1, t_2)$. For complex variables, the correlation function is defined as $R(t_1, t_2) \equiv \langle Y(t_1)Y(t_2)^*\rangle$, with the asterix denoting complex conjugate. If a time record is sampled at N points, $t_1, t_2, \ldots, t_N$ not necessarily equidistant, a correlation matrix can be defined with elements $R_{ij} \equiv R(t_i, t_j)$, where evidently $R_{ij} = R_{ji}$.

A *coherence function* can be defined as

$$C(t_1, t_2) \equiv \frac{\langle Y(t_1)Y(t_2)\rangle - \langle Y(t_1)\rangle\langle Y(t_2)\rangle}{\sqrt{\left(\langle Y(t_1)^2\rangle - \langle Y(t_1)\rangle^2\right)\left(\langle Y(t_2)^2\rangle - \langle Y(t_2)\rangle^2\right)}}, \qquad (19.29)$$

with the property $C(t_1, t_1) = C(t_2, t_2) = 1$. The coherence function is a normalized equivalent of the correlation function.

Considering n different times $t_1, t_2, \ldots, t_n$ in the ensemble of records of the signal $Y(t)$ we can construct $a_1 Y(t_1) + a_2 Y(t_2) + \cdots + a_n Y(t_n)$, where a_i are the elements of an arbitrary, non-zero, n-dimensional vector $\mathbf{a}$. We have evidently

$$\left\langle \left(\sum_i a_i Y(t_i) \right)^2 \right\rangle = \sum_i \sum_j a_i \langle Y(t_i) Y(t_j) \rangle a_j \\ \equiv \sum_i \sum_j a_i R(t_i, t_j) a_j \geqslant 0, \tag{19.30}$$

for any non-trivial real vector, $\mathbf{a}$. For complex $\mathbf{a}$, we use the complex conjugate a_j^* in (19.30). The correlation function $R(t_i, t_j)$ is said to be *non-negative definite*. This property of correlation functions will be important in discussions of their Fourier transforms If only the inequality sign applies in (19.30), the correlation function is said to be *positive definite*. The distinction between positive definite and non-negative definite is related to the notion of linear independence (Papoulis 1991). The correlation function $R(t_i, t_j)$ can be interpreted as the i, j-th element of a symmetric tensor, since evidently $R(t_i, t_j) = R(t_j, t_i)$

- **Example:** Consider the simple matrix with elements as stated

$$M_{\{i,j\}} = \begin{vmatrix} 1 & 1 & 0 \\ 1 & 1 & 1 \\ 0 & 1 & 1 \end{vmatrix}, \tag{19.31}$$

and let $\{a_i\} = \{1, -1, 1\}$. It is easily demonstrated that for this particular case $\sum_{i,j} a_i M_{\{i,j\}} a_j = -1$, i.e. it is not sufficient for a matrix to have positive elements only, in order to ensure it to be positive definite. The matrix M given by (19.31) cannot therefore represent a correlation matrix.

19.2.4 Conditional averages

Dealing with stochastic processes, it is a rather obvious question to ask how much are we actually able to predict when facing a given problem. For instance, without any prior knowledge, what is our best guess for the value of a function $\Phi(t)$? To answer this we define an error in the prediction

$$e \equiv \left\langle \left(\check{\Phi} - \Phi(t) \right)^2 \right\rangle,$$

stating the mean square average of the difference between our guess, or estimate, $\check{\Phi}$ and the *true* value $\Phi(t)$. (Note that the definition of e is not unique: we *might* have used $e \equiv \langle |\check{\Phi} - \Phi(t)| \rangle$ (Johnsen *et al* 1987).)

To minimize e we require $\mathrm{d}e/\mathrm{d}\check{\Phi} = 0$, which here gives $\check{\Phi} = \langle \Phi(t) \rangle$. This means that our best *a priori* guess for a variable $\Phi(t)$ is the average value of that variable! The uncertainty on our prediction is readily found to be $e = \langle \Phi^2(t) \rangle - \langle \Phi(t) \rangle^2$. We would like to be able to do better, but this requires that we have some *a priori* knowledge, for instance we may know with certainty the functional value $\Phi(t_0)$ and would like to predict it at a later time $t > t_0$. Alternatively, we might like to know what the variation was prior to the reference time. An obvious example for conditioned prediction is weather forecast: we know the weather today and would like to come up with a good guess on what it is tomorrow!

We propose an estimator in the form $\check{\Phi} = \sum_{n=0}^{N} a_n(t)\Phi_1^n$, where the a_js are for the time being unknown (Adrian 1979). In principle, we might let the summation go to infinity, but usually we have to truncate the series at a relatively low value of N. By Φ_1 we denote the presumed value of $\Phi(t_0)$, where we here take $t_0 = 0$ without loss of generality. Again, we want to minimize the mean square error for our prediction, and require $\partial e/\partial a_n = 0$ for $n = 0, 1, 2, \ldots, N$. Implicit in our analysis we have that the Φ_1s are taken from the ensemble of realizations of the process, and the statistics of Φ_1 must therefore be identical to that of $\Phi(t = 0)$. We find a set of equations, which can be expressed as

$$\sum_{n=0}^{N} a_n(t)\langle\Phi^{n+m}(0)\rangle = \langle\Phi^m(0)\Phi(t)\rangle \equiv R_m(t) \qquad \text{for} \qquad m = 0, 1, \ldots, N, \quad (19.32)$$

using the definition $R_m(t) \equiv \langle\Phi^m(0)\Phi(t)\rangle$. For some realistic models, see section 20.3.1, it is possible to obtain these averages analytically.

Conceptually we can describe the logic behind the result (19.32) by arguing that when we have found a prescribed event (here, Φ_1) in the time record at some time, we want to predict the most likely future (or, the most likely *past* as well, for that matter!). It is evident that, since Φ_1 is an element in the signal, its occurrence must follow the same statistics as the signal itself. We want the prediction to cover all conceivable values of Φ_1 and therefore minimize the error e for the average over *all* Φ_1 as indicated by the $\langle\rangle$-signs. The prediction will of course be different (at least it *can* be), depending on the magnitude of Φ_1, and for this reason linear as well as nonlinear terms enter the estimator $\check{\Phi}$.

The result (19.32) is of little value as it stands, since we are unable to give a general solution of the N equations. We consider here some illustrative cases, for different values of N (Pécseli and Trulsen 1989).

N = 0: For this case we obtain the trivial result $a_0 = \langle\Phi\rangle =$ const. i.e. $\check{\Phi} = \langle\Phi\rangle$, which is the simple *unconditioned* result, with associated error $e = \langle\Phi^2\rangle - \langle\Phi\rangle^2$, where these quantities are presumed to be known. We note that when Φ denotes an electrostatic potential, it is trivial to take the reference value $\langle\Phi\rangle = 0$, at least as far as homogeneous systems are concerned, but on the other hand this need not always be a smart assumption (Pécseli 2000).

N = 1: For this particularly simple case the linear estimate results in

$$\check{\Phi} = \langle\Phi\rangle + \frac{R_1(t) - \langle\Phi\rangle^2}{\langle\Phi^2\rangle - \langle\Phi\rangle^2}(\Phi_1 - \langle\Phi\rangle).$$

For the special case where $\langle\Phi\rangle = 0$, this result is particularly simple: $\check{\Phi} = \Phi_1 R_1(t)/\langle\Phi^2\rangle$. The function $R_1(t)$ is the correlation function for the (time stationary) process. The error on the estimate is trivially obtained as

$$e(t) = \langle\Phi^2\rangle - R_1^2(t)/\langle\Phi^2\rangle.$$

It is easily demonstrated that $e(0) = 0$, as expected, since we presume to know the functional value at $t = 0$, without uncertainty. For large times, larger than the correlation time where $R_1 \to 0$, we find on the other hand $e = \langle \Phi^2 \rangle$, and in this limit our prediction is the same as we would have obtained *without* any *a priori* knowledge. For short times, the error in the estimate depends critically on $\langle (\mathrm{d}\Phi/\mathrm{d}t)^2 \rangle$ as might be expected, and seen by use of results given later in section 19.2.5.

N = 2: To simplify the notation, we consider here the case $\langle \Phi(t) \rangle = 0$ from the outset, and find

$$\begin{aligned} a_0(t) &= \left(R_1(t)\langle \Phi^3 \rangle - R_2(t)\langle \Phi^2 \rangle \right) \frac{\langle \Phi^2 \rangle}{A} \\ a_1(t) &= \left(\langle \Phi^4 \rangle - \langle \Phi^2 \rangle^2 \right) \frac{R_1(t)}{A} - \langle \Phi^3 \rangle \frac{R_2(t)}{A} \\ a_2(t) &= \left(-R_1(t)\langle \Phi^3 \rangle + R_2(t)\langle \Phi^2 \rangle \right) \frac{1}{A} = -\frac{a_0}{\langle \Phi^2 \rangle}, \end{aligned} \tag{19.33}$$

with $A \equiv \langle \Phi^2 \rangle \langle \Phi^4 \rangle - \langle \Phi^3 \rangle^2 - \langle \Phi^2 \rangle^3$. Note that in general $a_0(t) \neq 0$, but $a_0(0) = 0$ as readily verified. For this case we find, after some standard manipulations the mean square error to be

$$\begin{aligned} e = a_0^2 + \langle \Phi^2 \rangle \left(1 + a_1^2 + 2a_0 a_2 \right) + 2a_1 a_2 \langle \Phi^3 \rangle + a_2^2 \langle \Phi^4 \rangle - 2a_1 R_1(t) \\ - 2a_2 R_2(t), \end{aligned}$$

which for $\langle \Phi^3 \rangle = 0$ simplifies to $e = \langle \Phi^2 \rangle - R_1^2(t)/\langle \Phi^2 \rangle$, i.e. the result we had for $N = 1$.

N = 3: This result is also relatively straight forward to obtain, but the expressions become very lengthy, so we refer to the paper by Pécseli and Trulsen (1991), where they are given explicitly, and tested with data from a numerical simulation. The agreement is convincing, indicating that the series representation for $\check{\Phi}$ is rapidly converging.

We would like to be able to impose more general conditions, for instance predict the time evolution (in an average sense) given the functional value at the reference time, together with the time derivative at the same time, for instance. To generalize the foregoing analysis, it is an advantage to introduce the formal vector $\mathbf{U} = \{U_1, U_2\} \equiv \{\Phi, \mathrm{d}\Phi/\mathrm{d}t\}$. The straight forward generalization of the estimator $\check{\Phi}$ is then

$$\check{\Phi}(t) = a_0(t) + \sum_j^2 b_j(t) U_j + \sum_j^2 \sum_k^2 c_{jk}(t) U_j U_k + \cdots \tag{19.34}$$

The criterion for the mean square errors on the prediction $e_\Phi = \langle (\check{\Phi}(t) - \Phi(t))^2 \rangle$ being minimum are (again) that the derivatives of e_Φ with respect to a_0, b_j and c_{jk}

vanish. Differentiating and carrying out the indicated averages gives algebraic equations for a_0, b_j and c_{jk}, which can be readily solved by standard techniques.

For more general conditions, the definition of the formal vector **U** can be generalized accordingly to arbitrary dimensions (Pécseli and Trulsen 1989).

As compared to the unconditional average, the conditional average gives a better approximation to features in the individual realizations. One can readily determine also a conditional variance (or conditional reproducibility (Øynes *et al* 1998)) and from this estimate the deviations from the conditional average found by inspection of an ensemble of individual realizations, all selected so that they fulfill the imposed condition.

For Gaussian random processes we know that all the statistical information can be deduced from the correlation function, see for instance case $N = 1$ before. In particular, also the conditional averages can be expressed in terms of the correlation function, as discussed later.

19.2.5 Time-stationary stochastic processes

Particularly important are *stationary* random processes (Davenport and Root 1958, Yaglom 1962, Cox and Miller 1965) defined by the requirement that all P_N depend on time separations only in the sense that

$$P_N\big(y_1,\, t_1 + \tau;\, y_2,\, t_2 + \tau;\, \ldots;\, y_N,\, t_N + \tau\big) = P_N\big(y_1,\, t_1;\, y_2,\, t_2;\, \ldots;\, y_N,\, t_N\big), \qquad (19.35)$$

for all N and τ. In particular $P_1(y_1)$ is independent of time. Shot noise is a particular stationary process defined by

$$P_N\big(y_1,\, t_1;\, y_2,\, t_2;\, \ldots;\, y_N,\, t_N\big) = P(y_1,\, t_1)^N,$$

for all N, i.e. the events at t_1, t_2, …, t_N are independent. Hence, the statistical properties of shot noise are described by one single probability density. An alternative name is *Poisson process*.

- **Example:** From the definition (19.35) of a stationary stochastic process we find for $N = 1$ that $P_1(y_1, t_1) = P_1(y_1, t_1 + \tau)$. This relation remains valid also for $\tau = -t_1$. Since the origin for the time-axis $t = 0$ is chosen arbitrarily, we argue that $P_1(y_1, t_1) = P_1(y_1)$, independent of time. Similarly, for $N = 2$ we have $P_2(y_1, t_1; y_2, t_2) = P_2(y_1, t_1 + \tau; y_2, t_2 + \tau)$, e.g. also for $\tau = -t_1$. Consequently, $P_2(y_1, t_1; y_2, t_2) = P_2(y_1; y_2; \tau)$, with $\tau = |t_2 - t_1|$.

Note that the analysis of section 19.2.4 did *not* assume stationary processes, but the results are of course trivially modified to cover these cases.

A stationary random process can be *ergodic*, in which case ensemble averages equal time averages, e.g. the correlation function can be written as

$$\langle Y(t_1)\, Y(t_1 + \tau)\rangle = \lim_{T\to\infty} \frac{1}{T} \int_0^{T^k} Y_x(t_1)^k\, Y(t_1 + \tau)\mathrm{d}t_1, \qquad (19.36)$$

no matter which member of the ensemble (labeled by k) is chosen. More generally (the notations used in the literature are unfortunately not unambiguous) it is required that all conceivable averages associated with the process should come out the same, no matter whether ensemble or time averages are used. For these stationary processes the random nature of the signal manifests itself in all detail if we wait long enough. Ergodic processes are important by allowing averages to be obtained from one (long) time series, i.e. one sample function or realization, rather than an ensemble of many such realizations (Khinchine 1949, Yaglom 1962). As summarized here, the concept of ergodicity seems innocent maybe, but we emphasize that statistical averages can have little relevance for observations of one realization. Ergodicity is an important concept: only for such cases will ensemble averages have value for predictions on spatial or temporal averages obtained from *one* realization.

- **Exercise:** Give an example for a *non*-stationary stochastic process where $P_1(y_1)$ is nevertheless independent of time.
- **Exercise:** Demonstrate by examples that stationary random processes need not be ergodic.
- **Exercise:** Obtain an explicit expression for the correlation function of a signal (a *random telegraph wave*) which alternates between +1 and −1 at irregular intervals, where the probability for change of sign in a small time interval Δt is $\nu \Delta t$ with ν being a given constant. Give the result for the case where the signal alternates between 0 and +1.

Although it can be defined quite generally, the *correlation function* given by (19.28) is particularly useful for characterizing *stationary* random processes, where it is a function of the time separation $\tau \equiv t_1 - t_2$ only, $R(t_1, t_2) = R(\tau)$. A triple correlation, as well as higher order correlations, can be defined similarly. Some random processes are stationary in a *wide sense*, meaning that their average is constant and their correlation function depends on time separations only, while this may not be true for other averages as triple correlations etc. Random processes satisfying (19.35) are *stationary in a strict sense* (Davenport and Root 1958). Time stationary processes are important for turbulence modeling and are central in what is called the 'random phase approximation'. We meet this in, for instance, chapter 28.

Cross-correlations

The foregoing discussions implicitly assumed that the variable Y refers to *one* particular quantity, say the fluctuating output from a mis-tuned radio receiver. Each realization of the ensemble contains one record for the time varying quantity, Y.It is customary to denote the correlation function the *auto-correlation* function for this case. More generally we consider joint probability densities from two signals of different origin, and let x_1 and y_1 refer to different time records, say $X(t)$ and $Y(t)$, i.e. each realization of the ensemble contains *two* records, one for X and one for Y. The correlation between these two records is called the *cross-correlation* and we indicate this by subscripts as R_{XY}

$$R_{XY}(t_1, t_2) \equiv \langle X(t_1) Y(t_2) \rangle = \iint_{-\infty}^{\infty} x_1(t_1) y_1(t_2) P_2(x_1, t_1; y_1, t_2) \mathrm{d}x_1 \mathrm{d}y_1, \tag{19.37}$$

where $P_2(x_1, t_1; y_1, t_2)\mathrm{d}x_1\mathrm{d}y_1$ is the probability that the variables X at time t_1 and Y at time t_2 assume a value in the narrow intervals x_1, $x_1 + \mathrm{d}x_1$ and y_1, $y_1 + \mathrm{d}y_1$. The notation given before will be unaffected but clearly we will have $R_{XY}(t_1, t_2) \neq R_{XY}(t_2, t_1)$, in general. For time stationary processes, X and Y, the cross-correlation is a function of time separations only, just as the auto-correlation, i.e. $R_{XY}(\tau)$. The arguments are readily generalized to the joint probability density of any number, N, of different signals and their cross-correlations.

- **Exercise:** Express the correlation function in terms of the appropriate bivariate characteristic function.

If two signals X and Y are statistically independent their joint probability density $P_2(x_1, t_1; y_1, t_2) = P(x_1, t_1)P(y_1, t_2)$ in (19.37) and consequently $R_{XY}(t_1, t_2) = \langle X(t_1) \rangle \langle Y(t_2) \rangle$, and the coherence function (19.29) both vanish. Statistically independent records are uncorrelated. It can on the other hand *not* be argued that uncorrelated records should be independent! It is quite easy to imagine cases where particular forms for the joint probability $P_2(x_1, t_1; y_1, t_2)$ happens to give $R_{XY}(t_1, t_2) = 0$, at least for some selected times t_1, t_2. The following chapters will give such examples. There is however one special case, signals having Gaussian statistics, where it will be demonstrated that uncorrelated events are also statistically independent. This example will be given particular attention.

Properties of correlation functions

Some basic properties of correlation functions can readily be noted. Thus the auto-correlation function for time-stationary random processes is symmetric,

$$R_{XX}(\tau) = R_{XX}(-\tau).$$

For the definition (19.37) of the *cross*-correlation we find that the time symmetry is lost, i.e. $R_{XY}(\tau) \neq R_{XY}(-\tau)$, in general.

Next, it can be demonstrated that for stationary ergodic processes, the auto-correlation function assumes its maximum value at $\tau = 0$, i.e.

$$R_{XX}(0) \geqslant R_{XX}(\tau). \tag{19.38}$$

The proof follows from Schwarz inequality (19.20).

Considering an auto-correlation as a measure of the ability of a function to interfere with itself, it can be argued that, physically, the result (19.38) means that any signal shows maximum interference with itself at zero time-delay. One exception is a harmonic signal, $\cos(2\pi t/T)$, where the auto-correlation function reaches the same peak value at regular times, $R_{XX}(\tau = nT) = R_{XX}(0)$, for $n = 1, 2, \ldots$, exemplifying the equality sign in (19.38).

As mentioned before, an auto-correlation function is non-negative definite. Not any symmetric function which is maximized by its value at the origin will do as an auto-correlation function. A trivial counter-example is a box-function which is constant for $|t| \leqslant \tau$ and vanishes elsewhere. The reader may verify that a power-spectrum obtained by its Fourier transform assumes also negative values. A less trivial example is a postulated unacceptable correlation function $\exp(-|\tau|^{\beta})$ with $\beta > 2$.

A *cross*-correlation function can assume its maximum value anywhere. One particular case deserves a remark. Often, the signal Y is just a delayed version of the signal X, and in this case the cross-correlation is just a shifted copy of the auto-correlation. Thus, if $Y(t) = X(t + \Delta t)$, we have $R_{XY}(\tau) = R(\tau - \Delta t)$. A measurement of a cross-correlation between a signal at its emitter and at a distant receiver can thus serve to measure a time-delay by the shift in maximum and thus, for instance, a speed of propagation.

Statistics of time-derivatives

Given a set of realizations of a differentiable random process, new ones can be generated by differentiation. It might then be desired to determine for instance

$$\left\langle Y(t)\frac{\mathrm{d}^n Y(t)}{\mathrm{d}t^n}\right\rangle,$$

assuming the derivatives to exist. To analyze this expression, it is first noted that differentiation and statistical averaging commute. This can be seen by noting that differentiation can be written as a difference $\mathrm{d}Y(t)/\mathrm{d}t \approx [Y(t + \Delta t) - Y(t)]\Delta t$ and that the average of a sum (difference) is the sum (difference) of the individual averages. We may write, for instance

$$\left\langle Y(t)\frac{\mathrm{d}^n Y(t)}{\mathrm{d}t^n}\right\rangle = \frac{\mathrm{d}}{\mathrm{d}t}\left\langle Y(t)\frac{\mathrm{d}^{n-1} Y(t)}{\mathrm{d}t^{n-1}}\right\rangle - \left\langle \frac{\mathrm{d}}{\mathrm{d}t}Y(t)\frac{\mathrm{d}^{n-1} Y(t)}{\mathrm{d}t^{n-1}}\right\rangle. \tag{19.39}$$

For time-stationary random processes the first term is vanishing and by induction it is easily demonstrated that

$$\left\langle Y(t)\frac{\mathrm{d}^n Y(t)}{\mathrm{d}t^n}\right\rangle = (-1)^p\left\langle \frac{\mathrm{d}^p Y(t)}{\mathrm{d}t^p}\frac{\mathrm{d}^{n-p} Y(t)}{\mathrm{d}t^{n-p}}\right\rangle, \tag{19.40}$$

for all $p < n$. In particular we find that

$$\left\langle \frac{\mathrm{d}^{2n+1} Y(t)}{\mathrm{d}t^{2n+1}}\frac{\mathrm{d}^{2m} Y(t)}{\mathrm{d}t^{2m}}\right\rangle = 0. \tag{19.41}$$

As a special case $\langle Y(t)\mathrm{d}Y(t)/\mathrm{d}t\rangle = 0$ for a stationary process, i.e. a signal is uncorrelated with its time-derivative (but of course not statistically independent, in general). These foregoing results evidently assume that all the derivatives exist *at all times*. Define now $\langle \Delta Y^2(t) \rangle$ as the mean-square change in value over time t of a real stationary function $Y(t)$. Then

$$\begin{aligned}\langle \Delta Y^2(t) \rangle &= \lim_{T\to\infty} \int_{-T}^{T} [Y(\tau + t) - Y(\tau)]^2 \, d\tau \\ &= \lim_{T\to\infty} \int_{-T}^{T} \left[\left[Y^2(\tau + t) + Y^2(\tau) - 2Y(\tau + t)Y(\tau)\right] d\tau\right] \\ &= 2\langle Y^2(t) \rangle - 2R(t).\end{aligned} \tag{19.42}$$

For sufficiently small times we may write $\langle \Delta Y^2(t) \rangle \approx \langle \Delta(Y(t_1 + t) - Y(t))^2 \rangle \approx \langle (t\,dY(t_1)/dt)^2 \rangle$ and consequently approximate

$$R(t) \approx \langle Y(t)^2 \rangle - \frac{1}{2} t^2 \left\langle \left(\frac{dY(t)}{dt} \right)^2 \right\rangle, \tag{19.43}$$

giving an important interpretation of the curvature of the correlation function at vanishing time-delay (Champeney 1973). Higher order terms in this expansion can of course be introduced, but they are seldom used. Note that for functions which are not differentiable we can still define a correlation function, this will still be symmetric $R(t) = R(-t)$, but will most likely be characterized by a cusp ('infinite curvature') at the origin. Singularities in the derivatives of a signal can be represented by δ-functions, and when we take the mean square value of the signal, we expect to find that this average diverges, giving a divergent second term in (19.43).

Characteristic time-scales

Usually, three different time-scales are associated with the normalized auto-correlation function $R(\tau) \equiv \langle Y(t)Y(t + \tau) \rangle / \langle Y^2 \rangle$ for time stationary processes.

- The *integral time-scale*, $t_I \equiv \int_0^\infty R(\tau) d\tau$.
- The notation *micro time-scale* is often seen for $t_m \equiv \sqrt{\langle Y^2 \rangle / \langle (dY/dt)^2 \rangle}$, which can be obtained from the curvature of the normalized auto-correlation function at zero time delay, see (19.43).
- Finally, a *correlation time* can be defined as the time-value t_C where $R(t) \approx 0$ when $t > t_C$. It is readily realized that $t_C \geqslant t_I$, in general.

Evidently t_I and t_m are well defined, but their physical interpretation is not entirely obvious. Although it is possible to construct cases where $R(t) = 0$ for some $t > t_C$, we will usually find $R(t) \neq 0$ for all t, although $R(t)$ generally becomes very small for large time separations t. The value of t_C is therefore not well defined in the general case. Its physical interpretation for ergodic Gaussian random signals is on the other hand quite explicit; two signal values obtained with time separations $t > t_C$ in a given record are statistically independent, see section 19.2.5. It is often assumed (and often erroneously) that two values of a signal can be considered statistically independent, when they are obtained at time separations larger than a value where the correlation has become very small.

'Almost' stationary processes
The stationary random process is an important concept, but evidently it can never be realized in practice; there will always be variations in one or all of the many parameters which usually determine the actual signal. Fluctuations in electric circuits depend on the temperature, etc and after all we have to turn on the device at some time, and it cannot be operative for all times either, so non-stationary conditions are inevitable in practice. Usually one introduces the idea of *locally* time stationary processes, or slowly varying stationary processes, in spite of the concept being just as meaningless as that of a slightly bent straight line, strictly speaking. Here we shall attempt to give at least *some* meaning to the idea (Pécseli 2000).

Consider first a non-stationary random process, $Y(t)$. It is still possible to obtain a correlation function $R(t_1, t_2) \equiv \langle Y(t_1)Y(t_2)\rangle$, but this function will now explicitly depend on t_1 and t_2, rather than just the separation $t_1 - t_2$. Introducing the average time $\mathcal{T} = \frac{1}{2}(t_1 + t_2)$ and separation time $\tau = t_1 - t_2$, the correlation function is rewritten in terms of these new variables as $R(\mathcal{T}, \tau)$, with $t_1 = \mathcal{T} + \frac{1}{2}\tau$ and $t_2 = \mathcal{T} - \frac{1}{2}\tau$. If $R(\mathcal{T}, \tau)$ varies much faster with τ than with $\mathcal{N}$ we can argue that the process is locally time stationary, at least in the wide sense discussed before in this section. (We use $\mathcal{N}$ here, instead of the more natural T, which is reserved for temperature.)

To complete the discussion we also consider the correlation function for the derivative process, i.e.

$$
\begin{aligned}
R_D(t_1, t_2) &\equiv \left\langle \frac{\mathrm{d}}{\mathrm{d}t_1} Y(t_1) \frac{\mathrm{d}}{\mathrm{d}t_2} Y(t_2) \right\rangle \\
&= \frac{\partial^2}{\partial t_1 \partial t_2} \langle Y(t_1) Y(t_2) \rangle \\
&\equiv \frac{\partial^2}{\partial t_1 \partial t_2} R(t_1, t_2).
\end{aligned}
\tag{19.44}
$$

Introducing again $R_D(\mathcal{T}, \tau)$ and using

$$
\frac{\mathrm{d}^2}{\mathrm{d}t_1 \mathrm{d}t_2} = \frac{1}{4}\frac{\partial^2}{\partial \mathcal{T}^2} - \frac{\partial^2}{\partial \tau^2},
$$

we find

$$
R_D(\mathcal{T}, \tau) = \frac{1}{4}\frac{\partial^2}{\partial \mathcal{T}^2} R(\mathcal{T}, \tau) - \frac{\partial^2}{\partial \tau^2} R(\mathcal{T}, \tau),
\tag{19.45}
$$

with $R(\mathcal{T}, \tau)$ given before.

Assuming now that we *do* have a stationary random process, we find in particular that

$$R_D(\mathcal{T}, \tau) = R_D(\tau) = -\frac{\mathrm{d}^2}{\mathrm{d}\tau^2} R(\tau), \tag{19.46}$$

which is a particularly simple relationship between the correlation functions for a stationary random process $Y(t)$ and for its time derivative process, $\mathrm{d}Y(t)/\mathrm{d}t$. The result (19.46) has evident generalizations for higher derivatives. For an almost stationary process we expect the first term in (19.45) to be much smaller than the second.

The n-th derivative of $R(\tau)$ at $\tau = 0$ determine $\langle(\mathrm{d}Y(t)/\mathrm{d}t)^{2n}\rangle$ in much the same way as the n-th derivative of the characteristic function at the origin determines $\langle Y^n(t)\rangle$.

Gaussian random processes

Gaussian random processes are particularly important because they adequately describe a large number of natural phenomena. By a Gaussian random process we understand one with the property that for every integer N and every set of points $Y(t_1)$, $Y(t_2)$, $Y(t_3)$, ..., $Y(t_N)$, its N-th order probability density function has an N-dimensional normal form

$$P_N(y_1, t_1; \ldots; y_N, t_N) = (2\pi)^{-N/2}\, |A|^{-1/2} \exp\left(-\frac{1}{2|A|} \sum_{i,j=1}^{N} A_{ij} y_i y_j\right), \tag{19.47}$$

where A_{ij} is the co-factor of $R(t_i, t_j)$ in the correlation matrix

$$\|A\| = \left\| \begin{matrix} R(t_1, t_1) & R(t_1, t_2) & \ldots & R(t_1, t_N) \\ R(t_2, t_1) & R(t_2, t_2) & \ldots & R(t_2, t_N) \\ \vdots & \vdots & \vdots & \vdots \\ R(t_N, t_1) & R(t_N, t_2) & \ldots & R(t_N, t_N) \end{matrix} \right\|, \tag{19.48}$$

and $|A|$ is the determinant of the matrix. Evidently, we are here dealing with square matrices, i.e. with the same number of rows and columns. The co-factor A_{ij} of any element $R(t_i, t_j)$ is defined to be the determinant of order $N - 1$ formed by omitting the i-th row and j-th column of $\|A\|$ and then multiplied by $(-1)^{i+j}$. Without loss of generality, we assumed from the outset that the signal has zero mean, $\langle Y(t)\rangle = 0$, since this can always be achieved by a proper change of variables for stationary processes.

Generally, a Gaussian random process is completely described by its average value and its auto-correlation function $R(t_i, t_j)$, since it completely specifies the joint probability density in this case. For stationary processes, as mentioned, $R(t_i, t_j) = R(|t_i - t_j|)$. The auto-correlation function is obtained from the joint probability density as

$$R(t_i, t_j) \equiv \left\langle Y(t_i)Y(t_j)\right\rangle = \iint_{-\infty}^{\infty} y_i y_j P_2\left(y_i, t_i; y_j, t_j\right)\mathrm{d}y_i\mathrm{d}y_j, \tag{19.49}$$

with $y_i = y(t_i)$ and $y_j = y(t_j)$. In this case, P_2 is the bivariate Gaussian obtained from (19.47) with $N = 2$.

- **Example:** For illustration consider first $N = 1$

$$P_1(y_1, t_1) = (2\pi)^{-1/2}\,|A|^{-1/2}\exp\left(-\frac{A_{11}}{2|A|}y_1^{\,2}\right), \tag{19.50}$$

where for this simple case $\|A\| = \|R(t_1, t_1)\|$, and $|A| = R(t_1, t_1)$, while $A_{11} = 1$.
A less trivial case is $N = 2$, where

$$P_2\left(y_1, t_1; y_2, t_2\right) = \frac{1}{2\pi\sqrt{|A|}}\exp\left(-\frac{1}{2|A|}\left(A_{11}y_1^{\,2} + A_{22}y_2^{\,2} + A_{12}y_1y_2 + A_{21}y_2y_1\right)\right). \tag{19.51}$$

Now the correlation matrix becomes

$$\|A\| = \left\|\begin{matrix} R(t_1, t_1) & R(t_1, t_2) \\ R(t_2, t_1) & R(t_2, t_2)\end{matrix}\right\|, \tag{19.52}$$

and $|A| = R(t_1, t_1)R(t_2, t_2) - R(t_1, t_2)R(t_2, t_1)$ is the determinant of the matrix. We have $R(t_1, t_2) = R(t_2, t_1) \equiv R^2$ independent of t_1 and t_2, and $|A| = R^2 - R(t_1, t_2)$. Then $A_{12} = A_{21} = -R(t_1, t_2)$. For time stationary processes $A_{11} = A_{22} = R(t_1, t_1) = R(t_2, t_2)$. The arguments are readily generalized to the joint probability density associated with any number, N.

Assume now that for some reason the signals obtained at different times are uncorrelated. In that case all $R(t_i, t_j) = 0$ for $i \neq j$ and the Gaussian joint probability density becomes

$$P_N(y_1, t_1; \ldots; y_N, t_N) = (2\pi)^{-N/2}\,|A|^{-1/2}\exp\left(-\frac{1}{2}\sum_{i=1}^{N}\frac{y_i^{\,2}}{R(t_i, t_i)}\right). \tag{19.53}$$

The correlation matrix becomes diagonal for this case

$$\|A\| = \left\|\begin{matrix} R(t_1, t_1) & 0 & \ldots & 0 \\ 0 & R(t_2, t_2) & \ldots & 0 \\ \vdots & \vdots & \vdots & \vdots \\ 0 & 0 & \ldots & R(t_N, t_N)\end{matrix}\right\|. \tag{19.54}$$

Then $|A| = \Pi_{i=1}^{N} R(t_i, t_i)$, and $A_{ij} = 0$ if $i \neq j$ and $A_{ii} = |A|/R(t_i, t_i)$ if $i = j$. Hence

$$P_N\left(y_1, t_1; \ldots; y_N, t_N\right) = P_1(y_1)P_1(y_2)\cdots P_1(y_N).$$

Thus, in this case, the signals y_i and y_j are independent for $i \neq j$, emphasizing an important property, namely that uncorrelated Gaussian variables are also statistically independent. Intuitively we expect that for a given ensemble of time records, the signals at two times t_1 and t_2 tend to become uncorrelated as $|t_1 - t_2| \to \infty$, and consequently for Gaussian processes, the two events also tend to be independent.

These latter expressions appear rather simple since they were obtained for the case where the variable Y refers to *one* particular quantity, i.e. the individual records of the ensemble are sampled at N different times. The arguments are readily generalized to the case where N signals of different origin are analyzed. The symmetry relation for the cross-correlation matrix becomes

$$\|A\|(\tau) = \|A\|^t(-\tau),$$

where the superscript ^{t} denotes the transposed matrix. If the signals constitute a joint Gaussian random process, we can again argue that lack of correlation is a sign of statistical independence of the signals.

- **Exercise:** Demonstrate that $\langle Y(t_1)Y(t_2)Y(t_3)\rangle = 0$ and

$$\begin{aligned}\langle Y(t_1)Y(t_2)Y(t_3)Y(t_4)\rangle &= \langle Y(t_1)Y(t_2)\rangle\langle Y(t_3)Y(t_4)\rangle \\ &+ \langle Y(t_2)Y(t_3)\rangle\langle Y(t_1)Y(t_4)\rangle + \langle Y(t_2)Y(t_4)\rangle\langle Y(t_1)Y(t_3)\rangle,\end{aligned} \tag{19.55}$$

  for a Gaussian random process $Y(t)$ with zero mean. Does this relation assume *stationary* random processes?

- **Exercise:** We have just stated that for a stationary Gaussian random process *any* statistical information is contained in the average (which often is zero) and the correlation function. Demonstrate that the conditional averages discussed in section 19.2.4 can indeed be obtained from the correlation function.
- **Example:** We can introduce the second-order *structure functions* $\Psi_2 \equiv \langle (y_1 - y_2)^2\rangle$ for characterizing fluctuating quantities. (The terminology is, strictly speaking, mostly used for spatial not temporal variations, but we leave that discussion to later.) The second order structure function is directly related to the correlation functions for the signal, since $\Psi_2 = \langle y_1^2\rangle + \langle y_2^2\rangle - 2\langle y_1 y_2\rangle$. For Gaussian random functions of zero mean, the correlation function contains all the available information (Bendat 1958). In particular, for two scalar variables, say y_1 and y_2, we thus have

$$\begin{aligned}P(y_1, y_2)\mathrm{d}y_1\mathrm{d}y_2 &= \frac{1}{2\pi\sigma_1\sigma_2\sqrt{1-\rho^2(1,2)}} \\ &\times \exp\left(\frac{-1}{2(1-\rho^2(1,2))}\left[\left(\frac{y_1}{\sigma_1}\right)^2 + \left(\frac{y_2}{\sigma_2}\right)^2 - 2\rho(1,2)\frac{y_1 y_2}{\sigma_1\sigma_2}\right]\right)\mathrm{d}y_1\mathrm{d}y_2,\end{aligned} \tag{19.56}$$

  where we have $\sigma_{1,2}^2 \equiv \langle y_{1,2}^2\rangle$. By $\rho(1,2) \equiv \langle y_1 y_2\rangle/(\sigma_1\sigma_2)$ we denote the correlation function for the two times t_1 and t_2 or, if spatial variations are considered, the two positions $\mathbf{r}_1$ and $\mathbf{r}_2$. For stationary or homogeneous conditions, respectively, we have $\sigma_1 = \sigma_2 \equiv \sigma$, as assumed in the following. For this case the normalized structure function $1 - \rho(1, 2)$ depends only on the separation of the two sampling positions, and not their absolute values. By inspection of (19.56) we readily find that it is natural to use σ for normalizing y, and this will be used in the following, to simplify the notation.

Introducing the difference and sum, $\Delta \equiv y_1 - y_2$ and $\Sigma \equiv y_1 + y_2$, we can readily rewrite (19.56) and have

$$P(\Delta, \Sigma)\mathrm{d}\Delta\mathrm{d}\Sigma = \frac{1}{\pi\sqrt{1-\rho^2}} \exp\left(\frac{-1}{1-\rho^2}(\Sigma^2 + \Delta^2 - \rho\Sigma^2 + \rho\Delta^2)\right)\mathrm{d}\Delta\mathrm{d}\Sigma,$$

where we used $\mathrm{d}y_1\mathrm{d}y_2 = 2\mathrm{d}\Delta\mathrm{d}\Sigma$.

After integration with respect to Σ, we obtain the probability density for Δ

$$P(\Delta) = \frac{1}{\sqrt{\pi(1-\rho)}} \exp\left(\frac{-1}{1-\rho^2}(\Delta^2 + \rho\Delta^2)\right) = \frac{1}{\sqrt{\pi(1-\rho)}} \exp\left(-\frac{\Delta^2}{1-\rho}\right),$$

where we recall that $\rho \leqslant 1$. The probability density $P(\Delta)$ will reduce to a δ-function, for the particular case of $\rho = 1$.

It is now a simple matter to obtain

$$\frac{\langle|\Delta|^n\rangle}{\langle\Delta^2\rangle^{n/2}} = \frac{2^{n/2}}{\sqrt{\pi}}\Gamma\left(\frac{1+n}{2}\right),$$

which is independent of $\rho(1, 2)$ for all n. It is here perfectly feasible to let n be a continuous variable. For Gaussian random processes, the ratio $\langle|\Delta|^n\rangle/(\langle\Delta^2\rangle)^{n/2}$ is thus *scale invariant* by being independent of the separation of the two sampling positions, here labeled 1 and 2.

For Gaussian processes we find

$$\langle|\Delta|^n\rangle = \frac{1}{\sqrt{\pi}}(1-\rho)^{n/2}\Gamma\left(\frac{1+n}{2}\right).$$

When $1 - \rho$ has a power-law dependence on the separation coordinate, e.g. $\gamma \equiv t_1 - t_2$, so that $1 - \rho \sim \gamma^\alpha$ in a non-trivial subrange, we then evidently find $\langle|\Delta|^n\rangle \sim \gamma^{n\,\alpha/2}$ in that same subrange. The power-law exponent for $\langle|\Delta|^n\rangle$ is consequently directly proportional to n for this case. We can obtain a *compensated* exponent by division with n, i.e. $\alpha/2$. A plot of the compensated exponent for varying n offers a test for Gaussian statistics. Be aware, though, that the larger the n value, the more data you will usually need in order to get a reliable estimate.

- **Example:** It is tempting to argue that two uncorrelated processes with Gaussian projections (or 'marginal distributions') are also independent. In most realistic cases it will indeed be so but at least one counterexample might be mentioned. Assume we have a random process $X(t)$ and another one

$Y(t) = X(t)$. The two processes are trivially correlated and not independent even though their projections are Gaussian. Now modify the process Y so that $Y(t) = A\,X(t)$ where A is a random number that can take values +1 and −1 with equal probability. The two processes are no longer correlated, but clearly not independent. Their joint probability density is not the product of the marginal probabilities. In a phase space spanned by $X - Y$ the joint probability density will look like an $\times$.

19.2.6 Spatial variables

Most of what has been said about functions of temporal variables can readily be 'translated' to cover functions of *spatial variables*. We can thus consider a quantity X depending on the spatial position $\mathbf{r}$, this may then be in one, in two or in three spatial dimensions, i.e. $X = X(\mathbf{r})$. At a certain position, say $\mathbf{r}_1$, the quantity X can assume some value, say x_1. In particular the discussion of temporal stationarity will then correspond to spatially *homogeneous* systems, in which case a probability density has the property

$$P_N(x_1, \mathbf{r}_1 + \mathbf{r}; x_2, \mathbf{r}_2 + \mathbf{r}; \ldots; x_N, \mathbf{r}_N + \mathbf{r}) = P_N(x_1, \mathbf{r}_1; x_2, \mathbf{r}_2; \ldots; x_N, \mathbf{r}_N), \qquad (19.57)$$

for all N and $\mathbf{r}$. In particular $P_1(x_1)$ is then independent of position. Note the functional dependence of the *directions* of the vectors $\mathbf{r}_j$, in particular, their *relative* directions. A homogeneous system is evidently one where it does not matter *where* we place the set of position vectors $\mathbf{r}_1, \mathbf{r}_2, \ldots, \mathbf{r}_N$ in the individual realization when we perform the ensemble averaging. If a *rotation* of the entire set of mutually fixed vectors turns out to be irrelevant as well, then we also call the system *isotropic*.

Correlations of vectors in three-dimensional isotropic turbulence

As far as *scalar* quantities are concerned, the correlation function with respect to spatial (rather than temporal) variables is a trivial generalization of the forgoing results; we simply replace the temporal variables with a spatial variables, and have for instance for $X = X(r)$ the correlation function $R(\mathbf{r}_1, \mathbf{r}_2) = \langle X(r_1)X(\mathbf{r}_2)\rangle$. A quantity closely related to the spatial correlation is the *structure function* defined as

$$\Psi_n(\mathbf{r}_1, \mathbf{r}_2) \equiv \langle (X(\mathbf{r}_1) - X(\mathbf{r}_2))^n \rangle, \qquad (19.58)$$

which can be defined for arbitrary powers n. The relation to the correlation function for $n = 2$ is trivial. Structure functions can of course also be defined for *temporal* variables, but that terminology is rarely used, with the term 'structure' referring to *spatial* variations.

If we are dealing with *vector* variables we have one extra freedom when correlating quantities in two different points separated by a vector $\mathbf{r}$: i.e. we can correlate the two vector components parallel to $\mathbf{r}$, the two perpendicular to $\mathbf{r}$, or mix one parallel and one perpendicular vector component. Correlations of vector

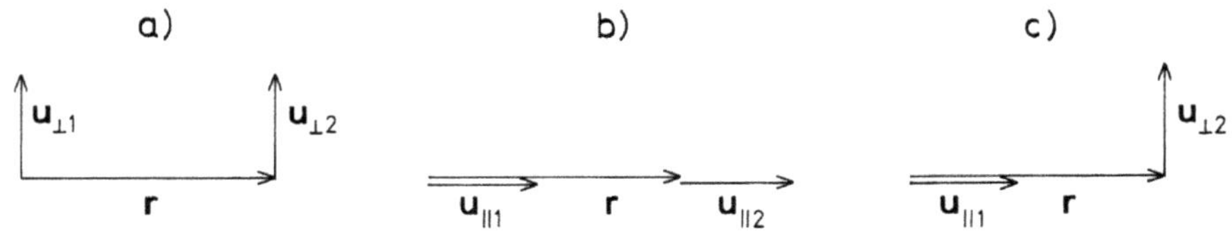

Figure 19.4. Illustration of combinations for two vector component correlations. One component could point out of the paper as well (not shown). The combinations shown here will apply for discussions of two-dimensional turbulence as well (now without including the case with a vector pointing out of the plane).

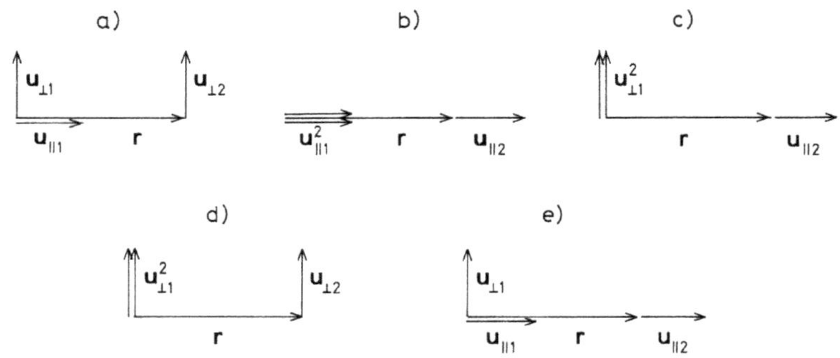

Figure 19.5. Illustration of combinations for three vector component correlations. One or two components could point out of the paper as well (not shown).

components referring to different orientations are illustrated in figure 19.4. We can write for instance for the case with $n = 2$

$$R_{i,j}(\mathbf{r}_1 = \mathbf{x}, \mathbf{r}_2 = \mathbf{x} + \mathbf{r}) \equiv \langle u_i(\mathbf{x}) u_j(\mathbf{x} + \mathbf{r}) \rangle,$$

where u_i and u_j refer for instance to different velocity components in a turbulent flow. For homogeneous flows, $R_{i,j}$ will depend on the separation $\mathbf{r}$ only. For three-dimensional homogeneous and isotropic flows it is found (Batchelor 1953, Tennekes and Lumley 1972) that the results can be expressed as

$$R_{i,j}(\mathbf{r}) = u^2 \left(\frac{f(r) - g(r)}{r^2} r_i r_j + g(r) \delta_{i,j} \right), \tag{19.59}$$

with $g = f + \frac{1}{2} r df/dr = (1/2r) d(r^2 f)/dr$, where

$$f(r) \equiv \frac{\langle u_\parallel(\mathbf{x}) u_\parallel(\mathbf{x} + \mathbf{r}) \rangle}{\langle u_\parallel^2 \rangle} \qquad \text{and} \qquad g(r) \equiv \frac{\langle u_\perp(\mathbf{x}) u_\perp(\mathbf{x} + r) \rangle}{\langle u_\perp^2 \rangle},$$

and $\langle u_\perp^2 \rangle = \langle u_\parallel^2 \rangle \equiv u^2 = \frac{1}{3} \langle u_i u_i \rangle$ with a summation convention understood. We explicitly write f and g as functions of $r \equiv |r|$, since the direction of this vector has been specified already by the definition, see figure 19.4, which also explains the various velocity components. Note that $f(0) = g(0) = 1$. In particular we find that $\langle u_{\|}(\mathbf{x})u_{\perp}(\mathbf{x}+\mathbf{r})\rangle = 0$, corresponding to c) in figure 19.4, as could be anticipated by symmetry arguments.

Correlation of *three* vector components obtained at *two* positions can be defined as well, see for instance figure 19.5. These results can be seen as a special case of a more general three point correlation. We can generalize the definitions of correlations to a signal obtained at arbitrarily many positions or at arbitrarily many different times. In particular we consider $S_{i,j,k}(\mathbf{r}) \equiv \langle u_i(\mathbf{x})u_j(\mathbf{x})u_k(\mathbf{x}+\mathbf{r})\rangle$.

In terms of the function

$$k(r) = \frac{1}{u^3}\left\langle u_{\|}^2(\mathbf{x})u_{\|}(\mathbf{x}+\mathbf{r})\right\rangle,$$

we have (Batchelor 1953)

$$S_{i,j,k}(\mathbf{r}) = u^3\left[\left(\frac{k(r) - rk'(r)}{2r^3}\right)r_i r_j r_k + \left(\frac{2k(r) + rk'(r)}{4r}\right)(r_i\delta_{j,k} + r_j\delta_{i,k}) - \frac{k(r)}{2r}r_k\delta_{i,j}\right], \tag{19.60}$$

where $k'(r) \equiv \mathrm{d}k(r)/\mathrm{d}r$, recalling that for homogeneous isotropic turbulence $k(\mathbf{r}) = k(|\mathbf{r}|)$. Note that $S_{i,j,k}(0) = 0$, and also $S_{1,1,k} + S_{2,2,k} + S_{3,3,k} = \langle u^2(\mathbf{x})u_k(\mathbf{x}+\mathbf{r})\rangle$. It can be demonstrated (Batchelor 1953), that other triple correlations like

$$h(r) = \frac{1}{u^3}\left\langle u_{\perp}^2(\mathbf{x})u_{\|}(\mathbf{x}+\mathbf{r})\right\rangle \qquad \text{or} \qquad q(r) = \frac{1}{u^3}\left\langle u_{\perp}(\mathbf{x})u_{\|}(\mathbf{x})u_{\perp}(\mathbf{x}+\mathbf{r})\right\rangle$$

can be expressed in terms of $k(r)$.

'Almost' homogeneous systems

The arguments outlined for 'almost' time stationary systems can evidently readily be generalized to *spatial* coordinates, in discussing 'almost' homogeneous systems. In this case we have spatial correlations $R(\mathbf{r}_1, \mathbf{r}_2)$. We introduce the separation variable $\mathbf{r} \equiv \mathbf{r}_2 - \mathbf{r}_1$ and average variable $\mathbf{R} \equiv \frac{1}{2}(r_2 + \mathbf{r}_1)$, and write $R(\mathbf{r}_1, \mathbf{r}_2) = R(\mathbf{R}, r)$. If the variation with $\mathbf{R}$ is slow, we call the system locally homogeneous. We might introduce a similar discussion of almost isotropic systems, but this concept will probably be of little use.

For homogeneous systems, we have $R(\mathbf{r}_1, \mathbf{r}_2) = R(\mathbf{r}_1 - r_2) = R(\mathbf{r})$, and similarly for the structure functions $\Psi_n = \Psi_n(\mathbf{r})$ from (19.58). For homogeneous and also *isotropic* systems we find $R = R(|\mathbf{r}|)$ and similarly $\Psi_n = \Psi_n(|\mathbf{r}|)$.

Mixed spatial-temporal variables

Assume for instance that N-signals are detected by obtaining a space-time varying signal at N spatially distributed detectors located at positions $\mathbf{r}_j$ for $j = 1, 2, \ldots, N$.

The joint probability density can then be specified as $P(r_1, \mathbf{r}_2, \ldots, \mathbf{r}_N; t_1, t_2, \ldots, t_M)$ with $N \neq M$ in general. Generalizing the notion of time-stationarity we call the space-time varying signal spatially *homogeneous* if

$$P_N(r_1, \mathbf{r}_2, \ldots, \mathbf{r}_N; t_1, t_2, \ldots, t_M) = P_N(r_1 + \mathbf{r}_0, \mathbf{r}_2 + \mathbf{r}_0, \ldots, r_N + \mathbf{r}_0; t_1, t_2, \ldots, t_M))$$

for all N and all combinations of positions. The signal is spatially *isotropic* if P_N is independent of the *direction* of $\mathbf{r}_0$. The correlation function, in particular, becomes $R(\mathbf{r}_i, \mathbf{r}_j; t_i, t_j) = R(\mathbf{r}_i - \mathbf{r}_j; t_i - t_j)$ for time-stationary and spatially homogeneous signals. The symmetry relation of the correlation function for time stationary, spatially homogeneous processes becomes

$$R\left(\mathbf{r}_i - \mathbf{r}_j; t_i - t_j\right) \equiv R\left(\boldsymbol{\xi}_i, \tau\right) = R\left(-\boldsymbol{\xi}_j, -\tau\right),$$

with $\boldsymbol{\xi}_i = \mathbf{r}_i - \mathbf{r}_j$ and $\tau = t_i - t_j$. Note that a signal can, of course, be spatially homogeneous and isotropic without being time-stationary and vice versa.

19.2.7 The Wiener–Khinchine theorem

The importance attributed to the auto-correlation function is due to the interpretation of its Fourier transform as a power spectral density. In fact, in many ways it might be argued that the wisest thing to do was simply to *define* the (auto) power spectral density of a stochastic process as the Fourier transform of the auto-correlation function. (Similarly, the *cross-spectral* density or *cross-power spectrum* of *two* processes is the Fourier transform of their cross-correlation function.) Such a definition, no matter how appropriate it might be, does not provide much physical insight. In the following the concept of the auto-power-spectrum is discussed in some detail.

Before proceeding, it might be worthwhile to recall that we in Fourier transform distinguish *finite energy* spectra where

$$F_e = 2\pi \int_{-\infty}^{\infty} f(t) f^*(t) \mathrm{d}t$$

is finite, and *finite power* spectra where

$$F_P = \lim_{\mathcal{T}\to\infty} \frac{1}{2\mathcal{T}} \int_{-\mathcal{T}}^{\mathcal{T}} F(t) F^*(t) \mathrm{d}t$$

exists (Champeney 1973). This distinction will be relevant in the following.

Consider an arbitrary real-valued function of time, ${}^k x(t)$, with $-\infty < t < \infty$, where ${}^k x(t)$ is a sample function from a (possibly non-stationary) random process as labeled by ^{k}. Following Kampe de Feriet (1954-1955) or Bendat (1958) we define

$$ {}^k x_{\mathcal{T}}(t) = \begin{cases} {}^k x(t) & |t| \leqslant \mathcal{T} \\ 0 & \text{otherwise.} \end{cases} \tag{19.61}$$

Assume that the Fourier transform of ${}^k x_{\mathcal{T}}(t)$ exists and is denoted by

$$A_{\mathcal{T}}(f,{}^k x) = \int_{-\infty}^{\infty} {}^k x_{\mathcal{T}}(t) \mathrm{e}^{-\mathrm{i}2\pi f t} \mathrm{d}t = \int_{-\mathcal{T}}^{\mathcal{T}} {}^k x(t) \mathrm{e}^{-\mathrm{i}2\pi f t} \mathrm{d}t.$$

Assume also that the power spectral density function $S_{XX}(f)$ exists, where

$$S_{XX}(f) = \lim_{\mathcal{T}\to\infty} S_{XX}(f,\mathcal{T}), \tag{19.62}$$

with

$$S_{XX}(f,\mathcal{T}) = \left\langle S_{\mathcal{T}}\left(f,{}^k x\right)\right\rangle,$$

where

$$S_{\mathcal{T}}\left(f,{}^k x\right) = \frac{\left|A_{\mathcal{T}}\left(f,{}^k x\right)\right|^2}{\mathcal{T}}.$$

Define

$${}^k J_{\mathcal{T}}(\tau) = \frac{1}{2\mathcal{T}} \int_{-\infty}^{\infty} {}^k x_{\mathcal{T}}(t) {}^k x_{\mathcal{T}}(t+\tau) \mathrm{d}t. \tag{19.63}$$

Then, from (19.61), it is clear that

$${}^k J_{\mathcal{T}}(\tau) = \frac{1}{2\mathcal{T}} \int_{-\mathcal{T}}^{\mathcal{T}} {}^k x(t) {}^k x(t+\tau) \mathrm{d}t + \mathcal{O}(1/\mathcal{T}) \qquad \text{as} \qquad \mathcal{T}\to\infty, \tag{19.64}$$

where the notation $\mathcal{O}(1/\mathcal{T})$ indicates an error term that approaches zero as $\mathcal{T}\to\infty$.

Assuming all mathematical operations in the following to be legitimate, we obtain starting with the Fourier transform of ${}^k J_{\mathcal{T}}(\tau)$

$$\begin{aligned} \int_{-\infty}^{\infty} {}^k J_{\mathcal{T}}(\tau) \mathrm{e}^{-\mathrm{i}2\pi f \tau} \mathrm{d}\tau &= \int_{-\infty}^{\infty} \mathrm{e}^{-\mathrm{i}2\pi f \tau} \left[\frac{1}{2\mathcal{T}} \int_{-\infty}^{\infty} {}^k x_{\mathcal{T}}(t) \right. \\ &\quad \left. {}^k x_{\mathcal{T}}(t+\tau) \mathrm{d}t \right] \mathrm{d}\tau \\ &= \frac{1}{2\mathcal{T}} \int_{-\infty}^{\infty} {}^k x_{\mathcal{T}}(t) \mathrm{e}^{\mathrm{i}2\pi f t} \mathrm{d}t \\ &\quad \times \int_{-\infty}^{\infty} {}^k x_{\mathcal{T}}(t+\tau) \mathrm{e}^{-\mathrm{i}2\pi f (t+\tau)} \mathrm{d}\tau \\ &= \frac{1}{2} S_{\mathcal{T}}\left(f,{}^k x\right), \end{aligned} \tag{19.65}$$

where we have in one place multiplied by $\mathrm{e}^{\mathrm{i}2\pi f t}\mathrm{e}^{-\mathrm{i}2\pi f t} = 1$, and in another place recognized that

$$\int_{-\infty}^{\infty} {}^k x_{\mathcal{T}}(t+\tau)\mathrm{e}^{-\mathrm{i}2\pi f(t+\tau)}\mathrm{d}\tau = \int_{-\infty}^{\infty} {}^k x_{\mathcal{T}}(t)\mathrm{e}^{-\mathrm{i}2\pi ft}\mathrm{d}t = A_{\mathcal{T}}\left(f, {}^k x\right).$$

Thus, for a single arbitrary time record ${}^k x(t)$, with k fixed, we have

$$S_{\mathcal{T}}\left(f, {}^k x\right) = 2\int_{-\infty}^{\infty} {}^k J_{\mathcal{T}}(\tau)\mathrm{e}^{-\mathrm{i}2\pi f\tau}\mathrm{d}\tau. \tag{19.66}$$

An average over all records k in the ensemble yields

$$S_{XX}(f, \mathcal{T}) = 2\int_{-\infty}^{\infty} \langle {}^k J_{\mathcal{T}}(\tau)\rangle \mathrm{e}^{-\mathrm{i}2\pi f\tau}\mathrm{d}\tau. \tag{19.67}$$

From (19.64) we have

$$\begin{aligned}\left\langle {}^k J_{\mathcal{T}}(\tau)\right\rangle &= \frac{1}{2\mathcal{T}}\int_{-\mathcal{T}}^{\mathcal{T}} \left\langle {}^k x(t) {}^k x\left(t+\tau\right)\right\rangle \mathrm{d}t + \mathcal{O}\left(1/\mathcal{T}\right)\\ &= \frac{1}{2\mathcal{T}}\int_{-\mathcal{T}}^{\mathcal{T}} \gamma_{XX}(t, t+\tau)\mathrm{d}t + \mathcal{O}\left(1/\mathcal{T}\right),\end{aligned} \tag{19.68}$$

as $\mathcal{T}\to\infty$, where we introduced

$$\gamma_{XX}(t, t+\tau) = \left\langle {}^k x(t) {}^k x\left(t+\tau\right)\right\rangle, \tag{19.69}$$

to represent the general non-stationary form of the auto-correlation function for the random process $X(t)$ at the times t and $t+\tau$. Thus, for any random process, as $\mathcal{N}$ becomes large, we derive the result

$$S_{XX}(f) = \lim_{\mathcal{T}\to\infty} 2\int_{-\infty}^{\infty} \mathrm{e}^{-\mathrm{i}2\pi f\tau}\left[\frac{1}{2\mathcal{T}}\int_{-\mathcal{T}}^{\mathcal{T}} \gamma_{XX}(t, t+\tau)\mathrm{d}t\right]\mathrm{d}\tau. \tag{19.70}$$

The amount of work involved in using this general formula is in practice quite significant.

If the random process is stationary, a considerable simplification occurs because, as discussed before

$$\gamma_{XX}(t, t+\tau) = \gamma_{XX}(0, \tau) = R_{XX}(\tau). \tag{19.71}$$

Then

$$\frac{1}{2\mathcal{T}}\int_{-\mathcal{T}}^{\mathcal{T}} \gamma_{XX}(t, t+\tau)\mathrm{d}t = R_{XX}(\tau) \quad \text{independent of } \mathcal{T}. \tag{19.72}$$

Hence, in this special case (19.70) becomes

$$S_{XX}(f) = 2\int_{-\infty}^{\infty} R_{XX}(\tau)\mathrm{e}^{-\mathrm{i}2\pi f\tau}\mathrm{d}\tau, \tag{19.73}$$

showing a simple relationship between the power spectral density function and the auto-correlation function of a stationary random process. Except for a factor of two, $S_{XX}(f)$ is the Fourier transform of $R_{XX}(\tau)$. Since $R_{XX}(\tau)$ is an even function of τ, the above is equivalent to

$$S_{XX}(f) = 4\int_{0}^{\infty} R_{XX}(\tau)\cos(2\pi f\tau)\mathrm{d}\tau. \tag{19.74}$$

Also, from (19.73), the inverse Fourier transform of $S_{XX}(f)$ is

$$R_{XX}(\tau) = \frac{1}{2}\int_{-\infty}^{\infty} S_{XX}(f)\mathrm{e}^{\mathrm{i}2\pi f\tau}\mathrm{d}f, \tag{19.75}$$

and using the fact that $S_{XX}(f)$ is an even function of f,

$$R_{XX}(\tau) = \int_{0}^{\infty} S_{XX}(f)\cos(2\pi f\tau)\mathrm{d}f. \tag{19.76}$$

From (19.76) is easily demonstrated that $R_{XX}(\tau) = R_{XX}(-\tau)$, $R_{XX}(0) \geqslant |R_{XX}(\tau)|$ for all τ, and $\mathrm{d}R_{XX}(\tau)/\mathrm{d}\tau|_{\tau=0} = 0$ for time stationary random processes, with the possible exception of processes with discontinuous derivatives.

In terms of angular frequency ω, in radians per second, (19.74) and (19.76) become

$$G_{XX}(\omega) = \frac{2}{\pi}\int_{0}^{\infty} R_{XX}(\tau)\cos(\omega\tau)\mathrm{d}\tau \tag{19.77}$$

$$R_{XX}(\tau) = \int_{0}^{\infty} G_{XX}(\omega)\cos(\omega\tau)\mathrm{d}\omega. \tag{19.78}$$

These last two formulas are known as the *Wiener–Khinchine relations*, named after two mathematicians, one American and the other Russian, who first noted the correspondence. (Due to the ambiguity in translating Russian names, different spellings of the latter name can be seen.) The definition of the Fourier transform and its corresponding inverse transform is not unique; a factor $1/2\pi$ can to some extent be placed after convenience (Champeney 1973). All that *can* be required is that the transform followed by an inverse transform must reproduce the original function.

Introducing the normalized version $D_{XX}(\omega) = G_{XX}(\omega)/\int_{0}^{\infty} G_{XX}(\omega)\,\mathrm{d}\omega$, with $\int_{0}^{\infty} G_{XX}(\omega)\,\mathrm{d}\omega = R_{XX}(0)$, the quantity $D_{XX}(\omega)$ is sometimes interpreted as a probability density for finding a particular frequency ω, although this interpretation may be too simplistic.

Instead of the power spectral density functions $G_{XX}(\omega)$ and $S_{XX}(f)$, as defined before, some writers define one-half of these particular values to be the 'power spectral density functions', namely by

$$W_{XX}(\omega) = \frac{1}{2} G_{XX}(\omega)$$
$$\mathcal{W}_{XX}(f) = \frac{1}{2} S_{XX}(f).$$

This notation makes the new functions $W_{XX}(\omega)$ and $\mathcal{W}_{XX}(f)$ the exact Fourier transforms of the auto-correlation function $R_{XX}(\tau)$, see (19.73). These slightly different definitions for the same quantity appearing in the literature should not be confused (Bendat 1958). Many formulas are unaffected by this change since the same constant multiplying factor may appear on both sides of the equation.

Equations (19.74) and (19.76) or, equivalently, (19.77) and (19.78) are well-known relations between power spectral density functions and auto-correlation functions for stationary random processes. These formulas do not involve any "ergodic hypothesis" as such, although this may actually be present in the manner by which the stationary auto-correlation function $R_{XX}(\tau)$ is calculated. By first determining $R_{XX}(\tau)$ and then using (19.74), we have a valid technique for computing $S_{XX}(f)$. This method was actually the one most often used until the fast Fourier transform (FFT, see e.g. Brigham (1974)) method was developed.

- **Example:** It is important to emphasize that in general $G(\omega = 0)$ is *not* the power in the dc-value of a signal. For instance a signal X which is a constant varying randomly over the realizations, is time-stationary and has an auto-correlation function $R(\tau) = \langle x^2 \rangle$, which is constant for all τ. By the Wiener–Khinchine relations we find the corresponding power spectrum $G(\omega) = \langle x^2 \rangle \delta(\omega)$. A dc-value is thus associated with δ-functions in the power spectra. On the other hand, the value $G(\omega = 0)$ for processes with *zero mean* is related to the integral of the auto-correlation function, see (19.77).

  More generally, for time-stationary ergodic signals, two signal samples taken at times with large separation are likely to be independent, implying that at large time-separations $\langle X(t)X(t+\tau)\rangle \to \langle X(t)\rangle\langle X(t+\tau)\rangle = \langle X(t)\rangle^2$. In other words, the auto-correlation function approaches a constant value at large time separations for signal with non-zero averages, $R(\tau \to \pm\infty) = \langle X(t)\rangle^2$. A non-vanishing average will also here imply a δ-function in the power spectrum.

  By choosing a proper reference value one might in principle always ensure that the average value of a given signal vanishes. In practice, this might not necessarily be a smart thing to do.
- **Example:** For a stationary random process we found a simple relation between the auto-correlation function for the process and the one for its time derivative process, see (19.46). Using the Wiener–Khinchine relations we find that the power spectrum, $G(\omega)$, associated with the time-series $Y(t)$ itself

and the spectrum, $G_D(\omega)$ for the series obtained by its time-derivative $\mathrm{d}Y(t)/\mathrm{d}t$ are related as $G_D(\omega) = \omega^2 G(\omega)$.

The auto-correlation function is non-negative definite as already mentioned, see (19.30). We generalize the result (19.30) by letting the index refer to continuous variables and obtain the often used, but slightly restrictive form

$$\iint_{-\infty}^{\infty} a(t_1)\langle Y(t_1)Y(t_2)\rangle a(t_2)\mathrm{d}t_1\mathrm{d}t_2 \equiv \iint_{-\infty}^{\infty} a(t_1)R(t_1, t_2)a(t_2)\mathrm{d}t_1\mathrm{d}t_2 \geqslant 0. \qquad (19.79)$$

Assuming now stationary processes, $R(t_1, t_2) = R(t_1 - t_2) \equiv R(\tau)$, and taking the special case of $a(t_{1,2}) = \exp(-\mathrm{i}2\pi f t_{1,2})/\sqrt{2\mathcal{T}}$ for $|t_{1,2}| \leqslant \mathcal{T}$ and $a(t_{1,2}) = 0$ otherwise, we find in the limit of $\mathcal{T} \to \infty$ that $S_{XX}(f) \geqslant 0$. This is a rather important result; it would be hardly meaningful to have a power spectral density which could assume negative values! Actually it can be demonstrated that any non-negative definite correlation function can be written as $R(\tau) = \int_{-\infty}^{\infty} S(\omega)\exp(-\mathrm{i}\omega\tau)\mathrm{d}\omega$ with $S(\omega) \geqslant 0$, this is *Bochner's theorem*.

- **Example:** Note that for instance a 'box-function' or a 'top hat' $f(\tau) = A$ for $|\tau| \leqslant \mathcal{T}$ and $f(\tau) = 0$ otherwise, cannot be acceptable as a correlation function even if it only assumes positive values; its Fourier transform is $2A\sin(\omega\mathcal{T})/\omega$. A box-function is not non-negative definite. Similarly $f(\tau) = A\cos(\tau/\mathcal{T})$ for $|\tau| \leqslant \pi\mathcal{T}/2$ and $f(\tau) = 0$ otherwise, is not non-negative definite either.
- **Exercise:** Express the spectral density of the product of two real Gaussian (or *normal*) processes $Z(t) = X(t)Y(t)$ in terms of the individual spectral densities by the relation

$$G_{ZZ}(\omega) = \int_{-\infty}^{\infty} G_{XX}(\omega - \omega')G_{YY}(\omega')\mathrm{d}\omega' + \int_{-\infty}^{\infty} G_{XY}(\omega - \omega')G_{YX}(\omega')d\omega' + \langle X\rangle^2 G_{XX}(\omega) + \langle Y\rangle^2 G_{YY}(\omega).$$

Alternatively, express the previous result in terms of correlation functions

$$R_{ZZ}(\omega) = R_{XX}(\tau)R_{YY}(\tau) + R_{XY}(\tau)R_{YX}(\tau) + \langle X\rangle^2 R_{XX}(\omega) + \langle Y\rangle^2 R_{YY}(\omega).$$

Write out the explicit expression for the special case of $X = Y$. Compare also the results with (19.55).
- **Exercise:** Find the spectral density for the process $Z(t) = X(t)\mathrm{d}X(t)/\mathrm{d}t$ if X is a Gaussian random process and $R_{XX}(\tau) = A\exp(-\alpha|\tau|)\,[\cos(\beta\tau) + (\alpha/\beta)\sin(\beta|\tau|)]$, with $\langle X\rangle = 0$.

19.2.8 Wavenumber spectra

The derivation of the Wiener–Khinchine theorem used a temporal variable for simplicity. We might have used a *spatial* variable as well, i.e. by Fourier transform of a correlation function $R(\mathbf{r})$, depending on a spatial separation, we obtain a power spectrum in terms of wavenumbers

$$F(\mathbf{k}) = \int R(\mathbf{r}) \cos(\mathbf{k} \cdot \mathbf{r}) \mathrm{d}\mathbf{r}.$$

The only difference is that we will be dealing with a *vector* variable $\mathbf{k}$, rather than a scalar ω. By the notation $\int \mathrm{d}\mathbf{r}$ we imply integration over all space here and in the following.

For inhomogeneous media we can define a local wavenumber spectrum from the correlation function $R(\mathbf{r}_1, \mathbf{r}_2) = R(\mathbf{r}, \mathbf{R})$, with the separation variable $\mathbf{r} \equiv \mathbf{r}_2 - \mathbf{r}_1$ and average variable $\mathbf{R} \equiv \frac{1}{2}(\mathbf{r}_2 + \mathbf{r}_1)$, see section 19.2.6. We can make a Fourier transform with respect to the separation variable, and obtain

$$F(\mathbf{k}, \mathbf{R}) = \int R(\mathbf{r}, \mathbf{R}) \cos(\mathbf{k} \cdot \mathbf{r}) \mathrm{d}\mathbf{r},$$

to be interpreted as a 'local' power spectrum, centered around the position $\mathbf{R}$. In cases where the variation with the variable $\mathbf{R}$ is slow, the definition is useful. The interpretation remains valid for wavenumbers $k \gg 1/\mathcal{L}$, where $\mathcal{L}$ is the characteristic length scale for variations with $\mathbf{R}$.

Generally, we have *anisotropic* conditions, where $R(\mathbf{r})$ depends on the magnitude as well as the direction of the position vector $\mathbf{r}$, and consequently $F(\mathbf{k})$ is anisotropic as well in wavenumber space. The expressions can be simplified somewhat for the special isotropic case, where $R = R(r)$. The results depend on the dimensionality of the problem (Champeney 1973), where evidently one-dimensional cases are trivial.

Two spatial dimensions

We write the basic transform in complex notation, since the imaginary part vanishes anyway for real R. For isotropic turbulence in two spatial dimensions we have

$$\begin{aligned} F(k) &= \int R(r) \exp(\mathrm{i}\mathbf{k} \cdot \mathbf{r}) \mathrm{d}\mathbf{r} \\ &= \int_0^\infty \int_0^{2\pi} R(r) \exp(\mathrm{i}kr \cos\theta) r \mathrm{d}r \mathrm{d}\theta \\ &= 2\pi \int_0^\infty R(r) J_0(kr) r \mathrm{d}r, \end{aligned} \tag{19.80}$$

obtained by use of the standard result

$$\int_0^{2\pi} \exp(\mathrm{i}x \cos\theta) \mathrm{d}\theta = 2\pi J_0(x),$$

where J_0 is the Bessel function (Champeney 1973).

Three spatial dimensions
The result in three spatial dimensions is somewhat different from that in two dimensions

$$\begin{aligned} F(k) &= \int R(r)\exp(\mathrm{i}\mathbf{k}\cdot\mathbf{r})\mathrm{d}\mathbf{r} \\ &= \int_0^\infty \int_0^\pi \int_0^{2\pi} R(r)\exp(ikr\cos\theta)r^2\sin\theta \mathrm{d}r\mathrm{d}\theta\mathrm{d}\phi \\ &= 2\pi\int_0^\infty \int_{-1}^1 R(r)(\cos(kr\gamma) + i\sin(kr\gamma))r^2\mathrm{d}r\mathrm{d}\gamma \\ &= 4\pi\int_0^\infty R(r)\frac{\sin(kr)}{kr}r^2\mathrm{d}r, \end{aligned} \tag{19.81}$$

where we one place used the substitution $\gamma \equiv \cos\theta$.

19.2.9 Reduced wavenumber spectra

For homogeneous but anisotropic conditions we can obtain a power spectrum of some fluctuating scalar quantity (say ζ) as, say, $F(k_x, k_y, k_z)$ so that $\langle\zeta^2\rangle = \iiint_{-\infty}^{\infty} F(k_x, k_y, k_z)\mathrm{d}k_x\mathrm{d}k_y\mathrm{d}k_z$. From this it is trivial to obtain reduced spectra by integration to give component spectra defined as $S_2(k_x, k_y) \equiv \int_{-\infty}^{\infty} F(k_x, k_y, k_z)\mathrm{d}k_z$ or $S_1(k_x) \equiv \iint_{-\infty}^{\infty} F(k_x, k_y, k_z)\mathrm{d}k_y\mathrm{d}k_z$. A two-dimensional wavevector spectrum is included by writing $F(k_x, k_y, k_z) = F(k_x, k_y)\delta(k_z)$ with Dirac's δ-function. Often we can assume local isotropy of the fluctuations and have $F(k_x, k_y, k_z) \approx F(|\mathbf{k}|)$, at least in a range of wavenumbers. For such a case we have e.g. $S_1(k_x) = \iint_{-\infty}^{\infty} F\left(\sqrt{k_x^2 + k_y^2 + k_z^2}\right)\mathrm{d}k_y\mathrm{d}k_z$ where now the x-direction is arbitrary. In figure 19.6 we illustrate the so called 'aliasing' effect, where we integrate wavevector

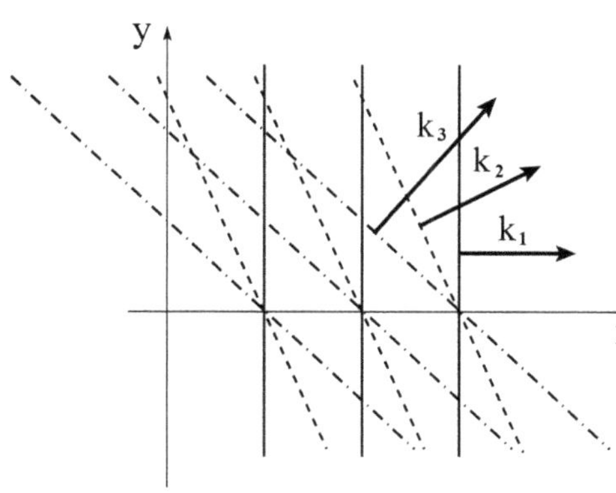

Figure 19.6. Illustration of 'aliasing' by integration over wavevector components perpendicular to a selected direction. All three wavevectors in the figure have the same k_x-component.

components perpendicular to a selected direction. The k_x-component of the wavevectors shown is the same, but the wavelengths of the three individual waves is very different. When integrated over k_y all these waves contribute to the same k_x.

Provided we can assume that the spectrum is isotropic in three-dimensional wavevector space, i.e. $F(k_x, k_y, k_z) = F(|\mathbf{k}|)$, we have a useful relation in the form

$$\begin{aligned} S_1(k_x) &\sim \iint_{-\infty}^{\infty} F(k_x, k_y, k_z)\mathrm{d}k_y\mathrm{d}k_z = \iint_{-\infty}^{\infty} F(k_x, \mathbf{k}_\perp)\mathrm{d}\mathbf{k}_\perp \\ &= 2\pi \int_0^{\infty} F\left(\sqrt{k_x^2 + k_\perp^2}\right)k_\perp \mathrm{d}k_\perp = 2\pi \int_{k_x}^{\infty} F(k')k'\mathrm{d}k', \end{aligned} \tag{19.82}$$

giving

$$F(k) = -\frac{1}{2\pi k}\frac{S_1(k)}{\mathrm{d}k}. \tag{19.83}$$

The expression (19.83) holds for any isotropic three-dimensional form of $S_1(k)$, it need not be a power law. A similarly nice relation for two-dimensional turbulence is not found, as the reader may verify.

Often it is found that a power spectrum can be characterized (at least in a range of wavenumbers) by a spectral power-law like $F(|\mathbf{k}|) \sim 1/k^\alpha$. Turbulence in neutral incompressible fluids offers such an example, the Kolmogorov–Oubokhov spectrum. If we have $F(|\mathbf{k}|) \sim 1/k^\alpha$ then it follows from (19.82) that for the three-dimensional case

$$S_1(k_x) \sim \iint_{-\infty}^{\infty} \frac{1}{\left(k_x^2 + k_y^2 + k_z^2\right)^{\alpha/2}}\mathrm{d}k_y\mathrm{d}k_z \sim \frac{1}{k_x^{\alpha-2}}.$$

For the related case with two-dimensional spectra we find

$$S_1(k_x) \sim \iint_{-\infty}^{\infty} \frac{\delta(k_z)}{\left(k_x^2 + k_y^2\right)^{\alpha/2}}\mathrm{d}k_y\mathrm{d}k_z \sim \frac{1}{k_x^{\alpha-1}}.$$

Although it seems like a trivial observation, it is important to make clear which version of the power spectra the exponent refers to, i.e. $F(k_x, k_y, k_z)$ or $S_1(k_x)$ for instance (Fredricks and Coroniti 1976). In particular in studies of plasma turbulence, several examples can unfortunately be found where experimentally obtained component spectra are compared with analytically argued two- or three-dimensional spectra without any comments on the differences.

For the Kolmogorov–Oubokhov spectrum mentioned before we have $S_1(k_x) \sim 1/k_x^{5/3}$. With the reasonable assumption of homogeneity and isotropy of fully developed turbulence in incompressible fluids we have $F(k_x, k_y, k_z) \sim 1/(k_x^2 + k_y^2 + k_z^2)^{11/6} \sim 1/|k|^{11/3}$.

Experiments often give the reduced power spectrum, and without additional information, *a priori* given isotropy for instance, it is not possible to obtain the full three-dimensional power spectrum, not even when the experiment gives solid indications for a power-law spectrum. Notice that a power-law spectrum can also be anisotropic, i.e. have $F(k_x, k_y, k_z) \sim \delta(k_z)/(ak_x^\alpha + bk_y^\beta)$, or similar, with the proper dimensions of a and b.

- **Exercise:** Try to carry out an analysis like the one giving (19.82), but now for two-dimensional turbulence where $F(k_x, k_y, k_z) = F(k_x, k_y)\delta(k_z) = F(|\mathbf{k}_\perp|)\delta(k_z)$.

Reduced wavenumber spectra with angular variables

We illustrated before how we can obtain a reduced wavenumber spectrum by projection on one wavevector direction. This form is important by being suitable for discussing Taylor's hypothesis, see also section 23.4.1. We can also write the spectrum in angular variables and integrate over angle as

$$S(k)\mathrm{d}k \sim k\int_0^{2\pi} F(k, \theta)\mathrm{d}\theta \mathrm{d}k,$$

using two-dimensional turbulence as an illustration where $F(k_x, k_y) \to F(k, \theta)$ with $k_x = k\cos\theta$ and $k_y = k\sin\theta$. In this form we have $\langle X^2\rangle = \int_0^\infty S(k)\mathrm{d}k$, with $X(\mathbf{r})$ being some random quantity varying homogeneously in space, but not necessarily isotropically.

19.3 Intermittency

One of the most often used (one might even say *abused*) words in turbulence is 'intermittency'. The term is here used to describe that turbulent energy dissipation occurring in localized spatial regions, or 'hot spots'. More generally, the word 'intermittent' is supposed to cover anything of bursty nature. Meteorologists use the word *gust* for sudden increases in wind speed, but the meaning is more or less the same. The concept is thus rather vague, and in reality what we need is a definition of what is *not* an intermittent signal. A somewhat pragmatic definition of intermittency was given by Rollefson (1978), stating that 'a variable with zero mean will be called intermittent if it has a probability distribution such that extremely small and extremely large excursions are more likely than in a normally distributed variable'.

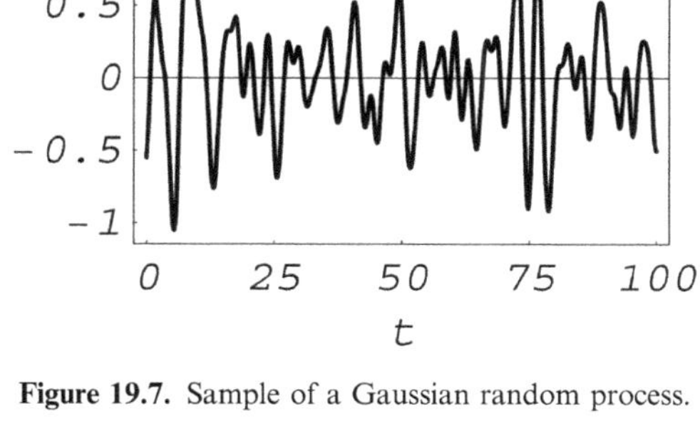

Figure 19.7. Sample of a Gaussian random process.

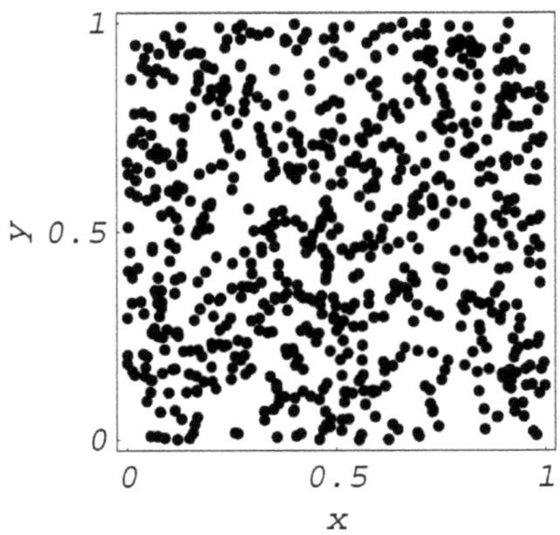

Figure 19.8. Randomly distributed points on a 1×1-plane. The $\{x, y\}$-coordinates of each point are chosen by a random number generator, so that the positions of all points are statistically independent.

In this sense, a random Gaussian process serves as the reference non-intermittent case. It is of course possible to find many types of signals with distributed large and small amplitude regions. This author will risk the bold statement that *any* random signal can be seen as intermittent unless we define a reference non-intermittent case, e.g. a Gaussian random process! To substantiate this idea, figure 19.7 shows a sample of such a Gaussian random process. Clearly, from time to time the signal has a localized large amplitude excursion, but by the foregoing definition, the signal will not be called intermittent.

19.3.1 Spatial intermittency

A discrete process can also exhibit intermittency. As an illustration we show in figure 19.8 results from a numerical experiment where 1000 points are distributed randomly in a 1×1 square. the x and y positions of each point are determined by random number generators, so that each point is placed independently of all the others, Nonetheless, we notice clear signs of 'clustering' and even appearance of 'channels' in certain regions. Incidentally, such channels were observed also in early photographs of the planet Mars, causing some turmoil by being taken as indications of life on the planet. As illustrated by figure 19.8, these and similar observations are purely artificial, and often the term 'Martian channels' are used for such observations. It is readily demonstrated that the spatial distribution of points, with conditions as described here, follows a Poisson distribution, i.e. the probability distribution of the number of points N in an area $\mathcal{A}$ is given by $P(N, \mathcal{A}) = (\mu\mathcal{A})^N \exp(-\mu\mathcal{A})/N!$, with μ being the density of points per unit area. The actual shape of the area element $\mathcal{A}$ is immaterial. As a simple test of the distribution of points we can divide the plane into many small identical boxes and count the number of points in each box. By this procedure it is possible to obtain an estimate for the probability density of points in the boxes which can subsequently be compared to what a Poisson distribution would give. For points distributed in three spatial dimensions it is possible to follow the same procedure.

19.3.2 Temporal intermittency

In the following we give attention to continuous and at least one time differentiable random processes (Gaussian processes in particular), this being the most relevant for turbulence applications.

A complete discussion of the statistical distribution of excursions in a random signal (such as the velocity or the electric field magnitude) is lengthy and complicated. Analytical forms for the duration of times spent by such a signal at amplitudes exceeding some selected levels were discussed by Rice 1945), but the practical applicability of the results is restricted to short times. Some useful estimates can, however, be obtained by simpler means (Barakat 1980, Kristensen *et al* 1991).

Considering a stationary random process $\Phi(t)$, which can represent the electric field magnitude or any other similar scalar process as well, we assume that the amplitude probability density is given as $P(\Phi)$. There are no restriction such as $\langle\Phi\rangle = 0$. Given a long time interval of duration $\mathcal{N}$, the time spent above a given reference level A is then

$$\Theta(A) = \mathcal{T}\int_{A}^{\infty} P(\Phi)\mathrm{d}\Phi. \tag{19.84}$$

This result can, however, be obtained in many different ways, by many short excursions above A or just a few, but long ones, for instance. To find the average number of excursion in the interval $\mathcal{N}$ it turns out that we need to know also the distribution of the time derivative $\mathrm{d}\Phi(t)/\mathrm{d}t \equiv \Phi'$ of the signal (Rice 1945, Bendat 1958, Lee 1967, Yura and Hanson 2010), i.e. the joint probability density $P(\Phi, \Phi')$. The assumed time-stationarity of the process becomes essential here by implying that $P(\Phi, \Phi')$ is independent of time.

First we note that the time τ it takes for the signal to cross a small interval $\mathrm{d}\Phi$ is given as $\mathrm{d}\Phi/\Phi'$. This result is most easily understood if we for the sake of argument

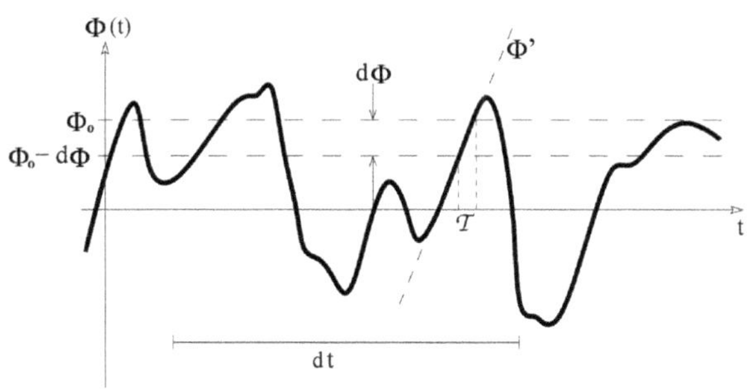

Figure 19.9. Schematic illustration of a signal $\Phi(t)$ with time differential $\Phi' \equiv \mathrm{d}\Phi(t)/\mathrm{d}t$ and $\tau \equiv \mathrm{d}\Phi/\Phi'$. The duration of the indicated time interval $\mathrm{d}t$ is of no concern for time stationary random processes.

consider the signal being a spatial displacement and its derivative then being the velocity. See figure 19.9 for an illustration.

The average fraction of time spent by the signal in an interval $\mathrm{d}t$ around any set of amplitudes and time derivatives $\{\Phi, \Phi'\}$ within a narrow interval $\mathrm{d}\Phi\mathrm{d}\Phi'$ is $P(\Phi, \Phi')\mathrm{d}\Phi\mathrm{d}\Phi'\mathrm{d}t$. For a given Φ' the average number of crossings of some signal level Φ within $\{t; t + \mathrm{d}t\}$ is consequently (Bendat 1958, Lee 1967) the ratio of the expected amount of time spent in the interval $\mathrm{d}\Phi$ for a given Φ' in a time $\mathrm{d}t$ and the time $\tau \equiv \mathrm{d}\Phi/\Phi'$ required to cross once for a given Φ' in the interval $\mathrm{d}\Phi$, implying

$$\frac{P(\Phi, \Phi')\mathrm{d}\Phi\mathrm{d}\Phi'\mathrm{d}t}{\mathrm{d}\Phi/\Phi'} = \Phi' P(\Phi, \Phi')\mathrm{d}\Phi'\mathrm{d}t.$$

The average number $\mathcal{N}$ of upward crossings of the level $\Phi = A$ in the time-interval $\mathcal{N}$ is then obtained by integration with respect to time, which becomes simple since $P(\Phi, \Phi')$ is independent of time, and with respect to all positive values of Φ' since we are interested in any level-crossing irrespective of the actual value of Φ'. The result of these two integrations is

$$\mathcal{N}(A) = \mathcal{T} \int_0^{\infty} \Phi' P(A, \Phi')\mathrm{d}\Phi', \tag{19.85}$$

which gives the average level crossing frequency $\langle f \rangle \equiv \mathcal{N}(A)/\mathcal{T} = \int_0^{\infty} \Phi' P(A, \Phi')\mathrm{d}\Phi'$ for the given value of A.

We would also like to know the statistical distributions, or at least the average values, for the time intervals from an up-crossing to the following down-crossing. This turns out to be a very difficult problem, even for a Gaussian random processes. A special case of the problem was studied by Rice 1945), by considering zero crossings of a signal. Assuming that the signal is known to pass through zero at $\tau = 0$ with a positive time derivative, what is the probability of passing through zero with a negative slope in the interval τ, $\tau + \mathrm{d}\tau$? The answer can be found (Rice 1945) but in reality it is of little use (Yura and Hanson 2010) since there may be other unaccounted zero crossings in between. In addition we need to extend the analysis beyond the zero-crossing analysis to an arbitrary signal level, but this is a minor concern.

An estimate for the average duration between an up and a down-crossing of a signal $\Phi(t)$ can be found by considering the following equality: for a long time record of duration $\mathcal{N}$ we let the fraction of time spent above a certain signal level A to be $Ta(A)$. We can write the identity $Ta(A) = \frac{1}{\mathcal{T}}\sum_{j=1}^{K}\Delta t_j$ where Δt_j is the time interval between the j-th up-crossing to the following down-crossing, while K is the number of such time intervals in the record. At the end of the record we can have an 'un-compensated' up-crossing, i.e. the following down-crossing falls outside the record, but if $\mathcal{N}$ is much larger than any Δt_j we need not worry because of this. Taking the ensemble average we have for the left side $\langle Ta(A)\rangle = \Theta(A)/\mathcal{T} = \int_A^{\infty} P(\Phi)\mathrm{d}\Phi$. The average of the right-hand side is not simple since in general the number $K = K(A)$ of time intervals in a record varies with the durations. We can however write an estimate as a product of averages as $\frac{1}{\mathcal{T}}\langle\sum_{j=1}^{K}\Delta t_j\rangle \approx \langle\Delta t_j\rangle\langle K\rangle/\mathcal{T}$ where $\langle K\rangle = \mathcal{N}(A)$. For time stationary processes we have $\langle\Delta t_j\rangle$ to be independent of j. The postulated average written as a product of averages would be correct for K and Δt_j being statistically independent. Generally this is not so, but it is used here as an approximation. An estimate (and not an accurate result) (Lee 1967, Barakat 1980, Kristensen *et al* 1991) for the average time spent in excess of a selected signal level $\Phi = A$ is then found by dividing all the time spent with $\Phi \geqslant A$ with the average number of upward crossings giving

$$\frac{\Theta(A)}{\mathcal{N}(A)} = \frac{\int_A^\infty P(\Phi)\mathrm{d}\Phi}{\int_0^\infty \Phi' P(A, \Phi')\mathrm{d}\Phi'}, \tag{19.86}$$

independent of $\mathcal{N}$. In the limit of $A \to -\infty$ we find $\Theta(A)/\mathcal{N}(A) \to \infty$ since we have $P(A \to -\infty, \Phi') \to 0$. For $A \to -\infty$ the entire (infinitely long) record will exceed the reference level.

- **Example:** Considering the importance of the result (19.86), it might be worthwhile to present an alternative derivation.

  The time it takes for the signal to cross a small interval $\mathrm{d}\Phi_x$ is given as $\mathrm{d}\Phi_x/\Phi'_x$. We then note that $P(\Phi_x, \Phi'_x)\mathrm{d}\Phi_x\,\mathrm{d}\Phi'_x\mathrm{d}t$ is the average fraction of time spent by the signal in an interval $\mathrm{d}t$ around any set of amplitudes and time derivatives $\{\Phi_x, \Phi'_x\}$ within a narrow interval $\mathrm{d}\Phi_x\mathrm{d}\,\Phi'_x$. The average number $\mathcal{N}$ of upward crossings of the level $\Phi_x = A$ in the time-interval $\mathcal{N}$ is then obtained by two integrations. First an integration with respect to time over the interval $\{0, \mathcal{T}\}$, which becomes simple since $P(\Phi_x, \Phi'_x)$ is independent of time, and then an integration with respect to all positive values of Φ'_x, implying that we include all upward crossing irrespective of the numerical value of Φ'_x. The result of these two integrations is

$$\mathcal{N}(A) = \mathcal{T}\int_0^\infty \Phi'_x P\left(A, \Phi'_x\right)\mathrm{d}\Phi'_x.$$

  As the next step, an estimate for the time spent in excess of a selected signal level $\Phi_x = A$ is then found again by dividing the total time spent with $\Phi_x \geqslant A$ with the average number of upward crossings (Kristensen *et al* 1991). This gives

$$\frac{\Theta(A)}{\mathcal{N}(A)} = \frac{\int_A^\infty P(\Phi_x)\mathrm{d}\Phi_x}{\int_0^\infty \Phi'_x P\left(A, \Phi'_x\right)\mathrm{d}\Phi'_x},$$

  i.e. again the result (19.86).

The problem addressed here has an interesting application for estimating the average number of real roots of a random algebraic equation (Kac 1943).

For an arbitrary process we do not in general know $P(\Phi_x, \Phi'_x)$, but for some models it is possible to obtain it analytically, as illustrated in section 20.2.

19.3.3 The Gaussian limit

In general we have no *a priori* knowledge of $P(\Phi, \Phi')$, except from the fact that Φ and Φ' are uncorrelated, since $\langle(\Phi(t) - \langle\Phi\rangle)\mathrm{d}\Phi/\mathrm{d}t\rangle = \frac{1}{2}\mathrm{d}\langle\Phi(t)^2\rangle/\mathrm{d}t = 0$, since $\langle\Phi'\rangle = 0$. This lack of correlation does not generally imply that Φ and Φ' are statistically independent.

In the non-intermittent reference Gaussian limit, the signal amplitude Φ and its time derivative Φ' are statistically independent for stationary random processes. We can write the joint probability density as a product $P(\Phi, \Phi') = P(\Phi)P(\Phi')$, where

$$P(\Phi) = \left(\frac{1}{\sqrt{2\pi\sigma^2}}\right)\exp\left(-\frac{1}{2}\frac{(\Phi - \langle\Phi\rangle)^2}{\sigma^2}\right)$$

and

$$P(\Phi') = \left(\frac{1}{\sqrt{2\pi\sigma'^2}}\right)\exp\left(-\frac{1}{2}\frac{\Phi'^2}{\sigma'^2}\right).$$

For this particular case we have Φ and Φ' to be statistically independent. The standard deviations σ^2 and σ'^2 of the two signals Φ and Φ' are most easily related by the power spectrum $G(\omega)$ of $\Phi(t)$, where we have $\sigma^2 = \int_0^\infty G(\omega)\mathrm{d}\omega$ and $\sigma'^2 = \int_0^\infty \omega^2 G(\omega)\mathrm{d}\omega$ for stationary Gaussian random processes. In this 'non-intermittent limit' we find the relatively simple result

$$\frac{\Theta(A)}{\mathcal{N}(A)} = \pi\frac{\sigma}{\sigma'}\exp\left(\frac{1}{2}(A - \langle\Phi\rangle)^2/\sigma^2\right) \times \operatorname{erfc}\left(\frac{1}{\sqrt{2}}(A - \langle\Phi\rangle)/\sigma\right). \tag{19.87}$$

where $\operatorname{erfc}(x) \equiv (2/\sqrt{\pi})\int_x^\infty \exp(-\xi^2)\mathrm{d}\xi$ is the complementary error function. The result (19.87) is illustrated in figure 19.10.

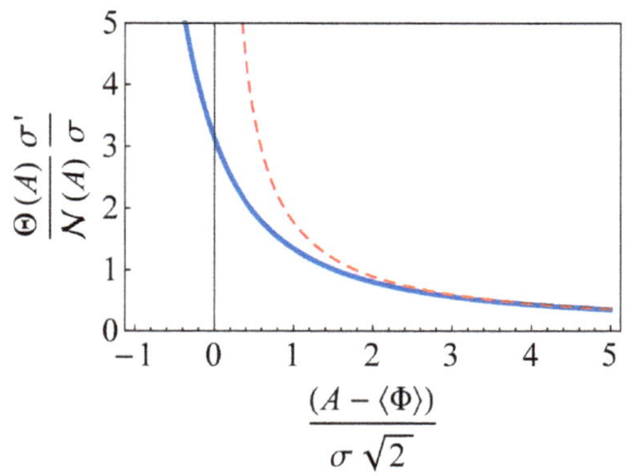

Figure 19.10. Illustration of the Gaussian limit (19.87) in terms of the normalized variables indicated. The asymptotic limit is shown by a dashed line.

For $A \to -\infty$ we also find here that $\Theta(A)/\mathcal{N} \to \infty$. We can define a mean square frequency as

$$\langle \omega^2 \rangle \equiv \int_0^\infty \omega^2 G(\omega) \mathrm{d}\omega \Big/ \int_0^\infty G(\omega) \mathrm{d}\omega$$

in which case $\sigma/\sigma' = 1/\sqrt{\langle \omega^2 \rangle}$.

For very large positive excursions, $A - \langle \Phi \rangle \gg \sigma$ the result (19.87) becomes particularly simple (Kristensen *et al* 1991), giving

$$\frac{\Theta(A)}{\mathcal{N}(A)} \approx \sqrt{2\pi} \frac{\sigma}{\sigma'} \frac{\sigma}{A - \langle \Phi \rangle},$$

implying that the average duration of an excursion is inversely proportional to deviations from the mean value. This asymptotic limit is shown with a dashed line in figure 19.10. The results from figure 19.10 are useful for comparison with results from data analysis.

We can give at least one example where the present results have applications. They can be particularly useful for studies of confinement of hot plasmas: it can here be important to distinguish many short plasma bursts from a few long ones. Although the accumulated time in the bursts can be the same, their consequences will be different as far as, for instance, the heat load on a confining wall is concerned. With many short bursts there can be time for the wall to cool down between bursts, while it need not be so for long bursts even when they are few (Fattorini *et al* 2012). The flux signal obtained from experiments or numerical simulations can be analyzed along the lines discussed here.

It should be mentioned that the statistical analysis of large excursions can be formulated in a slightly different way in terms of extremum distributions, here in terms of the distribution of local maxima (Gumbel 1958, Bendat 1958, Kristensen *et al* 1991). Intermittent features can also be revealed by a conditional data analysis that has also been successfully applied in studies of plasma turbulence (Johnsen *et al* 1987).

References

Abramowitz M and Stegun I A 1972 *Handbook of Mathematical Functions with Formulas, Graphs, and Mathematical tables* (New York: Dover)

Adrian R J 1979 Conditional eddies in isotropic turbulence *Phys. Fluids* **22** 2065–70

Barakat R 1980 The level-crossing rate and above-level duration time of the intensity of a Gaussian random process *Inf. Sci.* **20** 83–7

Batchelor G K 1953 *The Theory of Homogeneous Turbulence* (Cambridge: Cambridge University Press)

Bendat J S 1958 *Principles and Applications of Random Noise Theory* (New York: Wiley)

Brigham E O 1974 *The Fast Fourier Transform* (Englewood Cliffs, NJ: Prentice-Hall)

Champeney D C 1973 *Fourier Transforms and their Physical Applications* (London: Academic)

Cox D R and Miller H D 1965 *The Theory of Stochastic Processes* (London: Chapman and Hall)

Cramér H 1946 *Mathematical Methods of Statistics* (Princeton, NJ: Princeton University Press)

Davenport W B and Root W L 1958 *An Introduction to the Theory of Random Signals and Noise* (New York: McGraw-Hill)

Fattorini L, Fredriksen Å, Pécseli H L, Riccardi C and Trulsen J K 2012 Turbulent transport in a toroidal magnetized plasma *Plasma Phys. Control. Fusion* **54** 085017

Feller W 1968 *An Introduction to Probability Theory and its Applications* vol 1 (New York: Wiley)

Fredricks R W and Coroniti F V 1976 Ambiguities in the deduction of rest frame fluctuation spectrums from spectrums computed in moving frames *J. Geophys. Res.* **81** 5591–5

Gumbel E J 1958 *Statistics of Extremes* (New York: Columbia University Press)

Hansen F and Pedersen G K 1982 Jensen's inequality for operators and Löwner's theorem *Math. Ann.* **258** 229–41

Hogg R V and Craig A T 1970 *Introduction to Mathematical Statistics* 3rd edn (London: Macmillan)

Johnsen H, Pécseli H L and Trulsen J 1987 Conditional eddies in plasma turbulence *Phys. Fluids* **30** 2239–54

Kac M 1943 On the distribution of values of trigonometric sums with linearly independent frequencies *American J. Mathematics* **65** 609–15

Kampe de Feriet J 1954-1955 Introduction to the statistical theory of turbulence, parts I-IV *J. Soc. Indust. Appl. Math.* **2**(2,3,4,5) 1–9 143–174, 244–271, 90–117

Khinchine A I 1949 *Mathematical Foundations of Statistical Mechanics* (New York: Dover)

Kristensen L, Casanova M, Courtney M S and Troen I 1991 In search of a gust definition *Boundary-Layer Meteorol* **55** 91–107

Labit B, Furno I, Fasoli A, Diallo A, Müller S H, Plyushchev G, Podestà M and Poli F M 2007 universal statistical properties of drift-interchange turbulence in TORPEX plasmas *Phys. Rev. Lett.* **98** 255002

Lee W C- Y 1967 Statistical analysis of the level crossing and duration of fades of a signal from an energy density mobile radio antenna *Bell System Techn. J.* **46** 417–48

Mole N and Clarke E D 1995 Relationships between higher moments of concentration and of dose in turbulent dispersion *Boundary-Layer Meteorol* **73** 35–52

Oppenheim A V and Schafer R W 1975 *Digital Signal Processing* (New York: McGraw-Hill)

Øynes F J, Olsen O- M, Pécseli H L, Fredriksen Å and Rypdal K 1998 Experimental study of low-frequency electrostatic fluctuations in a magnetized toroidal plasma *Phys. Rev.* E **57** 2242–55

Papoulis A 1991 *Probability, Random Variables and Stochastic Processes* 3rd edn (New York: McGraw-Hill)

Pearson K 1929 Editorial note to Inequalities for moments of frequency functions and for various statistical constants *Biometrika* **21** 361–75

Pécseli H L 2000 *Fluctuations in Physical Systems* (Cambridge: Cambridge University Press)

Pécseli H L and Trulsen J 1989 A statistical analysis of numerically simulated plasma turbulence *Phys. Fluids* B **1** 1616–36
Erratum: Pécseli H L and Trulsen J 1990 *Phys. Fluids* B **2** 454

Pécseli H L and Trulsen J 1991 Analytical expressions for conditional averages: a numerical test *Phys. Scripta* **43** 503–7

Rice S O 1945 Mathematical analysis of random noise, II *Bell System Tech. J.* **24** 46–156 reprinted by
Wax N 1954 *Selected Papers on Noise and Stochastic Processes* (New York: Dover)

Rollefson J P 1978 On Kolmogorov's theory of turbulence and intermittency *Can. J. Phys.* **56** 1426–41

Tennekes H and Lumley J L 1972 *A First Course in Turbulence* (Cambridge, MA: The MIT Press)

Titchmarsh E C 1950 *The Theory of Functions* 2nd edn (London: Oxford University Press)

Wilkins J E 1944 A note on skewness and kurtosis *Annals Math. Stat.* **15** 333–5

Yaglom A M 1962 *An Introduction to the Theory of the Stationary Random Functions* (New York: Prentice-Hall)

Yura H T and Hanson S G 2010 Mean level signal crossing rate for an arbitrary stochastic process *J. Opt. Soc. Am.* A **27** 797–807

IOP Publishing

Low Frequency Waves and Turbulence in Magnetized Laboratory Plasmas and in the Ionosphere

Hans Pécseli

Chapter 20

A reference model

It is often a great advantage to have a simple model for random processes as a basis for discussions. Such a model should be based on some elementary statistical properties, and at least in principle allow relevant results to be obtained analytically. As an example we here consider a simple synthetic signal generated by a random superposition of a certain *a priori* given basic structure $\psi(t)$. In particular we shall use the model to illustrate the uncertainties associated with experimentally obtained average values, for instance when finite length records are used. Some basic results, averages, mean squares, etc can be obtained in generally useful expressions. More complicated results like probability densities, joint probability densities, etc are more complicated and can also give problems for numerical integrations. Some special choices for $\psi(t)$ can give results allowing simple and useful analytical results to be found (Garcia 2012).

With the model discussed in the following a time record becomes

$$\Phi(t) = \sum_{k=1}^{N} a\psi(t - t_k), \tag{20.1}$$

where the N pulse positions labeled t_k with $k = 1, 2, \ldots, N$ are randomly distributed in a large interval $\mathcal{T}$. In the following t_k is denoted 'pulse arrival-time' because t is most often a time-like variable, although this need not necessarily be the case. In most of the following analysis, the corresponding distribution is taken to be uniform, meaning that the probability $P(t_k) = 1/\mathcal{T}$ for $0 < t_k < \mathcal{T}$ and vanishing otherwise for all k. Any two arrival times are assumed to be statistically independent. A pulse amplitude a is included explicitly in (20.1) for later convenience. There are very few restrictions on the shape of the pulses given by $\psi(t)$. The basic feature of the model (20.1) is that everything is deterministic, except for the t_ks that have some known random distribution. With this information it is possible to describe the statistics of the signal $\Phi(t)$ completely.

The time series (20.1) constitutes a stationary random process and it will be demonstrated that it is ergodic by construction, for large $\mathcal{T}$. Examples from actual realization are shown in figure 20.1. The basic pulse was here taken to be $\psi(t) = \exp(-t^2)$ in all cases, with unit amplitude. The *shapes* of the structures entering are deterministic, so the statistical nature of the signal is manifested by the random distribution of positions. When we know the underlying statistical distributions of the 'pulse arrival times', we are in principle able to obtain analytically all the relevant statistical distributions associated with the signal.

The model (20.1) was analyzed in great detail by Rice (1944) and also by Davenport and Root (1958), see also summary by Pécseli (2000). It is general, and can refer to many possible physically relevant situations, and a few will be discussed later. Modeling physical processes, random flows in particular, in terms of self-consistently moving randomly distributed structures ('disturbance centers') was suggested by Wandel and Kofoed-Hansen (1962) and used also for simulating two-dimensional random flows (Pécseli and Trulsen 1997). Such models do not possess any cascade in the power spectrum and will thus not model what is properly denoted turbulence, but they can serve as interesting illustrations.

20.1 Campbell's theorem

Considering a time record as (20.1) a few basic results are readily obtained. Campbell's theorem (Campbell 1909) thus states that the average signal is given by

$$\langle \Phi \rangle = \mu a \int_{-\infty}^{\infty} \psi(t) \mathrm{d}t, \tag{20.2}$$

and the mean square value of the fluctuations about this average is

$$\langle (\Phi - \langle \Phi \rangle)^2 \rangle = \mu a^2 \int_{-\infty}^{\infty} \psi^2(t) \mathrm{d}t, \tag{20.3}$$

where μ is the average number of basic structures per time unit. The statement of the theorem is not precise until it is defined what is meant by 'average'. From the form of the equations one might be tempted to think of a time average; e.g. the value

$$\overline{\Phi(t)} \equiv \operatorname*{Lim}_{\mathcal{T}\to\infty} \frac{1}{\mathcal{T}} \int_0^{\mathcal{T}} \Phi(t) \mathrm{d}t. \tag{20.4}$$

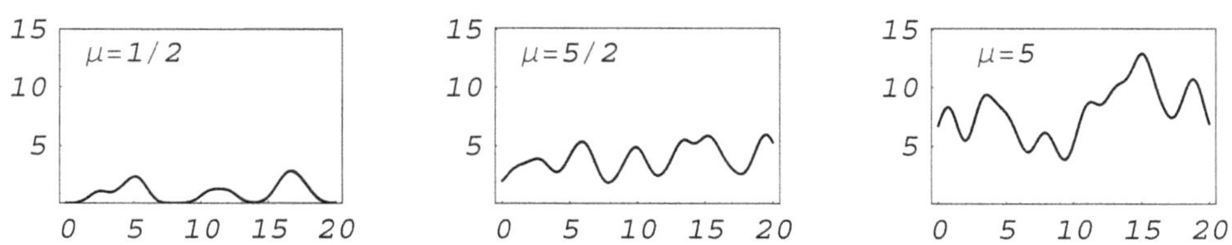

Figure 20.1. Examples of three realizations of the model for densities $\mu = N/\mathcal{T} = 1/2$, 5/2 and 5, respectively. Only a fraction of an extended sequence is shown.

However, in the proof of the theorem the average is generally taken over a great many intervals of length $\mathcal{T}$ with t held constant. This is the ensemble averaging discussed before. To make it clear in the present context we take the case of $\langle\Phi(t)\rangle$ for illustration. We observe $\Phi(t)$ for many, say M, intervals each of length $\mathcal{T}$, where $\mathcal{T}$ is large in comparison with the interval over which the effect $\psi(t)$ of a single basic structure, or pulse, is appreciable. Let ${}^{n}\Phi(t')$ be the value of $\Phi(t)$, at t' seconds after the beginning of the n-th interval. Thus, t' is equal to t plus a constant depending upon the beginning time of the interval. We place the subscript in front because we wish to reserve the usual place for another subscript later on. The value of $\langle\Phi(t')\rangle$ is then defined as

$$\langle\Phi(t')\rangle = \operatorname*{Lim}_{M\to\infty} \frac{1}{M}\left[{}^{1}\Phi(t') + {}^{2}\Phi(t') + \cdots + {}^{M}\Phi(t')\right]. \tag{20.5}$$

This limit is assumed to exist. The mean square value of the fluctuation of $\Phi(t')$ is defined in much the same way.

Actually, as the equations (20.2) and (20.3) of Campbell's theorem show, these averages and all the similar averages encountered later turn out to be independent of the time. When this is true and when the M intervals in (20.5) are taken consecutively, the time average (20.4) and the average (20.5) become the same; more generally the signal constitutes an ergodic random process. To show that the time average equals the ensemble average, we integrate both sides of (20.5) from 0 to $\mathcal{T}$ and obtain

$$\begin{aligned}\langle\Phi(t')\rangle &= \operatorname*{Lim}_{M\to\infty} \frac{1}{M\mathcal{T}} \sum_{m=1}^{M} \int_0^{\mathcal{T}} {}^{m}\Phi(t')\mathrm{d}t' \\ &= \operatorname*{Lim}_{M\to\infty} \frac{1}{M\mathcal{T}} \int_0^{M\mathcal{T}} \Phi(t)\,\mathrm{d}t = \overline{\Phi(t)}.\end{aligned} \tag{20.6}$$

This is the same as the time average (20.4), again assuming that the limit in (20.6) exists.

It is also possible here to choose the reference level so that $\langle\Phi(t)\rangle = 0$, but as the model construction indicates, this will not be a particularly smart thing to do!

20.1.1 Proof of Campbell's theorem

Consider the case in which exactly K pulses are present in the record of length $\mathcal{T}$. We imagine these K pulses as determined to arrive in the interval $\{0, \mathcal{T}\}$ but any particular pulse is just as likely to arrive at one time as any other time, i.e. the pulse-arrival times are uniformly distributed in the interval $\{0, \mathcal{T}\}$. These pulses are numbered from one to K for purposes of identification, but it is emphasized that the numbering need not have anything to do with the order of arrival. Thus, if t_k be the time of arrival of pulse number k, the probability that t_k lies in the interval $\{t, t + \mathrm{d}t\}$ is $\mathrm{d}t/\mathcal{T}$ irrespective of the arrival of any other pulse.

We take $\mathcal{T}$ to be very large compared with the range of values of t for which $\psi(t)$ is appreciably different from zero. In physical applications such a range usually exists and we shall call it Δ even though it need not be very definite. Then, when exactly K pulses arrive in the interval $\{0, \mathcal{T}\}$ the effect is given by (20.1) where we here include a subscript $_K$ to emphasize that the number of pulses is fixed in the given interval $\mathcal{T}$. In other words; the intervals are *conditionally* selected with respect to K.

Suppose that we examine a large number M of intervals of length $\mathcal{T}$. The number having exactly K arrivals will be, to a first approximation $M\ P(K)$ where $P(K)$ is a Poisson distribution. For a fixed value of t and for each interval having K arrivals, $\Phi_K(t)$ will have a definite value. As $M \to \infty$, the average value of the $\Phi_K(t)$ s, obtained by averaging over the intervals, is

$$\begin{aligned}\langle \Phi_K(t) \rangle &= \int_0^{\mathcal{T}} \frac{\mathrm{d}t_1}{\mathcal{T}} \cdots \int_0^{\mathcal{T}} \frac{\mathrm{d}t_K}{\mathcal{T}} \sum_{k=1}^{K} a\psi(t - t_k) \\ &= \sum_{k=1}^{K} \int_0^{\mathcal{T}} \frac{\mathrm{d}t_k}{\mathcal{T}} a\psi(t - t_k).\end{aligned} \tag{20.7}$$

If $\Delta < t < \mathcal{T} - \Delta$, we have effectively

$$\langle \Phi_K(t) \rangle = \frac{K}{\mathcal{T}} \int_{-\infty}^{\infty} a\psi(t)\mathrm{d}t. \tag{20.8}$$

If we now average $I(t)$ over all of the M intervals instead of only over those having K arrivals, we get by Bayes' rule, as $M \to \infty$,

$$\begin{aligned}\langle \Phi(t) \rangle &= \sum_{K=0}^{\infty} \langle \Phi_K(t) \rangle P(K) \\ &= \sum_{K=0}^{\infty} \frac{K}{\mathcal{T}} \frac{(\mu\mathcal{T})^K}{K!} \mathrm{e}^{-\mu\mathcal{T}} \int_{-\infty}^{\infty} a\psi(t)\,\mathrm{d}t \\ &= \mu a \int_{-\infty}^{\infty} \psi(t)\,\mathrm{d}t.\end{aligned} \tag{20.9}$$

This proves the first part of Campbell's theorem. We have used this rather elaborate proof on the relatively simple result (20.9) in order to illustrate a method which may also be used on more complicated problems. Of course, (20.9) could be obtained by noting that the integral is the average value of the effect produced by one pulse arrival, the average being taken over one time unit, and that μ is the average number of arrivals per time unit.

In order to prove the second part, (20.3), of Campbell's theorem calculate first $\langle \Phi^2(t) \rangle$ and use

$$\begin{aligned}\langle (\Phi(t) - \langle \Phi \rangle)^2 \rangle &= \langle \Phi^2 \rangle - 2\langle \Phi \langle \Phi \rangle \rangle + \langle \Phi^2 \rangle \\ &= \langle \Phi^2 \rangle - \langle \Phi \rangle^2.\end{aligned} \tag{20.10}$$

From the definition of $\Phi_K(t)$ we have

$$\Phi_K^2(t) = \sum_{k=1}^{K} \sum_{m=1}^{K} a^2 \psi(t - t_k)\psi(t - t_m).$$

Averaging this over all values of $t_1, t_2, \ldots, t_k$ with t held fixed as in (20.7),

$$\left\langle \Phi_K^2(t) \right\rangle = \sum_{k=1}^{K} \sum_{m=1}^{K} \int_0^{\mathcal{T}} \frac{\mathrm{d}t_1}{\mathcal{T}} \cdots \int_0^{\mathcal{T}} \frac{\mathrm{d}t_K}{\mathcal{T}}\, a^2 \psi(t - t_k) \psi(t - t_m).$$

The multiple integral has two principally different values. If $k = m$ its value is

$$\int_0^{\mathcal{T}} a^2 \psi^2(t - t_k) \frac{\mathrm{d}t_k}{\mathcal{T}},$$

and if $k \neq m$ its value is

$$\int_0^{\mathcal{T}} a\psi(t - t_k) \frac{\mathrm{d}t_k}{\mathcal{T}} \int_0^{\mathcal{T}} a\psi(t - t_m) \frac{\mathrm{d}t_m}{\mathcal{T}}.$$

Counting the number of terms in the double sum it is observed that of the altogether K^2 terms there are K having the first value and $K^2 - K$ having the second. Hence, if $\Delta < t < \mathcal{T} - \Delta$ we have

$$\left\langle \Phi_K^2(t) \right\rangle = \frac{K}{\mathcal{T}} \int_{-\infty}^{\infty} a^2 \psi^2(t)\, \mathrm{d}t + \frac{K(K-1)}{\mathcal{T}^2} \left[\int_{-\infty}^{\infty} a\psi(t)\, \mathrm{d}t \right]^2.$$

The foregoing analysis was conditional, i.e. it considered only realizations containing exactly K pulses. Averaging over all the realizations where K is statistically distributed we find, using that $\langle K(K-1) \rangle = \langle K \rangle^2$ for a Poisson distribution for K

$$\begin{aligned} \langle \Phi^2(t) \rangle &= \sum_{K=0}^{\infty} P(K) \left\langle \Phi_K^2(t) \right\rangle \\ &= \mu a^2 \int_{-\infty}^{\infty} \psi^2(t) \mathrm{d}t + \langle \Phi(t) \rangle^2, \end{aligned}$$

where the summation with respect to K is performed as in (20.9), and after summation the value (20.9) for $\langle \Phi(t) \rangle$ is used. Comparison with (20.10) establishes the second part of Campbell's theorem. The coefficient μ in front of the integral is obtained even when the distribution of K differs from a Poisson distribution, so this assumption for K is not strictly necessary.

20.1.2 Extension of Campbell's theorem

Consider now a slight generalization of the signal (20.1) in the form

$$\Phi(t) = \sum_{k}^{N} a_k \psi(t - t_k), \tag{20.11}$$

where $\psi(t)$ is an *a priori* given function as before, and where $a_1, \ldots a_k, \ldots$ are independent random variables all having the same distribution. It is assumed that all of the moments $\langle a^n \rangle$ exist, and that the events occur at random with a uniform distribution of t_k as before. In effect we *still* assume that all basic structures have the same form, but allow for a statistical distribution of their amplitudes.

The extension of Campbell's theorem states that the n-th semi-invariant of the probability density $P(\Phi)$ of Φ, where Φ is given by (20.11), is

$$\lambda_n = \mu\langle a^n\rangle \int_{-\infty}^{\infty} \psi^n(t)\,\mathrm{d}t, \tag{20.12}$$

where μ is the expected number of events per second. The semi-invariants (or *cumulants*) of a distribution are defined as the coefficients in the expansion

$$\ln\langle \mathrm{e}^{i\Phi u}\rangle = \sum_{n=1}^{N} \frac{\lambda_n}{n!}(iu)^n + \mathcal{O}(u^N), \tag{20.13}$$

i.e. as the coefficients in the expansion of the logarithm of the characteristic function. The λs are related to the moments of the distribution. Thus, if m_1, m_2, ... denote the first, second, ... moments about zero, we have

$$\langle \mathrm{e}^{\mathrm{i}\Phi u}\rangle = 1 + \sum_{n=1}^{N} \frac{m_n}{n!}(iu)^n + \mathcal{O}(u^N).$$

By combining this relation with the one defining the λs it may be shown that

$$\begin{aligned}\langle\Phi\rangle &= m_1 = \lambda_1,\\ \langle\Phi^2\rangle &= m_2 = \lambda_2 + \lambda_1 m_1,\\ \langle\Phi^3\rangle &= m_3 = \lambda_3 + 2\lambda_2 m_1 + \lambda_1 m_2.\end{aligned}$$

It follows that $\lambda_1 = \langle\Phi\rangle$ and $\lambda_2 = \langle(\Phi - \langle\Phi\rangle)^2\rangle$. Hence, (20.12) yields the original statement of Campbell's theorem when we set n equal to one. The symbol σ^2 is often used for λ_2, with σ being the *standard deviation*.

In principle we can use any form for $\psi(t)$ in (20.11), but for most relevant application we would like to see the average (20.2) and all the cumulants λ_n to be finite for all n. This requirement excludes, for instance, $\psi(t) \sim 1/(1+t)$ or $t/(1+t^2)$, etc.

- **Exercise:** Consider a process which is the sum of several independent processes. Demonstrate that the cumulants, or semi-invariants, associated with this process are given by the sum of cumulants for the individual processes.

- **Exercise:** Demonstrate the following relations for the cumulants λ_n in terms of the averages $m_n \equiv \langle\Phi^n\rangle$

$$\begin{aligned}\lambda_1 &= m_1\\ \lambda_2 &= m_2 - m_1^2 \equiv \sigma^2\\ \lambda_3 &= m_3 - 3m_1m_2 + 2m_1^3\\ \lambda_4 &= m_4 - 4m_1m_3 - 3m_2^2 + 12m_2m_1^2 - 6m_1^4.\end{aligned}$$

- **Exercise:** Consider a Gaussian random process, not necessarily with zero mean. Obtain the cumulants for this case.

- **Exercise:** A Gaussian probability density is completely characterized by its average and its standard deviation. More general distributions can also use the average and standard deviation, and include the skewness $\mathcal{S}$ and kurtosis $\mathcal{K}$ or excess of the distribution, where $\mathcal{S} \equiv \langle(\Phi - \langle\Phi\rangle)^3\rangle/\sigma^3$ and $\mathcal{K} \equiv \langle(\Phi - \langle\Phi\rangle)^4\rangle/\sigma^4$; as defined before. For a Gaussian distribution we have $\mathcal{S} = 0$ and $\mathcal{K} = 3$. Sometimes we find the kurtosis defined by subtracting the Gaussian value 3 from the definition used here.

  Consider a signal generated as in (20.1). Express the average, standard deviation, skewness and kurtosis in terms of first m_j and then λ_j, with $j = 1, 2, 3, 4$.

Further extensions of Campbell's theorem

Other extensions of Campbell's theorem may be made. The foregoing analysis might give the impression that the uniform distribution of pulse-arrival times t_k was essential. The following example, given by Rice 1944), demonstrates that the analysis can be extended to more general pulse-arrival time distributions. For example, suppose in the expression (20.11) for $\Phi(t)$ that $t_1, t_2, \ldots, t_k, \ldots$ while still random variables, are no longer uniformly distributed as assumed before. Suppose instead that a probability density $P(x)$ is given where $x_j > 0$ is the interval between two successive events labeled by j and $j + 1$,

$$\begin{aligned} t_2 &= t_1 + x_1, \\ t_3 &= t_2 + x_2 = t_1 + x_1 + x_2 \\ &\vdots \end{aligned} \tag{20.14}$$

We can use the arguments from section 3.2 and show that for the case treated before

$$P(x) = \mu e^{-\mu x}. \tag{20.15}$$

For other distributions more regular separations of pulses can be modeled. We can for instance illustrate a certain 'recovery time' for the pulse emission. Pulses appearing with fixed intervals are an extreme case. It is assumed in the following that the expected number of events per time unit is still μ.

Consider again the series (20.11). For a long time-interval extending from $t = t_1$ to $t = \mathcal{T} + t_1$ inside of which there are exactly K events we have by writing out the individual terms in (20.11) for clarity

$$\begin{aligned} \Phi(t) &= a_1\psi(t - t_1) + a_2\psi(t - t_1 - x_1) + \cdots \\ &\quad + a_{K+1}\psi(t - t_1 - x_1 \cdots - x_K) \\ &= a_1\psi(t') + a_2\psi(t' - x_1) + \cdots + a_{K+1}\psi(t' - x_1 - \cdots - x_K), \end{aligned}$$

and

$$\begin{aligned} \Phi^2(t) &= a_1^2\psi^2(t') + a_2^2\psi^2(t' - x_1) + \cdots + a_{K+1}^2\psi^2(t' - x_1 - \cdots - x_K) \\ &\quad + 2a_1a_2\psi(t')\psi(t' - x_1) + \cdots \\ &\quad + 2a_1a_{K+1}\psi(t')\psi(t' - x_1 - \cdots - x_K) \\ &\quad + 2a_2a_3\psi(t' - x_1)\psi(t' - x_1 - x_2) + \cdots + \cdots, \end{aligned}$$

where $t' = t - t_1$. If we integrate $\Phi^2(t)$ over the entire interval $0 < t' < \mathcal{T}$ and drop the primes we get approximately

$$\begin{aligned}\int_0^{\mathcal{T}} \Phi^2(t)\mathrm{d}t = \left(a_1^2 + \cdots + a_{K+1}^2\right)\varphi(0) \\ + 2a_1a_2\varphi(x_1) + 2a_1a_3\varphi(x_1 + x_2) + \cdots \\ + 2a_1a_{K+1}\varphi(x_1 + \cdots + x_k) \\ + 2a_2a_3\varphi(x_2) + \cdots + 2a_Ka_{K+1}\varphi(x_K),\end{aligned} \tag{20.16}$$

where we have defined the convolution

$$\varphi(x) = \int_{-\infty}^{\infty} \psi(t)\psi(t - x)\mathrm{d}t. \tag{20.17}$$

Now divide both sides of (20.16) by $\mathcal{T}$ and consider K and $\mathcal{T}$ to be very large. Then the following individual terms enter the expression

$$\frac{K}{\mathcal{T}}\frac{a_1^2 + \cdots + a_{K+1}^2}{K}\varphi(0) \approx \mu\varphi(0)\langle a^2\rangle,$$

and

$$\begin{aligned}\frac{1}{\mathcal{T}}\left[a_1a_2\varphi(x_1) + a_2a_3\varphi(x_2) + \cdots + a_Ka_{K+1}\varphi(x_K)\right] = \frac{K}{\mathcal{T}}\langle a_ka_{k+1}\varphi(x_k)\rangle \\ \approx \mu\langle a\rangle^2\int_0^{\infty} \varphi(x)P(x)\,\mathrm{d}x,\end{aligned}$$

and also

$$\begin{aligned}\frac{1}{\mathcal{T}}[a_1a_3\varphi(x_1 + x_2) + \cdots] = \frac{K-1}{\mathcal{T}}\langle a_ka_{k+2}\varphi(x_k + x_{k+1})\rangle \\ \approx \mu\langle a\rangle^2\int_0^{\infty}\mathrm{d}x_1\int_0^{\infty}\mathrm{d}x_2P(x_1)P(x_2)\varphi(x_1 + x_2).\end{aligned}$$

Consequently (20.16) can be reduced to

$$\begin{aligned}\langle\Phi^2(t)\rangle = \lim_{\mathcal{T}\to\infty}\frac{1}{\mathcal{T}}\int_0^{\mathcal{T}} \Phi^2(t)\,\mathrm{d}t \\ = \mu\langle a^2\rangle\varphi(0) + 2\mu\langle a\rangle^2\left[\int_0^{\infty} P(x)\varphi(x)\,\mathrm{d}x\right. \\ \left. + \int_0^{\infty}\mathrm{d}x_1\int_0^{\infty}\mathrm{d}x_2P(x_1)P(x_2)\varphi(x_1 + x_2) + \cdots\right].\end{aligned}$$

The foregoing results are rather general. We consider here the special, but important, case for which

$$\begin{aligned}\psi(t) &= 0, \qquad t < 0 \\ \psi(t) &= \mathrm{e}^{-\alpha t}, \qquad t > 0.\end{aligned} \tag{20.18}$$

For this particular exponential form for $\psi(t)$ we obtain the convolution (20.17) as

$$\varphi(x) = \frac{e^{-\alpha x}}{2\alpha}.$$

The multiple integrals appearing in the expression for $\langle \Phi^2(t) \rangle$ may be written in terms of powers of the integral

$$q \equiv \int_0^\infty P(x) e^{-\alpha x} dx. \tag{20.19}$$

Thus

$$2\alpha \langle \Phi^2(t) \rangle = \mu \langle a^2 \rangle + \langle a^2 \rangle \frac{\mu q}{1 - q}.$$

Since $\langle \Phi(t) \rangle = \mu \langle a \rangle \int_{-\infty}^{\infty} \psi(t) dt = \langle a \rangle \mu / \alpha$ with the present model for $\psi(t)$, we have

$$\langle \Phi^2(t) \rangle - \langle \Phi(t) \rangle^2 = \frac{\mu \langle a^2 \rangle}{2\alpha} + \left(\frac{\mu \langle a \rangle}{\alpha} \right)^2 \left[\frac{\alpha q}{\mu(1 - q)} - 1 \right]. \tag{20.20}$$

equations (20.19) and (20.20) give an extension of Campbell's theorem subject to the restrictions discussed in connection with equations (20.14) and (20.18). Other generalizations have been made by Rowland (1936).

- **Exercise:** Verify that (20.20) gives the correct answer when $P(x)$ is given by (20.15), and investigate also the case when the events are spaced equally.

20.2 Probability densities

The foregoing results for $\langle \Phi(t) \rangle$ and $\langle \Phi^2(t) \rangle$ could, as demonstrated, be obtained without the relevant probabilities being explicitly known. The one-point amplitude probability density for the model can however also be obtained by use of characteristic functions (Rice 1944, 1945). The problem is to obtain $P(\Phi)$ on the basis of the *a priori* given probability density of t_k i.e. $P(t_k)$.

The probability that a sum of K independent variables $x_1 + x_2 + \cdots + x_K$ lies in the range $[X, X + dX]$ is

$$dX \frac{1}{2\pi} \int_{-\infty}^{\infty} \exp(-iXu) \prod_{k=1}^{K} \langle \exp(ix_k u) \rangle du,$$

i.e. the characteristic function for a sum of independent variables is the product of their individual characteristic functions. These functions, $\langle \exp(ix_k u) \rangle$, associated with the distribution of x_k are obtained by averaging over the values of x_k.

In our case we want the characteristic function for the signal $\Phi(t)$ in (20.1), i.e.

$$\langle \exp(i\Phi u) \rangle \equiv \int \exp(i\Phi u) P(\Phi) d\Phi.$$

With $\Phi(t) = \sum_k^K a\psi(t - t_k)$ and $P(\psi)\mathrm{d}\psi = P(t_k)\mathrm{d}t_k$ we have

$$\langle \exp(\mathrm{i}a\psi(t - t_k)u)\rangle = \frac{1}{\mathcal{T}}\int_0^{\mathcal{T}} \exp(\mathrm{i}ua\psi(t - t_k))\mathrm{d}t_k,$$

noting that the time t_k is here the only statistically varying quantity. First we consider the conditional characteristic function, obtained for given K. All of the K characteristic functions are the same so

$$P_K(\Phi) = \frac{1}{2\pi}\int_{-\infty}^{\infty} \exp(-\mathrm{i}\Phi u)\left(\frac{1}{\mathcal{T}}\int_0^{\mathcal{T}} \exp(\mathrm{i}ua\psi(t - \tau))\,\mathrm{d}\tau\right)^K \mathrm{d}u.$$

This expression gives a *conditional* probability density. The averaging over all K is performed assuming the probability density $P(K) = \mathrm{e}^{-\mu\mathcal{T}}(\mu\mathcal{T})^K/K!$, i.e. a Poisson distribution for K. With $P(\Phi) = \sum_{K=0}^{\infty} P(K)P_K(\Phi)$ we perform the summation by use of $\mathrm{e}^{\mu S} = \sum_{K=0}^{\infty}(\mu S)^K/K!$ to obtain after some elementary manipulations

$$P(\Phi) = \frac{1}{2\pi}\int_{-\infty}^{\infty} \exp\left(-\mathrm{i}\Phi u + \mu\int_{-\infty}^{\infty}\left\{\exp[\mathrm{i}ua\psi(t)] - 1\right\}\mathrm{d}t\right)\mathrm{d}u, \tag{20.21}$$

where the density of structures $\mu = \langle K\rangle/\mathcal{T}$ was introduced in terms of the average number, $\langle K\rangle$, of structures in a record. Note that $P(\Phi)$ is real as it should be, since $P(\Phi) = P^*(\Phi)$.

Unfortunately, the result (20.21) is not as useful as we might have hoped it to be. There are quite few mathematical forms for $\psi(t)$ that allows an analytical solution. The 'box' or top-hat pulse is one, a half or double exponential is another (Garcia 2012). If we try to solve (20.21) numerically there will often be a 'spike' at the origin. For pulses $\psi(t)$ with finite support (i.e. some vanishing outside a finite time-interval) there will be a remnant of the δ-function from the small segments of no structures present in some short time intervals somewhere. These spiky contributions are numerically troublesome, see for instance a subsequent figure 20.8. This δ-function contribution is found only for basic structures or pulses with finite support.

Joint probability densities

The two-time amplitude, or joint probability density and also higher order densities can be obtained similarly. We imagine two records where in one we place a pulse $\phi(t - t_k)$ and in the other one $\psi(t - t_k)$, where the temporal form of the two structures need not have any relation at all. For the present application we assume however, one to be a time-shifted replica of the other, with time shift τ. The two-time probability density for this model becomes

$$\begin{aligned}P(\Phi, \Psi) = &\frac{1}{(2\pi)^2}\iint_{-\infty}^{\infty} \exp(-\mathrm{i}\Phi u - \mathrm{i}\Psi v)\\ &\times \exp\left(\mu\int_{-\infty}^{\infty}\left\{\exp[\mathrm{i}ua\psi(t) + \mathrm{i}va\psi(t + \tau)] - 1\right\}\mathrm{d}t\right)\mathrm{d}u\mathrm{d}v,\end{aligned} \tag{20.22}$$

where $\Phi = \Phi(t)$ and $\Psi = \Psi(t)$ in general corresponds to two records. For the present application we have $\Phi(t) = \sum_k^N a\psi(t - t_k)$ and $\Psi(t) \equiv \Phi(t + \tau) = \sum_k^N a\psi(t - t_k + \tau)$.

It is important to note that the two records Φ and Ψ are 'synchronized', i.e. every time we place a pulse in one we also place one in the other record. If the two records are completely independent (this would be the other extreme) we have $P(\Phi, \Psi) = P(\Phi)P(\Psi)$, where $P(\Phi)$ and $P(\Psi)$ are obtained as before. In this independence limit we place a number of, say, N_1 pulses of some sort in one record and after finishing that we place N_2 different pulses randomly in another record.

In section 19.3.2 we found use of the joint probability density $P(\Phi, \Phi')$ with $\Phi'(t) \equiv \mathrm{d}\Phi(t)/\mathrm{d}t$. To model this case we can place a pulse $\phi(t - t_k)$ in one of the records mentioned before and its time derivative $\phi'(t - t_k)$ in the other one. We find for this case

$$\begin{aligned} P(\Phi, \Phi') &= \frac{1}{(2\pi)^2} \iint_{-\infty}^{\infty} \exp(-\mathrm{i}\Phi u - \mathrm{i}\Phi' v) \\ &\quad \times \exp\left(\mu \int_{-\infty}^{\infty} \left\{\exp[\mathrm{i}ua\phi(t) + \mathrm{i}va\phi'(t)] - 1\right\} \mathrm{d}t\right) \mathrm{d}u \mathrm{d}v. \end{aligned}$$

- **Exercise:** Demonstrate that the result for the probability density becomes particularly simple where there are two pulse types having the same density and the same form but with opposite polarity.
- **Exercise:** Consider the particularly simple case where the basic pulse or structure is composed by one or more boxes of unit length as illustrated in figure 20.2. Demonstrate that the probability density for case I is

$$P(\Phi) = \sum_{N=0}^{\infty} \frac{\mu^N}{N!} \mathrm{e}^{-\mu} \delta(\Phi - N), \qquad \text{for} \qquad N = 0, 1, 2, \ldots$$

where μ is the density of pulses. Demonstrate that for case II as well as case III in figure 20.2 the probability density, in terms of the modified Bessel functions I_N of order N, is

$$P(\Phi) = \sum_{N=-\infty}^{\infty} I_N(\mu) \mathrm{e}^{-\mu} \delta(\Phi - N), \qquad \text{for} \qquad N = 0, \pm 1, \pm 2, \ldots$$

i.e. an even distribution with all odd semi-invariants vanishing, while all even semi-invariants are equal to μ. The *two-point* probabilities for the cases II and III are different but the one-point probability densities, or *projections*, are the

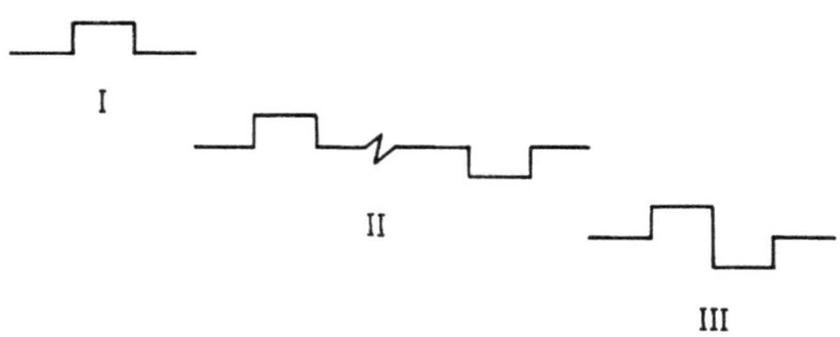

Figure 20.2. Illustration of particularly simple basic structures or pulses composed by one or more boxes or bins of unit amplitude and unit length.

same. Note that here μ becomes dimensionless because we used boxes of unit length.

What is the expression for the probability density for the limiting case where μ is small? Give a physical explanation of the results.

20.2.1 Extended model

The extension of Campbell's theorem follows almost at once from the generalization of the expression (20.21) for the probability density $P(\Phi)$. By proceeding as in the foregoing section and identifying x_k with $a_k\psi(t - t_k)$ we find that

$$\langle \exp(\mathrm{i}x_k u)\rangle = \frac{1}{T}\int_{-\infty}^{\infty} q(a) \int_0^T \exp\left[\mathrm{i}ua\psi(t - t_k)\right]\mathrm{d}t_k \,\mathrm{d}a,$$

where $q(a)$ is the probability density function for the as. It is readily demonstrated that the probability density $P(\Phi)$ of Φ as defined by (20.11) is

$$P(\Phi) = \frac{1}{2\pi}\int_{-\infty}^{\infty} \exp\left(-\mathrm{i}\Phi u + \mu\int_{-\infty}^{\infty} q(a)\,\mathrm{d}a \right. \\ \left. \times \int_{-\infty}^{\infty}\left[\exp(\mathrm{i}ua\psi(t)) - 1\right]\mathrm{d}t\right)\mathrm{d}u. \tag{20.23}$$

The logarithm of the characteristic function of $P(\Phi)$ is, from (20.23),

$$\mu\int_{-\infty}^{\infty} q(a)\,\mathrm{d}a\int_{-\infty}^{\infty}\left[\exp(\mathrm{i}ua\psi(t)) - 1\right]\mathrm{d}t \\ = \sum_{n=1}^{\infty}\frac{(iu)^n}{n!}\,\mu\int_{-\infty}^{\infty} q(a)da\, a^n \int_{-\infty}^{\infty} \psi^n(t)\,\mathrm{d}t.$$

Comparison with the series (20.13) defining the semi-invariants gives the extension of Campbell's theorem stated by (20.12).

The analysis given hitherto assumed one basic pulse shape only. It is straightforward to generalize the results to the case where many different pulses independently enter the construction of a signal. Thus, for instance (20.21) is generalized to

$$P(\Phi) = \frac{1}{2\pi}\int_{-\infty}^{\infty}\exp\left(-\mathrm{i}\Phi u + \sum_n \mu_n \int_{-\infty}^{\infty}\left\{\exp[\mathrm{i}ua\psi_n(t)] - 1\right\}\mathrm{d}t\right)\mathrm{d}u, \tag{20.24}$$

where the summation runs over all different pulse shapes ψ_n with densities μ_n. The result (20.24) is easily verified by recalling that the characteristic function for a sum of independent variables is the product of their respective, individual, characteristic functions.

20.2.2 Limit of low pulse densities

The probability densities and joint probability densities are generally complicated expressions that can be solved only in some special cases as shown in figure 20.2, although more complicated cases have also found solutions (Garcia 2012). The limit of sparse structure densities has interest and for $\mu \to 0$, we can simplify the expressions (20.21) and (20.22) somewhat. For (20.21) we find in this limit

$$P(\Phi) \to \delta(\Phi) + \frac{\mu}{2\pi}\int_{-\infty}^{\infty} \exp(-\mathrm{i}\Phi u)\int_{-\infty}^{\infty}(\exp[iua\phi(t)] - 1)\mathrm{d}t\mathrm{d}u. \qquad (20.25)$$

For the special case where the basic pulse $\phi(t)$ is a box-function with amplitude a and duration $\Delta\tau$, as in the first case of figure 20.2, we obtain the analytical result

$$P(\Phi) = (1 - \mu\Delta\tau)\delta(\Phi) + \mu\Delta\tau\delta(\Phi - a). \qquad (20.26)$$

We used $\int_{-\infty}^{\infty} \exp(iax)\mathrm{d}x = 2\pi\delta(a)$. The first δ-function term accounts for the zero-level between pulses. We note that for a long record, $\mu\Delta\tau$ is the probability of being inside one of the rectangular pulses. The first δ-function term can then be taken as the probability of being outside a pulse. In all its simplicity, the expression (20.26) and its generalizations has received wide interest for discussions of, for instance, studies of concentration fluctuations in turbulent environments (Mole and Clarke 1995, Yee and Chan 1997, Lewis and Chatwin 1997, Krane *et al* 2003, Mole *et al* 2008, Jørgensen *et al* 2010).

The general form of (20.22) becomes

$$\begin{aligned} P(\Phi, \Psi) \to \frac{1}{(2\pi)^2}\iint_{-\infty}^{\infty} \exp(-\mathrm{i}\Phi u - \mathrm{i}\Psi v) \\ \times\left(1 + \mu\int_{-\infty}^{\infty}\{\exp[iua\phi(t) + \mathrm{i}vb\psi(t)] - 1\}\mathrm{d}t\right)\mathrm{d}u\mathrm{d}v, \end{aligned} \qquad (20.27)$$

where the unity term in the brackets contribute with a product of δ-functions.

20.2.3 Transition to Gaussian distributions

It can be demonstrated analytically that $P(\Phi)$ approaches a Gaussian distribution when $\mu \to \infty$ as expected with reference to the central limit theorem since in this limit many pulses contribute to the signal at any time. To demonstrate this transition (Rice 1944, 1945), it is advantageous to express $P(\Phi)$ as

$$P(\Phi) = \frac{1}{2\pi}\int_{-\infty}^{\infty} \exp\left[-\mathrm{i}\Phi u + \sum_{n=1}^{\infty}(iu)^n\lambda_n/n!\right]\mathrm{d}u, \qquad (20.28)$$

using $\exp(x) = \sum_{n=0}^{\infty} x^n/n!$ and where the semi-invariants, λ_n, are defined as before by

$$\lambda_n = \mu\int_{-\infty}^{\infty}[a\psi(t)]^n\mathrm{d}t. \qquad (20.29)$$

Setting $\lambda_2 \equiv \sigma^2$ and introducing the new variable $x = (\Phi - \lambda_1)/\sigma$, the exponential $\exp \sum_{n=1}^{\infty}(iu)^n \lambda_n/n!$ is expanded as a power series in u. After term-wise integration the terms are collected according to their order in powers of $\mu^{-1/2}$ giving

$$P(\Phi) \sim f^{(0)}(x)/\sigma - \frac{\lambda_3}{3!\sigma^4} f^{(3)}(x) + \left[\frac{\lambda_4}{4!\sigma^5} f^{(4)}(x) + \frac{\lambda_3^2}{72\sigma^7} f^{(6)}(x)\right] + \cdots \tag{20.30}$$

The function $f(x)$ entering (20.30) is introduced through

$$\frac{1}{2\pi}\int_{-\infty}^{\infty} (iu\sigma)^n \exp\left([-iu\sigma x - u^2\sigma^2/2]\right) \mathrm{d}u = (-1)^n f^{(n)}(x)/\sigma, \tag{20.31}$$

where

$$f^{(n)}(x) = \frac{1}{\sqrt{2\pi}} \frac{\mathrm{d}^n}{\mathrm{d}x^n} \mathrm{e}^{-x^2/2}. \tag{20.32}$$

Since λ_n scales as μ^n we find as an order of magnitude that the first term in the series (20.30) is $\mathcal{O}(\mu^{-1/2})$ while the second term is $\mathcal{O}(\mu^{-1})$, and the term within the brackets $\mathcal{O}(\mu^{-3/2})$, etc. This is called the Edgeworth series (Rice 1944). The first term gives the normal distribution and the remaining terms show how it is approached as $\mu \to \infty$. The proof given here is based on a somewhat restrictive model and will not serve as a general proof of the central limit theorem, although it is closely related to it. Nonetheless we might argue on the basis of the results in the present section that a quantity being the sum of many independent contributions is likely to have a Gaussian distribution, at least to a good approximation.

The transition to multi-dimensional Gaussian, or normal, distributions for $\mu \to \infty$ is demonstrated similarly for multipoint distributions. In this Gaussian limit it will be argued that all the statistical information is contained in the correlation function.

- **Exercise:** As already mentioned, the transition to Gaussian statistics can be performed by keeping the mean-square fluctuation level constant, i.e. $\mu a^2 = \text{const.}$ while $\mu \to \infty$, implying $a \to 0$. Demonstrate that the triple-correlation function vanishes in this limit as $1/\sqrt{\mu}$, provided $\int_{-\infty}^{\infty} \psi(t)\,\mathrm{d}t = 0$. How about higher order correlations?

An apparent paradox

A paradox remains in the foregoing analysis for the case where the basic pulse in the time series takes discrete values only, i.e. the probability is non-zero for discrete values only. The question is how such a discrete distribution can be approximated by a continuous Gaussian probability distribution in the limit of $\mu \to \infty$. To study this particular question explicitly we consider a stochastic variable, X, that takes the value 0 and 1 with probability 1/2 each. Let Y be the sum of r such variables. Then Y takes the values $n = 0, 1, 2, 3, \ldots, r$ with probability

$$P_n = 2^{-r}\binom{r}{in}.$$

Noting the form of the average $\langle Y\rangle = r\langle X\rangle = \frac{1}{2}r$ and variance $\sigma_Y^2 = r\sigma_X^2 = \frac{1}{4}r$ it is found advantageous to introduce a new variable Z given through $Y = \frac{1}{2}r + \frac{1}{2}Z\sqrt{r}$. The probability that Z lies in the interval $(z; z + \Delta z)$ is

$$P_Z(z)\Delta z = \sum_{\frac{1}{2}r+\frac{1}{2}z\sqrt{r} \;<\; n \;<\; \frac{1}{2}r+\frac{1}{2}(z+\Delta z)\sqrt{r}} P_n.$$

For large r we have

$$\begin{aligned}
\log P_n = &- r\log 2 + \left(r + \frac{1}{2}\right)\log r - \left(n + \frac{1}{2}\right)\log n \\
&- \left(r - n + \frac{1}{2}\right)\log(r - n) - \frac{1}{2}\log 2\pi \\
= &\log 2 - \frac{1}{2}\log r - \left(\frac{1}{2}r + \frac{1}{2}r^{\frac{1}{2}}z + \frac{1}{2}\right)\left(r^{-\frac{1}{2}}z - \frac{1}{2}r^{-1}z^2 + \cdots\right) \\
&- \left(\frac{1}{2}r - \frac{1}{2}r^{\frac{1}{2}}z + \frac{1}{2}\right)\left(-r^{-\frac{1}{2}}z - \frac{1}{2}r^{-1}z^2 + \cdots\right) - \log 2\pi \\
= &\log 2 - \frac{1}{2}\log r - \frac{1}{2}z^2 - \log 2\pi + \mathcal{O}\left(r^{-\frac{1}{2}}\right).
\end{aligned}$$

Consequently

$$\begin{aligned}
P_Z(z)\Delta z &= \frac{1}{2}r^{\frac{1}{2}}\Delta z e^{\left(\log 2 - \frac{1}{2}\log r - \frac{1}{2}z^2 - \log 2\pi\right)} \\
&= \frac{1}{\sqrt{2\pi}}e^{-\frac{1}{2}z^2}\Delta z.
\end{aligned} \tag{20.33}$$

The implication of this result is that, for large r, we can use the approximation

$$P_Y(y) = \sqrt{\frac{2}{\pi r}}\exp\left[-\frac{\left(y - \frac{1}{2}r\right)^2}{\frac{1}{2}r}\right]. \tag{20.34}$$

This is evidently a coarse-grain distribution; it gives the probability of finding Y in an interval y, $y + \Delta y$ when $\Delta y \gg 1$. It is obviously incorrect in use when $\Delta y \leqslant 1$. It may also seem paradoxical that the range of (20.33) extends from $-\infty$ to ∞ while Y by construction can assume positive values only. The normalization $\int_{-\infty}^{\infty} P_Y(y)\mathrm{d}y = 1$ of the probability density will be violated if the range is reduced to $\{0, \infty\}$. Again it is emphasized that the distribution is only an approximation and when r is large it is actually exceedingly close to zero for all $y < 0$.

- **Exercise:** Consider the case (Papoulis 1991) where all x_js are uniformly distributed in the interval $\{0; 1\}$. Find $P(y)$ for this case.

20.3 Correlation functions

The two point correlation function associated with the model (20.1) can be calculated directly without use of the two point probability density. First select all records with K events and calculate

$$
\begin{aligned}
R_K(\tau) &\equiv \langle \Phi_K(t)\Phi_K(t+\tau)\rangle \\
&= \sum_{k=1}^{K}\sum_{m=1}^{K}\int_0^{\mathcal{T}}\frac{\mathrm{d}t_1}{\mathcal{T}}\cdots\int_0^{\mathcal{T}}\frac{\mathrm{d}t_K}{\mathcal{T}}\,a^2\psi(t-t_k)\psi(t+\tau-t_m).
\end{aligned}
$$

This expression contains two different types of terms; those where $k = m$, there are K of these, and then $K^2 - K$ terms where $k \neq m$. For the latter type t_k and t_m are independent and the average of the product becomes a product of averages. The result is

$$
R_K(\tau) = \frac{K}{\mathcal{T}}a^2\int_{-\infty}^{\infty}\psi(t)\psi(t+\tau)\mathrm{d}t + \frac{K(K-1)}{\mathcal{T}^2}\left[a\int_{-\infty}^{\infty}\psi(t)\,\mathrm{d}t\right]^2,
$$

where it was assumed that $\mathcal{T}$ is so large that it is safe to let the integration limits go to infinity. Assuming, as before, a Poisson distribution for K we have $\langle K(K-1)\rangle = \langle K\rangle^2$. With $\mu = \langle K\rangle/\mathcal{T}$ the final result becomes

$$
\begin{aligned}
\mathcal{R}(\tau) &= \langle \Phi(t)\Phi(\tau+t)\rangle \\
&= \mu a^2\int_{-\infty}^{\infty}\psi(t)\psi(\tau+t)\,\mathrm{d}t + \left[\mu a\int_{-\infty}^{\infty}\psi(t)\,\mathrm{d}t\right]^2.
\end{aligned}
\tag{20.35}
$$

In particular, the mean square fluctuation level is obtained from $\mathcal{R}(0)$ when the DC-level is subtracted. It may be worth noting that the assumption of a Poisson distribution for K was used explicitly in the second term of (20.35) only. This second term vanishes for many cases of practical interest. In these cases $\mathcal{R}(t)$ is unaffected by the transition to the Gaussian limit for $\mu \to \infty$, provided $a \to 0$ so that the mean square fluctuation level of the signal μa^2 is kept constant.

The correlation function, and thereby also the power spectrum, is insensitive to the sign of the basic structure, as well as to a 'mirroring', i.e. replacing of $\psi(t)$ by $\psi(-t)$. The power spectrum cannot distinguish the direction of time, and remains the same if you read the time series (20.1) 'backwards'. Note that *any* prescribed acceptable power spectrum (and by the Wiener–Khinchine theorem therefore any correlation function) can be modeled by proper choice of $\psi(t)$, but given a correlation function such as (20.35) alone, it is not possible uniquely to reconstruct the basic pulse which was used. The correlation function (20.35) is related to the function defined in (20.17), but different notations are used as the two functions have different meaning. To find information concerning phase relations in the basic structures we need triple correlations such as $\langle \Phi(t)\Phi(\tau_1+t)\Phi(\tau_2+t)\rangle$.

By Fourier transforming the correlation function (20.35) we obtain the corresponding power spectrum. By proper choice of the basic structures $\psi(t)$ in (20.1) we can generate time-series that can appear deceivingly similar to series obtained by sampling a turbulent wave field, and the resulting power spectra will appear similarly deceiving. This should serve as a warning that a continuous power spectrum is no evidence for turbulence.

- **Example:** A model like (20.1) can give an excellent representation for a variety of phenomena in systems not necessarily in thermal equilibrium. As an example we could model the noise caused by raindrops falling on a tin roof. A more common, but prosaic, application is for shot noise in diodes, see also section 9.2.2. In this case it can be argued that a model basic triangular current pulse is adequate as an illustration

$$i(t) = 2et/\tau_\alpha^2,$$

for $0 < t < \tau_\alpha$ and $i(t) = 0$ otherwise (Davenport and Root 1958). The auto-correlation function is readily obtained as

$$R_i(\tau) = \frac{4e\langle I\rangle}{3\tau_\alpha}\left(1 - \frac{3}{2}\frac{|\tau|}{\tau_\alpha} + \frac{1}{2}\frac{|\tau|^3}{\tau_\alpha^3}\right) + \langle I\rangle^2,$$

for $|\tau| < \tau_\alpha$ and $R_i(\tau) = \langle I\rangle^2$ otherwise, with $\langle I\rangle = e\mu$ where here μ is the number of electrons arriving at the cathode per second. By Fourier transforming the correlation function we obtain the frequency spectrum

$$\mathcal{S}_i(\omega) = e\langle I\rangle\frac{4}{(\omega\tau_\alpha)^4}\left\{(\omega\tau_\alpha)^2 + 2[1 - \cos(\omega\tau_\alpha) - \omega\tau_\alpha\sin(\omega\tau_\alpha)]\right\},$$

apart from a delta function at the origin due to the DC-current. The normalized correlation function and corresponding power-spectrum are shown in figure 20.3. In particular, the low-frequency limit of the spectral density of the shot noise current generated by a temperature limited diode is $\mathcal{S}_i(\omega) = e\langle I\rangle = e^2\mu$, known as the *Schottky formula*. Modifications to these results due to space-charge limitations are discussed in the literature (Davenport and Root 1958, MacDonald 1962). In particular, a space-charge-smoothing factor Γ^2 with $0 < \Gamma^2 \leqslant 1$ is sometimes introduced in the Schottky formula.

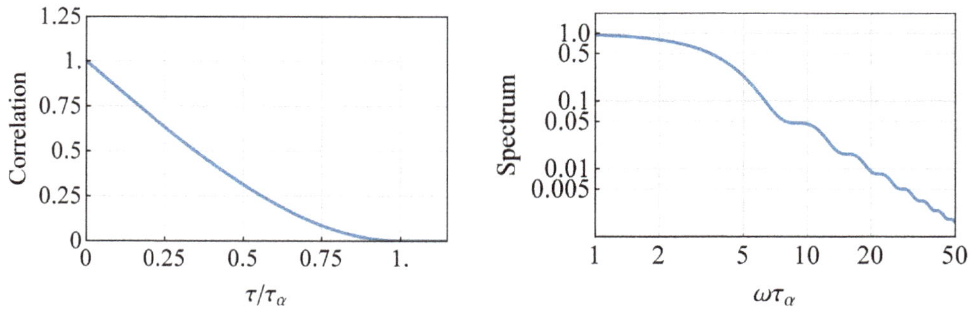

Figure 20.3. Correlation function, $(R_i(\tau) - \langle I\rangle^2)\tau_\alpha/e\langle I\rangle$, and power spectrum, $\mathcal{S}_i(\omega\tau_\alpha)/e\langle I\rangle$ (in double logarithmic representation), obtained by a random superposition of triangular pulses. Note that there are local maxima in the power spectrum, but no oscillations in the correlation function. This result may be somewhat counter-intuitive, and the model was mentioned here mostly for this property!

- **Example:** By incoherent radar scattering from the ionosphere (Bekefi 1966), we can in a first approximation assume that all electrons scatter the incoming radar beam independently. Some contributions cancel while others add-up, just like for a random superposition of wavepackets in a sequence as (20.1). The net scattering intensity will be proportional to the number of scatterers (i.e. electrons) in the scattering volume.
- **Exercise:** Obtain the analytical expression for the correlation function associated with a record generated by a random superposition of the pulses shown in figure 20.2.
- **Exercise:** Obtain the analytical expression for the triple-correlation function

$$S(\tau_1, \tau_2) = \langle \Phi(t)\Phi(t + \tau_1)\Phi(t + \tau_2)\rangle$$

by a similar analysis as for the model (20.1). Unlike the correlation function, the triple correlation changes sign with $\psi(t)$, so it is sensitive to the phase information.
- **Exercise:** Assume that it is *a priori* given that an ensemble of time series are generated by the model described by (20.1) with only one basic structure $\psi(t)$ entering the construction. Demonstrate that $\psi(t)$ can be determined if the correlation function, $\mathcal{R}(\tau)$, and the triple-correlation function, $S(\tau_1, \tau_2)$, are known. Can you, by proper choice of basic structure $\psi(t)$ in (20.1), generate an ensemble of time series which have *any* physically acceptable prescribed set of correlation and triple-correlation functions?
- **Exercise:** In this chapter all signals were assumed to be continuous functions of the independent variable, the time t in particular. In reality, the signal is often sampled, and available only at discrete times. Most of the analysis is readily generalized to discretized representations, with a wealth of literature being available on the subject.

  Demonstrate that a non-periodic signal which is sampled in time at regular intervals Δt, has a power spectrum which is continuous, with a periodicity $f_s = 1/\Delta t$, with f_s being the *sampling frequency*. Demonstrate that a periodic signal, with period $\mathcal{T}$, which is sampled in time at regular intervals Δt, has a power spectrum which is periodic with $f_s = 1/\Delta t$, and discrete with a frequency sampling interval of $1/\mathcal{T}$. It is implicit that $\mathcal{T} = N\Delta t$, with N being an integer.

An apparent paradox

Another paradox is concerned with the possible cases where the correlation function assumes a local zero value at a certain time-delay, say τ_*. This can happen when the basic pulse in (20.1) assumes positive as well as negative values, see (20.35). This will also remain so in the Gaussian limit where the density of pulses goes to infinity. Now this can appear puzzling; two uncorrelated events are also independent in the Gaussian limit. This means that for correlations of the particular sort mentioned here we start out for small time delays to have statistical *dependence*. For a certain time delay, τ_*, the correlation vanishes and the events are statistically independent.

Then, for time delays larger than τ_*, they become dependent again! This is however not really a paradox, as for any finite pulse density μ no matter how large, there will be small terms of importance in the series expansion in the exponents in the expression for the probability density also at the time delay where the correlation vanishes. These terms will ensure that the events with time separation τ_* retain their statistical dependence.

20.3.1 Cross-correlation functions

Assume that two different events, described by their respective time series $\Phi(t)$ and $\Psi(t)$, have a common cause. A certain random event at time t_k is thus assumed to excite pulses $\phi(t - t_k)$ and $\psi(t - t_k)$ in two different records. By a straightforward generalization of the foregoing analysis it is readily demonstrated that the correlation between the two records $\Phi(t)$ and $\Psi(t)$ can be obtained as

$$\begin{aligned}\mathcal{R}_{\Phi\Psi} &= \langle\Phi(t)\Psi(t)\rangle \\ &= \mu\int_{-\infty}^{\infty}\phi(t)\psi(t)\,\mathrm{d}t + \left[\mu\int_{-\infty}^{\infty}\phi(t)\,\mathrm{d}t\right]\left[\mu\int_{-\infty}^{\infty}\psi(t)\,\mathrm{d}t\right],\end{aligned} \tag{20.36}$$

where the densities, μ, of structures ϕ and ψ is the same in both records by construction. The result (20.36) is due to (Rowland 1936). It contains in particular (20.35) as a special case since it can be argued that the relation must also be valid if we identify $\psi(t)$ with $\phi(t + \tau)$. The result (20.36) can be generalized to other cases as well. Note that $\int\psi(t)\mathrm{d}t = 0$ and $\int\phi(t)\ \mathrm{d}t = 0$ for *orthogonal pulses*, $\int\phi(t)\psi(t)\ \mathrm{d}t = 0$, implies that $\mathcal{R}_{\Phi\Psi} = 0$, i.e. the events are uncorrelated, but evidently not *independent* since they are placed synchronously in the two records.

Again we can delay one time record with respect to the other and obtain the straightforward and almost trivial generalization of (20.36)

$$R_{\Phi\Psi}(\tau) = \mu\int_{-\infty}^{\infty}\phi(t)\psi(t+\tau)\,\mathrm{d}t + \left[\mu\int_{-\infty}^{\infty}\phi(t)\,\mathrm{d}t\right]\left[\mu\int_{-\infty}^{\infty}\psi(t)\,\mathrm{d}t\right].$$

Even if the basic pulses are orthogonal, $R_{\Phi\Psi}(0) = 0$, we generally have $R_{\Phi\Psi}(\tau) \neq 0$ for *some* values of τ. At first sight this seems surprising, since the statement apparently also holds for statistically independent records. Note, however, that the two records discussed here are by construction related; every time a pulse Φ is placed in one record, a pulse Ψ is placed in the other! The two records are synchronized, so to speak.

For analytical estimates of conditional averages in section 19.2.4 we found use of expressions like $\langle\Phi^m(0)\Phi(t)\rangle$. For the present model these can be found by application of Rowland's theorem as

$$\begin{aligned}\langle\Phi^m(0)\Phi(t)\rangle = \mu\int_{-\infty}^{\infty}\phi^m(t)\psi(t+\tau)\,\mathrm{d}t + \left[\mu\int_{-\infty}^{\infty}\phi^m(t)\,\mathrm{d}t\right] \\ \times\left[\mu\int_{-\infty}^{\infty}\psi(t)\,\mathrm{d}t\right].\end{aligned} \tag{20.37}$$

- **Example:** The cross-correlation is bounded by the *cross-correlation inequality* (Bendat and Piersol 1986) given as $R^2_{\Phi\Psi}(\tau) = |R_{\Phi\Psi}(\tau)|^2 \leqslant R_{\Psi\Psi}(0)R_{\Phi\Phi}(0)$. We prove this inequality in general by recalling that an expectation value

  $$E((a\phi(t) + b\psi(t+\tau))^2) \geqslant 0.$$

  This statement is equivalent to

  $$a^2 R_{\Phi\Phi}(0) + b2R_{\Psi\Psi}(0) + 2abR_{\Phi\Psi}(\tau) \geqslant 0.$$

  Consequently, for any $b \neq 0$ we have

  $$\left(\frac{a}{b}\right)^2 R_{\Phi\Phi}(0) + R_{\Psi\Psi}(0) + 2\left(\frac{a}{b}\right)R_{\Phi\Psi}(\tau) \geqslant 0.$$

  This is a quadratic equation in the ratio a/b without real different roots, since the expression is non-negative. Therefore the discriminant of this equation in a/b is negative, i.e.

  $$4R^2_{\Phi\Psi}(\tau) - 4R_{\Phi\Phi}(0)R_{\Psi\Psi}(0) \leqslant 0,$$

  giving the desired proof. This discussion is quite similar to the proof of the Cauchy–Schwarz inequality (19.20).

  The same proof gives a similar cross-covariance inequality (Bendat and Piersol 1986).
- **Exercise:** Consider a one-dimensional system, where pulses propagate without change in shape along the x-axis, with statistically distributed velocities, with given probability density for the velocities. Determine the cross-correlation of the two signals obtained by sampling the system at two positions with known separation.

20.4 Spectral representation

An alternative presentation of the signal (20.1) can be given in the form

$$\Phi(t) = \frac{a_o}{2} + \sum_{n=1}^{K}\left[a_n \cos\left(\frac{2\pi nt}{\mathcal{T}}\right) + b_n \sin\left(\frac{2\pi nt}{\mathcal{T}}\right)\right]. \tag{20.38}$$

The correspondence with (20.1) is obtained by the Fourier expansion of (20.1) over the interval $\mathcal{T}$, for a given N

$$\begin{aligned} a_n - ib_n &= \frac{2}{\mathcal{T}}\sum_{j=1}^{N}\int_0^{\mathcal{T}} a\psi(t-t_j)\mathrm{e}^{-\mathrm{i}2\pi nt/\mathcal{T}}\,\mathrm{d}t \\ &= \left[\frac{2}{\mathcal{T}}\int_{-\infty}^{\infty} a\psi(t)\mathrm{e}^{-\mathrm{i}2\pi nt/\mathcal{T}}\,\mathrm{d}t\right]\sum_{j=1}^{N}\mathrm{e}^{-\mathrm{i}n\theta_j}, \end{aligned} \tag{20.39}$$

with $\theta_j = 2\pi t_j/\mathcal{T}$. Since t_j for $j = 1, 2, \ldots N$ is uniformly distributed over $\mathcal{T}$ by assumption, the θ_js are random variables distributed uniformly over the interval

$\{0, 2\pi\}$. We can see a complex number $e^{-in\theta_j}$ for a given j as a vector in a plane, reaching from the origin to the position $(\cos(n\theta_j), -\sin(n\theta_j))$. The summation in (20.39) can then be interpreted as a sum over N randomly oriented vectors with a length given by the term in square brackets in (20.39). The angular bracket in (20.39) is deterministic, the coefficient sum is randomly varying. When N becomes large, as it does for instance when the density $\mu \to \infty$ with $\mathcal{T}$ fixed, the real and imaginary parts of this sum are random variables, which tend to become independent and normally distributed about zero (Rice 1944, 1945, Papoulis 1991).

For any pulse density μ it can be shown that

$$\langle a_n \rangle = \langle b_n \rangle = 0, \quad n \neq 0, \tag{20.40}$$

$$\left\langle a_n^2 \right\rangle = \left\langle b_n^2 \right\rangle = \sigma_n^2 \equiv \frac{2\mu}{\mathcal{T}} \left| \int_{-\infty}^{\infty} a\psi(t) e^{-i2\pi nt/\mathcal{T}} \mathrm{d}t \right|^2, \tag{20.41}$$

$$\langle a_n b_m \rangle = \langle a_n a_m \rangle = \langle b_n b_m \rangle = 0, \quad n \neq m. \tag{20.42}$$

The integration limits in (20.41) as well as (20.39) can be set to infinity when $\mathcal{T}$ is much larger than any timescale characterizing the individual pulse $\psi(t)$. The first one of these results (20.40) is trivially obtained by noting that $\langle \sum_{j=1}^{N} e^{-in\theta_j} \rangle \approx 0$ for large $\mathcal{T}$ and some fixed density μ. The other two results follow by some simple algebra from (20.39). The expressions (20.41) for the Fourier coefficients is consistent with the Wiener–Khinchine theorem expressing the power spectrum associated with the ensemble of records given in terms of the Fourier transform of the correlation function (20.32). The results (20.42) show that in the limit $\mathcal{T} \to \infty$, the coefficients corresponding to different Fourier components are uncorrelated (Davenport and Root 1958, Nicholson 1983); for Gaussian signals they are consequently also statistically independent. This observation is usually expressed as the statement that for Gaussian signals the phases of spectral components are uniformly distributed over the ensemble and statistically independent.

The phases of the Fourier transform of individual or isolated structures present in the signal will have some fixed relations. When many structures are randomly distributed in the record the phases of the different Fourier components of the entire time series become uncorrelated in the limit of $\mathcal{T} \to \infty$. In this limit the number of pulses or structures in the record becomes large.

The power spectrum obtained by Fourier transform of the correlation function is not sensitive to internal phase relations among the spectral components in the individual pulses. The bispectrum obtained by Fourier transform of the triple-correlation function will reveal these phase relations and is therefore of considerable interest for data analysis (Hasselman *et al* 1963, Lii *et al* 1976, Kim and Powers 1979). Model examples are given by Pécseli and Trulsen (1993). It is often an advantage to deal with normalized auto-correlations and triple-correlations. The auto-correlation is readily normalized by its value, σ^2, at zero time-delay. This is not quite so easy for the triple-correlation, which may be vanishing at the origin. A proper normalization is for instance

$$\langle\Phi(t)\Phi(t+\tau_1)\Phi(t+\tau_2)\rangle\Big/\sqrt{\langle\Phi^2(t)\Phi^2(t+\tau_1)\rangle\langle\Phi^2(t)\rangle},$$

with the result confined to the interval $\{-1, 1\}$ by the Schwartz inequality. (Other possible normalizations need not have this property.) The normalizing quantities can be given analytical expressions by use of higher order correlation functions as

$$\mathcal{Q}(\tau_1, \tau_2, \tau_3) = \langle\Phi(t)\Phi(t+\tau_1)\Phi(t+\tau_2)\Phi(t+\tau_3)\rangle,$$

and the correlation $\mathcal{R}(\tau)$. Also the normalized triple-correlation function is vanishing in the Gaussian limit where $\mu \to \infty$, with $a \to 0$.

- **Example:** The log-normal distribution. Assume that we are dealing with a multiplicative process, where we have for instance

$$Y = X_1 X_2 \cdots X_N,$$

or

$$Z \equiv \ln Y = \ln X_1 + \ln X_2 + \cdots + \ln X_N,$$

assuming all $X_j > 0$.

If now all X_js are statistically independent, we might argue by the central limit theorem that under relatively mild restrictions the variable Z is approximately normally distributed for large N (Papoulis 1991). Since $y = \exp z$, we then have the probability density

$$P(y) = \frac{1}{y\sigma\sqrt{2\pi}} \exp\left(-\frac{1}{2\sigma^2}(\ln y - \eta)^2\right), \qquad (20.43)$$

where

$$\eta \equiv \sum_i^N \langle \ln x_i\rangle \qquad \text{and} \qquad \sigma^2 \equiv \sum_i^N \langle(\ln x_i - \eta)^2\rangle.$$

The distributions of the individual x_js need not be the same. See figure 20.4 for an illustration.

- **Exercise:** Consider an amplitude and phase modulated narrow frequency band signal as

$$s(t) = R(t)\cos(\omega_c t + \theta(t)).$$

An equivalent representation is

$$s(t) = x(t)\cos(\omega_c t) - y(t)\sin(\omega_c t).$$

Suppose that the joint probability density $P(x, y)$ is known. Derive an expression for the joint probability density function $P(R, \theta)$, and write the expression for the marginal distributions $P(R)$ and $P(\theta)$.

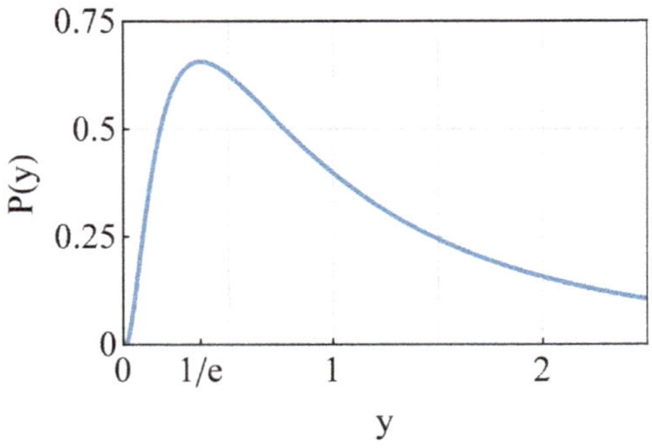

Figure 20.4. Illustration of a log-normal distribution, showing the normalized example $P(y) = \exp(-(\ln y)^2/2)/(y\sqrt{2\pi})$.

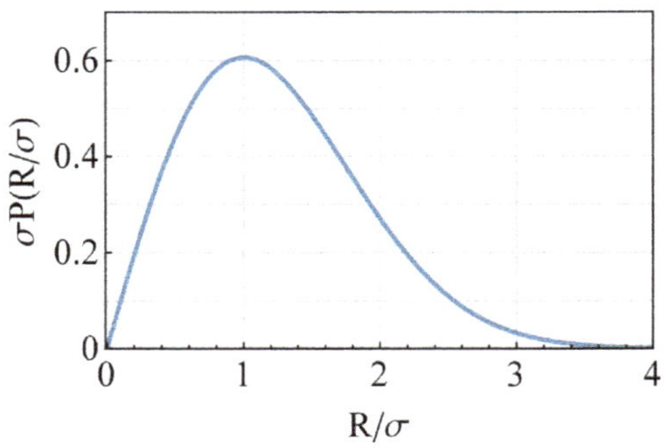

Figure 20.5. Probability density for the normalized Rayleigh distribution $\sigma P(R/\sigma)$, as given by (20.44).

Assume now that $s(t)$ is a normally distributed random signal

$$s(t) = \sum_{n=1}^{\infty} a_n \cos(n\omega t) + \sum_{n=1}^{\infty} b_n \sin(n\omega t),$$

where $\langle a_n b_m \rangle$ for all n, m, while $\langle a_n a_m \rangle = \langle b_n b_m \rangle = 0$ for $n \neq m$, and $\langle a_n^2 \rangle = \langle b_n^2 \rangle = \sigma^2$. Demonstrate that then also $x(t)$ and $y(t)$ are normally distributed. Given the foregoing results, demonstrate that $P(R)$ has a Rayleigh distribution

$$P(R) = \frac{R}{\sigma^2} e^{-\frac{1}{2}R^2/\sigma^2}, \tag{20.44}$$

for $R > 0$ and zero otherwise. Note that (20.44) is independent of ω_c. What is the distribution of θ in the expression for $s(t)$?

The Rayleigh distribution is shown in figure 20.5.

- **Exercise:** Demonstrate that the cumulative distribution of the Rayleigh distribution is $1 - \exp(-\frac{1}{2}R^2/\sigma^2)$ for $0 \leqslant R < \infty$ and vanishing for $R < 0$. Demonstrate that the mean value of the distribution is $\sigma\sqrt{\pi/2} \approx 1.25\,\sigma$, the variance is $\frac{1}{2}(4 - \pi)\sigma^2 \approx 0.43\,\sigma^2$, the skewness is $2\sqrt{\pi}(\pi - 3)/(4 - \pi)^{3/2} \approx 0.63$, the kurtosis is $(32 - 3\pi^2)/(4 - \pi)^2 \approx 3.25$. Show that the median value is $\sigma\sqrt{2\ln 2} \approx 1.18\,\sigma$ and the mode is σ.

- **Example:** By (20.1) we can describe a physical situation where two, or more, signal sources are present, each of them randomly emitting wave packets $\psi(t)$ the output being adequately represented by the expression (20.1). Assume for simplicity that the signal generators have the same statistical properties. Evidently, the resulting signal obtained by superposition of the outputs from the different sources will then have the same basic statistical properties as that of a single source, with the exception that the density μ has to be interpreted as $M\mu_s$, where M is the number of sources and μ_s is the density of pulses, or wave-packets, associated with one source, labeled by s. A basic consequence of this almost trivial observation is that, for instance, the power associated with the signal is proportional to the number of signal sources; five violins are five times as loud as one! (Sometimes it is experienced as being worse than that, but this is a subjective opinion without basis in the analysis.) This is actually not quite a trivial observation, as it is the amplitudes which are added by superposition of the signals, while the power is obtained by squaring the entire signal.

  The light from a group of statistically independent incoherent light sources can be discussed by entirely similar arguments. If the light sources have different spectra, the spectrum of their superimposed signal is readily obtained from the analysis as the weighted sum of the individual spectra.

- **Example:** In particular a signal with constant power spectrum, a *white noise* signal, can be realized by the model (20.1), by choosing the basic pulse to be a δ-function. In that case we readily find the correlation function by the convolution

$$R(\tau) = \mu a^2 \int \delta(t)\delta(t + \tau)\mathrm{d}t = \mu a^2 \delta(\tau),$$

  with a being an amplitude and μ the pulse density, as before. Fourier transform gives the constant power spectrum $G(\omega) = 2\mu a^2/\pi$. The total energy in this spectrum is infinite, as the correlation function is infinite at the origin, $\tau = 0$, implying that the mean square value of the signal diverges. Consistently, we find that $\int_0^\infty G(\omega)d\omega$ is also diverging.

20.5 Statistics of integrals

Consider a random signal ψ obtained by integration of another random signal $\phi(t)$. Assume also that the integration time can be subdivided into a large number N of shorter time intervals so that

$$\psi(T) = \int_0^T \phi(t)\mathrm{d}t = \int_0^T \phi(t)\mathrm{d}t + \int_T^{2T} \phi(t)\mathrm{d}t + \int_{2T}^{3T} \phi(t)\mathrm{d}t + \cdots \qquad (20.45)$$

We have $T = N\mathcal{T}$. We assume that the correlation time of $\phi(t)$ is finite. As T and then $\mathcal{T}$ increases, we expect the individual integrals in (20.45) become statistically independent when $\mathcal{T}$ is much larger than the correlation time. If this limit can be achieved it can be expected by the central limit theorem that $\psi(T)$ becomes Gaussian. The answer is not as obvious as might be expected, with more details given by for instance Tennekes and Lumley (1972). One special case should obviously be treated with care: assume that $\phi(t) = \mathrm{d}u(t)/\mathrm{d}t$. In that case $\psi(T) = u(T) - u(0)$, and ψ follows Gaussian statistics only if $u(t)$ does.

The ideas outlined here can be generalized. Take for instance a double integral

$$\psi(T) = \int_{-T}^{T}\int_{-t'}^{t'} \phi(t')\mathrm{d}t'\mathrm{d}t = 2T\int_{-T}^{T}\left(1 - \frac{|t|}{T}\right)\phi(t)\mathrm{d}t. \tag{20.46}$$

By arguments as before, it can be expected that $\psi(T)$ has Gaussian distribution under certain limiting conditions.

- **Exercise:** Demonstrate that if $\phi(t_1, t_2) = \phi(t_2, t_1)$ then we have

$$\int_0^t\int_0^t \phi(t_1, t_2)\mathrm{d}t_1\mathrm{d}t_2 = 2\int_0^t\int_0^{t_2} \phi(t_1, t_2)\mathrm{d}t_1\mathrm{d}t_2.$$

20.6 Consequences of finite record lengths

Given a finite set of data we are unable to obtain a probability density, or even an average: we can at best only obtain an *estimate* for a probability density or an average, this is not the same thing as the desired quantity. The actual estimates of, for instance, the correlations for a model as (20.1) from a given record are thus subject to uncertainties due to finite record lengths. The model discussed in this chapter also allows expressions to be obtained for these uncertainties (Bendat 1958, Pécseli and Trulsen 1993). The analysis can be generalized to other related problems with little effort.

20.6.1 Averages

For instance, assume that in the model we consider a case where the average value for the signal vanishes, $\langle\Phi(t)\rangle = \mu a\int_{-\infty}^{\infty}\psi(t)\,\mathrm{d}t = 0$. An actual estimate of this average, $\frac{1}{\mathcal{T}}\int_0^{\mathcal{T}}\psi(t)\,\mathrm{d}t$, based on a finite sample length, $\mathcal{T}$, will in general be different from zero. The estimate can itself be considered as a random variable and its standard deviation σ_{av} obtained as

$$\sigma_{av}^2 = \left\langle\left(\frac{1}{\mathcal{T}}\int_0^{\mathcal{T}}\Phi(t)\,\mathrm{d}t - \langle\Phi(t)\rangle\right)^2\right\rangle. \tag{20.47}$$

For the particular case $\int_{-\infty}^{\infty} \psi(t)\,\mathrm{d}t = 0$ this expression becomes

$$\sigma_{av}^2 = \frac{2}{\mathcal{T}^2}\int_0^{\mathcal{T}} (\mathcal{T} - \tau)\mathcal{R}(\tau)\mathrm{d}\tau, \qquad (20.48)$$

indicating that $\sigma_{av} \to 0$ as $\mathcal{T} \to \infty$, i.e. when $\mathcal{T}$ is much larger than the correlation time. On the other hand, for small $\mathcal{T}$ we have $\sigma_{av}^2 \approx R(0) = \langle \Phi^2 \rangle$.

20.6.2 Auto-correlation functions

The estimate $\frac{1}{\mathcal{T}}\int_0^{\mathcal{T}} \Phi(t)\Phi(t+\tau)\mathrm{d}t$ for the auto-correlation function is also randomly varying over the ensemble of many finite length records. The standard deviation for the error on this estimate is

$$\begin{aligned}\sigma_{\mathrm{corr}}^2(\tau) &= \left\langle \left(\frac{1}{\mathcal{T}}\int_0^{\mathcal{T}} \Phi(t)\Phi(t+\tau)\mathrm{d}t - \langle \Phi(t)\Phi(t+\tau)\rangle \right)^2 \right\rangle \\ &= \frac{2}{\mathcal{T}^2}\int_0^{\mathcal{T}} (\mathcal{T} - \nu)[\mathcal{Q}(\tau, \nu+\tau, \nu) - \mathcal{R}^2(\tau)]\mathrm{d}\nu,\end{aligned} \qquad (20.49)$$

where $\mathcal{Q}(\tau_1, \tau_2, \tau_3) = \langle \Phi(t)\Phi(t+\tau_1)\Phi(t+\tau_2)\Phi(t+\tau_3)\rangle$ is the quadruple correlation function. For the model given by (20.1) we obtain analytically

$$\begin{aligned}\mathcal{Q}(\tau_1, \tau_2, \tau_3) &= \mu a^4 \int_{-\infty}^{\infty} \psi(t)\psi(t+\tau_1)\psi(t+\tau_2)\psi(t+\tau_3)\mathrm{d}t \\ &+ \left\{ \mathcal{R}(\tau_1)\mathcal{R}(\tau_3 - \tau_2) + \mathcal{R}(\tau_2)\mathcal{R}(\tau_3 - \tau_1) + \mathcal{R}(\tau_3)\mathcal{R}(\tau_1 - \tau_2) \right\}.\end{aligned} \qquad (20.50)$$

We again explicitly used that $\langle \Phi(t) \rangle = 0$. The term $\mathcal{R}^2(\tau)$ cancels when (20.50) is inserted in (20.49). We note that $\sigma_{\mathrm{corr}}^2(\tau)$ contains $(\mu a^2)^2$ terms, originating from the second term in (20.50), and one varying with $(\mu a^2)a^2$, the latter vanishing in the Gaussian limit. The error on the mean-square estimate can be obtained as the special case with $\tau = 0$ in (20.49).

- **Exercise:** Demonstrate that

  $$\mathcal{Q}(\tau, \nu+\tau, \nu) = \mathcal{R}^2(\tau) + \mathcal{R}^2(\nu) + \mathcal{R}(\tau+\nu)\mathcal{R}(\tau-\nu),$$

  for a Gaussian random process described by its correlation function $\mathcal{R}(\tau)$.

- **Example:** Considering a Gaussian random process with the correlation function $R(\tau) = \exp(-b\tau)$ we can simplify the expression for $\mathcal{Q}$ in (20.50). It follows that

  $$\mathcal{Q}(\tau, \nu+\tau, \nu) - \mathcal{R}^2(\tau) = \mathrm{e}^{-2b|\nu|} + \mathrm{e}^{-b|\tau+\nu|}\mathrm{e}^{-b|\tau-\nu|}.$$

  Then (20.49) breaks up into two integrals

  $$\sigma_{\mathrm{corr}}^2(\tau) = \frac{2}{\mathcal{T}^2}\int_0^{\tau} (\mathcal{T} - \nu)(\mathrm{e}^{-b\nu} + \mathrm{e}^{-b\tau})\mathrm{d}\nu + \frac{4}{\mathcal{T}^2}\int_0^{\tau} (\mathcal{T} - \nu)\mathrm{e}^{-2b\nu}\mathrm{d}\nu.$$

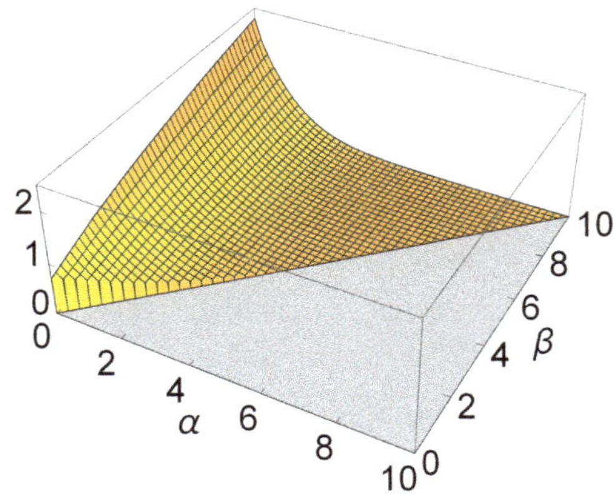

Figure 20.6. The 'signal-to-noise ratio' for an estimate of one particular correlation function, here given as the ratio $\exp(-\alpha)/\sigma_{\rm corr}(\alpha)$, see (20.51). The result is shown as a function of the normalized variables $\alpha = b\tau$ and the parameter $\beta = b\mathcal{T}$.

Introducing dimensionless variables $\alpha = b\tau$ and $\beta = b\mathcal{T}$, we find

$$\sigma^2_{\rm corr}(\alpha) = \frac{1}{2\beta^2}(2\beta - 1 + 2e^{-2\beta} + [(2\alpha + 1)(2\beta - 1) - 2\alpha^2]e^{-2\alpha}). \qquad (20.51)$$

This relation is exact for all α and $\beta > \alpha \geqslant 0$, that is for $\mathcal{T} > \tau \geqslant 0$. For a fixed value of β, the variance $\sigma_{\rm corr}$ indicates an uncertainty in the knowledge of $R(\tau)$ for all τ. For $\alpha = 0$, the result (20.51) reduces to

$$\sigma^2_{\rm corr}(0) = \frac{2\beta - 1 + e^{-2\beta}}{\beta^2}, \qquad (20.52)$$

giving $\sigma^2_{\rm corr}(0) \approx 2/\beta$ for large β. The 'signal-to-noise ratio' can be expressed as $R(\tau)/\sigma_{\rm corr}(\tau)$, which in the present case is $\exp(-\alpha)/\sigma_{\rm corr}(\alpha)$. This function is shown in figure 20.6. For large β we find using (20.51), that $\exp(-\alpha)/\sigma_{\rm corr}(\alpha) \approx \exp(-\alpha)\sqrt{\beta/k}$, with k being a number in the range $\{1; 2\}$ (Bendat 1958). For $\alpha = 0$, in particular, we find a $\sqrt{\beta}$ dependence as expected. The signal-to-noise as defined here should be large for reliable estimates, and figure 20.6 shows that this requires long, perhaps surprisingly long, time-series in particular if $\tau > 0$.

Although this example refers to some particular conditions, it serves as a useful illustration for more general cases.

20.6.3 Triple-correlation functions

The estimate for a triple-correlation function derived from a time series of finite length will deviate from the exact analytical result. An expression of the standard deviation on the error in such an estimate of the triple-correlation function can be obtained as

$$\begin{aligned}
&\sigma_{trp}^2(\tau_1, \tau_2) \\
&= \left\langle \left(\frac{1}{\mathcal{T}} \int_0^{\mathcal{T}} \Phi(t)\Phi(t+\tau_1)\Phi(t+\tau_2)\mathrm{d}t - \langle \Phi(t)\Phi(t+\tau_1)\Phi(t+\tau_2) \rangle \right)^2 \right\rangle \\
&= \frac{2}{\mathcal{T}^2} \int_0^{\mathcal{T}} (\mathcal{T} - \nu)\left[\mathcal{X}(\tau_1, \tau_2, \nu, \nu+\tau_1, \nu+\tau_2) - \mathcal{S}^2(\tau_1, \tau_2) \right] \mathrm{d}\nu,
\end{aligned} \tag{20.53}$$

expressed in terms of the sixth order correlation function

$$\begin{aligned}
&\mathcal{X}(\tau_1, \tau_2, \tau_3, \tau_4, \tau_5) \\
&= \mu a^6 \int_{-\infty}^{\infty} \psi(t)\psi(t+\tau_1)\psi(t+\tau_2)\psi(t+\tau_3)\psi(t+\tau_4)\psi(t+\tau_5)\mathrm{d}t \\
&\quad + \sum_{\mathrm{perm}(h,i,j,k,l)}^{15} \mathcal{R}(\tau_h)\mathcal{R}(\tau_i - \tau_j)\mathcal{R}(\tau_k - \tau_l) \\
&\quad + \sum_{\mathrm{perm}(h,i,j,k,l)}^{15} \mathcal{R}(\tau_h)\, \mathcal{Q}(\tau_i - \tau_j, \tau_k - \tau_j, \tau_l - \tau_j) \\
&\quad + \sum_{\mathrm{perm}(h,i,j,k,l)}^{15} \mathcal{S}(\tau_h, \tau_i)\, \mathcal{S}(\tau_j - \tau_l, \tau_k - \tau_l).
\end{aligned} \tag{20.54}$$

The summations are to be extended over all permutations of the τ indices, without double counting, just as with the last term in (20.50). Insertion of (20.54) in (20.53) results in a cancellation of the $\mathcal{S}^2(\tau_1, \tau_2)$ term. A term proportional to $(\mu a^2)^3$ appears in the error estimate $\sigma_{trp}^2(\tau_1, \tau_2)$. This term, which originates from the second term in (20.54), will remain in the Gaussian limit. In addition we find error contributions varying as $(\mu a^2)^2 a^2$ originating from the two last terms in (20.54).

20.6.4 Cross-correlation functions

The consequences of finite record lengths is analyzed for cross-correlations as well. The result is particularly interesting when two statistically independent records are analyzed. Ideally, the correlation should vanish, but in practice it will assume some non-zero value due to the finite length of the data sequence. Actually, it is rather difficult to find two records which give a vanishing cross-correlation, e.g. the birth-rate of babies and the population of storks in Denmark are found to be correlated. You can have a lot of fun with this and similar correlations!

The moral of the analysis in this section is that uncertainties on the estimation for a certain statistical moment, i.e. a correlation function of some order, depends on higher order correlations than the one investigated. It is not logically possible experimentally to determine these uncertainties; to express σ_{av}^2 we need $R(\tau)$ which can be measured only with some uncertainty, σ_{corr}^2, which can only be obtained by knowledge of $\mathcal{Q}(\tau_1, \tau_2, \tau_3)$, which is again uncertain, etc. From an experimental point of view, the uncertainties σ_{av}^2, σ_{corr}^2, etc are themselves also statistically uncertain quantities! Analytically, they can be obtained only by assuming certain models for the probabilities. These models build on some additional knowledge (or presumed knowledge) concerning the processes which may be supported, but never proven, by measurements carried out on the system.

20.6.5 Summary

The construction of synthetic data as discussed in this chapter is neat and interesting by representing solvable models; we do not experience too many cases like that! The idea of generating time series with known statistical properties can be carried one step further: if we purport to understand a process to all conceivable detail, we should then be able to produce an artificial time series with properties indistinguishable (in a statistical sense at least) from the original series. We can formulate our physical hypothesis and produce the time series with the postulated properties. We can then analyze this artificial series and compare with results from the true data and try to find differences. This sounds like a rather ambitious program, and indeed it is, but there are cases where solid insight has been gained by such a procedure (Kofoed-Hansen *et al* 1989).

In many ways the conclusions of the reference model described in the present chapter are self-evident, but a few comments might be in order. Basically, we can distinguish two cases: one where the fluctuation statistics are Gaussian, or so close to being Gaussian that we are unable to make a distinction on the basis of the available data-set. In that case all the available statistical information is exhausted by the correlation function. Alternatively, we can be dealing with *non*-Gaussian phenomena, and then there are no ends to what we can do.

After these introductory remarks, it may seem to be a fair question to ask where to begin, when faced with a given time record, or a sample of a spatial variation. The most obvious place to start, so it seems, is to obtain an estimate for the amplitude probability density of the signal. Note that we write *estimate*, because we will *never* be able to get the probability density, only a quantity which we hope will approximate it if the signal sequence is long enough. If the probability density seems to be significantly different from a Gaussian, then there is an obvious basis for detailed studies, involving higher order cumulants etc. If it *is* close to a Gaussian, there is still basis for detailed studies, but it is no longer obvious! In the latter case we have to consider two-point statistics, to see whether *these* are also bivariate Gaussians, if they are, then continue with three-point statistics, etc, but most likely we find out at this stage that the data sequence is not long enough to make a reliable estimate of higher order statistics. If we end up with the suspicion that the signal *is* a Gaussian random signal, all the available information is to be obtained by the correlation function, which we probably would like to estimate anyhow. Doing this, one should determine the necessary or desired accuracy. It is often a good idea to divide the signal into, say, three or more portions of equal length, and use each of these to get an estimate for the correlation function, and then take the average of these. This costs frequency resolution, but gives a better estimate for the points actually determined (Bendat and Piersol 1986).

20.7 A practical application: model analysis by synthetic data

As an illustration of the practical applicability of synthetic data we study the excess statistics outlined in section 19.3.2. The result (19.86) and its transition to the Gaussian limit (19.87) was studied by use of synthetic data (Sato *et al* 2012). Some results from this analysis can be illustrative and are summarized here. The data were constructed by random superposition of triangular double-pulses with zero average, the model being a special realization of (20.1), see figure 20.7. When generating the data $a^2\mu$ was kept constant as μ increased, so that the RMS signal was constant. For small densities of basic structures (meaning that only few structures overlap, on average) the signal is significantly non-Gaussian as evidenced by the signal amplitude probability densities. Figure 20.8 shows examples of probability densities (PDFs) for selected signals. The change from a non-Gaussian PDF for small μ to almost Gaussian distributions for large μ is clearly seen. The PDF for the strongly

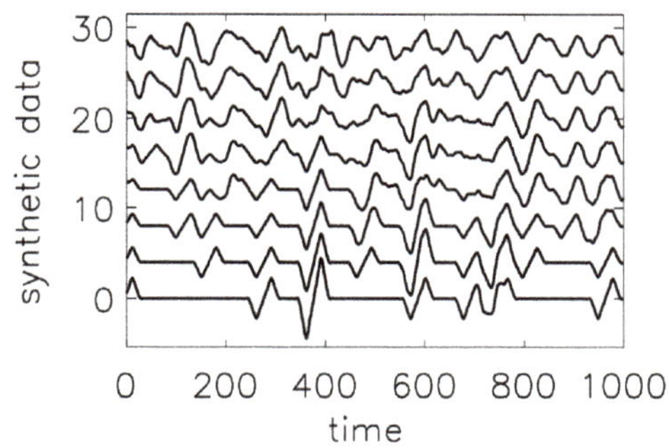

Figure 20.7. Illustration of a realization of synthetic data-sets. The density of structures increases from bottom to top, where the horizontal axes for each of the records are displaced for clarity. The average of the signals is vanishing in all cases. The densities are $\mu = 2, 2.5, \ldots, 5.5$ pulses per length unit of the basic pulse, here 55 computational units. The elementary pulse $\psi(t)$ in (20.1) is here a 'rounded off' triangular pulse that can be identified in the lowest trace, corresponding to $\mu = 2$. The figure is reproduced from Sato *et al* (2012) with permission from AGU.

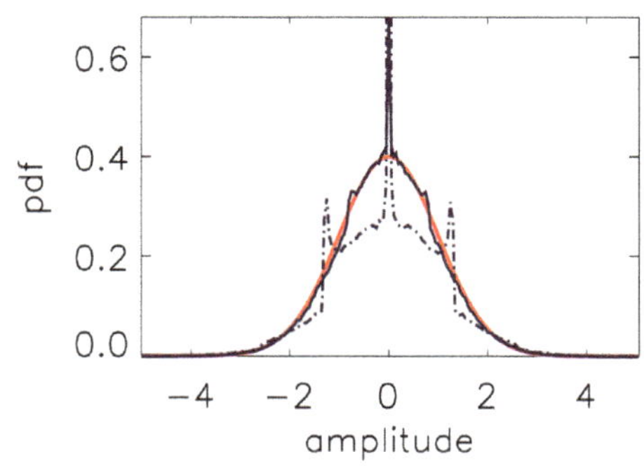

Figure 20.8. Examples of probability densities for two values of the pulse density, $\mu = 3$ and 5.5. The low density case shown with dashed lines is evidently far from a Gaussian, while the example for large μ (full line) is close to the Gaussian limit as given by a red curve. The figure is reproduced from Sato *et al* (2012) with permission from AGU. Note the 'spiky' contribution at the origin.

non-Gaussian case (pulse density $\mu = 2$) contains a δ-function that contributes significantly to the normalization. This δ-function is due to time intervals void of pulses: these are frequent for small μ, but a remnant can be found even for enhanced pulse densities. There are also two 'spikes' in the distribution corresponding to small μ: these originate from the flat tops of individual pulses that can still be recognized for small μ. For large pulse densities, these spikes disappear.

Note that the results from section 20.2 in principle give the analytical expression for the amplitude probability density of the signal. The expressions contain, however, some integrals in the exponential and all-in-all the results are very difficult to evaluate, even numerically, for general cases. The few analytical solutions that were presented in section 20.2 used very simple basic pulses in (20.1). One reason for these difficulties is the δ-function contributions mentioned before. In practice, it will be much simpler to make a synthetic data-set and then numerically obtain the required PDFs. This was done for obtaining figure 20.8. The practical value (except for the few analytically solvable simple cases) of the results in section 20.2 rests on the proof for the transition to Gaussian PDFs in the limit of large pulse-densities μ.

Using synthetic data-sets as those shown in figure 20.7 we are now in a position to study the expression (19.86) for Gaussian as well as non-Gaussian random signals. In figure 20.9 we show the variation of the average value for excess time intervals with varying threshold levels. For small μ we find clear deviations from the analytical results (19.87) based on the analysis of Kristensen *et al* (1991). In particular we also find an example demonstrating that the variation of the average excess time need not be a monotonically decreasing function for increasing threshold values! For large μ we find, on the other hand, a very good agreement, although as mentioned, (19.87) is not an exact result. In the present case we made the synthetic data-set so long that finite record effects were of negligible consequence. Examples illustrating such effects were shown by Pécseli and Trulsen (1993).

- **Exercise:** It is left to the reader as an exercise to explain why the average excess time need not be a monotonically decreasing function for increasing threshold values!

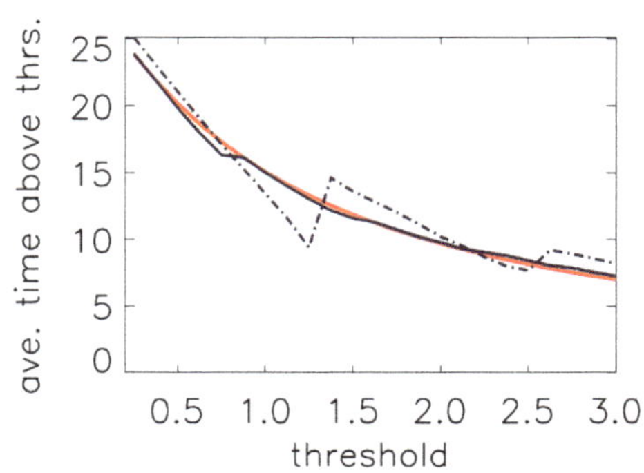

Figure 20.9. The average value for excess time intervals shown for varying levels of the threshold. Pulse densities are $\mu = 3$ and 5.5 shown with dash-dotted and solid lines, respectively. The full red line gives the result (19.87). The figure is reproduced from Sato *et al* (2012) with permission from AGU.

20.8 Signal modeling

On instrumented rockets, the fluctuations in electric fields are usually detected by measuring the difference signal between two spatially separated probe tips. By this method we essentially measure the force on an electron, and ideally the measurement can give the scalar as well as the vector potential associated with the electric field. For simplicity, we will in the following assume that the ionospheric fluctuations in question are electrostatic, and the measurements to be interpreted in terms of the electrostatic potential. The problem arises when a characteristic scale length of the plasma fluctuations becomes comparable to or smaller than the probe separation. This problem will be given particular attention.

To describe the relation between an actually measured probe-difference signal and the electrostatic fluctuations associated with ionospheric waves, we use first a simple model signal consisting of a superposition of identical electrostatic pulses with *a priori* known shapes $\psi(x)$, propagating in one spatial dimension, all with the same velocity V. With all pulses retaining their initial shape, we call this the frozen signal approximation. For simplicity we assume the pulses to have zero mean, $\int \psi(x)\,\mathrm{d}x = 0$. As they pass a detecting probe they are assumed to give rise to a temporally varying signal $\psi(Vt)$. The entire time record with assumed duration $\mathcal{T}$ is then

$$\Psi(t) = \sum_{j=1}^{N} \psi\big(x_j + Vt\big), \tag{20.55}$$

where x_j is the (random) initial position of the center of the actual structure. With $\psi(Vt)$ assumed known, the randomness in the signal arises solely due to the random distribution of x_j, here assumed uniform in space, and in the random number of pulses N entering a given record. Models of this type were analyzed in detail by Rice 1944), and were used to discuss ionospheric signals by, for instance, Pécseli and Trulsen (1993) and Pécseli *et al* (1993). Note that the signal is ill-posed when $V = 0$, since in this case the temporal signal has a constant contribution from those structures which happen to be detected by the probe and no contribution from the others, i.e. for $V = 0$ the signal has infinite energy.

From (20.55) we can by Rowland's theorem obtain the auto-correlation function for the potential $\mathcal{R}(\tau) = \mu \int_{-\infty}^{\infty} \psi(t)\psi(t+\tau)\mathrm{d}t$, where μ is the density of structures in the record, assuming the length of the record to be large. For simplicity, we omit the constant velocity V in the arguments.

However, with no suitable reference ground available, the correlation function $\mathcal{R}(\tau)$ for the potential is not directly accessible for measurements. What can be obtained is the potential difference $U(t) \equiv \Psi(t+s) - \Psi(t)$, with $s = \Delta/V$ being the time delay between the two signals due to the probe separation Δ, see also figure 20.10. It is again a standard problem to obtain the correlation function for the signal $U(t)$ with the result

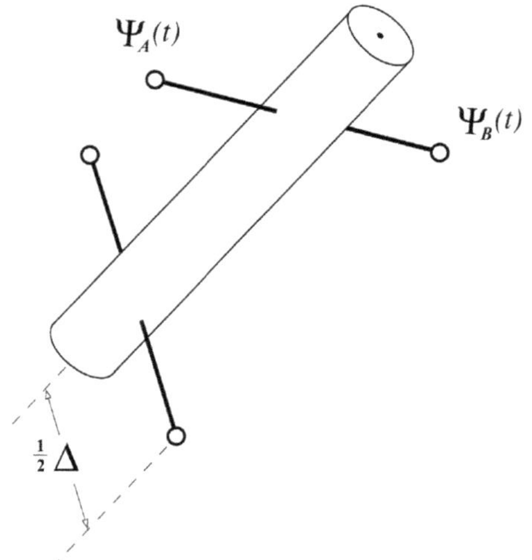

Figure 20.10. Schematic illustration of probe-pairs positioned on a rocket. Usually potential differences, such as $\Psi_A(t) - \Psi_B(t)$, are transmitted to ground. The probe positioning shown here is standard on many rocket flights (Rose *et al* 1992).

$$\begin{aligned} \mathcal{R}_U(\tau) &= \mu \int_{-\infty}^{\infty} [\psi(t) - \psi(t+s)] \times [\psi(t+\tau) - \psi(t+s+\tau)]\mathrm{d}t \\ &= 2\mathcal{R}(\tau) - \mathcal{R}(\tau+s) - \mathcal{R}(\tau-s), \end{aligned} \tag{20.56}$$

being expressed in terms of the potential auto-correlation. It is readily shown that if s is very small, we can approximate $\mathcal{R}_U(\tau) \approx -s^2\, \mathcal{R}''(\tau)$ which apart from a normalizing factor is the electric field auto-correlation. Assuming on the other hand that the pulses ψ have compact support, i.e. vanish for $|t| > t_c$ we readily find that $\mathcal{R}_U(\tau) \approx 2\mathcal{R}(\tau)$ for $|\tau| < t_c$ when $s > 2t_c$, since t_c is here also a representative correlation time. This limiting case corresponds to one where the spatial extent $2t_cV$ of the pulses $\psi(x)$ is smaller than the probe separation Δ. Here the correlation function of the detected signal, and thereby its power spectrum, is simply reflecting the properties of the potential fluctuations in the ionospheric plasma. In this limit, one probe is in a way acting as a reference for the other. The conditions are illustrated in figure 20.11 using a model correlation function $\mathcal{R}(t) = \exp(-(t/d_t)^2)$. For large separations, $s/d_t > 3$ we find a saturation, where we for small times have $\mathcal{R}_U(\tau) \approx 2\mathcal{R}(\tau)$. For $t \approx \pm s$ we recover the two last terms in (20.56), i.e. $-\mathcal{R}(\tau+s)$ and $-\mathcal{R}(\tau-s)$ which are here clearly separated from the rest.

Denoting the potential power spectrum by $S(\omega)$, as obtained by Fourier transforming $\mathcal{R}(\tau)$, we obtain trivially the corresponding power spectrum for the double probe signal as $S_U(\omega) = 4S(\omega)\sin^2(\omega s/2)$ by Fourier transform of (20.56). To obtain an estimate of the electric field spectrum $S_E(\omega)$, we divide $S_U(\omega)$ by the square of the probe separation, $\Delta = Vs$, and have

$$S_E(\omega) = S(\omega)\left(\frac{\omega}{V}\right)^2\left(\frac{\sin(\omega s/2)}{\omega s/2}\right)^2 \tag{20.57}$$

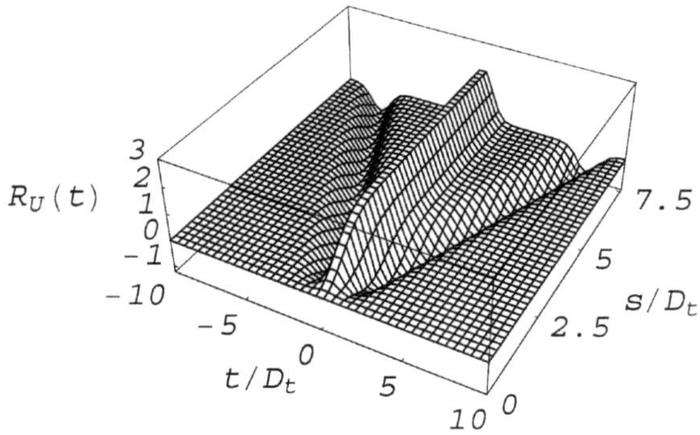

Figure 20.11. Model correlation function $\mathcal{R}_U(t)$ based on (20.56), with a model correlation function $\mathcal{R}(t) = \exp(-(t/d_t)^2)$.

containing the familiar sinc-squared function (Kelley and Mozer 1973, Krane *et al* 2000). Evidently, $S_E(\omega)$ measures the electric field spectrum only when ω s/2≪1. Interpreting the propagation velocity as a constant phase velocity, $\omega/k = V$, (after all this is the only consistent interpretation in the present context) we have $S_E(\omega) \to S(kV)k^2$ in the low frequency, long wavelength limit.

Assume now that the rocket is slowly spinning, with the consequence that the projection, Δ, of the probe separation vector on the direction of wave propagation, can be interpreted as varying on a slow time scale $\Delta = \Delta(t)$, allowing the spin phase to be considered constant during a time averaging. As a consequence we have also s being time varying, and with a constant spin period we have $s(t) = (\Delta/V)\sin(\Omega t)$ with Ω being the spin frequency. This feature can be used to determine the translation velocity V, provided the frozen signal approximation is fulfilled; by locating the frequency ω_n first ($n = 0$) or higher order ($n = 1, 2, \ldots$) zeroes of the sinc^2 function in (20.57) at $\sin(\Omega t) = 1$. This way we find e.g. $\omega_0 s/2 = \omega_0\Delta/2V\pi$, which determines V with given Δ. With $\omega_0/k_0 = V$ we may determine the wavenumber corresponding to ω_0. Note that there is no additional information in the higher order zeroes of the sinc-function; with the frozen signal approximation they all give the same estimate for V.

The observations of, for instance, Pfaff *et al* (1997) show a very clear spin modulation which is consistent with the assumption of structures propagating with only little change in shape, at least during the time it takes to pass from one probe-tip to another. In a general case, we must allow for the signal to be deformed at a faster rate and expect a smearing out of the signatures of the spin modulation. This effect will be included in a generalization of the analysis in the next section.

The analysis outlined here made reference to rocket measurements. The basic arguments apply equally well to several laboratory experiments using probe signal differences (Hamberger and Jancarik 1972). Many of the analytical results obtained before can be used directly for these cases as well.

20.9 Space–time varying signals

With the experience gained from the foregoing section, it is now straightforward to consider a space-time varying signal, which can simulate an actual physical process. In this case we assume that the structures are spatially varying, and at the same time growing and decaying in time, thus assuming that the structures are compact in time as well as spatial domains. A basic pulse is thus assumed to have a known spatial as well as temporal variation $\psi = \psi(\mathbf{r}, t)$, being sampled along the rocket trajectory. For simplicity, without loss of generality, we consider structures propagating in a simple space-time plane with some velocity, V. The positions in space as well as time, x_j, t_j, are randomly distributed as before. We can allow for the velocities V to be randomly distributed as well. The rocket trajectory with assumed constant velocity, V_R, is a straight line in this space-time diagram. The important modification by this generalized model is that the structures change as they propagate from one probe tip to the other, and this has consequences for the difference signal, as will be shown in the following.

The signal is assumed to be homogeneous in space. In the model, this is automatically achieved by the assumed statistically uniform distributions of structures in the space-time domain. The assumed spatial homogeneity implies that for instance the standard deviation as obtained by Campbell's theorem is independent of spatial positions, $\sigma^2 = \nu \int\int \psi^2(x, t)\mathrm{d}x\mathrm{d}t$, with the density ν having dimensions $(\textit{length} \times \textit{time})^{-1}$. The signal as such is varying in space and time, but when sampled by a moving rocket we obtain a signal varying with time only.

To obtain the correlation function associated with this signal, we first discuss the (ideal) space-time varying auto-correlation of the signal

$$\Psi(x, t) = \sum_{j=1}^{N} \psi(x_j + x, t_j + t). \tag{20.58}$$

The basic structures $\psi(x, t)$ positioned at (x_j, t_j) are here varying both in space and time and can be seen as one-dimensional 'blobs' with a finite spatial extent and a finite life time. Such a model has actually been found useful in modeling rocket data (Rinnert 1992).

By generalizing Rowland's theorem we obtain the auto-correlation associated with $\Psi(x, t)$ as

$$R(\xi, \tau) = \nu \int_{-\infty}^{\infty} \int_{-\infty}^{\infty} \psi(x + Vt, t)\psi(x + Vt + \xi, t + \tau)\mathrm{d}x\mathrm{d}t. \tag{20.59}$$

Referring to figure 20.10 we introduce the potential difference signal as

$$\Phi(x, t) = \sum_{j=1}^{N} \Big(\psi\big(x_j + x, t_j + t\big) - \psi\big(x_j + x + \Delta, t_j + t\big)\Big),$$

where we write the expression in one spatial dimension for simplicity. The signal $\Phi(x, t)$ represents what can actually be detected by the rocket. The auto-correlation function $\mathcal{R}_\Phi(\xi, \tau)$ for the process $\Phi(x, t)$ is expressed in terms of $R(\xi, \tau)$ as

$$\mathcal{R}_\Phi(\xi, \tau) = 2\, R(\xi, \tau) - R(\xi + \Delta, \tau) - R(\xi - \Delta, \tau).$$

The structures are all assumed to be moving with the same velocity V_R relative to the rocket. The signal transmitted to ground will be varying with time, where this variation is in part due to the time variation of the structures and in part due to the spatial variation sampled by the relative motion between structures and rocket.

A probe moving with velocity V_R and sampling the signal along a trajectory $\xi = \xi_0 + tV_R$ will ideally detect an auto-correlation function $R(\tau V_R, \tau)$ for large times where we can ignore the initial position ξ_0. For the special case with $V_R = 0$ this separation has to be retained. The case of propagating 'frozen' structures is a special limiting case. Note the distinction between the pulse propagation velocity V and the detector velocity V_R. As argued before, the auto-correlation function $R(\tau V_R, \tau)$ cannot be directly found since only potential differences from probes separated by Δ are available.

From (20.59) we obtain the correlation function for the difference signal obtained by use of two probes as in (20.56). We readily find

$$\mathcal{R}_{V_R}(\tau) = 2\,R(\tau V_R, \tau) - R(\tau V_R + \Delta, \tau) - R(\tau V_R - \Delta, \tau). \tag{20.60}$$

The time delay caused by the signal propagating from one probe to another enters only the first argument of R which refers to the spatial displacement, see (20.59).

Generally, we can assume the correlation function to have a finite correlation length ℓ_c and a correlation time t_c. When $\tau > t_c$ or $\tau V_R > \ell_c$ we have $\mathcal{R}(\tau V_R, \tau) \approx 0$. This implies that for large Δ the two last terms in (20.60) become negligible unless $\tau \approx \pm\Delta/V_R$.

To illustrate some basic properties of (20.60), we introduce $s \equiv \Delta/V_R$ and consider the simplified case where $R((\tau + s)V_R, \tau) = F((\tau + s)V_R)G(\tau)$. The power spectrum for the probe difference signal is obtained by Fourier transform of $\mathcal{R}_{V_R}(\tau) = G(\tau)(2F(\tau V_R) - F((\tau + s)V_R) - F((\tau - s)V_R))$ to give the convolution

$$S_E(\omega) = 4\int_{-\infty}^{\infty} \sin^2\left(\frac{\omega' s}{2}\right) S_F(\omega')S_G(\omega' - \omega)\mathrm{d}\omega', \tag{20.61}$$

where $S_F(\omega)$ is the spectrum obtained by Fourier transform of $F(\tau)$, and $S_G(\omega)$ is the similar power spectrum associated with $G(\tau)$. To simplify the notation we do not write V_R explicitly in the frequency variable ω/V_R. If we divide (20.60) by $\Delta^2 = (Vs)^2$ as in (20.56) we recover (20.57) in the limit of $G(\tau)$ being a constant, where we have $S_G(\omega' - \omega) \to \delta(\omega' - \omega)$.

- **Example:** As an illustrative model we consider a space-time varying correlation function of the form

$$R(x, t) = \exp\left(-\left(\frac{t}{D_t}\right)^2 - \left(\frac{x}{D_x}\right)^2\right)\cos\left(\frac{ax}{D_x}\right), \tag{20.62}$$

see figure 20.12. We use this model correlation function for illustrating $\mathcal{R}_{V_R}(t)$ from (20.60) with results shown in figure 20.13. Note the saturation observed for large values of the probe separation parameter $s \equiv \Delta/V$. In this limit, we have

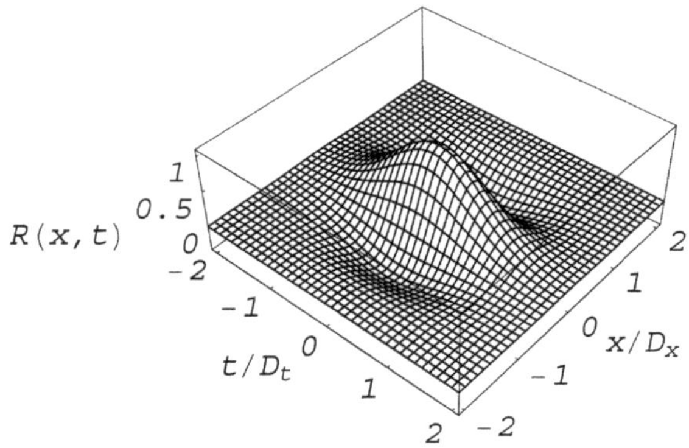

Figure 20.12. Space–time varying model correlation function $R(x, t)$ here shown for $a = 2.75$.

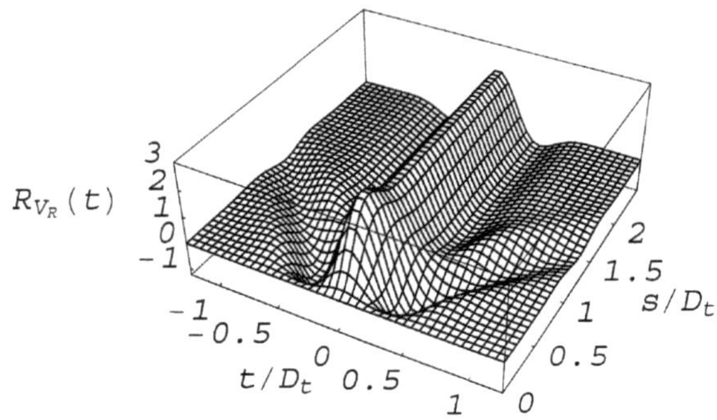

Figure 20.13. Model correlation function $\mathcal{R}_{V_R}(t)$ based on (20.62), here shown for $a/D_x = 2.75$, $V_R D_t = 2.25$. Note that $\mathcal{R}_{V_R}(t)$ no longer changes shape for $s/D_t > 1.75$.

$$\mathcal{R}_{V_R}(t) \approx 2\exp\left(-\left(\frac{t}{D_t}\right)^2 - \left(\frac{tV_R}{D_x}\right)^2\right)\cos\left(\frac{atV_R}{D_x}\right),$$

for the present model correlation function. The correlation time is here controlled by the smallest of the two quantities D_t or D_x/V_R.

- **Example:** Instruments on rockets and satellites usually detect fluctuations in plasma density. These signals can be considered as being obtained by 'point measurements'. Since no absolute ground is available, direct measurements of fluctuating potentials are not generally feasible and only potential differences between two probes can be obtained. When the two probes have a small separation, it can be argued that they to a good approximation give a component of the electric field corresponding to the direction given by the line connecting the two probes. For large separations the signal can be interpreted as being filtered, as best illustrated by considering a plane wave. Let the electrostatic potential be given as $\Psi(\mathbf{r}, t) = A\sin(\mathbf{k}_0 \cdot \mathbf{r} - \omega_0 t + \phi_0)$ where ϕ_0 is a phase. The two probes are separated by Δ and the difference signal divided by the probe separation becomes

$$
\begin{aligned}
\mathcal{S}_E(\mathbf{r}, t) &= \frac{k_\Delta}{\boldsymbol{\Delta} \cdot \mathbf{k}_0}(\Psi(\mathbf{r} + \boldsymbol{\Delta}, t) - \Psi(\mathbf{r}, t)) \\
&= Ak_\Delta \frac{\sin\left(\frac{1}{2}\boldsymbol{\Delta} \cdot \mathbf{k}_0\right)}{\frac{1}{2}\boldsymbol{\Delta} \cdot \mathbf{k}_0} \cos\left(\mathbf{k}_0 \cdot \mathbf{r} - \omega_0 t + \phi_0 + \frac{1}{2}\boldsymbol{\Delta} \cdot \mathbf{k}_0\right),
\end{aligned}
$$

where k_Δ is wavevector component along $\boldsymbol{\Delta}$, i.e. $k_\Delta \equiv \boldsymbol{\Delta} \cdot \mathbf{k}_0/|\Delta|$. For small $\boldsymbol{\Delta} \cdot \mathbf{k}_0$ we have $\sin(\frac{1}{2}\boldsymbol{\Delta} \cdot \mathbf{k}_0)/\frac{1}{2}\boldsymbol{\Delta} \cdot \mathbf{k}_0 \approx 1$ and $\mathcal{S}_E(\mathbf{r}, t)$ approximates the electric field derived from $\Psi(\mathbf{r}, t)$ when taken at the center-of-mass position, between the two detecting probes. For larger $\boldsymbol{\Delta} \cdot \mathbf{k}_0$ we can interpret $\sin(\frac{1}{2}\boldsymbol{\Delta} \cdot \mathbf{k}_0)/\frac{1}{2}\boldsymbol{\Delta} \cdot \mathbf{k}_0$ as a 'filter' that reduces the detected signal. If the rocket or satellite is spinning with some frequency Ω while detecting a plane wave with fixed phase velocity so that $\boldsymbol{\Delta} \cdot \mathbf{k}_0 = \Delta k_0 \sin(\Omega t)$, the filter characteristic will be observable (Kelley and Mozer 1973, Pfaff *et al* 1997).

For a wide spectrum of waves the situation is different. Using again a model in one spatial dimensional for simplicity we have

$$\Psi(x, t) = \sum_j a_j \phi(x - x_j, t - t_j),$$

where x_j, t_j is the random space-time position of a structure, here with randomly distributed amplitudes a_j. This model is standard and has been used many times. For a fully developed strong turbulence we can let some structures represent small scales and other structures intermediate or large scales. By Campbell's theorem and its extensions the correlation function of the signal $\Psi(x, t)$ can be determined analytically as $R_\Psi(\xi, \tau) = \nu\langle a^2\rangle \int\int \phi(x, t)\phi(x - \xi, t - \tau)\mathrm{d}x\mathrm{d}t$ where ν is the density of structures $\phi(x, t)$ in the space-time record, and ξ, τ represent spatial and temporal separations in the stationary and homogeneous space-time varying signal record. We denote the correlation length of $R_\Psi(\xi, \tau)$ by ℓ_{corr} and the correlation time by τ_{corr}.

For the difference signal we find the correlation function as when obtaining (20.56)

$$R_S(\xi, \tau) = 2R_\Psi(\xi, \tau) - R_\Psi(\xi + \Delta, \tau) - R_\Psi(\xi - \Delta, \tau).$$

We can divide $R_S(\xi, \tau)$ by Δ^2 and find for spatial separations $\Delta \ll \ell_{\mathrm{corr}}$ the result $R_S(\xi, \tau)/\Delta^2 \approx -\partial^2 R_\Psi(\xi, \tau)/\partial\xi^2$, i.e. the electric field correlation function, as expected.

We now introduce a relative velocity V_R between detector and structures. The signal sampled along this trajectory $\xi = V_R\tau$ will vary with time only. Basically, we find two cases as illustrated in figure 20.14: one is where $\Delta \ll \ell_{\mathrm{corr}}$ and we detect a temporal signal corresponding to the electric field,

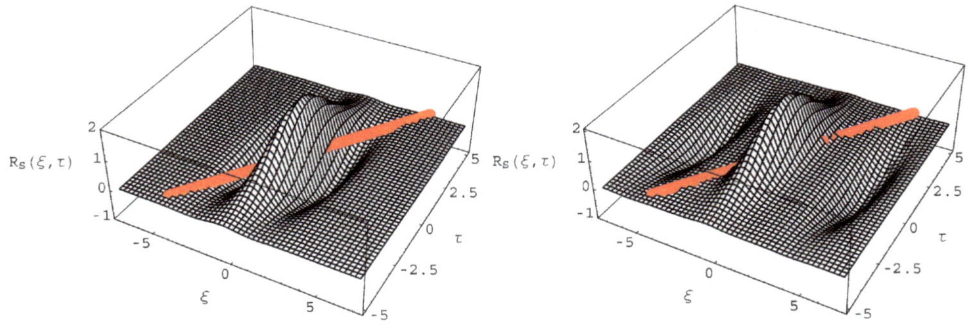

Figure 20.14. Two possibilities for sampling a signal along a trajectory $\xi = V_R\tau$ shown with a solid red line. In the first case (left with $\Delta \ll \ell_{\mathrm{corr}}$) we detect the correlation function for the electric field, in the other case (right with $\Delta \gg \ell_{\mathrm{corr}}$) we find the correlation function for the electrostatic potential. The figure uses an illustrative model with $R_\Psi(\xi, \tau) = \exp(-\xi^2/\ell_{\mathrm{corr}}^2 - \tau^2/\tau_{\mathrm{corr}}^2 + \zeta\xi\tau/(\ell_{\mathrm{corr}}\tau_{\mathrm{corr}}))$ with $\zeta = 0.75$.

and another limiting case $\Delta \gg \ell_{\mathrm{corr}}$ where the detector samples only the correlation function for the potential, provided the correlation time is smaller than Δ/V_R. In this latter limiting case one probe acts as a reference 'ground' for the other one (Krane *et al* 2000) and in this sense it is also possible to detect potential fluctuation by a rocket or satellite. For realistic conditions we have all scales represented at the same time, and the argument has to be rephrased: when the Taylor hypothesis is applicable we generally expect that when they are present, low frequency–long wavelength components represent electric fields, while high frequencies–short wavelengths will approximate the electrostatic potential variations in the power spectra.

The projection of the separation vector $\mathbf{\Delta}$ on the direction of propagation $\mathbf{V}_R$ can vary due to a rocket or satellite spin. In such cases it might be necessary to sample data in segments corresponding to a selected spin-phase in order to obtain spectra under uniform conditions (Krane *et al* 2000).

20.10 Models with non-overlapping structures

The model based on (20.1) assumes that any structure is placed in a record independently of all the others. It was outlined in section 20.1.2 how constraints on the positions can be introduced. Unfortunately, the general result is rather complicated and may not be useful for applications. Here we summarize a specific model, where structures or pulses are placed with no overlap (Bergsaker *et al* 2015). This model is also restrictive, but may have some use after all, and in particular it gives results that are easy to apply.

Consider first a simple two-level model where the signal assumes a constant positive value $\gamma > 0$ in a short time interval $\Delta\tau$ and is zero otherwise. The time variation of the event thus has a 'top-hat' shape. The random process is assumed to be stationary, and the probability for encountering a pulse at some position is the same for all times. At some fixed position we assume the probability for a pulse occurring in the time interval $\Delta\tau$ to be α with $0 \leqslant \alpha \leqslant 1$. It is essential for the following discussion that flux events do not overlap.

After averaging over different realizations, the probability density for the signal in this basic model is

$$P(\Phi) = (1 - \alpha)\delta(\Phi) + \alpha\delta(\Phi - \gamma), \tag{20.63}$$

where the first term accounts for the cases where the signal vanishes, corresponding to times where we are outside a pulse. The second term accounts for the signal encountered within a pulse.

For the case given by (20.63) we readily obtain $\langle\Phi^m\rangle = \alpha\gamma^m$ with $m = 1, 2, 3, \ldots$, giving the average value $\langle\Phi\rangle = \alpha\gamma$, the variance $\sigma^2 \equiv \langle(\Phi - \langle\Phi\rangle)^2\rangle = \alpha(1 - \alpha)\gamma^2$, the skewness $S = \alpha(1 - 3\alpha + 2\alpha^2)\gamma^3/\sigma^3$ and kurtosis $K = \alpha(1 - 4\alpha + 6\alpha^2 - 3\alpha^3)\gamma^4/\sigma^4$, implying here the exact relation $K = S^2 + 1$ for all α. For this particular signal we have $S = 0$ when $\alpha = 1/2$. For the simple two-level model discussed here it is not essential that all structures have the same duration (Mole and Clarke 1995, Sawford and Sullivan 1995, Schopflocher and Sullivan 2005), thus we can have $\Delta\tau$ to be statistically distributed. For generalizations of the model it is, however, an advantage to begin with $\Delta\tau$ being fixed.

For a long time record of duration $\mathcal{T}$, we have $\alpha = \mu\Delta\tau$ where $\mu \approx \langle N\rangle/\mathcal{T}$ is the number density of the appropriate flux-pulses in the record. Uncertainties arise due to end-effects, which can be made arbitrarily small by increasing $\mathcal{T}$. When μ is small, the distribution of the number N of pulses in a record can be approximated by a Poisson distribution, $P(N) = (\mu\mathcal{T})^N \exp(-\mu\mathcal{T})/N!$. More general distributions can be introduced where a minimum burst separation is ensured, i.e. a form where the positions of successive pulses are related as $x_{j+1} = x_j + \Delta\tau + X$, where $X > 0$ can be randomly distributed (Rice 1945, Pécseli 2000). The number of pulses or structures in any given record cannot exceed $N_m \equiv \mathcal{T}/\Delta\tau$, and it cannot be argued that a limiting case exists with $\mu \to \infty$, resulting in a Gaussian probability density, see section 20.2.3. The present model is thus inherently non-Gaussian for any α.

Generally it can be shown that $K \geqslant S^2 + 1$ holds for any probability distribution (Wilkins 1944), with the exception of PDFs with $\sigma = 0$, which are singular in the present context, see section 19.2.2. (As mentioned before, we use the definition where $K = 3$ for a Gaussian distribution.) The model PDF in (20.63) gives the equality sign, and it is the only model with this property. The simple analytical model (20.63) and its generalizations have been widely discussed in, for instance, studies of concentration fluctuations in turbulent environments (Mole and Clarke 1995, Yee and Chan 1997, Lewis and Chatwin 1997, Krane *et al* 2003, Mole *et al* 2008, Jørgensen *et al* 2010).

The model (20.63) can be generalized by allowing for a general temporal form of the burst event containing a random amplitude parameter a with some probability density $P_a(a)$, where in principle the amplitude can assume both signs even though the average is different from zero. Taking a as an amplitude factor we omit cases where the shape of the structure can depend on its magnitude, but the analysis can be extended to also include such cases. With a finite temporal duration $\Delta\tau$ of the flux event we exclude Gaussian or exponential pulses which in principle extend from $-\infty$ to ∞. We denote an acceptable, but otherwise general pulse form by $G_a(\tau)$ for $0 \leqslant \tau \leqslant \Delta\tau$ and $G_a(\tau) = 0$ otherwise, where we here take $\Delta\tau$ to be the same for all pulses, irrespective of a. One example could be $G_a(\tau) = a \sin^p(\pi\tau/\Delta\tau)$ for $0 \leqslant \tau \leqslant \Delta\tau$, where p is a deterministic fitting parameter. With this generalization we find the probability density of the signal to be

$$P(\Phi) = (1 - \alpha)\delta(\Phi) + \frac{\alpha}{\Delta\tau}\int_{-\infty}^{\infty}\int_0^{\Delta\tau} \delta(\Phi - G_a(\tau))\mathrm{d}\tau P_a(a)\mathrm{d}a. \tag{20.64}$$

For the 'top-hat' pulses assumed in e.g. (20.63) the amplitude probability density will be the same as $P_a(a)$, but in general the two PDFs are different. The general amplitude probability density can include pulses of both polarities.

The previous results for $\langle\Phi\rangle$, σ^2, S and K are now readily generalized (Jørgensen *et al* 2010, Sawford and Sullivan 1995). We find that, for instance, $\langle\Phi^m\rangle = (\alpha/\Delta\tau)\int_{-\infty}^{\infty}\int_0^{\Delta\tau} G_a^m(\tau)\mathrm{d}\tau P_a(a)\mathrm{d}a$. For the particular case emphasized here we have $G_a(\tau) = aG(\tau)$ and find the simple result $\langle\Phi^m\rangle = (\alpha/\Delta\tau)\int_0^{\Delta\tau} G^m(\tau)\mathrm{d}\tau$ with $\langle a^m\rangle = \int_{-\infty}^{\infty} a^m P_a(a)\mathrm{d}a$.

The expression (20.64) can be formulated differently, but it turns out that this form is the most convenient here. For completeness, we might add that (except for the top-hat signal) the distribution of amplitudes $P_a(a)$ is not simply related to the probability density of the signal amplitude.

With the model described here, a temporal signal will be composed of pulses with shape $G_a(\tau)$ placed at random positions (without overlap) and with randomly distributed amplitudes, and the ensemble average will have some value of the basic statistical averages, the skewness and kurtosis in particular, with values depending on the parameters used (Bergsaker *et al* 2015). For the special case where a enters only as an amplitude factor we can obtain some direct analytical results. We can use the general relation $\delta(bx) = \delta(x)/|b|$ in for instance (20.64) to give

$$P(\Phi) = (1 - \alpha)\delta(\Phi) + \frac{\alpha}{\Delta\tau}\int_0^{\Delta\tau} \frac{P_a\left(\dfrac{\Phi}{G(\tau)}\right)}{|G(\tau)|}\mathrm{d}\tau. \tag{20.65}$$

Taking a specific example $G_a(\tau) = a \sin^p(\pi\tau/\Delta\tau)$ and $G_a(\tau) = 0$ for $\tau < 0$ and $\tau > \Delta\tau$, we have

$$P(\Phi) = (1 - \alpha)\delta(\Phi) + \frac{\alpha}{\Delta\tau}\int_0^{\Delta\tau} \frac{P_a\left(\dfrac{\Phi}{\sin^p(\pi\tau/\Delta\tau)}\right)}{\sin^p(\pi\tau/\Delta\tau)}\mathrm{d}\tau. \tag{20.66}$$

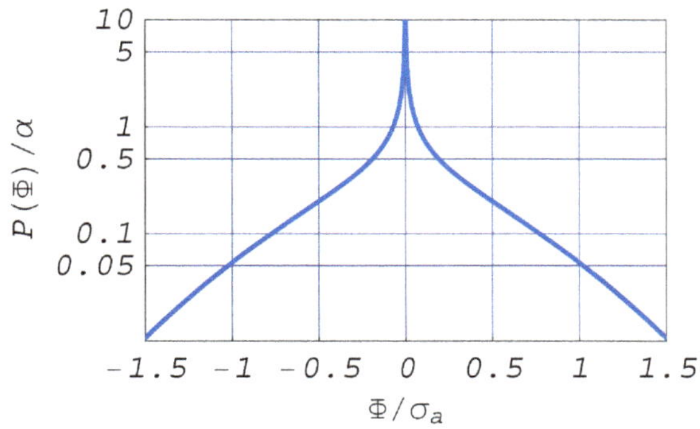

Figure 20.15. Example of a probability density derived from (20.66), assuming a distribution of pulses of the form $a\ \sin^p(\pi\tau/\Delta\tau)$, here with $p = 4$. The amplitude probability density is $P(a) = \exp(-a^2/\sigma_a^2)/\sqrt{\pi\sigma_a^2}$, where we for simplicity take $\langle a\rangle = 0$ here.

Since $P_a(a)$ is normalized, it must decrease faster than $1/a$ as $|a| \to \infty$. Except at $\Phi = 0$, there are thus no singularities associated with the limits where $\sin^p(\pi\tau/\Delta\tau) \to 0$.

Even with the present simplifying assumptions, the integral in the result (20.66) has to be solved numerically in most cases. Some cases can be solved analytically. If we, for instance, take $p = 2$ and the symmetric case $P_a(a) = \exp(-a^2)/\sqrt{\pi}$ we find (apart from the $\delta(\Phi)$-function contribution) $P(\Phi) = (\alpha/\Delta\tau)\exp(-\frac{1}{2}\Phi^2)\,K_{\frac{1}{4}}(\frac{1}{2}\Phi^2)/(\pi\sqrt{2})$ with a singularity for $\Phi = 0$, originating from the modified Bessel function of the second kind. In figure 20.15 we show an example for a solution of (20.66), based on a form $G_a(\tau) = a\ \sin^4(\pi\tau/\Delta\tau)$. A more general form will assume that the shape of a pulse or structure depends on the amplitude: such cases will probably have to rely on numerical solutions directly of (20.64).

In some studies of plasma turbulence (Labit *et al* 2007, Guszejnov *et al* 2013) much attention has been given to seemingly systematic relations between skewness and kurtosis of the form $K = aS^2 + b$, where $a \geqslant 1$ and $b \geqslant 1$ are some constants depending on experimental conditions. In meteorological studies such relation have been known for some time (Mole and Clarke 1995), as mentioned. The practical consequences of relations like these are probably limited, but a complete understanding of a signal requires that details like possible skewness–kurtosis relations can be accounted for.

It is interesting to note that the limits of low pulse densities discussed in section 20.2.2 reproduce probability densities as (20.63). For joint probability densities it might be that an expression like (20.27) can be used directly with relevance to the present section as well. Note that there is a non-trivial distinction here: the present section assumes finite support for the pulses (i.e. it is meaningful to distinguish being inside or outside a pulse) while the pulses assumed in section 20.2.2 have no such restriction.

References

Bekefi G 1966 *Radiation Processes in Plasmas* (New York: Wiley)

Bendat J S 1958 *Principles and Applications of Random Noise Theory* (New York: Wiley)

Bendat J S and Piersol A G 1986 *Random data, Analysis and Measurement Procedures* 2nd edn (New York: Wiley)

Bergsaker A S, Fredriksen Å, Pécseli H L and Trulsen J K 2015 Models for the probability densities of the turbulent plasma flux in magnetized plasmas *Phys. Scr.* **90** 108005

Campbell N 1909 The study of discontinuous phenomena *Proc. Phil. Soc.* **15** 117–36 and ibid. Discontinuities in light emission, 310-328.

Davenport W B and Root W L 1958 *An Introduction to the Theory of Random Signals and Noise* (New York: McGraw-Hill)

Garcia O E 2012 Stochastic modeling of intermittent scrape-off layer plasma fluctuations *Phys. Rev. Lett.* **108** 265001

Guszejnov D, Lazányi N, Bencze A and Zoletnik S 2013 On the effect of intermittency of turbulence on the parabolic relation between skewness and kurtosis in magnetized plasmas *Phys. Plasmas* **20** 112305

Hamberger S M and Jancarik J 1972 Experimental studies of electrostatic fluctuations in a turbulently heated plasma *Phys. Fluids* **15** 825–36

Hasselman K, Munk W and MacDonald G 1963 *Time Series Analysis* ed M Rosenblatt (New York: Wiley)

Jørgensen H E, Mikkelsen T and Pécseli H L 2010 Concentration fluctuations in smoke plumes released near the ground *Boundary-Layer Meteorol* **137** 345–72

Kelley M C and Mozer F S 1973 Electric field and plasma density oscillations due to the high-frequency Hall current two-stream instability in the Auroral E region *J. Geophys. Res.* **78** 2214–21

Kim Y C and Powers E J 1979 Digital bispectral analysis and its applications to nonlinear wave interactions *IEEE Trans. Plasma Sci.* **PS-7** 120–31

Kofoed-Hansen O, Pécseli H L and Trulsen J 1989 Coherent structures in numerically simulated plasma turbulence *Phys. Scripta* **40** 280–94

Krane B, Pécseli H L and Trulsen J 2003 Concentrations and concentration fluctuations in two-dimensional turbulence *Phys. Fluids* **15** 211–26

Krane B, Pécseli H L, Trulsen J and Primdahl F 2000 Spectral properties of low-frequency electrostatic waves in the ionospheric E region *J. Geophys. Res.* **105** 10585–601

Kristensen L, Casanova M, Courtney M S and Troen I 1991 In search of a gust definition *Boundary-Layer Meteorol* **55** 91–107

Labit B, Furno I, Fasoli A, Diallo A, Müller S H, Plyushchev G, Podestà M and Poli F M 2007 universal statistical properties of drift-interchange turbulence in TORPEX plasmas *Phys. Rev. Lett.* **98** 255002

Lewis D M and Chatwin P C 1997 A three-parameter PDF for the concentration of an atmospheric pollutant *J. Appl. Meteorol* **36** 1064–75

Lii K S, Rosenblatt M and Van Atta C 1976 Bispectral measurements in turbulence *J. Fluid Mech.* **77** 45–62

MacDonald D K C 1962 *Noise and Fluctuations: an Introduction* (New York: Wiley)

Mole N and Clarke E D 1995 Relationships between higher moments of concentration and of dose in turbulent dispersion *Boundary-Layer Meteorol* **73** 35–52

Mole N, Schopflocher T P and Sullivan P J 2008 High concentrations of a passive scalar in turbulent dispersion *J. Fluid Mech.* **604** 447–74

Nicholson D R 1983 *Introduction to Plasma Theory* (New York: Wiley)

Papoulis A 1991 *Probability, Random Variables and Stochastic Processes* 3rd edn (New York: McGraw-Hill)

Pécseli H L 2000 *Fluctuations in Physical Systems* (Cambridge: Cambridge University Press)

Pécseli H L and Trulsen J 1993 On the interpretation of experimental methods for investigating nonlinear wave phenomena *Plasma Phys. Contr. Fusion* **35** 1701–15

Pécseli H L and Trulsen J 1997 Eulerian and Lagrangian velocity correlation in two-dimensional random geostrophic flows *J. Fluid Mech.* **338** 249–76

Pécseli H L, Trulsen J, Primdahl F and Bahnsen A 1993 Propagation and nonlinear interaction of low-frequency electrostatic waves in the polar cap E region *J. Geophys. Res.* **98** 1603–12

Pfaff R F, Marionni P A and Swartz W E 1997 Wavevector observations of the two-stream instability in the daytime equatorial electrojet *Geophys. Res. Lett.* **24** 1663–6

Rice S O 1944 Mathematical analysis of random noise, I *Bell System Tech. J.* **23** 282–332
Rice S O 1954 *Selected Papers on Noise and Stochastic Processes* (New York: Dover)

Rice S O 1945 Mathematical analysis of random noise, II *Bell System Tech. J.* **24** 46–156
Rice S O 1954 *Selected Papers on Noise and Stochastic Processes* (New York: Dover)

Rinnert K 1992 Plasma waves observed in the Auroral E region ROSE campaign *J. Atmos. Terr. Phys.* **54** 683–92

Rose G *et al* 1992 The ROSE project — scientific objectives and discussion of I results', J *Atmos. Terr. Phys.* **54** 657–67

Rowland E N 1936 The theory of the mean square variation of a function formed by adding known functions with random phases, and applications to the theories of the shot effect and of light *Math. Proc. Cambridge Phil. Soc.* **32** 580–97

Sato H, Pécseli H L and Trulsen J 2012 Fluctuations in the direction of propagation of intermittent low frequency ionospheric waves *J. Geophys. Res.* **117** A03329

Sawford B L and Sullivan P J 1995 A simple representation of a developing contaminant concentration field *J. Fluid Mech.* **289** 141–57

Schopflocher T P and Sullivan P J 2005 The relationship between skewness and kurtosis of a diffusing scalar *Boundary-Layer Meteorol* **115** 341–58

Tennekes H and Lumley J L 1972 *A First Course in Turbulence* (Cambridge, MA: The MIT Press)

Wandel C F and Kofoed-Hansen O 1962 On the Eulerian-Lagrangian transformation in the statistical theory of turbulence *J. Geophys. Res.* **67** 3089–93

Wilkins J E 1944 A note on skewness and kurtosis *Annals Math. Stat.* **15** 333–5

Yee E and Chan R 1997 A simple model for the probability density function of concentration fluctuations in atmospheric plumes *Atmos. Environ.* **31** 991–1002

IOP Publishing

Low Frequency Waves and Turbulence in Magnetized Laboratory Plasmas and in the Ionosphere

Hans Pécseli

Chapter 21

Random walk and classical diffusion

By classical diffusion in fluids, gases and plasmas we usually understand collisional transport, as opposed to anomalous or turbulent transport which is mediated by fluid motions on many different scales. As an introduction to the problem we here first consider collisional transport.

21.1 Diffusion as a random walk

A very simple, yet illustrative, model for describing diffusion processes is called the *random walk* model (Cox and Miller 1965, Feller 1968). In its simplest version it is given as a discrete grid in one, two or three dimensions, where an object is moving one step at discrete times, the direction of each step, and possibly also its magnitude, being randomly selected. For rather self-evident reasons the process is sometimes called the *drunkard's walk*.

The origin of the model seems well established. It was introduced by Karl Pearson in a letter to *Nature* in 1905:

A man starts from a point 0 and walks ℓ yards in a straight line: he then turns any angle and walks another ℓ yards in a straight line. He repeats this process n times. I require[1] *the probability that after these n stretches he is at a distance between r and $r + \delta r$ from his starting point 0.*

The solution was provided in the same issue of *Nature* by Lord Rayleigh, who informed him that he had solved the problem some 25 years earlier, when studying the superposition of sound waves of equal frequency and amplitude, but with random phases.

As an illustration it is simplest to consider the problem in one spatial dimension, i.e. a line with discrete and equally spaced positions. Assume that an object, say a

[1] He forgot to say 'please'.

pawn from a chess game, is placed at the origin and then moved one step at regular time intervals, where the direction of the step is decided by the flip of a coin. The process can be visualized in a presentation like the one in figure 21.1 with points in a plane spanned by a displacement-axis and a time-axis. Often, we find the notation 'time' referring to a *time-duration*, i.e. $s\tau$ in the present case where τ is the interval between steps, while the term 'epoch' (Feller 1968, 1971) is used for a *position* on the time axis, which for the present case will be labeled by s.

The present problem is formally identical to a classical game of coin-tossing; if tails comes up player A wins, if it is heads then player B is the lucky one. The equivalent of the displacement-axis is then one for accumulated gain (or loss), taking for instance the positive direction to be the one favoring player A. This latter formulation of the problem has proved to be more appealing to the imagination and is one often found in the literature (Feller 1968, Mosteller 1987). Some of the basic results concerning random walks have been summarized by Codling *et al* (2008) or Pécseli (2000).

A discrete random walk of a free particle can in a limiting case be modeled by a Fokker-Planck equation (Ming and Uhlenbeck 1954, van Kampen 1981, Pécseli 2000). It is instructive to see how this transition from a discrete to a continuous case is brought about. Imagine a particle which moves along the x-axis in such a way that in each step it can move either Δx to the left or Δx to the right, and it does so at regular time intervals, τ. The step-length is determined by the physical properties of the model, as discussed for instance in section 3.1.1. For a free particle we can assume that a step in either of the two directions is equally probable, with probability $\frac{1}{2}$. Consider $P(n\Delta x|m\Delta x;\, s\tau)$ which is the probability density that the particle is at $m\Delta x$ at time $s\tau$, if at the beginning it was at $n\Delta x$. To arrive at position m at time $s + 1$, the particle has to be *either* at position $m - 1$ *or* at position $m + 1$ at the previous time step s, and jump from there to position m, both cases being equally probable in the present model. The conditional probability density then satisfies the difference equation (Kac 1946)

$$P(n|m;\, s+1) = \frac{1}{2}P(n|m+1;\, s) + \frac{1}{2}P(n|m-1;\, s), \tag{21.1}$$

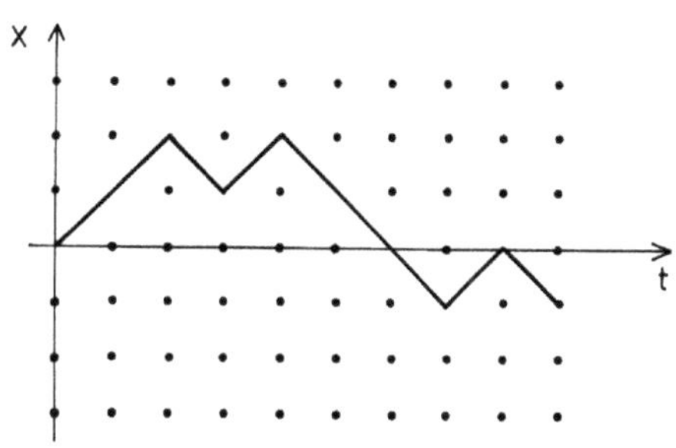

Figure 21.1. An illustration of space-time variation of a simple random walk. Points on the x-axis are labeled r, those on the time axis by n.

which, by adding and subtracting identical terms, can be rewritten in an equivalent form as

$$\begin{aligned}\frac{P(n\Delta x|m\Delta x;(s+1)\tau)-P(n\Delta x|m\Delta x;s\tau)}{\tau}\\ =\frac{\Delta x^2}{2\tau}\bigg(\frac{P(n\Delta x|(m+1)\Delta x;s\tau)+P(n\Delta x|(m-1)\Delta x;s\tau)}{\Delta x^2}\\ -\frac{2P(n\Delta x|m\Delta x;s\tau)}{\Delta x^2}\bigg).\end{aligned} \tag{21.2}$$

Suppose now that $\Delta x \to 0$ and $\tau \to 0$ in (21.2) in such a way that $\Delta x^2/2\tau = D =$ const. (having the consequence that $\Delta x/\tau \to \infty$). In this limit, with $n\Delta x \to x$ and $s\tau \to t$, we recover the diffusion equation from (21.2) as

$$\frac{\partial P}{\partial t} = D\frac{\partial^2 P}{\partial x^2}, \tag{21.3}$$

where D is a diffusion coefficient. The classical analysis assumes all step-lengths to be equal and the 'jumps' to be performed at regular intervals. In a more general analysis the jumps occur at random time intervals to be specified by some *waiting time distribution* (Balescu 1997).

The developments of the theoretical understanding of diffusion processes and the derivation of the diffusion equation is historically most interesting, see for instance the reviews by Narasimhan (1999) and Mazo (2002).

An important implication of the transition allowed to obtain this equation, is that the instantaneous particle velocity $\Delta x/\tau$ approaches infinity. This explains why the diffusion equation has a finite response at *any* position, no matter how far from the origin, at arbitrary short times, also for initial conditions with finite support. With the local particle velocity being infinite, we also find that the length of the path (not the net displacement) of a diffusing particle diverges in this diffusion limit.

The arguments outlined in this section contain two elements; first, time is made continuous by using infinitesimal time steps and, simultaneously, the spatial variable is also made continuous by using very small steps or jumps. The basic observation of the analysis outlined before is that a diffusion coefficient can be estimated by the ratio of the square of a characteristic step-length, divided by a characteristic time between steps. For classical collisional diffusion, with straight line orbits of length ℓ_c between collisions separated by a time interval $1/\nu_c$, with ν_c being the collision frequency, we find $D = \ell_c^2\nu_c$. In terms of a thermal speed $\sqrt{\kappa T/m}$, we have $D = \kappa T/(m\nu_c)$. This model might be applied to electrons colliding with a neutral background component, for instance.

For a magnetized plasma, we have that the step-length in the direction perpendicular to **B** is controlled by the Larmor radius, r_L. As an estimate we find $D_\perp = r_L^2\nu_c$, implying a classical scaling $D \sim 1/B^2$ with all other quantities constant, see also sections 3.1.1 and 3.5.1.

21.1.1 Random walk with persistence

The foregoing analysis explicitly assumed that in a one-dimensional model, the two directions of a random step are equally probable. It is possible to relax this assumption slightly, to obtain a somewhat more general model equation. We thus assume that the probability of a positive step, $+\Delta x$ is p and for $-\Delta x$ it is $1 - p$, where of course, the forgoing model is obtained by taking $p = \frac{1}{2}$. We still have the assumption that all steps are mutually independent. We now have, as a generalization of (21.1),

$$P(n|m; s + 1) = (1 - p)P(n|m + 1; s) + p\, P(n|m - 1; s). \tag{21.4}$$

With arguments like those giving (21.2), we here find a generalized diffusion equation ('advection diffusion equation' (Balescu 1997)) of the form

$$\frac{\partial P}{\partial t} = D\frac{\partial^2 P}{\partial x^2} - U\frac{\partial P}{\partial x}, \tag{21.5}$$

where we readily identify a *drift term*, with velocity U, and a *diffusion term*, with diffusion coefficient D. We find a drift velocity $U = \lim(2p - 1)\Delta x/\tau$ for $\Delta x \to 0$ and $\tau \to 0$, while we have $D = \lim\, p(1 - p)\Delta x^2/2\tau$. Both limits exist if we take

$$\Delta x = \sqrt{2D\tau}, \qquad \text{and} \qquad p = \frac{1}{2}\left(1 + \frac{U\tau}{\Delta x}\right) = \frac{1}{2}\left(1 + \sqrt{\frac{U^2\tau}{2D}}\right).$$

The important point concerning this relation is that for small τ, we have $\Delta x = \mathcal{O}(\sqrt{\tau})$, and that $p = \frac{1}{2} + \mathcal{O}(\sqrt{\tau})$. The displacement Δx must be of a considerably larger order of magnitude than the small time interval τ in which the step occurs, and p cannot deviate too much from $\frac{1}{2}$, if we are to avoid degeneracies (Cox and Miller 1965).

21.1.2 Forward and backward equations

The analysis leading to the diffusion equation in its two versions (21.3) and (21.5) were based on a 'forward' problem, one where we at a given epoch look forward in time to the next step. Alternatively, we could have looked *backward* in time, and asked how we reached that particular initial condition. The discussion will be made more clear when considering some illustrative special cases.

Simple birth process

First we consider a simple 'birth process', where the transition from a state E_i is possible only to state E_{i+1}. Thus, if we are a state E_n at an epoch t, and analyze what might happen in a short time interval h, we might either have a transition to E_{n+1} with an assumed probability $\lambda_n h$, *or* that nothing happens and we remain in the state E_n with probability $(1 - \lambda_n h)$. The probability for a transition *might* depend on the actual state, so we retain the index n on λ. We just *might* imagine that we had *two* or more transitions in the interval h, but we assume this probability to be negligibly small when h is small. To be in a state E_n at epoch $t + h$, we thus have to be in E_n at t and nothing happens, *or* be in state E_{n-1} and then experience a transition. The probability for being in a state labeled n is consequently

$$P_n(t + h) = P_n(t)(1 - \lambda_n h) + P_{n-1}(t)\lambda_{n-1}h. \tag{21.6}$$

With $\lim_{h\to 0}(P_n(t + h) - P_n)/h \to \mathrm{d}P_n/\mathrm{d}t$ we have the *forward equation*

$$\frac{\mathrm{d}}{\mathrm{d}t}P_n(t) = -\lambda_n P_n(t) + P_{n-1}(t)\lambda_{n-1},$$

with

$$\frac{\mathrm{d}}{\mathrm{d}t}P_0(t) = -\lambda_0 P_0(t).$$

We implicitly assumed that the system was at a state E_0 at time $t = 0$, but by interpreting P as transition probability, from state E_0 to E_n, we can write more generally $P_{i,n}$ with $n \geqslant i$ as the transition probability from any state E_i to E_n

$$\frac{\mathrm{d}}{\mathrm{d}t}P_{i,n}(t) = -\lambda_n P_{i,n}(t) + P_{i,n-1}(t)\lambda_{n-1}, \tag{21.7}$$

with condition $P_{i,i}(0) = 1$ and $P_{i,n}(0) = 0$.

We can now take a somewhat different approach: rather than extending the time interval t forwards, we might ask *how* we arrived at the initial state i, and extend the time interval *backwards*, but still with h. From an assumed state E_i at time $-h$ we can arrive at E_n at time t by having a transition to E_{i+1} in the interval h, again with probability $\lambda_i h$, and then have a transition from E_{i+1} to E_n in the time interval $\{0, t\}$, *or* we can have no transition in $\{-h, 0\}$, with probability $(1 - \lambda_i h)$, and then a transition from E_i to E_n in the time interval $\{0, t\}$. The important point is that the duration of the entire time interval is $h + t$ just as before. The probability of two transitions in the small interval h is also ignored here. With these arguments we find the *backward equation*

$$P_{i,n}(h + t) = P_{i,n}(t)(1 - \lambda_i h) + P_{i+1,n}(t)\lambda_i h, \tag{21.8}$$

with $P_{i,n}(t)$ being the transition probability $E_i \to E_n$ in the time interval t etc. With the same arguments as before we obtain a *backward time-differential equation* in the form

$$\frac{\mathrm{d}}{\mathrm{d}t}P_{i,n}(t) = -\lambda_i P_{i,n}(t) + P_{i+1,n}(t)\lambda_i, \tag{21.9}$$

with condition $P_{i,i}(0) = 1$ and $P_{i,n}(0) = 0$.

Birth and death processes

We now consider a slightly more general process, where we can have a transition $E_i \to E_{i+1}$ in a small time interval h with probability $\lambda_i h$, *or* have a transition $E_i \to E_{i-1}$ with probability $\mu_i h$. It is a simple matter to obtain the forward equation

$$P_{i,n}\big(t + h\big) = P_{i,n}(t)\big(1 - \lambda_n h - \mu_n h\big) + P_{i,n-1}(t)\lambda_{n-1}h + P_{i,n+1}(t)\mu_{n+1}h, \tag{21.10}$$

and then the forward differential equation

$$\frac{\mathrm{d}}{\mathrm{d}t}P_{i,n}(t) = -(\lambda_n + \mu_n)P_{i,n}(t) + P_{i,n-1}(t)\lambda_{n-1} + P_{n+1}(t)\mu_{n+1}.$$

For this case we can also derive a set of *backward equations*. It is a rather trivial generalization to obtain, for instance, the backward time differential equation for a birth–death process as

$$\frac{\mathrm{d}}{\mathrm{d}t}P_{i,n}(t) = -(\lambda_i + \mu_i)P_{i,n}(t) + P_{i+1,n}(t)\lambda_i + P_{i-1,n}(t)\mu_i. \tag{21.11}$$

The special simple case with $2\lambda_i = 2\mu_i \equiv \lambda$ for all i gives the forward equation

$$\frac{\mathrm{d}}{\mathrm{d}t}P_{i,n}(t) = -\lambda P_{i,n}(t) + P_{i,n-1}(t)\lambda$$

and the backward equation

$$\frac{\mathrm{d}}{\mathrm{d}t}P_{i,n}(t) = -\lambda P_{i,n}(t) + P_{i+1,n}(t)\lambda.$$

Both equations are satisfied by

$$P_{i,n}(t) = \mathrm{e}^{-\lambda t}\frac{(\lambda t)^{n-i}}{(n-i)!}$$

for $n \geqslant i$ and $P_{i,n}(t) = 0$ for $n < i$, characterizing the *Poisson process*. For the present process we have a finite probability to remain at the same position at each time step: for the random walk we *must* move in the forward *or* the backward direction, although not necessarily with the same probability.

Diffusion equations for the Wiener process

We consider a simple random walk with steps X_1, X_2, ..., each having a distribution $P(X_n = 1) = p$ and $P(X_n = -1) = q \equiv 1 - p$, see section 21.1.1. The forward equation is readily obtained as

$$P_{j,k}^{(n)} = P_{j,k-1}^{(n-1)}p + P_{j,k+1}^{(n-1)}q.$$

For fixed j, this is a difference equation of the first order in the time coordinate n and of second order in the space-like coordinate k (Cox and Miller 1965). We have $x_0 = j\Delta x$, $x = k\Delta x$, and $t = n\Delta t$. We assume that it is possible to make an expansion $P(x_0, x + \Delta x; t - \Delta t) \approx P(x_0, x; t) - \Delta t \partial P/\partial t + \Delta x \partial P/\partial x + \frac{1}{2}\Delta^2 x \partial^2 P/\partial x^2$. In the limit of $\Delta x \to 0$ and $\Delta t \to 0$, we then find a forward partial differential equation for the time and also the spatial coordinate as

$$\frac{\partial}{\partial t}P(x_0, x; t) = \frac{1}{2}\sigma^2\frac{\partial^2}{\partial x^2}P(x_0, x; t) - \mu\frac{\partial}{\partial x}P(x_0, x; t),$$

where we here have $4pq\Delta^2 x/\Delta t \to \sigma^2$ and $(p-q)\Delta x/\Delta t \to \mu$. See also section 21.1.1.

For the backward equation, we readily find

$$P_{j,k}^{(n-n_0+1)} = P_{j+1,k}^{(n-n_0)}p + P_{j-1,k}^{(n-n_0)}q,$$

where n_0 is the initial time label and n for the final time. The first term deals with the process of starting at j at t_0, going to $j+1$ in the first step, and from there transiting to k, while the second term accounts for the transition $j-1 \to j$ and from there to k. The corresponding differential equation is then found just as for the forward equation. We have

$$\frac{\partial}{\partial t}P(x_0, x; t) = \frac{1}{2}\sigma^2\frac{\partial^2}{\partial x_0^2}P(x_0, x; t) + \mu\frac{\partial}{\partial x_0}P(x_0, x; t).$$

For the particularly simple case where $p = q = \frac{1}{2}$, we find $\mu = 0$, giving the diffusion equation for the forward as well as for the backward equation. We could as well differentiate the backward equation with respect to t_0 as usually done, and have $\partial P/\partial t = -\partial P/\partial t_0$.

21.2 Imposed boundary conditions

Often, we are particularly interested in analyzing random walks with imposed boundary conditions. This is relevant if, for instance, we are to analyze the confinement time of particles in a hot magnetized plasma, i.e. study the statistical distribution of times elapsing from when such a particle is introduced in the plasma until it hits the walls confining the system.

In the following we will need the probability $P_{r,n}$ density, which refers to the case where the barrier is absent (Chandrasekhar 1954). The problem is to obtain, for instance, the probability of finding the moving object, e.g. a particle, at a position N steps from the origin after S intervals, or, if you like, the probability of the corresponding gain for player A. For definiteness, we introduce the notation of time being measured in multiples of Δ_t, i.e. $t = n\Delta_t$, and displacement measured in units of Δ_x, i.e. a position $X = r\Delta_x$ with n and r being integers. A position in space-time is thus given by (r,n). The starting point for a walk will always be $(0, 0)$. Assume that a walk at time n consists of p steps in the positive direction and q negative steps, $n = p + q$. The partial sum $s_k = \varepsilon_1 + \cdots + \varepsilon_k$ represents the difference between steps in the positive and the negative directions at time k, with ε_k taking the values $+1$ or -1 depending on the direction of the step with label k. Then $s_k - s_{k-1} = \pm 1$, with $k = 1, 2, \ldots, n$, $s_0 = 0$ and $s_n = p - q$. A path is defined as a polygonal line with vertices at abscissas at $0, 1, \ldots, n$ and ordinates $s_0, s_1, \ldots, s_n$, where $s_n = r$. We shall refer to n as the *duration* of a path. There are 2^n paths of duration n. If among all ε_k, p are positive and q are negative, then $n = p + q$ and $r = p - q$. A path from $(0, 0)$ to (r,n) only exist if n and r are of this form. In this case (Feller 1968) the p places for the positive ε_ks can be chosen from the $n = p + q$ available places in

$$N_{r,n} = \binom{p+q}{p} = \binom{n}{\frac{1}{2}(n+r)} \tag{21.12}$$

different ways, expressed in terms of the binomial coefficient. We define $N_{r,n} = 0$ whenever n and r are not of the form implied in (21.12), i.e. it is understood that the binomial coefficient is to be interpreted as zero, unless $\frac{1}{2}(n + r)$ is an integer between 0 and n, both inclusive. It requires at least $n = p$ steps to reach a position $r = p$, implying in particular that $N_{r,n} = 0$ if $r > n$. With this convention, there exists exactly $N_{r,n}$ different paths from the origin to a point (r,n). The probability that the position with label r is actually reached at time n is then

$$P_{r,n} = \binom{n}{\frac{1}{2}(n + r)} \frac{1}{2^n}, \tag{21.13}$$

recalling that there are 2^n paths available altogether for a given n, since at each time we have two choices for the direction of a step. As for $N_{r,n}$, we have that $P_{r,n} = 0$ for $r > n$.

- **Example:** The question of return to the origin at time t is a special case with $r = 0$ in (21.13). This particular probability is used frequently so it is given a special symbol

$$U_{2n} = \binom{2n}{n} \frac{1}{2^{2n}} = \frac{(2n)!}{2^{2n}(n!)^2}, \tag{21.14}$$

where, for this particular case, the argument in (21.13) is necessarily even, and therefore $2n$ is introduced instead of n. The variation of U_{2n} with n is shown in figure 21.2. Expressing the binomial coefficient in terms of factorials, and by use of Stirling's formula $n! \approx \sqrt{2\pi}\,\mathrm{e}^{-n} n^{n+\frac{1}{2}}$, we have that $U_{2n} \approx 1/\sqrt{\pi n}$ for large n. Note that we are not discussing the question of *first* return here: at step n, the origin may have been visited one or more times before.

- **Example:** From (21.14) we can obtain that the probability $\alpha_{2k,2n}$ that up to time t the last visit to the origin occurs at time $2k$. The result is obtained by arguing that U_{2k} is the number of favorable paths divided by the total number

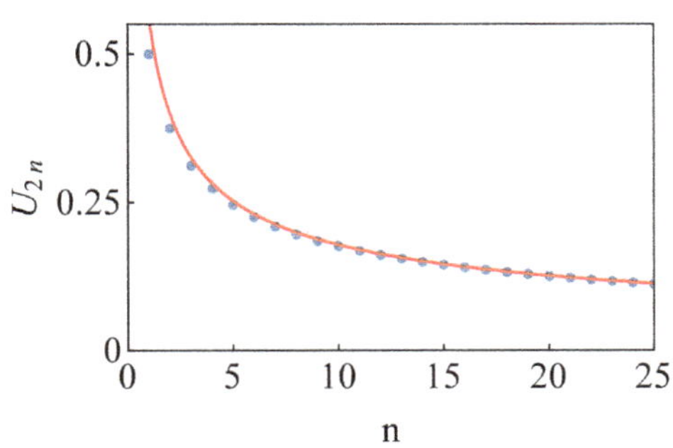

Figure 21.2. The variation of U_{2n} with n as given by (21.14). The full line gives the asymptotic approximation obtained by Stirling's formula. The formula is already a good approximation for modest n-values.

of paths up to (0, $2k$), the total number of paths being 2^n as before. The next $(2n - 2k)$ paths can be chosen in $U_{2n-2k}2^{2n-2k}$ different ways. Consequently

$$\alpha_{2k,2n} = U_{2k}U_{2n-2k}, \tag{21.15}$$

for $k = 0, 1, \ldots, n$. Note that the distribution is symmetric in the sense that $\alpha_{2k,2n} = \alpha_{2n-2k,2n}$. By use of Stirling's formula it is demonstrated that $\alpha_{2k,2n} \approx 1/[\pi\sqrt{k(n-k)}]$ with a negligible error, unless k is close to n or 0, which are the two points of maximum for $\alpha_{2k,2n}$. For $k = 0$ or $2n = 2k$ we have $\alpha_{0,2n} = \alpha_{2n,2n} = U_{2n}$ as given by (21.14).

With some algebra (Feller 1968) it can be demonstrated that the probability that in a time interval from 0 to $2n$ a random walk spends $2k$ time units on the positive side and $2n - 2k$ units on the negative side is $\alpha_{2k,2n}$. The time intervals are not necessarily continuous. (The total time spent on the positive side is necessarily an even number in time units.) It is evident that the probability for $2k$ being equal to n, i.e. half the time being spent on one side the rest on the other, is small, and in particular smaller than the probability for $2k = 2n$ or $k = 0$. These results have rather important implications; they mean that an object executing a random walk will most probably remain at the positive or the negative side of the origin for extended times, or if you like; in a long coin-tossing game it is quite likely that one of the players remain practically the whole time on the winning side, the other on the losing side (Feller 1968). This result may appear in contradiction with expectations, and it is presumably this failing intuition which causes many gambler's ruin.

21.2.1 Reflecting barriers

An important generalization of the foregoing results can be obtained by including the effects of barriers. Here we consider two types; a perfect reflector, and a perfect absorber.

Consider first an ideally reflecting barrier; by this we mean one having the property that if a particle undergoing a one-dimensional random walk arrives at the barrier at position m_r, it will with certainty be reflected one step back at its next time step, i.e. it will then arrive at the position with label $m_r - 1$. We now obtain an expression for the probability $P_{r,n;m_r}$ of finding a particle at the position labeled r at time labeled n, given that a reflecting barrier is placed at position m_r, as indicated by the subscript r. Two basically different situations may arise; either the particle has arrived at r without undergoing any reflections, or it has bounced off the barrier one or more times before arriving at r. In the latter case we can construct a virtual path by a mirror reflection of the actual path after reflection with respect to the position m_r as shown on figure 21.3.

Consider first a path undergoing only one reflection at m_r. By constructing the corresponding virtual path we find that this will at time n lead to the image position $2m_r - r$. Conversely, in the absence of a barrier, for every trajectory leading to the image point having crossed the line m_r only once, there is exactly one which leads to

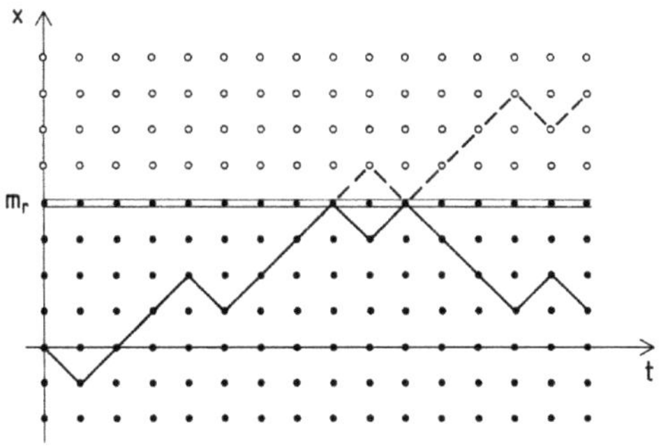

Figure 21.3. Illustration of the space-time variation of a simple one-dimensional random walk with the presence of a reflecting barrier at the position m_r. The dashed line indicates a virtual path explained in the text. Positions on the time axis have the label n.

m_r after a single reflection. In a similar way we can argue for a trajectory undergoing *two* reflections. In this case the actual path can, in the absence of a barrier, be completed with a segment of the mirror path between the two reflections, and this path will still end at (r, n). In the case of three reflections the actual path can, in the absence of a barrier, be replaced by one having a mirror segment between the two reflections, and then again a mirror segment from the third reflection until it reaches the point $(2m_r - r, n)$. In this way we can uniquely represent each path undergoing reflections in the case *with* a barrier, with another path in the *absence* of a barrier. All the relevant paths with zero or an even number of reflections end at (r,n), and these are represented with a probability $P_{r,n}$. All those with an uneven number of reflections end at $(2m_r - r, n)$ and these are represented by $P_{2m_r-r,n}$. Consequently we have (Chandrasekhar 1954)

$$P_{r,n:m_r} = P_{r,n} + P_{2m_r-r,n}, \quad r < m_r, \tag{21.16}$$

relating the probability *with* a reflecting barrier present at m to the corresponding probabilities *without* such a barrier. Note in particular that $P_{m_r,n:m_r} = 2P_{m_r,n}$, i.e. *with* a reflecting barrier we have twice the probability of finding a particle at the barrier position as compared to the case where no such barrier is present. This result can be understood on intuitive grounds.

21.2.2 Absorbing barriers

As an alternative to the discussion in the foregoing subsection we here consider a perfectly absorbing barrier at the position m_a. By a perfect absorber we here mean one with the property that if a particle undergoing a one-dimensional random walk arrives at the barrier, it will disappear from the system with certainty. The problem of 'gambler's ruin' mentioned before, can be formulated in this way; the absorbing barrier can represent the funds available for the gambler, and in the likely case where no credit is offered, it represents the maximum loss he can suffer (Cox and Miller 1965).

Also in the case of absorbing barriers we have two possibilities; a particle arrives at r at time n without ever reaching the barrier, or it can arrive at m_a at an earlier time and be lost from the system before reaching r. Once again we introduce the virtual path discussed in subsection 21.2.1 and figure 21.3. The present problem can be defined as one where we take *all* paths which arrive at r at time n without the presence of a barrier, and to remove from these all those which at some time encounter the barrier position m at least once. Out of these we note that to every path arriving *at* the barrier, there are uniquely defined paths which in the absence of the barrier leads to the point $2m_{a-r}$ at time n. The probability of finding a particle at the position r at time n with an absorbing barrier present at position m_a is subsequently

$$P_{r,n:m_a} = P_{r,n} - P_{2m_a-r,n}, \quad r < m_a. \tag{21.17}$$

The expression (21.17) relates the probability *with* an absorbing barrier present at m_a to the corresponding probabilities *without* such a barrier. Note in particular that $P_{m_a,n:m_a} = 0$, as expected.

Recall that $P_{r,n} = 0$ for $r > n$, implying that also $P_{2m_a-r,n} = 0$ for $r > n$ with $n < m_a$; within this time the particle has no possibility to reach the position r and the probability $P_{r,n:m_a} = 0$ as it should be.

21.2.3 Probable rate of arrival

The result for $P_{r,n:m_a}$ obtained in section 21.2.1 expresses the probability of actually finding a particle at the position r at time n with an absorbing barrier present at position m_a. In the derivation all orbits touching at some time the barrier itself were excluded, and the result $P_{m_a,n:m_a} = 0$ is therefore somewhat trivial. It is a different and just as meaningful question to ask for the rate of arrival $a_{m_a,n}$ of particles *at* the absorbing barrier. For this purpose we now assume that a particle *has* arrived at the screen position m_a at time n, without ever having touched that position before. Note that the meaning of $a_{m_a,n}$ is different from $P_{r=m_r,n:m_r}$ or $P_{r=m_a,n:m_a}$.

It is evident for the present analysis that n has to be even (odd) if m_a is even (odd), respectively. We assume that this is the case. We now attempt again to express $a_{m_a,n}$ in terms $P_{r,n}$ relating to the conditions without the presence of the barrier. Removing the barrier we note that a particle can arrive at (m_a, n) from either $(m_a - 1, n - 1)$ or $(m_a + 1, n - 1)$. The latter position is forbidden in the presence of the barrier, and we want to remove it from the allowed paths. This is however not sufficient, since some of the paths leading to $(m_a - 1, n - 1)$ are also forbidden, since they touch on the position m_a once or more at earlier times. To identify these, we note by considering the mirror image of the orbits with respect to the line $r = m_a$ in an r, n-plane, that to every orbit passing through $(m_a + 1, n - 1)$ we can uniquely associate another orbit of forbidden nature arriving at $(m_a - 1, n - 1)$, forbidden in the sense that it touches or crosses the line $r = m_a$ at some instant. Thus, all in all, the number of permitted ways to arrive at the position m_a for the first time after n steps is equal to all the possible ways of arriving at m_a after n steps in the absence of the absorber minus *twice* the number of ways of arriving at $m_a - 1$ after $n - 1$ steps again in the absence of the absorber. We therefore have

$$\begin{aligned}
&\frac{n!}{\left[\frac{1}{2}(n - m_a)\right]!\left[\frac{1}{2}(n + m_a)\right]!} - 2\frac{n!}{\left[\frac{1}{2}(n + m_a)\right]!\left[\frac{1}{2}(n - m_a - 2)\right]!} \\
&= \frac{n!}{\left[\frac{1}{2}(n - m_a)\right]!\left[\frac{1}{2}(n + m_a)\right]!}\left(1 - \frac{n - m_a}{n}\right) \\
&= \frac{n!}{\left[\frac{1}{2}(n - m_a)\right]!\left[\frac{1}{2}(n + m_a)\right]!}\frac{m_a}{n}
\end{aligned} \tag{21.18}$$

permitted paths. Recalling again that there are 2^n number of paths with duration n, we obtain the probability of a particle to arrive for the first time at m_a after exactly n steps to be

$$a_{m_a,n} = \frac{m_a}{n} P_{m_a,n}, \quad r < m_a, \tag{21.19}$$

with $P_{m_a,n}$ defined by (21.13). Evidently, $a_{m_a,n} < P_{m_a,n}$ for all $n > 1$.

- **Exercise:** The symmetry arguments in these latter three subsections may appear deceivingly simple and convincing; the reader is strongly urged to convince himself/herself that they are indeed complete.

21.2.4 Probability of returns

The random walk is a very simplified model of a physical process, but nonetheless it poses many interesting questions. One of them concerns the probability of return to a starting point (Mosteller 1987). In the simplest version of the problem we may ask 'if a random walker starts at the origin, when (on average) if ever will he/she return to this position?'. The question has many generalizations, and most of these will be left to the reader. If you want to pep up the question a little, you can turn it to a 'cliff hanger': if a random walker stands at the brink of a cliff, one step to the right will send him over the edge, one to the left to safety. What is the long term probability of surviving a random walk under these conditions? You can see this as a case of a perfectly absorbing barrier! In this formulation the question may appeal more to your imagination, so we start with this. It is clear that the dimensionality of the problem may matter: we can take a walk along the line perpendicular to the cliff-edge or consider a random walk on a plane where there is a well at some position, and we might like to know the probability of our 'walker' falling into this.

It is essential for the following arguments that each step in the random walk is independent of all others. We first consider the one-dimensional problem, using the illustration in figure 21.4. To make the analysis less morbid, let us use the term

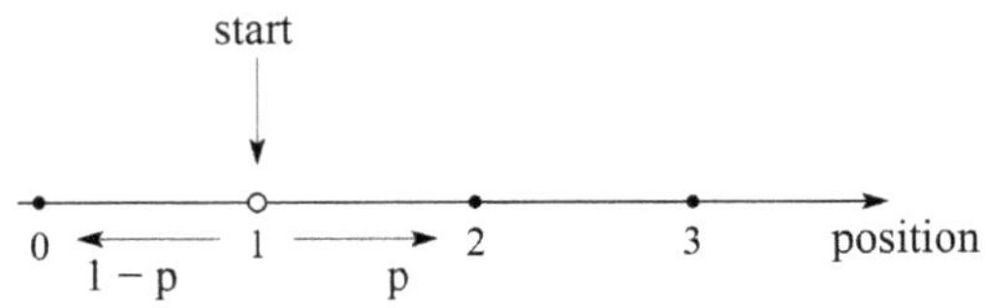

Figure 21.4. A particle starts a random walk at position 1 with a probability p for moving to the right, and $1-p$ for moving to the left. At position 0 it is absorbed.

'particle' instead of 'walker' and 'absorption' instead of sudden death. The probability of absorption at position 0 given a start at position 1 is denoted P_1, with start at position 2 it is called P_2, etc. If we consider the process step by step we find that at the first step we have a probability of absorption $1-p$, while arrival at position 2 has probability p. In that latter case, i.e. for starting point 2, we have the probability for absorption already defined as P_2. This means that we can write

$$P_1 = 1 - p + p\,P_2.$$

We note that we can arrive at position 0 only from position 1, and similarly position 1 can be reached only from position 2. The probability of arrival at position 1 from position 2 is completely analogous to arriving position 0 from position 1 and has the same probability. Arriving from position 0 from position 1 is already given as P_1, meaning that we can write $P_2 = P_1 \times P_1$. This implies

$$P_1 = 1 - p + p\,P_1^2$$

with two solutions $P_1 = 1$ or $P_1 = (1-p)/p$. One of these is false, but which one depends on p. For $p = 0$ we readily see that $P_1 = 1$, since the particle is absorbed with certainty after the first step. If $p = 1$ we see similarly that $P_1 = 0$ since the particle moves steadily to the right in this latter case. Clearly we require $P_1 \leqslant 1$, so the solution $P_1 = (1-p)/p$ has to be excluded for $p < 1/2$. We thus conclude that for $p < 1/2$ we have $P_1 = 1$. Requiring P_1 to be a continuous function of p we find $P_1 = (1-p)/p$ for $p > 1/2$, although admittedly the requirement of continuity is not completely obvious (Mosteller 1987), but we shall not go into further detail here. The result for $P_1(p)$ is shown in figure 21.5.

It is somewhat surprising that $p = 1/2$ gives absorption with certainty, but note that nothing is said about how long it takes! Another consequence of this result is that we can state that in this one-dimensional random walk, a particle visits its starting point with certainty at least once, which on the other hand is a much more reasonable formulation, agreeing with intuitive expectations. If on the other hand $p > 1/2$ there is a finite probability for the particle to wander off to infinity.

To discuss the two- and three-dimensional cases it turns out to be an advantage to first analyze the average number of returns to the origin of release (Mosteller 1987). We restrict the discussion to the classical random walk problem with a walk on a square grid, considering the plane first, see figure 21.6. We analyze this as a product of two random walks, where the flip of a coin determines the step in the north/south direction, and after this a similar operation for the completion of a step in the east/west direction. The components of the steps in the two directions are statistically

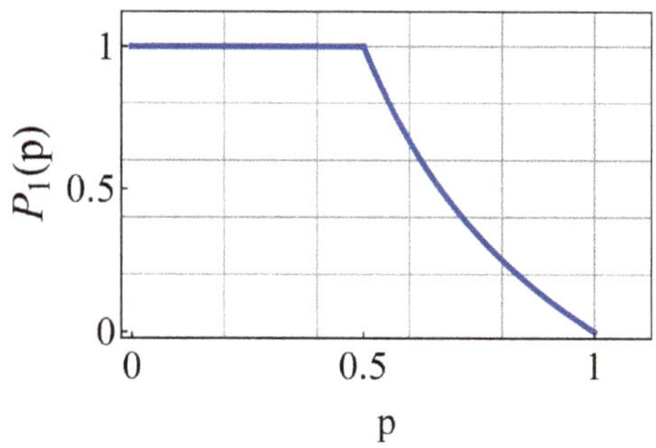

Figure 21.5. The probability $P_1(p)$ of absorption at position 0 for a particle starting a random walk at position 1 with a probability p for moving to the right, and $1 - p$ for moving to the left.

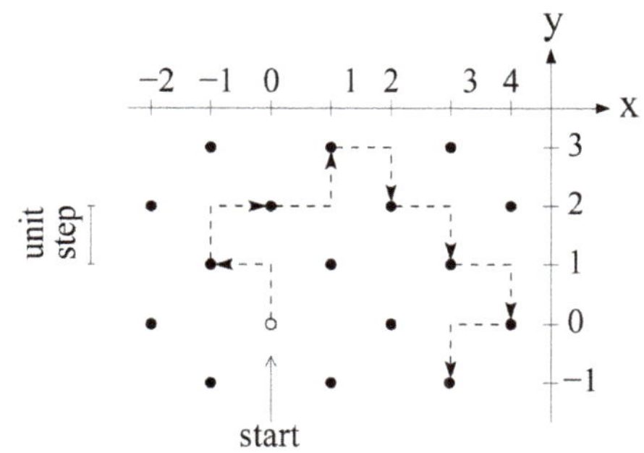

Figure 21.6. Illustration of a two-dimensional random walk.

independent by construction. The question of return to the origin at the $2n$-th step was found in (21.14): this result applies for the one-dimensional random walk. For the corresponding walk on a plane, we have the probability of a particle being at the origin of release after $2n$ steps to be

$$U_{2n}^2 = \left[\binom{2n}{n}\frac{1}{2^{2n}}\right]^2 = \left[\frac{(2n)!}{2^{2n}(n!)^2}\right]^2 \approx \frac{1}{\pi n},$$

where we introduced the approximation by Stirling's formula. The result we just found is the average number of times a particle is at the origin after exactly $2n$ steps. In between it can have been there many times, the average values being given by the expression shown evaluated for 2, 4, 6, 8, …, $2n$ steps. To get the total number of returns we have to sum all these since the average of a sum is the average of the individual terms We have, however, that $\sum_{n=1}^{N} 1/n \sim \ln N$, which diverges for N large. The total number of returns is unbounded and therefore the probability of return to the origin must be $P(return) = 1$. Particles not only return, but return infinitely often. Out of the infinitely many realizations you *might* find trajectories that go due north, or similar, but these have zero statistical weight.

We can repeat the same reasoning for the three-dimensional equivalent of the problem (Mosteller 1987) to find the probability of a particle being at the origin of release after $2n$ steps to be

$$U_{2n}^3 = \left[\binom{2n}{n}\frac{1}{2^{2n}}\right]^3 = \left[\frac{(2n)!}{2^{2n}(n!)^2}\right]^3 \approx \frac{1}{(\pi n)^{3/2}}.$$

The sum $\sum_{n=1}^{N} 1/n^{3/2}$ converges and the average number of returns is finite, numerically close to 0.32, although more generally the result depends on precisely how the grid and the steppings are defined (Feller 1968).

One conclusion from the outline in the present section is that we have to be careful in applying results from one- or two-dimensional random walk models to a fully three-dimensional problem. As a minimum the possible generalizations need some considerations.

- **Exercise:** How many times do you have to roll an honest die on average in order to get 6, or any other selected number? How is this question relevant for the discussion of the present section?

21.2.5 Limit of continuous distributions

By letting N be large we can apply a coarse-grained description where the relevant probabilities are considered as continuous functions. The results of the three sections 21.2.1, 21.2.2 and 21.2.3 will be considered independently.

No barriers

For later reference, it is advantageous first to consider the case where there are no barriers present, to obtain the results for the continuum limit. We first take the appropriate result (21.13) in the limit of large n and use Stirling's formula in the form

$$\log n! = \left(n + \frac{1}{2}\right)\log n - n + \frac{1}{2}\log(2\pi) + \mathcal{O}(n^{-1})$$

to approximate

$$P_{r,n} = \frac{n!}{\left[\frac{1}{2}(n-r)\right]!\left[\frac{1}{2}(n+r)\right]!}\left(\frac{1}{2}\right)^n, \tag{21.20}$$

and obtaining

$$\log P_{r,n} \approx -\frac{1}{2}\log n + \log 2 - \frac{1}{2}\log(2\pi) - \frac{r^2}{2n}. \tag{21.21}$$

With use of $\log(1 \pm r/n) \approx \pm(r/n) - (r^2/2n^2)$, we readily find (Chandrasekhar 1954)

$$P_{r,n} \approx \left(\frac{2}{\pi n}\right)^{\frac{1}{2}} \exp\left(-r^2/2n\right). \tag{21.22}$$

Introducing the net displacement as $x = r\Delta_x$, the actual time as $t = n\Delta_t$ and defining $D = \frac{1}{2}n\Delta_x^2$, we rewrite $P_{r,n}$ as a continuous function of new variables

$$P(x, t) \approx \frac{1}{2(\pi t D)^{\frac{1}{2}}} \exp\left(-x^2/4Dt\right). \tag{21.23}$$

The probability for a particle to be found in the interval $\{x, x + \Delta\}$ at a time t is expressed as $P(x, t)\Delta$. Note that (21.23) is a solution to the diffusion equation

$$\frac{\partial}{\partial t}P(x, t) = D\frac{\partial^2}{\partial x^2}P(x, t),$$

where D is a diffusion coefficient, and the initial condition is a δ-function, $P(x, t = 0) = \delta(x)$, i.e. the particle is known with certainty to be at the origin at time $t = 0$.

- **Example:** Since X at time $t = n\Delta_t$ is the sum of n independent variables x_i, with $i = 0, 1, \ldots, n$, where $\langle x_i \rangle = 0$ and $\langle x_i^2 \rangle = \Delta_x^2$, it is evident that the average is $\langle X \rangle = 0$ and the root mean square displacement is $\langle X^2 \rangle^{1/2} = \Delta_x\sqrt{n}$.

Reflecting barriers

Using the result (21.22), we find that (21.16) can be approximated as

$$P_{r,n:m_r} \approx \left(\frac{2}{\pi n}\right)^{\frac{1}{2}}\left(e^{-m^2/2n} + e^{-(2m_r - m)^2/2n}\right), \tag{21.24}$$

for $n \to \infty$. Introducing the net displacement as $x = r\Delta_x$, the actual time as $t = n\Delta_t$ and defining $D = \frac{1}{2}n\Delta_x^2$, we rewrite $P_{r,n:m_r}$ as a continuous function of new variables

$$P_r(x, t) \approx \frac{1}{2(\pi t D)^{\frac{1}{2}}}\left(\exp\left(-x^2/4Dt\right) + \exp\left(-(2x_r - x)^2/4Dt\right)\right), \tag{21.25}$$

with x_r being the position of the reflecting barrier.

From (21.25) we find $\partial P_r(x, t)/\partial x|_{x = x_r} = 0$, without conditions on the actual value of $P_r(x, t)$ at the boundary.

Absorbing barriers

Using the result (21.20)–(21.22), we find that (21.17) can be approximated by

$$P_{r,n:m_a} \approx \left(\frac{2}{\pi n}\right)^{\frac{1}{2}}\left(e^{-r^2/2n} - e^{-(2m_a - r)^2/2n}\right) \tag{21.26}$$

as $n \to \infty$. Introducing again the net displacement as $x = r\Delta_x$, the actual time as $t = n\Delta_t$ and D as defined before, we rewrite $P_{r,n:m_a}$ as a function of new variables

$$P_a(x, t) \approx \frac{1}{2(\pi t D)^{\frac{1}{2}}}\left(e^{-x^2/4Dt} - e^{-(2x_a - x)^2/4Dt}\right), \tag{21.27}$$

with x_a being the position of the absorbing barrier. From (21.27) we find $P_a(x = x_r, t) = 0$, without conditions on the value of the spatial derivative, $\partial P_r(x, t)/\partial x = 0$, at the boundary.

Probable rate of arrival

For the limiting case of large n we obtain from (21.19) and by use of the foregoing results

$$a_{m_a,n} \approx \frac{m_a}{n}\left(\frac{2}{\pi n}\right)^{\frac{1}{2}} \exp\left(-m_a^2/2n\right), \tag{21.28}$$

or in terms of the variables x_a and t introduced before

$$a(x_a, t) \approx \frac{x_a}{nt} \frac{1}{(\pi t D)^{\frac{1}{2}}} \exp\left(-x_a^2/4Dt\right). \tag{21.29}$$

Finally we ask for the probability, $q(x_a, t)\Delta$, that that the particle arrives at x_a in the time interval t, $t + \Delta$ for the first time. We have $q(x_a, t)\Delta = \frac{1}{2}a(x_a, t)n\Delta$, since $a(x_a, t)$ is the number of particles which arrive at x_a in the time it takes to traverse *two* steps, see section 21.2.3. We find

$$q(x_a, t) \approx \frac{x_a}{t} \frac{1}{2(\pi t D)^{\frac{1}{2}}} \exp\left(-x_a^2/4Dt\right). \tag{21.30}$$

We can interpret (21.30) as giving the fraction of a large number of particles released initially at $x = 0$ and which are deposited on an absorbing screen per unit time at the time t, assuming the particles to execute their random walk independently of each other.

It is easily verified that $q(x_a, t)$ satisfies the relation

$$q(x_a, t) = -D\frac{\partial}{\partial x}P_a(x, t)\bigg|_{x=x_a},$$

with $P_a(x, t)$ given by (21.27). This result can be understood by recalling that $(1/2\sqrt{\pi t D})\exp(-x^2/4Dt)$ is a solution to the diffusion equation

$$\frac{\partial}{\partial t}P(x, t) - \frac{\partial}{\partial x}D\frac{\partial}{\partial x}P(x, t) = 0, \tag{21.31}$$

where we placed the constant diffusion coefficient, D, under the first differentiation sign for later reference. Comparing this form of the diffusion equation with the equation of continuity written in one spatial dimension

$$\frac{\partial}{\partial t}\rho(x, t) + \frac{\partial}{\partial x}u(x, t)\rho(x, t) = 0,$$

for a fluid with local density ρ and velocity u, we see that $-D\partial P(x, t)/\partial x$ in the diffusion equation has the same role as the particle flux density $u(x, t)\rho(x, t)$ in the continuity equation, implying that we may interpret $-D\partial P(x, t)/\partial x$ as a 'probability flux'. For the drift-diffusion equation (21.5), the probability flux is $UP - D\partial P/\partial x$.

We might conclude that, in the limit where the diffusion model applies, the problem is fully determined when the diffusion coefficient is derived. This coefficient will depend critically on the parameters of the plasma. In the following, some particular problems will be considered placing the emphasis on *magnetized* plasma conditions.

21.3 Confinement times

One central problem in fusion plasma physics concerns the time energetic particles that can be confined by magnetic fields. Here we shall illustrate how this question can be analyzed in terms of simple diffusion equations, as discussed in sections 21.1, 3.5 and 3.6. For simplicity, the analysis will be restricted to one spatial dimension.

21.3.1 Forward and backward equations

In the following, we restrict the problem to the case where the diffusion equation addresses the time evolution of the probability density of particle positions subject to an initial condition

$$P(x, t_0|x_0, t_0) = \delta(x_0), \tag{21.32}$$

i.e. we know with certainty that the reference particle is at a reference position x_0, at the reference time t_0. To emphasize this, we write the probability density as $P(x, t|x_0, t_0)$ in the following. Discussing confinement times we want to gain information about the probability for this particle to arrive at a certain boundary at later times, $t > t_0$. The simplest way to do so is to assume that the particle *is* at the boundary, and analyze the probability evolution for *previous* times. This is however not the formulation we had for the diffusion equation! We need a *backward equation*, while the diffusion equation was derived for a *forward* problem, see section 21.1.2. Generally, the distinction between the two problems is not trivial (Feller 1968, 1971, Gardiner 1985, Paul and Baschnagel 1999). For the diffusion equation the situation is, fortunately, simple. This particular equation describes the continuous limit of a random walk as in figure 21.1. Here, for any trajectory traversed in the positive direction, there will one equally probable, which is traversed in the opposite direction. If we want an equation where the forward variables enter as boundary conditions, we have the backward diffusion equation (Cox and Miller 1965) as

$$-\frac{\partial P}{\partial t_0} = D\frac{\partial^2 P}{\partial x_0^2}, \tag{21.33}$$

where all derivatives are taken with respect to the initial coordinates x_0, t_0. The forward equation describes the evolution of future states, given the initial condition, here (21.32), while the backward equation can be interpreted as the one describing the distribution of *previous* states, given that we have arrived to (21.32) at $t = t_0$.

Problems concerning first passage times are often solved by defining a new stochastic process in which a boundary, or more generally a set of states, are absorbing; when such a state is entered, the process remains in that state thereafter. We usually denote these states or boundaries as *perfectly absorbing*. The idea is that when interested in first time arrivals only, we do not really care about the future of the particle after it has arrived at the selected stage, and we might just as well set it to being an absorber. There is one slight 'catch', which we will return to later on.

To add a little generality, which turns out to have non-trivial consequences, we can change the frame of reference to one moving with a constant velocity U, and rewrite the diffusion equation as

$$\frac{\partial P}{\partial t} = D\frac{\partial^2 P}{\partial x^2} - U\frac{\partial P}{\partial x}, \tag{21.34}$$

as easily demonstrated, see also (21.5). The corresponding backward equation (Cox and Miller 1965) becomes

$$-\frac{\partial P}{\partial t_0} = D\frac{\partial^2 P}{\partial x_0^2} + U\frac{\partial P}{\partial x_0}. \tag{21.35}$$

The forward and backward equations are equivalent to each other (Gardiner 1985). The basic difference is which set of variables is held fixed. In the case of the forward equations, we hold x_0, t_0 fixed, and the solution exists for $t > t_0$, so that (21.32) is an initial condition. For the backward equation, solutions exist for $t_0 < t$, so that x_0, t_0 should be termed *final condition* in this case, since the backward equation expresses the evolution in t_0. Since they are equivalent, the forward and backward equations are both useful. The forward equation gives more directly the distributions of measurable quantities as a function of observed time, t, and tends to be the one most commonly used in applications (Gardiner 1985). The backward equation is most often used in studies of *first passage problems*, where we analyze the probability of a particle leaving a fixed specified region in a given time. As already stated, the diffusion equation is the, presumably, simplest example of forward/backward equations. For time stationary problems as the present ones, we have that the probability densities depend on *time separations* like $t - t_0$ only, in which case we have a trivial relation between the time derivatives, $\partial P/\partial t = -\partial P/\partial t_0$. For more general cases, we refer to the literature (Feller 1968, 1971, Gardiner 1985, Paul and Baschnagel 1999).

In order to determine a particle confinement time, we now want to analyze the 'first passage problem', i.e. ask when a particle released at x_0 at the time t_0 passes for the first time an absorbing boundary, placed at a position a, where we here assume $x_0 < a$ for definiteness. It turns out that this particular question is analyzed best in terms of the backward equation. The first passage time is denoted $\mathcal{T}$ in the following, and defined by

$$X(t_0) = x_0, \quad X(t) < a \quad (t_0 < t < \mathcal{T}), \quad X(\mathcal{T}) = a.$$

In terms of the particle position $X(t)$ at a time t, we now note the following relation

$$\text{probability}\,(X(\tau) < a \quad \text{for} \quad t_0 < \tau < t, \;\; X(t) \leqslant x | X(t_0) = x_0) = \int_{-\infty}^{x} P(y, t|x_0, t_0)\mathrm{d}y \equiv \mathcal{P}(x, t|x_0, t_0), \tag{21.36}$$

where we distinguish the variable x and the particle positions $X(t)$. The relation (21.36) states the probability of finding a particle at a position $\leqslant x$, which was originally released at x_0 at time t_0, and has at no time up to the present at t been at the absorber at $x = a$. The function $P(x, t|x_0, t_0)$ satisfies the forward equation for $x < a$ and the backward equation for $x_0 < a$, subject to an imposed boundary condition for a perfect absorber at $x = a$.

In (21.36), we introduced by definition a new function $\mathcal{P}$, which satisfies the backward equations, just like $P(x, t|x_0, t_0)$, which can be seen by integrating the over the forward variable in (21.35). We readily find that $\mathcal{P}(a, t|x_0, t_0)$ denotes the probability that absorption has not yet occurred at time t, which implies

$$\mathcal{P}(a, t|x_0, t_0) = \text{probability}(\mathcal{T} \geqslant t). \tag{21.37}$$

The probability density of first passage times is introduced as $g(t|x_0, a)$. In general, we cannot be *certain* that a particle will be absorbed, so we have to accept the possibility that

$$\int_0^{\infty} g(t|x_0, a)\,\mathrm{d}t < 1.$$

It follows from (21.37) that

$$g(t|x_0, a) = -\frac{\partial}{\partial t}\mathcal{P}(a, t|x_0, t_0). \tag{21.38}$$

21.3.2 One boundary

First we consider the case where we have only one boundary in the system. We have to determine $\mathcal{P}(a, t|x_0, t_0)$, or $P(x, t|x_0, t_0)$ which is just as good. Both functions have to satisfy the backward equation (21.34) or (21.35) together with the absorbing boundary conditions. *Without* the boundary condition we have a solution to (21.34) in the form

$$P_1(x, t|x_0, t_0) = \frac{1}{2\sqrt{D\pi t}}\exp\left(-\frac{(x - x_0 - Ut)^2}{4Dt}\right). \tag{21.39}$$

The solution (21.39) fulfills the initial condition (21.32). The trick is now to also fulfill the *boundary condition* at $x = a$. This is done by *the method of images*. We introduce a fictitious *negative* probability at a suitable position at $x > a$: the negative probabilities are of course not acceptable, but we will manipulate the conditions so that the probability at $x < a$ is at all times acceptable, and this is all we need. In the present one-dimensional case, the problem is simple, and we introduce such a negative image at the position $x = 2a - x_0$ (Cox and Miller 1965). We find the full solution to be

$$P(x, t|0, 0) = \frac{1}{2\sqrt{D\pi t}}\left[\exp\left(-\frac{(x - Ut)^2}{4Dt}\right) - \exp\left(\frac{Ua}{D} - \frac{(x - 2a - Ut)^2}{4Dt}\right)\right]. \quad (21.40)$$

The initial condition is trivially fulfilled, and it is easily demonstrated that the boundary condition at $x = a$ is correct for (21.40) as well. For $U = 0$, the result (21.40) is consistent with (21.27) . To make life simpler, we took $x_0 = 0$ without loss of generality, and let the initial time be $t_0 = 0$. See figure 21.7 and 21.8 for an illustration of $P(x, t|0, 0)$.

We now find

$$\begin{aligned} g(t) \equiv g(t|0, a) &= -\frac{\mathrm{d}}{\mathrm{d}t}\int_{-\infty}^{a} P(x, t|0, 0) \\ &= -\frac{\mathrm{d}}{\mathrm{d}t}\left[\Phi\left(\frac{a - Ut}{\sqrt{2Dt}}\right) - \exp\left(\frac{Ua}{D}\right)\Phi\left(\frac{-a - Ut}{\sqrt{2Dt}}\right)\right] \end{aligned} \quad (21.41)$$

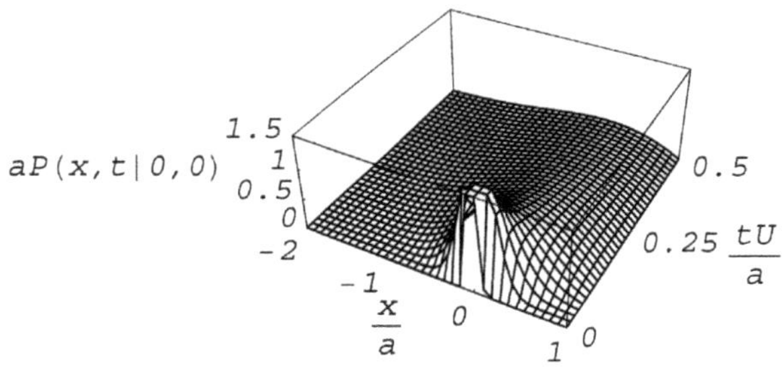

Figure 21.7. The normalized probability density $aP(x, t|0, 0)$, as a function of normalized variables, here illustrated for the case where $Ua/D = 1$. For $t \to 0$, $P(x, t|0, 0) \to \delta(x)$.

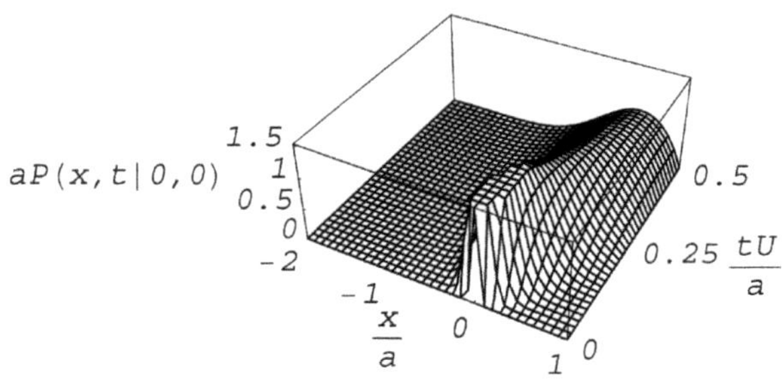

Figure 21.8. The normalized probability density $aP(x, t|0, 0)$, here illustrated for the case where $Ua/D = 4$. For $t \to 0$, $P(x, t|0, 0) \to \delta(x)$. Note the steepening of $P(x, t|0, 0)$ at the boundary $x/a = 1$.

in terms of the error integral

$$\Phi(x) \equiv \frac{1}{\sqrt{2\pi}} \int_{-\infty}^{x} e^{-y^2/2} dy.$$

For this case we find

$$g(t) = \frac{a}{\sqrt{\pi D t^3}} \exp\left(-\frac{(a - Ut)^2}{2Dt}\right), \tag{21.42}$$

being the probability density of confinement times. The *average* confinement time is readily obtained for $U > 0$, and only then, as

$$\langle t \rangle = \frac{a}{U}, \tag{21.43}$$

with variance $aD/(4U^3)$. Note in particular that with $U = 0$, we have $g(t) \sim 1/t^{3/2}$ asymptotically, i.e. the average confinement time is infinite in this case. This is to some extent a surprising result, since we found that a diffusion equation describes a random walk with very small steps. For the random walk (where the step lengths can be of *any* magnitude) we found, on the other hand, that any position on the axes of a one-dimensional system was visited infinitely often as $t \to \infty$. The solution to this seeming paradox lies in noting that the diffusion equation is accurate for the limit where the 'step-length' $\Delta x \to 0$, i.e. a finite fraction of the diffusing particles never make it to the boundary in a finite time. Measured in units of step-lengths, the absorbing boundary is 'infinitely far away'. The present analysis demonstrates how careful we should be in trusting diffusion equations too much when it comes to details in modeling the dynamics of particle distributions.

We have seen that the special case with diffusion but no flow, $U = 0$, in (21.40) and (21.42) is in principle simple, although it leads to singularities in the averages. The other limit *with* flow, but no diffusion, $D = 0$, is somewhat more intricate. *Physically* the problem is trivial, and the transit time is deterministic $\mathcal{T} = a/U$, but the boundary condition at $x = a$ becomes problematic in (21.40). The boundary condition is fulfilled for any $D > 0$, but the gradient at $x = a$ becomes steep, See figure 21.8, and in the limit of $D = 0$, the result (21.40) approaches a step function, and consequently the solution is not differentiable at $x = a$.

21.3.3 Two boundaries

We now consider the one-dimensional problem with *two* boundaries, one at $x = a$ and one at $x = -b$, with $b > 0$. One might think that solving the problem with two boundaries was a trivial extension of the foregoing analysis with one boundary. Alas, it is not so! We can still use the method of images to determine $P(x, t|x_0, t_0)$, but now we have one image in the 'mirror' at a having images in the mirror at b, which then has an image in a etc! The problem does not have a neat compact solution, as the one we encountered in the one-boundary case.

After some algebra, summarized for instance by Cox and Miller (1965), we find the solution

$$P(x, t|0, 0) = \frac{1}{2\sqrt{D\pi t}} \sum_{n=-\infty}^{\infty} \left[\exp\left(\frac{Ux_n'}{2D} - \frac{\left(x - x_n' - Ut\right)^2}{4Dt} \right) - \exp\left(\frac{Ux_n''}{2D} - \frac{\left(x - x_n'' - Ut\right)^2}{4Dt} \right) \right]. \tag{21.44}$$

where the abbreviations $x_n' \equiv 2n(a + b)$ and $x_n'' \equiv 2a - x_n'$ were introduced. The series is absolutely convergent, and differentiable, since each term satisfies (21.34), and so does the series. See figure 21.9 for an illustration of $P(x, t|0, 0)$, for a special choice of parameters.

Unfortunately, there seems no simple solution for $\langle t \rangle$ or similar quantities. It is however evident that with two absorbing boundaries, the confinement time is finite for any sign of U. For the purely diffusive case with $U = 0$, the result can be written in a more compact form.

A number of interesting applications of the diffusion equation, in particular also with imposed boundary conditions, can be found in the works of Carslaw and Jaeger (1947), Crank (1975), Crank *et al* (1981), and Seinfeld (1983).

21.3.4 A general method for determining average confinement times

The results of sections 21.3.2 and 21.3.3 gave, in a sense, too much information. Often we do not need the entire probability density for the particle distribution, but might be content with less information, the average transit times, for instance. Such special cases are often solved relatively easily. For instance, the illustrative cases discussed in sections 21.3.2 and 21.3.3 analyzed the full time evolution of an initial value problem: we are often content with an asymptotic steady state limit, for properly formulated conditions.

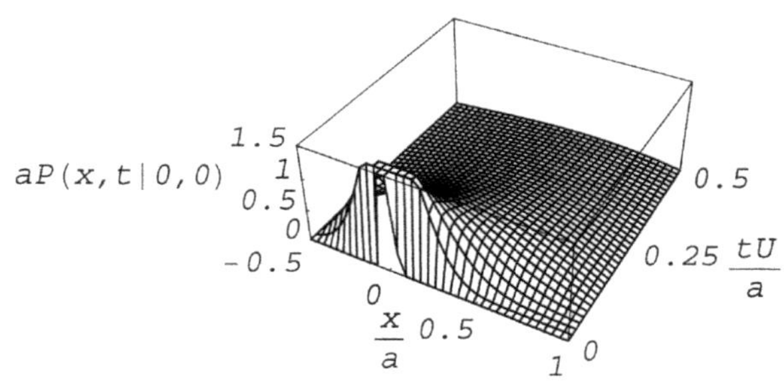

Figure 21.9. The normalized probability density $aP(x, t|0, 0)$ with *two* boundaries, one at $x = a$ and one at $x = -b$. The solution is illustrated for the case where $Ua/D = 1$ and $b = a/2$. Also here we have $P(x, t|0, 0) \rightarrow \delta(x)$ for $t \rightarrow 0$.

Consider first the almost trivial one-dimensional case, where particles propagate with constant velocities V, but statistically distributed velocities from one particle to the next. In this case we have the transit times (sometimes called occupation times) for particles released at x_0 being $t_t = (a - x_0)/V$ with $V > 0$, and $\langle t_t \rangle = (a - x_0)\int_0^\infty P(V)/V \, \mathrm{d}V$, with $P(V)$ being the probability density of particle velocities. We *might* anticipate that $\int_0^\infty P(V)/V \, \mathrm{d}V$ diverges, meaning that $\langle t_t \rangle$ is infinite. The contribution of a particle to the density in the system is proportional to the time it spends in the system, i.e. inversely proportional to its velocity. If we at a fixed point inject particles at a constant rate of J_0 particles per sec. and all with the same velocity V, we have the density to be $n = J_0/V$. The total number of particles is then $N = n(a - x_0) = J_0 t_t$ and $\langle N \rangle = J_0 \langle t_t \rangle$. This result hints that we might be able to relate the average transit time to the particle flux into (and therefore also out of) the system, and to the average number of particles in the system, under steady state conditions. This turns out to be correct. (It can be argued that the previous model is too simplistic to allow any expectations to be built upon it, but in reality it is not difficult to generalize it, with some rather mild restrictions. Let the particle velocity be time varying $V = V(t) = \mathrm{d}X(t)/\mathrm{d}t$, with $X(t)$ being the particle position. It is trivial to obtain the position at a given time $X(t) = \int_0^t V(\tau)\mathrm{d}\tau$, which can be averaged trivially, but this is actually not the result we need. Rather, we want the time it takes to reach a given position, $t(X)$, which is not always easy to determine.)

In a more general model, we assume that we inject particles into the system, so that we introduce $J_0 \, \mathrm{d}t$ particles in a short time interval $\mathrm{d}t$. Let now $F(t < t_t)$ denote the probability of transit times t_t being larger than the 'present' time, t, meaning that these particles are not (yet) lost from the system. By this definition we have $F(t < t_t)J_0\mathrm{d}t$ being the average number of the previously introduced particles remaining inside the system at a later time t. The *total* number of particles in the system is then all the particles that have not yet reached the boundary $\int_0^\infty F(t < t_t)J_0 \, \mathrm{d}t = N$. We *might* find that this integral diverges, meaning that particles remain, on average, for an infinitely long time in the system. In general we *can* let J_0 be time varying, but here we take it to be constant, and find $N = J_0 \int_0^\infty F(t < t_t)\mathrm{d}t$. This expression can be rewritten in terms of the transit time probability density $P(t_t) = -\mathrm{d}F/\mathrm{d}t_t$ as

$$
\begin{aligned}
N &= J_0 \int_0^\infty F(t < t_t)\mathrm{d}t = J_0 \int_0^\infty \int_t^\infty P(t_t)\mathrm{d}\, t_t \, \mathrm{d}t \\
&= J_0 \int_0^\infty \int_0^{t_t} P(t_t)\mathrm{d}t \, \mathrm{d}t_t = J_0 \int_0^\infty t_t P(t_t)\mathrm{d}t_t \\
&\equiv J_0 \, \langle t_t \rangle .
\end{aligned}
\tag{21.45}
$$

We used the integral operator identity $\int_0^\infty \mathrm{d}t \int_t^\infty \mathrm{d}t_t = \int_0^\infty \mathrm{d}t_t \int_0^{t_t} \mathrm{d}t$, which can easily be demonstrated. We have the general result

$$
\langle t_t \rangle = N/J_0, \tag{21.46}
$$

which can be used to determine the average transit time, when J_0 is known and N can be obtained by other means. Note that the result (21.46) is not based on any explicit assumption of diffusive motion (Boffetta and Sokolov 2002).

Consider, for instance, a simple one-dimensional model system, where particles are introduced at the origin at a constant rate J_0 and are absorbed at two boundaries placed at a and $-a$, respectively. Let the particle motion be governed by the diffusion equation with constant diffusion coefficient D. Under steady state conditions we have $Dd^2n/dx^2 = 0$ and easily find the spatially varying particle density to be $n(x) = A - B|x|$. The constant B is determined by equating the flux out of the system and into the system, i.e. $2DB = J_0$. Absorption at $|x| = a$ implies $n(x = \pm a) = 0$, giving $A = Ba$. The total number of particles in the system is then $N = \int_{-a}^{a} n(x)\mathrm{d}x = \frac{1}{2}a^2 J_0/D$. The average transit time becomes $\langle t_t \rangle = N/J_0 = \frac{1}{2}a^2/D$.

We can now relax the symmetry of the foregoing simple model, and assume the two absorbing boundaries to be placed at a and $-b$, respectively, with $a > 0$ and $b > 0$. The arguments are exactly the same as before, so we find the particle density to be given by $n = n_0(1 - Dx/a)$ for $x > 0$, and $n = n_0(1 - Dx/b)$ for $x < 0$. The density n_0 at $x = 0$ is determined by noting that the diffusive fluxes to the two boundaries are $J_a = n_0 D/a$ and $J_b = n_0 D/b$, respectively. The flux J_0 *into* the system is presumed known, so we have $J_a + J_b = J_0$ implying

$$n_0 = J_0 \frac{ab}{D(a+b)}.$$

For the present case we find the average transit, or occupation time to be

$$\langle t_t \rangle = \frac{\int_a^b n(x)\mathrm{d}x}{J_0} = \frac{n_0(a+b)/2}{n_0 D\left(\frac{1}{a} + \frac{1}{b}\right)} = \frac{ab}{2D}.$$

In particular, we find that $\langle t_t \rangle \to \infty$ if either a or $b \to \infty$, in agreement with the results from section 21.3.2.

References

Balescu R 1997 *Statistical Dynamics. Matter out of Equilibrium* (London: Imperial College Press)

Boffetta G and Sokolov I M 2002 Relative dispersion in fully developed turbulence: the Richardson law and intermittency corrections *Phys. Fluids* **14** 3224–32

Carslaw J C and Jaeger H S 1947 *Conduction of Heat in Solids* (Oxford: Clarendon)

Chandrasekhar S 1954 Stochastic problems in physics and astronomy ed N Wax *Selected Papers on Noise and Stochastic Processes* (New York: Dover) pp 3–91

Codling E A, Plank M J and Benhamou S 2008 Random walk models in biology *J. R. Soc. Interface* **5** 813–34

Cox D R and Miller H D 1965 *The Theory of Stochastic Processes* (London: Chapman and Hall)

Crank J 1975 *The Mathematics of Diffusion* 2nd edn (Oxford: Clarendon)

Crank J, N R McFarlane J C N, Paterson G D and Pedley J B 1981 *Diffusion Processes in Environmental Systems* (London: Palgrave MacMillan)

Feller W 1968 *An Introduction to Probability Theory and its Applications* vol 1 (New York: Wiley)
Feller W 1971 *An Introduction to Probability Theory and its Applications* vol 2 (New York: Wiley)
Gardiner C W 1985 *Handbook of Stochastic Methods for Physics, Chemistry and the Natural Sciences* 2nd edn (Berlin: Springer)
Kac M 1946 Random walk and the theory of Brownian motion *Am. Math. Monthly* **54** 369 Reprinted in
Kac M 1954 *Selected Papers on Noise and Stochastic Processes* ed N Wax (New York: Dover) pp 295–317
Mazo R M 2002 *Brownian Motion. Fluctuations, Dynamics and Applications* (Oxford: Oxford Science Publications)
Ming C W and Uhlenbeck G E 1954 On the theory of the brownian motion-II *Selected Papers on Noise and Stochastic Processes* ed N Wax (New York: Dover) pp 113–32
Mosteller F 1987 *Fifty Challenging Problems in Probability with Solutions* (New York: Dover)
Narasimhan T N 1999 Fourier's heat conduction equations: history, influence, and connections *Rev. Geophys.* **37** 151–72
Paul W and Baschnagel J 1999 *Stochastic Processes from Physics to Finance* (Berlin: Springer)
Pécseli H L 2000 *Fluctuations in Physical Systems* (Cambridge: Cambridge University Press)
Seinfeld J H 1983 Atmospheric diffusion theory *Advances in Chemical Engineering* vol 12 ed J Wei (New York: Academic Press) pp 209–99
van Kampen N G 1981 *Stochastic Processes in Physics and Chemistry* (Amsterdam: North-Holland)

IOP Publishing

Low Frequency Waves and Turbulence in Magnetized Laboratory Plasmas and in the Ionosphere

Hans Pécseli

Chapter 22

Turbulence in two and three spatial dimensions

Linearly unstable drift wave oscillations, treated as an initial value problem, will increase in amplitude at an exponential rate, until nonlinear effects ignored in the linear stability analysis ultimately become important. The final saturated stage of such linearized instabilities depends in general on many different effects, and no general statement can be made concerning the detailed nature of that stage. We might expect that the final stage can be dominated by a single high amplitude mode, or alternatively that many interacting modes can be excited. With particular reference to the drift wave instability, one might actually expect that at least for some conditions the saturated stage does not exist at all! After all, it would be natural to expect that as the drift wave amplitudes increase, the oscillations give rise to an increased anomalous diffusion rate which ultimately flattens out the density gradient, which originally gave rise to the instability. At this time the plasma becomes stable. In many cases of interest, the density gradient is, however, at least in some average sense, maintained by an external agency. The vertical density gradient of the ionosphere is one obvious example, and a saturated stage of the instability will exist in such cases. In such cases we might expect a stationary turbulent state to be maintained.

The description of turbulent drift waves relies on a number of basic methods, which are well developed for the study of turbulence in neutral fluids in three spatial dimensions. It seems to be a good starting point to give a summary of some of the basic methods and results obtained for that problem. A wealth of relevant literature exists, and no attempt is made here to give a full coverage. A few basic texts are provided by, for instance, Tennekes and Lumley (1972), Hinze (1975), and Pope (2000). A very readable introduction to the statistical theory of turbulence was given by Kampe de Feriet (1954–1955).

The results from neutral turbulence cannot directly be applied for magneto-hydrodynamic turbulence (Davidson 2001) where we have a characteristic velocity

doi:10.1088/978-0-7503-1251-6ch22

in the linearized set of equations. This means that information can be propagated by the Alfvén velocity. In neutral incompressible turbulence the linearized equations are close to trivial, and information is there propagated by nonlinear effects. Other models for plasma dynamics, in magnetized or unmagnetized conditions, are often even more different from the reference fluid case. The modeling and description of flows in fluids and plasmas can thus become rather different, in particular also when turbulent flows are considered. There are however a number of basic elements of models that are universal or general, and some of these will be discussed here.

22.1 Viscosity

One of the most important features of turbulent flows is their ability to mix constituents at a rate far exceeding that due to molecular mixing. One important constituent of a flow is its local momentum density, and this becomes mixed as any other constituent. Molecular momentum mixing is what we call 'classical viscosity', so it is appropriate to start out with a summary of this concept.

22.1.1 Molecular viscosity

In discussions of the Navier–Stokes equation we let the force on a fluid element be given by the gradient in the pressure. There are of course other contributions, gravity for instance, which may or may not be important, depending on conditions. We expect, however, the equivalent of *friction* to be important for a wide range of conditions. The force, $\mathbf{F}_U$, experienced by a fluid element due to the 'friction' with its surroundings is usually denoted *viscous interaction*. We expect these interactions to be dependent on the local velocity, $\mathbf{u}(\mathbf{r}, t)$ of the flow, and lead to an expansion of the form

$$\mathbf{F}_U = \mathbf{C}_0 + \mathsf{C}_1 \cdot \mathbf{u} + \mathsf{C}_2 \cdot \nabla\mathbf{u} + \mathsf{C}_{3a}\nabla^2\mathbf{u} + \mathsf{C}_{3b}\nabla\nabla \cdot \mathbf{u} + \cdots, \tag{22.1}$$

where $\mathbf{C}_0$ is a vector, C_1 is a tensor, etc. The two terms with scalar coefficients C_{3a} and C_{3b} represent the two possible combinations of a vector, where the ∇-operator acts twice on a vector $\mathbf{u}$. Recall that $\nabla \times \nabla\mathbf{u} = \nabla\nabla \cdot \mathbf{u} - \nabla^2\mathbf{u}$, so there are no rotational parts missing in (22.1).

Some of the coefficients in (22.1) can be discarded right away: it would thus be unphysical if friction was effective for a fluid at rest, and hence $\mathbf{C}_0 = 0$. Similarly, if $\mathsf{C}_1 \cdot \mathbf{u} \neq 0$ for a certain velocity profile; it would be strange that the viscous effects were to change sign just because the velocity gradient changed sign, i.e. we expect $\mathsf{C}_1 = 0$ and $\mathsf{C}_2 = 0$. The first effective term will therefore be the one with coefficient C_{3a} and C_{3b}. We will first consider *incompressible* flows, and take $\nabla \cdot \mathbf{u} = 0$. For isotropic flows, we then find the simplest lowest order form to be $\mathbf{F}_U = \mu\nabla^2\mathbf{u}$. This is the force density associated with the 'friction' of fluid with the surrounding flow. If we divide $\mathbf{F}_U$ with the fluid mass density we obtain a quantity having the dimension of acceleration. Intuitively, we expect viscosity to give a *deceleration*, but this *need* not be so: imagine a striated velocity field where we have a local *reduction* in velocity. For this case, the friction with the surrounding flow will *accelerate* a fluid element.

Physically the viscosity is expressing the internal momentum exchange mediated by atoms or molecules in the fluid when they cross an (imagined) boundary, or 'cut' $A - A'$, in the flow. Assume that the gradient is such that the horizontal velocity component *increases* with the vertical coordinate, z. Particles entering through the 'cut' $A - A'$ from *above* due to their thermal motion, have on average a larger horizontal velocity component compared to those crossing $A - A'$ from below. Hence, there must be a momentum exchange across $A - A'$, in general.

The *dynamic* viscosity is a measure for the fluids resistance towards deformations. Newton's relation gives for the shear stress τ, in the simple geometry outlined here

$$\tau = \mu \frac{\mathrm{d}}{\mathrm{d}z} u, \tag{22.2}$$

where μ is the *dynamic* viscosity. Usually, μ depends strongly on temperature, but not so much on pressure. In the oceans, μ depend also on the salinity of the water. We can define the *kinematic* viscosity, $\nu \equiv \mu/\rho$, where again, ρ is the mass density of the fluid. The dimension of τ is *force/area*, i.e. the same as *pressure*, but in this case the force acts *parallel* to the surface, so it is important to avoid confusion. The quantity μ in (22.2) has dimension *mass* × *length*$^{-1}$ × *time*$^{-1}$, and ν has dimensions *length*^{2} × *time*$^{-1}$.

An explicit calculation of the viscosity coefficient

By a simple model we are able to give a simple analytical expression for the viscosity of a fluid or a gas. We can calculate the net momentum carried from *above* the cut $A - A'$, *through* a unit area per second. Let the density of particles be $n_0 = \text{const.}$ and the directions of the particle velocity vectors $\mathbf{U}$ to be uniformly distributed over the spherical angle 4π. We use capital letters for particle velocities, in contrast to lower case letters for flow velocities. In this section we distinguish fluid velocity vectors (capital letters) with particle velocity vectors (lower case letters). The fraction of particles with velocities in the spherical angular element $\mathrm{d}\Omega$ is then $n_0 \mathrm{d}\Omega/(4\pi)$. We want to determine first the number of particles with velocity U that enter through a unit area within a directional angle in the narrow interval $\{\Theta, \Theta + \mathrm{d}\Theta\}$ within a time interval $\mathrm{d}t$. These particles fill a box with base being unit area, and height $U\,\mathrm{d}t \cos\Theta$. This expression only includes the *magnitude*, U, of the velocity. Only a fraction of the particles has the correct *direction* for the velocity vector, namely $n_0 U\,\mathrm{d}t \cos\Theta \mathrm{d}\Omega/(4\pi)$. We now use this expression for an angular interval of the form of two concentric conical segments with angle Θ and $\Theta + \mathrm{d}\Theta$, respectively. The corresponding angle is $2\pi \sin\Theta \mathrm{d}\Theta$, and the corresponding number of particles is

$$\frac{2\pi \sin\Theta \mathrm{d}\Theta}{4\pi} n_0 U\,\mathrm{d}t \cos\Theta.$$

Integration of this expression over all angles Θ in the interval $\{0, \pi/2\}$ gives $\frac{1}{4} n_0 U\,\mathrm{d}t$. We now integrate over all velocities, assuming a Maxwellian velocity distribution and obtain trivially $\frac{1}{4} n_0 \langle |U| \rangle$ particles per second (Reif 1965).

In order to find the *momentum* carried across the unit area of particles with mass m coming from the region above, we recall that the vertical momentum component of such a particle is $mu \cos\Theta$ and we have to integrate

$$\frac{2\pi \sin\Theta \mathrm{d}\Theta}{4\pi} n_0 U \cos^2\Theta m U.$$

Integrating again this expression over all angles Θ in the interval $\{0, \pi/2\}$ we find $\frac{1}{6} n_0 m U^2$. We now integrate over all velocities, assuming a Maxwellian velocity distribution and obtain trivially a momentum $\frac{1}{6} n_0 \langle U^2 \rangle$ per second. It is trivially demonstrated that $\langle U^2 \rangle = 3\kappa T/m$ for a Maxwellian distribution, while $\langle |U| \rangle^2 = 8\langle U^2 \rangle/(3\pi)$. If we are interested in the *pressure* exerted by such particles on a *solid* wall, we have to recall that the wall receives *twice* the momentum upon a completely elastic reflection, and the pressure, i.e. momentum received per area per second becomes $p = \frac{1}{3} n_0 \langle U^2 \rangle$.

We now consider the inhomogeneous velocity case where it is assumed that the average velocity is in all spatial positions much smaller than the thermal velocity. We can then, as an approximation, still use some of the previous results, when estimating the net momentum carried across the unit area as the difference between the momentum received from above minus the momentum received from below. We write the vertical variation of the horizontal average velocity component as $\langle u_y(z) \rangle$. A particle crossing $A - A'$ carries the information from the last time it collided with one of the other particles in the flow. For simplicity, and without much loss of generality, we assume that the mean free collision path, ℓ_c is identically the same for all particles.

To a good approximation we can assume, also for the present inhomogeneous flow, that the *vertical* momentum crossing $A - A'$ from above and below exactly cancels. Since, however, the *horizontal* flow component varies with z, there will be a net y-component of the momentum deposited at $A - A'$. To estimate this we write the net rate of momentum transfer across a unit area at the reference position z_0 of the cut $A - A'$ as

$$P_y(z_0, \Theta)\mathrm{d}\Theta = \frac{2\pi \sin\Theta \mathrm{d}\Theta}{4\pi} n_0 U \cos\Theta$$
$$m\Big[\big\langle u_y(z = z_0 - \ell_c \cos\Theta)\big\rangle - \big\langle u_y(z = z_0 + \ell_c \cos\Theta)\big\rangle\Big].$$

Making a Taylor expansion of the average velocity profile as

$$\big\langle u_y(z)\big\rangle \approx \frac{\mathrm{d}\langle u_y(z)\rangle}{\mathrm{d}z}\bigg|_{z=z_0} (z_0 + z) \equiv \big\langle u_y'(z_0)\big\rangle (z_0 + z),$$

we can write, without any explicit reference to the particular z_0 position,

$$P_y(z, \Theta)\mathrm{d}\Theta = -2\frac{2\pi \sin\Theta \mathrm{d}\Theta}{4\pi} n_0 m U\, \ell_c \cos^2\Theta \big\langle u_y''(z)\big\rangle.$$

After integration over all Θ in the range $\{0; \pi/2\}$ and average over all velocities assuming a local Maxwellian, we obtain

$$P_y(z) = -\frac{1}{3} n_0\, m\, \langle|U|\rangle \ell_c \frac{\mathrm{d}\langle u_y(z)\rangle}{\mathrm{d}z}, \qquad (22.3)$$

where, as mentioned before, we assume that $\langle U\rangle \ll \langle|U|\rangle$. This is the net *force* per unit *area* exerted of the fluid *above* z_0 on the fluid *below*. The dimension of $P_y(z)$ is *force* $\times$ *area*$^{-1}$. In order to obtain the net viscous drag we take a small box, or parcel of fluid, with base area A and height dz at the selected reference position, and calculate the difference in viscous forces between top and bottom, and find in terms of acceleration a_y in the y-direction that $a_y n_0 m A dz = AP_y(z - \mathrm{d}z/2) - AP_y(z + \mathrm{d}z/2)$. This gives

$$a_y = \frac{1}{3} \langle|U|\rangle \ell_c \frac{\mathrm{d}^2\langle u_y(z)\rangle}{\mathrm{d}z^2}$$

in the limit where $\mathrm{d}z \to 0$. By direct comparison with the first non-vanishing term in (22.1) we find the kinematic viscosity to be

$$\nu = \frac{1}{3} \langle|U|\rangle \ell_c, \qquad (22.4)$$

at least to the accuracy of the present calculations. As mentioned, for a Maxwellian distribution we have $\langle|U|\rangle = 2\sqrt{2\langle U^2\rangle/(3\pi)}$. The dynamic viscosity becomes $\mu = \frac{1}{3} n_0 m \langle|U|\rangle \ell_c$. This result is quite interesting since $\ell_c \approx 1/(n_0\, \sigma)$ in terms of a collisional cross-section σ, and μ is then, to the present approximation, independent of gas density! It varies with temperature, though, since $\langle|U|\rangle \sim \sqrt{T}$. If we have a *linear* gradient in velocity, the box or parcel of fluid, will be accelerated on the top by an amount equal to the deceleration at the bottom, and the net acceleration vanishes. To be effective, viscosity requires a non-vanishing second order derivative of the velocity gradient.

The foregoing analysis implicitly assumes that the density is constant, a requirement a bit stronger than the condition of incompressibility, which was made explicit. In realistic flows it is however rather unlikely to encounter density gradients which are comparable to the velocity gradients, at least if viscosity is to play any role at all. If required the analysis can readily be generalized. The problem of including *compressible* flows, is however a more important, but also more complicated problem. In this case *both* terms with C_{3a} and C_{3b} in (22.1) have to be retained, and the question is to determine these two coefficients. We will not enter these discussions here in any detail, only note the result, where the fluid equation is written as

$$\rho \frac{D}{Dt}\mathbf{u} = \rho\left(\frac{\partial}{\partial t}\mathbf{u} + \mathbf{u}\cdot\nabla\mathbf{u}\right) = \mathbf{F} + \mu\nabla^2\mathbf{u} + (\lambda + \mu)\nabla\nabla\cdot\mathbf{u}. \qquad (22.5)$$

For incompressible flows we have $\nabla\cdot\mathbf{u} = 0$. With $\mathbf{F} = -\nabla p$, the relation (22.5) is called the *Navier–Stokes* relation. In many cases it is possible to argue for Stokes relation $\lambda + \mu = \mu/3$ in (22.5). Often, it is possible to assume incompressibility for the flows, as already argued, but if we want to study, for instance, damping of sound waves in gases or liquids, the last term in (22.5) becomes important. Stokes relation is found to be good for mon-atomic gases, but often fails for liquids and poly-atomic gases. For gases, we generally find that viscosity *increases* with temperature, but for liquids, molecular liquids in particular, it might be the other way. For water, we thus find the viscosity to increase with decreasing temperature. Many liquids follow an approximate relation of the form $\mu \approx A + B/(\kappa T)$ for the dynamic viscosity (Kaye and Laby 1995), valid for a non-trivial temperature range with constant A and B.

A *Newtonian fluid* is one where there is a linear relationship between stress and strain (Aris 1962). This is generally given as a tensor relation, where (22.6) is a simple scalar version of this. Non-Newtonian fluids *do* exist, and can have quite peculiar properties.

- **Example:** Take a simple metal bar of length ℓ_0, and cross-section $\mathcal{A}$, fixed at one end, and subject to a small tensile force F at the other end. The force gives rise to an elongation $d\ell$, so that the *entire* length of the bar is $\ell_0 + d\ell$. *Hooke's law*, which relates the elongation to the force, has the form (Brekhovskikh and Goncharov 1985)

$$\mathcal{E}\frac{d\ell}{\ell_0} = \frac{F}{\mathcal{A}}. \tag{22.6}$$

If we double the length ℓ_0 of the bar, the relative elongation becomes only half of what is was before, for given $F/\mathcal{A}$. If we double the cross-sectional area, we need twice the force to produce the same elongation. The quantity $\mathcal{E}$ is *Young's modulus*, $F/\mathcal{A}$ is the *stress*, while the relative extension $\varepsilon \equiv d\ell/\ell_0$ is the *strain*. Hooke's law (22.6) is a *linear* relation: doubling the force gives a doubling of the elongation. These definitions can be readily generalized to cover forces in fluid or gases.

- **Exercise:** An elegant example suggested by Reif (1965) may illustrate the effect of viscosity. Consider two trains moving on parallel tracks, both of them loaded with sand-bags. Let the speed of the trains be somewhat different, and assume now that the crew on both trains constantly pick up sandbags on their own train, and throw them on board the other one. There will be a net transport of momentum between the trains, and the slower one will speed-up, while the faster one will be slowed down. Make a realistic model for the problem outlined here, and calculate the rate of acceleration of the two trains. Assume that at time $t = 0$ they have constant but different velocities. Generalize the problem to *three* trains with velocities $U_1 < U_2 < U_3$, where the workers on the middle one (nr. two) throw an equal number of sandbags to the right and to the left trains, the other two onto train nr. two only. Determine the parameters which give train nr. two a constant velocity.

22.1.2 Eddy viscosity

By adopting some of the basic arguments used in deriving expressions for molecular viscosity, we can obtain at least *some* insight into the basic properties of a turbulent flow.

In laminar incompressible flows, we assume the relation between the stress tensor $\tau_{i,j}$ and rate of strain as

$$\tau_{i,j} = -p\delta_{i,j} + \rho\nu\left(\frac{\partial u_i}{\partial r_j} + \frac{\partial u_j}{\partial r_i}\right),$$

where ν is the kinematic viscosity and ρ is the mass density of the fluid, assumed constant, see also (E.5) and (E.6). It was suggested, seemingly already by Boussinesq in 1877 (Beran 1968), that a more general relation could be obtained as

$$\overline{\tau}_{i,j} = -\overline{p}\,\delta_{i,j} + \rho(\nu + \varepsilon_\nu)\left(\frac{\partial \overline{u}_i}{\partial r_j} + \frac{\partial \overline{u}_j}{\partial r_i}\right), \tag{22.7}$$

where the overbar indicates time averaged quantities, and ε_ν is an additional viscosity term somehow related to the turbulence. We anticipate that ε_ν might depend on position (Beran 1968). The turbulence contribution can be estimated by arguments very similar to those used for deriving the expression for molecular viscosity in section 22.1.1 by replacing the particle collisional mean free path by a characteristic size, ℓ_t, of the turbulent eddies. Often we find the term 'mixing-length' used for ℓ_t, although its definition is not unique. Since ν was shown to be given as a product of the mean free path and a characteristic velocity, see (22.4), we expect

$$\varepsilon_\nu \approx \ell_t\langle|u|\rangle.$$

In general both ℓ_t and $\langle|u|\rangle$ will depend on position, in particular close boundaries, here taken to be along the x-direction. To estimate $\langle|U|\rangle$ we can argue that in a velocity gradient, taken to be in the y-direction, the variations in velocity occur when a fluid element traveling with the average velocity in the x-direction at a layer $y = y_0 - y'$ is moving, at some statistically distributed times, to a layer at $y = y_0$. Since the 'mean free path' for this estimate is ℓ_t, we expect

$$\langle|u|\rangle \approx \ell_t\left|\frac{\mathrm{d}\overline{u}}{\mathrm{d}y}\right|.$$

This approximation will be used for estimating the flow variation near the ground. As far as the mixing length ℓ_t is concerned we expect (Beran 1968, Tennekes and Lumley 1972) that close to a boundary it is given simply by the distance to the boundary: this will correspond to mixing by the largest eddies that can transfer momentum from the position y to the ground at $y = 0$. We thus find

$$\varepsilon_\nu \approx \ell_t^2\left|\frac{\mathrm{d}\overline{u}}{\mathrm{d}y}\right| = y^2\left|\frac{\mathrm{d}\overline{u}}{\mathrm{d}y}\right|, \tag{22.8}$$

apart from a numerical constant which is presumably of order unity. The model (22.8) has a certain appeal, but should be considered with some care, since unlike a particle or a molecule, a volume element of fluid does not travel a finite distance and then suddenly transfer its momentum, entirely or in part, to the surrounding fluid. It turns out, however, that (22.8) leads to physically plausible results.

In order to demonstrate the usefulness of (22.8) it may be instructive to consider a turbulent boundary layer flow, with a 'no-slip' condition at the wall, i.e. $u(y = 0, t) = 0$. The equation of motion for $\overline{u} = \overline{u}(y)$ under steady state conditions is

$$\frac{1}{\rho}\frac{\partial \overline{p}}{\partial x} = \frac{\partial}{\partial y}\left((\varepsilon_\nu(y) + \nu)\frac{\partial \overline{u}}{\partial y}\right), \tag{22.9}$$

where x is in the flow direction, $\overline{\mathbf{u}} \| \hat{\mathbf{x}}$, and $\hat{\mathbf{y}}$ is perpendicular to it. The partial derivative with respect to x has been set to zero for all terms except for the pressure term (after all, we expect the flow in the x-direction to be driven by a pressure gradient). We find with these assumptions together with (22.9) that the total shear, including both molecular viscosity and eddy viscosity is given by

$$\tau_{x,y} = \rho\left((\varepsilon_\nu(y) + \nu)\frac{\partial \overline{u}}{\partial y}\right).$$

In realistic turbulent flows, ℓ_t is much larger than any collisional mean free path, and $\varepsilon_\nu \gg \nu$ for such cases, except very close to the wall, where the turbulent eddies can be comparable in size to the mean free paths.

The ideas of mixing lengths are here introduced for discussions of boundary layers, but the basic elements of the arguments apply more widely. They have also been used for discussing particle diffusion, in particular also for magnetized plasmas (Kadomtsev 1965, Hatayama *et al* 1981) by using ℓ_t as a characteristic step-size for a turbulent random walk. When dealing with a linearly unstable plasma, resistive electrostatic drift waves for instance, we might within a weak turbulence approach expect the saturated spectrum to be dominated by the linearly most unstable wave component $\mathbf{k}^{(m)}$. In this case it can be argued that $\ell_t \sim 1/|k^{(m)}|$.

Logarithmic boundary layer

We assume now that the average pressure gradient is small, $\partial\overline{p}/\partial x \approx 0$, so that the flow is in an 'average' equilibrium. For a constant shear stress layer, characterized by τ_0 controlled solely by eddy viscosity, we can integrate

$$\tau_0 = \rho y^2\left(\frac{\mathrm{d}\overline{u}}{\mathrm{d}y}\right)^2 \tag{22.10}$$

to give

$$\overline{u}(y) = \frac{1}{A}\log(y)\sqrt{\frac{\tau_0}{\rho}} + B, \tag{22.11}$$

with B being an integration constant, which can be used to normalize the y-coordinate in (22.11), if you like. To compensate for all the numerical constants ignored in the analysis, we introduced an unknown numerical coefficient A, which we expect to be of order unity, though. Close to the wall, $y \approx 0$, we can no longer ignore molecular viscosity, as already mentioned, and it is no surprise that the expression (22.11) diverges there. The logarithmic flow variation given by (22.11) is well documented experimentally.

We note that the arguments for estimating ℓ_t are not unique: one might as well choose

$$\ell_t = k\frac{d^2\overline{u}}{d^2y}/\frac{d\overline{u}}{dy}, \tag{22.12}$$

as proposed by von Karman, with k being a universal constant (Beran 1968). The expression (22.12) ends up giving a logarithmic form just as (22.11). We see right away that for a logarithmic velocity profile, (22.12) also gives $\ell_t \sim y$.

The expression (22.11) can be generalized to also account for the region near the wall (Landahl and Mollo-Christensen 1992). The general expression, including *both* molecular and eddy diffusivities, is obtained as

$$\tau_0 = \rho y^2\left(\frac{d\overline{u}}{dy}\right)^2 + \mu\frac{d\overline{u}}{dy}, \tag{22.13}$$

replacing (22.10). The expression (22.13) is readily solved for $d\overline{U}/dy$, which can subsequently be integrated to give

$$\begin{aligned}\overline{u}(y) = \frac{1}{A}\Bigg[&\ln\left(2Ay\tau_0/\mu + \sqrt{1 + 4A^2y^2\tau_0^2/\mu^2}\right)\\ &+ \frac{\mu}{2Ay\tau_0}\left(1 - \sqrt{1 + 4A^2y^2\tau_0^2/\mu^2}\right)\Bigg].\end{aligned} \tag{22.14}$$

It is easily shown that in the limit of large y, the expression (22.14) reproduces the logarithmic law (22.11), while on the other hand we have $\overline{u} \to 0$ for $y \to 0$. The result is illustrated in figure 22.1.

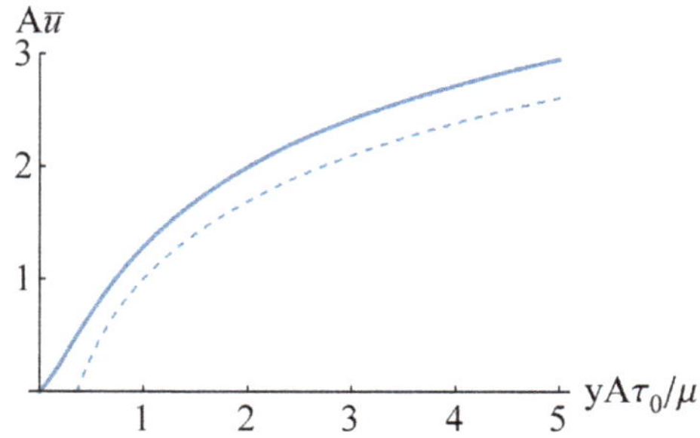

Figure 22.1. The general result (22.14) for the velocity distribution near a wall in normalized units. The dashed line shows $1 + \ln(yA\tau_0/\mu)$ for reference.

22.1.3 Heuristic model for a turbulent boundary layer

To give an additional discussion of the boundary layer model from section 22.1.2, we introduce a simple heuristic boundary layer model, illustrated in figure 22.2. We assume that close to the wall the flow is in general quiet, while it becomes increasingly probable to encounter turbulent regions as we move deeper into the flow, for increasing distances from the wall. We assume, heuristically, that there is a well defined boundary separating the turbulent region from the quiescent flow, with the positions of this boundary being fluctuating in time. We can define an *intermittency coefficient*, as the probability of being in the quiescent region (often the definition is the other way around, using the probability of being inside the turbulent region), see figure 22.2. The figure also defines the average flow, varying with distance from the wall, as well as the shear stress.

The logarithmic boundary layer can also be argued by dimensional reasoning which can also give some insight into the shortcomings of the model. For a half-open boundary layer (zero velocity at the bottom, no limits on the top) there are no length scales available, so the only quantity that can normalize the altitude y is the altitude itself: this is what the proportionality $\ell_t \sim y$ states. Physically this means that the turbulent eddies most efficient in transferring momentum from an altitude y to ground are the eddies of scale size y. The only scale-free function we have is a logarithm: a change in normalization gives merely a shift of the function as $\log(y) \rightarrow \log(y/a) + \log(a)$. Close to ground, typically a few hundred meters in normal atmospheres, the logarithmic profile serves as a very good approximation. The model also predicts however that the scaling increases linearly with altitude, and eventually very large velocities should be found (this would be a good reason to make very large windmills!). In reality it is not so, and a general study rests on fits based on the Weibull distribution

$$f(y;\, \lambda,\, k) = \frac{k}{\lambda}\left(\frac{y}{\lambda}\right)^{k-1} \exp\!\left(-(y/\lambda)^k\right)$$

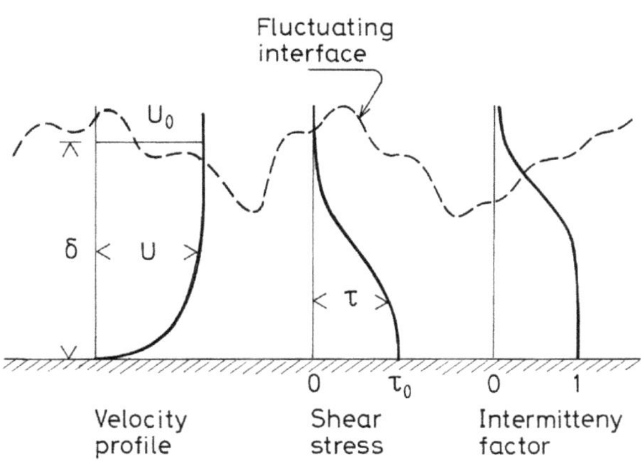

Figure 22.2. Schematic model for a turbulent boundary layer.

for $y \geqslant 0$ and $f(y; \lambda, k) = 0$ otherwise. Here $k > 0$ is the *shape parameter* and $\lambda > 0$ is the *scale parameter* of the distribution, the parameters usually fitted by experiments. For even larger altitudes, the Ekman spiral (Tritton 1988) becomes noticeable, but this topic is outside the present treatise.

Boundary layers have been studied in great detail in neutral flows, in the atmosphere or in the oceans. The ideas seem to be relevant for plasmas as well, but that problem usually involves recombination at the surfaces and the problem of turbulent plasma boundary layers in magnetized or unmagnetized plasmas has not been resolved.

- **Exercise:** Determine the moment generating function for the Weibull distribution and determine its variance, skewness and kurtosis.

22.2 Energy and enstrophy budgets in two- and three-dimensional turbulence

It may be worthwhile briefly to discuss energy and enstrophy cascades in three- as well as two-dimensional neutral fluid turbulence. In this case the description is usually based on the Navier–Stokes equation for the fluid velocity **u**

$$\frac{\partial}{\partial t}\mathbf{u} + \mathbf{u} \cdot \nabla\mathbf{u} = -\frac{1}{\rho}\nabla P + \nu\nabla^2\mathbf{u} \tag{22.15}$$

assuming incompressible flows, $\nabla \cdot \mathbf{u} = 0$, which for a uniform density ρ implies the following relation for the pressure $\frac{1}{2}\nabla^2 P/\rho + \nabla \cdot (\mathbf{u} \cdot \nabla\mathbf{u}) = 0$, having the form of a Poisson equations which can be solved for P in terms of $\mathbf{u}$. The assumption of incompressibility is equivalent to introducing an equation of state.

22.2.1 Energy budget expressed in wavenumbers

An equation for the fluid energy density $\frac{1}{2}\rho u^2$ with constant mass density ρ can be obtained from (22.15) as

$$\frac{1}{2}\frac{\partial}{\partial t}u^2 + \nabla \cdot \left(\mathbf{u}\frac{1}{2}u^2 + \mathbf{u}\frac{P}{\rho}\right) = \nu\nabla \cdot (\mathbf{u} \times \boldsymbol{\omega}) - \nu\omega^2. \tag{22.16}$$

The fluid vorticity was introduced as $\boldsymbol{\omega} \equiv \nabla \times \mathbf{u}$. (The notation using the symbol for angular frequency is a bit unfortunate, but is by now firmly established.)

The time evolution of the total volume integrated energy of the system, $E \equiv \frac{1}{2}\int \rho u^2 d\mathbf{r}$, is then given as

$$\frac{\partial}{\partial t}E = -\nu\rho\int \omega^2 d\mathbf{r}, \tag{22.17}$$

where we disregarded possible contributions from the bounding surface. It is tempting here to argue that in the limit $\nu \to 0$, this equation predicts energy conservation, but it should be emphasized that the argument relies on the additional assumption that $\int \omega^2 d\mathbf{r}$ remains bounded during the transition.

Expanding the spatial variation of the fluid velocity in Fourier components, we find

$$\frac{\partial}{\partial t}\mathbf{u}(\mathbf{k},\, t) = -i\sum_{\mathbf{k}'}\Big[\mathbf{u}(\mathbf{k}',\, t)\cdot\mathbf{k}\Big]\mathbf{u}(\mathbf{k}-\mathbf{k}',\, t) - i\frac{1}{\rho}P(\mathbf{k},\, t)\mathbf{k} - \nu k^2\mathbf{u}(\mathbf{k},\, t), \quad (22.18)$$

together with the incompressibility condition $\mathbf{u}(\mathbf{k}, t)\cdot\mathbf{k} = 0$, i.e. $\mathbf{u}$ and $\mathbf{k}$ are orthogonal. (Note that there are two different wavenumbers entering the angular brackets in (22.18), and $\mathbf{u}(\mathbf{k}', t)\cdot\mathbf{k} \neq 0$.) Multiplying (22.18) by $\mathbf{u}^*(\mathbf{k}, t)$ and using the orthogonality condition, we find

$$\begin{aligned}\frac{1}{2}\frac{\partial}{\partial t}|\mathbf{u}(\mathbf{k},\, t)|^2 = {}& -i\sum_{\mathbf{k}'}\Big[\mathbf{u}(\mathbf{k}',\, t)\cdot\mathbf{k}\Big]\mathbf{u}(\mathbf{k}-\mathbf{k}',\, t)\cdot\mathbf{u}(-\mathbf{k},\, t)\\ & - \nu k^2\,|\mathbf{u}(\mathbf{k},\, t)|^2,\end{aligned} \quad (22.19)$$

where we recall that $\mathbf{u}$ real implies $\mathbf{u}^*(\mathbf{k}, t) = \mathbf{u}(-\mathbf{k}, t)$. The important observation is that the *pressure* does not appear in (22.19). Pressure plays a role in exchanging energy between spectral velocity components belonging to the *same* wavenumber, but seemingly *not* in cascading energy from one wavenumber range to another. This spectral cascade is mediated by the first term on the right-hand side of (22.19).

If we complex conjugate (22.19), we obtain

$$\frac{1}{2}\frac{\partial}{\partial t}|\mathbf{u}(\mathbf{k},\, t)|^2 = i\sum_{\mathbf{k}'}\Big[\mathbf{u}(-\mathbf{k}',\, t)\cdot\mathbf{k}\Big]\mathbf{u}(\mathbf{k}'-\mathbf{k},\, t)\cdot\mathbf{u}(\mathbf{k},\, t) - \nu k^2\,|\mathbf{u}(\mathbf{k},\, t)|^2. \quad (22.20)$$

Adding (22.19) and (22.20), and summing over all $\mathbf{k}$ (integrating, if we have a continuous spectrum), we obtain

$$\frac{1}{2}\frac{\partial}{\partial t}\sum_{\mathbf{k}}|\mathbf{u}(\mathbf{k},\, t)|^2 = -\nu\sum_{\mathbf{k}}k^2\,|\mathbf{u}(\mathbf{k},\, t)|^2. \quad (22.21)$$

Since $\sum_{\mathbf{k}}|\mathbf{u}(\mathbf{k}, t)|^2 = 2\pi\int|\mathbf{u}(\mathbf{r}, t)|^2 d\mathbf{r}$ is proportional to the kinetic energy in the flow (we used Parseval's theorem), we see that (22.21) simply states that kinetic energy is dissipated by viscosity as in (22.17). (Surprise!) The dissipation term has k^2 as a multiplier, so the details of the long wavelength part of the spectrum $|\mathbf{u}(\mathbf{k}, t)|^2$ are immaterial in this context: viscosity acts on short wavelengths only.

The rate of specific energy dissipation ϵ (energy dissipation per unit mass) is then the right-hand side of (22.21), implying

$$\epsilon = \nu\sum_{\mathbf{k}}k^2\,|\mathbf{u}(\mathbf{k},\, t)|^2. \quad (22.22)$$

Note that the expressions so far do not imply any statistical averaging, and can as well refer to one realization of the ensemble. The distinction is not as trivial as it might seem: for any finite system, no matter how large, we will in general have that $|\mathbf{u}(\mathbf{k}, t)|^2$ depends explicitly on time for each individual realization. A constant value is obtained when averaging over all realizations of the ensemble, subject to the condition that the total energy input is the same in every realization, i.e. $|\mathbf{u}(\mathbf{k}, t)|^2 \to \langle|\mathbf{u}(\mathbf{k})|^2\rangle$ and the spectrum becomes a function of wavenumber only,

$S(k, t) \to \langle S(k)\rangle$. We usually omit the averaging brackets and imply that $S(k)$ refers to the ensemble averaged spectrum.

The ensemble averaged rate of specific energy dissipation ϵ is the average of (22.22), giving

$$\langle\epsilon\rangle = \nu \sum_{\mathbf{k}} k^2 \langle|\mathbf{u}(\mathbf{k}, t)|^2\rangle, \tag{22.23}$$

which is assumed to be a deterministic constant for time stationary conditions. We have

$$\langle\epsilon\rangle = \nu \int_0^\infty k^2 S(k, t) \mathrm{d}k$$

in terms of the energy spectrum $S(k, t)$ for the velocity fluctuations, the spectrum here defined in such a way that $\frac{1}{2}\langle u^2(\mathbf{r}, t)\rangle = \int_0^\infty S(k, t)\mathrm{d}k$. We can envisage the case where energy is fed continuously into the system, and have to introduce a source term in, for instance, (22.19), to have a steady state spectrum, where the energy input is balanced by the dissipation. Without such a term we have *decaying turbulence*, which is also an important topic.

With $\langle\mathbf{u}(\mathbf{r}, t) \cdot \nabla^2\mathbf{u}(\mathbf{r}, t)\rangle = \sum_{\mathbf{k}} k^2\langle|\mathbf{u}(\mathbf{k}, t)|^2\rangle$ we have the expression for the energy dissipation which can be obtained directly from the Navier–Stokes equation as well (Frisch 1995, Tsinober 2001)

$$\langle\epsilon\rangle = \nu\langle\mathbf{u}(\mathbf{r}, t) \cdot \nabla^2\mathbf{u}(\mathbf{r}, t)\rangle, \tag{22.24}$$

where the average is independent of time and position by the assumption of time stationary homogeneous turbulence (actually, the assumption of isotropy is not essential here). It should be mentioned that Serrin (1959) and Loitsyanskii (1966) demonstrated that the rate of which mechanical energy is dissipated into heat in incompressible Newtonian flows is (with the summation convention understood) $\epsilon = 2\nu s_{ij}s_{ij}$, where $s_{ij} = \frac{1}{2}(\partial u_i/\partial r_j + \partial u_j/\partial r_i)$ is the rate of strain tensor. In this latter formulation, the intermittent nature of the energy dissipation appears more evident. The expression (22.24) is an average, while the latter result is an exact local relation. The discussion can be extended to non-Newtonian compressible flows (Tsinober 2001).

Assume now that the energy input affects long wavelengths only. For the steady state conditions mentioned, ϵ is a constant, and has to equal the energy transported through the spectrum from long to short wavelengths where dissipation occurs. This is the basic argument for the energy cascade through the spectrum.

- **Exercise:** The expressions in this section have been formulated in terms of Fourier *series*. Rewrite as an exercise all expressions in terms of Fourier *integrals*.

22.2.2 Enstrophy budget

From (22.15) an equation for the fluid vorticity, $\boldsymbol{\omega} \equiv \nabla \times \mathbf{u}$, can be derived as

$$\frac{\partial}{\partial t}\boldsymbol{\omega} + \mathbf{u} \cdot \nabla\boldsymbol{\omega} = -\boldsymbol{\omega} \cdot \nabla\mathbf{u} + \nu\nabla^2\boldsymbol{\omega}, \tag{22.25}$$

where we note the absence of a pressure term.

In the same way as for the fluid energy density we find

$$\frac{1}{2}\frac{\partial}{\partial t}\omega^2 + \nabla\cdot\left(\frac{1}{2}\mathbf{u}\,\omega^2\right) = \boldsymbol{\omega}\cdot(\boldsymbol{\omega}\cdot\nabla\mathbf{u}) + \nu\nabla\cdot[\boldsymbol{\omega}\times(\nabla\times\boldsymbol{\omega})] - \nu(\nabla\times\boldsymbol{\omega})^2. \tag{22.26}$$

The time evolution of the volume integrated enstrophy, $W_\omega \equiv \frac{1}{2}\int\omega^2\mathrm{d}\mathbf{r}$, is obtained as

$$\frac{\partial}{\partial t}W_\omega = \int\boldsymbol{\omega}\cdot(\boldsymbol{\omega}\cdot\nabla\mathbf{u})\mathrm{d}\mathbf{r} - \nu\int(\nabla\times\boldsymbol{\omega})^2\mathrm{d}\mathbf{r}. \tag{22.27}$$

These expressions are derived for three-dimensional flows. By imposing some constraints they can be simplified to describe two-dimensional flows, where some additional simplifications can be found by introducing a stream function Φ so that $\boldsymbol{\omega} = -\nabla^2\Phi$.

The important observation is now that in the zero-viscosity limit energy is conserved (as expected!) also in the fully three-dimensional description, while enstrophy is not! The *physical* mechanism causing the time variation of W_ω for $\nu = 0$ is identified as vortex-line stretching (Tennekes and Lumley 1972), with results from numerical simulations summarized for instance by Aref and Kambe (1988).

In two-dimensional turbulence, however, all vortex tubes necessarily have to be parallel, and perpendicular to the plane of the analysis. The vortex stretching mechanism is consequently deactivated since in this case $\boldsymbol{\omega}\cdot\nabla\mathbf{u} = 0$, and the first term on the right-hand side of (22.27) is identically zero giving (Batchelor 1969)

$$\frac{\partial}{\partial t}W_{2\mathrm{d}\omega} = -\nu\int(\nabla\times\boldsymbol{\omega})^2\mathrm{d}\mathbf{r}. \tag{22.28}$$

For $\nu = 0$, enstrophy as well as energy is conserved in two-dimensional turbulence.

22.3 Reynolds' decomposition

In studies of for instance the Navier–Stokes equation it was found that it gives a significant improvement of our understanding of the inner dynamics of the model to decompose the flow into an average and a fluctuating component. The idea was seemingly first suggested by Reynolds (Tennekes and Lumley 1972). We thus write the velocity vector as $\mathbf{u}(\mathbf{r}, t) \equiv \mathbf{U}(\mathbf{r}) + \tilde{\mathbf{u}}(\mathbf{r}, t)$ where $\mathbf{U}$ is an average velocity vector. We use here time averages $\mathbf{U} \equiv \overline{\mathbf{u}}$ where the overline implies time averages, i.e. $\overline{\mathbf{u}} \equiv \frac{1}{\mathcal{T}}\int_0^{\mathcal{T}}\mathbf{u}\mathrm{d}t$ with $\mathcal{T}\to\infty$, this is the standard use in Reynolds' decomposition. Note that a time averaged quantity can have a spatial variation. This decomposition is always possible, where in some cases $\overline{\mathbf{u}}$ can be a trivial null-vector. Other averages can be used, ensemble averages, for instance. Inserting the velocity decomposition into the Navier–Stokes equation and averaging each term over time as indicated by the overline, we obtain

$$\mathbf{U}\cdot\nabla\mathbf{U} + \overline{\tilde{\mathbf{u}}\cdot\nabla\tilde{\mathbf{u}}} = -\frac{1}{\rho}\nabla\overline{P} + \nu\nabla^2\mathbf{U}, \tag{22.29}$$

where we again take ρ to be constant. The relation (22.29) demonstrates that the fluctuating component affects the mean flow. At this place the summation convention allows a particularly elegant formulation of the equation, obtained by writing

$$U_i \frac{\partial U_\ell}{\partial r_i} + \overline{\widetilde{u}_i \frac{\partial \widetilde{u}_\ell}{\partial r_i}} = -\frac{1}{\rho}\frac{\partial \overline{P}}{\partial r_\ell} + \nu \frac{\partial^2 U_\ell}{\partial r_i^2},$$

with $\partial^2/\partial r_i^2$ interpreted as a product and therefore implying summation over the i-s. Using the incompressibility condition explicitly, $\partial U_i/\partial r_i = 0$, we can write

$$U_i \frac{\partial U_\ell}{\partial r_i} + \frac{\partial(\overline{\widetilde{u}_\ell \widetilde{u}_i})}{\partial r_i} = -\frac{1}{\rho}\frac{\partial \overline{P}}{\partial r_\ell} + \nu \frac{\partial^2 U_\ell}{\partial r_i^2}.$$

We can express this relation in a more compact form (Tennekes and Lumley 1972) by writing

$$U_i \frac{\partial U_\ell}{\partial r_i} = \frac{1}{\rho}\frac{\partial}{\partial r_i}\left(\Sigma_{\ell,i} - \rho\overline{\widetilde{u}_\ell \widetilde{u}_i}\right), \tag{22.30}$$

with

$$\Sigma_{\ell,i} \equiv -\overline{P}\delta_{\ell,i} + \mu\left(\frac{\partial U_i}{\partial r_\ell} + \frac{\partial U_\ell}{\partial r_i}\right).$$

The result (22.30) is *Reynolds momentum equation* for the mean flow. The fluctuating velocity component contributes to the stress tensor by *Reynolds stress* $\tau_{\ell,i} \equiv -\rho\overline{\widetilde{u}_\ell \widetilde{u}_i}$, which is a symmetric tensor. The diagonal components are normal stresses or pressures. The off diagonal elements of $\tau_{\ell,i}$ are *shear stresses*, which play a dominant role in the momentum transfer by turbulent motions.

22.4 Homogeneous turbulence

We consider now the equivalent of Reynolds' decomposition, but now with ensemble averaged velocities, rather than time averages. With the summation convention implied we have

$$\frac{\partial U_j}{\partial r_j} = 0 \tag{22.31}$$

$$\frac{\partial U_j}{\partial t} + U_k \frac{\partial U_j}{\partial r_k} = -\frac{1}{\rho}\frac{\partial p}{\partial r_j} + \nu \frac{\partial^2 U_j}{\partial r_k \partial r_k} - \frac{\partial \langle \widetilde{u}_j \widetilde{u}_k \rangle}{\partial r_k} \tag{22.32}$$

$$\frac{\partial \widetilde{u}_j}{\partial r_j} = 0 \tag{22.33}$$

$$\frac{\partial \tilde{u}_j}{\partial t} + U_k \frac{\partial \tilde{u}_j}{\partial r_k} = -\frac{1}{\rho}\frac{\partial \tilde{p}}{\partial r_j} + \nu \frac{\partial^2 \tilde{u}_j}{\partial r_k \partial r_k} - \tilde{u}_k \frac{\partial U_j}{\partial r_k} - \frac{\partial\left(\tilde{u}_j\tilde{u}_k - \left\langle \tilde{u}_j\tilde{u}_k \right\rangle\right)}{\partial r_k}, \tag{22.34}$$

with $\tilde{u}_j$ denoting fluctuating quantities and $U \equiv \langle u \rangle$ here being the ensemble averaged velocity. We explicitly used the incompressibility condition $\partial \tilde{u}_j / \partial r_j = 0$ in the second equation. Implicitly we assumed ρ constant in agreement with incompressible motions. This is a minor restriction: in principle we *can* have varying fluid density with incompressibility, as already mentioned.

Due to the nonlinear advection term, the evolution of the averaged and the fluctuating motions are coupled. In the mean momentum equation the turbulent fluctuations appear through the Reynolds stresses gradient, while the mean field appears in the fluctuating momentum equation. The mean velocity field can be independent of the fluctuating velocity field if the Reynolds stresses are independent of position, meaning that the fluctuating motions are invariant with respect to translations. The fluctuations thus have to be homogeneous, though not necessarily isotropic. For homogeneous turbulence, we have the mean field decoupled from the turbulence.

22.4.1 Constraints of the mean field for homogeneous turbulence

The continuity equation (22.31) for incompressible motion has to be fulfilled.

From (22.34) and (22.34) a transport equation for Reynolds stresses can be obtained. Assuming these stresses to be invariant with respect to translations, we find that the mean velocity field must have constant velocity gradients (Hinze 1975). This implies in particular that the viscosity term $\nu\nabla^2\mathbf{U}$ vanishes.

From (22.32) we obtain

$$\frac{\partial}{\partial t}\mathbf{U} + \mathbf{U}\cdot\nabla\mathbf{U} = -\frac{1}{\rho}\nabla p + \nu\nabla^2\mathbf{U}.$$

Using the $\nabla\times$ operator on this equation together with the previous constraints, we obtain

$$\frac{\partial}{\partial t}\nabla\times\mathbf{U} + \nabla\times(\mathbf{U}\cdot\nabla\mathbf{U}) = 0.$$

For a steady state incompressible mean flow we thus find that either the mean flow is irrotational, *or* that the rotation axis is normal to the plane of strain everywhere.

22.4.2 Equations for the fluctuation averages

We can derive a set of transport equations for the mean squared average quantities by (22.34). Take two equations for U_j and U_ℓ, and multiply them by U_ℓ and U_j respectively. Then add and average the resulting equations to obtain

$$\frac{\partial\langle\tilde{u}_\ell\tilde{u}_j\rangle}{\partial t} + U_k\frac{\partial\langle\tilde{u}_\ell\tilde{u}_j\rangle}{\partial r_k} = -\frac{1}{\rho}\left\langle\tilde{u}_\ell\frac{\partial\tilde{p}}{\partial r_j} + \tilde{u}_j\frac{\partial\tilde{p}}{\partial r_\ell}\right\rangle + \nu\left(\left\langle\tilde{u}_\ell\frac{\partial^2\tilde{u}_j}{\partial r_k\partial r_k}\right\rangle + \left\langle\tilde{u}_j\frac{\partial^2\tilde{u}_\ell}{\partial r_k\partial r_k}\right\rangle\right)$$
$$- \langle\tilde{u}_j\tilde{u}_k\rangle\frac{\partial U_\ell}{\partial r_k} - \langle\tilde{u}_\ell\tilde{u}_k\rangle\frac{\partial U_j}{\partial r_k} - \left\langle\tilde{u}_\ell\frac{\partial(\tilde{u}_j\tilde{u}_k)}{\partial r_k}\right\rangle - \left\langle\tilde{u}_j\frac{\partial(\tilde{u}_\ell\tilde{u}_k)}{\partial r_k}\right\rangle.$$

After a little algebra (Hinze 1975), this equation can be rewritten as

$$\begin{aligned}\frac{\partial\langle\tilde{u}_\ell\tilde{u}_j\rangle}{\partial t} + U_k\frac{\partial\langle\tilde{u}_\ell\tilde{u}_j\rangle}{\partial r_k} &= \frac{1}{\rho}\left\langle\tilde{p}\left(\frac{\partial\tilde{u}_\ell}{\partial r_j} + \frac{\partial\tilde{u}_j}{\partial r_\ell}\right)\right\rangle - \frac{1}{\rho}\left(\frac{\partial\langle\tilde{p}\,\tilde{u}_\ell\rangle}{\partial r_j} + \frac{\partial\langle\tilde{p}\,\tilde{u}_j\rangle}{\partial r_\ell}\right) \\ &+ \nu\frac{\partial^2\langle\tilde{u}_\ell\tilde{u}_j\rangle}{\partial r_k\partial r_k} - 2\nu\left\langle\frac{\partial\tilde{u}_\ell}{\partial r_k}\frac{\partial\tilde{u}_j}{\partial r_k}\right\rangle - \langle\tilde{u}_j\tilde{u}_k\rangle\frac{\partial U_\ell}{\partial r_k} - \langle\tilde{u}_\ell\tilde{u}_k\rangle\frac{\partial U_j}{\partial r_k} - \frac{\partial\langle\tilde{u}_\ell\tilde{u}_j\tilde{u}_k\rangle}{\partial r_k},\end{aligned} \tag{22.35}$$

still assuming ρ to be constant. We recognize advection terms, dissipation due to viscosity and production terms due to gradients in the mean velocity field together with pressure-strain correlations.

Taking the trace (contraction) of the tensors in (22.35) we find

$$\begin{aligned}\frac{1}{2}\frac{\partial q^2}{\partial t} + \frac{1}{2}U_k\frac{\partial q^2}{\partial r_k} &= -\frac{\partial}{\partial r_\ell}\left\langle\left(\frac{\tilde{p}}{\rho} + \frac{\tilde{u}_j\tilde{u}_j}{2}\right)\tilde{u}_\ell\right\rangle \\ &+ \frac{\nu}{2}\frac{\partial^2 q^2}{\partial r_k\partial r_k} - \nu\left\langle\frac{\partial\tilde{u}_\ell}{\partial r_k}\frac{\partial\tilde{u}_\ell}{\partial r_k}\right\rangle - \langle\tilde{u}_\ell\tilde{u}_k\rangle\frac{\partial U_\ell}{\partial r_k},\end{aligned} \tag{22.36}$$

with $\langle\tilde{u}_\ell\tilde{u}_\ell\rangle \equiv q^2$. For steady state homogeneous turbulence we have the left side of (22.36) vanishing together with several terms on the right-hand side. Statistical homogeneity of the fluctuating velocity field then necessarily implies that $\partial U_\ell/\partial r_k$ is a constant, $U_0/\mathcal{L}$, for instance. We find

$$\nu\left\langle\frac{\partial\tilde{u}_\ell}{\partial r_k}\frac{\partial\tilde{u}_\ell}{\partial r_k}\right\rangle = -\langle\tilde{u}_x\tilde{u}_y\rangle\frac{U_0}{\mathcal{L}},$$

stating that the dissipated energy (left side) equals the generated energy (right-hand side) due to the average velocity gradients, where we introduced $\partial U_x/\partial y \equiv U_0/\mathcal{L}$, with y and z components of U_j assumed to vanish.

Obviously, we cannot have the macroscopic shear $\partial U_\ell/\partial r_k$ to be a constant in all space, but in wind-tunnels, for instance, it has been possible to make it a good approximation for large parts of the cross-sections. In such cases we can argue for locally homogeneous and isotropic turbulence to be a good approximation.

In a turbulent flow we deal with shear flows on all scales until we reach the dissipation length scale. In the inertial subrange you expect the local shear to be much larger than the mean shear, $\partial U_\ell/\partial r_k$ for all k and ℓ.

The results of this section were presented for a three-dimensional turbulent velocity field, but several of the results are readily rewritten for a two-dimensional case, which will often serve as an appropriate simplification for electrostatic turbulence in magnetized plasmas. For two spatial dimensions we take $U_z = 0$ and $\tilde{u}_z = 0$ so that all spatial derivatives also vanish in this direction. The rotation vector $\nabla \times \mathbf{u}$ will everywhere be parallel to the **z**-axis. The reader is advised to rewrite the results for this spatially two-dimensional case.

22.5 Structure functions and spectra

Assuming homogeneous and isotropic three-dimensional turbulence, we want to discuss structure functions and turbulent spectra of an incompressible fluid, which can adequately be described by the Navier–Stokes equation. In this model, the fluid itself is described by the kinematic viscosity, ν. The geometry of the system is described by one or more large length scales, $\mathcal{L}$. The turbulence is maintained by constant stirring at these large scales, they can represent rotating fans, or something else. By this, we assume to supply an energy ϵ per gram fluid per second, which must then also be the energy dissipated due to the viscosity, again per gram per second for a turbulence which is stationary in a statistical sense.

22.5.1 Spatial structure functions: inertial subrange

First we consider the second order structure function $\Psi_2(\mathbf{r}) = \langle(\mathbf{u}(\mathbf{r}_1, t) - \mathbf{u}(\mathbf{r}_1 + \mathbf{r}, t))^2\rangle$ for this problem, assuming without going into details that it exists as a smooth function. More often we use the structure function for a velocity component, see summary in figure 19.4. For instance the longitudinal structure function can be obtained as $\Psi_{2\|}(\mathbf{r}) = \langle((\mathbf{u}(\mathbf{r}_1, t) - \mathbf{u}(\mathbf{r}_1 + \mathbf{r}, t)) \cdot \hat{\mathbf{r}})^2\rangle$, where $\hat{\mathbf{r}} \equiv \mathbf{r}/r$ is a unit vector along $\mathbf{r}$. For time stationary and homogeneous conditions where $\langle u_1^2\rangle = \langle u^2(\mathbf{r}_1)\rangle \equiv \langle u^2(\mathbf{r}_2)\rangle$ the structure function is related to the un-normalized velocity correlation function $R(\mathbf{r}) \equiv \langle \mathbf{u}(\mathbf{r}_1, t) \cdot \mathbf{u}(\mathbf{r}_1 + \mathbf{r}, t)\rangle$ by $\Psi_2(\mathbf{r}) = 2\langle u^2\rangle - 2\langle \mathbf{u}(\mathbf{r}_1, t) \cdot \mathbf{u}(\mathbf{r}_1 + \mathbf{r}, t)\rangle = 2(\langle u^2\rangle - R(\mathbf{r}))$.

With reference to the discussion in section 22.2.1, we expect that at least for small scales the distribution of energy over the *directions* of wavenumbers is uniform, i.e. the turbulence has become *isotropic*, i.e. $\Psi_2(\mathbf{r}) = \Psi_2(r)$. In this limit, the turbulence can be considered as adequately described by ϵ. If the length scale $\mathcal{L}$ is very large, there are hopes that its actual value does not play a role for small wavelengths, corresponding to small separations $\mathbf{r}$. We have $[\Psi_2] = L^2T^{-2}$, and find the dimension matrix being

	ν	ϵ	r	Ψ_2
T	− 1	− 3	0	− 2
L	2	2	1	2

With a little algebra, we find

$$f\left(\frac{\Psi_2^2}{\epsilon\,\nu}, \frac{r^4\epsilon}{\nu^3}\right) = 0,$$

which we write as

$$\Psi_2 = (\epsilon\,\nu)^{1/2}\,\mathcal{G}\left(\frac{r\epsilon^{1/4}}{\nu^{3/4}}\right). \tag{22.37}$$

It is now a common experience that viscosity only affects the very small scales in a flow, and we can expect, still provided that $\mathcal{L}$ is large enough, that an interval for r can exist, where the structure function is independent of ν (Chandrasekhar 1957). For this to be the case, we require ν to cancel and find $\mathcal{G}(x) \sim x^{2/3}$, giving

$$\Psi_2 = C_K(\epsilon\,r)^{2/3}, \tag{22.38}$$

where C_K is the Kolmogorov constant, where experiments indicate that $C_K \approx 2.1-2.5$ when we consider $\Psi_{2\parallel}(\mathbf{r})$. The numerical constant will be the only distinction between $\Psi_{2\parallel}$ and $\Psi_{2\perp}$ for homogeneous isotropic turbulence.

For very short distances, we can obviously not rely on (22.38), but unfortunately, we cannot trust it for large separations either: it seems unacceptable that Ψ_2 can increase without limit. On the other hand, we were actually prepared for *some* problems when r became comparable to $\mathcal{L}$, so at least for the time being we accept the result (22.38) as referring to a *subrange*, usually denoted the 'inertial subrange'.

22.5.2 Wavenumber spectra: inertial subrange

Before discussing (22.38) in more detail, we derive the equivalent result for the wavenumber spectrum. By the Wiener–Khinchine theorem we obtain the wavenumber spectrum by a Fourier transform of the correlation function $R(r) = \langle(u(r_1) - u(r_1 + r))^2\rangle$, where we have $\Psi_2(r) = 2\langle u^2\rangle - 2R(r)$ for homogeneous isotropic turbulence. For time stationary turbulence there is no need to include time as a variable explicitly since we sample the two velocities at the same time. As far as the r-variation is concerned, we need not distinguish $R(r)$ and $\Psi_2(r)$, since $\langle u^2\rangle$ is a constant for homogeneous turbulence. For the velocity power spectrum for homogeneous and isotropic turbulence in three spatial dimensions we have (Champeney 1973)

$$S(k) = \int_0^\infty R(r)\frac{\sin(kr)}{kr}\mathrm{d}r = \frac{1}{k}\int_0^\infty R(\gamma/k)\frac{\sin(\gamma)}{\gamma}\mathrm{d}\gamma \equiv \frac{1}{k}F(k), \tag{22.39}$$

introducing $\gamma \equiv kr$, where F has dimension $length^2 \times time^{-2}$. We will re-express F as a coefficient having the correct dimension multiplied on a dimensionless function of a dimensionless variable, see also (19.81). To determine the power spectrum from dimensional arguments we have the dimension matrix

	ν	ϵ	k	$k\,S$
T	-1	-3	0	-2
L	2	2	-1	2

and find

$$S(k) = \frac{1}{k}(\epsilon\nu)^{1/2}\,\mathcal{G}\left(\frac{k\,\nu^{3/4}}{\epsilon^{1/4}}\right), \tag{22.40}$$

where $F(k) = (\epsilon\nu)^{1/2}\,\mathcal{G}(k\,\nu^{3/4}/\epsilon^{1/4})$. A universal numerical constant is assumed to be included in $\mathcal{A}$.

Our experience tells us once more that viscosity is effective only for small scales, and we expect that a spectral subrange may exist, where the spectrum is independent of viscosity. To have (22.40) independent of the kinematic viscosity ν, in such a postulated *inertial subrange*, we obviously require $\mathcal{G}(x) \sim x^{-2/3}$ giving

$$S(k) \sim \frac{\epsilon^{2/3}}{k^{5/3}}. \tag{22.41}$$

A numerical constant cannot be found by dimensional reasoning, but it is generally accepted that (22.41) can be completed by the Kolmogorov constant of approximately 1.5, which should not be confused with C_K in (22.38).

The result (22.41) is the *Kolmogorov–Oubokhov spectrum*, which has been verified to high accuracy in large scale turbulent flows (Hinze 1975). In a way, this might be a little surprising, since we made one implied assumption by treating ϵ as a constant. In reality it is a statistically varying quantity: in some regions of the flow we can have a large dissipation, in others a somewhat smaller one. We can interpret ϵ as the *average* of this energy dissipation, but in that case we have 'cheated' by taking it the actual value of ϵ outside the averaging sign, which is only permissible for statistically independent quantities. The local wavenumber spectrum *must* on the other hand depend strongly on the local energy dissipation: was it to be zero by chance, we would not have any local spectrum at all! This is a short summary of the so called 'intermittency problem' in turbulence, which has given (and will, at least for a while, continue to give) rise to many discussions in the literature. We will not go deeper into the problem here. *Very* heuristically it can be argued that if we consider ϵ as a randomly varying quantity, the dimensional reasoning outlined before will treat $\langle\epsilon^{2/3}\rangle$ and $\langle\epsilon\rangle^{2/3}$ equally, but obviously the two quantities are different in magnitude. If the spatial and temporal fluctuations in ϵ are large, they are likely to be *very* different! If we, for a moment consider a higher order structure function $\Psi_n(\mathbf{r}) = \langle(\mathbf{u}(\mathbf{r}_1, t) - \mathbf{u}(\mathbf{r}_1 + r, t))^n\rangle$, we readily find that dimensional reasoning gives $\Psi_n(\mathbf{r}) \sim (\epsilon r)^{n/2}$, where again we have the ambiguity between $\langle\epsilon\rangle^{n/2}$ and $\langle\epsilon^{n/2}\rangle$. The exception is the third order structure function, $n = 3$, where $\Psi_3(\mathbf{r}) \sim \epsilon r$. This particular result, interpreting the energy dissipated as an average, $\langle\epsilon\rangle$, is expected to be exact in the limit of infinite Reynolds numbers, where the distinction of the smallest and the largest scales is unambiguous. (Even this argument is somewhat ambiguous, after all, since $\langle\epsilon^{1/m}\rangle^m$ has the dimensions of ϵ for any m so we could have $\Psi_3(\mathbf{r}) \sim \langle\epsilon^{1/m}\rangle^m r$ by dimensional reasoning alone.) Intermittency corrections are likely to show up in higher order structure functions, but you need a *lot* of data to calculate these accurately, only few experimentalists are that fortunate!

Summary of basic scales:
$\eta = (\nu^3/\epsilon)^{1/4}$, the *Kolmogorov length scale*, or *inner scale*
$\tau = (\nu/\epsilon)^{1/2}$, the *Kolmogorov time scale*
$v = (\nu\epsilon)^{1/4}$ is a characteristic velocity at the inner scale.
Note that the Reynolds number constructed from η, τ and v is unity (Tennekes and Lumley 1972).

With reference to (22.23) or (22.24) we can also write

$$S(k) \sim \frac{\left(\nu \int_0^{k_{\max}} k'^2 S(k', t)\mathrm{d}k'\right)^{2/3}}{k^{5/3}}, \tag{22.42}$$

which can be interpreted as a self-consistency requirement for the time stationary power spectrum, $S(k)$. Note that we have to invoke some maximum wavenumber in order have a convergent integral: the Kolmogorov–Oubokhov spectrum in itself is integrable in the limit of large wavenumbers, but the integral diverges when the spectrum is multiplied by k^2. There is nothing unphysical in this, since this universal spectrum was derived for the spectral range independent of viscosity, which excludes the shortest wavelengths. As an order of magnitude estimate we expect the range of validity to stop at $k_{\max} \approx (\epsilon/\nu^3)^{1/4}$. For large wavenumbers, i.e. shorter wavelengths, we expect the spectrum to fall off exponentially, but no satisfactory theory is (yet?) available for describing details for this range although the analysis by Heisenberg (1948) attempted to extend the range of the Kolmogorov–Oubokhov spectrum. The spectrum fails also for the large non-universal scales at small k, but the k^2 multiplier in the integrand of (22.42) remedies this problem by giving this part of the spectrum a negligible weight.

Our *physical* understanding of the inertial spectral subrange can be outlined by stating that, as previously mentioned, viscous dissipation is effective at small wavelengths, the 'inner' scales, while we imagine that energy is fed into the turbulent motion by 'stirring' at large scales. The universal subrange is one where scale-lengths are so large that they are unaffected by viscosity, but on the other hand much smaller than the 'outer' scales. The universal subrange consists of the scales through which energy is cascaded from outer scales into dissipation at the small scales (Tennekes and Lumley 1972, She and Leveque 1994). The 'critical' scale-length, η, where viscosity begins to compete with the energy cascade can be estimated by dimensional arguments to be $\eta = (\nu^3/\epsilon)^{1/4}$, often called the *Kolmogorov scale*. Similarly we can obtain expressions for the time-scale characterizing the inner scales as $\tau = (\nu/\epsilon)^{1/2}$, and a velocity as $v = (\nu\epsilon)^{1/4}$. It might be emphasized that an inertial, or universal, subrange *need* not develop for small or moderate Reynolds numbers. There can be cases where the viscosity is effective in making an overlap between the inner and outer scales.

- **Exercise:** Prove

$$S(k) = \int_0^{\infty} R(r/\mathcal{L})\frac{\sin(kr)}{kr}\mathrm{d}r$$

in (22.39) for the three-dimensional homogeneous isotropic case. Find the similar relation for homogeneous isotropic turbulence in two spatial dimensions (Champeney 1973).

22.5.3 Viscous spectral subrange

The Kolmogorov–Oubokhov structure function for the universal subrange has a cusp at vanishing separation. This particular result is unphysical in the sense that for very small separations we expect that viscosity should have an effect and the analysis that forms the basis for a Kolmogorov–Oubokhov result ceases to be valid. It can be argued from first principles (Pécseli 2000) that the velocity correlation function, and thus the structure function should have a vanishing derivative at the origin $r = 0$. It is reasonable to assume that for small r we have a variation like $\Psi_2 \sim r^2$. (Strictly speaking all we can argue for is an even exponent in Ψ_2 for small r, it *could* be $\Psi_2 \sim r^4$, for instance.) A detailed analysis (Davidson 2004) confirms that the assumption of $\Psi_2 \sim r^2$ is correct, and an exact expression can be found for this subrange, but we found here how easily the basic form can be obtained by dimensional reasoning. This conjecture implies that $\mathcal{G} \sim r^2$, giving by the general result (22.37) that

$$\Psi_2 \sim \frac{\epsilon}{\nu} r^2. \tag{22.43}$$

By definition we had $\Psi_{2\|}(\mathbf{r}) = \langle((\mathbf{u}(\mathbf{r}_1, t) - \mathbf{u}(\mathbf{r}_2 + \mathbf{r}, t)) \cdot \hat{\mathbf{r}})^2\rangle$. We can then take

$$\frac{1}{r_\|^2}\left\langle\left(\left(\mathbf{u}(\mathbf{r}_1, t) - \mathbf{u}(\mathbf{r}_2 + \mathbf{r}, t)\right) \cdot \hat{\mathbf{r}}\right)^2\right\rangle \approx \left\langle\left(\frac{\partial U_\|}{\partial r_\|}\right)^2\right\rangle \approx C_\nu \frac{\epsilon}{\nu},$$

in agreement with previous results. We introduced a numerical coefficient C_ν. By assuming homogeneity and isotropy, the relation can be used to determine C_ν via (22.22) or (22.24). The result gives $C_\nu = 1/15$ for $\Psi_{2\|}$ in this viscous subrange (Davidson 2004).

By dimensional arguments we estimated the scale η where viscosity begins to compete with the energy cascade. These dimensional arguments cannot account for a numerical constant, and by simple arguments we can do a somewhat better job. We can ask where the structure function for the universal subrange is crossing the structure function characterizing the viscous subrange (Pécseli and Trulsen 2007). We denote this transition length scale, the modified Kolmogorov scale, as η_0 and readily find $\eta_0 = (C_K/C_\nu)^{3/4}(\nu^3/\epsilon)^{1/4} \equiv (C_K/C_\nu)^{3/4}\eta$. With the given numerical constants we find $\eta_0 \approx 13\eta$ which after all demonstrates a non-trivial difference between η and η_0.

A short note of warning concerning terminology seems appropriate here: the present use of the words 'inertial subrange' is almost universally applied, but sometimes one might find the same part of the structure function or power law spectrum called 'universal subrange'. In a way it is understandable that a wording differing from 'inertial' is attractive, since the relevance of inertia is poorly argued for the subrange bearing its name, but we should keep in mind that the viscous subrange for the structure function is also 'universal' so the former terminology is to be preferred after all.

By its definition the Kolmogorov–Oubokhov turbulent spectrum is characterizing the fluctuations in fluid velocity. Other quantities also have their characteristic spectra, temperature, for instance. This is a *scalar* spectrum. Temperature, as well as many other scalar quantities are passively carried around by the turbulent motions in the flow. For very small scale sizes, diffusivity, here *thermal diffusivity*, begins to be important, but the characteristic length scale where this occurs is in general different from the Kolmogorov micro-scale.

For very small scales, much smaller than η, molecular diffusivity becomes important, as modeled by a diffusion coefficient κ.

- $(\kappa^3/\epsilon)^{1/4}$ was introduced by Corrsin and Oubokhov as the length scale where conduction 'cuts-off' temperature fluctuations.
- $(\nu\kappa^2/\epsilon)^{1/4}$ was introduced by Batchelor (1959) as the length scale where conduction 'cut-offs' temperature fluctuations. The two scales mentioned seem to be in contradiction, but they apply for different parameter relations. The latter expression holds when $\nu \gg \kappa$ and vice versa. A detailed discussion is given by Batchelor (1959).
- $\eta/\sqrt{Sc} \equiv (\nu D^2/\epsilon)^{1/4}$ was introduced by Batchelor (1959) to describe the size of a droplet of some passively diffusing material that will diffuse in the same time it takes the energy in an eddy of size η to dissipate. The Schmidt number was introduced as $Sc \equiv \nu/D$, where D is the molecular mass diffusion coefficient. The same number turns out to have a role in separating the length scale for Brownian diffusion and turbulent diffusion in the viscous subrange of turbulence (Pécseli *et al* 2014). This is not so surprising, since the basic elements of the problem are the same.

One might be interested in studies of the turbulent spectra for short wavelengths, i.e. scales shorter than the Kolmogorov length. Unfortunately, this is a rather complicated problem (even worse than that for the universal subrange), and only relatively sparse analytical and experimental results are available. It was suggested that the extension of the Kolmogorov–Oubokhov law was a *steeper* power law, but there are good reasons to be skeptical on such results. As long as the Navier–Stokes equation is valid (i.e. down to atomic or molecular scales), there will not be any more length scales entering to normalize with. Therefore, a power law spectrum k^{-m} with some $m > 5/3$, will supposedly continue as long as the basic equation is valid. Now this does not seem likely, because when $\mathcal{E}(k)$ is the power spectrum for the velocity fluctuations, we have $k^2\mathcal{E}(k)$ being the power spectrum of the velocity derivatives, $k^4\mathcal{E}(k)$ for the second derivatives etc. We find that for a large enough derivative, the integral will diverge for any power law spectrum, $\mathcal{E}(k) \sim k^{-m}$. We have, however, all reason to believe that initially smooth (infinitely many times differentiable) conditions lead to time evolutions which are smooth in the same sense at all later times. Due to the viscosity term, the Navier–Stokes equation does not develop discontinuities in a function, nor its derivatives in three dimensions, unless they were there to begin with (Sulem *et al* 1983, Constantin 1991). (It *might* be different in two spatial dimensions.) We therefore require the integral $\int_0^\infty k^m \mathcal{E}(k)\mathrm{d}k$ to be finite for all m, implying that $\mathcal{E}(k)$ has to fall off faster than any power law for $k \to \infty$, for instance exponentially.

22.5.4 Temporal structure functions

Assume that we want to determine the *temporal* variation for the structure function taken in one position, i.e. $\Psi_2(t) \equiv \langle(\mathbf{u}(0, t_1) - \mathbf{u}(0, t_1 + t))^2\rangle$. We find by a dimensional analysis similar to that used before that

$$\Psi_2 = (\epsilon\nu)^{1/2} f\left(t\epsilon^{1/2}/\nu^{1/2}\right). \tag{22.44}$$

Presuming the existence of a temporal subrange, where $\Psi_2(t)$ is independent of ν, we find

$$\Psi_2 \sim \epsilon t. \tag{22.45}$$

By its derivation, the result (22.45) may appear just as convincing as (22.38), but there is a problem: it is not obvious what problem we have solved! A temporally varying velocity function can be obtained in two different ways (Tennekes 1973, 1975): either we sample the fluid velocity vector in a fixed position (Eulerian sampling), or along a self-consistent trajectory (Lagrangian sampling). We can write the two corresponding vectorial time-records as $\mathbf{u}_E(t) = \mathbf{u}(\mathbf{r}_0, t)$ and $\mathbf{u}_L(t) = \mathbf{u}(\mathbf{r}(t), t)$. These two cases will suffice for the discussion, although others can also be envisaged, sampling along a prescribed curved trajectory, for instance. In the first case the fluid elements are continuously being replaced, while in the second case we follow a fluid element as it is being deformed by the turbulent stretching and folding. There is no corresponding distinction for the structure function in terms of spatial variables: there it does not make any difference how we arrived at the two sampling positions with separation $\mathbf{r}$.

As we have seen, there will inevitably be a non-universal part of the spectrum for large scales, i.e. small wavenumbers. Since the intensity in the power spectrum is increasing with decreasing k, most of the energy is actually presumed to reside in these large scales. This large scale motion is, however, carrying the small scales with it and for a fixed observer attempting to obtain $\Psi_2(t)$, the signal will be 'contaminated' by the non-universal Doppler shifts. The $\Psi_2 \sim \epsilon t$ subrange in (22.45) is likely to apply only for the case where we follow a fluid element along its Lagrangian trajectory. The stationary observer will obtain a temporally varying signal, where the frequency spectrum at large frequencies will have its largest contribution from the Doppler shifts of the *wavenumber* spectrum (22.41). This is actually a 'good thing', by offering a simple means of obtaining the wavenumber spectrum experimentally. The implied idea is often called *Taylor's hypothesis*, or the *frozen turbulence approximation*. For isotropic homogeneous fully developed turbulence a fixed observer is likely to find an $\omega^{-5/3}$ spectral subrange at high frequencies.

- **Exercise:** Carry out the analysis giving (22.44) and (22.45).

Following the analysis of section 22.5.2 we can attempt to obtain a frequency spectrum as well. For the frequency power spectrum we have by dimensional reasoning

$$\omega S(\omega) = (\epsilon\nu)^{1/2} f\left(\omega\sqrt{\nu/\epsilon}\,\right),$$

which becomes independent of ν by $S(\omega) \sim \epsilon\omega^{-2}$. The objections regarding the temporal structure function apply to the frequency spectrum as well, so the present result will not be discussed any further, and should be seen mostly as an exercise.

22.5.5 A summary of Heisenberg's derivation

The Kolmogorov–Oubokhov power-law spectrum was independently obtained by Heisenberg (1948), and it may be instructive to see a summary of his derivation. Basically, he visualized the formation of the spectrum as a transport in wavenumber space. The energy transfer was assumed to be local, in the sense that energy from a wavenumber interval was transferred to a neighboring interval of larger wavenumbers. The transfer was assumed to be mediated by a process very much akin to a viscous process, attributed to a sort of turbulent viscosity ν_T. Inspection of (22.19) shows that 'ordinary' viscosity depletes energy from the interval $\{0, k\}$ by

$$\epsilon_k = \nu \int_0^k k'^2 \langle |\mathbf{u}(\mathbf{k}', t)|^2 \rangle \mathrm{d}k', \tag{22.46}$$

see also (22.22), so Heisenberg generalized the integral of (22.19) for the interval $\{0, k\}$ as

$$\mathcal{F}_k = -\nu_T(k) \int_0^k k'^2 \langle |\mathbf{u}(\mathbf{k}', t)|^2 \rangle \mathrm{d}k' = -\nu_T(k) \int_0^k k'^2 S(k')\, \mathrm{d}k'. \tag{22.47}$$

The question is now to model the turbulent viscosity $\nu_T(k)$. Heisenberg suggested the expression

$$\nu_T(k) = C \int_k^\infty \sqrt{\frac{S(k')}{k'^3}}\, \mathrm{d}k', \tag{22.48}$$

where C is some constant, presumed universal. Apart from being dimensionally correct, the expression (22.48) can be argued physically by noting that large wavelengths (small k) *convect* structures of the size $2\pi/k$, so we expect any turbulent viscosity effects to be associated with smaller scale-lengths, i.e. wavenumbers larger than k. Also, we would intuitively expect wavelengths smaller than, but comparable to $2\pi/k$ to be the most effective by remembering collisional transport by long mean free paths. On the basis of these arguments we accept (22.48) as a working hypothesis, but are prepared for criticism of details in the derivation. We note that hidden in the postulate (22.47) there is an assumption of $\nu_T(k)$ being placed outside an ensemble average $\langle\ \rangle$.

In order to have a constant transfer of energy through the spectrum, we require $\mathcal{F}(k)$, with $\nu_T(k)$ inserted, to be a constant. This constant must equal the total energy dissipation ϵ. This gives $S(k) \sim \epsilon^{2/3}k^{-5/3}$ in complete agreement with Kolmogorov's result. Heisenberg's analysis was extended and discussed in detail by Chandrasekhar (1949) and also summarized by Stanišić (1988).

More generally, Heisenberg included molecular dissipation for the short wavelength part of the spectrum, and wrote

$$\frac{1}{2}\frac{\partial}{\partial t}\int_0^k S(k', t)\mathrm{d}k' = -(\nu + \nu_T(k))\int_0^k k'^2 S(k', t)\mathrm{d}k', \tag{22.49}$$

with ν again being the kinematic viscosity of the fluid. Assuming a time stationary spectrum, we can differentiate (22.49) with respect to k and find a complicated relation for the spectrum $S(k)$. Heisenberg provided an interpolation formula for the result

$$S(k) \approx C_1 \frac{\epsilon^{2/3}k^{-5/3}}{\left(1 + C_2\left(\nu^3/\epsilon\right)k^4\right)^{4/3}},$$

which correctly reproduces the $\epsilon^{2/3}k^{-5/3}$ law for long wavelengths, but also contains a short wavelength dissipation subrange $\sim(\epsilon/\nu^2)^2k^{-7}$. The latter part seems to be in good agreement with observations, but as already mentioned, this power law cannot be expected to extend into infinitely large wavenumbers since averages of the form $\langle k^n\rangle$ will diverge for some sufficiently large n. The unphysical properties of the result are usually attributed to an oversimplified modeling of the turbulent viscosity. One might be tempted to argue that a k^{-7} could continue until a new short scale is reached, but such a scale-length is not present in the basic equations from which the results are supposed to be obtained. Then it could be expected that it actually *is* so that averages like $\langle k^n\rangle$ diverge for sufficiently large n in turbulence described by the Navier–Stokes equation (implying a limitation of this equation as a model for turbulence), but no evidence for that has been found either: this would imply that some higher order derivatives became discontinuous and this has not been observed, neither in experiments nor in numerical simulations. In spite of some supporting experimental evidence, a k^{-7} dissipative spectral subrange has not gained much attention, probably because this particular form only moves the problem to divergences of higher n-values in $\langle k^n\rangle$. The intermittency problem involves localized dissipation regions, and modifications of the short wavelength part of the velocity power spectrum will be influenced by these intermittency effects (She and Leveque 1994).

Heisenberg also discussed decaying turbulence by use of (22.49), i.e. a case where the external stirring has been turned off. In this case viscous dissipation will 'win' ultimately, and we have $S = S(k, t)$, with an explicit time dependence. Heisenberg proposed a similarity of the spectrum as it decayed, and searched for solutions of the form

$$S(k, t) \sim \frac{1}{\sqrt{t}}F(k\sqrt{t}).$$

Two limiting cases were identified, one for small values of the argument, where $F \sim k\sqrt{t}$ and one for large arguments, where $F \sim (k\sqrt{t})^{-5/3}$. The small argument limit is in variance with observations, which indicate that a variation $\sim k^4$ seems more appropriate there.

Chandrasekhar (1949) re-analyzed Heisenberg's work and corrected some of his numerical results, also obtaining an exact analytical expression for the stationary velocity power spectrum $S(k)$, although none of Heisenberg's results were changed qualitatively. According to Chandrasekhar we can study decaying turbulence by starting with an equilibrium spectrum and then suddenly stopping the external energy source. The turbulence will then decay in three stages.

a) In the early stage, the largest eddies adjust to the lack of energy input.
b) An intermediate stage, where the remaining energy from the larges eddies maintains the Kolmogorov–Oubokhov spectrum at small scales.
c) In the final stage, the energy from the largest eddies is exhausted, and the Reynolds number falls to zero.

Within this scheme, Heisenberg's results seem to describe stage b) best. Extensive numerical studies of Heisenberg's results were carried out by Proudman and Batchelor (1951).

The analysis outlined here, and indeed the one found in most expositions on the subject explicitly assumes homogeneous and isotropic conditions by assuming that the structure functions and thereby correlation functions depend on position variable separations only, and not their absolute values. Fourier transform of the correlation function with respect to this separation variable gives a wavenumber spectrum. Energy cascade in a strongly stratified fluid has also been analyzed with results found in the literature (Lindborg 2006).

22.5.6 Extended models

The $\epsilon^{2/3}k^{-5/3}$ spectral law applies only for scales in the inertial subrange. The problems concerning even shorter scales, the viscous subrange, have already been mentioned. For large scales it could be expected that details would depend on details in the forcing of the turbulence and other non-universal conditions, but it has been found that a universal form with an extra length scale, an outer non-universal scale, L seems to accommodate a wide range of observations. It was thus proposed by von Kármán (1948) that a spectrum of the form

$$S(k) = \alpha\epsilon^{2/3}L^{5/3}\frac{(Lk)^4}{\left(1 + (Lk)^2\right)^{17/6}}, \tag{22.50}$$

could be used for most observations of homogeneous isotropic turbulence, with α being a numerical constant of the same order of magnitude as the spectral Kolmogorov constant. It is readily found that (22.50) reproduces the Kolmogorov–Oubokhov spectrum in the limit of large $k \gg 1/L$ as it should. The result (22.50) is illustrated in figure 22.3. We have

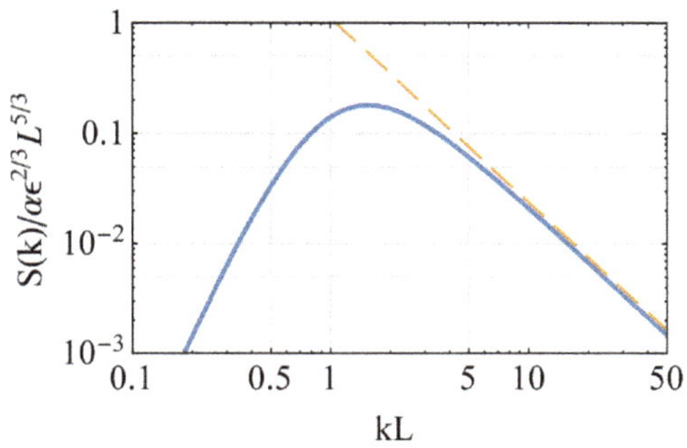

Figure 22.3. Illustration of the von Karman spectrum (22.50) as a function of normalized wavenumber kL. A dashed line gives the Kolmogorov–Oubokhov $k^{-5/3}$ law for reference. The outer scale L can depend on experimental conditions.

$$\int_0^{\infty} S(k)\mathrm{d}k = \frac{3}{8}\alpha\epsilon^{2/3}L^{2/3}\sqrt{\pi}\,\frac{\Gamma\left(\frac{1}{3}\right)}{\Gamma\left(\frac{17}{6}\right)}$$

for (22.50) in terms of the standard Γ-function. Note, however, that $\int_0^{\infty} k^2 S(k)\mathrm{d}k$ diverges for (22.50) and the integral (which is needed for expressing ϵ, see (22.23) or (22.24)) has to be cut-off at some maximum wavenumber $k_{\,\mathrm{max}}$, where we usually take $k_{\mathrm{max}} \approx 1/\eta$ in terms of the Kolmogorov inner length scale.

22.5.7 Discussions of the universal spectral subrange

Given the universal spectral subrange

$$S(k) \sim \frac{\epsilon^{2/3}}{k^{5/3}},$$

see (22.41), we can discuss some of its properties, and quantities which can be derived from it. We can define (Orszag 1977) a characteristic time scale associated with a wavenumber k as

$$\tau(k) = a\epsilon^{-1/3}k^{-2/3}, \qquad (22.51)$$

with a being some dimensionless constant, presumably of the order unity. There are several interpretations of this time scale. We can for instance obtain the energy associated with a wave-packet built-up by wavenumber components around the selected wavenumber by

$$\frac{1}{2}\Delta_k^2 U = \int_{k/2}^{2k} S(k')\mathrm{d}k' \sim \frac{\epsilon^{2/3}}{k^{2/3}},$$

where there is an evident arbitrariness in *how* we define the wave-packet: for a given central wavenumber k we here took a range from $k/2$ to $2k$ for composing the selected wave-packet. We have several possibilities here:

- We can consider $\Delta_k U$ as a measure of the typical velocity differences across distances $\sim 2\pi/k$. It is then reasonable to interpret $\tau(k)$ in (22.51) as the time it takes an eddy with Fourier components centered around k to rotate around itself, or 'turn over', i.e. take $\tau(k) \sim 1/(k\Delta_k U)$.

- As an alternative (Orszag 1977), we can interpret $\tau(k)$ as the time it takes for energy transfer occurring at a rate ϵ to deplete the energy $\frac{1}{2}\Delta_k^2 U$, in the case of the input to $\Delta_k U$ from lower k-modes was to be turned off, for some reason. This time is clearly $\frac{1}{2}\Delta_k^2 U/\epsilon$, which turns out to be the same as $\tau(k)$ obtained before.
- Finally, as a third possible interpretation (Orszag 1977), we can consider $\tau(k)$ to be the time it takes to de-correlate the relative phases of the *simultaneous* Fourier amplitudes of distinct wave-vectors of magnitude k.

We might try to use these results to obtain a deeper insight into the nature of the spectral cascade, mediating the Kolmogorov–Oubokhov law. We *postulate* that this transport through the spectrum is of a diffusive nature. By 'diffusive' we imply a similarity to the diffusion equation, but now written in wavenumber space as

$$\frac{\partial}{\partial t}S(k, t) - \frac{\partial}{\partial k}D(k, t)\frac{\partial}{\partial k}S(k, t) = 0, \qquad (22.52)$$

where $D(k, t)$ is a general, and for the moment unspecified, diffusion coefficient, noting that actually we just *might* have a time dependence also in D. The diffusion flux is $F(k, t) \equiv -D(k, t)\partial S(k, t)/\partial k$. To have a constant power spectrum we require F to be independent of time as well as k. We now try to argue for a diffusion coefficient. First we note that the dimension of D must be K^2/T, where $[K]$ is *length*$^{-1}$, as for wavenumbers. The characteristic time is evidently $\tau(k)$ given by (22.51). As far as the wavenumber K is concerned, the most obvious choice is k itself, since we expect the interactions between various components in the spectrum to be strongest when the wavenumbers entering the interaction are of approximately the same magnitude. Consequently we have $D(k) \sim k^2(\epsilon^{-1/3}k^{-2/3})^{-1}$. The constant flux must be equal to the energy dissipated per kg fluid per unit time, which means that

$$k^2\epsilon^{1/3}k^{2/3}\frac{\partial}{\partial k}S(k) \sim \epsilon.$$

This expression can readily be integrated to give $S(k) \sim \epsilon^{2/3}k^{-5/3}$, again the Kolmogorov–Oubokhov result. Now, this does not really *prove* anything, but the result is at least not in *contradiction* with the hypothesis of diffusive transport through the spectrum.

In a fully developed turbulent flow we deal with shear flows on all scales until we reach the dissipation length scale. In most cases found in nature and in laboratory experiments, the turbulence is generated by large scale velocity shears. In the inertial subrange you expect the local shear to be much larger than the mean shear.

The discussion of this section serves to emphasize one difference between what we might term a truly turbulent spectrum as distinguished from those discussed in, say, section 20.3. One possible, though not unique, definition of a truly turbulent process is one with a cascade in wavenumber space: this is lacking in spectra derived from (20.35). The question is now how to distinguish? Since the various Fourier components are coupled somehow (although they remain uncorrelated in the sense discussed in section 20.4), it is plausible that methods that are sensitive to relative phases can be useful. One such candidate is a bispectrum, see section 20.4. If the power spectrum is due to a random superposition of coherent structures as in (20.1), we would expect that Fourier components are strongly phase-coupled over a wide range. For a spectral cascade as the one discussed in the present section, it will be argued that only nearby Fourier components are phase-coupled in a more or less uniform manner over all of Fourier space. Experimental studies of incompressible fluid turbulence by Lii *et al* (1976) support these ideas. A distinction of multimode and turbulent regimes in plasma turbulence have been demonstrated (Kim and Powers 1978, Kim *et al* 1980).

22.5.8 Intermittency and the velocity probability densities

Section 19.3 discussed intermittency in a somewhat abstract way as a property of a signal. One might like to know the *physical* origins of the intermittency effects, i.e. the reasons for deviations from a strictly Gaussian random process. There is no universal answer to this question; it depends on the problem we consider. As an illustration we can take (again) incompressible fluid turbulence as a reference case. The probability density for the fluctuations in absolute velocity is very close to a Gaussian: the velocity at a given position can be seen as the result of many widely separated disturbances in the sense of, for instance, Wandel and Kofoed-Hansen (1962). At any given position in space, the velocity is a result of a superposition of many vorticity centers in the surrounding, at least when the Reynolds number is large. By the central limit theorem, we then expect the velocity to have Gaussian distribution, at least to a good approximation. The argument seems quite safe for the Eulerian problem, and it turns out that the conclusion applies for the Lagrangian velocity as well under relatively general conditions, to be discussed in chapter 25.

Higher order statistics, the second order structure function for instance, exhibits non-Gaussian features. For small separations, in particular, the second order structure function approximates the square of the spatial derivative by $\langle (u_{\|}(\xi) - u_{\|}(\xi + r))^2\rangle \approx r^2\langle(\partial u_{\|}/\partial r)^2\rangle$. For fluid turbulence, these short separations are in the viscous subrange.

Intermittency effects in neutral fluid turbulence are manifested, for instance, by a skewness in the probability density of the fluid velocity derivatives (Batchelor 1953). In particular, for three-dimensional turbulence, it is found that

$$\left\langle \left(\frac{\partial u_i}{\partial r_i}\right)^3 \right\rangle \left\langle \left(\frac{\partial u_i}{\partial r_i}\right)^2 \right\rangle^{-3/2} \approx -0.3.$$

Small positive values of $\partial u_i/\partial r_i$ are slightly more probable than small negative values, but this positive contribution to the skewness is more than balanced by the higher frequency of occurrence of large negative values of $\partial u_i/\partial r_i$ (Batchelor 1953). In spite of the single point velocity field having a Gaussian distribution, the velocity

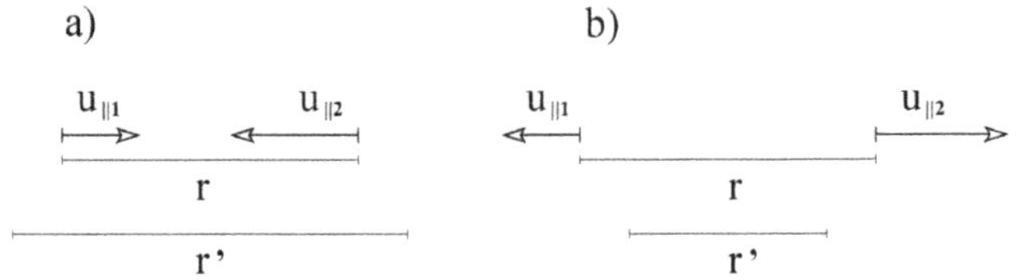

Figure 22.4. Illustration of velocity vector components for discussing intermittency effects in incompressible fluid turbulence. Sketch showing the intrinsic origin of the skewness. The fluid particles corresponding to case a) being at separations r at time t, were at a separation $r' > r$ at a time τ_c before, where randomization occurred. Thus negative values of $u_{\|}(\xi + r) - u_{\|}(\xi)$, correspond to larger r' values, i.e. a larger variance σ of the velocity difference distribution.

derivatives are not Gaussian, and the turbulent velocity fluctuations do not constitute a Gaussian random process.

The physical origin of the skewness of the velocity difference probability density can be explained by reference to, for instance, figure 22.4(a) and (b). The two cases illustrated might at first sight be expected to be equally probable, and indeed in large parts of the turbulent flows it is so. The energy cascade, which is manifested by vortex stretching (see Tennekes and Lumley (1972) and section 22.2), breaks this symmetry in regions close to vortex lines. Here the velocity difference between the two points 1 and 2 in figure 22.4 is $\Delta u \equiv u_{\|2} - u_{\|1}$. If $\Delta u > 0$ the two points 1 and 2 will originate from closer positions, $r' < r$. As the variance of the velocity fluctuations increase with increasing r, the negative wing of the distribution of Δu will be wider than the positive one, as illustrated by an example by Batchelor (1953) or Castaing *et al* (1990).

The situation for incompressible turbulent flows in *two* spatial dimensions is somewhat different. Here the vortex stretching mechanism is in effect arrested, as mentioned before in section 22.2, so the same arguments do not apply there. The velocity components entering $\Delta u \equiv u_{\|2} - u_{\|1}$ can in principle come from points which were previously larger $r' > r$ or smaller $r' < r$ separations. As the variance of the fluctuations increase with r, the negative wing will be wider than the positive one.

For short separations, we expect the skewness of the distribution of velocity separations to vanish in two-dimensional flows (Castaing *et al* 1990) where the vortex stretching mechanism is absent.

The tools described in section 19.3 can be used to quantify intermittency effects, but their physical origin is not always so easy to identify.

22.5.9 Conditional averages

We might use the results from section 19.2.4 and follow Adrian (1979) to answer the following question: if we go through all realizations of the flow and find a velocity in a narrow velocity interval around **u** in a narrow spatial interval around a reference position **x** at a reference time set to $t = 0$, then what is the average velocity found in the surroundings of this reference position at this or other times (this may be earlier times, or later)? The results are somewhat different in two and three spatial dimensions, so the two cases are considered separately, with the two-dimensional results presented in section 23.1.2.

Using the summation convention for repeated indices we write here for the time $t = 0$ the estimate for the three-dimensional velocity field as

$$\check{u}_i(\mathbf{r} + \mathbf{x}) = A_{ij}u_j(\mathbf{x}) + B_{ijk}u_j(\mathbf{x})u_k(\mathbf{x}),$$

where u_j is a component of the imposed velocity condition.

For three spatial dimensions (Adrian 1979) it is found that

$$A_{ij} = \frac{\langle u_j(\mathbf{x})u_i(\mathbf{r} + \mathbf{x})\rangle}{\mathcal{U}},$$

and

$$B_{ijk} = 3\frac{\langle u_j(\mathbf{x})u_k(\mathbf{x})u_i(\mathbf{r} + \mathbf{x})\rangle}{2K\mathcal{U}^4},$$

where $\mathcal{U} \equiv \frac{1}{3}\langle u_i u_i\rangle$ and K is the kurtosis. We can then write $\check{U}_i(\mathbf{r} + \mathbf{x}) = \check{u}_i^{(1)}(\mathbf{r} + \mathbf{x}) + \check{u}_i^{(2)}(\mathbf{r} + \mathbf{x})$ with

$$\check{u}_i^{(1)} \equiv \left(-\frac{1}{2}f'(r)\frac{r_i r_j}{r} + \left(f + \frac{1}{2}rf'\right)\delta_{ij}\right)u_j(\mathbf{x}),$$

and

$$\check{u}_i^{(2)} \equiv \frac{3}{2K\mathcal{U}}\left(\left(\frac{k - rk'}{2r^3}\right)r_j r_k r_i + \left(\frac{2k + rk'}{4r}\right)\left(r_j\delta_{ki} + r_k\delta_{ji}\right) - \frac{k}{2r}r_i\delta_{jk}\right)u_j(\mathbf{x})u_k(\mathbf{x}),$$

where we used the expressions for the correlation functions for homogeneous and isotropic turbulence in three dimensions (Batchelor 1953) where two-dimensional equivalents are given in section 23.1.

The previous results can be expressed most conveniently in terms of stream functions $\check{\Psi} = \check{\Psi}^{(1)} + \check{\Psi}_i^{(2)}$ where $\check{\Psi}^{(1)} = \frac{1}{2}u(\mathbf{x})r^2 f(r)\sin\theta$ and $\check{\Psi}^{(2)} = (3u^2(\mathbf{x})/4K\mathcal{U})r^2 k(r)\sin^2\theta\cos\theta$, introducing $\theta \equiv \arccos(\mathbf{r}\cdot\mathbf{u}(\mathbf{x})/ru(\mathbf{x}))$ being the polar angle of $\mathbf{r}$ with respect to $\mathbf{u}(\mathbf{x})$ and $r \equiv |\mathbf{r}|$. The stream function is cylindrically symmetric with respect to the direction of the imposed condition vector $\mathbf{u}$. The functions $f(r)$ and $k(r)$ can be obtained by measurements, in general. Extensions of the analysis to include even higher order terms have been given by Tung and Adrian (1980).

References

Adrian R J 1979 Conditional eddies in isotropic turbulence *Phys. Fluids* **22** 2065–70

Aref H and Kambe T 1988 Report on the IUTAM symposium: fundamental aspects of vortex motion *J. Fluid Mech.* **190** 571–95

Aris R 1962 *Vectors, Tensors, and the Basic equations of Fluid Mechanics* (London: Prentice-Hall)

Batchelor G K 1953 *The Theory of Homogeneous Turbulence* (Cambridge: Cambridge University Press)

Batchelor G K 1959 Small-scale variation of convected quantities like temperature in turbulent fluid. Part 1. General discussion and the case of small conductivity *J. Fluid Mech.* **5** 113–33

Batchelor G K 1969 Computation of the energy spectrum in homogeneous two-dimensional turbulence *Phys. Fluids Supplement* **12** II-233–9

Beran M J 1968 *Statistical Continuum Theories* Vol 9 *of Monographs in Statistical Physics and Thermodynamics* (New York: Interscience)

Brekhovskikh L and Goncharov V 1985 *Mechanics of Continua and Wave Dynamics (Series on Wave Phenonmena)* (Berlin: Springer)

Castaing B, Gagne Y and Hopfinger E J 1990 Velocity probability density functions of high Reynolds number turbulence *Physica D* **46** 177–200

Champeney D C 1973 *Fourier Transforms and their Physical Applications* (London: Academic)

Chandrasekhar S 1949 On Heisenberg's elementary theory of turbulence *Proc. Roy. Soc. (London)* **A 200** 20–33

Chandrasekhar S 1957 The theory of turbulence *J. Madras Univ.* **B 27** 251–75

Constantin P 1991 *New Perspectives in Turbulence* chapter 8. Remarks on the Navier–Stokes equations (Berlin: Springer)

Davidson P A 2001 *An Introduction to Magnetohydrodynamics* (Cambridge: Cambridge University Press)

Davidson P A 2004 *Turbulence. An Introduction for Scientists and Engineers* (Oxford: Oxford University Press)

Frisch U 1995 *Turbulence: The Legacy of A. N. Kolmogorov* (Cambridge: Cambridge University Press)

Hatayama A, Koshi Y and Ogasawara M 1981 Mixing length hypothesis in the dissipative trapped ion instability *J. Phys. Soc. Japan* **50** 976–9

Heisenberg W 1948 Zur statistische theorie der turbulenz *Z. Phys* **124** 628–57 An English translation of Heisenberg's paper is available for instance as Nato Advisory Committee for Aeronautics, Technical Memorandum 1431

Hinze J O 1975 *Turbulence* 2nd edn (New York: McGraw Hill)

Kadomtsev B B 1965 *Plasma Turbulence* (New York: Academic)

Kampe de Feriet J 1954–1955 Introduction to the statistical theory of turbulence, parts I-IV *J. Soc. Indust. Appl. Math.* **2** 1–9 143–174, 244–271, 90–117

Kaye G W C and Laby T H 1995 *Tables of Physical and Chemical Constants* 16th edn (London: Longmans)

Kim Y C, Beall J M, Powers E J and Miksad R W 1980 Bispectrum and nonlinear wave coupling *Phys. Fluids* **23** 258–63

Kim Y C and Powers E J 1978 Digital bispectral analysis of self-excited fluctuation spectra *Phys. Fluids* **21** 1452–3

Landahl M T and Mollo-Christensen E 1992 *Turbulence and Random Processes in Fluid Mechanics* 2nd edn (Cambridge: Cambridge University Press)

Lii K S, Rosenblatt M and Van Atta C 1976 Bispectral measurements in turbulence *J. Fluid Mech.* **77** 45–62

Lindborg E 2006 The energy cascade in a strongly stratified fluid *J. Fluid Mech.* **550** 207–42

Loitsyanskii L G 1966 *Mechanics of Liquids and Gases* (Oxford: Pergamon)

Orszag S A 1977 *Lectures on the Statistical Theory of Turbulence* Fluid Dynamics—Dynamique des Fluides, Les Hauches, Juillet 1973, Cours de l'Ecole d'Eté de Physique Théorique Organe d'intérèt commun de l'U.S.M.G. et I.N.P.G., subventionné par le Commissariat à l'Energie Atmique (London: Gordon and Breach)

Pécseli H L 2000 *Fluctuations in Physical Systems* (Cambridge: Cambridge University Press)

Pécseli H L and Trulsen J 2007 Turbulent particle fluxes to perfectly absorbing surfaces: a numerical study *J. Turbul.* **8** N42

Pécseli H L, Trulsen J K and Fiksen Ø 2014 Predator-prey encounter and capture rates in turbulent environments *Limnol. Oceanog.: Fluids Environments* **4** 85–105

Pope S B 2000 *Turbulent Flows* (Cambridge: Cambridge University Press)

Proudman I and Batchelor G K 1951 A comparison of Heisenberg's spectrum of turbulence with experiment *Math. Proc. Cambridge Phil. Soc.* **47** 158–76

Reif F 1965 *Fundamentals of Statistical and Thermal Physics* (Blacklick, OH: McGraw-Hill College)

Serrin J 1959 *Mathematical Principles of Classical Fluid Mechanics* Vol VIII/1 *of Handbuch der Physik* (Berlin: Springer) chapter Ströhmungsmechanik pp 125–263

She Z-S and Leveque E 1994 Universal scaling laws in fully developed turbulence *Phys. Rev. Lett.* **72** 336–9

Stanišić M M 1988 *The Mathematical Theory of Turbulence, Universitext* 2nd edn (Berlin: Springer)

Sulem C, Sulem P-L and Frisch H 1983 Tracing complex singularities with spectral methods *J. Comp. Phys.* **50** 138–61

Tennekes H 1973 Intermittency of the small-scale structure of atmospheric turbulence (Part II) *Boundary-Layer Meteorol* **4** 241–50

Tennekes H 1975 Eulerian and Lagrangian–time microscales in isotropic turbulence *J. Fluid Mech.* **67** 561–7

Tennekes H and Lumley J L 1972 *A First Course in Turbulence* (Cambridge, MA: The MIT Press)

Tritton D J 1988 *Physical Fluid Dynamics* 2nd edn (Oxford: Oxford Science Publications)

Tsinober A 2001 *An Informal Introduction to Turbulence, Fluid Mechanics and its Applications* (Dordrecht: Kluwer)

Tung T C and Adrian R J 1980 Higher order estimates of conditional eddies in isotropic turbulence *Phys. Fluids* **23** 1469–70

von Kármán T 1948 Progress in the statistical theory of turbulence *Proc. National Acad. Sci. USA* **34** 530–9

Wandel C F and Kofoed-Hansen O 1962 On the Eulerian–Lagrangian transformation in the statistical theory of turbulence *J. Geophys. Res.* **67** 3089–93

Chapter 23

Low frequency turbulence in magnetized plasmas

In magnetized plasmas, we expect the time averaged plasma parameters to be constant along magnetic field lines, but can in general have inhomogeneous conditions in the directions transverse to **B**. Drift waves are associated with such inhomogeneities. However, it turns out that even a homogeneous magnetized plasma has interesting properties.

23.1 Homogeneous plasmas

The simplest, yet relevant, problem we can consider deals with homogeneous plasmas, i.e. the case where we have no density or temperature gradients.

By considering two-dimensional random or turbulent flows, in particular, we can directly derive an important relation for the power spectral tensor. We assume that an incompressible flow is described through a velocity potential Φ, so that $\mathbf{u} = \nabla\Phi \times \hat{\mathbf{z}}$, i.e. $u_x = \partial\Phi/\partial y$ and $u_y = -\partial\Phi/\partial x$. For these two-dimensional flows we have in particular the vorticity to be $\boldsymbol{\omega} \equiv \nabla \times \mathbf{u} = -\nabla^2\Phi$. In neutral flows the velocity potential is an auxiliary function: for the equivalent two-dimensional homogeneously magnetized plasma problem where $\mathbf{u}_\perp = \mathbf{E}_\perp \times \mathbf{B}/B^2 = -\nabla_\perp\Phi \times \hat{\mathbf{z}}/B$ giving $\nabla \times \mathbf{u} = 0$, the electrostatic potential (with a sign opposite of the velocity potential) is a measurable quantity that offers a significant experimental simplification.

We have the following expression for the correlation tensor for time stationary, and homogeneous, but not necessarily isotropic, random flows

$$\mathsf{R}(\mathbf{r}, t) = \begin{vmatrix} \left\langle u_y(\mathbf{r}_1, t_1) u_y(\mathbf{r}_1 + \mathbf{r}, t_1 + t) \right\rangle & -\left\langle u_x(\mathbf{r}_1, t_1) u_y(\mathbf{r}_1 + \mathbf{r}, t_1 + t) \right\rangle \\ -\left\langle u_y(\mathbf{r}_1, t_1) u_x(\mathbf{r}_1 + \mathbf{r}, t_1 + t) \right\rangle & \left\langle u_x(\mathbf{r}_1, t_1) u_x(\mathbf{r}_1 + \mathbf{r}, t_1 + t) \right\rangle \end{vmatrix}$$

doi:10.1088/978-0-7503-1251-6ch23

or, in terms of the correlation function for the velocity potential,

$$\mathsf{R}(\mathbf{r}, t) = \begin{vmatrix} \frac{\partial^2}{\partial y_1 \partial y_2}\langle \Phi(\mathbf{r}_1, t_1)\Phi(\mathbf{r}_1 + \mathbf{r}, t_1 + t)\rangle & -\frac{\partial^2}{\partial x_1 \partial y_2}\langle \Phi(\mathbf{r}_1, t_1)\Phi(\mathbf{r}_1 + \mathbf{r}, t_1 + t)\rangle \\ -\frac{\partial^2}{\partial y_1 \partial x_2}\langle \Phi(\mathbf{r}_1, t_1)\Phi(\mathbf{r}_1 + \mathbf{r}, t_1 + t)\rangle & \frac{\partial^2}{\partial x_1 \partial x_2}\langle \Phi(\mathbf{r}_1, t_1)\Phi(\mathbf{r}_1 + \mathbf{r}, t_1 + t)\rangle \end{vmatrix}$$

where $\langle u_x(\mathbf{r}_1, t_1)u_y(\mathbf{r}_1 + \mathbf{r}, t_1 + t)\rangle = \langle u_y(\mathbf{r}_1, t_1)u_y(\mathbf{r}_1 + \mathbf{r}, t_1 + t)\rangle$ for the assumed conditions. Introducing the (scalar) correlation function for the velocity potential as $R_\Phi(\mathbf{r}, t) = \langle \Phi(\mathbf{r}_1, t_1)\Phi(r_1 + \mathbf{r}, t_1 + t)\rangle$, we can evidently write

$$\mathsf{R}(\mathbf{r}, t) = \begin{Bmatrix} \frac{\partial^2}{\partial y_1 \partial y_2} & -\frac{\partial^2}{\partial x_1 \partial y_2} \\ -\frac{\partial^2}{\partial x_2 \partial y_1} & \frac{\partial^2}{\partial x_2 \partial x_1} \end{Bmatrix} R_\Phi(\mathbf{r}, t).$$

For isotropic turbulence, where $R_\Phi(\mathbf{r}, t) = R_\Phi(r, t)$, this relation can be made much simpler and more elegant. With $r = \sqrt{(x_1 - x_2)^2 + (y_1 - y_2)^2}$, we have

$$\frac{\partial}{\partial x_1}R_\Phi(r) = -\frac{x_1 - x_2}{r}\frac{\partial}{\partial r}R_\Phi(r)$$
$$\frac{\partial^2}{\partial x_2 \partial x_1}R_\Phi(r) = \frac{(x_1 - x_2)^2}{r^3}\frac{\partial}{\partial r}R_\Phi(r) - \frac{(x_1 - x_2)^2}{r^2}\frac{\partial^2}{\partial r^2}R_\Phi(r) - \frac{1}{r}\frac{\partial}{\partial r}R_\Phi(r)$$
$$\frac{\partial^2}{\partial y_1 \partial x_2}R_\Phi(r) = \frac{(x_1 - x_2)(y_1 - y_2)}{r^3}\frac{\partial}{\partial r}R_\Phi(r) - \frac{(x_1 - x_2)(y_1 - y_2)}{r^2}\frac{\partial^2}{\partial r^2}R_\Phi(r)$$

and similar relations obtained by permuting indices. We wrote the expressions for zero time-delay for simplicity only. We find after simple manipulations

$$\mathsf{R}(\mathbf{r}, t) = \frac{1}{r^2}\left(\frac{1}{r}\frac{\partial}{\partial r}R_\Phi(\mathbf{r}, t) - \frac{\partial^2}{\partial r^2}R_\Phi(\mathbf{r}, t)\right)\begin{Bmatrix} r_2^2 & -r_1 r_2 \\ -r_1 r_2 & r_1^2 \end{Bmatrix} - \frac{1}{r}\frac{\partial}{\partial r}R_\Phi(\mathbf{r}, t)\begin{Bmatrix} 1 & 0 \\ 0 & 1 \end{Bmatrix},$$

or for any tensor component denoted by the subscript i, j

$$\{R\}_{i,j}(\mathbf{r}) = -\frac{\partial^2}{\partial r^2}R_\Phi(r)\delta_{i,j} + \frac{r_i r_j}{r^2}\left(r\frac{\partial}{\partial r}\frac{1}{r}\frac{\partial}{\partial r}R_\Phi(r)\right). \qquad (23.1)$$

The form $R_{i,j} \sim a\delta_{i,j} + b\, r_i r_j / r^2$ can be argued on general principles (Batchelor 1953, Beran 1968).

With the often used notation

$$g(r) \equiv -\frac{\partial^2}{\partial r^2}R_\Phi(r) \qquad \text{and} \qquad f(r) \equiv -\frac{1}{r}\frac{\partial}{\partial r}R_\Phi(r)$$

we have

$$\{R\}_{i,j}(\mathbf{r}) = g(r)\delta_{i,j} + (f(r) - g(r))\frac{r_i r_j}{r^2},$$

where $g(r) = f + r\partial f/\partial r = \partial(rf)/\partial r$ in two-dimensional isotropic turbulence (Pécseli and Trulsen 1995, Pécseli and Trulsen 1997). It is evident that for a correlation function $R_\Phi(r)$ maximized at the origin and vanishing for $r \to \infty$, both g and f take on negative values at least for *some* r, and in particular $\int_0^\infty g(r)\mathrm{d}r = 0$. This latter relation is a consequence of the necessity of back-flow somewhere in the plane. *If* we had a positive correlation, e.g. $R_{x,x}(y) > 0$ for all y, this would imply that we were likely to find parallel velocity vectors in the x-direction as we moved along the y-direction, but this would correspond to a net flow in the y-direction, in variance with the assumed isotropy. For small y-separations, we *are* likely to find parallel velocity vectors, since a realistic flow is unlikely to change direction over very short distances. On larger scales, however, we *must* have a back-flow. This observation indicates that two-dimensional turbulence is likely to consist of large vortices. Similar observations will of course also apply to *three*-dimensional isotropic turbulence.

To obtain the power spectral tensor $\mathsf{F}(\mathbf{k})$, we recall that when Fourier transforming a derivative we multiply with the transform variable and obtain trivially

$$\begin{aligned} \mathsf{F}(\mathbf{k}) &= F_\Phi(\mathbf{k})\begin{vmatrix} k_y^2 & -k_x k_y \\ -k_x k_y & k_x^2 \end{vmatrix}, \\ \{\mathsf{F}\}_{i,j}(\mathbf{k}) &= k^2 F_\Phi(\mathbf{k})\left[\delta_{i,j} - \frac{k_i k_j}{k^2}\right] \end{aligned} \tag{23.2}$$

in terms of the Kronecker $\delta_{i,j}$, which is unity for $i = j$ and zero otherwise. We introduced the (scalar) wavenumber power spectrum $F_\Phi(\mathbf{k})$ for the velocity potential, which can be obtained by Fourier transform of $R_\Phi(r)$. It turns out that a relation somewhat like (23.2) is valid for homogeneous and isotropic turbulence in three spatial dimensions (Batchelor 1953).

The wavenumber spectrum $F_\Phi(\mathbf{k})$ is defined so that $\langle\phi^2\rangle \equiv \iint F_\Phi(\mathbf{k})\mathrm{d}k_x\mathrm{d}k_y$. In the case of isotropic turbulence, we can introduce another spectrum so that $\langle\phi^2\rangle \equiv \int_0^\infty \mathcal{E}_\Phi(k)\mathrm{d}k$. Evidently we have $2\pi k F_\Phi(\mathbf{k}) = \mathcal{E}_\Phi(k)$. Actually, we can define such a 'one-dimensional' spectrum also for *anisotropic* turbulence, but its purpose is then no longer so obvious, since it can depend on a reference direction. For homogeneous and isotropic turbulence we have

$$\mathsf{F}(\mathbf{k}, t) = \frac{k}{2\pi}\mathcal{E}_\Phi(k, t)\left[\delta_{i,j} - \frac{k_i k_j}{k^2}\right]. \tag{23.3}$$

The properties of correlation functions and power spectra discussed in this section made no reference to the *nature* of the phenomena, but in order for the description to be of any use, the application to *random* flows is self evident. The randomness of the flow can, however, have different origins. We might have a condition where the energy is constantly fed into the flow at some characteristic scale lengths, to be dissipated at some other scale lengths. This is what we will usually consider a proper turbulent flow, i.e. the driven-dissipative situation. The Kolmogorov–Oubokhov spectrum discussed in section 22.5 is an example of such a model. Alternatively, we might consider a dissipationless system, and assume that some energy is released initially, to be subsequently distributed over all modes, with distributed amplitudes. This flow will eventually reach a statistical equilibrium, though not necessarily in thermal equilibrium. In a real physical condition we will not expect to encounter the latter type of phenomena, but they make perfect sense in, for instance, numerical simulations of plasma turbulence. We might also be dealing with decaying turbulence, where the energy input has been terminated, and the turbulence decays, subject to viscous dissipation. This problem has also received considerable interest.

- **Exercise:** Show that for an (1) incompressible and (2) irrotational flow it is possible to find a potential function $\phi(\mathbf{r}, t)$ as different from the velocity potential, so that $\mathbf{u} = \nabla\phi$. Deduce that the flow is a unique solution when you specify the normal velocity of a closed boundary (Aris 1962). Is this result restricted to 2 spatial dimensions?

23.1.1 Two-dimensional random flows generated by superposition of line vortices

In section 17.1.1 we found that line charge, distributed along homogeneous magnetic field lines gives rise to a vortex-like velocity field. The form of this vortex depends on the description used for the electrons, where the assumption of isothermally Boltzmann distributed electrons implies a shielding of the line charge, see section 17.1.2. As outlined in sections 17.1.3 and 17.1.4, we can imagine an approximation to the actual charge distribution to be obtained by a distribution of line charges, implying that the actual velocity field is approximated by the sum of the contributions from the individual vortices. The vortices define the velocity field at any instant, and the velocity field determines the motion of the individual line vortices. If we now make the additional statement that the positions at any time can be assumed to be randomly distributed on the plane, we can obtain the correlation function, spectra and potential amplitude distributions by reference to the analysis of chapter 20, the only difference being that we here are dealing with two spatial variables, instead of one temporal variable. We write

$$\Phi(\mathbf{r}) = \sum_{j}^{N} \phi(\mathbf{r} - \mathbf{r}_j) - \sum_{\ell}^{N} \phi(\mathbf{r} - \mathbf{r}_\ell),$$

where ϕ denotes the potential variation associated with an individual vortex, all assumed equal, and by Φ we denote the resulting potential variation. The sum consists of two parts, taking into account that we have an equal number of positive and negative line sources, with a total number of $2N$. The model might seem restrictive by taking all line vortices to have the same numerical value of the charge, but this is just assumed to simplify the notation, and the full case, with distributed charges, is trivially obtained by use of section 20.1.2.

The correlation function for the potential is now readily obtained to be

$$R_\Phi(r) = 2\frac{\langle N\rangle}{\mathcal{A}} \iint \phi(\mathbf{r}')\phi(\mathbf{r} + \mathbf{r}')\mathrm{d}\mathbf{r}',$$

where $\mathcal{A}$ is the (large) area of the system. As evident, we explicitly used that the average is vanishing, with the present model, see also section 20.3. The power spectrum is subsequently obtained by the Fourier transform of $R_\Phi(r)$.

Taking the shielded limit, we have $\phi(r) = (A/2\pi)K_0(r/\lambda_D)$, giving

$$S_\Phi(k) = C\big/\big(1 + (k\lambda_D)^2\big)^2, \tag{23.4}$$

where we have a constant C depending on the strength of the individual vortices (Pécseli and Trulsen 1997). The velocity power spectral tensor is then

$$S_{i,j}(\mathbf{k}) = k^2 S_\Phi(k)\left(\delta_{i,j} - \frac{k_i k_j}{k^2}\right). \tag{23.5}$$

The trace of the tensor becomes $S_{j,j}(k) = Ck^2/(1 + (k\lambda_D)^2)^2$.

The present model for the fluctuating velocity spectrum corresponds to the case where energy is released into a dissipationless system, where it is 'sloshing around' for all later times. It is essential for the model that the electric field is generated by *line* charges. If we have a finite, though narrow, spatial charge distribution, there will be a finite possibility for overlap between two vortex cores, and the model is no longer applicable. The point charge gives rise to a singularity of the flow by giving infinite velocities as we approach a vortex center. This divergence is manifested in the divergence of the integral $\int_0^\infty S_{j,j}(k)2\pi k\mathrm{d}k$. Note, though, that $\int_0^\infty S_\Phi(k)2\pi k\mathrm{d}k$ is finite.

The *unshielded* vortex limit discussed in section 17.1.3 can be obtained from the foregoing results by letting $\lambda_D \to \infty$, giving $S_\Phi(k) = C_1 k^{-4}$ and $S_{j,j}(k) = C_2 k^{-2}$. In this case we have a divergence of the spectral integrals in the large as well as the small wavenumber limits. The small wavenumber divergence (i.e. at large scales) led Taylor and McNamara (1971) to conclude that the physical size of the system (or rather the 'outer' non-universal scale of the spectrum) was important for the analysis, in particular for the expression of the spectral energy entering the turbulent diffusion coefficient.

The analysis of the present section assumed implicitly that the positions of the vortices were independent. This is of course not strictly correct, because of the Hamiltonian property of the system, see section 17.1.4. If we have already placed $N - 1$ vortices, the position of the last one is *not* arbitrary if we want to obtain a certain value of the Hamiltonian! When H is large and positive, as is the case if we initially placed two of the vortices with the same sign close together, it will at all later times be slightly more probable to find like-signed vortices (but of course not always the same ones) close to each other. If we, however, have $H \approx 0$, then it seems more likely that relative vortex positions can be weakly coupled, and results like (23.4) and (23.5) may apply. Indeed, some numerical simulations support this interpretation

(Lynov *et al* 1991, Pécseli and Trulsen 1997), and the model was used to discuss turbulent plasma diffusion across magnetic field lines (Pécseli and Mikkelsen 1985).

23.1.2 Conditional averages

Following the discussion in section 19.2.4 (see also section 22.5.9), we can attempt to derive an analytical expression for the average velocity distribution in stationary, homogeneous and isotropic, two-dimensional turbulence, subject to the condition that the direction and magnitude of the velocity vector is given as $\mathbf{U}(0, 0) = \mathbf{U}_0$ in one point, say the origin at a reference time, say $t = 0$. This analysis was carried out by Pécseli and Mikkelsen (1986), assuming that the local fluid velocity in the direction was well represented by the $\mathbf{E} \times \mathbf{B}/B_0^2$-velocity, possibly with the lowest order finite Larmor radius corrections to the velocity included. They gave the result in terms of a stream function

$$\check{\psi}(r, \theta) = rf(r)\sin(\theta),$$

with

$$f(r) = -\frac{1}{r}\frac{\mathrm{d}}{\mathrm{d}r}\left(1 + \frac{r_L^2}{4}\left(\frac{1}{r}\frac{\mathrm{d}}{\mathrm{d}r} + \frac{\mathrm{d}^2}{\mathrm{d}r^2}\right)\right)^2 R_\psi(r),$$

where $R_\psi(r)$ is the spatial correlation for the electrostatic potential associated with the electric field fluctuations. Finite ion Larmor radius corrections to the $\mathbf{E} \times \mathbf{B}/B_0^2$-velocity were included also, by the term with $\mathrm{r}_L{}^2$ as coefficient. From $\check{\psi}$ we obtain the conditionally averaged velocity field from

$$\check{\mathbf{U}}(x, y) = \left\{\check{U}_x, \check{U}_y\right\} = U_0\left\{\frac{\partial}{\partial y}\check{\psi}, -\frac{\partial}{\partial x}\check{\psi}\right\},$$

with $r^2 = x^2 + y^2$. The direction of the conditional velocity vector is taken to be along the x-axis, which is not restrictive for homogeneous turbulence. We assumed that U is normalized by $\sqrt{\langle E^2\rangle}/B_0$. The results quoted correspond to retaining the first term in the series expansion discussed in section 19.2.4. Pécseli and Mikkelsen (1986) gave results based on experimentally determined correlation functions $R_\psi(r)$, using data from Pécseli *et al* (1983), although the assumptions of local homogeneity and isotropy were certainly questionable in that experiment. The basic feature of a large double vortex came out, however, quite clearly in the results. As an illustration we show in figure 23.1 the case where $R_\psi(r)$ is modeled as a normalized Gaussian, $\exp(-r^2)$. The bottom figure shows the Eulerian vector field associated with a conditional vortex in this model case. The figure illustrates, in particular, the back-flow mentioned before, when we observe the flow along the x-axis for $y = 0$.

23.2 Spectral cascade

The Hasegawa–Mima equation describes weakly nonlinear electrostatic fluctuations, in particular also wave interactions. For this analysis, the equations are rewritten in a spectral representation by introducing

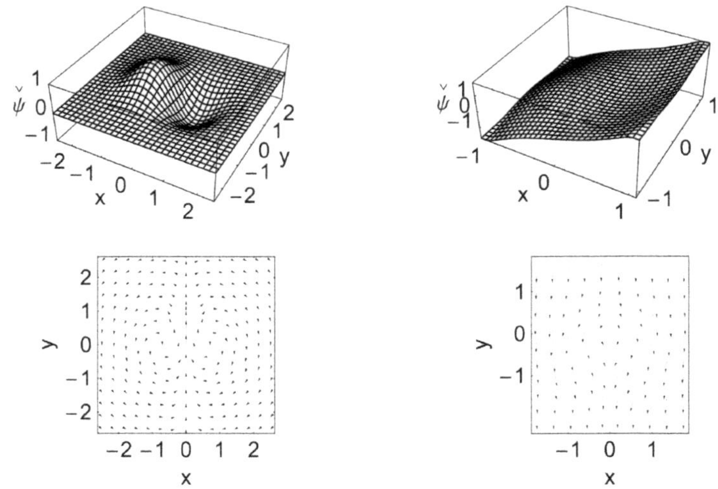

Figure 23.1. Illustration of a conditional vortex obtained by modeling $R_\psi(r) = \exp(-r^2)$. The top figure shows the conditional stream function, $\check{\psi}(r, \theta)$, while the bottom figure illustrates the conditional velocity field derived from the model. The conditionally selected velocity vector is at the origin, in the positive y-direction. The left-hand side shows the results in the rest frame, the right-hand side is in the co-moving frame.

$$\phi(\mathbf{r}, t) = \sum_k \phi_\mathbf{k}(t) \mathrm{e}^{\mathrm{i}\mathbf{k}\cdot\mathbf{r}}, \tag{23.6}$$

where $\phi_\mathbf{k}(t) = \phi^*_{-\mathbf{k}}(t)$ since $\phi(\mathbf{r}, t)$ is a real-valued function. If (23.6) is inserted into the *linearized* version of (18.7) the coefficients are obtained as $\phi_\mathbf{k}(t) = \exp(\mathrm{i}\omega_\mathbf{k} t)$ where $\omega_\mathbf{k} = \beta k_y/(1 + k^2)$ is the linear dispersion relation for (18.7). Note that the linear properties of (18.7) are trivial in the case where the density gradient can be ignored, i. e. $\beta = 0$. In this case the result is $\nabla_\perp^2 \phi = \phi$, giving the linearized version of Debye-shielding. The presence of a gradient, or rather the electron diamagnetic drift, gives rise to the linear dynamics of the waves. The nonlinear terms induce an additional time-variation of the wave amplitudes $\phi_\mathbf{k}$, to be obtained by writing the Fourier transform of the nonlinear terms in (18.7) as a convolution sum. In this representation we have (18.7) in the form

$$\left(1 + k^2\right)\frac{\mathrm{d}}{\mathrm{d}t}\phi_\mathbf{k}(t) + i\beta k_y \phi_\mathbf{k}(t) = \sum_{\mathbf{k}=\mathbf{k}'+\mathbf{k}''} k''^2 (\hat{\mathbf{z}} \times \mathbf{k}') \cdot \mathbf{k}''\, \phi_{\mathbf{k}'}(t)\phi_{\mathbf{k}''}(t). \tag{23.7}$$

By simple manipulations this equation is rewritten as

$$\frac{\mathrm{d}}{\mathrm{d}t}\phi_\mathbf{k}(t) + i\omega_\mathbf{k}\phi_\mathbf{k}(t) = \sum_{\mathbf{k}'+\mathbf{k}''-\mathbf{k}=0} \Lambda^{-\mathbf{k}}_{\mathbf{k}',\mathbf{k}''}\left(k''^2 - k'^2\right)\phi_{\mathbf{k}'}(t)\phi_{\mathbf{k}''}(t), \tag{23.8}$$

with

$$\Lambda^{-\mathbf{k}}_{\mathbf{k}',\mathbf{k}''} = \frac{1}{2}\frac{(\mathbf{k}' \times \mathbf{k}'') \cdot \hat{\mathbf{z}}}{\left(1 + k^2\right)},$$

and the wavenumber matching condition

$$\mathbf{k}' + \mathbf{k}'' - \mathbf{k} = 0.$$

Physically, the relation (23.8) demonstrates the interaction of three wavenumber components labeled by $\mathbf{k}$, $\mathbf{k}'$ and $\mathbf{k}''$, where a matching condition is fulfilled. The strength of the interaction is measured by the coupling coefficient $\Lambda^{-\mathbf{k}}_{\mathbf{k}',\mathbf{k}''}$. The coupling is large for 'fat' triangles composed by the three vectors $\mathbf{k}'$, $\mathbf{k}''$ and $\mathbf{k}$, while $\Lambda^{-\mathbf{k}}_{\mathbf{k}',\mathbf{k}''}$ is small for 'slim' triangles, and vanishes when the three vectors are on a straight line. A more detailed and somewhat different analysis of wave-vector triad interactions was given by Bustamante *et al* (2014).

The sum in (23.8) is over all possible triplets in the actual perturbation. In general, a large number, N, of different terms interact simultaneously, assuming a numerical simulation with a finite number of Fourier modes. We may however imagine that during a small time interval $\Delta t/N$ only one term acts, however with the N-fold intensity with all other terms 'switched-off'. During the next small time interval $\Delta t/N$, another term acts and so on, until all terms on the right-hand side of (23.8) have been activated (Knorr *et al* 1990). This procedure is known in numerical methods as 'the method of weak approximation' or as 'the method of fractional steps' or simply 'splitting'. It can be proved that the approximation converges to the true solution if the time interval Δt is reduced. Seen in this light, the basic nonlinearity is adequately described by considering one triplet only. This interaction is described by

$$\frac{\mathrm{d}}{\mathrm{d}t}\phi_{\mathbf{k}}(t) + i\omega_{\mathbf{k}}\phi_{\mathbf{k}}(t) = \Lambda^{-\mathbf{k}}_{\mathbf{p},\mathbf{q}}\left(p^2 - q^2\right)\phi_{\mathbf{p}}(t)\phi_{\mathbf{q}}(t). \tag{23.9}$$

By permutation of indexes, i.e. $p \to q$, $q \to -k$, and $-k \to p$ we find

$$\frac{\mathrm{d}}{\mathrm{d}t}\phi_{\mathbf{p}}(t) + i\omega_{\mathbf{p}}\phi_{\mathbf{p}}(t) = \Lambda^{\mathbf{p}}_{\mathbf{q},-\mathbf{k}}\left(q^2 - k^2\right)\phi^*_{\mathbf{q}}(t)\phi_{\mathbf{k}}(t), \tag{23.10}$$

where we used the reality condition $\phi_{-\mathbf{p}} = \phi^*_{\mathbf{p}}$ and $\omega_{-\mathbf{p}} = -\omega_{\mathbf{p}}$ and recalling that for the present case $\Lambda^{-\mathbf{k}}_{\mathbf{p},\mathbf{q}}$ is real. Similarly, by $p \to -k$, $q \to p$, and $-k \to q$, we find

$$\frac{\mathrm{d}}{\mathrm{d}t}\phi_{\mathbf{q}}(t) + i\omega_{\mathbf{q}}\phi_{\mathbf{q}}(t) = \Lambda^{\mathbf{q}}_{-\mathbf{k},\mathbf{p}}\left(k^2 - p^2\right)\phi_{\mathbf{k}}(t)\phi^*_{\mathbf{p}}(t), \tag{23.11}$$

where $\mathbf{p} + \mathbf{q} - \mathbf{k} = 0$. Note that no explicit assumption on frequency matching has been imposed, i.e. $\omega_{\mathbf{p}} + \omega_{\mathbf{q}} - \omega_{\mathbf{k}} \neq 0$ in general. The equations (23.9)–(23.11) can be solved exactly. This solution will not be obtained here, where only a qualitative analysis is attempted.

The conservation laws from the foregoing section imply

$$\left(1 + k^2\right)\left|\phi_{\mathbf{k}}(t)\right|^2 + \left(1 + p^2\right)\left|\phi_{\mathbf{p}}(t)\right|^2 + \left(1 + q^2\right)\left|\phi_{\mathbf{q}}(t)\right|^2 = \text{const}. \tag{23.12}$$

and

$$\left(1 + k^2\right)k^2\left|\phi_{\mathbf{k}}(t)\right|^2 + \left(1 + p^2\right)p^2\left|\phi_{\mathbf{p}}(t)\right|^2 + \left(1 + q^2\right)q^2\left|\phi_{\mathbf{q}}(t)\right|^2 = \text{const}. \tag{23.13}$$

- **Exercise:** Prove the relation

$$\Lambda^{-\mathbf{k}}_{\mathbf{p},\mathbf{q}}\left(p^2 - q^2\right) + \Lambda^{\mathbf{p}}_{\mathbf{q},-\mathbf{k}}\left(q^2 - k^2\right) + \Lambda^{\mathbf{q}}_{-\mathbf{k},\mathbf{p}}\left(k^2 - p^2\right) = 0,$$

for the coupling coefficients.
- **Exercise:** Prove the relations (23.12) and (23.13) directly from (23.9)–(23.11).

Introducing the energy of a mode k as $(1 + k^2)|\phi_{\mathbf{k}}(t)|^2 \equiv w_{\mathbf{k}}(t)$, and similarly for $\mathbf{p}$ and $\mathbf{q}$, we have from (23.12) and (23.13)

$$w_{\mathbf{k}}(t) + w_{\mathbf{p}}(t) + w_{\mathbf{q}}(t) \equiv 2W = \text{const.} \tag{23.14}$$

and

$$k^2 w_{\mathbf{k}}(t) + p^2 w_{\mathbf{p}}(t) + q^2 w_{\mathbf{q}}(t) \equiv 2E = \text{const.} \tag{23.15}$$

Introducing a Cartesian coordinate system, with axes in directions $w_{\mathbf{k}}$, $w_{\mathbf{p}}$ and $w_{\mathbf{q}}$, the relations (23.14)–(23.15) represent two intersecting planes. Any change in energy must satisfy $\delta w_{\mathbf{k}} + \delta w_{\mathbf{p}} + \delta w_{\mathbf{q}} = 0$ and $k^2 \delta w_{\mathbf{k}} + p^2 \delta w_{\mathbf{p}} + q^2 \delta w_{\mathbf{q}} = 0$, or $\delta \mathbf{w} \cdot \mathbf{e} = 0$ and $\delta \mathbf{w} \cdot \mathbf{s} = 0$, where $\delta \mathbf{w} = (\delta w_{\mathbf{k}}, \delta w_{\mathbf{p}}, \delta w_{\mathbf{q}})$, $\mathbf{e} = (1, 1, 1)$ and $\mathbf{s} = (k^2, p^2, q^2)$ in the coordinate system defined before. A system point can only move in the direction $\mathbf{e} \times \mathbf{s}$, so the changes become explicitly

$$\begin{aligned}
\delta w_{\mathbf{k}} &= \left(q^2 - p^2\right)\frac{\mathrm{d}}{\mathrm{d}t} f(t)\delta t \\
\delta w_{\mathbf{p}} &= \left(k^2 - q^2\right)\frac{\mathrm{d}}{\mathrm{d}t} f(t)\delta t \\
\delta w_{\mathbf{q}} &= \left(p^2 - k^2\right)\frac{\mathrm{d}}{\mathrm{d}t} f(t)\delta t.
\end{aligned} \tag{23.16}$$

The results (23.16) are easily verified by insertion into (23.14)–(23.15). The function $\mathrm{d}f(t)/\mathrm{d}t$ with initial condition $\mathrm{d}f(t)/\mathrm{d}y|_{t=0} = 0$, can be determined by (23.9)–(23.11). We do not need this explicit result here, but rather follow Fjørtoft (1953) in considering the ratios

$$\frac{\delta w_{\mathbf{k}}}{\delta w_{\mathbf{p}}} = \frac{\left(q^2 - p^2\right)}{\left(k^2 - q^2\right)} \tag{23.17}$$

$$\frac{\delta w_{\mathbf{p}}}{\delta w_{\mathbf{q}}} = \frac{\left(k^2 - q^2\right)}{\left(p^2 - k^2\right)} \tag{23.18}$$

$$\frac{\delta w_{\mathbf{q}}}{\delta w_{\mathbf{k}}} = \frac{\left(p^2 - k^2\right)}{\left(q^2 - p^2\right)}. \tag{23.19}$$

Assuming for instance that $p < k < q$, we have for example $\delta w_{\mathbf{k}}/\delta w_{\mathbf{p}} < 0$ and $\delta w_{\mathbf{q}}/\delta w_{\mathbf{k}} < 0$, while $\delta w_{\mathbf{p}}/\delta w_{\mathbf{q}} > 0$. From this result it is readily seen that no single of the three wavenumber components can represent a source or a sink for *both* two remaining components, unless it is represented by a wavenumber (i.e. inverse scale size) intermediate between the wavenumbers (or inverse scales) of the two other components (Fjørtoft 1953).

Using the wavenumber matching condition $\mathbf{p} + \mathbf{q} = \mathbf{k}$, it is readily shown that $(k^2 - q^2)/(p^2 - k^2) > 1$ for parameters p, q inside the area bounded by $p < q$, $p/q = -2\cos(\mathbf{q}, \mathbf{p}) - \sqrt{4\cos^2(\mathbf{q}, \mathbf{p}) - 1}$, $q/p = -2\cos(\mathbf{q}, \mathbf{p})$ and $\cos(\mathbf{q}, \mathbf{p}) > -1$, with $(\mathbf{q}, \mathbf{p})$ being the angle between the vectors $\mathbf{q}$ and $\mathbf{p}$. Therefore $\delta w_{\mathbf{p}}/\delta w_{\mathbf{q}} > 1$ in this region with $p < k < q$. From this result together with (23.19), it is evident that the numerical value of the change in energy will be largest for the component with the intermediate wavenumber, and smallest for the one with the largest wavenumber. This is actually a remarkable result: in contrast to incompressible fluid turbulence in three spatial dimensions, the energy will not cascade mainly from large scales towards small scales. The simultaneous conservation of energy as well as generalized enstrophy implies that if some energy is cascaded toward small scales in a triplet interaction *more* has to be cascaded towards larger scales (i.e. smaller wavenumbers) for wave vectors in the given region. This is the *inverse cascade*, being a characteristic of many two-dimensional flows (Kraichnan and Montgomery 1980), not only drift waves. For such systems, in the case of an initial perturbation being concentrated on small scales, i.e. large wavenumbers, eventually a significant part of the energy will be spreading to the long wavelength regime (Fjørtoft 1953, Merilees and Warn 1975). The proper interpretation of the inverse cascade has been subject to some controversy and the reader is urged to consult the literature, e.g. Tung and Welch (2001).

A byproduct of the inverse cascade turns out to be most disturbing in numerical simulations; In actual simulations it is difficult to avoid introducing an error at small scales, where the spatial resolution of the numerical code is limited. This deviation between the numerical solution and the 'true' solution would usually be of little concern, in other types of problems. This deviation can, however, in the analysis of for instance drift waves, be seen as an (unwanted) contribution to the spectral energy distribution, and the associated error will consequently spread to the long wavelength regime as well, due to the inverse cascade!

23.2.1 Simple argument for the inverse cascade

The simultaneous conservation of energy, E, and enstrophy, W, in two-dimensional flow systems has interesting and important consequences, outlined here with heuristic arguments, based on the energy power spectrum $S(\mathbf{k}, t) \equiv S(k_x, k_y, t)$, see also (18.15). First it is useful to introduce the one-dimensional spectra defined by rewriting the spectrum S in cylindrical wavenumber coordinates

$$\mathcal{E}(k, t)\mathrm{d}k \equiv k\mathrm{d}k \int_0^{2\pi} S(\mathbf{k}, \theta, t)\,\mathrm{d}\theta. \tag{23.20}$$

This relation does *not* necessarily imply that the spectrum is spatially isotropic, i.e. $S = S(|\mathbf{k}|, t)$ is *not* assumed. The total spectral energy is obtained as $E = \int_0^\infty \int_0^{2\pi} S(\mathbf{k}, \theta, t) k \mathrm{d}k \mathrm{d}\theta \equiv \int_0^\infty \mathcal{E}(k, t)\, \mathrm{d}k$. For an undriven system without dissipation, E is a constant even though the shape of the spectrum can vary with time.

From $\mathcal{E}(k, t)$ we can define a possibly time varying average wavenumber as

$$\overline{K_0}(t) \equiv \int_0^\infty k\mathcal{E}\mathrm{d}k \bigg/ \int_0^\infty \mathcal{E}\mathrm{d}k,$$

and a time varying width of the spectrum as

$$\Delta K^2(t) \equiv \int_0^\infty \left(k - \overline{K_0}\right)^2 \mathcal{E}\mathrm{d}k \bigg/ \int_0^\infty \mathcal{E}\mathrm{d}k.$$

These definitions can be argued by considering a power spectrum as a probability density for finding a certain wavenumber, and $\int_0^\infty \mathcal{E}\mathrm{d}k$ then enters as a means of normalizing this probability density.

Following e.g. Rhines (1975) or Pedlosky (1987), we assume now that initially the spectrum is peaked around $\overline{K_0}$, but otherwise smooth and integrable. Our experience then tells us that in general the spectrum will widen as time increases, implying $\mathrm{d}\Delta K^2/\mathrm{d}t > 0$. An exception would be that we by chance found an exact stationary nonlinear solution, but we exclude this case.

The expression for ΔK^2 is now rewritten as

$$\begin{aligned} \Delta K^2(t) &= \left[\int_0^\infty k^2\mathcal{E}\mathrm{d}k - 2\overline{K_0}\int_0^\infty k\mathcal{E}\mathrm{d}k + \overline{K_0}^2\int_0^\infty \mathcal{E}\mathrm{d}k\right] \bigg/ \int_0^\infty \mathcal{E}\mathrm{d}k \\ &= \left[\int_0^\infty k^2\mathcal{E}\mathrm{d}k - \overline{K_0}^2(t)\int_0^\infty \mathcal{E}\mathrm{d}k\right] \bigg/ \int_0^\infty \mathcal{E}\mathrm{d}k. \end{aligned} \tag{23.21}$$

The first term in (23.21) is readily identified as the integrated enstrophy spectrum giving the mean square enstrophy $W = \int_0^\infty k^2\mathcal{E}(k, t)\, \mathrm{d}k / \int_0^\infty \mathcal{E}\mathrm{d}k$, see also (18.16), while the second term in (23.21) contains the mean square energy as before. Since both these quantities are conserved we have

$$\frac{\mathrm{d}}{\mathrm{d}t}\overline{K_0}^2 = -\frac{\mathrm{d}}{\mathrm{d}t}\Delta K^2(t) < 0. \tag{23.22}$$

Thus, under the influence of nonlinear interactions which conserve both energy and enstrophy, the mean square wavenumber must decrease; a somewhat unexpected result when it was first obtained. The derivation summarized here has a certain appeal by its simplicity, but leaves much of the underlying physics obscured.

The present illustrative summary was referring to an initial value problem, but as a concept, we can imagine a similar process for a driven dissipative problem, where energy is continuously injected at some intermediate wavenumbers, and dissipated by sinks at large as well as small wavenumbers. Also other more detailed discussions on the proper interpretation of the inverse cascade are published (Tung and Welch 2001), but these will not be included here.

Experiments on soap films (this comes as close to truly two-dimensional systems as we can imagine), have given support for the energy-enstrophy cascade (Rutgers 1998).

23.2.2 Enstrophy spectrum

In section 23.2 we learned that enstrophy is likely to cascade towards short wavelengths. From (22.27) we found that in two dimensions, enstrophy is dissipated at short wavelengths.

Thus, for homogeneous and isotropic conditions, we can apply dimensional arguments very much along the lines presented in section 22.5, in order to discuss power spectra in two-dimensional turbulence (Batchelor 1969). As mentioned already, one basic difference between turbulence in two and three spatial dimensions is the role of vorticity, $\nabla \times \mathbf{u}(\mathbf{r}, t)$, see section 22.2. For the present two-dimensional conditions, enstrophy will be conserved in the absence of viscosity, see (22.28). We can propose a 'vorticity wavenumber power spectrum', $S_\omega(k)$, which has the dimension *length* $time^{-2}$. Assuming that, for short wavelengths, vorticity is dissipated by molecular viscosity at a rate of ϵ_ω pr. kg fluid, where the dimension of ϵ_ω is $time^{-3}$, we can argue (Pedlosky 1987) that the contribution to the mean square vorticity density from a wavenumber interval $\{k; k + \mathrm{d}k\}$ is $S_\omega(k)\mathrm{d}k$ so that

$$S_\omega(k) = \frac{\epsilon_\omega^{2/3}}{k} \mathcal{G}\left(k\nu^{1/2}/\epsilon_\omega^{1/6}\right),$$

just as in (22.40), with $\mathcal{A}$ being some unknown function. For a universal subrange independent of viscosity ν, we require $\mathcal{G}$ = constant, giving the vorticity power spectrum (Batchelor 1969)

$$S_\omega(k) \sim \frac{\epsilon_\omega^{2/3}}{k}. \qquad (23.23)$$

By similar arguments we can determine the second order structure function for the vorticity, and find

$$\Psi_\omega(r) = \epsilon_\omega^{2/3} \mathcal{F}\left(r\epsilon_\omega^{2/3}/\nu^{1/2}\right).$$

For the expression to be independent of viscous dissipation, we require here that $\mathcal{F}$ = const. implying that

$$\Psi_\omega(r) \sim \epsilon_\omega^{2/3},$$

in the corresponding separation range.

From, for instance, the vorticity spectrum (23.23) we can trivially obtain the velocity power spectrum as $S(k) = S_\omega(k)/k^2$ from the definition of vorticity. We consequently find

$$S(k) \sim \frac{\epsilon_\omega^{2/3}}{k^3}. \qquad (23.24)$$

Stimulated by the results of section 22.5.7, we can try to repeat the arguments for diffusive transport now for the enstrophy spectrum. According to (22.52), we again seek an expression for the diffusion coefficient D. Again we find the wavenumber k to be the most appropriate scale, while the relevant time scale is argued to be $\tau_\omega(k) \sim \epsilon_\omega^{-1/3}$, since there are no other time scales around. Consequently, in order to have a stationary enstrophy spectrum, we require

$$k^2 \epsilon_\omega^{1/3} \frac{\partial}{\partial k} S_\omega(k) \sim \epsilon_\omega.$$

This expression is readily integrated to give $S_\omega(k) \sim \epsilon_\omega^{2/3}/k$, which is nothing but the previous result (23.23). Also in this case, we find that the result lends some weight for the arguments in favor of diffusive transport through the power spectrum, in this case for the vorticity spectrum. If we now try to use the very same arguments for the power spectrum of velocity, $S(k)$, we find an inconsistent result! The reason is (evidently) that the transport in the velocity power spectrum is for the present case *not* diffusive, and the spectral amplitude is in practice vanishing in the corresponding subrange (Pedlosky 1987), consistent with the steeply decreasing k^{-3} spectrum (23.24).

Simple interpretations of experimentally observed energy cascade

Based on observations in a modified Penning discharge, Roth (1971) offered a simple argument for determining the direction of energy cascade. In a number of his experimental data on turbulent spectra in his device he identified cases like those shown in figure 23.2, where a 'peak' can be found in the power spectrum. He argued that it is close to self-evident that this peak represents injection of wave energy in the spectrum and that an enhancement on one side of the power indicates that energy is cascaded in that direction. Thus, in figure 23.2(a) it could be argued that energy is cascaded towards larger frequencies, which we by Taylor's hypothesis will expect to correspond to large wavenumbers. In figure 23.2(b) the same arguments will imply that energy cascades to smaller wavenumbers. Although very convincing, the arguments by Roth (1971) rely on a particular feature observed in a spectrum, and in reality this is not found so often, although it seemed to be a characteristic of his device. The ideas have no universal applicability.

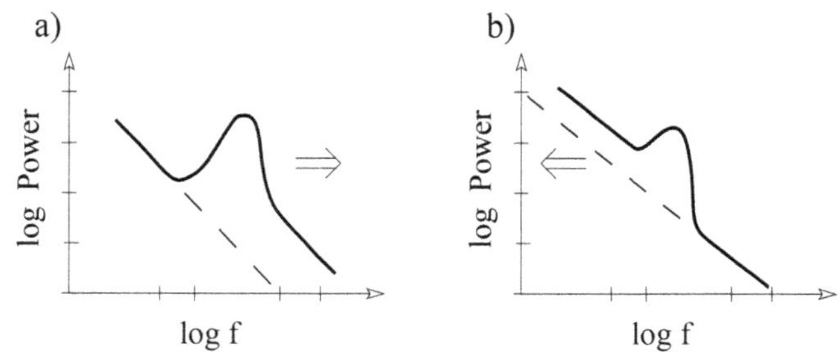

Figure 23.2. Schematic illustration of spectra observed by Roth (1971) in a modified Penning discharge, with a 'common sense' argument for the direction of a spectral cascade indicated.

In other experiments (Huld *et al* 1988) it was tried to inject a localized large amplitude disturbance into a turbulent magnetized plasma and then follow the space-time evolution of the non-stationary, inhomogeneous structure function of the electrostatic potential. The idea was that a broadening or narrowing of the normalized structure function could be interpreted as a sign of the energy cascade. The results were not conclusive.

23.3 Inhomogeneous plasmas

The discussion in the foregoing section 23.1 implicitly assumed that we *have* a fully developed strongly turbulent state. For three dimensions it can be demonstrated that the origin of the turbulence is a fluid instability (Hinze 1975, Landahl and Mollo-Christensen 1992), where the free energy for driving the turbulence is an inhomogeneous mean flow. For tidal flows in channels, and similar conditions, this energy can indeed drive the flow into a strongly turbulent state, as amply demonstrated by many observations, in nature as well as in laboratory experiments. It is not quite obvious that drift waves will necessarily reach a strongly turbulent state. Also in this case we have an instability driving the fluctuations, but this need not be a *strong* instability! We can of course postulate conditions where the condition is met, but it is worthwhile to discuss whether these conditions are met sufficiently often in nature to justify a major effort for our understanding of such condition. If we do *not* have strong turbulence, we might anticipate that no universal subrange exists, i.e. there are no wavenumbers where a spectral cascade is determined solely by the properties of drift waves as such, independent of the finer details of the driving instability and the overall geometry.

23.3.1 Correlation functions

Although formally they contain the same information, it is often convenient to distinguish correlation functions and power spectra. For inhomogeneous conditions, as those typically met in studies of drift waves, we can possibly hope for *locally* homogeneous conditions, and be able to define a correlation function of the form $R(\mathbf{R}, \mathbf{r}, \mathcal{T}, t)$, as in section 19.2.6, where the variation with the average coordinate $\mathcal{R}$ is slow, and fast with the separation variable $\mathbf{r}$. As far as the *time* variation of the signal is concerned, it can be considered a stationary random process for many relevant conditions, and $R(\mathbf{R}, \mathbf{r}, t)$ is a function of time separations, t, only. In a streaming plasma, we can obtain an estimate for the time variation of the correlation function for vanishing spatial separation, while a two-point correlation can readily give the $\mathbf{r}$ variation of the correlation function for zero time-lag. Examples of such results are given by Pécseli and Mikkelsen (1986), as also discussed in section 23.4.1.

As already mentioned, we can obtain a local power spectrum by Fourier transforming $R(\mathbf{R}, \mathbf{r}, t)$ with respect to time as well as the separation coordinate $\mathbf{r}$. The result will subsequently depend on the selected value for $\mathbf{R}$, but hopefully in such a way that it takes a large variation to see a small change.

23.3.2 Turbulent spectra

In section 18.2 we discussed some plausible equations for describing the nonlinear equations for the evolution of resistive drift waves, and their nonlinear saturation, the so-called Hasegawa–Wakatani equations (18.44) and (18.45). These are for the time being the best equations we have, at least as long as a manageable set of equations are required. They are not quite satisfactory by giving a fluctuation level of the fully developed saturated turbulent stage, which is unexpectedly high, but we accept nevertheless the equations as a basis for discussions of drift wave turbulence. We found that the basic dimensionless quantities characterizing the model can be argued as $\beta \equiv a_i n_0'/n_0$, $\eta \equiv n/n_0$, $\chi \equiv e\phi/\kappa T_e$, $t\Omega_c$, $r_\perp/a_i$, z/ℓ_c and $\mu/(a_i^2\Omega_c)$, with $a_i^2 = \kappa T_e/M\Omega_c^2$, while ℓ_c is the collisional mean free path and finally $C_c \equiv \kappa T_e/(n_0 e^2 \Omega_c \xi \ell_c^2) = \kappa T_e/(\Omega_c \nu_{ei} m \ell_c^2) \approx \nu_{ei}/\Omega_{ci}$, which measures the effect of plasma resistivity. To simplify the notation, we here let μ denote the kinematic viscosity with physical dimension *length*$^2 \times$ *time*$^{-1}$. The fluid literature uses ν for kinematic viscosity, but we frequently use that notation for collision frequencies.

By dimensional reasoning we now propose the form

$$G\left(\frac{n}{n_0}, \frac{e\phi}{\kappa T_e}, t\Omega_c, \frac{r_\perp}{a_i}, \frac{z}{\ell_c}, a_i\frac{n_0'}{n_0}, \frac{\mu}{a_i^2\Omega_c}, \frac{\kappa T_e}{\Omega_c \nu_{ei} a_i^2}, \theta\right) = 0$$

containing the parameter dependence of, for instance, the spatial correlation function. We also introduced the angle θ between $\mathbf{k}_\perp$ and ∇n_0. Emphasizing the properties of the wavenumber spectrum of the density fluctuations, we argue that the electrons can be taken so close to being in local isothermal equilibrium that the information in $\chi \equiv e\phi/\kappa T_e$ and $\eta \equiv n/n_0$ is practically the same. Following section 22.5, we write for the second order structure function

$$\begin{aligned}\left\langle (\tilde{n}(\boldsymbol{\xi}, t) - \tilde{n}(\boldsymbol{\xi} + \mathbf{r}, t))^2 \right\rangle \frac{1}{n_0^2} &\equiv \left\langle \left(\frac{\Delta\tilde{n}}{n_0}\right)^2 \right\rangle \\ &= g\left(\frac{r_\perp}{a_i}, C_c\left(\frac{z}{\ell_c}\right)^2, \frac{\mu}{a_i^2\Omega_c}, \beta, \theta\right).\end{aligned} \tag{23.25}$$

Our main argument for writing $C_c(z/\ell_c)^2$ as *one* quantity is that this is the most obvious combination suggested from the basic equations (18.44) and (18.45). To obtain a corresponding expression for the wavenumber power spectrum for the relative density fluctuations, we write

$$S(k_\perp, k_\parallel, \theta) = G\left(k_\perp a_i, \frac{1}{C_c}(k_\parallel \ell_c)^2, \frac{\mu}{a_i^2\Omega_c}, \beta, \theta\right)\frac{1}{k_\perp k_\parallel}, \tag{23.26}$$

defined in such a way that

$$\left\langle \left(\frac{\tilde{n}}{n_0}\right)^2 \right\rangle = \int S(k_\perp, k_\parallel, \theta) \mathrm{d}k_\perp \mathrm{d}k_\parallel \mathrm{d}\theta.$$

There is not much we can do in general with these expressions. We can have some expectations, though. For instance, we know that for $\mathcal{C}_c = 0$, the plasma is linearly stable in the present model, and hence the turbulence vanishes in this case. Since the drift waves remain stable when formally $\mathcal{C}_c < 0$, we can argue that there may exist a range of $\mathcal{C}_c$ values where the fluctuation level is linearly proportional to $\mathcal{C}_c$. This then implies that, with the other quantities taken constant, we expect

$$S(k_\perp, k_\parallel, \theta) \sim \frac{\mathcal{C}_c}{k_\parallel^3 \ell_c^2},$$

a result which should be easy to check, for instance by numerical simulations. For the structure function, this result will correspond to

$$\left\langle \left(\frac{\Delta \tilde{n}}{n_0} \right)^2 \right\rangle \sim \mathcal{C}_c \left(\frac{z}{\ell_c} \right)^2,$$

again with other quantities kept constant.

We can apply similar arguments involving β. We can argue that the steeper the gradient (larger β) the stronger we expect the turbulence to be. For steep density gradients it becomes less likely to find locally homogeneous turbulence conditions in the plane $\perp \mathbf{B}_0$, but we assume that at least some interval exists where these conditions can be found. Now, the *large* scales of the turbulence are expected to be independent of the length-scale characterizing the *smallest* scales. These are expected to be characterized by a_i, as also argued by Chen (1965). We cannot offhand predict the scaling of turbulence level with the density gradient, but intuitively we expect it unlikely that it scales linearly, since the sign of β should not enter the expressions, but β^2 seems acceptable, or any other even exponent. *If* we assume that a subrange in the $k_\perp$-spectrum exists, which is proportional to β^2, and at the same time is independent of a_i, this implies that the a_i scaling of β must cancel that of $k_\perp a_i$. Consequently, we expect $S(k_\perp, k_\parallel) \sim (|n_0'|/n_0)^2/k_\perp^3$ there, all other quantities assumed constant.

Chen (1965) also used dimensional arguments to obtain a power spectrum for the density fluctuations in drift wave turbulence, but his results are somewhat suspect by not including any dimensional coefficient accounting for the turbulence intensity, i.e. corresponding to a Kolmogorov–Oubokhov spectrum (22.41) independent of dissipated energy, ε. (Related results were obtained also by Vedenov (1968), but again the same objections can be made.) A different derivation (Fowler 1966) of the k^{-3} power spectrum for electric field fluctuations tried in part to remedy this problem by introducing the macroscopic radius of the plasma column as an outer scale, but the result remains independent of the nature of the density gradient. Also the relevance of the model to a slab-geometry was not evident. Nonetheless, the results concerning a k^{-5} spectrum for the fluctuations in electrostatic potential found some experimental support by e.g. Mikkelsen and Pécseli (1978), Pécseli (1982) and Pécseli *et al* (1983), see also summary by Tchen *et al* (1980).

Some more detailed analytical works (Tchen *et al* 1980) obtained the k^{-5} spectral characteristics for a short wavelength spectral subrange, albeit with more complicated dimensional coefficients. It was found that a characteristic frequency

$$\omega_0 \equiv \left(\Omega_{ci} \int_0^\infty k^2 G(k) \mathrm{d}k\right)^{1/3} \tag{23.27}$$

controls the inertial spectral range of drift-wave turbulence, and with this frequency it is possible to recover the k^{-5} spectrum $G(k)$ for the normalized electrostatic potential, where we note that the spectrum is defined so that $\langle\psi^2\rangle = \int_0^\infty G(k)\mathrm{d}k$, where ψ is the normalized electrostatic potential. The drift wave power spectrum for the potential fluctuations was found to be

$$G(k) \sim \frac{\omega_0^4}{\lambda_c \Omega_{ci}} \frac{1}{k^5}, \tag{23.28}$$

where $\lambda_c \equiv a_D(\kappa T_e/M)/\Omega_{ci}$ with an undetermined numerical coefficient a_D. The result (23.28) together with (23.27) can be seen as a self-consistency relation for the power spectrum $G(k)$ just like (22.42) for the Kolmogorov–Oubokhov spectrum.

Since the inertial range of drift-wave turbulence concerns short wavelengths, we can assume that the phase difference between density and potential to be small. For this range we can then assume locally Boltzmann distributed electrons and argue that fluctuation amplitude of density and potential are proportional for all wavelengths, and that the k^{-5} spectrum applies for both potential and density, albeit with different coefficients. The power spectrum of the electric field will consequently be proportional to k^{-3}. Many observations give support for power spectra of the form k^{-5} for fluctuating density and potential in drift wave turbulence (D'Angelo and Enriques 1966, Mikkelsen and Pécseli 1978, Pécseli *et al* 1983), and a $1/k^3$ spectrum for fluctuations in electric fields as summarized after a literature search by Chen (1965) and Tchen *et al* (1980).

It is interesting that the study of Tchen *et al* (1980) predicted a long wavelength part of the spectrum (the 'production subrange') to scale as

$$S\left(k_\perp, k_\parallel, \theta\right) \sim \left(\frac{|n_0'|}{n_0}\right)^2 \frac{1}{k_\perp^3},$$

i.e. precisely the spectral subrange obtained by dimensional reasoning before. The variation with θ for this subrange is presumably significant, and unaccounted for here. The production subrange has a counterpart in neutral velocity shear-driven turbulence (Tchen 1953, Tchen and Pierson 1988).

A work by Gürcan *et al* (2009) based on a 'shell model' derives power spectra for density and potential in the form

$$\langle |n|_k^2 \rangle \sim \left\langle |\phi|_k^2 \right\rangle \sim \frac{k^{-3}}{(1 + k^2)^2}, \tag{23.29}$$

which has the form k^{-7} for large k. The spectra for density and potential are proportional because the electrons are allowed to follow a local Boltzmann equilibrium in the analysis. The spectrum (23.29) is obtained so that e.g. $\langle n^2(\mathbf{r})\rangle = \int_0^\infty \langle |n|_k^2\rangle \, \mathrm{d}k$. These results correspond to much steeper spectra than other works and have so far found little support in laboratory observations and numerical simulations.

A study of an energy and enstrophy conserving, undriven equilibrium model (Fyfe and Montgomery 1979), without dissipation gave a power law spectrum following $\langle |\phi(\mathbf{k}|^2\rangle \sim k^{-6}$ for large k. This example reduces to the k^{-5} component law if we assume the turbulence to be locally homogeneous in two spatial dimensions. The relation of these equilibrium models to physically relevant driven-dissipative turbulence remains unclear, however.

Numerical studies by Camargo *et al* (1995) based on the Hasegawa–Wakatani equations, see section 18.2 (although restricted to one wavenumber component along **B**) demonstrated a significant dependence of the turbulent spectra on their 'adiabaticity parameter' $\mathcal{C}$ controlling the plasma resistivity. Increasing $\mathcal{C}$ gave systematically steeper power-law spectra, with the steepest density spectra reported there having $k^{-4.2}$ which is not too different from the k^{-5} advocated earlier. It is conceivable that the increase in $\mathcal{C}$ brings their simulation system from a weak to a strongly turbulent state in the sense discussed in chapter 16. Two stages, a weak and a strong turbulence limit, for drift wave turbulence were anticipated by Ottaviani and Krommes (1992), in their case for the forced, or driven, Hasegawa–Mima model.

23.4 Detection of power spectra

Detection of power spectra is in principle simple: you obtain a long data sequence and use a fast Fourier transform (FFT) routine (Brigham 1974) to get the Fourier components and then construct the power spectrum. This spectrum refers to a single realization of the ensemble, and if a full ensemble average is required, the process has to be repeated many times, with a subsequent average. The result is in principle easy to interpret, but on second thought it might not be the form wanted: for a time record of duration $\mathcal{T}$, the frequency resolution is $\Delta f = 1/\mathcal{T}$ and often it is an advantage to be content with a smaller frequency resolution, and reduce the number of realizations by averaging the spectral coefficients in an interval (Bendat and Piersol 1986). Ultimately, we might expect the process to be ergodic, so one long time record should be sufficient for obtaining the statistical average. Given one record we can cut it up into shorter sequences, obtain the power spectrum for each and then take an average. It is an interesting optimization problem to determine the best number of sub-sequences of data to be obtained from one long record (Kristensen and Kirkegaard 1986).

23.4.1 Taylor's hypothesis

Obtaining a frequency spectrum is relatively straightforward. Many (most?) of the analytical results refer, however, to wavenumber spectra. These can be simple to obtain in, for instance, numerical simulations. In nature or in laboratory experiments wavenumber spectra can be difficult to measure directly. An often used hypothesis is the 'Taylor hypothesis' or the 'frozen turbulence approximation' that relates readily obtained frequency spectra to wavenumber spectra (Taylor 1938, Granatstein 1967, Shkarofsky 1969, Fredricks and Coroniti 1976). Basically, the theorem assumes that under rather general conditions it is possible to identify the frequency by the Doppler shift, i.e. assume $\omega \approx \mathbf{k} \cdot \mathbf{V}$, where $\mathbf{V}$ is some relevant bulk drift velocity. The hypothesis should, however, be implemented with some care, so a brief outline of its implications may be in order.

As a part of the most primitive argument supporting Taylor's hypothesis, we can relate a frequency ω' as detected by a moving observer to the wavevector and the rest frame frequency ω as $\omega' = \omega + \mathbf{k} \cdot \mathbf{V}$, where $\mathbf{k}$ is the wavevector associated with the frequency via some dispersion relation, this relation assumed to be unique, while $\mathbf{V}$ is the velocity of the observer. Taylor's hypothesis then states $\omega' \approx \mathbf{k} \cdot \mathbf{V}$, i.e. we identify the observed frequency with the Doppler shift. The assumption can be justified for an observer moving at high velocity compared to the phase velocity of the wave. The relation $\omega' \approx \mathbf{k} \cdot \mathbf{V}$ does not give any one-to-one relation between observed frequency and wavenumber, due to the aliasing illustrated in figure 23.3, see also section 19.2.8. At first sight it might seem that the hypothesis works best for large k, but that depends on the dispersion relation: generally we might have $\omega = \omega(\mathbf{k})$, a relation that need not be *a priori* known.

The foregoing discussion related frequencies to wavenumber components, i.e. observed frequency spectra to wavenumber component spectra. Another closely related argument relates temporal to spatial correlation functions (or structure functions). It is thus close to trivial to argue that when Taylor's hypothesis is applicable we have

$$R(0, t) \approx R(\mathbf{V}t, 0), \tag{23.30}$$

in terms of temporal and spatial auto-correlation functions $R(\mathbf{r}, t)$, which may or may not be isotropic.

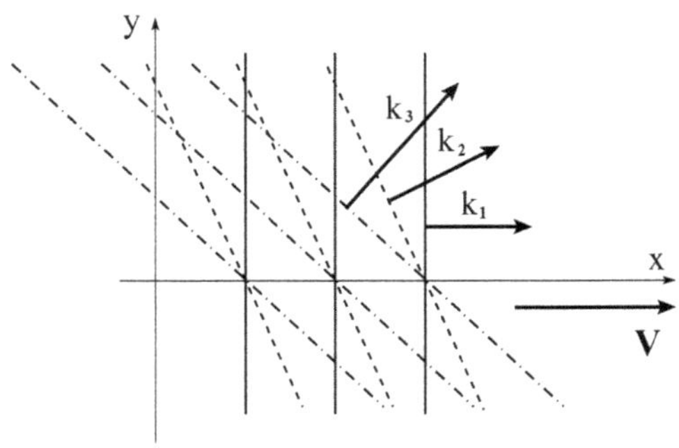

Figure 23.3. Illustration of 'aliasing' implied in Taylor's hypothesis. All three wavevectors in the figure have the same k_x-component. The observer is assumed to move with velocity $\mathbf{V}$ along the x-axis.

It is possible to give a more detailed discussion favoring Taylor's hypothesis. First we recall that the space-time varying auto-correlation function and the wave-number-frequency power spectrum form a Fourier transform pair

$$R(\mathbf{r}, t) = \frac{1}{(2\pi)^4} \iint S(\mathbf{k}, \omega)\exp(\mathrm{i}(\omega t - \mathbf{k}\cdot\mathbf{r})\mathrm{d}\mathbf{r}\mathrm{d}t,$$

and

$$S(\mathbf{k}, \omega) = \iint R(\mathbf{r}, t)\exp(-\mathrm{i}(\omega t - \mathbf{k}\cdot\mathbf{r})\mathrm{d}\mathbf{k}\mathrm{d}\omega.$$

For simplicity we write $\iiint \mathrm{d}x\mathrm{d}y\mathrm{d}z = \int \mathrm{d}\mathbf{r}$, etc here and in the following. There will be several Fourier transforms and inverse transforms, so we make an effort to retain the 2πs that enter. We have by definition

$$R(t) = \frac{1}{2\pi} \int S(\omega)\exp(\mathrm{i}\omega t)\mathrm{d}\omega,$$

with $R(t) \equiv R(t, 0)$ and similarly $R(\mathbf{r}) = R(\mathbf{r}, 0)$, although we are aware that the notation might be a bit confusing.

In addition to these two relations we can define the restricted forms, they will be useful later on

$$S(\mathbf{k}) \equiv \frac{1}{2\pi} \int S(\mathbf{k}, \omega)\mathrm{d}\omega, \tag{23.31}$$

which is the wavevector power spectrum. We can always write

$$S(\mathbf{k}, \omega) = S(\mathbf{k})\Psi(\mathbf{k}, \omega)$$

and have $\int \Psi(\mathbf{k}, \omega)\mathrm{d}\omega = 2\pi$ with the foregoing definition, since $\Psi(\mathbf{k}, \omega) = S(\mathbf{k}, \omega)/S(\mathbf{k})$. We can write a Fourier transform pair also for $\Psi(\mathbf{k}, \omega)$ as, e.g.

$$\Psi(\mathbf{r}, t) = \frac{1}{(2\pi)^3} \int \Psi(\mathbf{k}, t)\exp(-\mathrm{i}\mathbf{k}\cdot\mathbf{r})\mathrm{d}\mathbf{r},$$

and

$$\Psi(\mathbf{k}, t) = \iint \Psi(\mathbf{r}, t)\exp(\mathrm{i}\mathbf{k}\cdot\mathbf{r})\mathrm{d}\mathbf{k},$$

where the argument accounts for the variables, where the time variable enters through an ω-transform. The form Ψ will be a useful form for modeling later on. By the convolution theorem we have

$$R(\mathbf{r}, t) = \int R(\mathbf{r}')\Psi(\mathbf{r} - \mathbf{r}', t)\mathrm{d}\mathbf{r}'.$$

For a perfectly frozen flow moving with velocity $\mathbf{V}$ we have $\Psi(\mathbf{k}, \omega) = 2\pi\delta(\omega - \mathbf{k}\cdot\mathbf{V})$ and consequently $\Psi(\mathbf{r}, t) = \delta(\mathbf{r} - \mathbf{V}t)$ so that $R(\mathbf{r}, t) = R(\mathbf{r} - \mathbf{V}t) = R(\mathbf{r} - \mathbf{V}t, 0)$. This is the limiting case where Taylor's hypothesis is exactly fulfilled. In its exact form Taylor's hypothesis states that $R(0, t) = R(\mathbf{r} - \mathbf{V}t, 0)$, i.e. a measurement at a fixed position of the time variation of the normalized correlation function produces the same result as obtained by an observer moving with velocity $\mathbf{V}$, measuring the spatial variation of the correlation function for zero time-delay.

- We may illustrate this result in more detail: assume we have a realization of some randomly varying quantity, say plasma density n. Let this density be perfectly frozen, meaning that $n = n(\mathbf{r})$ with no time dependence. If this realization is in uniform translation with velocity $\mathbf{V}$, the signal will become space-time varying as $n(\mathbf{r} - \mathbf{V}t)$, i.e. a fixed observer will see the signal vary in time, and if you take a 'snap-shot' at a fixed time, the signal will vary in space. Assume that the process is ergodic in the sense that spatial averaging of one realization gives the same result as the ensemble averaging. (Usually, we reserve the idea of ergodicity to temporally varying signals, but the generalization is feasible here.) The system is homogeneous, but not necessarily isotropic. We have $\langle n^2 \rangle = \lim_{L^3\to\infty} \int_0^L n^2(\mathbf{r}')\,\mathrm{d}\mathbf{r}'$.

  The spatial correlation function of the signal in its rest frame (indicated by subscript r) is

$$R_r(\mathbf{r}) = \frac{1}{\langle n^2 \rangle} \lim_{L^3\to\infty} \int_0^L n(\mathbf{r}')n(\mathbf{r}' - \mathbf{r})\mathrm{d}\mathbf{r}',$$

  so that $R_r(0) = 1$. In this frame there is no time variation.

  The space-time varying correlation function of the signal seen from a moving frame (indicated by subscript m) is

$$R_m(\mathbf{r}, t) = \frac{1}{\langle n^2 \rangle} \lim_{\mathcal{T}, L^3\to\infty} \int_0^{\mathcal{T}} \int_0^L n(\mathbf{r}' - \mathbf{V}t')n(\mathbf{r}' - \mathbf{r} - \mathbf{V}(t' - t))\mathrm{d}\mathbf{r}'\mathrm{d}t'$$
$$= R_r(\mathbf{r} - \mathbf{V}t).$$

  With a slightly different terminology and notation, this is Taylor's hypothesis obtained before. If the signal is not isotropic, the correlation will depend not only on the magnitude of the velocity, but also its direction.

We now generalize $\Psi(\mathbf{k}, t)$ by writing $\Psi(\mathbf{k}, t) \equiv \Psi_L(\mathbf{k}, t)\exp(i\mathbf{k}\cdot\mathbf{V}t)$ where we by modeling Ψ_L can account for deviation from the perfectly frozen flow condition. A fairly general form can be

$$\Psi_L(\mathbf{k}, t) = \exp\left(-\frac{1}{2}k^2\Delta^2(t)\right)\exp(-\Gamma|t|),$$

where the temporal damping constant Γ and the time varying broadening in wavenumbers can be modeled through $\Delta^2(t)$, one such example being (Shkarofsky

1969) $\Delta^2(t) = t^2\langle\widetilde{U}^2\rangle + 2D|t|$ with constant D and fluctuating velocity $\widetilde{U}$. For such a case we find

$$\Psi(\mathbf{r}, t) = \frac{1}{\left(2\pi\Delta^2(t)\right)^{3/2}} \exp\left(-\frac{(\mathbf{r} - \mathbf{V}t)^2}{2\Delta^2(t)} - \Gamma|t|\right).$$

If we take Γ and D to be small, we can argue that for Taylor's hypothesis to be applicable the root-mean-square of the displacement due to velocity fluctuation in the medium $\widetilde{U}t$ should be small compared to characteristic spatial correlation distance L_c and also smaller than the translational displacement Vt. We can express these two inequalities as

$$\langle\widetilde{U}^2\rangle t^2 \ll L_c^2, \qquad \text{and} \qquad V^2 \gg \langle\widetilde{U}^2\rangle.$$

Given a long time-record obtained by a rapidly moving probe one may ask what do we know then? The proper answer is probably 'not much'. Application of Taylor's hypothesis can concern only one component of the wavenumber, the one parallel to the direction of the propagation velocity. In terms of the correlation functions, this amounts to $R(\mathbf{r} - \mathbf{V}t)$ depending on the direction of $\mathbf{V}$ in general. Additional information is essential if results concerning wavevectors in other directions are needed. If we have *a priori* knowledge of the turbulence being isotropic we might use (19.83) to obtain the isotropic spectrum. See also the discussion in section 19.2.9. Alternatively we may have some *a priori* information indicating that the turbulence is composed by structures or waves propagating in essentially one dimension. Many authors have discussed the conditions for one-, two- or three-dimensional turbulence, for instance Kintner (1976). The design and instrumentation of for instance the ROSE rockets (Rose *et al* 1992) demonstrate how additional information can be obtained and utilized to interpret the available signals (Krane *et al* 2000).

From one point measurements it is logically impossible to confirm the validity of Taylor's hypothesis. For cases where measurements from two points are available, it is possible to extend the analysis by comparing auto-correlations with cross-correlations, assuming the two positions $\mathbf{r}_1$ and $\mathbf{r}_2$ are placed along the streaming direction. For an ideally frozen flow the cross-correlation is merely a shifted copy of the auto-correlation, and the hypothesis can be tested by making the comparison. For studies of drift wave turbulence, results are given by Mikkelsen and Pécseli (1980) and Pécseli *et al* (1983). For examples taken from fluid turbulence, see Johnston *et al* (1969). One interesting observation can be made when this two point information is available: if we take the peak value of the cross-correlation at the time t where $\mathbf{r}_2 - \mathbf{r}_1 = \mathbf{V}t$ we can argue that the variation of this value represents the temporal de-correlation an observer would experience by following the flow. By varying the separation $\mathbf{r}_2 - \mathbf{r}_1$ it is thus feasible to obtain an approximation for the temporal auto-correlation an observer would find by following a fixed point moving with the flow, i.e. with the velocity $\mathbf{V}$. See for instance figure 24.4a).

The present outline emphasizes the case where Taylor's hypothesis applies due to rapid motion of a detector, space-craft or similar, along a deterministic orbit, i.e. a straight line, giving a Galileo transformation of the signal. This is the ideal case often used experimentally. The argument applies for fluid turbulence or for any other random field. The arguments can be valid however also for random translations: this case has received particular attention for fluid turbulence and the interpretation of the frequency power spectrum obtained by a detector at a fixed position. It is argued (Tennekes 1975, Tsinober *et al* 2001) that the advection of the small scales by the large turbulent scales can *also* be interpreted in terms of Taylor's hypothesis, and that the fixed point frequency power spectrum in effect reproduces the wavenumber spectrum, for this case the Kolmogorov–Oubokhov spectrum. It is not known to what extent similar arguments apply for turbulence in magnetized plasmas. For two-dimensional low frequency electrostatic turbulence in a plane perpendicular to a homogeneous magnetic field we would expect the turbulence properties to be similar to that of neutral incompressible fluids. The similarity is probably sufficient to allow application of Taylor's hypothesis in the general sense discussed before.

A model example

As an illustrative model we consider again a space-time varying correlation function of the form (20.62), see figure 20.12, which can also approximate some correlation functions found in nature. By definition we have $R(x, t) = \langle \phi(0, 0)\phi(x, t)\rangle$ assuming the signal $\phi(x, t)$ to be a spatially homogeneous stationary random process sampled at position x at a time t, multiplied by $\phi(0, 0)$ and ensemble averaging the resulting product. We can, however, also sample $\phi(x, t)$ at a moving point, for instance at $x = Ut$ where U is some constant velocity. The result gives a time varying correlation function $R(Ut, t)$ which depends also on U as shown in figure 23.4. For $U = 0$, in particular, we recover $R(0, t) = \exp(-(t/D_t)^2)$. For large velocities, on the other hand, we have $R(Ut, t) \approx \exp(-(Ut/D_x)^2)\cos(aUt/D_x)$ which represents the spatial variation of the correlation function: this is the limit giving Taylor's hypothesis and the transition is particularly clear in figure 23.4. The summary here merely assumes that the relative velocity between the medium and the detector is U, how this motion is realized in practice is immaterial for the argument.

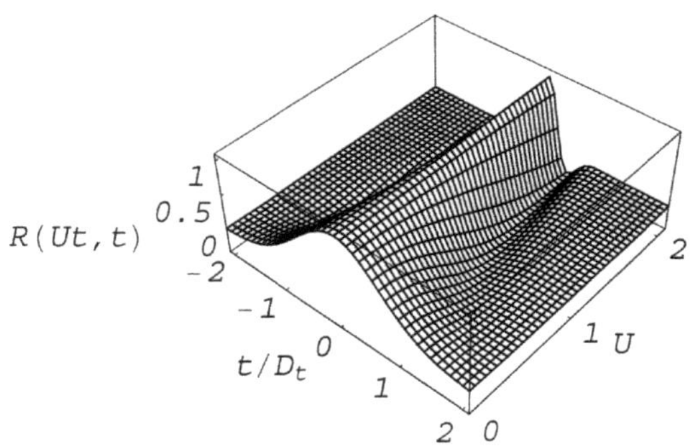

Figure 23.4. Time variation of the model correlation function $R(Ut, t)$ for varying velocities U.

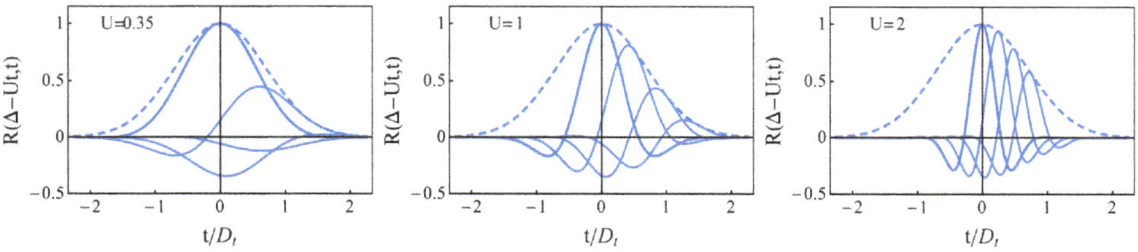

Figure 23.5. Time varying correlation functions obtained by correlating two signals originating from two probes with separation $\Delta/Dx = 0$, .5, 1, and 1.5 along the flow direction, here with $U = 0.5$, 1 and $U = 2$, shown in the left, middle and right hand figures respectively. Increasing spatial separations give (for fixed U) an increasing time-delay in direct proportion. The dashed line gives $R(0, t)$ for reference, see also Johnston *et al* (1969).

The signal $\phi(x, t)$ can be sampled in other ways, for instance by taking $\phi(0, 0)$ multiplied by $\phi(\Delta - Ut, t)$ which gives a time varying correlation function $R(\Delta - Ut, t)$ shown in figure 23.5. This detection can be realized by an experiment where the medium is flowing with velocity U and the signal is detected at two positions with separation Δ along the flow direction. If we follow the medium (it need not be a plasma) as it moves with the flow, we will find a decreasing temporal correlation and the peak value of the cross-correlation function will decrease accordingly (Johnston *et al* 1969). To illustrate this, we included by a dashed line the temporal correlation $R(0, t)$ in figure 23.5. Provided U is large enough, we can (at least in the present simple model) estimate the fixed position temporal cross-correlation function by the envelope of the cross-correlations for varying separation. The approximation of the envelope of the cross-correlation function to the temporal cross-correlation function is another representation of Taylor's hypothesis and should not be seen as an independent result. The use of the cross-correlation envelope as an approximation for the temporal correlation function will be illustrated by plasma examples later in sections 23.4.2 and 24.2.

A relatively simple and intuitively appealing possibility for testing Taylor's hypothesis can be mentioned. The idea assumes that a finite size probe can properly detect only wavelengths longer than the size of the probe. If the probe is elongated, the resolution will be anisotropic. The discussion can be made more detailed (Tolias *et al* 2016), but this is the main idea. Assume that a power-law spectrum as a function of frequency for fluctuations in density or floating potential has been measured by a probe that is as small as you can get. Now measure the same spectra with a larger size probe. According to the previous argument you would expect the spectral index to become steeper at frequencies corresponding to a wavelength determined by the probe size. The result can then be checked against what is expected by use of Taylor's hypothesis. It will be best to do this with several probes of varying size. The author actually tried this out in a laboratory plasma long ago, and found it to be a reasonable although not particularly accurate method, and the results were never published. One problem was that in a laboratory there are rather strict limits to how large probes you can accept. The idea might work better on a space experiment?

Unfortunately there are no general objective criteria for when Taylor's hypothesis is applicable, all we have given here and in previous sections were inequalities. All we can say with certainty is that a sufficiently large velocity allows the frozen turbulence approximation. The magnitude of the required velocity depends on the nature of the fluctuations and also on the required accuracy. The outline in the present section argues that it might be advisable to find an approximate model for the space-time varying correlation function, and then test various approximations on it for comparison with observations. Such a model can be constructed by random superposition of structures as outlined before, see for instance chapter 20, where each structure has an associated space-time variation.

23.4.2 Experimental observations of turbulent drift-wave spectra

A number of experiments have reported power spectra of turbulent plasma fluctuations, mostly fluctuating densities and electrostatic potentials since they are most easily measured, but in some cases also electric field fluctuations were measured. Several of these works were *a posteriori* interpreted in terms of drift wave turbulence. Some of these early results are summarized by Chen (1965) and Tchen *et al* (1980). Studies of drift wave turbulence were reported also by Buchel'nikova *et al* (1967), with emphasis on the turbulent diffusion, rather than the turbulent spectra.

A k^{-5} spectrum for the fluctuations in electrostatic potential found some experimental support by measurements in a Q-machine plasma by e.g. D'Angelo and Enriques (1966), Mikkelsen and Pécseli (1978), Pécseli (1982) and Pécseli *et al* (1983), see also summary by Tchen *et al* (1980). The drift wave spectrum was analyzed by microwave scattering in a toroidal plasma by Okabayashi and Arunasalam (1977), who found indications of a k^{-6} spectrum, albeit with a non-trivial experimental uncertainty, which *could* accommodate also the k^{-5} spectrum mentioned before. An interesting result from these experiments was an evidence of local homogeneity and isotropy of the smallest scales of the turbulence in the two dimensions perpendicular to the magnetic field (Okabayashi and Arunasalam 1977). Physically, this result is consistent with the expectation that small scales propagate on the gradients of the larger scales and thereby randomizing the local direction of propagation as the scales become smaller and smaller. Similar observations of local two-dimensional isotropy were made by TFR-Group and Truc (1984) in tokamak plasmas. Only the motions of the largest turbulent eddies are directly coupled to the direction given by the diamagnetic drift determined by the magnetic field and the average large scale density gradient, so that these largest scales will be anisotropic. This would correspond to the 'production subrange' predicted by Tchen *et al* (1980).

Some basic results from the Q-machine study of Pécseli *et al* (1983) are summarized here. That experiment for testing the results was conducted in a single ended Q-machine alkali plasmas (cesium), where an externally imposed radial electric field $\mathbf{E}_0 \sim \mathbf{r}$ gave rise to a near solid-body rotation of the plasma column. The conditions are similar to those studied analytically by Chu *et al* (1969). The hot spiral used for this is shown in figure 23.6. The magnetic field could be varied in the range of 1–2.6 kG. Plasma temperatures were approximately 2000 K, where $T_e > T_i$. Ion Larmor radii were of the order of 1 mm. Figure 23.7 shows the average density $\bar{n}$ and relative fluctuation level $\tilde{n}/\bar{n} \equiv \sqrt{\langle (n - \bar{n})^2 \rangle} / \bar{n}$ for varying radial positions in the plasma column. The relative fluctuation level was maximum in the region where the average density gradient is largest. The peak density in the central part of the plasma column was high for a Q-machine, of the order of 10^{16} m^{-3}. The fluctuation level decreased for increasing B, corresponding to a decrease of the bulk E_0/B plasma rotation, thus demonstrating that the rotation is important for the growth rate of the instability.

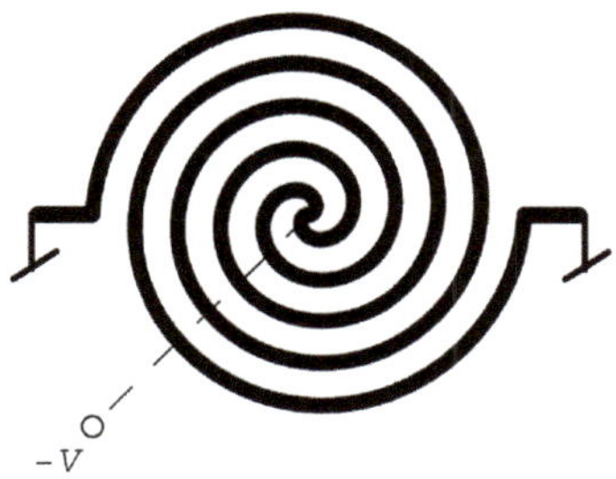

Figure 23.6. Schematic diagram of the hot spiral made by 2 mm tantalum wire and heated by ~100 A, used as 'hot-plate' in a Q-machine in order to create a rotating plasma column (Pécseli *et al* 1983). The center of the spiral is at a negative potential of the order of −5 V, the outer contacts grounded. The potential drop along the spiral is proportional to the radial position (this being a property of Archimedean spirals), thus giving a DC electric field that increases linearly with radius. With constant magnetic field this gives a solid body rotation of the plasma column. The magnetic field was variable up to 0.6 T.

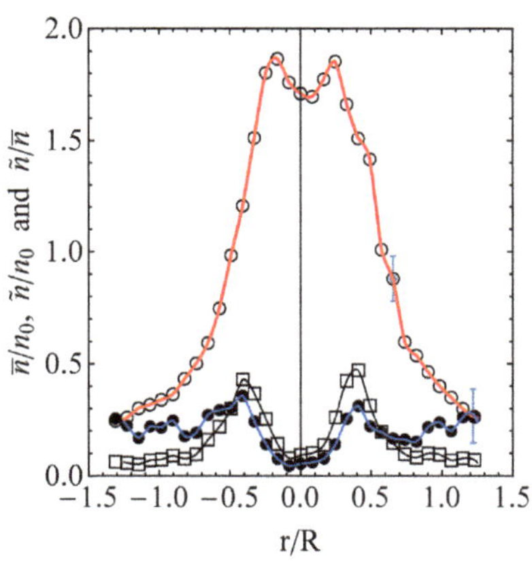

Figure 23.7. The variation with radial position of the average or DC density $\bar{n}/n_0$ normalized by a reference density n_0 is shown by ∘. The fluctuating RMS-density $\tilde{n}/n_0$ is given by open squares. The relative density fluctuation level $\tilde{n}/\bar{n}$ shown with filled circles maximizes at the steepest radial density gradient and is approximately constant for larger radial positions. In this case $n_0 = 10^{16}$ m^{-3}. Radial positions are normalized by the length-scale $R = 4$ cm. The figure is reproduced from Pécseli (2015).

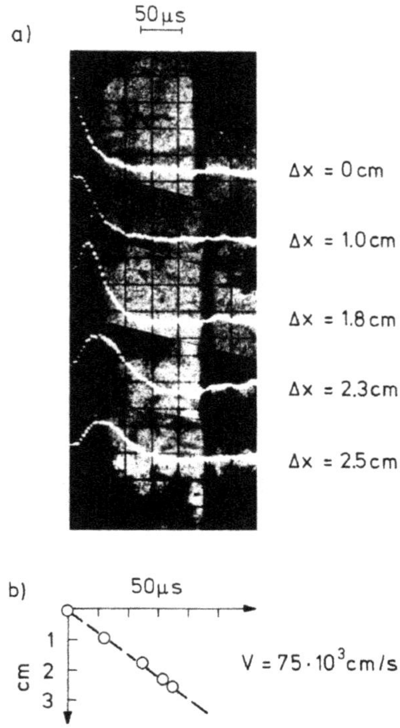

Figure 23.8. Cross-correlation of fluctuating plasma densities detected by two probes with separations Δx as indicated in (a). Both probes are at the same radial position. See figure 23.5 for comparison. The flow velocity deduced from the time delays is shown in (b). The figure is reproduced from Pécseli *et al* (1983),

Cross-correlations between two probes with variable separation Δx along the flow direction (i.e. both are at the same radial position but separated in the azimuthal direction) are shown in figure 23.8. The rotation velocity could be confirmed by these results, and also an interval of a temporal cross-correlation could be estimated as the envelope of the cross-correlations for varying Δx, see also Johnston *et al* (1969) and also figure 23.5. With a typical azimuthal velocity of 750 ms^{-1} we can argue that a perturbation of the order of an ion Larmor radius will correspond to a frequency of 400 kHz on the basis of Taylor's hypothesis. This estimate indicates that the dominant part of the observed power spectra correspond to subranges with scale-lengths much larger than the Larmor radii.

The fluctuations were characterized by a wide continuous spectrum with a shape that is robust for changes in plasma parameters, for instance the magnetic field, see figure 23.9. The various spectral parameters are defined in figure 23.9, where in particular the production and inertial subranges of Tchen *et al* (1980) are identified. For drift waves, we expect fluctuation in density and potential to be nearly proportional. For a linearized analysis this statement can be quantified as summarized in section 6.5, see for instance figure 6.3. If we can assume nearly Boltzmann distributed electrons, the potential-density proportionality can be argued for any frequency and wavelength, implying that the turbulent spectrum for density and

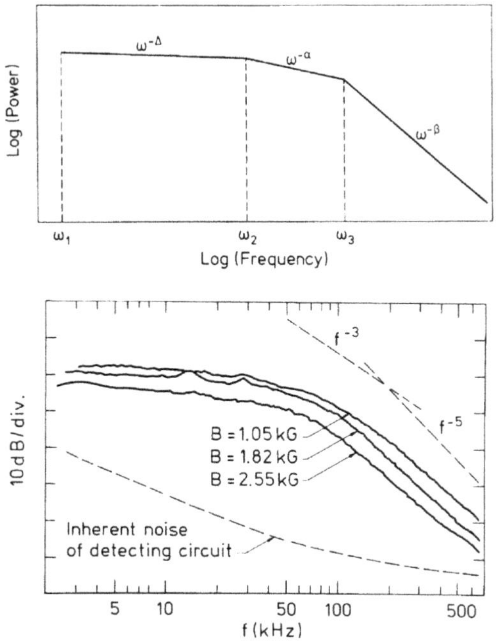

Figure 23.9. Top: definition of parameters for a model spectrum, i.e. spectral indexes and transition wavenumbers, as analyzed by Pécseli *et al* (1983). Bottom: power spectrum of potential fluctuations, shown for three values of the magnetic field. Dashed lines give some reference slopes. The figure is in double logarithmic presentation. In the terminology of Tchen *et al* (1980), the subrange with $\sim f^{-3}$ corresponds to a 'production subrange', followed by a 'coupling subrange' with a f^{-5} variation. The figure is reproduced from Pécseli *et al* (1983), © IOP Publishing. Reproduced with permission. All rights reserved.

potential should be the same, apart from a dimensional constant factor. As shown in figure 23.10 this is nearly so.

Due to an externally imposed rapid solid-body rotation of the plasma column (D'Angelo *et al* 1974, Mikkelsen and Pécseli 1980), see also figure 23.6, we can assume that Taylor's hypothesis, or the frozen turbulence approximation, is applicable at large radial positions. Experimentally observed frequency spectra can be compared to analytical results for wavenumber spectra. Particular attention was given to results by Tchen *et al* (1980), mainly because they contain not only discussions of spectral indexes but also for the parameter dependencies of the spectral coefficients. The variation of the spectral index for fluctuations in floating potential for different radial positions were analyzed as shown in figure 23.11. The assumed Taylor hypothesis is best fulfilled at large radial positions, as stated before.

Pécseli *et al* (1983) concluded that the agreement with a k^{-5} power law for the component spectrum was satisfactory although not perfect. The enhanced fluctuation level was found to dominate at the position of the steepest density gradient and the spectra for fluctuations in relative density and in potential were similar (albeit not identical), both features in agreement with drift wave driven turbulence. The mean free path for electron–ion collisions was in the cm-range, but even in this case the growth rate of a resistive drift wave is relatively low. The rotation of the plasma

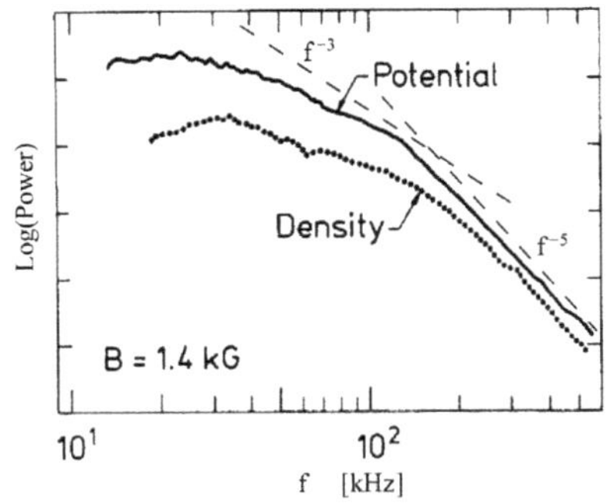

Figure 23.10. Simultaneously obtained turbulent spectra for fluctuations in density and electrostatic floating potential. Two dashed lines giving the slopes for f^{-3} and f^{-5} are inserted for reference. The figure is reproduced from Pécseli *et al* (1983), © IOP Publishing. Reproduced with permission. All rights reserved.

column is, however, giving rise to a centrifugal force that implies a different rotation frequency for electrons and ions (Odajima 1978). This azimuthal current comes in addition to the electron diametrical current, and the instability is enhanced to give a fully developed turbulent spectrum. The experimental results were compared with the analysis of Tchen *et al* (1980) which accounts not only for the spectral index but also the coefficient variations and for transition wavenumbers between spectral subranges. A sensitivity analysis was also carried out (Pécseli *et al* 1983), demonstrating that with the variation of plasma parameter accessible for the experimental conditions, it was possible to test at least some of these predictions from theory qualitatively as well as quantitatively. In those cases the agreement was satisfactory. Numerical simulations could offer possibilities for more detailed studies, but no such investigations seem to have been made so far.

An electric field power spectrum varying like k^{-3} corresponds to a similar $\mathbf{E} \times \mathbf{B}/B^2$-velocity spectrum. As noted by Misguich *et al* (1987) such a spectrum is particularly interesting for relative particle diffusion, i.e. for studies of a cloud of charged particles as they expand in the direction perpendicular to **B**, see the discussion in section 25.2.

23.5 Coherent structures

A linear plasma instability need not always saturate in a wide-range power spectrum. A wide wavenumber spectrum will generally develop for cases where the system can be considered locally homogeneous. In other cases we might find a saturated stage with large coherent structures embedded. In neutral turbulence it is for instance found that large coherent structures, often called 'hair-pin vortices' (Adrian 2007), can develop near walls where the flow and the turbulence are inhomogeneous. For such conditions we might suggest the use of a conditional averaging, or conditional sampling, in order to reveal the presence of such structures. Rather than using the analytical expression in sections 19.2.4 and 23.1.2, we might prefer to carry out the averaging directly. The analysis basically requires a reference signal available possibly together with other simultaneously obtained signals. These can be obtained all at the same time, or two-by-two, so that

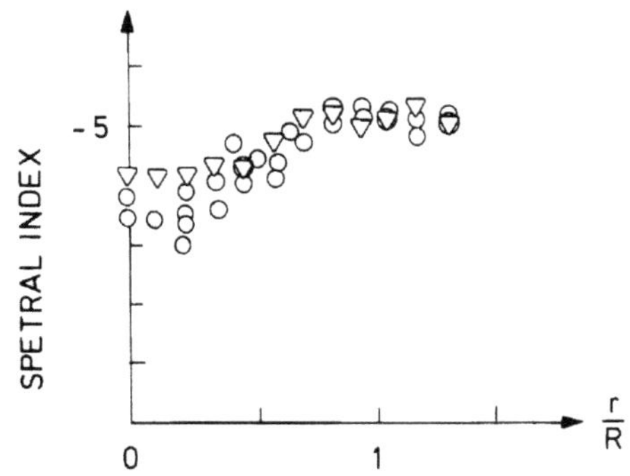

Figure 23.11. Spectral index for the 'coupling subrange' for varying radial positions for two values of the magnetic field as indicated by ∘ and ∇. In the central region near $r = 0$, where the spectral index is ~3–4 we have small fluctuation levels, see figure 23.7. The normalizing radius is $R \approx 4$ cm. The figure is reproduced from Pécseli *et al* (1983), © IOP Publishing. Reproduced with permission. All rights reserved.

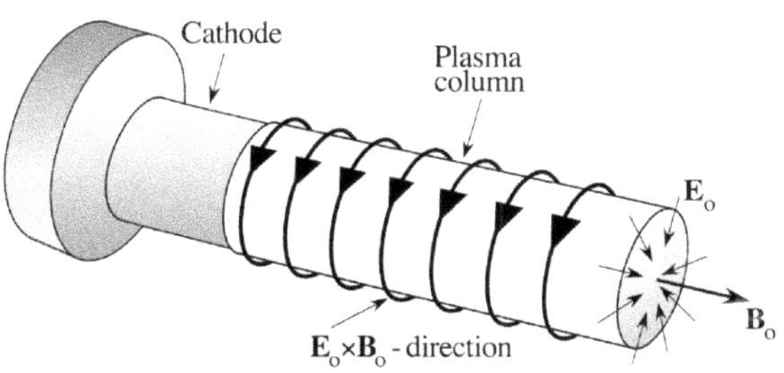

Figure 23.12. Schematic diagram of a Q-machine plasma column for single ended conditions. The localized radial electric field at the scrape-off layer gives rise to a strongly sheared azimuthal $\mathbf{E}_0 \times \mathbf{B}_0/B_0^2$-velocity.

the reference is always accompanied by at least one other dataset. Various conditions can be imposed on the reference data and the same or the other data can subsequently be analyzed, subject to the imposed condition. The problem is that if very detailed conditions are imposed, then these will be fulfilled rarely in the dataset, and the resulting signal-to-noise ratio on the averaging will be poor. The challenge with conditional averaging is to find a useful set of imposed conditions, which can give the desired information with an acceptable signal-to-noise ratio.

As an illustrative example we consider here the experiment by Huld *et al* (1991). The essential part of their experimental condition is illustrated in figure 23.12. The central part of the plasma column has a very low DC electric field (i.e. the part terminated by the hot plate of the Q-machine). At the edge of the column, the radial electric field increases so that the potential reaches ground at the confining vessel. The radial electric field varies significantly within a narrow radial interval, where also the plasma density decreases from the central high density (in the central Q-machine column) until it vanishes before the plasma recombines on the wall of confining vessel. The $\mathbf{E}_0 \times \mathbf{B}_0/B_0^2$-drifting plasma has features similar to an azimuthal 'jet' to be seen in for instance figure 23.13. The width of this layer is several ion Larmor radii, so we can safely use a simple characteristic $\mathbf{E}_0 \times \mathbf{B}_0/B_0^2$-velocity for both electrons and ions. The radial electric field could be controlled externally by

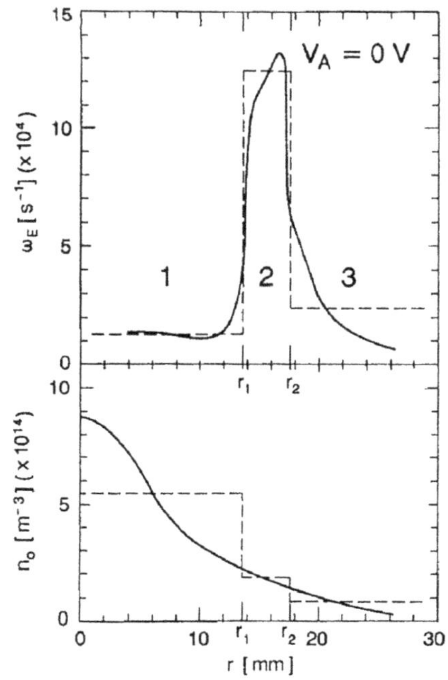

Figure 23.13. Radial variation of density and azimuthal velocity in the experimental set-up illustrated in figure 23.12. V_A denotes the bias of an aperture plate controlling the radial electric field. Note that the local $\mathbf{E} \times \mathbf{B}/B^2$-velocity is strongly peaked at the plasma edge (top frame, region 2), in the scrape-off region. The result can be described as an azimuthal 'jet'. The figure is reproduced from Huld *et al* (1991) with permission. Copyright 1991, AIP Publishing LLC.

varying the bias V_A of an aperture plate (Huld *et al* 1991). Applying a bias close to the plasma potential of the central plasma column gave a moderate electric field, while V_A close to ground potential (i.e. the potential of the confining vacuum vessel) gave large E_0. With these conditions we expect a strong Kelvin–Helmholtz instability excited in the region with large azimuthal $\mathbf{E}_0 \times \mathbf{B}_0/B_0^2$-velocity shears (Kent *et al* 1969, Huld *et al* 1991). Similar conditions can be produced in other devices (Shinohara *et al* 2001) and enhanced fluctuation levels were observed there as well.

The stability analysis of the present conditions is complicated, but it can be facilitated by approximating the radial variation of plasma density and azimuthal flow velocity by step-functions as indicated in figure 23.13. By such assumptions, appropriate basic equations can be solved in each of the intervals with constant parameter values, and the results fitted together by 'jump conditions' at the transition points (Chandrasekhar 1961, Kent *et al* 1969). The procedure works for any number of steps in the approximations but the resulting dispersion relation will become of increasingly higher order in frequency. Often only a few levels will be used, for instance 3 as shown for the present case in figure 23.13. The results usually indicate that relatively large mode-numbers are the linearly most unstable. For two reasons this might be surprising: first, the analysis uses results for the steady state conditions that originate from a case where the fluctuations are saturated, presumably due to some nonlinear effects, i.e. we deal with some smoothing out averaged steady state conditions that are approaching a marginally stable limit. Second, we note that the structures observed by Huld *et al* (1991), for instance, cover a rather large azimuthal fraction of the circumference of the plasma column. This means that the structures are associated with a small azimuthal mode number and are not directly excited by the linear instability. It can be argued that some nonlinear effects seem to cascade energy from the linearly most unstable short scales to the observed larger scales.

23.5.1 Conditional averaging, experimentally

Conditional averaging has been used as a diagnostic tool in describing fluid turbulence, wall bounded turbulence in particular. The basic ideas are reviewed by for instance Antonia (1981), but even earlier a simple so called 'method of superposed epoch' (at times also called 'Chree analysis' after Charles Chree (Chapman 1941)) was in use (Shapley and Beynon 1965), in the analysis of space data in particular.

Fluid turbulence is problematic in one respect by dealing with vector quantities, a scalar velocity potential is not accessible for experiments. This means that a full description has to obtain three vector components simultaneously at selected positions at selected times. Apart from all the other problems arising, this also imposes restrictions on the data storage and handling. Plasma turbulence is more fortunate in this respect: in this case a relevant scalar quantity, the electrostatic potential or the plasma density, is usually directly amenable for measurements, and the analysis is considerably simplified.

The method as used in the following is based on a two-probe diagnostic (Øynes *et al* 1998). One probe is here kept at a fixed reference position measuring an observable Φ_1, while the other probe scans selected points in a cross-section of the plasma column measuring Φ_2. The quantities Φ_1 and Φ_2 can individually represent any observable, such as density, potential or temperature. We use a condition, C, defining time-windows of selected size centered around a specific signature at the reference probe, for instance a certain signal amplitude. We define the *conditional time-windows* to be the sub-series at the movable probe corresponding to a fulfillment of C. At each position of the moving probe we average all acquired conditional time windows, getting the conditional averaged time series

$$\Phi_{ca} = \langle \Phi_2 | C \rangle. \tag{23.32}$$

In this manner we are able to determine the average time variation of an observable Φ_2 at any location in the plasma centered around an event at a fixed spatial position. By presenting the results of this analysis as contour or surface plots, we can describe the average time evolution of the structures causing, or being correlated with, the signature selected as our triggering condition. The simplest signature to look for is amplitude levels higher than some critical value, either absolute or relative to the average fluctuation level. The ideas are illustrated in more detail in figure 23.14,

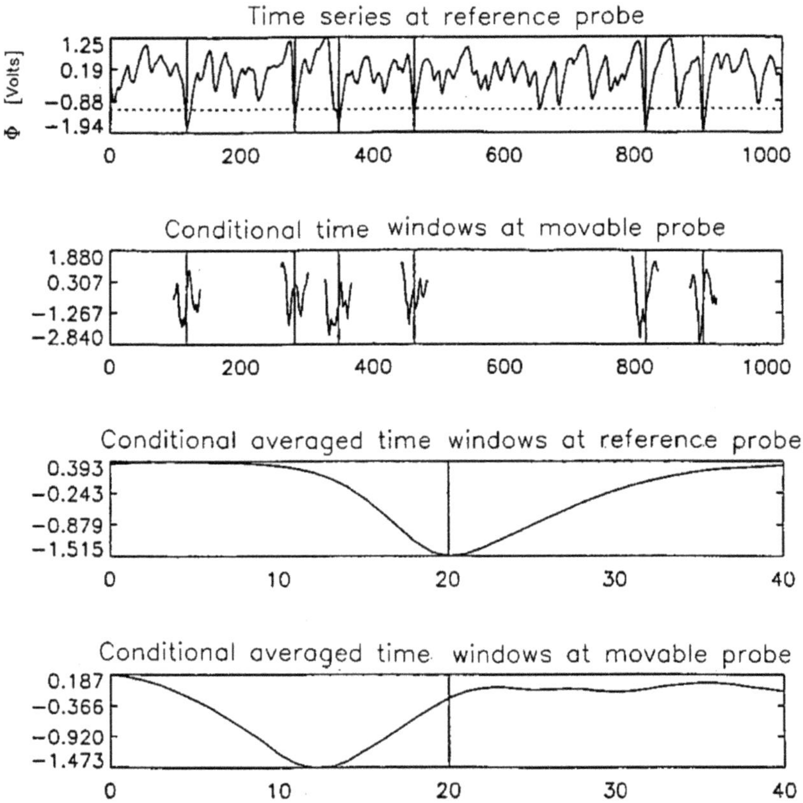

Figure 23.14. Illustration of conditional sampling as applied to actual data. From a reference signal (top frame), we select times where the signal fulfills a pre-defined condition, in the present case, $\Phi_1 < -2\sigma_1$, where σ_1 is the root-mean-square of the reference signal. The selection of the reference time is here done by locating a time interval between two level-crossings with opposite slopes of the signal. A conditional time interval is obtained between these two times. The reference time is then (for the present case) placed at the deepest local minimum, within this time interval. Given the reference time, conditional time-windows with given duration are selected symmetrically around the reference time in the other signal, here the one from the movable probe, see second frame. The signals from many such conditional time-intervals are averaged to give results as shown in the two lower frames, one for the reference signal and one for the movable probe. The figure is reproduced from Øynes *et al* (1998) with permission from APS.

showing actual data. The conditional average is the simplest analysis to implement in practice, but it was pointed out by Johnsen *et al* (1987) that a conditional median is in reality more relevant by giving the conditionally most probable value, assuming it is uniquely determined. The problem with this approach is that it requires the conditional probability density to be obtained at each position of the movable probe, which is a non-trivial extension of the data analysis. For the case studied by Johnsen *et al* (1987) the difference between the conditional average and the conditional median were not significant, but for strongly intermittent fluctuations it might be worthwhile to study and compare both cases.

While the conditional averaging method gives the space-time evolution of an *average* structure, it does not give any indication of how individual structures deviate from this average. By defining the *conditional variance* as

$$CV = \frac{\left\langle (\Phi_2 - \langle \Phi_2 | C \rangle)^2 \middle| C \right\rangle}{\left\langle (\Phi_2 | C)^2 \right\rangle}, \qquad (23.33)$$

where $0 \leqslant CV \leqslant 1$, we obtain a measure for the reproducibility of the *entire* event, i.e. the shape and position of a potential structure, and in particular also its trajectory. Small values of CV indicate a high degree of reproducibility, and $CV \approx 1$ means little or no reproducibility (Johnsen *et al* 1987, Øynes *et al.* 1998). For representation it might be advantageous to define a *conditional reproducibility* as

$$CR \equiv 1 - CV. \qquad (23.34)$$

The conditional analysis is different from a simple correlation. It is thus readily seen that the conditional average is sensitive for instance to the sign or polarity of structures, in contrast to a simple correlation which in effect squares the signal and is thus insensitive to a sign reversal.

We can also calculate the conditional variance given by (23.33). Typically we find that the regions near the extremum values of structures are highly reproducible. The statistical uncertainty of the result increases away from the extremum value. At present we cannot distinguish whether this uncertainty represents variations in the structures from one realization to another, or being alternatively a signature of a 'flapping' of the edges of the structure as it propagates.

23.5.2 Results from conditional averaging

Selected results from a conditional averaging in the study of Huld *et al* (1991) are shown in figure 23.15 and 23.16. We note the formation and propagation of large coherent vortex-like potential structures. By closer inspection (see figure 23.15) it was found that the structures had fine details, down to the ion Larmor radii scale. The details of the phenomena depend on external parameters: when E_0 is large we note the formation of a monopolar vortex, while for moderate E_0 double-vortices are formed. Note that the central art of the plasma is quiescent, the structures are seemingly confined to the region of large azimuthal $\mathbf{E}_0 \times \mathbf{B}_0/B_0^2$-velocity shears. The spatial resolution available limited the small details that the analysis could reveal. By focusing on a small area as illustrated in the lower frames in figure 23.15 and using all the available resolution there, it was found that the structures had fine details down to the ion Larmor radius scale. Since the spatial extent of the instability region is narrow we cannot expect any local homogeneity or isotropy. It could be argued that perturbations could propagate out of the region of linear instability, but for large distances from the plasma column, the plasma density is strongly reduced and perturbations will be damped there. In the central part of the plasma column we have a short circuiting affect of the hot plate and for the moderate plasma densities

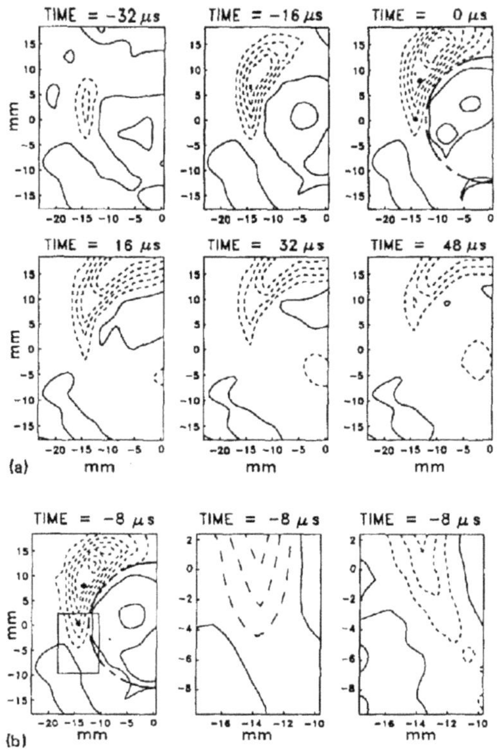

Figure 23.15. Results from a conditional averaging for a large radial electric field E_0. In figure (b) we have a part of figure (a) with enhanced resolution. Times before and after the reference time $t = 0$ are indicated above each frame. The projection of the hot plate (see figure 23.12) is indicated on one of the figures with a dashed line for reference. The figure is reproduced from Huld *et al* (1991) with permission. Copyright 1991, AIP Publishing LLC.

in this experiment we expect radial potential variations to be strongly reduced. Significant perturbations will be confined to the residual plasma, or scrape-off layer.

A detailed analysis demonstrated (Huld *et al* 1991) that a significant part of the anomalous transport in the radial direction of the device was associated with the presence of these coherent structures. The magnitude of this transport depended critically on the fluctuation level. The largest transport was observed when large scale double-vortex structures were present as in figure 23.16. It may not be proper to call this 'turbulent diffusion'. We will find that such a model for turbulent transport has a more restrictive definition. The experiment of Huld *et al* (1991) allowed also a study of the relative importance of the transport by large and small amplitude structures. The relevant analysis is made by selecting a threshold for the amplitudes of the structures, and time-integrate the averaged flux at a fixed position. The amplitude level is then varied and the integrated fluxes compared. The integrated flux increases when the amplitude level is decreased and for small amplitudes an 'accumulated' flux is obtained by this analysis. Differentiating the flux curve with respect to amplitude, the flux contribution for prescribed structure

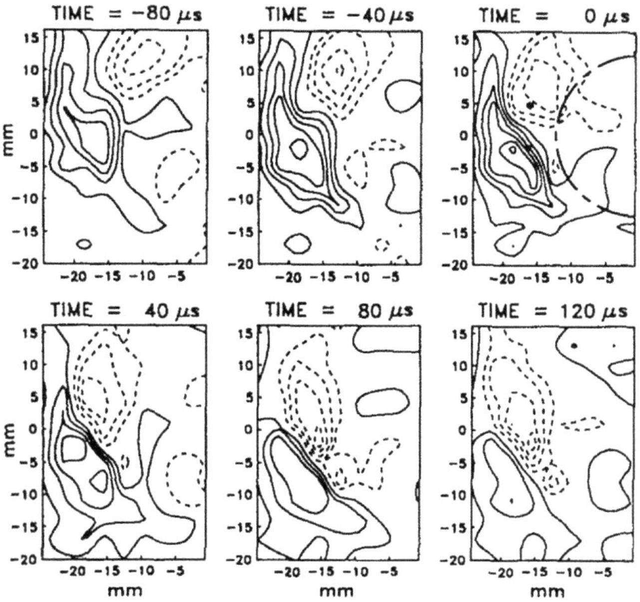

Figure 23.16. Results from a conditional averaging for a moderate radial electric field E_0, see figure 23.14 for comparison. The figure is reproduced from Huld *et al* (1991) with permission. Copyright 1991, AIP Publishing LLC.

amplitude levels as obtained (Huld *et al* 1991). It was found that the largest structures contributed individually more than smaller ones (this may hardly be surprising) but they were so few that the net effect of the many small structures dominated the transport. In fusion plasma experiments it is found that coherent structures, there termed 'blobs', often mediate a significant part of the plasma losses. The scalings of these structures have been studied in detail as summarized by D'Ippolito *et al* (2011).

By now many experiments have been carried out in search of coherent structures by use of conditional averaging. In some cases a reader may sense some disappointment: the results are at times found to differ only little from the correlation function associated with the turbulence (Filippas *et al* 1995). This observation should not come as a surprise: after all, we have no *a priori* guarantee that a turbulent flow (in neutral gases or fluids, or in plasmas) necessarily develop large scale coherent structures. For non-trivial conditions a near-Gaussian state may also develop, and this is well described solely by the space-time varying correlation function, as found also in section 19.2.4.

23.5.3 Bi-orthogonal decomposition

The present summary has emphasized conditional averaging as a means for identifying large coherent structures, in part because the method is easy to implement and also easy to visualize. Other methods exist, for instance bi-orthogonal decomposition (Benkadda *et al* 1994, Dudok de Wit *et al* 1994, Dudok de Witt 1995) that belong to a class of methods that includes proper orthogonal decomposition, Karhunen-Loeve expansion, and singular value decomposition, that were originally introduced in fluid turbulence (Lumley 1970). In its standard form the method assumes space-time separability, which need not always be a good assumption, necessitating extensions of the analysis.

Assume that we have given a data matrix D with elements $D(\ell, j)$ in the form

$$D(\ell, j) = \begin{vmatrix} \mathrm{d}(x_1, t_1) & \mathrm{d}(x_2, t_1) & \dots & \mathrm{d}(x_N, t_1) \\ \mathrm{d}(x_1, t_2) & \mathrm{d}(x_2, t_2) & \dots & \mathrm{d}(x_N, t_2) \\ \vdots & \vdots & \vdots & \vdots \\ \mathrm{d}(x_1, t_M) & \mathrm{d}(x_2, t_M) & \dots & \mathrm{d}(x_N, t_M) \end{vmatrix}, \tag{23.35}$$

representing a space-time varying signal, sampled at selected times and positions, t and x. Here we have j to be the temporal and ℓ to be the spatial index. The restriction in the chosen form is not serious when it comes to actual signals. The matrix need not be quadratic: we can have $M \neq N$.

The signal matrix is decomposed into a space and a time varying part, usually called 'topos' and 'chronos', respectively, expressed as

$$D(\ell, j) = \sum_k \lambda_k \Psi_k(\ell) \Phi_k(j), \tag{23.36}$$

with λ_k being a 'weight' coefficient to be determined. It is assumed that Ψ_k and Φ_k are orthonormal, i.e. $\sum_j \Psi_k(j)\Psi_\ell(j) = \delta_{k,\ell}$ and $\sum_j \Phi_k(j)\Phi_\ell(j) = \delta_{k,\ell}$ in terms of the Kronecker δ. Other normalizations can be used (van Milligen *et al* 2015). The covariance matrix of the signal is expressed as

$$Cov(j_1, j_2) = \frac{1}{M} \sum_\ell D(\ell, j_1) D(\ell, j_2).$$

Using the decomposition (23.36) we can also write

$$Cov(j_1, j_2) = \frac{1}{M} \sum_k \lambda_k^2 \Phi_k(j_1) \Phi_k(j_2).$$

Using the orthogonality already stated we find

$$\sum_{j_2} Cov(j_1, j_2) \Phi_k(j_2) = \frac{\lambda_k^2}{M} \Phi_k(j_1).$$

We thus found that the 'topos' are the eigenvectors of the covariance matrix, and λ_k^2/M are the corresponding eigenvalues. This observation offers a method for determining Φ_k and the corresponding λ_k's. With these functions and eigenvalues determined, it is possible to find all Ψ_k s by a simple matrix multiplication, i.e. $\lambda_k \Psi_k(\ell) = \sum_j D(\ell, j) \Phi_k(j)$. The reader is referred to the scientific literature for more details (Benkadda *et al* 1994, Dudok de Wit *et al* 1994, Dudok de Witt 1995, van Milligen *et al* 2015). It has been argued that the method is not always easy to use or implement (Davenport and Root 1958).

A draw back of the method in its basic form is that all elements $\mathrm{d}(x_\ell, t_j)$ for varying ℓ are assumed to be determined at the same time, which can be problematic if the analysis is based on probes that can disturb the plasma. The conditional averaging has a great advantage in this respect, although it will take more time. It is plausible that a two-step procedure can be advantageous: first the space-time variation of the phenomena are analyzed by the conditional averaging. The resulting reduced data-set often has a complicated space-time variation (Fredriksen *et al* 2003). A subsequent bi-orthogonal decomposition can resolve inherent details of this reduced signal. This two-step procedure has promise, but to the authors knowledge it has never been applied.

23.5.4 Comments of structures as a concept

Based on the foregoing discussions in sections 19.2.4 and 23.5.1 it may be worthwhile to discuss in more detail what we understand by a coherent structure or rather a coherent element in a signal. A realistic situation is one where we have a noisy signal from a receiver and some *a priori* information of a coherent signal, often in the form of some pulse. A radar can be an example: there will always be some noise received, sometimes with an airplane crossing the beam. We want to know if the signal record contains such an *a priori* known pulse, say $p(t)$, usually having some finite temporal support. The received signal is then $n(t) + p(t - t_0)$, where $n(t)$ represents the noise and t_0 is a reference time for the pulse event if it is present at all. The standard approach in such cases is to devise a 'matched filter'. The signal output for a filter with characteristic $h(t)$ becomes $s(t) = \int_0^T h(\tau)(n(t-\tau) + p(t - t_0 - \tau))\mathrm{d}\tau$. It can be demonstrated analytically (Bendat 1958, Davenport and Root 1958, Turin 1960) that provided we deal with white noise, the optimum filter characteristic is $h(t) = p(-t)$, i.e. the signal run backward. In lack of better, matched filters with this characteristic are then also used for noise characteristics other than white noise. The filter is not reproducing the signal, it merely gives a peak at a time where the signal or pulse is present. Basically we get a 'yes' or 'no', do we have a signal or not? Did an airplane pass by or not?

Another limit is found when the pulse-signals represent a significant constituent of the entire signal, i.e. when many pulses are present. In the signals discussed in chapter 20 the signal was *entirely* composed of structures, and given such a signal we would here ask for the shape of the structures and their density if we do not know this from the outset. The intermediate case with a signal composed by some noise superimposed on a large number of coherent structures is what seems most relevant for a large class of turbulent plasma phenomena.

The foregoing discussions in for instance section 23.5.1 were based on some semi-intuitive ideas on the interpretation of 'structures' as a concept. The implications of the conditional sampling is the expectation that coherent structures have large amplitudes and are embedded in a moderate or small amplitude noise. Selecting some large reference amplitude with some additional conditions on signal derivatives for instance, we are likely to pick out a subensemble that contains structures and by averaging we reduce or even eliminate the noise contribution. This idea will not work if the structures have a small amplitude compared to the RMS-noise level. In such cases it might be possible to make some advance by using some *a priori* knowledge, or assumed knowledge, to devise a matched filter to give reference times for the subensemble that forms the basis for the ensuing averaging. These ideas were tried out on data from a plasma experiment by Fredriksen *et al* (2008) with good results, including some test of the ideas by use of synthetic data. Related ideas were formulated by Teliban *et al* (2007).

References

Adrian R J 2007 Hairpin vortex organization in wall turbulence *Phys. Fluids* **19** 041301

Antonia R A 1981 Conditional sampling in turbulence measurement *Annual Rev. Fluid Mech.* **13** 131–56

Aris R 1962 *Vectors, Tensors, and the Basic equations of Fluid Mechanics* (London: Prentice-Hall)

Batchelor G K 1953 *The Theory of Homogeneous Turbulence* (Cambridge: Cambridge University Press)

Batchelor G K 1969 Computation of the energy spectrum in homogeneous two-dimensional turbulence *Phys. Fluids Supplement* **12** II–233–II–9

Bendat J S 1958 *Principles and Applications of Random Noise Theory* (New York: Wiley)

Bendat J S and Piersol A G 1986 *Random data, Analysis and Measurement Procedures* 2 edn (New York: Wiley)

Benkadda S, de Wit T D, Verga A, Sen A, ASDEX-team and Garbet X 1994 Characterization of coherent structures in tokamak edge turbulence *Phys. Rev. Lett.* **73** 3403–6

Beran M J 1968 *Statistical Continuum Theories* Vol 9 of *Monographs in Statistical Physics and Thermodynamics* (New York: Interscience)

Brigham E O 1974 *The Fast Fourier Transform* (New Jersey: Prentice-Hall)

Buchel'nikova N S, Salimov R and Eidel'man Y I 1967 Plasma turbulence in the presence of a drift instability *Sov. Phys. JETP* **25** 548–56

Bustamante M D, Quinn B and Lucas D 2014 Robust energy transfer mechanism via precession resonance in nonlinear turbulent wave systems *Phys. Rev. Lett.* **113** 084502

Camargo S J, Biskamp D and Scott B 1995 Resistive drift-wave turbulence *Phys. Plasmas* **2** 48–62

Chandrasekhar S 1961 *Hydrodynamic and Hydromagnetic Stability* (Oxford: Oxford University Press)

Chapman S 1941 Charles chree and his work on geomagnetism *Proc. Phys. Soc.* **53** 629–34

Chen F F 1965 Spectrum of low-β plasma turbulence *Phys. Rev. Lett.* **15** 381–3

Chu T K, Coppi B, Hendel H W and Perkins F W 1969 Drift instabilities in a uniformly rotating plasma cylinder *Phys. Fluids* **12** 203–8

D'Angelo N and Enriques L 1966 Turbulence of a low-β plasma *Phys. Fluids* **9** 2290–1

D'Angelo N, Pécseli H L and Petersen P I 1974 Farley instability - a laboratory test *J. Geophys. Res.* **79** 4747–51

Davenport W B and Root W L 1958 *An Introduction to the Theory of Random Signals and Noise* (New York: McGraw-Hill)

D'Ippolito D A, Myra J R and Zweben S J 2011 Convective transport by intermittent blob-filaments: comparison of theory and experiment *Phys. Plasmas* **18** 060501

Dudok de Wit T, Pecquet A-L, Vallet J-C and Lima R 1994 The biorthogonal decomposition as a tool for investigating fluctuations in plasmas *Phys. Plasmas* **1** 3288–300

Dudok de Witt T 1995 Enhancement of multichannel data in plasma physics by biorthogonal decomposition *Plasma Phys. Control. Fusion* **37** 117–35

Filippas A V, Bengston R D, Li G X, Meier M, Ritz C P and Powers E J 1995 Conditional analysis of floating potential fluctuations at the edge of the texas experimental tokamak upgrade (TEXT-U) *Phys. Plasmas* **2** 839–45

Fjørtoft R 1953 On the changes in the spectral distribution of kinetic energy for two dimesional, nondivergent flow *Tellus* **5** 225–30

Fowler T K 1966 Plasma fluctuations and the control of pumpout *Phys. Fluids* **9** 1437–41

Fredricks R W and Coroniti F V 1976 Ambiguities in the deduction of rest frame fluctuation spectrums from spectrums computed in moving frames *J. Geophys. Res.* **81** 5591–5

Fredriksen Å, Pécseli H L and Trulsen J 2008 Crash and recovery of the potential in a toroidal plasma column, as observed by generalized conditional sampling *New J. Phys.* **10** 033030

Fredriksen Å, Riccardi C, Cartegni L and Pécseli H 2003 Coherent structures, transport and intermittency in a magnetized plasma *Plasma Phys. Control. Fusion* **45** 721–33

Fyfe D and Montgomery D 1979 Possible inverse cascade behavior for drift-wave turbulence *Phys. Fluids* **22** 246–8

Granatstein V L 1967 Structure of wind-driven plasma turbulence as resolved by continuum ion probes *Phys. Fluids* **10** 1236–44

Gürcan Ö D, Garbet X, Hennequin P, Diamond P H, Casati A and Falchetto G L 2009 Wave-number spectrum of drift-wave turbulence *Phys. Rev. Lett.* **102** 255002

Hinze J O 1975 *Turbulence* 2nd edn (New York: McGraw Hill)

Huld T, Iizuka S, Pécseli H L and Rasmussen J J 1988 Experimental investigation of flute-type electrostatic turbulence *Plasma Phys. Control. Fusion* **30** 1297–318

Huld T, Nielsen A H, Pécseli H L and Juul Rasmussen J 1991 Coherent structures in two-dimensional turbulence *Phys. Fluids* B **3** 1609–25

Johnsen H, Pécseli H L and Trulsen J 1987 Conditional eddies in plasma turbulence *Phys. Fluids* **30** 2239–54

Johnston T W, Richards C, Ghosh A K, Carswell A I and Graf K 1969 *Turbulence in Fluids and Plasmas* (Brooklyn, NY: Polytechnic Press) chapter 8 "Correlation study in neutral-dominated plasma turbulence"

Kent G I, Jen N C and Chen F F 1969 Transverse kelvin-helmholtz instability in a rotating plasma *Phys. Fluids* **12** 2140–51

Kintner P M 1976 Observations of velocity shear driven plasma turbulence *J. Geophys. Res.* **81** 5114–22

Knorr G, Lynov J P and Pécseli H L 1990 Self-organization in three-dimensional hydrodynamic turbulence *Z. Naturforsch.* **45 a** 1059–73

Kraichnan R H and Montgomery D 1980 Two-dimensional turbulence *Rep. Prog. Phys.* **43** 547–619

Krane B, Pécseli H L, Trulsen J and Primdahl F 2000 Spectral properties of low-frequency electrostatic waves in the ionospheric E region *J. Geophys. Res.* **105** 10585–601

Kristensen L and Kirkegaard P 1986 Sampling problems with spectral coherence, Technical Report Risø-R-526, Risø National Laboratory, Denmark.

Landahl M T and Mollo-Christensen E 1992 *Turbulence and Random Processes in Fluid Mechanics* 2 edn (Cambridge: Cambridge University Press)

Lumley J L 1970 *Stochastic Tools in Turbulence* (London: Academic Press)
Lynov J P, Nielsen A H, Pécseli H L and Juul Rasmussen J 1991 Studies of the eulerian-lagrangian transformation in two-dimensional random flows *J. Fluid Mech.* **224** 485–505
Merilees P E and Warn H 1975 On energy and enstrophy exchanges in two-dimensional non-divergent flow *J. Fluid Mech.* **69** 625–30
Mikkelsen T and Pécseli H L 1978 Investigations of strong turbulence in a low-β plasma *Phys. Rev. Lett.* **41** 951
Mikkelsen T and Pécseli H L 1980 Strong turbulence in partially ionized plasmas *Phys. Lett.* A **77** 159–62
Misguich J H, Balescu R, Pécseli H L, Mikkelsen T, Larsen S E and Xiaoming Q 1987 Diffusion of charged particles in turbulent magnetoplasmas *Plasma Phys. Control. Fusion* **29** 825–56
Odajima K 1978 Effects of radial electric-field on drift wave instability in a weakly ionized plasma *J. Phys. Soc. Japan.* **44** 1685–93
Okabayashi M and Arunasalam V 1977 Study of drift-wave turbulence by microwave scattering in a toroidal plasma *Nucl. Fusion* **17** 497–513
Ottaviani M and Krommes J A 1992 Weak- and strong-turbulence regimes of the forced Hasegawa-Mima equation *Phys. Rev. Lett.* **69** 2923–6
Øynes F J, Olsen O-M, Pécseli H L, Fredriksen Å and Rypdal K 1998 Experimental study of low-frequency electrostatic fluctuations in a magnetized toroidal plasma *Phys. Rev.* E **57** 2242–55
Pécseli H L 1982 Drift-wave turbulence in low-β plasmas *Phys. Scripta* **T2/1** 147–57
Pécseli H L 2015 Spectral properties of electrostatic drift wave turbulence in the laboratory and the ionosphere *Ann. Geophysicae* **33** 875–900
Pécseli H L and Mikkelsen T 1985 Turbulent diffusion in two-dimensional strongly magnetized plasmas *J. Plasma Phys.* **34** 77–94
Pécseli H L and Mikkelsen T 1986 Clumps in drift wave turbulence *Plasma Phys. Control. Fusion* **28** 1025–41
Pécseli H L, Mikkelsen T and Larsen S E 1983 Drift wave turbulence in a low-β plasma *Plasma Phys.* **25** 1173–97
Pécseli H L and Trulsen J 1995 Velocity correlations in two-dimensional electrostatic turbulence in low-β plasmas *J. Plasma Phys.* **54** 401–30
Pécseli H L and Trulsen J 1997 Eulerian and lagrangian velocity correlation in two-dimensional random geostrophic flows *J. Fluid Mech.* **338** 249–76
Pedlosky J 1987 *Geophysical Fluid Dynamics* 2nd edn (New York: Springer)
Rhines P B 1975 Waves and turbulence on a β-plane *J. Fluid Mech.* **69** 417–43
Rose G *et al* 1992 The ROSE project - scientific objectives and discussion of 1st results *J. Atmos. Terr. Phys.* **54** 657–67
Roth J R 1971 Experimental study of spectral index, mode coupling, and energy cascading in a turbulent, hot-ion plasma *Phys. Fluids* **14** 2193–202
Rutgers M A 1998 Forced 2D turbulence: experimental evidence of simultaneous inverse energy and forward enstrophy cascades *Phys. Rev. Lett.* **81** 2244–7
Shapley A H and Beynon W J G 1965 Winter anomaly in the ionospheric absorption and stratospheric warmings *Nature* **206** 1242–3
Shinohara S, Matsuoka N and Matsuyama S 2001 Establishment of strong velocity shear and plasma density profile modification with associated low frequency fluctuations *Phys. Plasmas* **8** 1154–8
Shkarofsky I P 1969 *Turbulence in Fluids and Plasmas* (Brooklyn, NY: Polytechnic Press) chapter 21 Analytic forms for decaying turbulence functions

Taylor G I 1938 The spectrum of turbulence *Proc. R. Soc. London Ser.* A **164** 476–90
Taylor J B and McNamara B 1971 Plasma diffusion in two dimensions *Phys. Fluids* **14** 1492–9
Tchen C-M 1953 On the spectrum of energy in turbulent shear flow *J. Res. Bur. Stand.* **50** 51–62
Tchen C-M, Pécseli H L and Larsen S E 1980 Strong turbulence in low-β plasmas *Plasma Phys.* **23** 817–29
Tchen C-M and Pierson W J 1988 A group-kinetic theory with a closure by memory loss for modeling turbulence in the atmosphere and the oceans *Meteorol. Atmos. Phys.* **38** 95–105
Teliban I, Block D, Piel A and Greiner F 2007 Improved conditional averaging technique for plasma fluctuation diagnostics *Plasma Phys. Control. Fusion* **49** 485–97
Tennekes H 1975 Eulerian and Lagrangian time microscales in isotropic turbulence *J. Fluid Mech.* **67** 561–7
TFR-Group and Truc A 1984 Turbulent spectrum analysis in TFR tokamak plasmas *Plasma Phys. Control. Fusion* **26** 1045
Tolias P, Ratynskaia S, de Angelis U and Lazzaro E 2016 The finite probe size effect in fluctuation measurements; application to dusty plasmas *J. Plasma Phys.* **82** 615820201 (23 pages)
Tsinober A, Vedula P and Yeung P K 2001 Random Taylor hypothesis and the behavior of local and convective accelerations in isotropic turbulence *Phys. Fluids* **13** 1974–84
Tung K K and Welch W T 2001 Remarks on Charney's note on geostropic turbulence *J. Atmos. Sci.* **58** 2009–12
Turin G L 1960 An introduction to matched-filters *IRE Trans. Information Theory* **6** 311–29
van Milligen B P, Sánchez E, Alonso A, Pedrosa M A, Hidalgo C, de Aguilera A M and Fraguas A L 2015 The use of the biorthogonal decomposition for the identification of zonal flows at TJ-II *Plasma Phys. Control. Fusion* **57** 025005
Vedenov A A 1968 *Theory of Turbulent Plasma* (London: Ilife)

Hans Pécseli

Chapter 24

Turbulence in the ionosphere

Low frequency electrostatic fluctuations in the ionosphere are often observed to develop into a strongly turbulent state (Kintner and Seyler 1985). Strong turbulence, with a fully developed spectrum seems mostly to be associated with gradient driven instabilities, as discussed for instance in section 11.5. We might discuss ionospheric spectra on the basis of the relations (11.31)–(11.32). By inspection, we note that only dimensional parameters are the ambipolar diffusion coefficient D_α, the density gradient for the plasma $\mathcal{L}_n \equiv n_0/|\nabla n_0|$ and the drift velocity U_0. All other parameters are related to these three (Ott and Farley 1974, Sudan and Keskinen 1977). We *assume* now that the equations (11.31)–(11.32) indeed give an adequate description of the full nonlinear problem and, moreover, that they have solutions corresponding to a fully developed stationary turbulent stage. These questions may not have been fully explored yet. If our expectations are fulfilled, we may hope that the physical problem is adequately described by the dimensional quantities characterizing (11.31)–(11.32), and we expect that the dimensionless structure function for the fluctuations in relative density can be expressed as

$$\langle(\tilde{n}(\boldsymbol{\xi}, t) - \tilde{n}(\boldsymbol{\xi} + \mathbf{r}, t))^2\rangle \frac{1}{n_0^2} = g\left(\frac{r}{\mathcal{L}_n}, \frac{rU_0}{D_\alpha}, \theta, \beta_1, \beta_2\right), \tag{24.1}$$

where θ is the angle between $\mathbf{U}_0$ and $\mathbf{r}$, in the plane perpendicular to the magnetic field. Since the two dimensionless quantities β_1 and β_2 enter the basic equations in the given form, it is natural to retain them in the same form when carrying out the dimensional analysis. The structure function can be written in different dimensionless ways, but the following discussion will demonstrate why the present version is preferable.

As a direct alternative to the density structure function (24.1), we might consider a spectral representation of the correlation function obtained from the structure

doi:10.1088/978-0-7503-1251-6ch24

function, see section 22.5. Retaining the dimensionless quantities used in (24.1), we find

$$S(k,\theta) = G\left(k\mathcal{L}_n, \frac{U_0}{D_a k}, \theta, \beta_1, \beta_2\right)\frac{1}{k}, \tag{24.2}$$

defined in such a way that, for the present two-dimensional model, we have

$$\left\langle\left(\frac{\tilde{n}(\boldsymbol{\xi}, t)}{n_0}\right)^2\right\rangle = \int S(k,\theta)\mathrm{d}k\mathrm{d}\theta.$$

We note that (24.2) does not contain the energy dissipated per kg plasma explicitly, as one might have expected by comparison with (22.37). The reason is that we here believe that the fundamental equations implicitly or self consistently contain such a generating mechanism, when basic ionospheric parameters as U_0 etc are given. When deriving (22.37) we had no general basic equations and had to introduce the dissipated energy 'ad hoc'.

We note now that experimental observations by 6 m wavelength radar scattering seem to indicate that the scattered power increases approximately with the square of the drift velocity U_0 (Balsley 1969, Balsley and Farley 1973, Ott and Farley 1974). It is not evident from these experiments that U_0 was the only parameter that was changed, but we will assume that this is the case. Referring to (24.2) we consequently expect the spectral variation to be given by

$$S(k,\theta) \sim \frac{U_0^2}{D_a^2 k^3} \tag{24.3}$$

for the given θ direction. The spectrum does not depend on $\mathcal{L}_n$ in the given subrange, but there will presumably be a long wavelength part of the wavenumber spectrum where it enters. If the variation is linear with $\mathcal{L}_n$, we can conclude right away that the density power spectrum is constant in this subrange. It is interesting to note here the analytical studies of (Moiseev *et al* 1981), where a $1/k^3$ spectral subrange was derived for the gradient instability in the ionospheric plasma.

If we use the present results on the structure function, we might conclude that there is a separation range where we have

$$\left\langle(\tilde{n}(\boldsymbol{\xi}, t) - \tilde{n}(\boldsymbol{\xi} + \mathbf{r}, t))^2\right\rangle\frac{1}{n_0^2} \sim \frac{U_0^2\, r^2}{D_a^2}.$$

The present discussions explicitly deal with *density* fluctuations. The fluctuating electric field is related to fluctuating ion velocity by a constant mobility $\widetilde{\mathbf{U}} = \mu\widetilde{\mathbf{E}}$ in the present inertia-free model. The continuity equation then relates density to electric fields, via the velocity. To lowest order approximation, with non-dispersive waves, we expect a proportionality between $\tilde{n}/n_0$ and E. Such relations are amenable to experimental verification by comparison of spectral indices (Krane *et al* 2000). If, for instance, $\tilde{n}/n_0$ and $\widetilde{E}$ are to be proportional, then the spectral index for the turbulent spectra for the two quantities must be the same. The nice feature of this argument is that we need not rely on any fitted constants, coupling coefficients between probes and plasma, for instance.

For short wavelengths and high relative ion–electron drift velocities, the Farley–Buneman instability, see chapter 11, can directly excite short wavelength fluctuations, and these are not included in the present analysis. The present author (Pécseli 2015) found it interesting though that the observations seem to support the classical result (Kadomtsev 1965) found by weak turbulence theory for ion acoustic wave power spectra excited by an electron current where short wavelengths are also directly excited. The analytical results of Kadomtsev (1965) found experimental support in studies reported by Daughney *et al* (1970) and Machalek and Nielsen (1973). Some analytical results are summarized by Paul *et al* (1969).

By analyzing data from instrumented ROSE4 rocket (Rose *et al* 1992) flown over northern Scandinavia, Krane *et al* (2000) observed a significantly enhanced levels of low frequency wave activity in the ionospheric E region, as illustrated in figure 24.1. It was found that the high frequency part of the spectrum of density as well as potential fluctuations could be approximated by simple power laws. See examples shown in figure 24.2, for both upleg and downleg parts of the flight. In effect the rocket thus makes *two* measurements of the fluctuations in the ionosphere, and with the two parts of the flight being separated by more than 20 km, they can be considered as two *independent* experiments.

It is possible to make a convincing power-law fit to the high frequency part of the spectra at a given segment of the data, but it might be expected that the spectral index varies with altitude. The variations of the spectral indices are shown in figure 24.3. It is interesting that the index variation is systematic, and that by and large it is reproducible for the upleg and downleg conditions. Also it is important to note that the spectral index is significantly different for the fluctuations in density and in electrostatic potential. This observation rules out the possibility of local isothermal Boltzmann distribution of electrons, and assumption which would otherwise be tempting. The region where the two indices become similar is characterized by a very low level of fluctuations, see also figure 24.1. The high frequency part of the spectrum is expected to be influenced mainly by the Farley–Buneman instability, see chapter 11. The conditions for this instability were fulfilled by having an

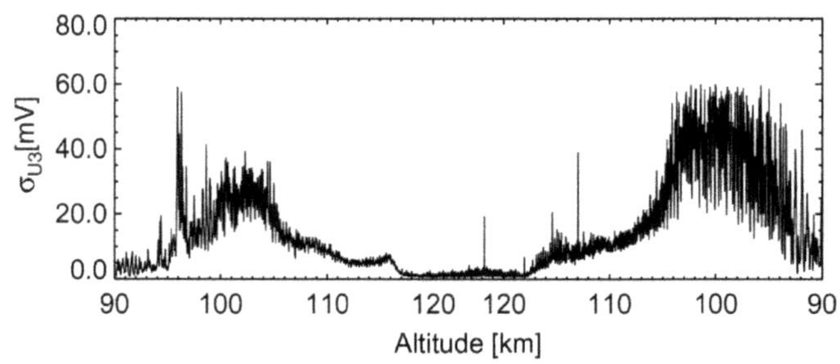

Figure 24.1. Standard deviation of the potential fluctuation levels for rocket flight. The figure covers both upleg and downleg parts of the flight. The figure is reproduced from Krane *et al* (2000) with permission from AGU.

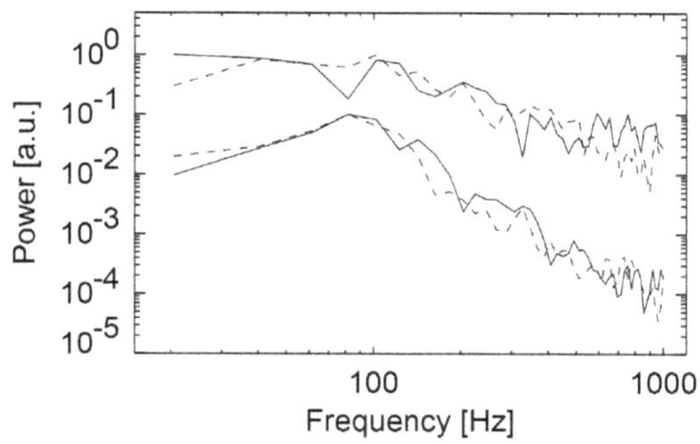

Figure 24.2. Spectra of ionospheric fluctuations as detected by instrumented rockets. The top curves are for fluctuations in density, the lower ones for potential. The upleg parts are shown with full lines, downleg with dashed lines. The two sets of data refer to the same altitude. Data are shown in double-logarithmic presentation. The figure is reproduced from Krane *et al* (2000) with permission from AGU.

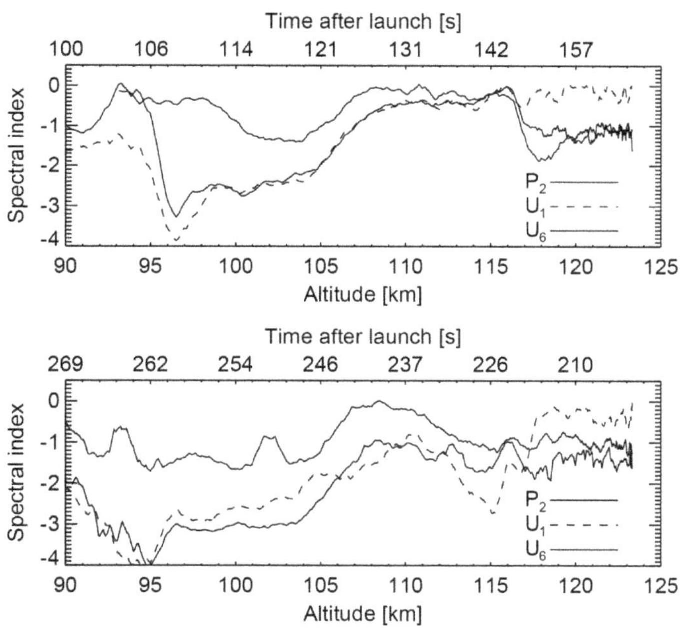

Figure 24.3. Spectral index variation of ionospheric fluctuations as detected by instrumented rockets. The full line is for density fluctuations, and the two dashed lines refer to potential differences for two probe-sets. The two figures refer to upleg and downleg parts of the flight, respectively. Note that the spectral index varies systematically with altitude, and that this variation is quite similar for upleg and downleg conditions. The figures are reproduced from Krane *et al* (2000) with permission from AGU.

$\mathbf{E} \times \mathbf{B}/B^2$-drift of the electrons significantly exceeding the ion acoustic sound speed. Since the rocket in effect makes a 'one point measurement' it was not possible to deduce the presence of a density gradient in the direction perpendicular to the magnetic field, and little can be said concerning the presence of density gradient instabilities for those conditions. See also the discussion in section 20.8 for relevant models of synthetic rocket signals.

Also Pécseli *et al* (1989) and Pécseli *et al* (1993) present power spectra from their rocket experiment, but these results are shown in a color coded representation that is not readily compared with the results from the ROSE rockets (Krane *et al* 2000). Within the uncertainty, their results are however not inconsistent with the observations of Krane *et al* (2000).

Since the Farley–Buneman instability can be seen as current driven instability, it is interesting to compare the observed spectra with those deduced for classical current driven instabilities as studied by for instance Kadomtsev (1965), predicting a spectrum

$$S(k) \sim \frac{1}{k^3} \ln\left(\frac{1}{k\lambda_{De}}\right)$$

for the electrostatic potential, giving a power spectrum in the form $k^2S(k)$ for the electric field. Experimental results by laser scattering from plasma shock waves (Machalek and Nielsen 1973) are in remarkable agreement with this spectrum, but the interpretation relies on an assumed relation between the scattered signal and the fluctuating plasma potential.

The analysis of Krane *et al* (2000) was continued by Larsen *et al* (2002) to analyze amplitude probability densities, demonstrating significantly non-Gaussian features. Phase coherence effects were studied by triple correlation methods (Pécseli *et al* 1993). The results indicate that there *are* non-trivial phase coherent coupling in the spectrum, but these were most apparent in the low frequency part where a spectral index fit was inaccurate. One interesting observation in this context was the close agreement between data from northern Scandinavia obtained in 1988–1989 and rocket observations from Greenland dating from 1972: in spite of different locations, different times, different methods for the triple correlation analysis, etc, the results look more or less the same!

The summary in this section assumes that the turbulence is generated by the instability of the ionospheric plasma. It can, however, easily be imagined that a formally stable ionosphere will be in a turbulent state, simply because it is embedded in a turbulent neutral atmosphere, which is strongly coupled to the ion dynamics because of the large ion–neutral collision frequency (Gurevich *et al* 1997). The lower we get in the ionosphere (lower parts of the E-region, upper parts of the D-region), the larger is the possibility for this effect to be present. On the other hand, it might be added that not much is known about neutral turbulence at those high altitudes, although a few rocket experiments have given at least *some* information (Wu and Widdel 1992). Indirect evidence for the presence of neutral turbulence at altitudes below 95 km was found much earlier by Prakash *et al* (1968), using instrumented rockets.

24.1 Observations

The foregoing section presented observations of turbulent spectral for fluctuations in density and electrostatic potential as obtained by an instrumented rocket. Several such spectral observations have been reported and some such results will be

Table 24.1. Observations made by instrumented rockets. If two values for a spectral index are listed, then the lowest frequency range is listed first. Note that the uncertainty on the spectral estimates is not always given in the literature. The data are sorted according to average altitudes, the lowest altitudes first.

Source	Region/ Altitude	Spectral indices, density	Spectral indices, potential	Spectral indices, electric field
Krane *et al* (2000)	100–110 km	1 ± 0.25	3 ± 0.3	–
Prakash *et al* (1971)	150 km	–	–	2.5 ± 0.5
LaBelle *et al* (1986)	300–500 km	2.5/4.5	–	–
Kelley *et al* (1982)	390–560 km	$2.2 \pm 0.2/5.5 \pm 0.5$	–	–
Earle *et al* (1989)	450–500 km	2.2 ± 0.5	–	2.9 ± 0.5

summarized here in tables 24.1 and 24.2. A somewhat more detailed summary, including laboratory results, is found elsewhere (Pécseli 2016). The scatter in spectral indexes is large, even within one experiment, and it is recommended that the original papers are consulted to see precisely what result they are being cited for. Note the correction of spectral indexes (Prakash *et al* 1971) that takes into account a varying filter bandwidth: this reference is predated by others (Prakash *et al* 1968, 1969) containing an erroneous value for spectral indexes. Values for positive spectral indices have been reported (Prakash *et al* 1971) but these apply only to some unspecified spectral subranges and they are not included in the present summary. Observations of turbulent electric field spectra have been made by instrumented balloons (Mozer 1971), but these will not be discussed further here, although they can be relevant for comparison with E-region turbulence results. An overview of ionospheric turbulent conditions was given by Hysell (2000), with emphasis on the F-region.

A vast number of numerical simulations contain partial results for spectral power laws: there are too many to be summarized here, in part also because it is not always evident which ionospheric parameter range (if relevant at all) the analysis refers to. We can mention an early summary of studies of the type II irregularities (see section 11.4.1) by McDonald *et al* (1974) giving an density fluctuation spectrum close to $k^{-3.5}$, albeit with a clear tendency for steepening at small wavelengths. It is stated explicitly that this result is obtained by averaging over all directions, so that a k-integration directly gives $\langle \tilde{n}^2 \rangle$. Gondarenko and Guzdar (2004) compare results of numerical simulations with experimental observations of Basu *et al* (1990) and emphasize a subrange that would correspond to the 'production subrange' in the terminology of Tchen *et al* (1980). This subrange is, however, in their simulations followed by a 'steeper' subrange that is not discussed in detail, but it might be noted that it has the characteristics of a 'coupling subrange', see figure 23.9.

Some reports seem to distinguish drift and transverse velocity shear driven turbulence (Kintner and Seyler 1985, Basu *et al* 1988). Indeed in what could be

Table 24.2. Observations made by instrumented satellites. The results are obtained by determining the exponent α in frequency spectra $\sim f^{-\alpha}$ as in table 24.1. For the data in Kintner (1976) only the high intensity cases are cited. The data are sorted according to average altitudes, the lowest altitudes first.

Source	Region/Altitude	Spectral indices, density	Spectral indices, potential	Spectral indices, electric field
Villain *et al* (1986)	200–400 km	2.1 ± 0.1 1.7/2.9	–	–
Shume and Hysell (2004)	325–450 km	1.6 ± 0.05	–	–
Basu *et al* (1990)	400 km	1.9 ± 0.1	–	1.9 ± 0.1, 2.3 ± 0.1, 3.3 ± 0.1
Rodrigues *et al* (2009)	500 km	5.5 ± 1	–	1 ± 1
McDaniel and Hysell (1997)	530 km	1.7/3.5	–	–
Cerisier *et al* (1985)	400–2.000 km	1.93	–	1.77
Golovchanskaya *et al* (2012)	700–2.500 km	–	–	2.2 ± 0.3
Kintner (1976)	500–125.000 km	–	–	2.8 ± 0.3
Gurnett and Frank (1977)	24.000–30.000 km	–	–	3 ± 0.3
Angelopoulos *et al* (2001)	5.700–54.000 km	–	–	1/2.5

called the production range, we might expect differences, but for an inertial range the precise nature of the energy input should be of little consequence and the spectral shape would be expected to be controlled by the cascade which will be the same for small scale electrostatic drift and velocity shear instabilities.

The results of Basu *et al* (1990) for the electric field in table 24.2 are obtained for different orbits, with a seemingly large variability. It might be that the variation is due to different subranges being sampled. This work includes an interesting comparison of spectra for different wavenumber component directions perpendicular to the magnetic field giving some indications for local isotropy in that plane.

The lists in tables 24.1 and 24.2 are by no means exhaustive, but already on this limited basis we can state that the results are not promising. There is a large scatter in the spectral indices reported. It seems unlikely that a universal power-law can be determined. The reason is not obvious. It might not be permissible to compare data from such diverse regions of space, but on the other hand, such a comparison should be whole point of universal properties. Can it be that such a universal law simply does not exist? This conclusion seems somewhat pessimistic, and in the opinion of the present author it is a question of sorting the data according to an effective

Reynolds number. universal laws will be found when the nonlinear effects dominate, and with the exception of the results of Kintner (1976) the data are not sorted according to wave intensity. In the one case mentioned, the distinction of low and high intensity spectra was significant. Most laboratory results refer to large intensity fluctuations and there we could find indications for universal power-law spectra. It can also be that the instrumentations of the rockets and satellites are insufficient to distinguish the nature of the turbulence. As mentioned, by applying Taylors hypothesis (or the 'frozen turbulence hypothesis', see section 23.4.1) to a 'one-point signal' alone, it is not logically possible to distinguish the dimensionality of the turbulence field. It is quite possible that some of the observations deal with plasma turbulence being close to two-dimensional and perpendicular to **B**, other cases may have a driving that includes components parallel to **B**, where insufficient instrumentation does not allow distinction of the separate cases. A part of the variability in spectral index found in the rocket data table 24.1 can however be resolved by a dedicated laboratory experiment summarized in the following section 24.2.

24.2 Laboratory studies

The results shown in figure 24.2 and 24.3 were obtained by instrumented rockets. The Hall current driven instabilities and their saturated turbulent stage have also been studied in laboratory experiments by, e.g. Mikkelsen and Pécseli (1980). The experimental set-up was also used in studies of electrostatic drift waves (Mikkelsen and Pécseli 1978, Pécseli *et al* 1983). Basically, the experiment consists of a homogeneously magnetized plasma column in a steady state parabolic potential profile. The resulting steady state electric field $\mathbf{E}_0$ increases linearly with radius, and it gives rise to a uniform $\mathbf{E}_0 \times \mathbf{B}/B^2$ rotation of the entire plasma column. The root-mean-square value of the relative density fluctuations were largest at a radial position where the steady state average density gradient was strongest.

The basic features of the cross-correlation are illustrated in figure 24.4. Two signals were obtained for varying azimuthal probe separation with both probes being at the same radial position. The cross-correlations were obtained by an analog device and shown in figure 24.4(*a*) for various separations, see also discussion of figure 23.5. The time delay of the maximum is plotted for varying separations as shown in figure 24.4(*b*), where the best linear fit estimates the average propagation velocity. As a cross-check, this velocity could be compared to the local $\mathbf{E}_0 \times \mathbf{B}/B^2$ velocity deduced from the known magnetic field and $\mathbf{E}_0$ at the radial probe position. The authors found the agreement to be satisfactory. For perfectly frozen turbulence, the propagation velocity would be the same for all frequencies. This hypothesis could also be tested, with results shown in figure 24.4(*c*) obtained by band-pass filtering the signal prior to the cross-correlation. (This method for frequency separation is not unique.) It was found that the highest frequencies actually propagated at the local $\mathbf{E}_0 \times \mathbf{B}/B^2$ velocity, while lower frequencies were somewhat slower.

The experimental conditions allowed injection of a neutral gas, and the effects of the turbulent spectra were studied, as illustrated in figure 24.5(*a*). Heuristically, it

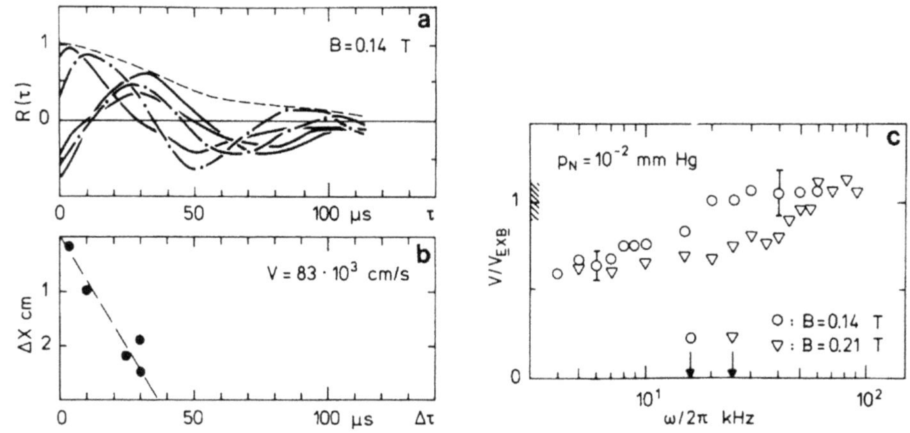

Figure 24.4. Cross-correlation function used for estimating propagation velocities at a reference neutral pressure, see also figure 23.5. The two symbols ○ and ▽ in (*c*) refer to different values of the magnetic field, as indicated. Vertical arrows with the corresponding symbols in (*c*) indicate the corresponding ion cyclotron frequencies. The data are reproduced from Mikkelsen and Pécseli (1980) with permission from Elsevier.

can be argued that increasing the neutral background pressure corresponds to moving down from the ionospheric F-region into the E-region. As the neutral drag on the ions due to the neutral background increase with increasing neutral pressure, the fluctuation level was seen to increase, consistently with an increasing Hall current giving the free energy for the instability see figure 24.5(*c*). For low neutral pressures the free energy is in the electron diamagnetic drift (that is at least one way to see it) and in the azimuthal current due to the difference in rotation velocity of the light electrons and the heavy ions (Pécseli 1982). For low neutral pressures we find the approximate ω^{-5} power law for both plasma density and electrostatic potential, as would be expected for a drift wave instability, with the electrons being close to an isothermal Boltzmann distribution at all times. At enhanced neural pressures the electron–neutral collisions prohibit the free motion of electrons and the density and potential spectra begin to deviate as shown by the difference between the two spectral indices in figure 24.5(*b*). The spectra also become flatter: this is consistent with energy input at large wavenumbers by the Farley–Buneman instability, see chapter 11. It is interesting to note that the rocket data in table 24.1 are at least not in contradiction with the variation in spectral index illustrated in figure 24.5 if we associate a decreasing neutral collision frequency with increasing ionospheric altitude.

The sample spectra shown in figure 24.3 agree well with those obtained for enhanced neutral pressures (corresponding to E-region conditions) shown in figure 24.5(*a*). The results summarized in figure 24.3 agree well with the appropriate values for relevant neutral pressures in figure 24.5(*b*). Fluctuations in the top-side high latitude F-region in the ionosphere were observed by the Aureol 3 satellite (Cerisier *et al* 1985). The spectra for the fluctuating density and electric fields were very similar, with a spectral index 1.9 for the density power spectrum and 1.8 for the electric field. The observations were by the observers related to low frequency

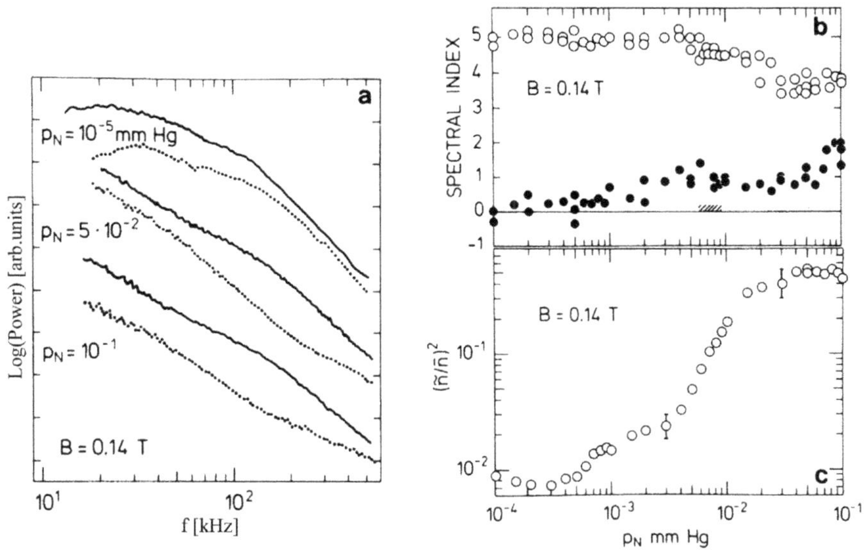

Figure 24.5. Variation of density and potential spectra for varying neutral pressure inlet. Fully developed turbulent spectra excited by the Hall current instabilities are obtained for high neutral pressures. Open circles ∘ in (*b*) give the spectral index for the potential fluctuations in the high frequency limit, while filled circles • give the difference between the spectral index for density and potential in the same frequency interval. In (*c*) we have the RMS-fluctuation level of the plasma density. The data are reproduced from Mikkelsen and Pécseli (1980) with permission from Elsevier.

electrostatic drift waves for wavelengths in the range 50–8 $\times 10^3$ m, deduced from a measured frequency and a known satellite velocity. The interplay between the laboratory experiment and the ionospheric rocket and satellite data is interesting and valuable, but illustrates also the superiority of the level of information available by laboratory conditions.

Results from other laboratory experiments have been reported (Saxena and John 1975, John and Saxena 1975). For conditions relevant here, when the criteria for exciting the Farley–Buneman instability were satisfied, potential power spectra with power-law index 2.8 ± 0.7 were reported (John and Saxena 1975). No systematic variation was made in those studies, but the results do not disagree with those of Mikkelsen and Pécseli (1978).

References

Angelopoulos V, Mozer F S, Bonnell J, Temerin M, Somoza M, Peterson W K, Collin H L and Giles B 2001 Wave power studies of cusp crossings with the polar satellite *J. Geophys. Res.: Space Phys.* **106** 5987–6006

Balsley B B 1969 Some characteristics of the non-two-stream irregularities in the equatorial electrojet *J. Geophys. Res.* **74** 2333–47

Balsley B B and Farley D T 1973 Radar observations of two-dimensional turbulence in the equatorial electrojet *J. Geophys. Res.* **78** 7471–9

Basu S, Basu S, MacKenzie E, Coley W R, Sharber J R and Hoegy W R 1990 Plasma structuring by the gradient drift instability at high latitudes and comparison with velocity shear driven processes *J. Geophys. Res.* **95** 7799–818

Basu S, Basu S, MacKenzie E, Fougere P F, Coley W R, Maynard N C, Winningham J D, Sugiura M, Hanson W B and Hoegy W R 1988 Simultaneous density and electric field fluctuation spectra associated with velocity shears in the auroral oval *J. Geophys. Res.* **93** 115–36

Cerisier J C, Berthelier J J and Beghin C 1985 Unstable density gradients in the high-latitude ionosphere *Radio Sci.* **20** 755–61

Daughney C C, Holmes L S and Paul J W M 1970 Measurement of spectrum of turbulence within a collisionless shock by collective scattering of light *Phys. Rev. Lett.* **25** 497–9

Earle G D, Kelley M C and Ganguli G 1989 Large velocity shears and associated electrostatic waves and turbulence in the auroral F region *J. Geophys. Res.: Space Physics* **94** 15321–33

Golovchanskaya I V, Kozelov B V and Despirak I V 2012 Investigation of the broadband ELF turbulence by observations of the FAST satellite *Geomagn. Aeronomy* **52** 474–81

Gondarenko N A and Guzdar P N 2004 Plasma patch structuring by the nonlinear evolution of the gradient drift instability in the high-latitude ionosphere *J. Geophys. Res.: Space Phys.* **109** A09301

Gurevich A V, Borisov N D and Zybin K P 1997 Ionospheric turbulence induced in the lower parts of the E region by the turbulence in the neutral atmosphere *J. Geophys. Res.* **102** 379–88

Gurnett D A and Frank L A 1977 A region of intense plasma wave turbulence on auroral field lines *J. Geophys. Res.* **82** 1031–50

Hysell D 2000 An overview and synthesis of plasma irregularities in equatorial spread F *J. Atmos. Solar-Terrestrial Phys.* **62** 1037–56

John P I and Saxena Y C 1975 Observation of Farley-Buneman instability in laboratory plasma *Geophys. Res. Lett.* **2** 251–4

Kadomtsev B B 1965 *Plasma Turbulence* (New York: Academic)

Kelley M C, Livingston R C, Rino C L and Tsunoda R T 1982 The vertical wave number spectrum of topside equatorial spread f: estimates of backscatter levels and implications for a unified theory *J. Geophys. Res.: Space Physics* **87** 5217–21

Kintner P M 1976 Observations of velocity shear driven plasma turbulence *J. Geophys. Res.* **81** 5114–22

Kintner P M and Seyler C E 1985 The status of observations and theory of high latitude ionospheric and magnetospheric plasma turbulence *Space Sci. Rev.* **41** 91–129

Krane B, Pécseli H L, Trulsen J and Primdahl F 2000 Spectral properties of low-frequency electrostatic waves in the ionospheric E region *J. Geophys. Res.* **105** 10585–601

LaBelle J, Kelley M C and Seyler C E 1986 An analysis of the role of drift waves in equatorial spread *F J. Geophys. Res.* **91** 5513–25

Larsen Y, Hanssen A, Krane B, Pécseli H L and Trulsen J 2002 Time-resolved statistical analysis of nonlinear electrostatic fluctuations in the ionospheric E region *J. Geophys. Res.* **107(A1)** SIA5–1–1

Machalek M D and Nielsen P 1973 Light-scattering measurements of turbulence in a normal shock *Phys. Rev. Lett.* **31** 439–42

McDaniel R D and Hysell D L 1997 Models and DE II observations of inertial-regime irregularities in equatorial spread *F J. Geophys. Res.* **102** 22233–46

McDonald B E, Coffey T P, Ossakow S and Sudan R N 1974 Preliminary reports of numerical simulation of type 2 irregularities in the equatorial electrojet *J. Geophys. Res.* **79** 2551–4
Mikkelsen T and Pécseli H L 1978 Investigations of strong turbulence in a low-β plasma *Phys. Rev. Lett.* **41** 951
Mikkelsen T and Pécseli H L 1980 Strong turbulence in partially ionized plasmas *Phys. Lett.* A **77** 159–62
Moiseev S S, Sagdeev R Z, Tur A V and Yanovskiĭ V V 1981 Energy transfer and turbulence spectrum when the gradient instability is excited in the ionospheric plasma *Zh. Eksp. Teor. Fiz.* **80** 597–607
Moiseev S S, Sagdeev R Z, Tur A V and Yanovskiĭ V V 1981 russian original *Sov. Phys. JETP* **53** 301–5
Mozer F S 1971 Power spectra of the magnetospheric electric field *J. Geophys. Res.* **76** 3651–67
Ott E and Farley D T 1974 The *k* spectrum of ionospheric irregularities *J. Geophys. Res.* **79** 2469–72
Paul J W M, Daughney C C and Holmes L S 1969 Study of turbulence within a collisionless shock using collective scattering of light, *in* Collision-Free Shocks in the Laboratory and Space, number ESRO SP-51 *in* Proceedings of a Study Group held at the European Space Research Institute (ESRIN), Frascati, 11-20 June 1969, Organisation Européenne de Recherches Spatiales, European Space Research Organisation pp 207–27
Pécseli H L 1982 Drift-wave turbulence in low-β plasmas *Phys. Scripta* **T2/1** 147–57
Pécseli H L 2015 Spectral properties of electrostatic drift wave turbulence in the laboratory and the ionosphere *Ann. Geophysicae* **33** 875–900
Pécseli H L 2016 Low frequency electrostatic waves propagating in plasmas with parameters varying along magnetic field lines *Plasma Sources Sci. Tech.* **25** 015010
Pécseli H L, Mikkelsen T and Larsen S E 1983 Drift wave turbulence in a low-β plasma *Plasma Phys.* **25** 1173–97
Pécseli H L, Primdahl F and Bahnsen A 1989 Low-frequency electrostatic turbulence in the polar cap E region *J. Geophys. Res.* **94** 5337–49
Pécseli H L, Trulsen J, Primdahl F and Bahnsen A 1993 Propagation and nonlinear interaction of low-frequency electrostatic waves in the polar cap E region *J. Geophys. Res.* **98** 1603–12
Prakash S, Gupta S P, Subbaraya B H and Jain C L 1971 Electrostatic plasma instabilities in the equatorial electrojet *Nature-Physical Sciences* **233** 56–8
Prakash S, Gupta S P and Subbaraya B H 1969 Irregularities in the equatorial E region over Thumba *Radio Sci.* **4** 791–6
Prakash S, Subbaraya B H and Gupta S P 1968 A study of the equatorial E-region during evening twilight using a langmuir probe *J. Atmos. Terrestrial Phys.* **30** 1193–202
Rodrigues F S, Kelley M C, Roddy P A, Hunton D E, Pfaff R F, de La Beaujardiére O and Bust G S 2009 C/NOFS observations of intermediate and transitional scale-size equatorial spread F irregularities *Geophys. Res. Lett.* **36** L00C05
Rose G *et al* 1992 The ROSE project - scientific objectives and discussion of 1st results *J. Atmos. Terr. Phys.* **54** 657–67
Saxena Y C and John P I 1975 Observation of the cross-field instability in laboratory plasma *Geophys. Res. Lett.* **2** 492–5
Shume E B and Hysell D L 2004 Spectral analysis of plasma drift measurements from the AE-E satellite: evidence of an inertial subrange in equatorial spread *F J. Atmospheric Solar-Terrestrial Phys.* **66** 57–65

Sudan R N and Keskinen M 1977 Theory of strongly turbulent two-dimensional convection of low-pressure plasma *Phys. Rev. Lett.* **38** 966–70

Tchen C-M, Pécseli H L and Larsen S E 1980 Strong turbulence in low-β plasmas *Plasma Phys.* **23** 817–29

Villain J P, Beghin C and Hanuise C 1986 ARCAD3-SAFARI coordinated study of auroral and polar F-region ionospheric irregularities *Ann. Geophys.* **4** 61–8

Wu Y-F and Widdel H-U 1992 Observational evidence of local turbulence production due to gravity wave saturation in the summer mesosphere *J. Atmospheric Terrestrial Phys.* **54** 143–51

IOP Publishing

Low Frequency Waves and Turbulence in Magnetized Laboratory Plasmas and in the Ionosphere

Hans Pécseli

Chapter 25

Turbulent diffusion

One of the most important properties of turbulent flows is their ability to disperse particles at a rate which by far exceeds transport by classical molecular diffusion. This property is the same for neutral fluids and for plasmas. We have seen (see section 6.7, in particular discussion of figure 6.4) that during the linear growth phase, the drift wave instability is inherently connected with particle transport along the direction of the density gradient. The question is now how the saturated turbulent fluctuating field (assuming that such a state exists) transports particles. First we make some general remarks on this problem, and present some of the basic results on turbulent diffusion. The outline is independent of the dimensionality of the problem and the ideas apply equally well for three-dimensional and two-dimensional turbulence, given the assumption of homogeneity and isotropy. Particle displacement will be expressed in terms of a velocity, but this velocity can be obtained from an electrostatic electric field in magnetized plasmas as $\mathbf{u}(\mathbf{r}, t) = \mathbf{E}(\mathbf{r}, t) \times \mathbf{B}_0/B_0$ for homogeneous magnetic fields, assuming low frequency turbulence where all relevant frequencies are $\omega \ll \Omega_{ci}$. This model will then assume two-dimensional turbulence in the plane $\perp\mathbf{B}_0$. Many experiments have demonstrated that turbulent transport is an important mechanism for limiting the plasma confinement in, for instance, fusion plasma experiments (Liewer 1985), and significant efforts have been made to understand this mechanism for anomalous transport in the context of drift wave turbulence (Dupree 1967, Misguich *et al* 1987, Horton 1990, 1999).

25.1 Single particle diffusion

Consider the simplest possible problem, namely the one where a single particle is released in a homogeneous and isotropic turbulent velocity field, $\mathbf{u}(\mathbf{r}, t)$, which is for the moment not further specified but for a two-dimensional model it can represent an $\mathbf{E}(\mathbf{r}, t) \times \mathbf{B}_0/B_0$ velocity. We assume that the particle is convected as a passive

doi:10.1088/978-0-7503-1251-6ch25

scalar by the flow, and want to determine its mean-square displacement with respect to the origin of release, which we without loss of generality take to be at the origin of the coordinate system (Roberts 1957).

Since, by assumption, $\mathrm{d}\mathbf{r}(t)/\mathrm{d}t = \mathbf{u}(\mathbf{r}(t), t)$, we have in a given realization of the flow the particle position to be

$$\mathbf{r}(t) = \int_0^t \mathbf{u}(\mathbf{r}(t'), t')\mathrm{d}t'. \tag{25.1}$$

The *average* position $\langle \mathbf{r}(t) \rangle$ vanishes, since we have assumed $\langle \mathbf{u}(\mathbf{r}(t), t) \rangle = 0$. This is actually not quite as self-evident as it might appear (Tennekes and Lumley 1972), since we can assume from the outset only that $\langle \mathbf{u}(\mathbf{r}, t) \rangle = 0$, but this concerns a function of a spatial as well as temporal variables, while $\langle \mathbf{u}(\mathbf{r}(t), t) \rangle$ is a function of time alone. We postpone the discussion of this question.

The mean square displacement is a positive quantity, and we find

$$\begin{aligned} \mathbf{r}(t) \cdot \frac{\mathrm{d}\mathbf{r}(t)}{\mathrm{d}t} = \frac{1}{2}\frac{\mathrm{d}r^2(t)}{\mathrm{d}t} &= \mathbf{u}(\mathbf{r}(t), t) \cdot \int_0^t \mathbf{u}(\mathbf{r}(t'), t')\mathrm{d}t' \\ &= \int_0^t \mathbf{u}(\mathbf{r}(t), t) \cdot \mathbf{u}(\mathbf{r}(t'), t')\mathrm{d}t'. \end{aligned} \tag{25.2}$$

Taking the ensemble average, we have

$$\frac{\mathrm{d}}{\mathrm{d}t}\langle r^2(t) \rangle = 2\int_0^t \langle \mathbf{u}(\mathbf{r}(t), t) \cdot \mathbf{u}(\mathbf{r}(t'), t') \rangle \mathrm{d}t'. \tag{25.3}$$

For time stationary turbulence we require $\langle \mathbf{u}(\mathbf{r}(t), t) \cdot \mathbf{u}(\mathbf{r}(t'), t') \rangle = R_L(t - t')\langle u^2 \rangle$, with the subscript L reminding us to sample the velocity field along a Lagrangian–orbit. We introduced R_L as the normalized Lagrangian velocity correlation function, $R_L(0) = 1$.

When integrating (25.2) as $\langle r^2(t) \rangle = 2\langle u^2 \rangle \int_0^t \int_0^{t''} R_L(t'' - t')\, \mathrm{d}t'\mathrm{d}t''$, it is an advantage to introduce the variables $\tau \equiv t'' - t'$ and $s \equiv t''$, giving the Jacobian

$$J = \begin{vmatrix} \dfrac{\partial \tau}{\partial t''} & \dfrac{\partial \tau}{\partial t'} \\ \dfrac{\partial s}{\partial t''} & \dfrac{\partial s}{\partial t'} \end{vmatrix} = \begin{vmatrix} 1 & -1 \\ 1 & 0 \end{vmatrix} = 1,$$

so that $\mathrm{d}t''\mathrm{d}t' = \mathrm{d}\tau\mathrm{d}s$. Making the variable transforms as indicated, we note that we can change the order of integration as $\int_0^t \int_0^s R_L(\tau)\, \mathrm{d}\tau\mathrm{d}s = \int_0^t \int_\tau^t R_L(\tau)\, \mathrm{d}s\mathrm{d}\tau$ to give

$$\langle r^2(t) \rangle = 2t\langle u^2 \rangle \int_0^t (1 - \tau/t)R_L(\tau)\mathrm{d}\tau. \tag{25.4}$$

Two relevant limiting cases of (25.4) can be distinguished here (Roberts 1957).

1. Very short times, where it can be assumed that $R_L(\tau) \approx 1$. In this limit we find $\langle \mathbf{r}^2(t)\rangle \approx \langle u^2\rangle t^2$, which is often called the *ballistic limit* since it is the result we would have obtained by assuming the particle to follow straight lines of orbit, $\mathbf{r}(t) = \mathbf{u}t$, and simply average over all velocities. We might expect to find auto-correlation functions that are not differentiable for $\tau = 0$, and in such cases the ballistic limit does not exist. Such a case requires that $\langle (\mathrm{d}\mathbf{u}/\mathrm{d}t)^2\rangle$ diverges. With finite particle inertia we would expect that $\mathrm{d}\mathbf{u}/\mathrm{d}t$ should be finite at all times. The absence of a ballistic limit simply indicates a very rapidly changing velocity field, and the interval where $R_L(\tau) \approx 1$ may be negligible.
2. Very large times, $t \to \infty$, where it can be assumed that $\int_0^t R_L(\tau)\mathrm{d}\tau \approx \tau_L$, introducing the Lagrangian integral correlation time

$$\tau_L \equiv \int_0^\infty R_L(\tau)\mathrm{d}\tau. \tag{25.5}$$

In this limit we find the important result

$$\langle r^2(t)\rangle \approx 2t\tau_L\langle u^2\rangle. \tag{25.6}$$

At least formally this looks like the result one obtains by using the classical diffusion equation with a diffusion coefficient $D \equiv 2\tau_L\langle u^2\rangle$ to obtain the mean square particle displacement. Such cases have $\langle \mathbf{r}^2(t)\rangle \sim t$. The limit of times much larger than the Lagrangian correlation time is consequently called the diffusion limit. This limiting case is consistent with a random walk with a typical length step $\tau_L\sqrt{\langle u^2\rangle}$ per time τ_L, consistent with a mixing length model as the one outlined in section 22.1.2.

Note that $\mathbf{r}(t)$ is not a time-stationary random process, although it is derived from $\mathbf{u}(\mathbf{r}, t)$ which can be assumed to be so! The initial time (the time of release of the particle) has a special role for $\mathbf{r}(t)$.

Introducing the Lagrangian power spectrum, $S_L(\omega) \equiv (1/2\pi)\int_{-\infty}^\infty R_L(t)\mathrm{e}^{-i\omega t}\mathrm{d}t$, with $R_L(t) = \int_{-\infty}^\infty S_L(\omega)\,\mathrm{e}^{i\omega t}\mathrm{d}\omega$, we can write the result (25.4) in the form (Lumley and Panofsky 1964)

$$\langle r^2(t)\rangle = t^2\,\langle u^2\rangle \int_{-\infty}^\infty \left(\frac{\sin(\omega t/2)}{\omega t/2}\right)^2 S_L(\omega)\mathrm{d}\omega. \tag{25.7}$$

The function $\sin^2(\omega t/2)/(\omega t/2)^2$ originates from the Fourier transform of the 'triangular' function $1 - \tau/t$ entering the convolution (25.4). We note that $\sin^2(\omega t/2)/(\omega t/2)^2$ (illustrated in figure 25.1) acts as a time-varying filter that emphasizes low frequencies as $t \to \infty$. For large times it is evident that dispersion of the test particle is primarily due to the low frequencies in the Lagrangian spectrum. We have $\lim_{t\to\infty}\sin(\omega t/2)/(\omega/2) = \pi\delta(\omega)$, so we recover $\tau_L = S_L(\omega = 0)$, see also figure 25.2. Note that oscillations in the Lagrangian spectrum with frequencies being multiples of $1/t$ do not contribute to the particle displacement.

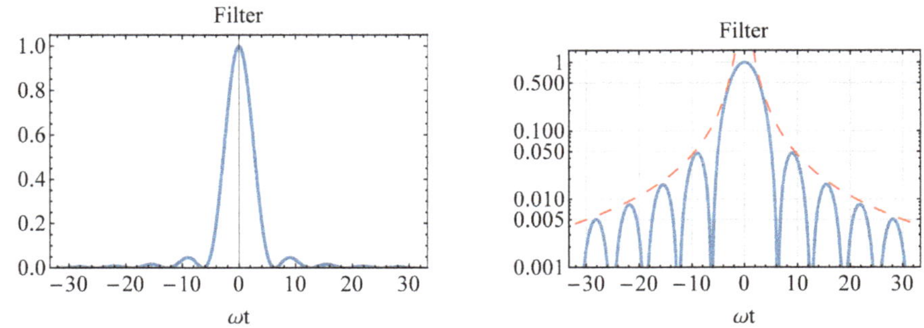

Figure 25.1. Illustration of the filter $\sin^2(\omega t/2)/(\omega t/2)^2$ as a function of ωt shown in linear as well as semi-logarithmic representation. The dashed lines in the latter figure give $\sim(\omega t)^{-2}$ for reference.

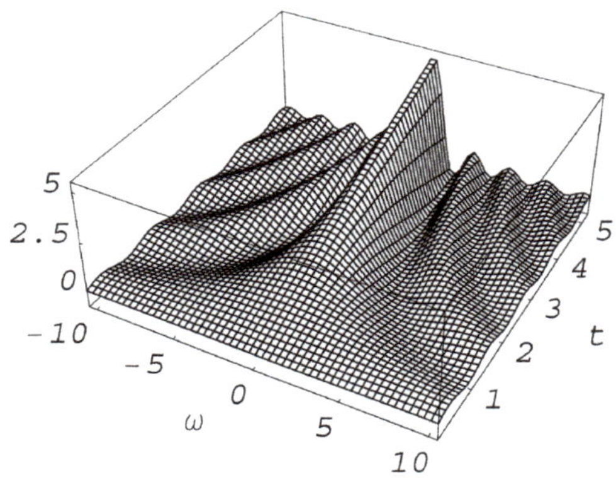

Figure 25.2. Illustration of $\sin(\omega t/2)/(\omega/2)$ as a function of ω and t, demonstrating the transition to a δ-function for $t \rightarrow \infty$.

This is because they return the particle to its starting point after a time t (Batchelor 1949). Often it is assumed that low frequencies in (25.7) correspond to large wavelengths (or rather large scales), but we should be aware that there is no *a priori* reason to believe this.

By (25.7) our problem seems to be solved once and for all, at least for the case of homogeneous isotropic turbulence! Alas, it is not so, since we do not know the spectrum $S_L(\omega)$, and it is very complicated to obtain it experimentally. It has been a major enterprise over the years to find ways of predicting $S_L(\omega)$ on the basis of the more readily measurable *Eulerian* correlation function. Amazingly good results can be obtained, at least as far as predictions of $\langle r^2(t)\rangle$ are concerned. However, this might as well imply that this is a very robust result, and that almost any reasonable guess on $R_L(t)$ will give acceptable results. After all, we *must* require that $R_L(0) = 1$ and that $R_L(t \rightarrow \infty) \rightarrow 0$, and with a little common sense all reasonable guesses of R_L tend to look more or less the same. Historically, one of the first attempts to relate the Eulerian and Lagrangian correlation functions was empirical and very easy to use in practice (Hay and Pasquill 1959). The hypothesis was given analytical support by Wandel and Kofoed-Hansen (1962), but it seems to serve only for qualitative modeling (Pécseli and Trulsen 1997). Also other related studies can be found (Weinstock 1976).

- **Example:** For anisotropic conditions (Crank 1975) we can write quite generally a diffusion equation in the form

$$\frac{\partial}{\partial t}P(\mathbf{r}, t) = D_{ji}(t)\frac{\partial}{\partial r_i}\frac{\partial}{\partial r_j}P(\mathbf{r}, t),$$

with summation convention implied. In general, the diffusion coefficient tensor can be time-dependent. A transformation to rectangular coordinates can be found that transforms this equation to

$$\frac{\partial}{\partial t}P(\mathbf{r}, t) = C_j(t)\frac{\partial^2}{\partial r_i^2}P(\mathbf{r}, t),$$

where $j = 1, \ldots, d$ with d being the dimensionality of the problem. By introducing $P(\mathbf{r}, t) = P(r_1, t)P(r_2, t)\ldots$, this equation can be solved by separation of variables. In the diffusion limit this equation will also account for turbulent diffusion in incompressible flows.

We now let all $C_1 = C_2 = \ldots$ be the same. For the probability density for one of the components, r_j, of the particle positions we have

$$P(r_j, t) = \frac{1}{\sigma_j\sqrt{2\pi}}\mathrm{e}^{-r_j^2/2\sigma_j^2}$$

with

$$\sigma_j^2(t) \equiv \left\langle r_j^2\right\rangle = \frac{1}{d}\langle \mathbf{r}^2\rangle \equiv \frac{\sigma^2(t)}{d},$$

where we do not assume the summation convention when writing r_j^2. We have d to be the dimensionality of the problem: we can have two- or three-dimensional turbulence ($d = 2$ or 3), or imagine a one-dimensional case ($d = 1$), although this will be rather trivial for incompressible flows. In the diffusion limit we have $\sigma_j^2(t) = 2t\tau_L\langle u_j^2\rangle$. The probability density satisfies the one-dimensional diffusion equation

$$\frac{\partial}{\partial t}P(r_j, t) = \frac{1}{2}\frac{\mathrm{d}\sigma_j^2(t)}{\mathrm{d}t}\frac{\partial^2}{\partial r_j^2}P(r_j, t), \tag{25.8}$$

again *without* summation convention.

More generally we can write $Q_2(\mathbf{r}, t) = P(x, t)P(y, t)$ in two, or $Q_3(\mathbf{r}, t) = P(x, t)P(y, t)P(z, t)$ in three spatial dimensions, and have

$$\frac{\partial}{\partial t}Q_d(\mathbf{r}, t) = \frac{K(t)}{d}\nabla^2 Q_d(\mathbf{r}, t),$$

again with d being the dimensionality of the problem, and

$$K(t) = \frac{1}{2}\frac{\mathrm{d}}{\mathrm{d}t}\langle r^2\rangle = \langle u^2\rangle \int_0^t R_L(\tau)\mathrm{d}\tau,$$

with the proper dimensionality implied in the ∇-operator. The point in introducing the notation (25.8) is that more generally we can define $K(t)$ by (25.4), and we then have a diffusion equation which correctly reproduces the mean-square particle displacement. However, the solution for the probability density P may very well be wrong, although it is interesting to note that it will be correct in the ballistic as well as in the diffusion limits ($K = \langle u^2\rangle t$ and $K = \langle u^2\rangle \tau_L$, respectively). In an intermediate range, it turns out to be incorrect, but the solution might be used nonetheless as an approximation, provided the problem does not require fine details to be reproduced accurately.

It turns out that many quite different models give a satisfactory account for the time evolution of $\langle r^2(t)\rangle$ in turbulent flows. The point may be that the initial ballistic and the asymptotic diffusion limits are described by two parameters, $\langle u^2\rangle$, which can be measured quite accurately, and τ_L. The latter scalar parameter is much easier to model than an entire function. The variation of $\langle r^2(t)\rangle$ for the intermediate times has then a limited range of variability, so you cannot go *completely* wrong with models for $R_L(\tau)$.

25.1.1 Eulerian and Lagrangian mean-square velocities

Before closing the present discussion, we should clear up an ambiguity hinted at already: we introduced $\langle u^2\rangle$ in a Lagrangian context. In principle we should distinguish ensemble averages of Eulerian and a Lagrangian velocities. These two are however simply related in cases relevant here. Consider first the Eulerian characteristic function for the fluctuating velocity. which we can define as the ensemble average $\langle \exp(i\mathbf{k}\cdot\mathbf{u}_E(\mathbf{r}, t))\rangle$. The corresponding structure function for the Lagrangian velocity field, as sampled by a self-consistently moving passive test particle is $\langle \exp(i\mathbf{k}\cdot\mathbf{u}_L(\mathbf{r}(t), t))\rangle$. We then integrate the two exponential functions entering the characteristic functions over a (large) volume, V. It is here essential that the volume is well-defined, but for incompressible flows we can assume that the fluid or gas fills the same volume for all times. Selecting a certain reference time, the volume is defined by the same boundaries for both integrals, i.e. we have

$$\frac{1}{V}\int_V \exp(\mathrm{i}\mathbf{k}\cdot\mathbf{u}_E(\mathbf{r}, t))\mathrm{d}\mathbf{r} = \frac{1}{V}\int_V \exp(\mathrm{i}\mathbf{k}\cdot\mathbf{u}_L(\mathbf{r}(t), t))\mathrm{d}\mathbf{r}. \tag{25.9}$$

We now take the ensemble average of both sides of this relation, but in order to have an expression involving the characteristic function, we would like to move the average signs *inside* the integral signs. As far as the Eulerian integral is concerned, we can retain the integration volume to be the same, without further ado, provided the flow is incompressible, as we have assumed from the outset here. Lagrangian integral is problematic, however. Here, the volume is deformed as the flow defining the volume evolves. The *relative* deformation is however, immaterial in the limit where $V \to \infty$, simply because the number of particles leaving the reference volume scales with the *surface* of V (Tennekes and Lumley 1972). Consequently, we can formally consider the surface of this volume as fixed also for the Lagrangian case. In the limit $V \to \infty$ we therefore have

$$\lim_{V\to\infty}\frac{1}{V}\left\langle\int_V \exp(i\mathbf{k}\cdot\mathbf{u}_E(\mathbf{r},t))d\mathbf{r}\right\rangle = \lim_{V\to\infty}\frac{1}{V}\int_V\langle\exp(i\mathbf{k}\cdot\mathbf{u}_E(\mathbf{r},t))\rangle d\mathbf{r}$$
$$=\lim_{V\to\infty}\frac{1}{V}\left\langle\int_V \exp(i\mathbf{k}\cdot\mathbf{u}_L(\mathbf{r}(t),t))d\mathbf{r}\right\rangle = \lim_{V\to\infty}\frac{1}{V}\int_V\langle\exp(i\mathbf{k}\cdot\mathbf{u}_L(\mathbf{r}(t),t))\rangle d\mathbf{r}. \tag{25.10}$$

For homogeneous and isotropic turbulence we have $\langle\exp(i\mathbf{k}\cdot\mathbf{u}_E(\mathbf{r},t)\rangle$ as well as $\langle\exp(i\mathbf{k}\cdot\mathbf{u}_L(\mathbf{r}(t),t)\rangle$ being independent of position (that is the whole idea with homogeneity and isotropy) and we can take these quantities outside the integration. The integrations are then trivially giving the volume, which cancels with the volume in the divisor and we and up with the relation

$$\langle\exp(i\mathbf{k}\cdot\mathbf{u}_E(\mathbf{r},t))\rangle = \langle\exp(i\mathbf{k}\cdot\mathbf{u}_L(\mathbf{r}(t),t))\rangle. \tag{25.11}$$

When the characteristic functions for the Eulerian and the Lagrangian velocities are equal, we trivially have $\langle u_E^2\rangle = \langle u_L^2\rangle$ and we did right in avoiding any subscript in (25.4). This is, however, correct *only* for homogeneous and incompressible turbulence.

The velocity field mediating the turbulent transport is *for absolute diffusion* a stationary random process. The particle displacement is of course steadily increasing in a mean square sense, and is therefore *not* a stationary random process.

- **Exercise:** A model considered by e.g. Carreras *et al* (1996) assumes that in a one-dimensional model the fluctuating fluid velocity and density are characterized by a joint Gaussian probability density

$$P(\tilde{n},\tilde{u}) = \frac{1}{2\pi}\frac{1}{\sigma_n\sigma_u\sqrt{1-\rho^2}}\times\exp\left(-\frac{1}{2(1-\rho^2)}\left(\frac{\tilde{n}^2}{\sigma_n^2}+\frac{\tilde{u}^2}{\sigma_u^2}-2\rho\frac{\tilde{n}\tilde{u}}{\sigma_n\sigma_u}\right)\right),$$

where ρ is a correlation coefficient. Demonstrate that by this model we have a turbulent flux $\tilde{\Gamma}\equiv\tilde{n}\tilde{u}$ with a probability density given by

$$P(\Gamma) = \frac{1}{\pi}\frac{1}{\sigma_n\sigma_u\sqrt{1-\rho^2}}\times K_0\left(\frac{|\Gamma|}{\sigma_n\sigma_u(1-\rho^2)}\right)\exp\left(\frac{\rho\,\Gamma}{\sigma_n\sigma_u(1-\rho^2)}\right),$$

where K_0 is the modified Bessel function of the second kind. Demonstrate that

$$\langle\Gamma\rangle \equiv \int_{-\infty}^{\infty} \Gamma\, P(\Gamma)\mathrm{d}\Gamma = \sigma_n\sigma_u\,\rho.$$

Demonstrate that the flux variance is

$$\sigma_\Gamma^2 \equiv \int_{-\infty}^{\infty} (\Gamma - \langle\Gamma\rangle)^2 P(\Gamma)\mathrm{d}\Gamma = (\sigma_n\sigma_u)^2(1 + \rho^2).$$

Discuss also the skewness and kurtosis of $P(\widetilde{\Gamma})$ and demonstrate that $S = -2\rho(3 + \rho^2)/(1 + \rho^2)^{3/2}$ and $K = 3(3 + 14\rho^2 + 3\rho^4)/(1 + \rho^2)^2$. Give a parametric plot of K versus S.

25.2 Relative diffusion

As an alternative problem to the question of *single* particle diffusion, we might consider *two* particles simultaneously released in a turbulent flow (Roberts 1957, 1961, Sawford 2001). This might sound as a rather trivial generalization, but as we will see this is a much more complicated problem as may have been expected. Single particle diffusion refers to the motion of one particle in a deterministic frame of reference (usually taken to be the fixed laboratory frame). *Relative* particle motion refers to the time variation of the separation of two particles, analyzed in a randomly moving frame of reference!

For the present analysis, we study the time varying separation of particles at positions $\mathbf{r}_1(t)$ and $\mathbf{r}_2(t)$. We introduce the separation coordinate as $\mathbf{y}(t) \equiv \mathbf{r}_1 - \mathbf{r}_2$, we have

$$\begin{aligned}\mathbf{y}(t) &= \mathbf{y}(0) + \int_0^t [\mathbf{u}(\mathbf{r}_1(t'), t') - \mathbf{u}(r_2(t'), t')]\mathrm{d}t' \\ &\equiv \mathbf{y}_0 + \int_0^t \Delta\mathbf{u}(t')\mathrm{d}t'\end{aligned} \tag{25.12}$$

for a given velocity field $\mathbf{u}(\mathbf{r}, t)$. With $\langle\mathbf{u}(t)\rangle = 0$ we have a constant average separation $\langle\mathbf{y}(t)\rangle = \langle\mathbf{y}_0\rangle$, while the mean-square separation can formally be expressed in dyadic form by

$$\begin{aligned}\frac{\mathrm{d}}{\mathrm{d}t}[\mathbf{y}(t)\mathbf{y}(t)] &= \Delta\mathbf{u}(t)\mathbf{y}(t) + y(t)\Delta\mathbf{u}(t) \\ &= \Delta\mathbf{u}(t)\mathbf{y}_0 + \mathbf{y}_0\Delta\mathbf{u}(t) + \int_0^t [\Delta\mathbf{u}(t)\Delta\mathbf{u}(t') + \Delta\mathbf{u}(t')\Delta\mathbf{u}(t)]\mathrm{d}t',\end{aligned} \tag{25.13}$$

giving

$$\frac{\mathrm{d}}{\mathrm{d}t}\langle\mathbf{y}(t)\mathbf{y}(t)\rangle = \int_0^t \langle\Delta\mathbf{u}(t)\Delta\mathbf{u}(t') + \Delta u(t')\Delta\mathbf{u}(t)\rangle\mathrm{d}t'. \tag{25.14}$$

In the short time limit we readily obtain

$$\begin{aligned}\left\langle y_i y_j\right\rangle &= y_{0i}y_{0j} + t^2\left\langle u^2\right\rangle\left(\delta_{ij} - 2\, R_{ij}^E\left(y_0, 0\right)\right) \\ &= y_{0i}y_{0j} + t^2\left\langle u^2\right\rangle\left(\delta_{ij} - 2\, g(y_0)\delta_{ij} - 2\left[f(y_0) - g(y_0)\right]\frac{y_{0i}\, y_{0j}}{y_0^2}\right),\end{aligned} \tag{25.15}$$

where R_{ij}^E is the normalized Eulerian velocity correlation tensor (Batchelor 1953, Roberts 1957), here taken at zero time delay and at spatial separation y_0. For the present two-dimensional case we have $g = \mathrm{d}(yf)/\mathrm{d}y$ and $f(0) = g(0) = 1/2$ for the longitudinal and lateral velocity correlations f and g (Pécseli and Trulsen 1997). In deriving (25.15) it was assumed that the time after release is so short that the two particle velocities can be considered constant, i.e. the ballistic limit. The mean square particle separation is conditional in the sense that $\mathbf{y}_0$ is considered given.

Taking the trace of $\langle \mathbf{y}(t)\mathbf{y}(t)\rangle$ we can write for short times

$$\left\langle y^2(t)\right\rangle = y_0^2 + 2\, t^2\left\langle u^2\right\rangle\left(1 - \left[f(y_0) + g(y_0)\right]\right). \tag{25.16}$$

For arbitrary times we have

$$\begin{aligned}\frac{1}{2}\frac{\mathrm{d}}{\mathrm{d}t}\left\langle y^2(t)\right\rangle &= \int_0^t \langle \Delta\mathbf{u}(t') \cdot \Delta\mathbf{u}(t)\rangle\, \mathrm{d}t' \\ &\equiv \left\langle \Delta u^2(t)\right\rangle \int_0^t \mathcal{R}(t, t')\mathrm{d}t',\end{aligned} \tag{25.17}$$

with $\mathcal{R}(t, t') \equiv \langle \Delta\mathbf{u}(t') \cdot \Delta\mathbf{u}(t)\rangle / \langle \Delta u^2(t)\rangle$. The initial condition for the differential equation (25.17) is $\langle y^2(t)\rangle = y_0^2$.

Introducing $t_{\mathcal{R}}(t) \equiv \int_0^t \mathcal{R}(t, t')\mathrm{d}t'$ and using the definition of $\Delta\mathbf{u}(t)$ we write

$$\frac{\mathrm{d}}{\mathrm{d}t}\left\langle y^2(t)\right\rangle = 4t_{\mathcal{R}}(t)\left(\left\langle u^2\right\rangle - \left\langle \mathbf{u}\left(r_1(t), t\right) \cdot \mathbf{u}\left(\mathbf{r}_2(t), t\right)\right\rangle\right), \tag{25.18}$$

for time stationary turbulence. The relative diffusion of two particles is, on the other hand, a non-stationary process since the initial time (where the particles are chosen to be close) has a special role.

In the limit where the Lagrangian velocity correlation vanishes in (25.18), i.e. for large particle separations when $\langle \mathbf{u}(\mathbf{r}_1(t), t) \cdot \mathbf{u}(\mathbf{r}_2(t), t)\rangle \approx 0$, we find that the relative particle separation proceeds at a rate *twice* the value we would find for a *single* particle dispersing in the same flow, compare with, for instance, (25.4) in the limit $t \to \infty$. We find that the term $\langle u^2\rangle - \langle \mathbf{u}(r_1(t), t) \cdot \mathbf{u}(\mathbf{r}_2(t), t)\rangle$ acts as a 'filter' (Csanady 1973, Misguich *et al* 1987, Mikkelsen *et al* 1987), which allows smaller and smaller frequencies in the Lagrangian velocity frequency spectrum to contribute as $t \to \infty$. Intuitively, we expect that the low frequencies correspond to large scales, or long wavelengths, in the Eulerian wavenumber spectrum, but the direct relation between the two quantities is by no means obvious. For short times, only a modest part of the power spectrum is effective, and the separation $\langle y^2(t)\rangle$ increases slowly. We can make this postulated 'filtering effect' more evident by reformulating (25.18). First we note that $\langle \mathbf{u}(r_1, t) \cdot \mathbf{u}(\mathbf{r}_2, t) \rangle$ is the Fourier transform of the power spectrum $\langle u^2 \rangle \iiint_{-\infty}^{\infty} S(\mathbf{k}) \exp(-\mathrm{i}\mathbf{k} \cdot (\mathbf{r}_1 - \mathbf{r}_2)) \mathrm{d}\mathbf{k}$. The problem here is that the separation $\mathbf{y}(t) \equiv \mathbf{r}_1(t) - \mathbf{r}_2(t)$ is randomly varying and differs from one realization to another even when taken at the same time. We deal with this problem by taking the ensemble average written as

$$\frac{\mathrm{d}}{\mathrm{d}t}\left\langle y^2(t)\right\rangle = 4\, t_R(t)\left\langle u^2\right\rangle \iiint_{-\infty}^{\infty} S(k)(1 - \langle \exp(-\mathrm{i}\mathbf{k} \cdot \mathbf{y}(t))\rangle)\mathrm{d}\mathbf{k}, \qquad (25.19)$$

where the power spectrum $S(k)$ is here normalized for simplicity so that $\iiint_{-\infty}^{\infty} S(k)\mathrm{d}\mathbf{k} = 1$, with the 'usual' assumptions of homogeneity and isotropy of the turbulence. We introduced again $\mathbf{y}(t) \equiv r_1(t) - \mathbf{r}_2(t)$ and implicitly assumed that we could make the approximation $\langle \mathbf{u}_\mathbf{k} \cdot \mathbf{u}_{-\mathbf{k}} \exp(-\mathrm{i}k \cdot (\mathbf{r}_1 - \mathbf{r}_2))\rangle \approx \langle \mathbf{u}_k \cdot \mathbf{u}_{-\mathbf{k}}\rangle \langle \exp(-\mathrm{i}\mathbf{k} \cdot (\mathbf{r}_1 - \mathbf{r}_2))\rangle \equiv S(\mathbf{k})\langle \exp(-\mathrm{i}\mathbf{k} \cdot (\mathbf{r}_1 - \mathbf{r}_2))\rangle$. We recognize $\langle \exp(-\mathrm{i}\mathbf{k} \cdot \mathbf{y}(t))\rangle$ as the characteristic function for the probability density of particle separations $\mathbf{y}(t)$. By the 'cumulant discard assumption' (Misguich *et al* 1987) we can make the further simplification $\langle \exp(-\mathrm{i}\mathbf{k} \cdot \mathbf{y}(t))\rangle \approx \exp(-k^2\langle y^2(t)\rangle)$, see also section 20.1.2. In effect, we keep the cumulant we would obtain for a Gaussian random process with zero mean, which seems to be a fair assumption for the non-stationary particle displacement in the limit where $\langle y^2(t)\rangle \gg y_0^2$. For this symmetric form we can rewrite (25.19) as

$$\frac{\mathrm{d}}{\mathrm{d}t}\left\langle y^2(t)\right\rangle = 16\pi\, t_R(t)\left\langle u^2\right\rangle \int_0^{\infty} k^2 S(k)\left(1 - \exp\left(-k^2\left\langle y^2(t)\right\rangle\right)\right)\mathrm{d}k, \qquad (25.20)$$

where $S_1(k) \equiv 4\pi k^2 S(k)$ is the effective one-dimensional wavenumber spectrum. For the equivalent two-dimensional problem we would have $2\pi k S(k)$ to the effective spectrum, see also the discussion in section 19.2.9.

By inspection we find that the parentheses in the integrands of (25.19) or (25.20) acts as a 'wavenumber filter' emphasizing only large wavenumbers, i.e. small wavelengths, the transition wavenumber being $k \sim 1/\sqrt{\langle y^2(t)\rangle}$. Physically, this is precisely what we would expect: large scales (i.e. small k) move both particles together, while separations are induced only by scales smaller than or comparable to the separation distance. Note that the filter is time-dependent! This is a not a time-stationary process. As the separation $|\mathbf{y}(t)|$ between the two particles increase, we find that scales that a little while ago were larger than the separation, and therefore moved both particles simultaneously, now become smaller and contribute to the separation instead.

We have no analytical exact results for the Lagrangian–time-scale $t_R(t)$ for relative motion, but it is amenable for modeling (Misguich *et al* 1987, Mikkelsen *et al* 1987). For very short times we have $\langle \Delta\mathbf{u}(t') \cdot \Delta\mathbf{u}(t)\rangle / \langle \Delta u^2(t)\rangle \approx 1$, and $t_R(t) \equiv \int_0^t \mathcal{R}(t, t')\mathrm{d}t' \approx t$ in this limit. More generally we can argue that Buckingham's Π determinant must have the form

	ν	ϵ	t_R	t	y_0
T	-1	-3	1	1	0
L	2	2	0	0	1

In separations in the inertial subrange we expect viscosity to be immaterial, and ν should not appear in the expression for t_R. Also we expect that after some time the initial separation y_0 should be of little importance for the variation of $\langle y^2(t)\rangle$. Consequently we have no terms to cancel the spatial dimension of ϵ, so this quantity cannot enter explicitly in the result for t_R either! For short times, $t \ll t_L$, the Lagrangian–time scale, we have only one quantity of dimension *time* available, namely time itself, and the only logical approximation in this time range is consequently

$$t_R(t) \approx t, \tag{25.21}$$

apart from a possible dimensionless constant (Mikkelsen *et al* 1987, Misguich *et al* 1987). We have argued, however, that for very small times we should have $t_R(t) = t$, so the numerical constant consequently equals unity! For very large times $t \gg t_L$, on the other hand, we expect that $t_R(t)$ approaches a constant value. A model encompassing a wide time-range was suggested by Misguich *et al* (1987).

As a short time simplification of (25.19) we find with these approximations that

$$\frac{\mathrm{d}}{\mathrm{d}t}\left\langle y^2(t)\right\rangle = 4\,t\left\langle u^2\right\rangle \int_0^{\infty} S_1(k)\left(1 - \exp\left(-k^2\left\langle y^2(t)\right\rangle\right)\right)\mathrm{d}k. \tag{25.22}$$

This result applies for two- and three-dimensional turbulence, with the assumptions of homogeneity and isotropy.

Results for special cases

We can solve (25.22) for some special cases (Mikkelsen *et al* 1987). We can postulate a power law spectrum $\langle u^2\rangle S_1(k) = Ak^{-p}$ and study various values for the exponent p. We are aware, of course, that a power law spectrum as the one indicated cannot cover all of k-space, in particular not the region $k \approx 0$, but we note that the filter function in (25.22) in effect quenches small wavenumbers, so there are reasons to hope that such details are of minor consequences at least for *some* range of p-values.

i) $1 < p < 3$. We find here $\langle y^2(t)\rangle = (ct^2 + \sigma_0^{1/q})^{2q}$.

ii) $p = 3$. For this case $\langle y^2(t)\rangle = \sigma_0^2 \exp(A\langle u^2\rangle t^2)$. There seems to be general consensus concerning $p \approx 3$ to be appropriate for the fluctuating electric fields in drift wave spectra (Tchen *et al* 1980, Pécseli *et al* 1983). If we in the plane perpendicular to the magnetic field approximate the fluctuating velocity by $\widetilde{\mathbf{E}} \times \mathbf{B}_0/B_0^2$, we note that this power law index $p = 3$ has special interest for two particle separations induced by drift wave turbulence.

iii) $p > 3$. For this case the integral in (25.22) will diverge, so we have to sharpen the filter function to become a Heaviside's step function $H(k - \langle y^2(t)\rangle^{-1/2})$. With this simplification we find the result $\langle y^2(t)\rangle = (\tilde{c}t^2 + \sigma_0^{1/q})^{2q}$.

In all cases we have $q = 1/(3-p)$, $\tilde{c} = -\langle u^2\rangle A(3-p)/(1-p)$, $c = \tilde{c}\,\Gamma(\frac{1}{2}(3-p))$ and $\sigma_0 = \sqrt{\langle y^2(t=0)\rangle}$. Case (iii) above has an explosive solution by diverging at $t = t_{max} \equiv \sigma_0^2/|\tilde{c}|$: no matter how interesting such a solution might be from a purely mathematical point of view, this result has little practical relevance since any realistic spectrum will be 'flat' for small k, and the integral in (25.22) has to be analyzed with greater care for times comparable to t_{max}. The importance of the spectral form for the time dependence of relative diffusion was noted by Roberts (1957).

As a test of the foregoing analysis, we can use the Kolmogorov–Oubokhov spectrum $\sim\epsilon^{2/3}k^{-5/3}$ i.e. $p = 5/3$ and $q = 3/4$, with $c \sim \epsilon^{2/3}$ in example (i). As a result we find $\langle y^2(t)\rangle = (C_0\epsilon^{2/3}t^2 + \sigma_0^{4/3})^{3/2}$ with C_0 being a constant related to the Kolmogorov constant in some unspecified manner. For large times this gives the well known Richardson law $\sim C_0^{3/2}\epsilon t^3$ for relative particle motion in incompressible turbulence (Richardson 1926). This is a comforting result: after all, we made several simplifying assumptions on the way and we have no guarantee that the final result is robust!

- **Exercise:** Derive the result (25.19).

The analysis outlined here assume the power spectrum of velocity fluctuations to be known. In neutral incompressible flows this will be a standard measurement but for plasma applications in general it will be a quantity not easily measured. For two-dimensional turbulence in magnetized plasmas we might associate a velocity component perpendicular to the magnetic field by assuming $\mathbf{u} = \mathbf{E}\times\mathbf{B}/B^2$ for frequencies well below the ion cyclotron frequency. By this approximation, we can analyze the turbulent transport in the plane $\perp\mathbf{B}$ using the previous expressions. This model has been widely applied, although it could be comforting to see a test of the importance of the ion polarization drift. The direction along $\mathbf{B}$ requires a different analysis which can become necessary in a transient period, if we consider release of a small cloud of contaminant in the magnetized turbulent plasma. This could be by release of a neutral barium cloud in the Earth's ionosphere (Goldman *et al* 1974) or by a small pellet of a deuterium-tritium mixture injected into a hot fusion plasma (Chang 1991). In both cases, the material will be ionized and expand rapidly along magnetic field lines to fill a magnetic flux tube. The turbulent transport will be dominating the direction $\perp\mathbf{B}$.

25.2.1 Model equations for relative diffusion

The analysis summarized before was restrictive by addressing only the variation of $\langle y^2(t)\rangle$. More generally, we would like a result for the entire probability density for the two-particle separations $P(y, t)$, denoting the distance-neighbor function by Richardson (1926, 1929). In his original work Richardson (1929) suggested a differential-equation, making the *a priori* assumption that it had to have the form of a diffusion equation, here written in spherical coordinates

$$\frac{\partial}{\partial t}P(y, t) = \frac{1}{y^2}\frac{\partial}{\partial y}y^2 K\frac{\partial}{\partial y}P(y, t), \tag{25.23}$$

where K is some diffusion coefficient. (Actually Richardson (1926) used a one-dimensional argument (Ott and Mann 2000), but that does not change his arguments in any significant way.) Richardson (1929) then argued that K is proportional to $y^{4/3}$. It was, however, noted by Batchelor (1952a) that the argument Richardson (1926) was restrictive by making K depend on the separation variable only. Relative diffusion is a non-stationary process, and a diffusion equation with a *time varying* diffusion coefficient could be justified as well! A model with K being proportional to t^2 could be argued, and by this we could place K outside the spatial differentiations making the diffusion equation much simpler! It can then readily be solved to give a Gaussian form for $P(y, t)$. The controversy between these two models was not properly resolved before an experimental study by Ott and Mann (2000).

Generally it can argued by dimensional reasoning that for the inertial subrange we must have

$$K = \epsilon^{1/3} y^{4/3} F\left(t\epsilon^{1/3} y^{-2/3}\right),$$

where F is some dimensionless function of a dimensionless variable. This expression contains Richardson's argument as a special limit with $F \approx$ const, while the limit suggested by Batchelor (1952a) is found when $F(s) \sim s^2$ so that the y-dependence of K cancels. With some algebra (Ott and Mann 2000) it can be shown that (25.23) has the general solution

$$P(y, t) = \frac{1}{\epsilon^{3/2} t^{9/2} N} G\left(t\epsilon^{1/3} y^{-2/3}\right),$$

where

$$G(s) = \exp\left(\frac{9}{4} \int_1^s \frac{\mathrm{d}x}{x^2 F(x)}\right).$$

We introduced a normalizing constant N. If we now calculate the mean square particle separation from this $P(y, t)$, we find the Richardson law

$$\langle y^2(t) \rangle = C_R \epsilon t^3,$$

with $C_R = (6\pi/N) \int_0^\infty s^{-17/2} G(s) ds$ being some universal numerical constant. The precise value for this Richardson constant was for a long time unknown, but recent detailed laboratory experiments by Ott and Mann (2000) indicated that $C_R = 0.5 \pm 0.2$.

We thus found that it does not matter which one of the model equations (Richardson (1926, 1929) or Batchelor (1952a)) is used, the mean-square separations obtained come out the same! To distinguish the two model equations, the full space-time variation of the distance-neighbor function has to be determined, and this is a much greater challenge than finding $\langle y^2(t) \rangle$. Experimental results by Ott and Mann (2000) clearly favor the model suggested by Richardson (1926) (see also Richardson (1929)), but the agreement is best for large values of the normalized variable $y^{2/3}/(\epsilon t^3)^{1/3}$. The model suggested by Batchelor (1952a) is, however, clearly inapplicable.

25.3 Models for center-of-mass diffusion

For studying relative motion of two particles we introduced the separation coordinate $\mathbf{y}(t) \equiv \mathbf{r}_1 - \mathbf{r}_2$. Modeling a release of a cloud of contaminants we can let the separation of a pair of particles represent the temporal evolution of the width of the cloud. This information is important by allowing at least a crude estimate of the local average cloud concentration since we will argue that for incompressible flows the cloud volume is conserved and hence the average concentration should scale as the diameter of the cloud to the power −1/3.

The question of the center-of-mass $\mathbf{z}(t) \equiv \frac{1}{2}(\mathbf{r}_1 + \mathbf{r}_2)$ can, however, be just as interesting. The information of $\mathbf{z}(t)$ allows an estimate for the position of a cloud. We have trivially that $\langle \mathbf{z}(t) \rangle = 0$ for isotropic homogeneous turbulence. For the mean square average position we find

$$\langle z^2(t) \rangle = \frac{1}{2}\langle r^2(t) \rangle + \frac{1}{2}\langle \mathbf{r}_1(t) \cdot \mathbf{r}_2(t) \rangle .$$

If the two particles move independently, i.e. $\langle \mathbf{r}_1(t) \cdot \mathbf{r}_2(t) \rangle = 0$, we have their mean square separation to be half the mean square displacement of the single randomly moving particle. The correction due to correlated particle motions, i.e. the term $\langle \mathbf{r}_1(t) \cdot \mathbf{r}_2(t) \rangle$, is found in principle by precisely the same analysis as the one summarized in section 25.2. The discussion will be relevant for contaminants released in the neutral atmosphere but the problem also has plasma applications as mentioned before.

25.4 Analytical expressions

Given the basic properties and expressions for turbulent transport as summarized in sections 25.1 and 25.2, we can attempt to find analytical expressions for the diffusion coefficients. For homogeneous and isotropic conditions in two spatial dimensions, a general summary was given by Misguich *et al* (1987). It is not possible to predict anything 'out of the blue', so the problem is essentially to find out what we need to know, in order to make a prediction of the diffusion coefficient discussed in e.g. Section 25.1. This *a priori* information should be obtainable experimentally, in order for the problem to make sense. The simplest information is of course $\langle u^2 \rangle$, which can obtained by relatively simple measurements. We can argue that the $k^{-5/3}$ power spectrum can be assumed to be known from first principles, but unfortunately, these pieces of information alone are of little help. In part because the energy dissipation rate ϵ cannot be deduced from $\langle u^2 \rangle$ alone, but also because the universal, short wavelength, part of the spectrum is unlikely to contain *all* the information we need. Thus, the diffusion limit of turbulent transport is controlled by the smallest frequencies of the Lagrangian spectrum, which we expect to be related to *long* wavelengths, in some possibly not-so-simple manner. We could argue that knowing the Lagrangian integral time scale would do the trick at least in the diffusion limit, but in order to obtain this we had to follow a large number of particles being transported by the turbulent velocity fluctuations, and then obtain the velocity correlation function along their orbit. Although this is in principle possible by modern techniques (Ott and Mann 2000), the idea misses the point: having done this we *have* solved the entire problem of turbulent diffusion, at least for that given flow, and the point was to make a usable prediction on the basis of some reduced information. We can try to discuss what is easily available from an experimental point of view: we have already mentioned $\langle u^2 \rangle$, and we can claim that as long as Taylor's hypothesis (or the assumption of a frozen flow) is applicable, we can also obtain the Eulerian wavenumber spectrum. The Eulerian frequency spectrum might also be available, but probably not quite as easily. On the basis of this information, one might like to know to what accuracy we can predict a mean square particle displacement? This means in reality that we want to predict some properties of the Lagrangian averages on the basis of the measured Eulerian averages, so the problem is often formulated in a more general way, as the question of the Eulerian–Lagrangian transformation.

25.4.1 Corrsin's hypothesis

One of the most widely made assumptions concerning the relation between some Eulerian and some Lagrangian averages is the so called Corrsin hypothesis (Corrsin 1959). Its starting point is an exact tensorial relation for the Lagrangian velocity components for fluid particles taken to have their position at the origin at time t'

$$\begin{aligned} R_{i,j}^L(t) &= \left\langle u_i\left(\mathbf{r}(t'), t'\right) u_j\left(\mathbf{r}(t'+t), t'+t\right)\right\rangle \\ &= \left\langle \int u_i(\mathbf{r}(t_0), t_0) u_j(\mathbf{r}, t)\delta(\mathbf{r} - \mathbf{R}(t))\mathrm{d}\mathbf{r} \right\rangle \\ &= \int \left\langle u_i\left(\mathbf{r}(t_0), t_0\right) u_j\left(\mathbf{r}, t\right)\delta\left(\mathbf{r} - \mathbf{R}(t)\right)\right\rangle \mathrm{d}\mathbf{r}, \end{aligned} \tag{25.24}$$

with $\mathbf{R}(t = t_0) = \mathbf{r}_0$, and the integral is over all space. After averaging, the initial position does not enter for homogeneous turbulence. It was argued by Corrsin (1959) that at least for large times, we might expect it to be possible to let the average of the product in (25.24) to be approximated by the product of two averages as

$$\int \left\langle u_i\left(\mathbf{r}(t_0), t_0\right) u_j\left(\mathbf{r}, t\right)\delta\left(\mathbf{r} - \mathbf{R}(t)\right)\right\rangle \mathrm{d}\mathbf{r} \approx \int \left\langle u_i\left(\mathbf{r}(t_0), t_0\right) u_j\left(\mathbf{r}, t\right)\right\rangle \langle \delta(\mathbf{r} - \mathbf{R}(t))\rangle \mathrm{d}\mathbf{r}.$$

We have the identity $P(\mathbf{r}, t) = \langle \delta(\mathbf{r} - R(t))\rangle$, with $P(\mathbf{r}, t)$ being the probability density of particle positions at t. We can then write

$$R_{i,j}^L(t) = \int \left\langle u_i\left(\mathbf{r}', t'\right) u_j\left(\mathbf{r}' + \mathbf{r}, t' + t\right)\right\rangle P(\mathbf{r}, t)\mathrm{d}\mathbf{r}. \tag{25.25}$$

This seemingly reasonable argument (presented in a few sentences in an also otherwise short paper) has had a great significance. It has been widely applied (although not always given the proper reference!), and the results tested to give no little confidence in the applicability of the approximation (Peskin 1974). Often the argument is carried one step further: since the approximation is expected to be valid for large times only, we might as well approximate the probability density of particle positions with a Gaussian, since we know this to be a generic form for diffusion processes, which we expect to account for the diffusion limit at large times (that is why it is called the diffusion limit!). This additional assumption thus implies $P(\mathbf{r}, t) \approx (2\pi Dt)^{-d/2} \exp(-r^2/(2Dt)^2)$, with $D = 2\tau_L\langle u^2\rangle$, and d being the dimensionality of the problem.

Given (25.25), we can sum the diagonal tensor elements (using the summation convention) to have $R^L_{i,i}(t)$ expressed in terms of $R_E(\mathbf{r}, t) \equiv R^E_{i,i}(r, t) \equiv \langle u_i(\mathbf{r}', t')u_i(\mathbf{r}' + \mathbf{r}, t' + t)\rangle$, and introduce the velocity power spectrum in wavenumbers by $R_E(\mathbf{r}, t) = \int S_E(\mathbf{k}, t)\exp(-i\mathbf{k} \cdot \mathbf{r})dk$. We then find

$$R_L(t) = \int S_E(\mathbf{k}, t) \int P(\mathbf{r}, t)e^{-i\mathbf{k}\cdot\mathbf{r}}\mathrm{d}\mathbf{r}\mathrm{d}\mathbf{k}. \tag{25.26}$$

In (25.26) we recognize the characteristic function for the particle position probability density. With the simple Gaussian model advocated before we can further simplify (25.26) as

$$R_L(t) = \int S_E(\mathbf{k}, t)e^{-\frac{1}{2}\mathbf{k}^2\langle\mathbf{r}^2(t)\rangle}\mathrm{d}\mathbf{k}, \tag{25.27}$$

where $\langle\mathbf{r}^2(t)\rangle$ is the mean square particle displacement, assuming $\langle\mathbf{r}(t)\rangle = 0$ consistent with the simple Gaussian form. The version (25.27) is probably the form most often used for representing Corrsin's hypothesis, but we emphasize that it is somewhat more restrictive than the original version (25.25).

Using (25.3), we can introduce $\mathrm{d}^2\langle\mathbf{r}^2(t)\rangle/\mathrm{d}t^2 = 2R_L(t)$ and obtain one closed equation for $\langle\mathbf{r}^2(t)\rangle$. This is the most important consequence of Corrsin's hypothesis!

It seems generally agreed that for *long* times, Corrsin's hypothesis has to be applicable, but unfortunately, it is still not clear *how* long this time has to be! Numerical simulations by Peskin (1974) carried out in two spatial dimensions, demonstrated a remarkable agreement between the directly obtained Lagrangian correlation function and the one derived by (25.25). In particular it was maintained that a slight improvement of the agreement could be noticed if the 'true' probability density $P(\mathbf{r}(t))$ was used, instead of its limiting Gaussian model. Experimental results (Ott and Mann 2005) on the other hand indicate that Corrsin's hypothesis is not applicable for times of interest, and that the good results obtained by its use appear coincidental! The question seems not to be fully resolved yet.

Various extensions of the Corrsin hypothesis were reviewed by Misguich *et al* (1987), in particular also the renormalized model suggested by Dupree (1967). The basic starting point in almost all models is the separation of the average of the product by a product of averages: this being the basic form for Corrsin's hypothesis. The models differ in the modeling of the probability density for the particle displacements. For turbulence modeling by a random superposition of self-consistently moving coherent forms or structures it was argued by Pécseli and Trulsen (1997) that in two spatial dimensions special cases can be found where the hypothesis may be exact.

An example of a result obtained by Corrsin's hypothesis

Taylor and McNamara (1971) investigated turbulent plasma transport in a plane perpendicular to an externally imposed strong magnetic field. They identified the cross-field plasma velocity by the $\mathbf{E} \times \mathbf{B}/B^2$-velocity and found by use of Corrsin's hypothesis

$$\frac{1}{2}\frac{\mathrm{d}^2}{\mathrm{d}t^2}\langle r^2(t)\rangle = \int_a^b \left\langle \left|\frac{E(k)}{B}\right|^2 \right\rangle \mathrm{e}^{-\frac{1}{2}k^2\langle r^2(t)\rangle} k\,\mathrm{d}k \tag{25.28}$$

where a and b are introduced to limit the wavenumber range of electric field fluctuations. The power spectrum of the velocity fluctuation is written as $\langle |E(k)/B|^2\rangle$, and the problem is not fully defined before this spectrum is given. Taylor and McNamara (1971) considered an energy conserving random flow in statistical equilibrium, rather than a driven-dissipative turbulent condition. The result by Taylor and McNamara (1971) is closely related to Corrsin's result, but somewhat more restrictive by ignoring the possible time-variation of the wavenumber spectrum (most generally we have the Eulerian spectrum being of the form $S_E(\mathbf{k}, \omega)$). From (25.28) one readily finds

$$\frac{1}{2}\left(\frac{\mathrm{d}\langle r^2(t)\rangle}{\mathrm{d}t}\right)^2 = \int_a^b \left\langle \left|\frac{E(k)}{B}\right|^2 \right\rangle \left(1 - \mathrm{e}^{-\frac{1}{2}k^2\langle r^2(t)\rangle}\right)\frac{\mathrm{d}k}{k}. \tag{25.29}$$

With the definition $D = \lim_{t\to\infty} \mathrm{d}\langle r^2(t)\rangle/\mathrm{d}t$ and with $\exp(-\frac{1}{2}\langle r^2(t)\rangle k^2) \to 0$ for $t \to \infty$ we readily have

$$D = \frac{\sqrt{2}}{B}\sqrt{\int_a^b \langle |E(k)|^2\rangle \frac{\mathrm{d}k}{k}}. \tag{25.30}$$

Provided the power spectrum of the electric field fluctuations is independent of the magnetic field, we readily find that (25.30) contains the *Bohm scaling* of the diffusion coefficient with B, see also (3.18). For plasmas in thermal equilibrium Taylor and McNamara (1971) found that $\langle |E(k)|^2\rangle \sim \kappa T/(1 + (k\lambda_D)^2)$, independent of B, giving the important observation that the seemingly anomalous Bohm diffusion was actually the *natural* one for plasmas in thermal equilibrium. This was unpleasant news indeed, but fortunately (for fusion plasma confinement) many experiments have found ways around this scaling.

Before the results of Taylor and McNamara (1971) the Bohm-scaling for the anomalous turbulent diffusion was not taken seriously by everybody: after all, it was argued, in its original version the result was based on nothing but a dimensional analysis of a simplified parameter set, and a numerical factor was unaccounted for, in particular. The first analytical basis for the Bohm-scaling gave rise to a series of analytical, numerical and experimental investigations. The studies of, in particular, Dawson *et al* (1971) and Okuda and Dawson (1973) confirmed the Bohm-scaling and outlined also its restrictions. It was thus found that there are three regions: for sufficiently weak magnetic fields the diffusion coefficient is the classical one with $D_\perp$ scales like $1/B^2$. For moderate magnetic fields characterized by $\omega_{ce} \approx \omega_{pe}$ the diffusion rate is enhanced and $D_\perp$ is almost independent of B. Finally, for very large fields, where $\Omega_{ci} > \Omega_{pi}$, the diffusion coefficient scales like $1/B$. The enhanced diffusion at moderate and high magnetic fields is dominated by collective modes; i.e. by thermally excited convective modes (Shukla *et al* 1984). Theory and simulations were in good agreement. It was also shown that the diffusion coefficient behaves essentially the same way for a three-dimensional plasma when the magnetic field lines are closed.

25.5 Elongation of a contour

Turbulent diffusion problems can be formulated in many different ways: we have already met the problem of single particle and two-particle relative diffusion. One might also consider the process where an infinitely flexible *string* is released in the turbulence. Formally, this might be realized by taking an initial condition, where particles are placed densely along a contour (not necessarily a straight line) initially at $t = 0$. This will be the way the problem would be treated in a numerical study (Krane *et al* 2003). As long as the relative particle separations are *all* smaller than the inner scale (let this be in two or three spatial dimensions) we might treat the line as continuous. In particular, we can consider the elongation of a vortex-line as a special case of this problem (in three spatial dimensions).

A phenomenological model for the problem of the elongation of a vortex line in three spatial dimensions was suggested by Chorin (1994). The essential point of the argument is that vorticity is most likely to change in places where the vorticity is strong already: vortex lines stretch vortex lines. One can model the basic ideas outlined here by

$$\frac{\mathrm{d}|\boldsymbol{\xi}(t)|}{\mathrm{d}t} = b(t)|\boldsymbol{\xi}(t)|, \tag{25.31}$$

where $\boldsymbol{\xi}$ is the vorticity vector and $b(t)$ is some random coefficient. We can write the solution of (25.31) as

$$\log|\boldsymbol{\xi}(t)| - \log|\boldsymbol{\xi}_0| = \int_0^t b(\tau)\mathrm{d}\tau. \tag{25.32}$$

For large time separations $t_1 - t_2$ it can be assumed that $b(t_1)$ and $b(t_2)$ are independent. By the central limit theorem we would then expect that the right-hand side of (25.32) should be a Gaussian random variable when t is large. This will imply that $\log|\boldsymbol{\xi}|$ is a Gaussian random variable, or consequently that $|\boldsymbol{\xi}|$ is log-normal variable. Since $|\boldsymbol{\xi}|$ is inversely proportional to the cross-section of the vortex filament, this cross-section should then *also* have a log-normal distribution. These results, obtained here by rather basic arguments, seem to have experimental support (Chorin 1994).

Elongation of a line-element in a turbulent flow was studied analytically by Batchelor (1952b). An exponential elongation $\sim\exp(\zeta t)$ was obtained, with $\zeta = c\sqrt{\epsilon/\nu}$ where c is a universal constant of order unity, implying that ζ is proportional to and of the same order of magnitude as the root-mean-square of the average Eulerian spatial velocity derivatives (Batchelor 1952b). Numerical simulations based on two different realizations of two-dimensional turbulence (Krane *et al* 1996, 2003) demonstrated a very clear asymptotic exponential elongation of a macroscopic closed contour: this is a problem *related* to that analyzed by Batchelor (1952b), but not the same.

25.6 Eulerian and Lagrangian dynamics in shear flows

Here we will illustrate some consequences of relaxing the assumption of homogeneity of the turbulent fluctuations by considering shear flows (Hinze 1975). The velocity field then becomes in general anisotropic and inhomogeneous as well, and this has significant implications for description of the turbulent transport in such flows. For inhomogeneous flows we may have, for instance, $\langle \mathbf{U}(\mathbf{r}, t)\rangle \neq 0$ and depending explicitly on position $\mathbf{r}$. We might have $\langle \mathbf{U}(\mathbf{r}, t)\rangle = 0$ also for such conditions, and the inhomogeneity may be manifested in different ways, for instance by $\langle U^2(\mathbf{r}, t)\rangle$ depending explicitly on position and also more generally time. Here we assume the Eulerian flow to be a possibly anisotropic, inhomogeneous but time stationary random process, where $\langle \mathbf{U}(\mathbf{r}, t)\rangle \neq 0$ implying that we have a velocity shear flow.

We let the i-th component of the Eulerian velocity field be described by

$$u_i(\mathbf{r}, t) = \overline{u_i(\mathbf{r}, t)} + \tilde{u}_i(\mathbf{r}, t), \tag{25.33}$$

with $\overline{u}_i$ denoting the time average of u_i. The time averaged flow $\overline{u}_i$ is independent of time, but may depend on position $\mathbf{r}$. We may indicate this by writing $\overline{u_i(\mathbf{r}, t)} = \overline{u}_i(\mathbf{r})$. An arbitrary component of the velocity vector $\mathbf{u}$ is denoted by u_i. Evidently we have $\overline{u}_i = 0$. Ensemble averages will be denoted by $\langle\ \rangle$ as usual. With the given assumptions, we allow for $\mathbf{u}(\mathbf{r}, t)$ being a spatially inhomogeneous and anisotropic random process.

We also introduce the Lagrangian velocity component $V_i(\mathbf{r}_0, t)$ as detected by a particle that was at the position $\mathbf{r}_0$ at time $t = 0$, and use the subscript $_0$ as a label. At a later time t, this particle will be at a position vector with components $\mathbf{r}$, where the Eulerian velocity component is $u_i(\mathbf{r}, t)$. We can express this by writing

$$V_i(\mathbf{r}_0, t) = u_i(\mathbf{r}, t). \tag{25.34}$$

The distance traveled by the particle in the time interval $\{0; t\}$ is then

$$Y_k(\mathbf{r}_0, t) \equiv x_k - x_{0k} = \int_0^t V_k(\mathbf{r}_0, \tau)\mathrm{d}\tau. \tag{25.35}$$

There are two possibilities for the time variation of $v_i(\mathbf{r}_0, t)$. One is because at each point the Eulerian velocity varies with time. The other is because $\overline{u}_i$ as well as u_i varies due to the inhomogeneity of the time-averaged flow field (Hinze 1975).

We can recall here the analysis and comparison of $\langle \exp(\mathrm{i}\mathbf{k} \cdot \mathbf{u}_E(\mathbf{r}, t))\rangle$ and $\langle \exp(\mathrm{i}\mathbf{k} \cdot \mathbf{u}_L(\mathbf{r}(t), t))\rangle$ for Eulerian and Lagrangian velocities in section 25.1.1. As evident from that analysis (Tennekes and Lumley 1972), it was essential that the flow was homogeneous in order to obtain $\langle u_E^2\rangle = \langle u_L^2\rangle$. This will in general not be the case for shear flows, as we have seen.

25.6.1 Turbulent diffusion in a linear shear flow

As discussed before we have a fairly good basic understanding of turbulent transport in homogeneous isotropic turbulence, see chapter 25. It is evident that one of the basic features of these turbulent conditions is that the turbulence is time stationary no matter whether it is sampled in an Eulerian or a Lagrangian frame: if it is non-stationary in the Lagrangian frame, then we cannot uniquely define the Lagrangian–time-scale (25.5) because both times t and t' will appear explicitly and not merely the difference $t - t'$. In shear flows we have found that generally the turbulence is inhomogeneous and anisotropic, and seemingly we can give up all of our previous understandings of turbulent diffusion! There is however a special case, already discussed in section 22.4.2. We found there that for a linear shear flow we *can* have that the turbulence is homogeneous and isotropic. Turbulent diffusion under these conditions can be assigned a diffusion coefficient as in chapter 25 and the resulting transport will be due to a combination of the turbulent diffusion and the bulk shear flow. It is of course also evident that an ideal linear shear flow cannot be realized, there *must* be boundaries somewhere, but we can hope for locally linear shear flows and locally homogeneous isotropic turbulence! This idealization is not so different from what we have to assume anyhow.

To illustrate turbulent transport in homogeneous turbulence in linear shear flows with homogeneous turbulence (Tennekes and Lumley 1972) we consider the Eulerian test particle displacements

$$Y_1(\mathbf{r}_0, t) - x_{01} = \int_0^t \left(Y_2(\mathbf{r}_0, t)\frac{u_0}{\mathcal{L}} + \tilde{v}_1(\mathbf{R}(\mathbf{r}_0, \tau), \tau) \right) \mathrm{d}\tau \tag{25.36}$$

$$Y_2(\mathbf{r}_0, t) - x_{02} = \int_0^t \tilde{v}_2(\mathbf{r}_0, \tau)\mathrm{d}\tau \tag{25.37}$$

$$Y_3(\mathbf{r}_0, t) - x_{03} = \int_0^t \tilde{v}_3(\mathbf{r}_0, \tau)\mathrm{d}\tau. \tag{25.38}$$

The Lagrangian velocities $\tilde{v}_j$ constitute a stationary random process by assumption. The scale-length for the linear shear flow $\mathbf{U} = (y_2 u_0/\mathcal{L}, 0, 0)$ is denoted $\mathcal{L}$ as before. We have introduced $\mathbf{r}_0 \equiv (x_{01}, x_{02}, x_{03})$ and $\mathbf{R} \equiv (Y_1, Y_2, Y_3)$. Two of the equations, (25.37) and (25.38), have a structure similar to the one we encountered already for homogeneous isotropic turbulence, and we can write down the asymptotic expressions similar to (25.6) for the mean square particle separation for the two relevant directions as

$$\left\langle (Y_2(\mathbf{r}_0, t) - x_{02})^2 \right\rangle = 2t\, \tau_{L2} \left\langle \tilde{v}_2^2 \right\rangle \quad \text{and} \quad \left\langle (Y_3(\mathbf{r}_0, t) - x_{03})^2 \right\rangle = 2t\, \tau_{L3} \left\langle \tilde{v}_3^2 \right\rangle,$$

with Lagrangian integral time scales τ_{L2} and τ_{L3}, and mean square velocities $\langle \tilde{v}_3^2 \rangle$ and $\langle \tilde{v}_3^2 \rangle$, respectively, so as to serve as a reminder that even though the turbulence was assumed to be homogeneous, it need not be isotropic.

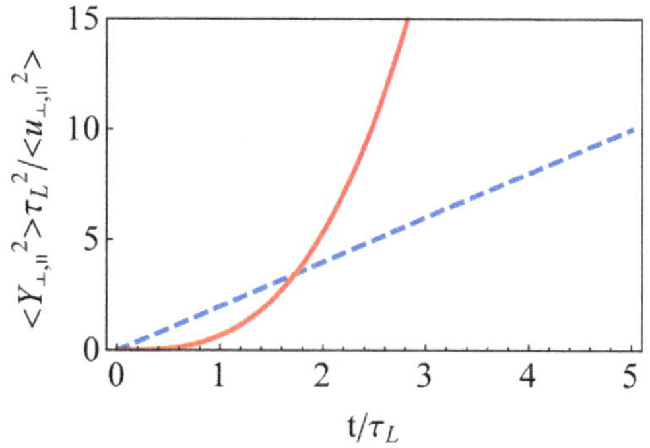

Figure 25.3. Illustration of the diffusion limit for a turbulent shear flow. We normalized the displacement to be $\langle Y^2\rangle/(\tau_L^2\langle\tilde{v}_1^2\rangle)$ and time as t/τ_L. This normalization leaves us with a dimensionless parameter $(u_0\tau_L/\mathcal{L})^2$ in the expression for the dispersion of the component along the flow. This parameter is here taken to be unity for simplicity. Full line shows the longitudinal, $\langle Y_\parallel^2\rangle$, while a dashed line gives the transverse diffusion, $\langle Y_\perp^2\rangle$. All Lagrangian–time-scales are taken equal.

The third component of the particle displacement requires a little more work. First we note that

$$\langle Y_1(\mathbf{r}_0, t) - x_{01}\rangle = \langle Y_1(\mathbf{r}_0, t)\rangle - x_{01} = t\, u_0\frac{x_{02}}{\mathcal{L}}.$$

We then find

$$\langle (Y_1(\mathbf{r}_0, t) - \langle Y_1(\mathbf{r}_0, t)\rangle)^2\rangle = \left\langle \left(\int_0^t \left(\frac{u_0}{\mathcal{L}}\int_0^\tau \tilde{v}_2(\mathbf{r}_0, \tau')d\tau' + \tilde{v}_1(\mathbf{r}_0, \tau)\right)d\tau\right)^2\right\rangle. \quad (25.39)$$

After some algebra we find in the limit of large times

$$\langle (Y_1(\mathbf{r}_0, t) - \langle Y_1(\mathbf{r}_0, t)\rangle)^2\rangle = \frac{2}{3}t^3\left(\frac{u_0}{\mathcal{L}}\right)^2 \tau_{L2}\left\langle \tilde{v}_2^2\right\rangle. \quad (25.40)$$

If you release a small drop of ink in a linear shear flow supporting homogeneous (but not isotropic) turbulence, then the drop will on average expand linearly with time in the direction transverse to the flow, but as t^3 in the direction *along* the flow. The result will apply equally well to a two-dimensional situation, with the flow direction being the plane considered. We illustrate the results in figure 25.3 with normalized units for both transverse $Y_\perp^2$ and longitudinal $Y_\parallel^2$ components, where we note that strictly speaking the diffusion limit does not apply for the region at $t/\tau_L \ll 1$.

- **Exercise:** Find $\langle (Y_1(\mathbf{r}_0, t) - \langle Y_1(\mathbf{r}_0, t)\rangle)(Y_2(\mathbf{r}_0, t) - x_{02})\rangle$, and also the correlation coefficient between $(Y_1(\mathbf{r}_0, t) - \langle Y_1(\mathbf{r}_0, t)\rangle)$ and $(Y_2(\mathbf{r}_0, t) - x_{02})$.

References

Batchelor G K 1949 Diffusion in a field of homogeneous turbulence *Aust. J. Sci. Res.* **2** 437–50

Batchelor G K 1952a Diffusion in a field of homogeneous turbulence II. The relative motion of particles *Proc. Cambridge Philos. Soc. (London)* **48** 345–62

Batchelor G K 1952b The effect of homogeneous turbulence on material lines and surfaces *Proc. R. Soc. (London)* A **213** 349–66

Batchelor G K 1953 *The Theory of Homogeneous Turbulence* (Cambridge: Cambridge University Press)

Carreras B A, Hidalgo C, Sánchez E, Pedrosa M A, Balbín R, Garcia-Cortés I, van Milligen B, Newman D E and Lynch V E 1996 Fluctuation-induced flux at the plasma edge in toroidal devices *Phys. Plasmas* **3** 2664–72

Chang C T 1991 Pellet-plasma interactions in tokamaks *Phys. Reports* **206** 143–96

Chorin A J 1994 *Vorticity and Turbulence* number 103 Applied mathematical Sciences (Berlin: Springer)

Corrsin S 1959 Progress report on some turbulent diffusion research *Adv. Geophys.* **6** 161–4

Crank J 1975 *The Mathematics of Diffusion* 2nd edn (Oxford: Clarendon)

Csanady G T 1973 *Turbulent Diffusion in the Environment* (Dordrecht: Reidel)

Dawson J M, Okuda H and Carlile R N 1971 Numerical simulation of plasma diffusion across a magnetic field in two dimensions *Phys. Rev. Lett.* **27** 491–4

Dupree T H 1967 Nonlinear theory of drift-wave turbulence and enhanced diffusion *Phys. Fluids* **10** 1049–55

Goldman S R, Ossakow S L and Book D L 1974 On the nonlinear motion of a small barium cloud in the ionosphere *J. Geophys. Res.* **79** 1471–7

Hay J S and Pasquill F 1959 Diffusion from a continuous source in relation to the spectrum and scale of turbulence *Adv. Geophys.* **6** 345–65

Hinze J O 1975 *Turbulence* 2nd edn (New York: McGraw Hill)

Horton W 1990 Nonlinear drift waves and transport in magnetized plasma *Phys. Reports* **192** 1–77

Horton W 1999 Drift waves and transport *Rev. Modern Phys.* **71** 735–78

Krane B, Pécseli H L and Trulsen J 1996 Concentration fluctuations in two-dimensional turbulence *Europhys. Lett.* **36** 99–104

Krane B, Pécseli H L and Trulsen J 2003 Concentrations and concentration fluctuations in two-dimensional turbulence *Phys. Fluids* **15** 211–26

Liewer P C 1985 Measurements on microturbulence in Tokamaks and comparison with theories of turbulence and anomalous transport *Nucl. Fusion* **25** 543–621

Lumley J L and Panofsky H A 1964 *The Structure of Atmospheric Turbulence* Vol XII *of Monographs and Texts in Physics and Astronomy, Interscience* (New York: Interscience)

Mikkelsen T, Larsen S E and Pécseli H L 1987 Diffusion of Gaussian puffs *Quart. J. Roy. Meteorol. Soc.* **113** 81–105

Misguich J H, Balescu R, Pécseli H L, Mikkelsen T, Larsen S E and Xiaoming Q 1987 Diffusion of charged particles in turbulent magnetoplasmas *Plasma Phys. Contr. Fusion* **29** 825–56

Okuda H and Dawson J M 1973 Theory and numerical simulation on plasma diffusion across a magnetic field *Phys. Fluids* **16** 408–26

Ott S and Mann J 2000 An experimental investigation of the relative diffusion of particle pairs in three dimensional turbulent flow *J. Fluid Mech.* **422** 207–23

Ott S and Mann J 2005 An experimental test of Corrsin's conjecture and some related ideas *New J. Phys.* **7** 142

Pécseli H L, Mikkelsen T and Larsen S E 1983 Drift wave turbulence in a low-β plasma *Plasma Phys.* **25** 1173–97

Pécseli H L and Trulsen J 1997 Eulerian and Lagrangian velocity correlation in two-dimensional random geostrophic flows *J. Fluid Mech.* **338** 249–76

Peskin R L 1974 Numerical simulation of Lagrangian turbulent quantities in two and three dimensions; Frenkiel F N and Munn R E Turbulent Diffusion in Environmental Pollution. Proceedings of a Symposium held at Charlottesville, Virginia, April 8–14, 1973 vol 18A *Advances in Geophysics* (New York: Academic) pp 141–63

Richardson L F 1926 Atmospheric diffusion shown on a distance-neighbour graph *Proc. R. Soc. London* A **110** 709–37

Richardson L F 1929 A search for the law of atmospheric diffusion *Beitr. Phys. Frei. Atmos.* **15** 24

Roberts P H 1957 On the application of a statistical approximation to the theory of turbulent diffusion *J. Math. and Mech.* **6** 781–99

Roberts P H 1961 Analytical theory of turbulent diffusion *J. Fluid Mech.* **11** 257–83

Sawford B 2001 Turbulent relative dispersion *Annu. Rev. Fluid Mech.* **33** 289–317

Shukla P K, Yu M Y, Rahman H U and Spatschek K H 1984 Nonlinear convective motion in plasmas *Phys. Reports* **105** 229–328

Taylor J B and McNamara B 1971 Plasma diffusion in two dimensions *Phys. Fluids* **14** 1492–9

Tchen C- M, Pécseli H L and Larsen S E 1980 Strong turbulence in low-β plasmas *Plasma Phys.* **23** 817–29

Tennekes H and Lumley J L 1972 *A First Course in Turbulence* (Cambridge, MA: The MIT Press)

Wandel C F and Kofoed-Hansen O 1962 On the Eulerian-Lagrangian transformation in the statistical theory of turbulence *J. Geophys. Res.* **67** 3089–93

Weinstock J 1976 Lagrangian-Eulerian relation and the independence approximation *Phys. Fluids* **19** 1702–11

Part VII

Analytical tools in turbulence

IOP Publishing

Low Frequency Waves and Turbulence in Magnetized Laboratory Plasmas and in the Ionosphere

Hans Pécseli

Chapter 26

Langevin's model for Brownian motion

Thermal motions of small, light particles like pollen or dust suspended in a suitable liquid are observable by microscopes. It is thus experienced that pollen of sizes 50 μm is large enough to be discriminated in even relatively simple microscopes, but light enough to exhibit an observable thermal motion. Observations of this motion were reported as early as 1785, by the Dutch physician Jan Ingenhausz, who examined powdered charcoal on an alcohol surface (Klafter *et al* 1996). The phenomenon was however named later after a biologist Robert Brown, who in 1827 made an extensive report of his detailed investigations of the motion of tiny pollen grains in water as observed by a microscope. He eventually got carried away by his studies, investigated all kinds of materials (including, apparently, dust from a fragment of the Sphinx) and found that this sort of motion persisted for every kind of substance he examined. Brown's own conclusion was seemingly that he had encountered the existence of some elementary form of life in all organic and inorganic matter. For the history of Brownian motion, see for instance the interesting and informative notes by Fürth in Einstein's collected papers on Brownian motion (Einstein and Fürth 1956) and also MacDonald (1962). It took rather a long time before the present day explanation of the phenomenon was accepted, although it was already formulated by several authors by the end of the 19th century (MacDonald 1962). Systematic research made it evident that the explanation was to be sought in the collisions with molecules of the surrounding fluid being in thermal motion. In 1906 Einstein and Smoluchowski both derived essentially the same formula from considerations of kinetic theory although by different methods, see for instance Lee *et al* (1973). Ironically, it is now being argued (Deutsch 1992) that what Brown observed was *not* what we today call Brownian motion! His microscopes were not sufficiently good (some of them are still in existence and can be tested). He might have seen convection motion that was generated by small temperature gradients in his water droplets.

doi:10.1088/978-0-7503-1251-6ch26

Shortly after Einstein's work was published, Langevin (1908) presented a different approach which, according to him, was 'infinitely more simple'. Although the scope of this first work was limited, it turns out that Langevin's ideas have much wider applicability, in particular for turbulence modeling.

Langevin's original analysis can be summarized as follows (Lee *et al* 1973). The force on an individual particle (e.g. pollen) at a position $x = x(t)$ was assumed to be composed of an average and a fluctuating part, $F(t)$, giving

$$M\frac{\mathrm{d}^2x}{\mathrm{d}t^2} = -6\pi\eta a\frac{\mathrm{d}x}{\mathrm{d}t} + F(t), \tag{26.1}$$

where Langevin assumed that the coefficient to the average part $-\mathrm{d}x/\mathrm{d}t$ could be identified with the viscous drag given by the same formula as in hydrodynamics, i.e. $-6\pi\eta a U$, with η being the dynamic viscosity of the liquid surrounding the particle, U is the particle velocity and a the diameter of the particle, assumed to be spherical. The dimension of η is $[\eta] = \text{mass(length} \times \text{time)}^{-1}$. Langevin did not argue in any greater detail for the separation between the fluctuating and average forces in (26.1), but the criticism faded by the elegance of the analysis and the beauty of the results it gave.

Multiplying by x, the equation (26.1) can be written as

$$\frac{1}{2}M\frac{\mathrm{d}^2\,x^2}{\mathrm{d}t^2} - M\left(\frac{\mathrm{d}x}{\mathrm{d}t}\right)^2 = -3\pi\eta a\frac{\mathrm{d}\,x^2}{\mathrm{d}t} + xF(t). \tag{26.2}$$

As the initial condition it is assumed that the particle is released at the origin. Now, we take the average value of each term in (26.2). By physical reasoning we argue as follows; $F(t)$ is caused by the thermal agitation of the medium surrounding the particle, i.e. an irregular bombardment of molecules. Due to its large mass the observed particle changes its velocity very little by each impact. For time intervals of relevance for the movement of the heavy particle we may assume $F(t)$ to be statistically independent of x and, since $\langle F(t)\rangle = 0$ by assumption, $\langle xF(t)\rangle \approx 0$ can be argued. It is assumed that the initial position (taken to be at the origin of the coordinate system) of a particle (pollen, dust-grain or similar) is known with certainty, implying $\langle x\rangle = 0$ and also $\langle x^2\rangle = 0$ at time $t = 0$. The initial *velocity* of the particle is on the other hand unknown. Since the particle is in thermal equilibrium with the surrounding medium we know on the other hand that $\langle(\mathrm{d}x/\mathrm{d}t)^2\rangle \equiv \langle U^2\rangle = \kappa T/M$ for this system in one spatial dimension. Consequently

$$\frac{1}{2}M\frac{\mathrm{d}^2\langle x^2\rangle}{\mathrm{d}t^2} + 3\pi\eta a\frac{\mathrm{d}\langle x^2\rangle}{\mathrm{d}t} = \kappa T. \tag{26.3}$$

The equation (26.3) has the solution

$$\frac{\mathrm{d}\langle x^2\rangle}{\mathrm{d}t} = \frac{\kappa T}{3\pi\eta a} + Ce^{-6\pi\eta a t/M}, \tag{26.4}$$

where $C = -\kappa T/3\pi\eta a$ is a constant determined by the initial conditions. For large times the last term is negligible (in reality for $t > 10^{-8}$ s or so) and Einstein's result

$$\langle x^2 \rangle = 2Dt, \tag{26.5}$$

is obtained with

$$D = \frac{\kappa T}{6\pi\eta a}. \tag{26.6}$$

It is important to note that the diffusion coefficient D here is independent of the mass M of the macroscopic particle, it enters only in the short transient time where the second term in (26.4) is important. This particular result was very surprising when it was first obtained. Nevertheless, this prediction of the theory was later shown by Perrin (1916) and coworkers to be satisfied to high precision for the substances they investigated.

For the large particles considered in this section, the diffusion coefficient D in (26.6) depends explicitly on the radius of the particle; this can be used to determine the size of small particles suspended in a fluid by measuring the diffusion coefficient. The method works well for spherical objects, for elongated or irregular ones an average scale size is obtained. The method has been used for measuring the magnitude of vira, for instance. The analysis predicts $\langle x^2 \rangle$ proportional to T. Since, however, viscosity usually decreases rapidly with increasing temperature, the thermal variation of the diffusion coefficient is not easy to verify directly.

The last term in (26.4) can be traced back to the inertia term $M\mathrm{d}^2x/\mathrm{d}t^2$ in (26.1). Since this last term is ignored, we find that Brownian motion is described as *inertialess*, meaning that a particle is supposed to be able to change its direction of motion abruptly. When observing Brownian motion in a microscope, this is actually how it appears; the tiny rounding-off of the corners of their trajectories due to the last term in (26.4) are hardly visible.

At first sight it seems surprising that $F(t)$ has disappeared completely from the results, apparently it might have been put to zero from the outset. However it appears indirectly by the identification $M\langle U^2 \rangle = \kappa T$. A random force is necessary to mediate the relaxation of the fluctuating velocities to the universal form given by thermodynamic arguments.

A mobility can be defined as the ratio of an average velocity and a deterministic external force, although this definition is not universally applied. From (26.1) we readily find a velocity $U = F/(6\pi\eta a)$, where F can denote a force of gravity Mg, for instance. The mobility is thus $\mu = 1/(6\pi\eta a)$ and the ratio $D/\mu = \kappa T$, i.e. the Einstein relation is again recovered.

Langevin's equation was presumably the first example of a stochastic differential equation to be analyzed, i.e. one where a random term appears explicitly and thus having random functions as solutions (Risken 1984).

26.1 Limitations of the Langevin equation

It was noted by Lorentz (1921) that the Langevin equation in its original form can be applied within certain limits only. His arguments were essentially based on a dimensional analysis which is quite instructive. Introduce first a time-scale τ_B for relaxation time of the velocity correlation function for the Brownian particle. In terms of the viscosity η we have as an order of magnitude $\tau_B \sim M/(\eta R)$, with R being a typical scale size of this particle. Also another time-scale can be introduced, namely one characterizing the relaxation of a change in fluid velocity as a response to the motion of the particle. The time it takes a hydrodynamic perturbation to relax over a length scale R is $\tau_H \sim R^2/\nu$, where ν is here the kinematic viscosity, $\eta \approx \nu\rho_F$ where ρ_F is the fluid mass density. A diffusive relaxation was assumed for the hydrodynamic velocity, for the present condition this is the most reasonable hypothesis. In order to have a well defined fluid velocity which is the same all over the surface of the Brownian particle, as implicitly assumed in (26.1), it is evidently necessary to require $\tau_H \ll \tau_B$, i.e. $R^2 \ll \nu M/(\eta R)$, or

$$\rho_F \ll M/R^3 \sim \rho_B, \tag{26.7}$$

where ρ_B is the mass density of the Brownian particle, implying that the Langevin equation applies for heavier particles in a lighter fluid. The inequality (26.7) is not always satisfied and the analysis has been extended to more general cases by e.g. Hauge and Martin-Löf (1973).

It might be noted that the inequality (26.7) need not imply that the heavier particles condense at the bottom of the fluid. The particles participate in the thermal motion while influenced by gravity as well as the lift of the surrounding fluid, and will obtain a height distribution with a scale length $\mathcal{L} = (\kappa T/Mg)/(1 - \rho_F/\rho_B)$, see for instance Chandrasekhar (1954) as also summarized by Pécseli (2000). Provided this scale length is macroscopic, it is meaningful to expect the thermal fluctuations to be well described by the Langevin equation. The problem of thermal motion of a heavy particle subject to a central force is, however, probably the most relevant case for the Langevin equation in its basic original form, see section 26.2. An example for this is the so called *galvanometer fluctuations*, which have been previously studied in detail (MacDonald 1962, Pécseli 2000).

26.2 Brownian motion with central force

The analysis of the foregoing section can be made somewhat more general by including a central force (Risken 1984), giving a Langevin type equation

$$M\left[\frac{\mathrm{d}^2}{\mathrm{d}t^2}x + \omega_0^2 x + \gamma\frac{\mathrm{d}}{\mathrm{d}t}x\right] = F(t), \tag{26.8}$$

where a friction coefficient γ is introduced. This equation is capable of describing a variety of phenomena, e.g. it is formally equivalent to $L\ddot{q} + R\dot{q} + q/C = e(t)$ with $\omega_0^2 = 1/(LC)$ describing a parallel connection of a resistor R, self-inductance L and capacitor C while $e(t)$ is a fluctuating voltage and q, as before, the charge on the condenser. Equation (26.8) might as well describe the motion of a mechanical system, such as a simple pendulum. For $F(t)$ it is again assumed that $\langle F(t)\rangle = 0$ and $\langle F(t_1)F(t_2)\rangle = R(t_1 - t_2)$. With $x = x_0$ and $u \equiv \mathrm{d}x/\mathrm{d}t = u_0$ at $t = 0$ one finds by integration following Uhlenbeck and Ornstein 1930)

$$
\begin{aligned}
u = &-\frac{2\omega_0 x_0 + \gamma u_0}{2\omega_1} \mathrm{e}^{-\gamma t/2} \sin(\omega_1 t) + u_0 \mathrm{e}^{-\gamma t/2} \cos(\omega_1 t) \\
&+ \frac{1}{\omega_1} \int_0^t F(\tau) \mathrm{e}^{\gamma(\tau - t)/2} \left\{ \omega_1 \cos[\omega_1(t - \tau)] - \frac{\gamma}{2} \sin[\omega_1(t - \tau)] \right\} \mathrm{d}\tau,
\end{aligned} \tag{26.9}
$$

and

$$
\begin{aligned}
x = &\frac{2u_0 + \gamma x_0}{2\omega_1} \mathrm{e}^{-\gamma t/2} \sin(\omega_1 t) + x_0 \mathrm{e}^{-\gamma t/2} \cos(\omega_1 t) \\
&+ \frac{1}{\omega_1} \int_0^t F(\tau) \mathrm{e}^{\gamma(\tau - t)/2} \sin[\omega_1(t - \tau)] \mathrm{d}\tau,
\end{aligned} \tag{26.10}
$$

where $\omega_1^2 = \omega_0^2 - \gamma^2/4$. Note that there is no *a priori* reason that x_0 should correspond to the equilibrium position. From the foregoing expressions one obtains immediately

$$
\langle x \rangle = \frac{2u_0 + \gamma x_0}{2\omega_1} \mathrm{e}^{-\gamma t/2} \sin(\omega_1 t) + x_0 \mathrm{e}^{-\gamma t/2} \cos(\omega_1 t). \tag{26.11}
$$

demonstrating that the average displacement of the harmonically bound particle goes to zero with the time-constant $\gamma/2$. In an individual realization of the ensemble, this displacement is of course non-zero in general, as evidenced by the standard deviation which can be obtained as

$$
\begin{aligned}
\langle x^2 \rangle = &\left(\frac{2u_0 + \gamma x_0}{2\omega_1} \mathrm{e}^{-\gamma t/2} \sin(\omega_1 t) + x_0 \mathrm{e}^{-\gamma t/2} \cos(\omega_1 t) \right)^2 \\
&+ \frac{\tau_1}{2\omega_1^2 \gamma} (1 - \mathrm{e}^{-\gamma t}) \\
&- \frac{\tau_2}{8\omega_0^2 \omega_1^2} \left\{ \gamma \left[1 - \mathrm{e}^{-\gamma t} \cos(2\omega_1 t) \right] + 2\omega_1 \mathrm{e}^{-\gamma t} \sin(2\omega_1 t) \right\},
\end{aligned} \tag{26.12}
$$

introducing $\tau_1 = \int_{-\infty}^{\infty} R(\tau) \cos(\omega_1 \tau) \mathrm{d}\tau$ and $\tau_2 = \int_{-\infty}^{\infty} R(\tau) \mathrm{d}\tau$. In agreement with the foregoing arguments we expect that $R(\tau)$ is sharply peaked around zero, which means that $\tau_1 \approx \tau_2$. Calculating the velocity of the particle and using the fact that for large times thermal equilibrium is reached, i.e. use $\langle (\mathrm{d}x/\mathrm{d}t)^2 \rangle = \kappa T/M$, it is demonstrated that $\tau_1 = \tau_2 = 2\gamma\kappa T/M$, and by insertion the problem is solved. This result would have been obtained from the outset by assuming $\langle F(t_1)F(t_2) \rangle \approx 2D\delta(t_1 - t_2)$ with $D = \kappa T/\gamma$ as before. In particular, the mean square particle displacement becomes $\langle x^2 \rangle = \kappa T/(M\omega_0^2)$ for large times where the particle has reached equilibrium with the surroundings. The particle is thus confined to a finite region, in contrast to the free Brownian motion discussed before, but if the restoring force is made weaker, i.e. $\omega_0 \to 0$, the mean square displacement increases.

- **Exercise:** Obtain the expressions for $\langle u \rangle$, $\langle u^2 \rangle$ and $\langle ux \rangle$ from (26.9) and (26.10).

If we now assume that the particle is taken from actual realizations of the ensemble, it is evident that the initial velocity has the same statistics as that caused by the thermal fluctuations. The problem has a natural reduction obtained by an average over all realizations of u_0, subject to this condition. Physically, a sub-ensemble is selected by choosing realizations subject only to the condition on the initial particle position, and arbitrary, randomly distributed, initial velocities. With the present assumption we now have $\langle u^2 \rangle = \kappa T / M$ at all times.

Other averages can be calculated than those discussed previously. It may for instance be instructive to also consider

$$\langle xu \rangle = \frac{1}{\omega_1 \omega_2}\left(\frac{\kappa T}{M\omega_0^2} - x_0^2\right)\left(\cos(\omega_1 t) + \frac{\gamma}{2\omega_1}\sin(\omega_1 t)\right)\mathrm{e}^{-\gamma t}\sin(\omega_1 t), \tag{26.13}$$

with the initial velocities being statistically distributed as discussed before. The correlation between position and velocity thus starts out being zero, then oscillates and eventually vanishes.

- **Exercise:** Discuss the damped harmonic oscillator subject to thermal agitation as described by the foregoing analysis for the overdamped, critically damped and underdamped case.

26.3 Stochastic differential equations

The present summary emphasizes the applications of the Langevin equation for the fluctuation–dissipation theorem. Historically, this is the proper approach. The basic idea had applications probably way beyond Langevin's own imagination: a deterministic equation driven by a random forcing. Such models found applications in many branches of physics, in particular for turbulence modeling (Nicholson 1983). A particularly valuable general reference is van Kampen (1976). There are a vast number of relevant studies where one starting point can be the book by Gardiner (1985) or by Lawrence (2013). The problem of stochastic differential equations can be formulated in many different ways, where basically two different approaches have emerged: the Ito and the Stratonovich calculus. The latter one is considered easiest to deal with analytically, the former best suited for numerical schemes.

A basic problem has to do with our concept of randomness: A time varying signal is deterministic, i.e. completely predictable, provided it is infinitely many times differentiable and *a priori* known in a small time interval. Then, by a Taylor expansion, it can in principle be described with arbitrary accuracy to arbitrary later times. Unpredictability is therefore here related to discontinuities either in the functional values or in time derivatives at some order invalidating this expansion. Quite formally this is so, and it also agrees with intuitive expectations of random functions looking ragged.

Mathematics can only derive the probabilities of outcomes of experiments or trials from *a priori given* probability densities or distributions. The only restrictions imposed on these probabilities is that they have to be positive definite and normalizable (and even this condition can formally be relaxed somewhat). There is no mathematical requirement of averages to exist, although it is sometimes hard to imagine a physical process to be described without. From basic statistical requirements, we cannot argue for restrictions on the variability of a random signal. An example for a probability density without finite averages was mentioned in chapter 19. When derivatives of first or some higher order of a temporally varying signal are diverging frequently, we expect this to be reflected in the probability density by corresponding divergences of some averages.

With these scattered remarks in mind we go back to the definition of an integral

$$\int_0^t f(t)\mathrm{d}t = \lim_{n\to\infty} \sum_{j=1}^{n} f(\tau_j)(t_{j+1} - t_j),$$

where τ_j is in the interval $[t_j,\ t_{j+1}]$, written as $\tau_j \in [t_j,\ t_{j+1}]$. More generally, we have the Riemann-Stieltjes integral

$$\int_0^t f(t)dg(t) = \lim_{n\to\infty} \sum_{j=1}^{n} f(\tau_j)\big(g(t_{j+1}) - g(t_j)\big).$$

For a smooth measure $g(t)$, the limit converges to a unique value regardless of where τ_j taken in interval $[t_j,\ t_{j+1}]$. The key word here is 'smooth'. A random function satisfying our intuitive ideas is *not* smooth, on the contrary, the forcing of the Langevin equation for the Brownian diffusion is a series of δ-functions, statistically independent in the sense that the arrival of one is independent of any of the others.

To illustrate the problem we consider a simplified version of (26.1) giving

$$\frac{\mathrm{d}x}{\mathrm{d}t} = -\frac{x}{\tau} + \gamma\frac{\mathrm{d}F(t)}{\mathrm{d}t},$$

where $F(t)$ represent the noise. The solution is

$$x(t) = x_0\mathrm{e}^{-t/\tau} + \gamma \int_0^t \mathrm{e}^{-(t-t')/\tau}\frac{\mathrm{d}F(t')}{\mathrm{d}t'}\mathrm{d}t'.$$

The integral is undefined for white noise that changes infinitely rapidly (that is why the frequency spectrum is constant at all frequencies) and its variance is infinite (the integral of the power spectrum diverges). This description corresponds to forcing with white noise, but the basic idea applies also to more general random forcings. For such cases the limit in the definition of the integral depends on where we place τ_j in the corresponding integral. We can take two choices (out of many): either $\tau_j = t_j$ or $\tau_j = (t_j + t_{j+1})/2$. The latter leads to what is called the Stratonovich calculus, the former to the Ito calculus (Risken 1984). A detailed comparison of the different approaches is given by, for instance, van Kampen (1981).

- **Example:** We can illustrate the difference between the Ito and the Stratonovich results by considering a very simple case where $x(t) = \alpha t$ and want to estimate the integral $\int_0^T \alpha t'\mathrm{d}t'$. Sampling the function $x(t)$ in the interval $\{0;\ T\}$ at $N + 1$ equidistant positions labeled 0, 1, ..., N we have by the Ito approach $\alpha\sum_0^{N-1} n = \frac{\alpha}{2}N(N-1)$, while on the other hand the Stratonovich approximation gives $\alpha\sum_0^{N-1}(n+(n+1))/2 = \frac{\alpha}{2}N^2$, for all N. In both cases we take the sampling interval to be of unit length. The exact result for the integral in this simple case is $N^2\alpha/2$ with the given definitions. According to this example, we can expect that the 'normal' laws of integration apply to the Stratonovich calculation, but not to the Ito approach.

The problem is physically relevant in general for many reasons, one deals with linearly unstable systems. Naively one might expect that the initial conditions are irrelevant, since the nonlinearities will eventually smear out any memory of what conditions were imposed at say $t = 0$. There is, however, a class of such systems being described by nonlinear model equations that are highly dependent on the initial conditions (Breizman and Malkin 1980, Pécseli *et al* 1984). A natural time evolution starts from thermal noise. Modeling of such a noise level can be important in a correct numerical model, for instance. Pioneering numerical simulations of weakly turbulent Langmuir waves by Guio and Forme (2006) took care to model this thermal noise correctly.

References

Breizman B N and Malkin V M 1980 Dynamics of the modulational instability of a broad spectrum of Langmuir waves *Sov. Phys. JETP* **52** 435–41

Breizman B N and Malkin V M 1980 *Russian original Zh. Eksper. Teor. Fiz.* **79** 857–69

Chandrasekhar S 1954 Stochastic problems in physics and astronomy ed N Wax *Selected Papers on Noise and Stochastic Processes* (New York: Dover) pp 3–91

Deutsch D H 1992 Brownian motion *Nature* **357** 354

Einstein A 1956 Reprint of translations of the original papers: *On the movement of small particles suspended in a stationary liquid demanded by the molecular-kinetic theory of heat, On the theory of the Brownian movement, A new determination of molecular dimensions, Theoretical observations of the Brownian motion and Elementary theory of the Brownian motion* ed Fürth R (New York: Dover) transl. Cowper A D

Gardiner C W 1985 *Handbook of Stochastic Methods for Physics, Chemistry and the Natural Sciences* 2nd edn (Berlin: Springer)

Guio P and Forme F 2006 Zakharov simulations of Langmuir turbulence: Effects on the ion-acoustic waves in incoherent scattering *Phys. Plasmas* **13** 122902

Hauge E H and Martin-Löf A 1973 Fluctuating hydrodynamics and Brownian motion *J. Stat. Phys.* **7** 259–81

Klafter J, Schlesinger M F and Zumofen G 1996 Beyond Brownian motion *Phys. Today* **49** 33

Langevin P 1908 Sur la théorie du mouvement brownien *C. R. Acad. Sci. (Paris)* **146** 530–3

Lawrence C E 2013 *An Introduction to Stochastic Differential equations* (Washington, DC: American Mathematical Society)

Lee J F, Sears F W and Turcotte D L 1973 *Statistical Thermodynamics* 2nd edn (Reading, MA: Addison-Wesley)

Lorentz H A 1921 *Lessen over Theoretische Natuurkunde* vol 5 Kinetische Problemen (Leiden: E J Brill)

MacDonald D K C 1962 *Noise and Fluctuations: an Introduction* (New York: Wiley)

Nicholson D R 1983 *Introduction to Plasma Theory* (New York: Wiley)

Pécseli H L 2000 *Fluctuations in Physical Systems* (Cambridge: Cambridge University Press)

Pécseli H L, Trulsen J and Armstrong R 1984 Formation of ion phase-space vortexes *Phys. Scripta* **29** 241–53

Perrin J 1916 *Atoms* (London: Constable)

Risken H 1984 *The Fokker-Planck equation: Methods of Solution and Applications* (Berlin: Springer)

Uhlenbeck G E and Ornstein L S 1930 On the theory of the Brownian motion *Phys. Rev.* **35** 823–41 Reprinted in
Uhlenbeck G E and Ornstein L S 1954 *Selected Papers on Noise and Stochastic Processes* ed N Wax (New York: Dover) pp 93–111

van Kampen N 1976 Stochastic differential equations *Phys. Rep.* **24** 171–228

van Kampen N 1981 Itô versus Stratonovich *J. Stat. Phys.* **24** 175–87

IOP Publishing

Low Frequency Waves and Turbulence in Magnetized Laboratory Plasmas and in the Ionosphere

Hans Pécseli

Chapter 27

Markov processes

A particular set of random processes, Markov processes play an important role for natural phenomena, serving in particular as a limiting case for some turbulence phenomena. These processes can be described by a lack of memory; knowledge of the actual state of systems described by these processes determines the distribution of future states (Risken 1984, Papoulis 1991). For Markovian systems the present separates the past from the future; knowledge of the system at one instant, the present, permits a prediction of the 'average' future and indeed the probability of various possible futures (Lax 1960, Cox and Miller 1965). Information of the past is not needed, and is not relevant even when available. Here the basic properties of such systems will be discussed without particular reference to actual physical details. In particular it is demonstrated that the relaxation to thermal equilibrium of Markovian systems in response to an initial perturbation can be given a general formalism.

It is advantageous to discuss Markov processes in terms of probability densities (Cox and Miller 1965, Risken 1984, Balescu 1997). Let $Y_x(t)$ be a stochastic process, i.e. an ensemble of functions of t labeled by the parameter x with given distribution $D(x)$. As before, we use the notation Y for the variable and lower case y for the values it can assume.

Let $P_1(y_1, t_1)\mathrm{d}y_1$ be the probability that $Y(t_1)$ lies in the interval $[y_1; y_1 + \mathrm{d}y_1]$. Then we have

$$P_1(y_1, t_1) = \int_{-\infty}^{\infty} \delta\big[Y_x(t_1) - y_1\big] D(x)\mathrm{d}x. \tag{27.1}$$

Similarly, we let $P_2(y_1, t_1; y_2, t_2)\mathrm{d}y_1\mathrm{d}y_2$ be the joint probability that $Y(t_1)$ lies is the interval $[y_1; y_1 + \mathrm{d}y_1]$ and simultaneously $Y(t_2)$ in the interval $[y_2; y_2 + \mathrm{d}y_2]$. By these arguments a hierarchy of probabilities P_1, P_2, P_3, … can be defined to constitute a possible way of describing the stochastic process. Conditional probability densities can also be defined in terms of these joint probability density functions

$$P_{n\,|\,n-1}(y_n,\, t_n|y_1,\, t_1;\, \dots,\, y_{n-1},\, t_{n-1}) = \frac{P_n(y_1,\, t_1;\, \dots;\, y_n,\, t_n)}{P_{n-1}(y_1,\, t_1;\, \dots;\, y_{n-1},\, t_{n-1})}, \tag{27.2}$$

by use of Bayes' rule, where the time ordering $t_n \geq t_{n-1} \geq \cdots \geq t_2 \geq t_1$ was assumed. In the following we omit the index on conditional probabilities for simplicity.

The Markov property is expressed in terms of the conditional probabilities as

$$P(y_n,\, t_n|y_1,\, t_1;\, \dots;\, y_{n-1},\, t_{n-1}) = P(y_n,\, t_n|y_{n-1},\, t_{n-1}), \tag{27.3}$$

with a self-evident interpretation, i.e. the state y_{n-1}, t_{n-1} determines the probability density of states y_n, t_n irrespective of the states y_m, t_m with $m < n - 1$. An arbitrary joint probability density for a Markov process can then be expressed as

$$P_n(y_1,\, t_1;\, \dots,\, y_n,\, t_n) = P(y_n,\, t_n|y_{n-1},\, t_{n-1})\cdots P(y_2,\, t_2|y_1,\, t_1)P_1(y_1,\, t_1). \tag{27.4}$$

For Markov processes all information is contained in the joint probability density function, P_2, since P_1 can be derived from it together with the conditional probability density P_2/P_1 by use of Bayes' rule, (19.14). In particular we also have

$$P(y_2,\, t_2;\, y_3,\, t_3|y_1,\, t_1) = P(y_3,\, t_3|y_2,\, t_2;\, y_1,\, t_1)P(y_2,\, t_2|y_1,\, t_1). \tag{27.5}$$

With the Markov property this relation becomes

$$P(y_2,\, t_2;\, y_3,\, t_3|y_1,\, t_1) = P(y_3,\, t_3|y_2,\, t_2)P(y_2,\, t_2|y_1,\, t_1). \tag{27.6}$$

In terms of conditional probabilities we have

$$\begin{aligned} P_1(y_2,\, t_2) &= \int_{-\infty}^{\infty} P_2(y_1,\, t_1;\, y_2,\, t_2)\mathrm{d}y_1 \\ &= \int_{-\infty}^{\infty} P(y_2,\, t_2|y_1,\, t_1)P_1(y_1,\, t_1)\mathrm{d}y_1, \end{aligned} \tag{27.7}$$

and also

$$\begin{aligned} P(y_3,\, t_3|y_1,\, t_1) &= \int_{-\infty}^{\infty} P(y_2,\, t_2;\, y_3,\, t_3|y_1,\, t_1)\mathrm{d}y_2 \\ &= \int_{-\infty}^{\infty} P(y_3,\, t_3|y_2,\, t_2;\, y_1,\, t_1)P(y_2,\, t_2|y_1,\, t_1)\mathrm{d}y_2. \end{aligned} \tag{27.8}$$

Introducing the Markov assumption this latter relation becomes

$$P(y_3,\, t_3|y_1,\, t_1) = \int_{-\infty}^{\infty} P(y_3,\, t_3|y_2,\, t_2)P(y_2,\, t_2|y_1,\, t_1)\mathrm{d}y_2, \tag{27.9}$$

known as the Smoluchowski equation or as the Chapman–Kolmogorov equation, see also figure 27.1 for a schematic illustration. The relation states that the probability of going from y_1 to y_3 via y_2 is the product of the probability for going

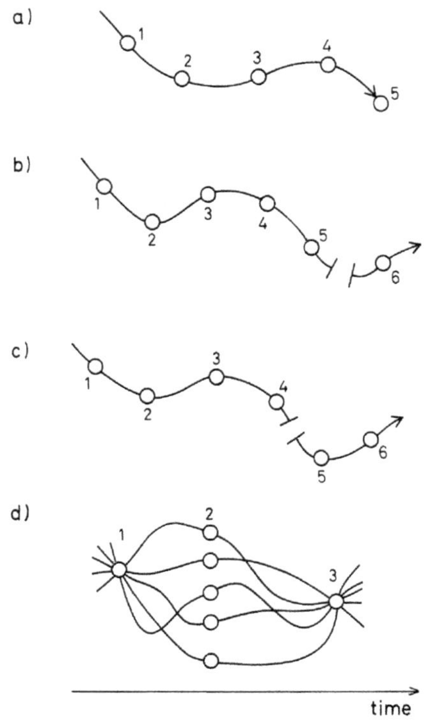

Figure 27.1. Schematic illustration of the Chaman—Kolmogorov equation in (d) and the one or two step Markov property in (b) and (c). The general conditional probability that a particle passes through positions 2, 3, 4, 5 subject to the condition that it passed through position 1 is illustrated in (a), or passing through 3, 4, and 5, given conditions 1 and 2, etc.

from y_1 to y_2 times that for going subsequently from y_2 to y_3; that is, successive transitions are statistically independent. The time ordering $t_1 < t_2 < t_3$ is essential. It can*not* be assumed that an arbitrary joint probability density can describe a Markov-process; it has to satisfy (27.9) also.

27.1 Master equation

A more manageable form of (27.9) is obtained by taking $t_3 = t_2 + \Delta t$ with small Δt. For physically acceptable processes it is expected that the probability that Y makes a jump during Δt will be proportional to Δt apart from small corrections. This is expressed by

$$P(y_3, t_3|y_2, t_2) = \delta(y_3 - y_2)[1 - w(y_2)\Delta t] + \Delta t\, W(y_3|y_2) + \mathcal{O}(\Delta t^2), \tag{27.10}$$

where $W(y_3|y_2)$ is the transition probability per unit time from y_2 to y_3. The term $w(y_2)\Delta t$ is the total probability that a transition took place, *from* y_2 to *any* y_3,

i. e. $w(y_2) = \int W(y_3|y_2)\mathrm{d}y_3$ while $1 - w(y_2)\Delta t$ is the probability for nothing happening in the time interval Δt. Ignoring terms $\mathcal{O}(\Delta t^2)$ in (27.10), direct substitution gives

$$\frac{\partial P(y,\, t|y_1,\, t_1)}{\partial t} = \int_{-\infty}^{\infty} \{W(y|y')P(y',\, t|y_1,\, t_1) - W(y'|y)P(y,\, t|y_1,\, t_1)\}\mathrm{d}y', \tag{27.11}$$

with use of the approximation

$$\frac{\partial P(y,\, t|y_1,\, t_1)}{\partial t} \approx \frac{P(y,\, t|y_1,\, t_1) - P(y,\, t - \Delta t|y_1,\, t_1)}{\Delta t}.$$

As an alternative, more customary formulation; select a subensemble of the functions $Y_x(t)$ by prescribing a distribution $P(y_1,\, t_1)$ at some initial time t_1. Then the distribution in this subensemble at $t \geq t_1$ obeys the master equation

$$\frac{\partial P(y,\, t)}{\partial t} = \int_{-\infty}^{\infty} \left\{W(y|y')P(y',\, t) - W(y'|y)P(y,\, t)\right\}\mathrm{d}y'. \tag{27.12}$$

This differential version of the Chapman–Kolmogorov equation, valid for the probability density associated with any stationary Markov process obeying (27.10), is called the master equation. For a discrete sample space but still a continuous time variable, this equation takes the form

$$\frac{\partial P_n(t)}{\partial t} = \sum_{n'} \{ W_{n,n'}\, P_{n'}(t) - W_{n',n}\, P_n(t)\}. \tag{27.13}$$

These equations have an intuitively plausible appearance; they are gain-loss equations for the probability of states y or, in the discrete case, n. The first term is the gain due to transitions from other states y', and the second term is the loss due to transitions into other states. It is an advantage to make further simplifications of the master equations. The Fokker–Planck equation is a particularly useful example.

27.2 Fokker–Planck equations

As a first step the transition probability is rewritten in terms of the size r of the jump and of the starting point y as $W(y|y') = W(y';\, r)$ with $r = y - y'$. The master equation can then be written as (e.g. van Kampen (1981))

$$\frac{\partial P(y,\, t)}{\partial t} = \int_{-\infty}^{\infty} W(y - r;\, r)P(y - r,\, t)\mathrm{d}r - P(y,\, t)\int_{-\infty}^{\infty} W(y;\, -r)\mathrm{d}r. \tag{27.14}$$

The basic assumption is now that only small jumps occur, that is $W(y';\, r)$ is a sharply peaked function of r but varies slowly with y. The second assumption is that $P(y,\, t)$ varies only slowly with y for cases of interest. Introducing a Taylor expansion up to second order it is possible to shift from $y - r$ to y in the first integrand

$$\frac{\partial P(y,t)}{\partial t} = \int_{-\infty}^{\infty} W(y;r)P(y,t)\mathrm{d}r - \int_{-\infty}^{\infty} r\frac{\partial}{\partial y}\{W(y;r)P(y,t)\}\,\mathrm{d}r$$
$$+\frac{1}{2}\int_{-\infty}^{\infty} r^2\frac{\partial^2}{\partial y^2}\{W(y;r)P(y,t)\}\,\mathrm{d}r \qquad (27.15)$$
$$- P(y,t)\int_{-\infty}^{\infty} W(y;-r)\mathrm{d}r.$$

The first and last terms cancel, as demonstrated by a change of integration variable, while the remaining two terms are expressed by introducing the jump moments

$$a_n(y) = \int_{-\infty}^{\infty} r^n W(y;r)\mathrm{d}r, \qquad (27.16)$$

to give the Fokker–Planck equation

$$\frac{\partial P(y,t)}{\partial t} = -\frac{\partial}{\partial y}[a_1(y)P(y,t)] + \frac{1}{2}\frac{\partial^2}{\partial y^2}[a_2(y)P(y,t)], \qquad (27.17)$$

where only a_1 and a_2 enter. For an n-dimensional process equation (27.17) is generalized to

$$\frac{\partial P}{\partial t} = -\sum_l \frac{\partial}{\partial y_l}\left[A_l(\mathbf{y})P\right] + \frac{1}{2}\sum_{l,m}\frac{\partial^2}{\partial y_l \partial y_m}\left[B_{lm}(\mathbf{y})P\right], \qquad (27.18)$$

where now A_l and B_{lm} are the components of a vector and a tensor, respectively (van Kampen 1981).

The Fokker–Planck equation allows a study of non-stationary conditions, i. e. the time evolution of the probability density function, in particular the relaxation to thermal equilibrium. The diffusion equation is a particularly simple version. Often the Fokker–Planck equation is solved with the initial condition $P(y,0) = \delta(y)$ although strictly speaking this function violates the assumption of P being slowly varying with y.

In principle one might include all terms in a Taylor expansion of (27.14) giving

$$\frac{\partial P(y,t)}{\partial t} = \sum_{j=1}^{\infty}\frac{(-1)^j}{j!}\frac{\partial^j}{\partial y^j}\left[a_j(y)P(y,t)\right],$$

known as the Kramers–Moyal expansion, with a detailed discussion given for instance by van Kampen (1981). The Fokker–Planck equation assumes that all terms with $j > 2$ are negligible in the Kramers–Moyal expansion. Note that this assumption need not always be justified (van Kampen 1981).

- **Exercise:** Demonstrate that the form

$$\frac{\partial P}{\partial t} = -\sum_l C_l\frac{\partial}{\partial y_l}P - \sum_l A_l\frac{\partial}{\partial y_l}\left[y_l P\right] + \frac{1}{2}\sum_{l,m} B_{lm}\frac{\partial^2}{\partial y_l \partial y_m}P, \qquad (27.19)$$

can be reduced to (27.18) by proper change of variables (van Kampen 1981).

27.2.1 Transition from discrete to continuous case

A discrete random walk of a free particle can in a limiting case be modeled by a Fokker–Planck equation. It is instructive to see how this transition from a discrete to a continuous case is brought about. Imagine a particle which moves along the x-axis in such a way that in each step it can move either Δx to the left or Δx to the right. For a free particle we can assume that a step in either of the two directions is equally probable, with probability $\frac{1}{2}$. Consider $P(n\Delta x|m\Delta x;\, s\tau)$ which is the probability density that the particle is at $m\Delta x$ at time $s\tau$, if at the beginning it was at $n\Delta x$. The conditional probability density satisfies the difference equation (Kac 1946)

$$P(n|m;\, s+1) = \frac{1}{2}P(n|m+1;\, s) + \frac{1}{2}P(n|m-1;\, s), \tag{27.20}$$

which can be rewritten in an equivalent form as

$$\begin{aligned} &\frac{P(n\Delta x|m\Delta x;\, (s+1)\tau) - P(n\Delta x|m\Delta x;\, s\tau)}{\tau} \\ &= \frac{\Delta x^2}{2\tau}\left(\frac{P(n\Delta x|(m+1)\Delta x;\, s\tau) + P(n\Delta x|(m-1)\Delta x;\, s\tau)}{\Delta x^2} \right. \\ &\left. - \frac{2P(n\Delta x|m\Delta x;\, s\tau)}{\Delta x^2}\right). \end{aligned} \tag{27.21}$$

Suppose now that $\Delta x \to 0$ and $\tau \to 0$ in (27.21) in such a way that $\Delta x^2/2\tau = D=$ const. In this limit, with $n\Delta x \to x$ and $s\tau \to t$, we recover the diffusion equation from (27.21) as

$$\frac{\partial P}{\partial t} = D\frac{\partial^2 P}{\partial x^2},$$

where D is a diffusion coefficient, see also chapter 21.

An important implication of the transition allowed to obtain this equation is that the instantaneous particle velocity $\Delta x/\tau$ approaches infinity. This explains why the diffusion equation has a finite response at any position, no matter how far from the origin, at arbitrary short times, also for initial conditions with finite support. With the local particle velocity being infinite, we also find that the length of the path (not the net displacement) of a diffusing particle diverges in this diffusion limit.

The arguments outlined in this section contain two elements; first time is made continuous by using infinitesimal time steps and, simultaneously, the spatial variable is also made continuous by using very small steps or jumps. This limit can be obtained by 'brute force' using (27.16) with the transition probability $W_{n,n'} = \frac{1}{\tau}(\frac{1}{2}\Delta_{n',n+1} + \frac{1}{2}\Delta_{n',n-1})$ per time interval τ for the symmetric random walk. Clearly, $a_1 = 0$ while $a_2 = \frac{1}{2}$ $\Delta x^2/\tau \equiv D$ and the corresponding Fokker–Planck equation becomes again the diffusion equation, $\partial P/\partial t = D\partial^2 P/\partial x^2$. With some care, the expression (27.16) can thus also be used to obtain a Fokker–Planck equation for processes which are by nature discrete, with some self-evident restrictions imposed on the interpretation of the results.

The limiting process of a simple random walk is called the *Wiener process*, or the *Brownian motion process*. For such a process, the increment $\Delta X(t) = X(t + \Delta t) - X(t)$ in a small time interval Δt is independent of $X(t)$ and has a mean and a variance proportional to Δt. The process is a Markov process, since, if we are given that $X(t) = a$, the distribution of $X(u)$ with $u > t$ is fixed, namely it is the distribution of the quantity $a + \sigma(u - t)^{1/2} Y$, where Y is a standard normal variable and σ is the variance of the process (Cox and Miller 1965).

The developments of the theoretical understanding of diffusion processes and the derivation of the diffusion equation is historically interesting, see for instance the review by Narasimhan (1999).

- **Example:** An elastically bound particle undergoing a random walk with a finite step-size in one spatial dimension can be described by a random walk model. Here the step-probabilities were assigned as $\frac{1}{2}(1 + k/R)$ and $\frac{1}{2}(1 - k/R)$ for moving one step to the left or to the right, respectively, provided the particle is at the position labeled k, see e. g. Kac (1946). Here, R is a certain integer, and possible positions of the particle are limited by the condition $-R \leq k \leq R$. The transition probability can consequently be written as $W(k; m) = \frac{1}{2}(1 + k/R) \delta_{m,k-1} + \frac{1}{2}(1 - k/R)\delta_{m,k+1}$ in terms of Kronecker's delta.

  The generalization of the equation (27.21) for the probability density becomes

$$P(n|m; s + 1) = \frac{R + m + 1}{2R} P(n|m + 1; s) + \frac{R - m + 1}{2R} P(n|m - 1; s), \tag{27.22}$$

  to be solved with the initial condition $P(n|m; 0) = \delta_{m,n}$. Here, $P(n|m; s + 1)$ denotes the probability for the particle being at the position with label m at time labeled s, provided it was at position n initially.

  It is readily demonstrated (e.g. Wang and Uhlenbeck (1945)) that average values are, for instance

$$\langle m(s) \rangle = \sum_m m P(n|m; s) = \left(1 - \frac{1}{R}\right) \langle m(s - 1) \rangle.$$

  Since $\langle m(s) \rangle = n$ at $s = 0$ we have

$$\langle m(s) \rangle = n \left(1 - \frac{1}{R}\right)^s,$$

  which shows that the average position of the particle approaches zero. Similarly one finds

$$\langle m^2(s) \rangle = n^2 \left(1 - \frac{2}{R}\right)^s + \frac{R}{2}\left[1 - \left(1 - \frac{2}{R}\right)^s\right],$$

  demonstrating that $\langle m^2(s) \rangle \to R/2$ as $s \to \infty$.

In the continuum limit as before, in particular with $R \to \infty$, we find

$$\frac{\partial}{\partial t} P = \gamma \frac{\partial}{\partial x}(xP) + D\frac{\partial^2}{\partial x^2} P, \tag{27.23}$$

with the diffusion coefficient D found before, and $1/(R\tau) \to \gamma$ as $\tau \to 0$. The equation (27.23) is valid for the overdamped case.

- **Example:** As an illustrative example, we consider the dichotomic Markov process. Assume that the particles are moving with uniform, finite velocity $\pm u$ and have a probability α of direction change per unit time. Denoting the probability densities that a particle moves in the positive and in the negative direction by $p^{\pm}$ respectively, the master equation for this process becomes

$$\frac{\mathrm{d}p_n^+}{\mathrm{d}t} = \frac{1-\alpha}{\tau} p_{n-1}^+ + \frac{\alpha}{\tau} p_{n+1}^-, \tag{27.24}$$

and

$$\frac{\mathrm{d}p_n^-}{\mathrm{d}t} = \frac{1-\alpha}{\tau} p_{n+1}^- + \frac{\alpha}{\tau} p_{n-1}^+, \tag{27.25}$$

with a self-evident derivation and interpretation. With a Taylor expansion $p_{n\pm1}^{\pm} \approx p^{\pm} \pm \Delta x \, \mathrm{d}p_n^{\pm}/\mathrm{d}x$ and defining particle probability fluxes and densities as $F = u(p^+ - p^-)$ and $C = (p^+ + p^-)/\Delta x$, respectively, we find an equation for F and C. Eliminating F we find

$$\frac{\partial^2 C}{\partial t^2} + \frac{2\alpha}{\tau}\frac{\partial C}{\partial t} + (2\alpha - 1)u^2 \frac{\partial^2 C}{\partial x^2} = 0. \tag{27.26}$$

When a velocity-change frequency $a \equiv \alpha/\tau$ is introduced, a limiting case can be considered, where $\tau \to 0$, $\Delta x \to 0$, $\alpha \to 0$, $\alpha/\tau = a$ and $\Delta x/\tau = u$. The result is the telegraph equation

$$\frac{\partial^2 C}{\partial t^2} + 2a\frac{\partial C}{\partial t} - u^2 \frac{\partial^2 C}{\partial x^2} = 0. \tag{27.27}$$

The diffusion equation results in the limit where $u \to \infty$, $a \to \infty$, in such a way that $u^2/a \to D$. The generalization obtained by describing the classical Brownian motion using the telegraph equation rather than the diffusion equation is immaterial, but (27.27) can be used with advantage for a number of other problems. It has been applied to diffusion caused by random, turbulent, motions in fluids, for instance.

The random walk model may be generalized by introducing a statistical correlation between two successive steps, in such a way that the probability α for a step in the same direction as the previous step differs from the probability β for a step back; 'random walk with persistence'. In this case the position variable, Y, no longer constitutes a Markov process, because its probability density at time t_n

depends not only on its value at this time but also on its value at t_{n-1}. However, the Markov property can be restored by introducing the previous position explicitly as an additional variable. The resulting two-component process, (Y_n, Y_{n-1}), is again Markovian, and this model can again be treated in full detail. A stochastic process that can be made Markovian by means of one additional variable is called 'Markovian of second degree', and if more variables are needed it is Markovian of some higher degree (van Kampen 1981, Risken 1984).

- **Example:** Consider the Brownian motion of a simple pendulum in contact with a surrounding gas in thermal equilibrium. The fluctuations in the position of the pendulum can be assumed to be driven by Gaussian random noise. This problem serves to illustrate a case where the displacement $X(t)$ is *not* a Markov process. It is the *projection* of the two-dimensional (Gaussian) Markov process $X(t)$, $p(t)$, where $p(t) \equiv \mathrm{d}X(t)/\mathrm{d}t$. In this case the correlation function has to be extended to a correlation matrix

$$\|R\| = \left\| \begin{matrix} \langle x(t)x(t+\tau)\rangle & \langle x(t)p(t+\tau)\rangle \\ \langle p(t)x(t+\tau)\rangle & \langle p(t)p(t+\tau)\rangle \end{matrix} \right\|, \tag{27.28}$$

where the elements are determined by the appropriate Langevin equation (26.8), see section 26.2.

- **Exercise:** Write (27.20) in the symmetric form $P(m; s) = \frac{1}{2}[P(m+1; s+1) + P(m-1; s-1)]$ and demonstrate that the solution is

$$P(n|m, s) = \frac{s!}{\left(\frac{1}{2}(\nu+s)\right)!\left(\frac{1}{2}(\nu-s)\right)!}\left(\frac{1}{2}\right)^s,$$

with $P(m, 0) = \delta_{n,m}$ and $\nu = |n - m|$. See e.g. Wang and Uhlenbeck 1945).

27.3 Correlation functions and spectra

The master equation, or the Fokker–Planck equation, allows calculation of the correlation function and the fluctuation spectrum for time stationary processes. It is advantageous to approach the problem by the two-time correlation function defined as $\mathcal{R}(\tau) = \langle y(t_1)y(t_1+\tau)\rangle$ where $y(t)$ is the appropriate variable for the problem. For time stationary processes we may as well take $t_1 = 0$, i.e. $\mathcal{R}(\tau) = \langle y(0)y(\tau)\rangle$. The correlation function is determined by first obtaining the conditional probability density subject to the condition that $y = y_0$ at $t = 0$. The conditional average is then $\langle y, \tau|y_0, 0\rangle = \int_{-\infty}^{\infty} yP(y, \tau|y_0, 0)\,\mathrm{d}y$ which is explicitly a function of τ. In the next step we find

$$\mathcal{R}(\tau) = \int_{-\infty}^{\infty} \langle y, \tau|y_0, 0\rangle y_0 P(y_0)\mathrm{d}y_0,$$

where $P(y_0)$ is the equilibrium solution of the probability density. The fluctuation spectrum can then, if desired, be obtained by Fourier transforming $\mathcal{R}(\tau)$. By a procedure similar to the one outlined here the entire two point probability density can be determined, and also N-point probability densities.

27.3.1 Correlation functions for Gaussian Markov processes

A particular example of a Markov process is one where the statistics are Gaussian, in which case the entire statistical information is contained in the correlation function. Being a Markov process on the other hand, it is completely defined by a joint probability distribution from which the correlation function can be derived so it seems natural to expect that the Markov property imposes some restrictions of $R(t_i, t_j)$. In order to determine the nature of these restrictions we first consider the joint probability density for a Gaussian random process

$$P_2(y_1, t_1; y_2, t_2) = (2\pi)^{-1}\,|A|^{-1/2}\exp\left(-\frac{1}{2|A|}\sum_{i,j=1}^{2} A_{ij}y_iy_j\right), \tag{27.29}$$

where A_{ij} is the co-factor of

$$R(t_i, t_j) \equiv \left\langle Y(t_i)Y(t_j)\right\rangle = \iint_{-\infty}^{\infty} y_1y_2P_2(y_1, t_i; y_2, t_j)\mathrm{d}y_1\mathrm{d}y_2,$$

in the correlation matrix

$$\|A\| = \begin{Vmatrix} R(t_1, t_1) & R(t_1, t_2) \\ R(t_2, t_1) & R(t_2, t_2) \end{Vmatrix}, \tag{27.30}$$

and $|A|$ is the determinant of the matrix, see also section 19.2.5. Also

$$P_3(y_1, t_1; y_2, t_2; y_3, t_3) = (2\pi)^{-3/2}\,|B|^{-1/2}\exp\left(-\frac{1}{2|B|}\sum_{i,j=1}^{3} B_{ij}y_iy_j\right), \tag{27.31}$$

where B_{ij} is the co-factor of $R(t_i, t_j)$ in the correlation matrix

$$\|B\| = \begin{Vmatrix} R(t_1, t_1) & R(t_1, t_2) & R(t_1, t_3) \\ R(t_2, t_1) & R(t_2, t_2) & R(t_2, t_3) \\ R(t_3, t_1) & R(t_3, t_2) & R(t_3, t_3) \end{Vmatrix}, \tag{27.32}$$

and $|B|$ is the determinant of the matrix. Bayes' rule relates the conditional probability $P(y_3, t_3|y_1, t_1; y_2, t_2)$ to (27.29) and (27.31) by

$$\begin{aligned} P(y_3, t_3|y_1, t_1; y_2, t_2) &= \frac{P_3(y_1, t_1; y_2, t_2; y_3, t_3)}{P_2(y_1, t_1; y_2, t_2)} \\ &= \sqrt{\frac{|A|}{2\pi|B|}}\exp\left(-\frac{1}{2|B|}\sum_{i,j=1}^{3} B_{ij}y_iy_j + \frac{1}{2|A|}\sum_{i,j=1}^{2} A_{ij}y_iy_j\right) \end{aligned} \tag{27.33}$$

Note that the summation index in the first term in the exponent goes to 3 according to (27.31) but only to 2 in the second term according to (27.29).

In order for $P(y_3, t_3|y_1, t_1; y_2, t_2)$ to be independent of y_1, t_1 as required for a Markov process, it is necessary that

$$B_{13} = 0 = B_{31} \tag{27.34}$$

and

$$\frac{B_{11}}{A_{11}} = \frac{B_{12}}{A_{12}} = \frac{B_{21}}{A_{21}} = \frac{|B|}{|A|}. \tag{27.35}$$

The first condition $B_{13} = 0$ implies

$$R(t_2, t_2)R(t_3, t_1) = R(t_3, t_2)R(t_2, t_1). \tag{27.36}$$

As yet, no use has been made of the assumption that the signal is time stationary in a statistical sense, implying for instance $R(t_i, t_j) = R(t_i - t_j)$. With this hypothesis, together with normalizing the auto-correlation function so that $R(0) = 1$ which is no loss of generality, the relation (27.36) becomes

$$R(t_3 - t_1) = R(t_3 - t_2)R(t_2 - t_1). \tag{27.37}$$

This last equation, in addition to satisfying $B_{13} = 0$, also makes the various parts of (27.35) valid. The problem now is to solve the functional relation (27.37). This is done by introducing $\tau = t_3 - t_2$ and $\nu = t_2 - t_1$. Then $\tau + \nu = t_3 - t_1$ and (27.37) states

$$R(\tau + \nu) = R(\tau)R(\nu) \quad \text{for all} \quad \tau, \nu > 0. \tag{27.38}$$

Trivial solutions are offered by $R(\tau)$ identically equal to 0 or 1. If $R(\nu_0) = 0$ at one point ν_0, then $R(\tau + \nu_0) = 0$ for all $\tau > 0$ implying $R(\tau) = 0$ for $\tau > \nu_0$. Hence $R(\tau)$ cannot equal zero at any point without obtaining the trivial result that $R(\tau)$ equals zero for all τ to the right of that point. Since, by construction, $R(0) = 1$, this means that $R(\tau) \geq 0$ for all τ. Excluding the equality sign one has $R(\tau) > 0$ for all τ. It is then legitimate to take the logarithm of both sides of (27.38) since all quantities are positive. Thus

$$\ln R(\tau + \nu) = \ln R(\tau) + \ln R(\nu). \tag{27.39}$$

If $R(\tau)$ is analytic at $\tau = 0$, that is, has a power series expansion about $\tau = 0$, the same is true of $\ln R(\tau)$. Assume

$$\ln R(\tau) = \sum_{n=0}^{\infty} c_n \tau^n, \tag{27.40}$$

where the coefficients c_n are to be determined. Then, from (27.39)

$$\sum_0^{\infty} c_n(\tau + \nu)^n = \sum_0^{\infty} c_n \tau^n + \sum_0^{\infty} c_n \nu^n. \tag{27.41}$$

If at all, this relation is satisfied for all (τ, ν), provided $c_n = 0$ for $n = 0$ and for all $n \geq 2$. Hence $\ln R(\tau)$ must be

$$\ln R(\tau) = c_1 \tau. \tag{27.42}$$

It is *a priori* known that $R(\tau)$ is an even function of τ which approaches zero as $\tau \to \infty$. This implies that the coefficient c_1 is negative. Therefore, letting $c_1 = -\beta$ where $\beta > 0$, the final result is

$$\ln R(\tau) = -\beta|\tau|.$$

Solving for $R(\tau)$, we find

$$R(\tau) = e^{-\beta|\tau|}, \qquad \text{with} \qquad \beta > 0. \tag{27.43}$$

This completes the desired proof (Bendat 1958) and also demonstrates the fundamental importance of exponential auto-correlation functions. From this theorem a strong theoretical justification exists to expect many random noise phenomena to exhibit an exponential auto-correlation function or something close to it.

The spectrum for a Gaussian Markov process is obtained by Fourier transforming the correlation function (the Wiener–Khinchine theorem) and it is found to have the universal, Lorentzian, form

$$S(\omega) = \frac{2\beta}{\beta^2 + \omega^2}. \tag{27.44}$$

The micro time-scale defined as $[-\mathrm{d}^2 R(\tau)/\mathrm{d}\tau^2|_{\tau=0}]^{-1/2}$ vanishes for correlation functions of the form (27.43). This is consistent with the Markov property, i.e. the lacking memory of the process. Correspondingly, we have a divergence in the derivative process, $\langle (\mathrm{d}Y(t)/\mathrm{d}t)^2 \rangle = \int \omega^2 S(\omega)\, \mathrm{d}\omega \to \infty$, indicating that Gaussian Markov processes usually have singularities in their time derivatives.

When modeling actual spectra, the result obtained here is often considered too restrictive. Deviations from Gaussianity are allowed for and correlation functions of the form $R(\tau) = A\mathrm{e}^{-\beta|\tau|}\cos \Omega t$ are assumed, where the constants β and Ω are obtained by fits to results based on actual data. The corresponding frequency spectrum becomes

$$S(\omega) = \frac{2A\beta}{\pi}\left[\frac{\omega^2 + \left(\beta^2 + \Omega^2\right)}{\omega^4 + 2\left(\beta^2 - \Omega^2\right)\omega^2 + \left(\beta^2 + \Omega^2\right)^2}\right]. \tag{27.45}$$

When $3\Omega^2 > \beta^2$, this spectrum has a peak for $\omega_1 = (\beta^2 + \Omega^2)^{1/4}[2\Omega - (\beta^2 + \Omega^2)^{1/2}]^{1/2}$ with a peak power $S(\omega_1) = (2A\beta/\pi)/(\Omega^2 + \beta^2)$. Surprisingly many experimentally obtained power spectra can be approximated by (27.45) by proper choice of Ω and β.

- **Example:** As an important special case we consider an amplitude and phase modulated harmonic oscillation, where one record, or one realization, labeled k is given as

$${}^k s(t) = {}^k a(t)\sin\left(\Omega t + {}^k\theta\right),$$

where Ω is constant, while a and θ are mutually independent random variables, where in particular θ is uniformly distributed over the interval $\{0; 2\pi\}$. Suppose that the auto-correlation of a is given by $R_{aa}(\tau) \equiv \langle a(t)a(t+\tau)\rangle = A\exp(-\beta|\tau|)$. The auto-correlation of $s(t)$ is then readily obtained as

$$
\begin{aligned}
R_{ss}(\tau) &= \langle a(t)\sin(\Omega t + \theta(t))a(t+\tau)\sin(\Omega(t+\tau) + \theta)\rangle \\
&= R_{aa}(\tau)\frac{1}{2\pi}\int_0^{2\pi} \sin(\Omega t + \theta)\sin(\Omega(t+\tau) + \theta)\mathrm{d}\theta \\
&= \frac{1}{2}A\mathrm{e}^{-\beta|\tau|}\cos(\Omega\tau),
\end{aligned}
$$

demonstrating a possible way of obtaining an exponential-cosine correlation function.

27.4 Relevance for turbulence

A Markov process is one with no or limited memory, and it may be appropriate to ask to what extent such a model can be applied to turbulence. It is easily argued that as a general model, the Markov approach is not relevant for turbulence modeling, but it may be more constructive to ask what it can do right. Does a limiting case exist where a Markov process is useful? One such case is found in section 25.1, where a diffusion limit was demonstrated for turbulent transport. A diffusion equation is a simple example of a Fokker–Planck equation and in the following we discuss one particular application of this sort.

27.4.1 One test of the Markov assumption in turbulence

We consider a diffusion problem that has relevance in nature, where details are discussed by Kiørboe (2008) and with special attention to the experimental conditions by Mann *et al* (2005). We want an estimate for the turbulent flux to a spherical surface moving with the flow in such a way that its center at all times assumes the local fluid velocity. The fluid (or air when we discuss atmospheres) has an initially uniform distribution of a contaminant. With radius in this center we have a spherical surface that absorbs all of the contaminating particles that arrive at the surface. We want to model the time variation of the particle flux, in particular its time stationary asymptotic value. The problem has relevance for small organisms (plankton) in turbulent environments, where a predator can be seen as a perfect absorber of prey that enters its ‘sphere of interception’. In a more general sense the problem is relevant to absorbing closed surfaces moving in turbulent media.

Dimensional arguments

The present problem is characterized by a few dimensional quantities. With the viscosity, ν, being immaterial for the flow dynamics on scale lengths larger than the Kolmogorov length scale $\eta \equiv (\nu^3/\epsilon)^{1/4}$, we only have one quantity characterizing the flow, namely ϵ with dimension *length*^{2}*time*$^{-3}$, and the length scale $\mathcal{R} < \mathcal{L}_{\rm int}$ characteristic for the particular problem, here a moving sphere of interception. Out of these quantities the only combination giving a quantity with dimension *time* is $\mathcal{R}^{2/3}/\epsilon^{1/3}$, while $\epsilon^{1/3}\mathcal{R}^{7/3}$ gives *length*^{3}/*time*. The physical dimension of the averaged normalized particle flux $\langle J\rangle/n_0$ is *length*^{3}/*time*.

Quite generally it can then be argued, by purely dimensional reasoning, that the turbulent flux for given reference density n_0 must have the form

$$\frac{\langle J\rangle}{n_0} = \epsilon^{1/3}\mathcal{R}^{7/3}f\left(t\epsilon^{1/3}/\mathcal{R}^{2/3}\right), \tag{27.46}$$

with f being a dimensionless function of a dimensionless variable. The actual form of f can only be determined by a more detailed model analysis, for instance by a laboratory experiment. If we can argue that a steady state asymptotic limit exists where $f(\xi \to \infty)\to$ const., then this limit must scale as

$$\frac{\langle J\rangle}{n_0} \sim C_\infty\, \epsilon^{1/3}\mathcal{R}^{7/3}, \tag{27.47}$$

where C_∞ is a universal constant.

- **Example:** By the reasoning before we implicitly assumed that the turbulent flux *has* a non-trivial asymptotic limit. This need not be so as the following discussion will show.

  An interesting problem described by the diffusion equation with imposed boundary conditions deals with absorbing surfaces in two and three spatial dimensions.

1) Consider first a three-dimensional problem with a perfectly absorbing spherical surface with radius R. This condition is specified by requiring $n(r = R, t) = 0$. The sphere can itself participate in the diffusive motion (Chandrasekhar 1954). With the given symmetry we write the ∇^2-operator in the form that gives the diffusion equation

$$\frac{\partial}{\partial t}[r\, n(r, t)] = D\frac{\partial^2}{\partial r^2}[r\, n(r, t)], \tag{27.48}$$

where radius, r, and time, t, are the only independent variables (Chandrasekhar 1954, Carslaw and Jaeger 1947). For $r > R$ the solution of this equation, with the given boundary conditions, is

$$n(r, t) = n_0\left[1 - \frac{R}{r} + \frac{2R}{r\sqrt{\pi}}\int_0^{(r-R)/2\sqrt{Dt}} \exp(-\xi^2)\mathrm{d}\xi\right]. \tag{27.49}$$

The particle flux through a sphere with radius R is then $4\pi R^2 D\ \partial n/\partial r|_{r=R}$. We find

$$4\pi r^2 D\frac{\partial}{\partial r}n(r, t)\bigg|_{r=R} = 4\pi DRn_0\left(1 + \frac{R}{\sqrt{\pi Dt}}\right), \tag{27.50}$$

in units of number of particles per unit time, with n_0 being the particle concentration at $r \to \infty$, where n_0 is independent of t. We note an asymptotic saturation level for the diffusive particle flux to be $4\pi D R n_0$, i.e. a constant. The asymptotic result can be made equivalent with (27.47) if we choose $D = (C_\infty/4\pi)\epsilon^{1/3}\mathcal{R}^{4/3}$. The choice is not unreasonable: an effective turbulent diffusion coefficient must depend on ϵ somehow and the scaling with $\mathcal{R}$ is reasonable as well. Eddies much larger than $\mathcal{R}$ will simply move the sphere and its near surroundings as an entity, and will therefore not contribute to the relative motion between the sphere and particles around it. Turbulent eddies with scale sizes much smaller than $\mathcal{R}$ will be ineffective since they contain very little energy according to the Kolmogorov–Oubokhov power spectrum, so the most effective eddies are those with scale-sizes $\sim\mathcal{R}$. The only dimensionally correct combination of $\mathcal{R}$ and ϵ giving a diffusion coefficient is the one given, apart as usual from a numerical constant.

2) For two spatial dimension the result is quite different. Here the diffusion equation becomes

$$\frac{\partial}{\partial t}[r\, n(r, t)] = D\frac{\partial}{\partial r}r\frac{\partial}{\partial r}\, n(r, t), \tag{27.51}$$

assuming cylindrical symmetry. After some algebra (Carslaw and Jaeger 1947) we find the space-time varying particle density to be

$$n(r, t) = n_0\,\frac{2}{\pi}\int_0^\infty \frac{J_0(\lambda R)Y_0(\lambda r) - Y_0(\lambda R)J_0(\lambda r)}{J_0^2(\lambda\mathcal{R}) + Y_0^2(\lambda\mathcal{R})}\frac{\mathrm{e}^{-\lambda^2 Dt}}{\lambda}\mathrm{d}\lambda$$

fulfilling the imposed condition $n(r = R, t) = 0$. By differentiation of $n(r, t)$ we obtain the diffusive particle flux through a cylinder segment of length $\mathcal{L}$ and radius R to be

$$\mathcal{L}\,2\pi r D\frac{\partial}{\partial r}n(r, t)\bigg|_{r=R} = \mathcal{L}\mathcal{R}Dn_0\,4\int_0^\infty \frac{J_1(\lambda R)Y_0(\lambda R) - Y_1(\lambda R)J_0(\lambda R)}{J_0^2(\lambda\mathcal{R}) + Y_0^2(\lambda\mathcal{R})}\,\mathrm{e}^{-\lambda^2 Dt}\mathrm{d}\lambda,$$

in terms of Bessel functions J_0, Y_0, J_1 and Y_1. It turns out (Carslaw and Jaeger 1947) that this result can be written as

$$\mathcal{L}\,2\pi r D\frac{\partial}{\partial r}n(r, t)\bigg|_{r=R} = \mathcal{L}\mathcal{R}Dn_0\frac{8}{\pi}\int_0^\infty \frac{\mathrm{e}^{-\lambda^2 Dt}}{J_0^2(\lambda\mathcal{R}) + Y_0^2(\lambda\mathcal{R})}\frac{\mathrm{d}\lambda}{\lambda}. \tag{27.52}$$

The interesting point here is that the asymptotic limit of the particle flux is zero! A perfect absorber in cylindrical symmetry eventually sucks up all the particles in the surrounding and n_0 is here the initial particle concentration. In the three spatial dimensions version of the problem it is not so, there the spatially asymptotic concentration remains at its initial value.

For proving the cylindrical case you might need the identity

$$1 = \frac{2}{\pi}\int_0^\infty \frac{J_0(\lambda R)Y_0(\lambda r) - Y_0(\lambda R)J_0(\lambda r)}{J_0^2(\lambda R) + Y_0^2(\lambda R)}\frac{d\lambda}{\lambda},$$

in terms of the Bessel functions J_0 and Y_0 of first and second kinds, respectively (Abramowitz and Stegun 1972).

A model diffusion equation

The particle flux to a perfectly absorbing sphere, which is moving with the flow has been modeled by, for instance, a simple diffusion equation with a properly chosen diffusion coefficient, depends on the simultaneous mean square velocity differences obtained at given spatial separations, but independent of time (Osborn 1996). Essentially, the argument is based on the second order structure function

$$\Psi_2(r) \equiv \langle (u_r(0, t) - u_r(r, t))^2 \rangle \approx C_K(\epsilon r)^{2/3}, \tag{27.53}$$

with the approximation being valid for separations r smaller than the length scale of the turbulence, i.e. the outer scales. A diffusion coefficient is constructed from a characteristic velocity and a characteristic length. The velocity is taken to be $\sqrt{\Psi_2(r)}$. For the limiting form expressed in (27.53), the only length characterizing the two particles is their separation r. The resulting diffusion coefficient is consequently $K(r) \sim r^{4/3}\epsilon^{1/3}$. The proposed diffusion equation for the density n is actually identical to the one suggested by Richardson in his study of distance-neighbor functions (Richardson 1926)

$$\frac{\partial}{\partial t}n(r, t) = C\frac{\epsilon^{1/3}}{r^2}\frac{\partial}{\partial r}r^{10/3}\frac{\partial}{\partial r}n(r, t). \tag{27.54}$$

The result is written for spherically symmetric geometry, with r being the radial coordinate, measured from the position of the center of the reference sphere, and C is a numerical constant, assumed to be universal. While (27.54) was here argued by dimensional reasoning, it has also an analytical derivation (Roberts 1961). As a consequence of (27.54) we have the well known result for the mean-square separation of two initially close particles $\langle r^2 \rangle = C_R\epsilon t^3$, with the Richardson's constant being $C_R \approx 0.5$ (Ott and Mann 2000). We have the relation $C = (3/2)(3C_R/143)^{1/3}$, giving $C \approx 0.33$.

For a diffusion equation of the form $\partial n/\partial t = \nabla \cdot D\nabla n$, possibly with $D = D(\mathbf{r})$, we have the flux density to be $-D\nabla n$, see also the discussion of (21.31). In the present model, the time varying diffusion flux of particles to a perfectly absorbing sphere with radius $\mathcal{R}$ is given by $4\pi C\epsilon^{1/3}\mathcal{R}^{10/3}\mathrm{d}n(r, t)/\mathrm{d}r|_{r=\mathcal{R}}$, with $n(r, t)$ obtained from (27.54).

The derivation of (27.54) assumes that ϵ is a deterministic constant, and thereby ignores intermittency corrections (Boffetta *et al* 1999). Although the relation (27.54) had some experimental support from the time when it was first proposed (Richardson 1926), and also supported more recently (Ott and Mann 2000), its general validity has been criticized (Batchelor 1952, Roberts 1961), see also the summary by (Ott and Mann 2000). In neutral flows the range of validity of (27.54) is not fully explored. For large separations, a simple diffusion equation, with constant diffusion coefficient, is expected to apply, as indicated for instance by experimental results (Virant and Dracos 1997) for initial conditions having scales larger than the integral length scale. These cases (Ott and Mann 2000, Virant and Dracos 1997) referred to particle releases considered as initial value problems. It seems that a diffusion equation as (27.54) can indeed be applied for analyzing relative two-particle diffusion in certain variable ranges (Ott and Mann 2000). On the other hand, one cannot expect a diffusion coefficient depending solely on relative times or spatial separations to be universally applicable for this problem (Roberts 1961). In general, a Fokker–Planck equation, with (27.54) being one special example, describes a Markov process, where the probabilities of future states depend solely on the present, and not past ones. Modeling of turbulent displacements as a simple Markov process is known to be rather inaccurate, and a study of the limits of applicability of models like (27.54) is therefore worthwhile.

From (27.54) it is easy to derive (Osborn 1996) a steady state flux to a sphere with radius $\mathcal{R}$ as

$$\frac{J_0}{n_0} = \frac{28\pi}{3} C\epsilon^{1/3}\mathcal{R}^{7/3}, \tag{27.55}$$

where n_0 is the constant particle density at $r \to \infty$. The result is consistent with the previous result obtained by dimensional reasoning, but this *has* to be so. The time variation of the turbulent flux to a perfectly absorbing sphere has been studied by data from a laboratory experiment (Mann *et al* 2005) with results shown in figure 27.2, and also by numerical simulations (Boffetta *et al* 2006). The results from these two studies agree very well, but both differ from the flux solution obtained by (27.54), particularly for short times after release.

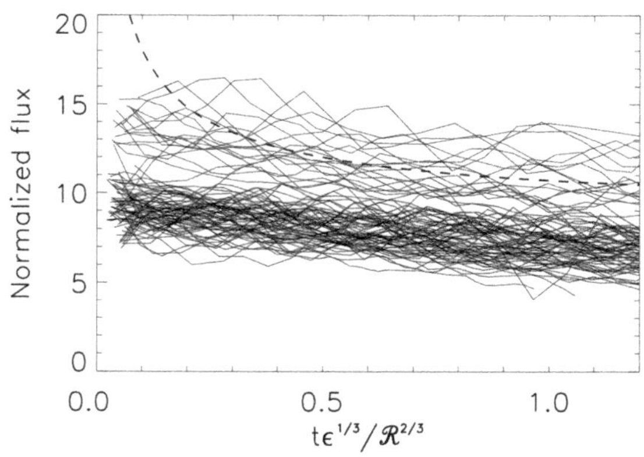

Figure 27.2. Normalized flux, $\langle J\rangle/(n_0\epsilon^{1/3}\mathcal{R}^{7/3})$, to a moving sphere as a function of the normalized temporal variable, $t\epsilon^{1/3}/\mathcal{R}^{2/3}$. Curves are shown for eight realizations with different turbulence conditions, each with curves for different $\mathcal{R}$. The dashed line gives a theoretical result, obtained by solution of the diffusion equation (27.54). The (unphysical) singularity at $t = 0$ for the dashed line solution is due to the assumed infinite initial gradient at $r = \mathcal{R}$. The figure is reproduced from Mann *et al* (2005) with permission from the Cambridge University Press.

The experimental uncertainty on tracer particle positions in the experiment (Mann *et al* 1999, Ott and Mann 2000, Mann *et al* 2005) is approximately 0.02 mm. The idealized step function in particle density is therefore not necessarily a correct representation for our experimental conditions. For the numerical simulations (Boffetta *et al* 2006) the uncertainty is much smaller. In spite of this minor uncertainty we can learn two things by the results in figure 27.2. First of all, a diffusion equation is completely wrong for describing the initial phase, up to a turn over time for eddies of scale size $\mathcal{R}$. Asymptotically, the results seems OK, but parts of these we could have guessed by dimensional reasoning almost as well. Another lesson is that the numerical constant does not look right either, although it is reasonable within a factor of two or so. Considering that we started out with an empirically well known Kolmogorov constant, we could have expected a better agreement, but then again, the modeling of the diffusion constant has some unspecified shortcomings as a concept. The conclusion seems to be that a diffusion equation, or more generally a Fokker–Planck equation, can serve to describe asymptotic large scale limits, but cannot be a general model for turbulence phenomena.

References

Abramowitz M and Stegun I A 1972 *Handbook of Mathematical Functions with Formulas, Graphs, and Mathematical tables* (New York: Dover)

Balescu R 1997 *Statistical Dynamics. Matter out of Equilibrium* (London: Imperial College Press)

Batchelor G K 1952 Diffusion in a field of homogeneous turbulence II. The relative motion of particles *Proc. Cambridge Philos. Soc. (London)* **48** 345–62

Bendat J S 1958 *Principles and Applications of Random Noise Theory* (New York: Wiley)

Boffetta G, Celani A, Crisanti A and Vulpiani A 1999 Pair dispersion in synthetic fully developed turbulence *Phys. Rev.* E **60** 6734–41

Boffetta G, Pécseli H L and Trulsen J 2006 Numerical studies of turbulent particle fluxes into perfectly absorbing spherical surfaces *J. Turbul.* **7** N22

Carslaw J C and Jaeger H S 1947 *Conduction of Heat in Solids* (Oxford: Clarendon Press)

Chandrasekhar S 1954 Stochastic problems in physics and astronomy *Selected Papers on Noise and Stochastic Processes* N Wax (New York: Dover) pp 3–91

Cox D R and Miller H D 1965 *The Theory of Stochastic Processes* (London: Chapman and Hall)

Kac M 1946 Random walk and the theory of Brownian motion *Am. Math. Monthly* **54** 369 Reprinted in: Kac M 1954 *Selected Papers on Noise and Stochastic Processes* N Wax (New York: Dover) pp 295–317

Kiørboe T 2008 *A Mechanistic Approach to Plankton Ecology* (Princeton, NJ: Princeton University Press)

Lax M 1960 Fluctuations from the nonequilibrium steady state *Rev. Modern Phys.* **32** 25–64

Mann J, Ott S and Andersen J S 1999 Experimental study of relative, turbulent diffusion Technical Report Risø-R-1036(EN), Risø National Laboratory, DK-4000 Roskilde, Denmark

Mann J, Ott S, Pécseli H L and Trulsen J 2005 Turbulent particle flux to a perfectly absorbing surface *J. Fluid Mech.* **534** 1–21

Narasimhan T N 1999 Fourier's heat conduction equations: history, influence, and connections *Rev. Geophys.* **37** 151–72

Osborn T 1996 The role of turbulent diffusion for copepods with feeding currents *J. Plankton Res.* **18** 185–95

Ott S and Mann J 2000 An experimental investigation of the relative diffusion of particle pairs in three dimensional turbulent flow *J. Fluid Mech.* **422** 207–23

Papoulis A 1991 *Probability, Random Variables and Stochastic Processes* 3rd edn (New York: McGraw-Hill)

Richardson L F 1926 Atmospheric diffusion shown on a distance-neighbour graph *Proc. R Soc. London* A **110** 709–37

Risken H 1984 *The Fokker-Planck equation: Methods of Solution and Applications* (Berlin: Springer)

Roberts P H 1961 Analytical theory of turbulent diffusion *J. Fluid Mech.* **11** 257–83

van Kampen N G 1981 *Stochastic Processes in Physics and Chemistry* (Amsterdam: North-Holland)

Virant M and Dracos T 1997 3D PTV and its application on Lagrangian motion *Meas. Sci. Technol.* **8** 1529–52

Wang M C and Uhlenbeck G E 1945 On the theory of Brownian motion II *Rev. Mod. Phys.* **17** 323–42

Reprinted in: Wang M C and Uhlenbeck G E 1954 *Selected Papers on Noise and Stochastic Processes* N Wax (New York: Dover) pp 113–32

IOP Publishing

Low Frequency Waves and Turbulence in Magnetized Laboratory Plasmas and in the Ionosphere

Hans Pécseli

Chapter 28

The quasi-normal approximation

The pioneering approach for describing a turbulent field was essentially based on dimensional arguments. In modern physics such methods may not be held in the same high esteem they once had, but one should keep in mind that it was a bold and novel idea of making predictions not of individual flow realizations but of statistical ensemble averages. The road for this approach may have been paved by the statistical mechanics of gases and the novel approach (at that time) to quantum physics in terms of the wave-function as described by the Schrödinger equation.

Some of the first approaches to analytical models for the space time evolution of turbulent flows (or more generally: random vector fields) were based on the observation that the statistical averages of the velocity fields were close to Gaussians. The general ideas are best illustrated by considering some model equations, and this has been done by many. Here we mostly follow expositions by Orszag (1970), Kono and Ichikawa (1973), Cook (1972), Nicholson (1983) and Diamond *et al* (2010), taking the basic equations to be

$$\frac{\partial}{\partial t}A_k + i\beta_k A_k = \epsilon \sum_{j,\ell} M_{j,\ell}^{\{-k\}} A_j A_\ell, \tag{28.1}$$

where the As can represent a wave amplitude of a Fourier component specified by their subscript, with ϵ being a parameter measuring the strength of the nonlinear interaction. In the given form we have that (28.1) includes the Fourier transformed version of the Navier–Stokes equation (16.3). The interaction matrix should properly be written by vectors as $M_{\mathbf{j},\mathbf{l}}^{\{\mathbf{k}\}}$, but for simplicity we omit the bold-face vector notation in the following. The model will be assumed to have some simplifying properties.

1) We assume that β_k is real for all k and as we argue later also $\beta_k = -\beta_{-k}$.
2) We choose to consider cases where we have symmetry in the indices as $M_{j,\ell}^{\{k\}} = M_{\ell,j}^{\{k\}}$.

doi:10.1088/978-0-7503-1251-6ch28

3) With A_k being the spatial Fourier transform of the space-time varying scalar wave-field we have

$$A_k = \frac{1}{\mathrm{Vol}} \int A(r, t) \exp(ikr) \mathrm{d}r.$$

Since only real $A(r, t)$ will be relevant we have the reality condition $A_{-k} = A_k^*$ and consequently $\beta_k = -\beta_{-k}$ and $M_{j,\ell}^{\{k\}*} = M_{-j,-\ell}^{\{-k\}}$. One implication of this assumption together with assumption **1**) is that when the linearized version of (28.1) has a solution $A_k(t) = A_{k0} \exp(-i\beta_k t)$ with A_{k0} being some initial value then also $A_k(t)^* = A_{k0}^* \exp(i\beta_k t) = A_{-k0} \exp(-i\beta_{-k} t)$ is a solution for that linear equation.

4) We assume the spatial average field to be vanishing, i.e. $\frac{1}{\mathrm{Vol}} \int A(r, t) \mathrm{d}r = 0$ giving $A_0 = 0$ and $M_{\ell,j}^{\{0\}} = 0$, where we here and in the following implicitly assume the limit where the volume extends to infinity here denoted by Vol $\to \infty$.

5) We assume the problem to be time-reversible, i.e. without losses: this is a non-trivial restriction consistent with assumption **1**), and it simplifies the analysis. For nonlinear wave interaction related to the Vlasov equation the assumption seems reasonable but here we make it without specific references. Assuming $A_k(t)$ to be a solution of (28.1) we define the criterion for time reversibility to be that $A_{-k}(-t)$ is a solution as well. By insertion we can determine the condition for time reversibility in this sense as the condition $M_{j,\ell}^{\{-k\}*} = -M_{j,\ell}^{\{-k\}}$, meaning that $M_{j,\ell}^{\{-k\}}$ is imaginary.

6) As a simplification of the notation we can introduce a new variable to replace A_k by taking $E_k(t) \equiv A_k(t) \exp(i\beta_k t)$. The dynamic equation for E_k is then found as

$$\frac{\partial}{\partial t} E_k = \epsilon \sum_{j,\ell} M_{j,\ell}^{\{-k\}} \mathrm{e}^{i(\beta_k - \beta_j - \beta_\ell)t} E_j E_\ell \equiv \epsilon \sum_{j,\ell} L_{j,\ell}^{\{-k\}} E_j E_\ell, \tag{28.2}$$

where we introduced the abbreviation $L_{j,\ell}^{\{-k\}} \equiv M_{j,\ell}^{\{-k\}} \mathrm{e}^{\mathrm{i}(\beta_k - \beta_j - \beta_\ell)t}$. Note that we had $M_{j,\ell}^{\{-k\}}$ constant in time, but that $L_{j,\ell}^{\{-k\}} = L_{j,\ell}^{\{-k\}}(t)$ is time-varying due to the $\mathrm{e}^{i(\beta_k - \beta_j - \beta_\ell)t}$-exponential. Implied in our assumption **5**) was that $M_{j,\ell}^{\{-k\}}$ are all imaginary, but $L_{j,\ell}^{\{-k\}}$ are complex.

Assuming that we have a really large number of wave modes interacting, we give up the hope of obtaining exact results for individual realizations, and will be content with some statistical averages. The simplest one is trivial with the given assumptions: $\langle E_k \rangle = 0$. A non-trivial average is

$$\langle E_k E_j \rangle = \frac{1}{\mathrm{Vol}^2} \int\int \mathrm{e}^{ikr} \mathrm{e}^{ijr'} \langle E(r) E(r') \rangle \mathrm{d}r \mathrm{d}r', \tag{28.3}$$

where we first write the double Fourier transform and then take an ensemble average. We now assume that the system is spatially homogeneous: by this we mean that the E-correlation depends on spatial separation only and not on the absolute positions i.e. $\langle E(r)E(r')\rangle = Q(r - r')$. Introducing a new variable $\xi \equiv r - r'$ we can introduce the Fourier transform $Q_k = \frac{1}{\mathrm{Vol}}\int Q(\xi)\exp(\mathrm{i}k\xi)d\xi$ and rewrite the double integral in (28.3) to give

$$\langle E_k E_j \rangle = Q_k \frac{1}{\mathrm{Vol}} \int e^{\mathrm{i}(k+j)r}\mathrm{d}r = Q_k \delta(k + j), \tag{28.4}$$

where we used basic relations for δ-functions (Pécseli 2000). The relation (28.4) implies that $Q_k = \langle E_k E_{-k}\rangle = \langle |E_k|^2\rangle$ because of the reality condition mentioned before. By Parseval's theorem (Champeney 1973) we have

$$\sum_k |E_k|^2 = \frac{1}{\mathrm{Vol}} \int E^2(r, t)\mathrm{d}r, \tag{28.5}$$

relating Q_k to the mean-square average of the spatially varying field. By the conditions listed before we have $|E_k|^2 = |A_k|^2$, but in general the relation between higher order averages of $E_k(t)$ and $A_k(t)$ can become lengthy, although they are formally easy to write down.

We can now repeat the entire procedure and demonstrate that $\langle E_k E_j E_\ell\rangle = R_{k,j}\delta(k + j + \ell)$ and similarly for higher order moments. We have seen similar results in section 20.4 but there they referred to a specific model for synthetic data.

We now proceed to obtain a set of equations for statistical averages. It turns out that there are several ways to do this, but we start with the seemingly most obvious one and use (28.2) to find

$$\frac{\partial}{\partial t} Q_k = \epsilon \sum_{j,\ell} L_{j,\ell}^{\{-k\}} \langle E_k E_j E_\ell \rangle + \text{complex conjugate} \tag{28.6}$$

found by multiplying (28.2) with E_{-k}, ensemble averaging and then adding the result to its complex conjugate. To find an equation for Q_k we need information concerning $\langle E_k E_j E_\ell\rangle$. No problem: we repeat the previous procedure, i.e. multiply (28.2) by $E_j E_\ell$ and take the average to find

$$\frac{\partial}{\partial t}\langle E_k E_j E_\ell \rangle = \overset{\text{perm}}{\{k, j, \ell\}} \left\langle E_k E_j \frac{\partial}{\partial t} E_\ell \right\rangle = \epsilon \overset{\text{perm}}{\{k, j, \ell\}} \sum_{m,n} L_{m,n}^{\{-\ell\}} \langle E_k E_j E_m E_n \rangle,$$

where we used (28.2) and introduced the notation $\overset{\text{perm}}{\{k, j, \ell\}}$ to indicate permutations of the indices in the { } parentheses. We learn that an equation for $\langle E_k E_j E_\ell\rangle$ requires information concerning $\langle E_k E_j E_m E_n\rangle$ and we are correct in suspecting that this will continue for any higher order moment equation we attempt to derive. We have seen this in many other problems, for instance when attempting to derive the standard fluid equation from the Vlasov equation, etc. We can truncate the series of equations by some assumption or postulate, for instance simply throw away some of the higher order terms Here we will illustrate another approach in terms of characteristic functions, see section 19.1.2. We therefore define the time depending characteristic function

$$\mathcal{G}(\{\mu_k\}, t) = \left\langle \exp\left(i \sum_k \mu_k E_k(t)\right)\right\rangle, \tag{28.7}$$

where we introduced the notation $\{\mu_k\}$ as a reminder that G depends on *all* of the variables μ_k for all k. The advantage of characteristic functions is, as we saw in section 19.1.2, that it offers an easy, elegant way of obtaining statistical moments, such as $\partial\mathcal{G}(\{\mu_k\}, t)/\partial\mu_k|_{\{\mu_k\}=0} = i\langle E_k\rangle$, etc demonstrating that the statistical moments contain all the information needed for a Taylor series expansion of the characteristic function at the origin. In this sense the series $\mathcal{G}(\{\mu_k\}, t)$ contains all the statistical information, but there are other ways to make a series expansion. Here we expand the exponent in (28.7) to find

$$\begin{aligned}\mathcal{G}(\{\mu_k\}, t) &\equiv \exp\left(\ln\left\langle \exp\left(i\sum_k \mu_k E_k(t)\right)\right\rangle\right)\\ &= \exp\left(i\sum_k A_k\mu_k - \frac{1}{2!}\sum_{k,j} B_{k,j}\mu_k\mu_j\right.\\ &\quad \left. - \frac{i}{3!}\sum_{k,j,\ell} C_{k,j,\ell}\mu_k\mu_j\mu_\ell + \frac{1}{4!}\sum_{k,j,\ell} D_{k,j,\ell,m}\mu_k\mu_j\mu_\ell\mu_m + \cdots\right),\end{aligned} \tag{28.8}$$

in terms of the variables μ_k for the characteristic function.

To determine the coefficients A_k, $B_{k,j}$ etc in the series expansion we differentiate (28.8) and find

$$A_k = \langle E_k\rangle = 0 \tag{28.9}$$

$$B_{k,j} = \langle E_k E_j\rangle = Q_k \delta(k + j) \tag{28.10}$$

$$C_{k,j,\ell} = \langle E_k E_j E_\ell\rangle = R_{k,j}\delta(k + j + \ell). \tag{28.11}$$

No matter how tempting, the next term is not found by continuing the hierarchy in what seems to be the obvious way. Instead we find

$$\langle E_k E_j E_\ell E_m\rangle = B_{k,j}B_{\ell,m} + B_{k,\ell}B_{j,m} + B_{k,m}B_{j,\ell} + D_{k,j,\ell,m}. \tag{28.12}$$

The coefficients A_k, $B_{k,j}$, $C_{k,j,\ell}$, $D_{k,j,\ell,m}$ etc are the *cumulants* or *semi-invariants*, see also section 20.1.2. One important feature of these cumulants is that they offer a measure of the deviations of a given process from a Gaussian process: for a Gaussian random process with zero mean, all cumulants vanish with the exception of $B_{k,j}$.

Assume for the moment that the process $E_k(t)$ was Gaussian. In that case (28.6) tell us that $\partial Q_k/\partial t = 0$ and similarly for the temporal derivatives of all higher moments: we have no dynamics at all! The best we can hope for is that the process $E_k(t)$ is in some sense close to a Gaussian, so at least some of the cumulants can be taken to be negligible.

Expressed in terms of cumulants we can write

$$\frac{\partial}{\partial t} Q_k(t) = \epsilon \sum_{j,\ell} L_{j,\ell}^{\{k\}} C_{k,j,\ell}(t). \tag{28.13}$$

We now use

$$\frac{\partial C_{k,j,\ell}}{\partial t} = \epsilon \overset{\text{perm}}{\{k, j, \ell\}} \sum_{m,n} L_{j,\ell}^{\{-k\}} \left(B_{k,j} B_{\ell,m} + B_{k,\ell} B_{j,m} + B_{k,m} B_{j,\ell} + D_{k,j,\ell,m} \right),$$

with $B_{k,j} = \langle E_k(t) E_j(t) \rangle = Q_k \delta(k + j)$ to obtain

$$\begin{aligned} \frac{\partial C_{k,j,\ell}}{\partial t} &= 2\epsilon \left(L_{-k,-j}^{\{-\ell\}} Q_k Q_j + L_{-j,-\ell}^{\{-k\}} Q_j Q_\ell + L_{-\ell,-k}^{\{-j\}} Q_\ell Q_k \right) \\ &\quad + \overset{\text{perm}}{\{k, j, \ell\}} \epsilon \sum_{m,n} L_{m,n}^{\{-\ell\}} D_{k,j,m,n}, \end{aligned} \tag{28.14}$$

where a term $L_{m,n}(0)$ is ignored since $M_{m,n}(0) = 0$.

If we were dealing with a Gaussian random process we would have $C_{k,j,\ell}(t) = 0$, as already mentioned, as well as $D_{k,j,m,n} = 0$, giving $Q_k(t)$ constant by (28.13). We might hope that the process, albeit non-Gaussian, is so close to one that we have $C_{k,j,\ell}(t) \neq 0$ thus retaining a time variation through (28.13), but being allowed to assume $D_{k,j,m,n} \approx 0$ with negligible error. In all its simplicity, this is the basis for the quasi-normal approximation. It states, with different wording, that the fourth moment is related to the second moment as it would be for a Gaussian process, and the equation for the time evolution of the third moment is retained without further assumptions. The two coupled equations (28.13) and (28.14) with the last term ignored form the basis of the quasi-normal model of turbulence. Note that ignoring $D_{k,j,m,n}$ will not by itself make the equations (28.13) and (28.14) irreversible with time.

Before discussing the quasi-normal model in further detail, it might be appropriate to emphasize that we have so far not assumed that ϵ is a small parameter. Neither have we provided any arguments for expecting the fluctuations to have near Gaussian statistics. The assumption of small ϵ gives problems for turbulence in incompressible fluids.

Assume first that the linearized version of (28.1) represents waves propagating according to some dispersion relation, implying that $\beta_k = \omega(\mathbf{k})$ is the dispersion relation for this case. Assuming the dispersion relation to be isotropic, i.e. $\omega(\mathbf{k}) = \omega(k)$ in two or three spatial dimensions an observer at some position will in general at any time expect perturbations arriving from widely separate regions in space. In a linear model all these contributions add up to represent the perturbation detected by this observer. If the origins of these perturbation are widely separated we might as a working hypothesis assume that they are uncorrelated, and by pushing the argument a little also assume that they are statistically independent. By the central limit as discussed in section 20.2.3 we can then argue that for homogeneous isotropic conditions the resulting wave-field will at any position represent a Gaussian random process (Benney and Saffman 1966), implying that not only the amplitude probability density is Gaussian but so are all joint probability densities we can construct. Note that the argument does not distinguish non-dispersive and dispersive wave-types. If consider a problem in one spatial dimension the argument no longer applies since non-dispersive perturbation arrive *at the same time* only from two positions, one behind and one in front of the detector. If the waves are strongly dispersive (Langmuir waves are one example) perturbations spread out as they propagate and the argument also becomes applicable for this case. Note that many plasma wave-types found in laboratory or plasma are neither dispersive nor having isotropic dispersion relation (Krane *et al* 2010), see also section 11.4.1, so the arguments outlined here are more restrictive than they may seem.

Now, assuming that the linear arguments support a locally Gaussian wave-field, then a small value of the coupling coefficient, i.e. $\epsilon \ll 1$ in (28.1) will allow us to assume that the near Gaussian assumption in the sense that we can assume $D_{k,j,m,n} \approx 0$ in (28.14). Hydrodynamic turbulence as illustrated by (16.3) has no small parameter, and the quasi-normal model fails miserably by predicting negative power spectra, for instance. It took some time to discover these shortcomings since solution of the basic equations needed modern computers. It could be argued that by choosing conditions where $D_{k,j,m,n} \approx 0$ initially, there might be a set of sub-problems that could be approached by the quasi-normal model, but even this hope is in vain, since $D_{k,j,m,n}$ will in general develop to reach significant magnitudes at later times. Hydrodynamic turbulence is entirely dominated by nonlinearities. For isotropic plasmas the quasi-normal approximation has been more successful.

The quasi-normal model can be developed in more detail by a few additional assumptions.

- We assume that $Q_k(t)$ changes slowly, in the sense that the time variation of $C_{k,j,\ell}(t)$ can be modeled by a function of $Q_k(t)$, i.e. $C_{k,j,\ell}(t)$ evolves in a quasi-stationary manner.
- We assume also that $C_{k,j,\ell}(t \to -\infty) = 0$.

Using (28.14) with these assumptions we find for large times

$$C_{k,j,\ell}(t) = 2\epsilon \int_{-\infty}^{t} \Big(L_{k,j}^{\{\ell\}*}(t')Q_k(t)Q_j(t) + L_{j,\ell}^{\{k\}*}(t')Q_j(t)Q_\ell(t) \\ + L_{\ell,k}^{\{j\}*}(t')Q_\ell(t)Q_k(t)\Big)\mathrm{d}t',$$

where we used $L_{-k,-j}^{\{-\ell\}} = L_{k,j}^{\{\ell\}*}$ by reference to $M_{j,\ell}^{\{k\}*} = -M_{j,\ell}^{\{k\}}$. This result for $C_{k,j,\ell}(t)$ is substituted into (28.13) to give

$$\frac{\partial}{\partial t}Q_k(t) = 2\epsilon^2 \sum_{j,\ell} \int_{-\infty}^{t} L_{j,\ell}^{\{k\}}\Big(L_{k,j}^{\{\ell\}*}(t')Q_k(t)Q_j(t) \\ + L_{j,\ell}^{\{k\}*}(t')Q_j(t)Q_\ell(t) + L_{\ell,k}^{\{j\}*}(t')Q_\ell(t)Q_k(t)\Big)\mathrm{d}t'. \tag{28.15}$$

Recalling that we defined $L_{j,\ell}^{\{k\}}(t) = M_{j,\ell}^{\{k\}} \mathrm{e}^{-i(\beta_k+\beta_j+\beta_\ell)t}$ with $M_{j,\ell}^{\{k\}}=$ const. we can solve the integrations in (28.15) to be in the form

$$\int_{-\infty}^{t} L_{j,\ell}^{\{k\}}(t')L_{k,j}^{\{\ell\}*}(t')\mathrm{d}t' = M_{j,\ell}^{\{k\}}M_{k,j}^{\{\ell\}*}\int_{-\infty}^{t} \mathrm{e}^{-i(\beta_k+\beta_j+\beta_\ell)(t-t')}$$
$$= M_{j,\ell}^{\{k\}}M_{k,j}^{\{\ell\}*}\int_{0}^{\infty} \mathrm{e}^{-i(\beta_k+\beta_j+\beta_\ell)s}\mathrm{d}s.$$

We have the relation

$$\int_0^\infty \mathrm{e}^{-i(\beta_k+\beta_j+\beta_\ell)s}\mathrm{d}s = P\frac{i}{\beta_k+\beta_j+\beta_\ell} + \pi\delta(\beta_k+\beta_j+\beta_\ell),$$

where P means that the principal value should be taken when the expression is integrated (Pécseli 2000). Recalling that the $M_{j,\ell}^{\{k\}}$s are imaginary for the cases considered here, we find as a final result a closed expression for $Q_k(t)$ in the form

$$\frac{\partial}{\partial t}Q_k(t) = 4\pi\epsilon^2\sum_{j,\ell} M_{j,\ell}^{\{k\}}\Big(M_{k,j}^{\{\ell\}*}Q_k(t)Q_j(t) + M_{j,\ell}^{\{k\}*}Q_j(t)Q_\ell(t)$$
$$+ M_{\ell,k}^{\{j\}*}Q_\ell(t)Q_k(t)\Big)\delta(\beta_k+\beta_j+\beta_\ell). \qquad (28.16)$$

Without more knowledge concerning $M_{j,\ell}^{\{k\}}$ we cannot arrive at any final solution for (28.16) but some general comments can be made. First of all, the equation is not time reversible although the problem in its original formulation *was* time reversible. This breaking of the symmetry for $t \to +\infty$ and $t \to -\infty$ was introduced when it was assumed that $C_{k,j,\ell}(t)$ was a function of the function $Q_k(t)$ at the same time t, not including the past. Second, we note that the conditions implied in the δ-functions have the form

$$\mathbf{k}+\mathbf{j}+\mathbf{l}=0 \qquad \text{and} \qquad \beta_k+\beta_j+\beta_\ell=0.$$

These are equivalent to momentum and energy conservations, indicating that the right-hand side of (28.16) can be interpreted as a 'wave-collision' integral. For applications to collisionless plasmas a quasi-normal model has to be extended by including the interaction of waves and particles. It turns out that such a generalization can be made by terms accounting for energy and momentum conserving wave-particle interactions that involves the particles as well.

There are two cases of physical relevance (Cook 1972, Nicholson 1983) where we can give some general comments:

1) $M_{j,\ell}^{\{k\}} + M_{\ell,k}^{\{j\}} + M_{k,j}^{\{\ell\}} = 0$
2) $M_{j,\ell}^{\{k\}} + M_{\ell,k}^{\{j\}} + M_{k,j}^{\{\ell\}} \propto \beta_k+\beta_j+\beta_\ell$.

For these cases we have a solution $Q_k = \alpha =$ const., this is readily shown. It takes a little more work to demonstrate that the solution actually approaches this state, i.e. to demonstrate $Q_k(t\to\infty) = \alpha$. It can also be shown that this solution has a characteristic function

$$\mathcal{G}(t \to \infty) = \exp\left(-\frac{\alpha}{2}\sum_k \mu_k \mu_{-k}\right).$$

It should be emphasized that the formulation discussed in this section refers to what is properly a 'decaying turbulence' problem. A more relevant and maybe also more interesting problem will have a driving term accounting for a constant energy input in (28.1), but then we also need energy sinks in order to have possibilities for a steady state solution to develop. All this will complicate the analysis considerably. The aim of the present summary is to account for the basic features of the quasi-normal approximation and this can be done with simpler systems.

In theoretical plasma physics the weak turbulence theory was historically developed first for describing Langmuir turbulence in kinetic models of collisionless plasmas (Vedenov *et al* 1961, Vedenov *et al* 1962, Al'tshul' and Karpman 1966), and in its most general form the analysis includes wave-particle interactions (Kadomtsev 1965, Sagdeev and Galeev 1969). A nice and detailed textbook summary was given by Nicholson (1983), containing in particular a detailed reference list. A quasi-particle model that can be best understood in terms of propagating wave-packets was introduced by Vedenov *et al* (1967) although some early works (Camac *et al* 1962) contain some of the same ideas. Some of the results obtained by detailed models based on the quasi-normal approximation has been very successful in describing, for instance, weak turbulence in the solar wind (Kontar 2001a, 2001c) but it should be added that many of these models depend heavily on time consuming numerical solutions even in one spatial dimension (Kontar 2001b).

- **Example:** It is instructive to derive a differential equation for the characteristic function. We have

$$\begin{aligned}\frac{\partial \mathcal{G}}{\partial t} &= \mathrm{i}\sum_k \mu_k \left\langle \frac{\partial E_k}{\partial t} \exp\left(\mathrm{i}\sum_k \mu_k E_k\right)\right\rangle \\ &= \mathrm{i}\sum_k \mu_k \left\langle \epsilon \sum_{j,k} L_{j,\ell}^{\{-k\}} E_j E_\ell \exp\left(\mathrm{i}\sum_k \mu_k E_k\right)\right\rangle \\ &= -\,\mathrm{i}\epsilon \sum_{k,j,\ell} \mu_k L_{j,\ell}^{\{-k\}} \frac{\partial^2 \mathcal{G}}{\partial \mu_j \partial \mu_\ell}.\end{aligned} \tag{28.17}$$

  The relation (28.17) has the form of a Liouville equation. Attempts have been made to use equations like this as a basis for a theory of incompressible fluid turbulence (Edwards 1969).

- **Example:** The 'random phase approximation' is a term we often encounter in discussions of the quasi-normal approximation. Often it is identified (Cook 1972, Nicholson 1983) by the expression (28.4)

$$\langle E_j E_k \rangle = Q_k \delta(k + j),$$

meaning that two different Fourier components of the signal (this may actually be spatial as well as temporal Fourier transforms) are uncorrelated. If they represent Gaussian random processes, then they are also statistically independent. We have seen that (28.4) is nothing but a consequence of assuming homogeneous turbulence, where expressions like $\langle E(r_j)E(r_k)\rangle$ depend on the separation $r_j - r_k$ rather than r_j and r_k individually and the result that two different Fourier components are uncorrelated is merely a consequence of this. Recall that also in section 20.4 we learned that for a Gaussian random process the phases of different Fourier components are uncorrelated and statistically independent.

To discuss this question in more detail we take two plane waves with different wavenumbers $A_1 \cos(k_1 x)$ and $A_2 \cos(k_2 x + \psi)$, where ψ is some phase difference at the position $x = 0$. We have here

$$\begin{aligned}\langle E_{k_1} E_{k_2}\rangle &= \langle A_1 A_2 \cos(k_1 x)\cos(k_2 x + \psi)\rangle \\ &= \frac{1}{2}\langle A_1 A_2 \cos((k_1 + k_2)x + \psi) + A_1 A_2 \cos((k_1 - k_2)x - \psi)\rangle.\end{aligned}$$

To carry out the averaging we have to specify the random variable. We assume the amplitudes A_1 and A_2 to be deterministic but ψ to be uniformly distributed in the interval $\{0; 2\pi\}$. In any one realization the relative phase is given, but it varies over the ensemble. Since, e.g., $\frac{1}{2\pi}\int_0^{2\pi} \cos((k_1 - k_2)x - \psi)\mathrm{d}\psi = 0$ we have $\langle E_{k_1} E_{k_2}\rangle = 0$ when $k_1 \neq k_2$. If the two waves have the *same* wavenumber we are in reality dealing with only one plane wave $A\cos(kx + \psi)$ so that $\langle E_{k_1} E_{k_2}\rangle \to \langle E_k^2\rangle$, giving $\langle E_k^2\rangle = \frac{1}{2}A^2$ upon averaging over all uniformly distributed ψ. We could have introduced *two* phase values ψ_1 and ψ_2, one for each cosine, but this would only complicate the notation, without contributing to the physics.

More generally we have for this example $\langle E_{k_1}(x_1)E_{k_2}(x_2)\rangle = \langle A_1 A_2 \cos(k_1 x_1) \times \cos(k_2 x_2 + \psi)\rangle = \frac{1}{2}A_1 A_2\langle(\cos(k_1 x_1 + k_2 x_2 + \psi) + \cos(k_1 x_1 - k_2 x_2 - \psi)\rangle$. Also with this slight generalization we have $\langle E_{k_1}(x_1)E_{k_2}(x_2)\rangle = 0$ for $k_1 \neq k_2$, while $\langle E_k(x_1)E_k(x_2)\rangle = \frac{1}{2}A^2 \cos(k(x_1 - x_2)) = \frac{1}{2}A^2\cos(k(x_2 - x_1))$ for $k_1 = k_2 = k$. With the random phases we have $\langle E_k(x_1)E_k(x_2)\rangle$ to depend on the position separations $x_1 - x_2$ and not the positions x_1 and x_2 independently.

We emphasize once more that the distinction of ensemble and time averages. Naively one might argue that we could let the phases ψ vary randomly with time in each realization to make ensemble and time averages the same, but then the problem is changed since the temporal phase-modulation gives rise to a frequency broadening and we no longer have two well defined frequencies. The two models can be made formally equivalent only for very long records (in terms of the oscillation periods) and the phases very slowly varying.

As seen, the statement of 'random phase approximation' in its strict sense covers a somewhat trivial statistical assumption, inviting for instance Marshall Rosenbluth renaming it to the 'random phrase approximation'. In reality, however, it is often found that implicit in the stated approximation lies *also* the assumption of $D_{k,j,m,n} \approx 0$ in (28.14). This is not at all the same thing, and the reader should check carefully the meaning of the approximations made when consulting the literature (Cook 1972).

References

Al'tshul' L M and Karpman V I 1966 Theory of nonlinear oscillations in a collisionless plasma *Soviet Phys JETP-USSR* **22** 361–9

Al'tshul' L M and Karpman V I 1966 *Russian original J. Exptl. Theoret. Phys. (USSR)* **49** 515–28

Benney D J and Saffman P G 1966 Nonlinear interactions of random waves in a dispersive medium *Proc. Royal Soc. London. Ser. A, Math. Phys. Sci.* **289** 301–20

Camac M, Kantrowitz A R, Litvak M M, Patrick R M and Petschek H E 1962 Shock waves in collision-free plasmas *Nuclear Fusion Suppl.* **2** 423–45

Champeney D C 1973 *Fourier Transforms and their Physical Applications* (London: Academic)

Cook I 1972 The non-linear theory of plasmas UKEA Research group, Culham Laboratory Report CLM-L 17. Summer School Lectures, ed B E Keen (Abingdon: Culham Laboratory)

Diamond P H, Itoh S- I and Itoh K 2010 *Modern Plasma Physics* Vol I *Physical Kinetics of Turbulent Plasmas* (Cambridge: Cambridge University Press)

Edwards S F 1969 *Non-linear Effects in Plasmas* (New York: Gordon and Breach) ch Strong Turbulence

Kadomtsev B B 1965 *Plasma Turbulence* (New York: Academic)

Kono M and Ichikawa Y H 1973 Renormalization of the wave-particle interaction in weakly turbulent plasmas *Prog. Theor. Phys.* **49** 754–63

Kontar E P 2001a Dynamics of electron beams in the inhomogeneous solar corona plasma *Solar. Phys.* **202** 131–49

Kontar E P 2001b Numerical consideration of quasi-linear electron cloud dynamics in plasma *Comp. Phys. Commun* **138** 222–33

Kontar E P 2001c Propagation of a fast electron cloud in a solar-like plasma of decreasing density *Plasma Phys. Control. Fusion* **43** 589–601

Krane B, Pécseli H L, Sato H, Trulsen J and Wernik A W 2010 Low-frequency electrostatic waves in the ionospheric E region *Plasma Sources Sci. Technol.* **19** 034007

Nicholson D R 1983 *Introduction to Plasma Theory* (New York: Wiley)

Orszag S A 1970 Analytical theories of turbulence *J. Fluid Mech.* **41** 363–86

Pécseli H L 2000 *Fluctuations in Physical Systems* (Cambridge: Cambridge University Press)

Sagdeev R Z and Galeev A A 1969 *Nonlinear Plasma Theory* Frontiers in Physics (Amsterdam: Benjamin)

Vedenov A A, Gordeev A V and Rudakov L I 1967 Oscillations and instability of a weakly turbulent plasma *Plasma Phys.* **9** 719–35

Vedenov A A, Velikhov E P and Sagdeev R Z 1961 Stability of plasma *Soviet Phys. Uspekhi* **4** 332–69
Vedenov A A, Velikhov E P and Sagdeev R Z 1961 Russian original Usp. Fiz. Nauk. **73** 701–66
Vedenov A A, Velikhov E P and Sagdeev R Z 1962 Quasi-linear theory of plasma oscillations *Nucl. Fusion Suppl.* **2** 465–75

IOP Publishing

Low Frequency Waves and Turbulence in Magnetized Laboratory Plasmas and in the Ionosphere

Hans Pécseli

Chapter 29

The direct interaction approximation

One of the most successful approaches in describing the statistical properties of turbulence are based on the direct interaction approximation (DIA) introduced by Robert Kraichnan (Kraichnan 1958a, 1958b, 1959) and developed further by himself and many other authors. The relations between mixing length theories and the DIA, in particular, were discussed by Sudan and Pfirsch (1985). The amount of literature on the subject is overwhelming, so reference is here given only to the book by Leslie (1973) which offers one of the most detailed introductions. More details can be found in the specialized literature (Betchov 1966, 1967, Orszag 1970, Krommes and Smith 1987, Stanišić 1988, Similon and Sudan 1990, Krommes 2002, 2015). Most of the following summary is based on the authors scattered notes from a few inspired seminars by Dwight Nicholson that he gave around 1979 when he visited us at the Risø National Laboratory in Denmark. His ambition was high: we should be introduced not only to the DIA, but also to the general notation in the more advanced literature. The author is deeply indebted to Dwight Nicholson for many discussion on the subject, the direct interaction approximation in particular.

29.1 Introduction to functional differentiation

We can have a function $f(t)$ of a variable t. This can be implicit in the sense that f depends on another function x that depends on t, this we write as $f(x(t))$. In this sense we have already found a rather trivial example for a function depending on another function, but this dependence refers to one selected value of the variable t. We can have a functional dependence of *all* variables in an interval if, for instance, we have

$$f[x(t)] = \int_a^b k(t)x(t)\mathrm{d}t, \tag{29.1}$$

where $k(t)$ is some weight function. To distinguish the two examples we denote one with ()-brackets, the other with []-brackets. We can differentiate $f(x(t))$ with respect to t to find $\mathrm{d}f/\mathrm{d}t = (\mathrm{d}f/\mathrm{d}x)\mathrm{d}x/\mathrm{d}t$.

We can give a general expression for differentiation of $f(x)$ mentioned before by writing $f(x_1, x_2, x_3, \ldots, x_N)$ which gives the differential

$$\mathrm{d}f = \sum_{j=1}^{N} \frac{\mathrm{d}f}{\mathrm{d}x_j}\mathrm{d}x_j.$$

In view of this expression we generalize functional differentiation to give the form

$$\delta f = \int_a^b \frac{\delta f}{\delta x(t)}\delta x(t)\,\mathrm{d}t, \tag{29.2}$$

taking care to write $\delta f/\delta x(t)$ and not just $\delta f/\delta x$. We introduced δ replacing the standard differentiation symbol *d*. To indicate the fact that $\delta f/\delta x(\xi)$ is a functional that also depends on the parameter ξ, we shall often write $f'[x(t), \xi] = \delta f/\delta x(\xi)$ to make this explicit. In contrast to the ordinary derivative notation, where $\mathrm{d}y/\mathrm{d}x$ is also the quotient of differentials, we have here that $(\delta f/\delta x(\xi))\delta x(\xi)$ is *not* equal to δf.

We can also differentiate $f[x(t)]$ by using the basic definition in terms of small increments, giving

$$f[x(t) + \delta x(t)] = \int_a^b k(t)(x(t) + \delta x(t))\mathrm{d}t,$$

and thus

$$f[x(t) + \delta x(t)] \approx f[x(t)] + \delta f = \int_a^b k(t)(x(t) + \delta x(t))\mathrm{d}t$$

or

$$\delta f[x(t)] = \int_a^b k(t)\delta x(t)\mathrm{d}t.$$

By comparing the two results for δf we find that $f'[x(t), \xi] = k(\xi)$. We can take a slightly more advanced problem by considering

$$f[x(t)] = \int_a^b k(t)x^2(t)\mathrm{d}t,$$

and argue that

$$\begin{aligned} f[x(t) + \delta x(t)] &= \int_a^b k(t)(x(t) + \delta x(t))^2\mathrm{d}t \\ &\approx \int_a^b k(t)x^2(t)\mathrm{d}t + \int_a^b k(t)2x(t)\delta x(t)\mathrm{d}t, \end{aligned}$$

giving $\delta f/\delta x(t) = 2k(t)x(t)$.

Two rules for differentiation will be particularly useful later on:

Rule 1: We can write

$$x(t) = \int_a^b x(t')\delta(t - t')\mathrm{d}t'$$

as an identity. It is a bit of a cheat to write the expression directly, since we presuppose δ-functions, where strictly speaking they should emerge from the functional analysis, but for the present purpose we can allow this. We note that $\delta(t - t')$ takes the role of $k(t)$ in (29.1). By the previous differentiation rule we find directly

$$\frac{\delta x(t_1)}{\delta x(t')} = \delta(t_1 - t'). \tag{29.3}$$

Rule 2: By dividing (29.2) with $\delta y(t'')$ we postulate

$$\frac{\delta f[x(t)]}{\delta y(t'')} = \int_a^b \frac{\delta f}{\delta x(t')}\frac{\delta x(t')}{\delta y(t'')}\mathrm{d}t'. \tag{29.4}$$

We can interpret this as the 'chain-rule' for functional differentiation. To be correct with $x(t) = x(y(t))$ this should give

$$\begin{aligned}
\delta f[x(t)] &= \int_a^b \frac{\delta f}{\delta y(t'')}\delta y(t'')\mathrm{d}t'' \\
&= \int_a^b\int_a^b \frac{\delta f}{\delta x(t')}\frac{\delta x(t')}{\delta y(t'')}\delta y(t'')\mathrm{d}t'\mathrm{d}t'' \\
&= \int_a^b \frac{\delta f}{\delta x(t')}\int_a^b \frac{\delta x(t')}{\delta y(t'')}\delta y(t'')\mathrm{d}t''\mathrm{d}t' \\
&= \int_a^b \frac{\delta f}{\delta x(t')}\delta x(t')\mathrm{d}t',
\end{aligned}$$

QED! where we used the rule (29.3) on $\delta y(t)$ in the last integral.

An excellent summary of generalized function theory is given in a small book by Lighthill (1964) and also by Berz (1967), Beran (1968) or Krommes and Smith (1987). For more details, the reader is referred to these and other related works.

29.2 Basic equations for a reference model

To discuss the basic ideas we choose first a simple, well known problem, the harmonic oscillator as expressed in terms of a potential $V(x)$ as

$$\frac{\mathrm{d}^2x}{\mathrm{d}t^2} = -\frac{\mathrm{d}V(x)}{\mathrm{d}x}.$$

We choose first the basic oscillator form with $V(x) = \frac{1}{2}x^2$ and have

$$\frac{\mathrm{d}^2 x}{\mathrm{d}t^2} = -x + \eta(t),$$

where $\eta(t)$ is some general forcing term, here introduced for later convenience. We will use this simple oscillator form to illustrate a general notation that will be used later on. To obtain a compact formulation we introduce two functions $x_1(t) \equiv x(t)$ and $x_2(t) \equiv \mathrm{d}x(t)/\mathrm{d}t$ and have $\mathrm{d}x_2/\mathrm{d}t = -x_1 + \eta$. This we can write as a formal expression that has general applicability

$$\begin{pmatrix} \mathrm{d}x_1(t)/\mathrm{d}t \\ \mathrm{d}x_2(t)/\mathrm{d}t \end{pmatrix} = \begin{pmatrix} 0 & 1 \\ -1 & 0 \end{pmatrix} \begin{pmatrix} x_1(t) \\ x_2(t) \end{pmatrix} + \begin{pmatrix} 0 \\ \eta(t) \end{pmatrix}. \tag{29.5}$$

This expression can be written in a different way as

$$\begin{pmatrix} \mathrm{d}x_1(t_1)/\mathrm{d}t_1 \\ \mathrm{d}x_2(t_1)/\mathrm{d}t_1 \end{pmatrix} = \int_{-\infty}^{\infty} \begin{pmatrix} 0 & \delta(t_1 - t_2) \\ -\delta(t_1 - t_2) & 0 \end{pmatrix} \begin{pmatrix} x_1(t_2) \\ x_2(t_2) \end{pmatrix} \mathrm{d}t_2 + \begin{pmatrix} 0 \\ \eta(t_1) \end{pmatrix}. \tag{29.6}$$

This result can be compressed to give

$$\dot{\Psi}(1) = X_1(12)\Psi(2) + \eta(1). \tag{29.7}$$

The terms can be identified one by one, but some explanation may be in order: the 'dot' indicated time-differentiation as in the standard Newton notation. By writing (1) we mean (t_1), etc. The standard summation convention has been extended to imply that repeated variables are integrated. Finally, in (29.7) we let $\eta(1)$ represent the last vector in (29.6). The notation given here is the standard one (DuBois and Espedal 1978). The present author is tempted to insert a 'comma' to distinguish 'one-two' from 'twelve', but the misunderstanding will never occur in the following applications, so it is left to stand as it is. It is standard in many texts.

The model is now extended to include a quadratic term so the oscillator becomes an-harmonic by taking $V(x) = -\frac{1}{2}x^2 - \frac{1}{3}x^3$. It may be instructive for the reader to make a small sketch of $V(x)$. We now have $\dot{x}_2 = -x_1 - x_1^2 + \eta$. Following the foregoing prescription we obtain with a little effort

$$\dot{\Psi}(1) = X_1(12)\Psi(2) + \frac{1}{2}X_2(123)\Psi(2)\Psi(3) + \eta(1). \tag{29.8}$$

where we have either $X_2(123) = -2\delta(t_1 - t_2)\delta(t_1 - t_3)$ or $X_2(123) = 0$, depending on the variable-values. It might be instructive for the reader to write up the equation (29.8) in its full matrix version to see the proper signs of the terms in X_2. The present example has by construction only a time-variable, but the formalism can be extended to include space-time varying phenomena by generalizing the meaning of the variables.

Often, we see the notation

$$\dot{\Psi}(1) = J(1) + \eta(1), \tag{29.9}$$

where $J(1) \equiv X_1(12)\Psi(2) + \frac{1}{2}X_2(123)\Psi(2)\Psi(3)$ is denoted a 'current' in analogy to Maxwell's equations. The expression (29.9) will be used as a general reference in the following, where the particular problem will be introduced by defining $J(1)$ where the present case (29.8) will serve as an illustration later on. The nonlinear an-harmonic oscillator as such may not have a great appeal, but the structure of the equations derived from it can be seen as universal for quadratically nonlinear systems, where the Navier–Stokes equation is one important example (Martin *et al* 1973). Plasma dynamics offer many more.

We will now discuss ensemble averages of (29.9). Doing so, we have a goal namely to analyze response functions. One reason for this is that the classical fluctuation–dissipation theorem discussed at great length in the literature (Callen and Welton 1951, MacDonald 1962, Bekefi 1966, Pécseli 2000) basically concerns thermal fluctuations and response functions. If we let $\eta(1) = 0$ in e.g. (29.9) then the system is at least formally deterministic. In its most general form a system might have so many degrees of freedom that it is amenable only for a statistical description. Also systems with a small number of degrees of freedom can exhibit chaotic time-variations (Berge *et al* 1984) but a significant number of relevant systems do not have such properties, in particular not the harmonic oscillator. Forced systems can on the other hand exhibit random behavior and the response function is central for their modeling. We here define a response function as

$$R(12) \equiv \frac{\delta\langle\Psi(1)\rangle}{\delta\eta(2)}.$$

Note that we might have $\langle\Psi\rangle = 0$ but a perturbation η will induce a response $\delta\langle\Psi(1)\rangle \neq 0$. First we take the ensemble average of (29.9) to give

$$\langle\dot{\Psi}(1)\rangle = \langle J(1)\rangle + \eta(1).$$

Since we are in control of η we do not write an explicit ensemble average of the last term. We now differentiate this expression in the sense discussed in section 29.1 to give

$$\frac{\partial}{\partial t_1}\frac{\delta\langle\Psi(1)\rangle}{\delta\eta(1')} = \frac{\delta\langle J(1)\rangle}{\delta\eta(1')} + \frac{\delta\eta(1)}{\delta\eta(1')}.$$

The left side is the time differential of the response function, this is what we wanted. The last term on the right-hand side can be identified with $\delta(1-1')$ by the rule (29.3). The first term on the right-hand side can be rewritten by the chain rule (29.4) so that we have

$$\begin{aligned}\frac{\partial}{\partial t_1}R(11') &= \frac{\delta\langle J(1)\rangle}{\delta\langle\Psi(2)\rangle}\frac{\delta\langle\Psi(2)\rangle}{\delta\eta(1')} + \delta(1-1')\\ &\equiv \frac{\delta\langle J(1)\rangle}{\delta\langle\Psi(2)\rangle}R(21') + \delta(1-1').\end{aligned}$$

In quantum field theory $\delta\langle J(1)\rangle/\delta\Psi(2)$ would be called the self-energy. At closer inspection we find that there is an implicit assumption that all of the η-dependence of

$\langle J(1)\rangle$ can be expressed through $\langle\Psi\rangle$. With this reasonable assumption, we are allowed to use the chain rule (29.4). We rewrite the result in the form

$$\left(\frac{\partial}{\partial t_1}\delta(1-2) - \frac{\delta\langle J(1)\rangle}{\delta\langle\Psi(2)\rangle}\right)R(21') = \delta(1-1'). \tag{29.10}$$

This expression is one of two forming the basis of the analysis. It has the form of a Green's function equation, giving the response to a δ-pulse at some time t_1, where R (21) is the response to this perturbation at a time t_2. We use the boundary condition $R(21) = 0$ if $t_2 < t_1$ for the response function. The relation (29.10) is exact as it stands, with all the relevant physics entering through $\delta\langle J(1)\rangle/\delta\Psi(2)$. The analysis is, as such general, and we have so far not used the actual form of J.

So far so good! We have an equation accounting for the evolution of $R(12)$. It is about time to discuss what we need. A central quantity in studies of turbulence is the power spectrum, or what (by the Wiener-Khinchine theorem, section 19.2.7) is just as good, the correlation function. This is what we want. We define the correlation function for the process Ψ as

$$C(12) \equiv \langle\Psi(1)\Psi(2)\rangle - \langle\Psi(1)\rangle\langle\Psi(2)\rangle. \tag{29.11}$$

To determine $C(12)$ we need one more equation and start with (29.9). Subtracting the same term on both sides, we have

$$\dot{\Psi}(1) - \frac{\delta\langle J(1)\rangle}{\delta\langle\Psi(2)\rangle}\Psi(2) = J(1) - \frac{\delta\langle J(1)\rangle}{\delta\langle\Psi(2)\rangle}\Psi(2) + \eta(1)$$

or

$$\left(\frac{\partial}{\partial t_1}\delta(1-2) - \frac{\delta\langle J(1)\rangle}{\delta\langle\Psi(2)\rangle}\right)\Psi(2) = \Delta J(1) + \eta(1), \tag{29.12}$$

with the definition

$$\Delta J(1) \equiv J(1) - \frac{\delta\langle J(1)\rangle}{\delta\langle\Psi(2)\rangle}\Psi(2).$$

The term η in (29.12) can be turned on and off at will, so we turn it off at the moment. The similarity between (29.12) and (29.10) is then evident: the operator on the left sides is the same indicating that $R(12)$ is the Green's function that propagates the source $\Delta J(1)$ forward in time to produce $\Psi(2)$. Realizing this we can write

$$\Psi(2) = R(12)\Delta J(1) + \Psi_0(2),$$

where Ψ_0 is the initial value of $\Psi(2)$. We assume the initial time to be somewhere in the infinite past and make the bold, but not unrealistic, assumption that the turbulence has mixed up things to such a degree that the initial value is forgotten and no longer matters and take $\Psi_0 \to 0$. We then have

$$\Psi(2) = R(12)\Delta J(1). \tag{29.13}$$

We are now in a position to express the correlation function $C(12)$ in the form

$$C(11') \equiv R(12)R(1'2')(\langle \Delta J(2)\Delta J(2')\rangle - \langle \Delta J(2)\rangle\langle \Delta J(2')\rangle). \qquad (29.14)$$

In the formalism of DuBois and Espedal (1978) we would call (29.14) a 'non-equilibrium fluctuation dissipation theorem' and $\langle \Delta J(2)\Delta J(2')\rangle - \langle \Delta J(2)\rangle\langle \Delta J(2')\rangle$ would be the 'generalized thermal noise correlator'.

We have now a complete set of equations (29.10) and (29.14) expressed in terms of known forms for the currents $J(1)$ and $\Delta J(1)$. In the present form the set of equations are not in a directly applicable form, but they serve to illustrate the relevance and importance of the response function $R(12)$. It is not evident that time-stationarity has been assumed, and indeed it is not made explicit yet (although there is a hint in the discard of the initial value $\Psi_0(2)$). Formulation of a set of equations in a more useful form still remains, and also more technical discussions of time stationary processes have to be explained.

29.3 The direct interaction approximation

The expressions (29.10) and (29.14) were derived in a general form, and to solve an explicit problem we need to specify $J(1)$ and thereby $\Delta J(1)$. As an illustration of such a 'practical' procedure we return to the problem of the an-harmonic oscillator (29.8), being aware of the limited practical interest of the problem. It can, however, serve well as a guide to a general approach. The analysis is very lengthy and placed in appendix J, so here we summarize only the results, mentioning that at some place the quasi-normal assumption is also made here. It is also assumed that $\langle \psi \rangle = 0$, although it is important to note that this does not imply that the variation $\delta\langle \psi \rangle$ vanishes! We have

$$C(11') = \frac{1}{2}R(12)R(12')X_2(234)X_2(2'3'4')C(33')C(44') \qquad (29.15)$$

$$\begin{aligned}\frac{\partial}{\partial t_1}R(11') = \delta(1-1') &+ X_1(12)R(21')\\ &+ X_2(123)X_2(1''2'3')R(21'')R(3'1')C(2'3).\end{aligned} \qquad (29.16)$$

These two coupled equations form the basic model for the DIA.

It is more common to present the equations in the form of two coupled partial differential equations. We thus note that the index '1' appears only in one place on the right-hand side of (29.15). Differentiating (29.15) with respect to t_1 we have

$$\begin{aligned}\frac{\partial}{\partial t_1}C(11') &= \frac{1}{2}\left(\frac{\partial}{\partial t_1}R(12)\right)R(12')X_2(234)X_2(2'3'4')C(33')C(44')\\ &= \frac{1}{2}(\delta(1-2) + X_1(12'')R(2''2)\\ &\quad + X_2(12^\dagger 3^\dagger)X_2(1^{\dagger\dagger}2''3'')R(2^\dagger 1^{\dagger\dagger})R(2''2)C(3^\dagger 3''))\\ &\quad \times R(12')X_2(234)X_2(2'3'4')C(33')C(44'),\end{aligned} \qquad (29.17)$$

where 29.16 was used. To avoid ‴ I had to introduce †, etc. to make the expressions look nicer. Using (29.15) once again we can rewrite (29.17) in the form

$$\begin{aligned}\frac{\partial}{\partial t_1}C(11') = &\frac{1}{2}R(1'2')X_2(134)X_2(2'3'4')C(33')C(44')\\ &+ X_1(12'')C(2''1')\\ &+ X_2(12^\dagger 3^\dagger)X_2(1^{\dagger\dagger}2^\dagger 3')R(2^\dagger 1^{\dagger\dagger})C(3^\dagger 3')C(2^\dagger 1').\end{aligned} \tag{29.18}$$

This latter equation can be cleaned up a little by a change of variables. The two equations (29.16) and (29.18) constitute basic equations for DIA. Note the structure

$$\dot{R} \sim CRR$$

$$\dot{C} \sim RCC,$$

which turns out to be a characteristic of the DIA in its general from.

The DIA-equations involve several integrations, and these will be troublesome for solutions of realistic problems even when done numerically. Although it may not seem evident from the present rather formal summary, the development of DIA was a great step forward in the analysis of turbulence. The advantages of the method will be more evident by considering a simple example. This is done in the next section.

29.3.1 DIA by example

The DIA is by many considered to be the most developed self-consistent analytical turbulence theory yet developed. Many other independently developed theories can be shown to be contained in it. Given its importance, it might be worthwhile to discuss it once more by a different approach, considering a set of quadratically nonlinear equations

$$\frac{\mathrm{d}}{\mathrm{d}t}y_i(t) \equiv \dot{y}_i(t) = A_{ijk}y_j(t)y_k(t) + \eta_i(t), \tag{29.19}$$

with $i = 1, 2, \ldots, N$, with summation over repeated indexes understood. The term $\eta_i(t)$ accounts for an external driving force that we can use at will, just like before in (29.12). We choose to consider a set of initial conditions so that the ensemble average is

$$\left\langle y_i(t=0)y_j(t=0)\right\rangle = \delta_{ij},$$

using Kronecker's delta. The coefficients describe the physical system, and we assume that

$$A_{ijk} + A_{kij} + A_{jki} = 0 \tag{29.20}$$

$$A_{ijk} = A_{ikj} \tag{29.21}$$

$$A_{ijk} = 0 \qquad \text{if} \qquad j = k \tag{29.22}$$

$$A_{ijk}A_{ijk} = C = \text{const.} \tag{29.23}$$

The coupling coefficients A_{ijk} are assumed to be the same in each realization, but this is not a strictly necessary requirement. Given (29.20) it can be shown that 'energy' defined as $E \equiv y_i(t)y_i(t)$ is conserved. The auto-correlation function is $R(t - t') \equiv \langle y_i(t)y_i(t')\rangle$ with $R(0) = 1$.

The first step in developing a model with the properties characterizing the DIA consists of separating y_i into two parts

$$y_i(t) = y_i^{\{0\}}(t) + \tilde{y}_i(t), \tag{29.24}$$

where $y_i^{\{0\}}(t)$ denotes the part of $y_i(t)$ we would find with $\eta = 0$, while $\tilde{y}_i(t)$ is the correction induced by $\eta \neq 0$, assuming that η is turned on at $t = 0$, implying that $\tilde{y}_i(t = 0) = 0$. The idea is illustrated in figure 29.1. We have

$$y_i^{\{0\}}(t) = A_{ijk} y_j^{\{0\}}(t) y_k^{\{0\}}(t).$$

To find a similar expression for $\tilde{y}_i(t)$ we assume η to be small and ignore second order terms to give

$$\begin{aligned} \dot{\tilde{y}}_i(t) &\approx A_{ijk}\left(y_j^{\{0\}}(t)\tilde{y}_k(t) + y_k^{\{0\}}(t)\tilde{y}_j(t)\right) + \eta_i(t) \\ &= 2A_{ijk} y_j^{\{0\}}(t)\tilde{y}_k(t) + \eta_i(t), \end{aligned} \tag{29.25}$$

where we used $A_{ijk} = A_{ikj}$. In a compact tensor notation we have

$$\dot{\tilde{\mathbf{y}}} = 2\mathbf{A} : \mathbf{y}^{\{0\}}\tilde{\mathbf{y}} + \eta.$$

This is an ordinary inhomogeneous differential equation and its solution can be written in terms of a Green function as

$$\dot{\tilde{\mathbf{y}}} = \int_0^t \mathbf{G} \cdot \eta \, \mathrm{d}t'.$$

We will simplify the problem by ignoring all possible off diagonal terms in $\mathbf{G}$ and write $G_{ij=i} \equiv G_i$ so that we have $\mathbf{G}(t, t') \cdot \eta(t') = G_i(t, t')\eta_i(t')$. We find that $\tilde{\mathbf{y}}(t)$ is given by its *direct interaction* with $\eta_i(t)$. This is an approximation: from (29.19) we see that any change in one component $\tilde{y}_I(t)$ induced by $\eta_I(t)$ will induce a change in another component, say $\tilde{y}_{II}(t)$ which will have a feedback on $\tilde{y}_I(t)$. This is an *indirect interaction*, which is ignored.

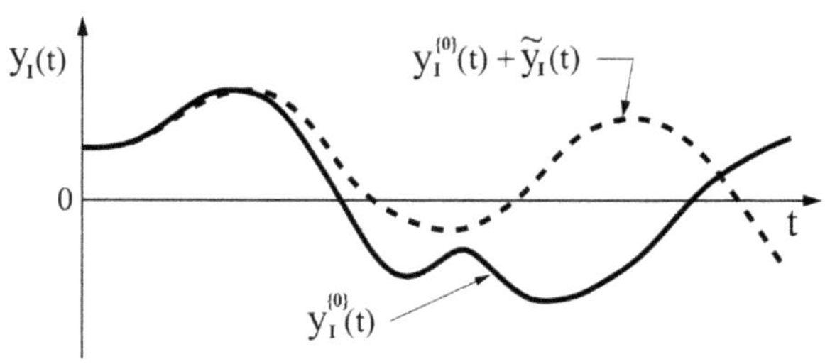

Figure 29.1. Illustration of $y_I^{\{0\}}(t)$ and $\tilde{y}_I(t)$. The label is taken as I to emphasize that it deals with one particular, representative function.

Introducing the averaged Green function into the analysis of turbulence was Kraichnan's first major step (Leslie 1973).

The Green function $\mathbf{G}$ follows

$$\dot{\mathbf{G}}(t, t') = 2\mathbf{A} : \mathbf{y}^{\{0\}}(t)\mathbf{G}(t, t'),$$

with the condition $\mathbf{G}(t, t' = t) = 1$ and $\mathbf{G}(t, t') = 0$ for $t < t'$. The function $\mathbf{G}(t, t')$ changes from one realization to another. We seek the ensemble averaged value $\langle \mathbf{G}(t, t') \rangle$. When analyzing homogeneous turbulence we might assume that $\langle G_i(t, t') \rangle$ is independent of the index i for symmetry reasons, but for studies of turbulence in magnetized plasma it is not so. To simplify the notation we here assume isotropy nonetheless and use $\langle G(t, t') \rangle$ without subscript. The next step is to write the approximation

$$\dot{\tilde{y}}_i = \int_0^t \langle G(t, t') \rangle \eta_i \, dt'. \tag{29.26}$$

This approximation implies that we consider only the average propagator in each realization, and ignore the coupling to the fluctuations. This can be seen by writing $G(t, t') \equiv \langle G(t, t') \rangle + \widetilde{G}(t, t')$, which is merely a definition. The approximation consists in ignoring the coupling between fluctuations, i.e. terms like $\widetilde{G}(t, t')\eta_i$. In some general sense, it can be argued that this corresponds to the so-called first order smoothing approximation. The relation (29.26) is applicable for any component η_i. In particular, we can select one of the elements in the sum in (29.19) and consider it as a noise driving the others, i.e. let $\eta_i(t) \to A_{ijk}\, y_j(t) y_k(t) \approx A_{ijk}\, y_j^{\{0\}}(t) y_k^{\{0\}}(t)$. We can do this for any choice of j and k so we can write

$$\dot{\tilde{y}}_i = A_{ijk} \int_0^t \langle G(t, t') \rangle y_j^{\{0\}}(t') y_k^{\{0\}}(t') \, dt', \tag{29.27}$$

summing over all indices. Implied in this argument is the assumption that we have many components interacting. The analysis will work very poorly for three coupled nonlinear oscillators.

The goal of the analysis is to find expressions for the correlation function $(t, t') = R(t - t') = \frac{1}{N}\langle y_i(t) y_i(t') \rangle$ defined so that $R(0) = 0$. For this purpose we write

$$\dot{R}(t - t') = \frac{1}{N}\left\langle \dot{y}_i(t) y_i(t') \right\rangle = \frac{1}{N} A_{ijk} \left\langle y_j(t) y_k(t) y_i(t') \right\rangle, \tag{29.28}$$

where we used (29.19) with $\eta = 0$. The triple correlation can be written as

$$\begin{aligned}
&\left\langle y_i(t) y_j(t) y_k(t') \right\rangle \\
&= \left\langle \left(y_i^{\{0\}}(t) + \tilde{y}_i(t)\right)\left(y_j^{\{0\}}(t) + \tilde{y}_j(t)\right)\left(y_k^{\{0\}}(t) + \tilde{y}_k(t')\right) \right\rangle \\
&= \left\langle y_i^{\{0\}}(t) y_j^{\{0\}}(t) y_k^{\{0\}}(t') \right\rangle + \left\langle \tilde{y}_i(t) y_j^{\{0\}}(t) y_k^{\{0\}}(t') \right\rangle \\
&\quad + \left\langle y_i^{\{0\}}(t) \tilde{y}_j(t) y_k^{\{0\}}(t') \right\rangle + \left\langle y_i^{\{0\}}(t) y_j^{\{0\}}(t) \tilde{y}_k(t') \right\rangle,
\end{aligned} \tag{29.29}$$

where terms of the order $\tilde{y}\tilde{y}$ are ignored. The first term is vanishing for a Gaussian random process. We have no *a priori* guarantee that we *are* dealing with such a process, but assume that we can ignore this first term nonetheless. With this assumption in mind, we now insert (29.27) into (29.29) to be used in (29.28). We have three terms contributing, here denoted T_1, T_2 and T_3, where the first one is

$$T_1 \equiv A_{ipq} \int_0^t \langle G(t,\, t'')\rangle \left\langle y_p^{\{0\}}(t'')y_q^{\{0\}}(t'')y_k^{\{0\}}(t)y_i^{\{0\}}(t')\right\rangle \mathrm{d}t''.$$

At this stage we make the quasi-normal approximation (which is consistent with our ignoring the first triple correlation in (29.29)) and write

$$\begin{aligned}&\left\langle y_p^{\{0\}}(t'')y_q^{\{0\}}(t'')y_k^{\{0\}}(t)y_i^{\{0\}}(t')\right\rangle \\ &\quad= \left\langle y_p^{\{0\}}(t'')y_q^{\{0\}}(t'')\right\rangle\left\langle y_k^{\{0\}}(t)y_i^{\{0\}}(t')\right\rangle + \left\langle y_p^{\{0\}}(t'')y_k^{\{0\}}(t)\right\rangle\left\langle y_q^{\{0\}}(t'')y_i^{\{0\}}(t')\right\rangle \\ &\qquad+ \left\langle y_p^{\{0\}}(t'')y_i^{\{0\}}(t')\right\rangle\left\langle y_k^{\{0\}}(t)y_q^{\{0\}}(t'')\right\rangle.\end{aligned} \tag{29.30}$$

Each of these three terms are recognized as the correlation function to $^{\{0\}}$ order. Using the symmetry $R(t,\, t') = R(t',\, t)$ we can therefore write

$$T_1 \equiv 2A_{ipq} \int_0^t \langle G(t,\, t'')\rangle R(t,\, t'')R(t',\, t'')\mathrm{d}t''.$$

We can now proceed in exactly the same manner with the two other terms T_2 and T_3 and find results that can be obtained by simple renaming of variables from T_1. The final result is a nonlinear differential equation for the correlation function

$$\begin{aligned}\dot{R}(t-t') = -\frac{2C}{N}\bigg(&\int_0^t \langle G(t,\, t'')\rangle R(t-t'')R(t'-t'')\mathrm{d}t'' \\ &- \int_0^t \langle G(t',\, t'')\rangle R^2(t-t'')\mathrm{d}t''\bigg),\end{aligned} \tag{29.31}$$

recalling that $C \equiv A_{ijk}A_{ijk}$. This is an equation for determining $R(t-t')$ once the average propagator $\langle G(t,\, t')\rangle$ is known. We now need an independent equation for $\langle G(t,\, t')\rangle$. We found before that

$$\dot{G}_{i\ell}(t,\, t') = 2A_{ijk}y_j^{\{0\}}(t)G_{k\ell}(t,\, t'), \tag{29.32}$$

when $t > t'$. We now recall the expression (29.25) we found for $\tilde{y}(t')$. The two differential equations (29.25) and (29.32) are quite similar, which is not so surprising since the Green function was derived to give the solution for $\tilde{y}(t)$. We can also here take one of the terms from (29.32) and let it represent a driving 'noise' term denoted $f_i(t)$. In the same spirit as when obtaining (29.26), we postulate

$$G_{i\ell}(t,\, t') = \int_0^t G_{k\ell}(t,\, t'') f_i(t'') \mathrm{d}t''$$
$$= 2A_{ijk} \int_0^t G_{k\ell}(t,\, t') y_j^{\{0\}}(t'') G_{k\ell}(t'',\, t') \mathrm{d}t'',$$

with the same arguments as used for obtaining (29.27). The expression for $G_{i\ell}(t, t')$ can now be used on the right hand side of (29.32) to give

$$\dot{G}_{i\ell}(t,\, t') = 2A_{ijk}\, y_j^{\{0\}}(t) G_{k\ell}(t,\, t')$$
$$= 2A_{ijk}\, y_j^{\{0\}}(t) 2A_{kpq} \int_0^t G(t,\, t'') y_p^{\{0\}}(t'') G_{q\ell}(t'',\, t') \mathrm{d}t''.$$

In principle this expression holds for any realization. We now take the ensemble average and assume again isotropy with $\langle G_{ii} \rangle = N\langle G \rangle$, where N was introduced in (29.19). This gives

$$\langle \dot{G}(t,\, t') \rangle = \frac{4}{N} A_{ijk} A_{kpq} \int_0^t \langle G(t,\, t'') \rangle \left\langle y_j^{\{0\}}(t) y_p^{\{0\}}(t'') G_{q\ell}(t'',\, t') \right\rangle \mathrm{d}t''.$$

We again make an assumption similar to the quasi-normal approximation (not exactly the same this time) and write an average of a product as a product of averages

$$\left\langle y_j^{\{0\}}(t) y_p^{\{0\}}(t'') G_{q\ell}(t'',\, t') \right\rangle = \left\langle y_j^{\{0\}}(t) y_p^{\{0\}}(t'') \right\rangle \left\langle G_{q\ell}(t'',\, t') \right\rangle \equiv \delta_{jp} R(t - t'') \delta_{qi} \left\langle G(t'',\, t') \right\rangle,$$

using that $y_j^{\{0\}}$ and $y_p^{\{0\}}$ are assumed statistically independent for $j \neq p$ and introducing the Kronecker deltas. With these approximations we find a differential equation for $G(t, t')$ in the form

$$\langle \dot{G}(t,\, t') \rangle = \frac{4}{N} A_{ijk} A_{kpq} \delta_{jp} \delta_{qi} \int_{t'}^t \langle G(t,\, t'') \rangle \langle G(t'',\, t') \rangle R(t - t'') \mathrm{d}t''$$
$$= \frac{4}{N} A_{ijk} A_{kji} \int_{t'}^t \langle G(t,\, t'') \rangle \langle G(t'',\, t') \rangle R(t - t'') \mathrm{d}t''.$$

The lower limit of integration is set to t' since $\langle G(t'', t') \rangle = 0$ for $t'' < t'$. It can be demonstrated (here left for the reader as an exercise) that $A_{ijk} A_{kji} = -\frac{1}{2} A_{ijk} A_{ijk} = -C/2$. With this simplification we can write

$$\langle \dot{G}(t,\, t') \rangle = -\frac{C}{2N} \int_{t'}^t \langle G(t,\, t'') \rangle \langle G(t'',\, t') \rangle R(t - t'') \mathrm{d}t''. \tag{29.33}$$

The relation together with (29.31) constitute the full set of equations for the direct interaction approximation. Noting the resemblance of the two equations we make a bold guess: $\langle G(t - t') \rangle = R(t - t')$ for $t > t'$ giving

$$\langle \dot{G}(t,\, t') \rangle = -\frac{C}{2N} \int_{t'}^t \langle G(t,\, t'') \rangle^2 \langle G(t'',\, t') \rangle \mathrm{d}t''. \tag{29.34}$$

Albeit obtained by a number of assumptions, this equation nonetheless represents the result of a step-by-step derivation of a nonlinear turbulence model equation without free adjustable parameters. Numerical solutions of (29.34) have been found by Betchov (1966) with applications for a particular model problem. The model received convincing support.

The direct interaction approximation has been applied to a number of problems relevant for low frequency plasma turbulence (Mattor and Terry 1992), for ionospheric conditions (Sudan and Keskinen 1977) in particular, and also for describing anomalous transport in tokamak plasma (Thoul *et al* 1994). A general summary is given by, for instance, Krommes (2002) and Krommes (2015).

As said before, I owe Dwight Nicholson a lot for the presentation in this chapter.

References

Bekefi G 1966 *Radiation Processes in Plasmas* (New York: Wiley)

Beran M J 1968 *Statistical Continuum Theories* Vol 9 of *Monographs in Statistical Physics and Thermodynamics* (New York: Interscience)

Berge P, Pomeau Y and Vidal C 1984 *Order within Chaos* (New York: Wiley)

Berz E 1967 *Verallgemeinerte Funktionen und Operatoren*, Vol 122/122a of *Hochschultaschenbücher*, Bibliographisches Institut, Mannheim

Betchov R 1966 *Dynamics of Fluids and Plasmas, Proceedings of the Symposium held in honor of Professor Johannes M. Burgers, 7-9 October, 1965 at University of Maryland* (New York: Academic) pp 215–37 chapter Introduction to the Kraichnan theory of turbulence,

Betchov R 1967 Review of Kraichnan's theory of turbulence *Phys. Fluids* **10** S17–24

Callen H B and Welton T A 1951 Irreversibility and generalized noise *Phys. Rev.* **83** 34–40

DuBois D F and Espedal M 1978 Direct interaction approximation and plasma turbulence theory *Plasma Phys.* **20** 1209–39

Kraichnan R H 1958a Higher order interactions in homogeneous turbulence theory *Phys. Fluids* **1** 358–9

Kraichnan R H 1958b Irreversible statistical mechanics of incompressible hydromagnetic turbulence *Phys. Rev.* **109** 1407–22

Kraichnan R H 1959 The structure of isotropic turbulence at very high Reynolds numbers *J. Fluid Mech.* **5** 497–543

Krommes J A 2002 Fundamental statistical descriptions of plasma turbulence in magnetic fields *Phys. Reports* **360** 1–352

Krommes J A 2015 A tutorial introduction to the statistical theory of turbulent plasmas, a half-century after Kadomtsev's plasma turbulence and the resonance-broadening theory of Dupree and Weinstock *J. Plasma Phys.* **81** 205810601

Krommes J A and Smith R A 1987 Rigorous upper bounds for transport due to passive advection by inhomogeneous turbulence *Annals Phys.* **177** 246–329

Leslie D C 1973 *Developments in the Theory of Turbulence* (Oxford: Oxford University Press)

Lighthill M J 1964 *An Introduction to Fourier Analysis and Generalized Functions* (Cambridge: Cambridge University Press)

MacDonald D K C 1962 *Noise and Fluctuations: an Introduction* (New York: Wiley)

Martin P C, Siggia E D and Rose H A 1973 Statistical dynamics of classical systems *Phys. Rev.* A **8** 423–37

Mattor N and Terry P W 1992 Frequency spectrum in drift wave turbulence *Phys. Fluids* B **4** 1126–38

Orszag S A 1970 Analytical theories of turbulence *J. Fluid Mech.* **41** 363–86

Pécseli H L 2000 *Fluctuations in Physical Systems* (Cambridge: Cambridge University Press)

Similon P L and Sudan R N 1990 Plasma turbulence *Annual Rev. Fluid Mech.* **22** 317–47

Stanišić M M 1988 *The Mathematical Theory of Turbulence, Universitext* 2nd edn (Berlin: Springer)

Sudan R N and Keskinen M 1977 Theory of strongly turbulent two-dimensional convection of low-pressure plasma *Phys. Rev. Lett.* **38** 966–70

Sudan R N and Pfirsch D 1985 On the relation between 'mixing length' and 'direct interaction approximation' theories of turbulence *Physics of Fluids (1958-1988)* **28**

Thoul A A, Similon P L and Sudan R N 1994 Anomalous perturbative transport in tokamaks due to drift-wave turbulence *Phys. Plasmas* **1** 579–600

IOP Publishing

Low Frequency Waves and Turbulence in Magnetized Laboratory Plasmas and in the Ionosphere

Hans Pécseli

Chapter 30

Diagram methods

Although it might be difficult to reach consensus on the opinion, it can be postulated that differential equations are short hand for physics and diagrams are shorthand for differential equations. Diagram methods have found wide applications for many body systems, or rather systems with many degrees of freedom, statistical models in particular. A complete summary is outside the scope of the present study. As in many other cases, diagram methods are most easily approached by trying to work out some practical problems. Some can be found in books, for instance in Mattuck (1967), Orszag (1970), Yeh and Liu (1972) or Nayfeh (1973), with a historical summary given by Kaiser (2005). Diagram methods have been used for fluctuating systems in general (Martin *et al* 1973) and also in formulating the kinetic quasi-linear analysis of weakly unstable electron plasma instabilities (Al'tshul' and Karpman 1966). The method has been widely used in studies of wave propagation in randomly varying media.

As an illustration we here have a simple equation

$$\frac{\partial}{\partial t}G(t, t') = -ib(t)G(t, t'), \tag{30.1}$$

with $G(t, t) = 1$ and $G(t, t') = 0$ for $t < t'$, while $G(t, t')$ is a Green's function or 'propagator'. Generally the function $b(t)$ is a random variable with a normal or Gaussian distribution. A classical example will be one where $\langle b(t)b(t')\rangle = q(t, t')$, for stationary random processes in particular we will have $\langle b(t)b(t')\rangle = q(t - t')$. For most of the analysis we will here be content with an even simpler case where b is a constant in each realization, but varying over the ensemble so that $\langle b\rangle = 0$ while $\langle b(t)b(t')\rangle = b_0^2$ is a constant. The exact solution of (30.1) is then readily obtained as

$$G(t, t') = \mathrm{e}^{-ib(t-t')} \tag{30.2}$$

doi:10.1088/978-0-7503-1251-6ch30

for $t > t'$ while $G(t, t') = 0$ for $t < t'$. Since b is a constant in each realization it is easy to take the ensemble average to give

$$\langle G(t, t') \rangle = \mathrm{e}^{-\frac{1}{2} b_0^2 (t-t')^2}. \tag{30.3}$$

Before proceeding, it might here be appropriate to insert a short note on the distinction between (30.2) and (30.3). Their definition is clear, but the two results serve to emphasize that statistical averages can have little relevance for observations of one realization (see also section 19.2.5). Ergodicity is an important concept giving ensemble averages value for predictions on spatial or temporal averages obtained from one realization. The present model is *not* ergodic, but we use it for illustration only, so we will not let this trouble us here.

In (30.3) we now have the exact solution for this simple problem and will use it to illustrate the accuracy of various approximate solutions, also using the process to illustrate the use of diagrams. We thus return to the original equation (30.1) and write it in a different way

$$G(t, t') = \begin{cases} 1 - i \int_{t'}^{t} b(t_1) G(t_1, t') \mathrm{d}t_1 & \text{for} \quad t > t' \\ 0 & \text{for} \quad t \leqslant t'. \end{cases} \tag{30.4}$$

Introducing Heaviside's step-function $\mathcal{H}(t, t')$ we can express this equation in a more compact way as

$$G(t, t') = \mathcal{H}(t, t') - i \int_{t'}^{t} \mathcal{H}(t, t') b(t_1) G(t_1, t') \mathrm{d}t_1, \tag{30.5}$$

for all t. You might argue that the $\mathcal{H}(t, t')$ function in the integral is redundant, but its purpose will become evident in what follows. We have $\mathcal{H}(t, t') = \mathcal{H}(t - t')$ since $\mathcal{H}(t, t') = 1$ for $t > t'$ and $\mathcal{H}(t, t') = 0$ for $t \leqslant t'$, but by retaining the form used here we obtain expressions that are more easily generalized, so we keep this slight cosmetic improvement of the formulation.

- **Example:** We have that any implicit equation of the form

$$f(x) = g(x) + \epsilon \int_a^b K(x, y) f(y) \mathrm{d}y$$

can be expanded in a Neumann series (or Born expansion) (Nayfeh 1973) provided we can consider ϵ as a small parameter. To lowest (or zeroth) order, where we set $\epsilon = 0$, we evidently have $f(x) = g(x)$. The first order solution (ϵ^1) is obtained by iteration as

$$f(x) = g(x) + \epsilon \int_a^b K(x, y) g(y) \mathrm{d}y.$$

The second order solution (ϵ^2) is readily found as

$$f(x) = g(x) + \epsilon \int_a^b K(x, y)\left(g(y) + \epsilon \int_a^b K(y, y_1)g(y_1)\right)\mathrm{d}y_1\mathrm{d}y \tag{30.6}$$

$$= g(x) + \epsilon \int_a^b K(x, y)g(y)\mathrm{d}y + \epsilon^2 \int_a^b \int_a^b K(x, y)K(y, y_1)g(y_1)\mathrm{d}y_1\mathrm{d}y. \tag{30.7}$$

We can continue and obtain an infinite series, although without any *a priori* guarantee for convergence.

To illustrate the nature of the problems that may arise we can consider the simple case (Nayfeh 1973) where

$$\frac{\partial^2}{\partial t^2}u(t) + k^2(\epsilon)u(t) = 0,$$

having the exact solution $u(t) = \exp(\mathrm{i}k(\epsilon)t)$. We now specify $k(\epsilon) \equiv \sum_{n=0}^{\infty}\epsilon^n k_n$. The Born (or von Neumann) expansion of $u(t)$ gives

$$u(t) = \exp(\mathrm{i}k_0 t)\left(1 + \mathrm{i}\epsilon k_1 t + \epsilon^2\left(\mathrm{i}k_2 t - \frac{1}{2}k_1^2 t^2\right)\right. \tag{30.8}$$

$$\left. + \epsilon^3\left(\mathrm{i}k_3 t - k_1 k_2 t^2 - \frac{i}{6}k_1^3 t^3\right)\ldots\right). \tag{30.9}$$

This expansion is valid only for short times and breaks down for $k_1 t \geqslant \mathcal{O}(\epsilon^{-1})$. To increase the range of validity we collect all terms containing $\mathrm{i}k_1 t$ to form $1 + \mathrm{i}k_1 t + \cdots(\mathrm{i}k_1 t)^n/n!\cdots = \sum_{n=0}^{\infty}(\mathrm{i}k_1 t)^n/n! = \exp(\mathrm{i}k_1 t)$ taking care to note all terms where we have mixtures like $k_1 k_2$ etc. We can then rewrite (30.9) as

$$u = \exp(\mathrm{i}(k_0 + \epsilon k_1)t)\left(1 + \mathrm{i}\epsilon^2 k_2 t + \mathrm{i}\epsilon^3 k_3 t - \frac{1}{2}\epsilon^4 k_2^2 t^2 + \cdots\right). \tag{30.10}$$

This series expansion has a wider range of applicability and breaks down $k_2 t \geqslant \mathcal{O}(\epsilon^{-2})$. We could now continue and collect all terms containing $k_2 t$, etc. The root of all these problems can be identified in the observation that a series expansion of $\exp\psi = 1 + \psi + \psi^2\cdots$ converges uniformly and absolutely for all values of ψ, but any finite number of terms from the series cannot represent $\exp\psi$ to any accuracy for all values of ψ (Nayfeh 1973). Fortunately, we are often content with a limited range of convergence, the important thing is to know how large it is!

- **Example:** One of our students gave me a nice example demonstrating how careful we have to be when manipulating series. He took $a \equiv \sum_{n=1}^{\infty} 9 \times 10^{-n} = 0.99999\ldots$ and then wrote $10a = 9.99999\ldots = 9 + 0.99999\ldots$ The last figure is a, so we have $10a = 9 + a$ giving $9a = 9$. The rest is silence.

Stated in the form (30.7) this a quite general result. We can now apply it to the present problem given by (30.5) and find to zeroth order $G(t, t') = \mathcal{H}(t, t')$, which we call the 'bare propagator' obtained for $b = 0$.

To higher orders we have

$$G(t, t') = \mathcal{H}(t, t') - i\int_{t'}^{t} \mathcal{H}(t, t')b(t_1)\mathcal{H}(t_1, t')\mathrm{d}t_1 + (-i)^2\int_{t'}^{t} \mathcal{H}(t, t_1)b(t_1)\int_{t'}^{t_1} b(t_2)\mathcal{H}(t_1, t_2)\mathcal{H}(t_2, t')\mathrm{d}t_1\mathrm{d}t_2 + \cdots. \tag{30.11}$$

The solution is formal in the sense that we have no *a priori* well defined small expansion parameter unless we assume b to be small in any realization.

It is evident that the series (30.11) soon becomes messy, and it is an advantage to look for simplifications in the presentation. Diagram methods present such a simplification. We introduce a notation where the bare propagator $\mathcal{H}$ is indicated by a thin horizontal line labeled with the times where it begins and ends. The coefficient $(-i)$ is denoted a 'vertex' and indicated by a small filled circle while $b(t)$ is denoted by a vertical dashed line, see (30.12)

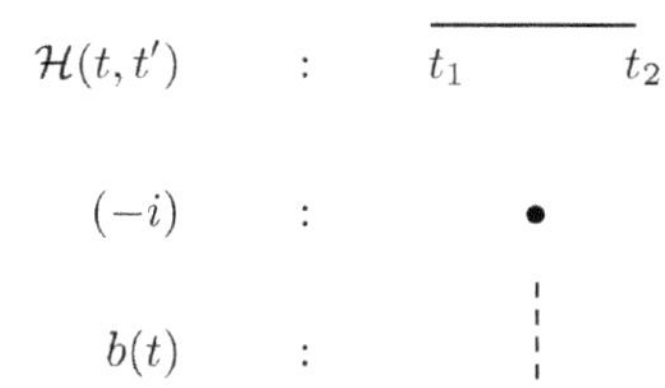

(30.12)

The expression (30.11) can be expressed in terms of diagrams by use of (30.12) as

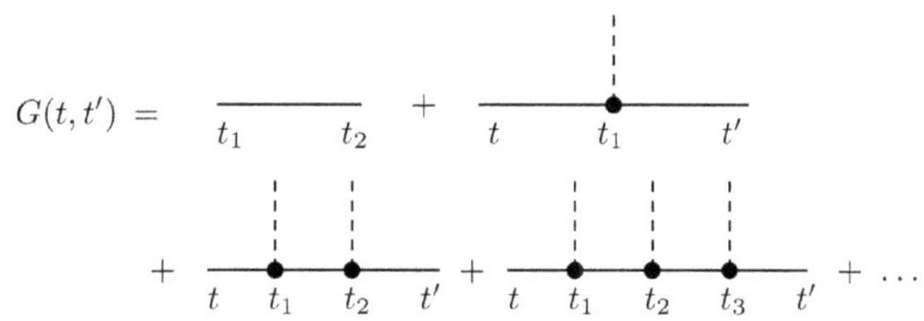

(30.13)

We note how much the presentation is improved: it is compact and more transparent and maybe most of all, much faster to write down.

So far we have discussed the exact propagator. The goal of the present analysis is a presentation of the ensemble average $\langle G(t, t')\rangle$. We will assume from the outset that the process $b(t)$ is Gaussian with zero mean so all odd moments (such as $\langle b(t)\rangle$, $\langle b^3(t)\rangle$, etc) will vanish when we take averages. By averaging (30.11) we find

$$\langle G(t, t')\rangle = \mathcal{H}(t, t') + (-i)^2\int_{t'}^{t}\int_{t'}^{t_1} \mathcal{H}(t, t_1)\mathcal{H}(t_1, t_2) \langle b(t_1)b(t_2)\rangle\mathcal{H}(t_2, t')\mathrm{d}t_1\mathrm{d}t_2 \tag{30.14}$$

$$+(-i)^4 \int_{t'}^{t} \int_{t'}^{t_1} \int_{t'}^{t_2} \int_{t'}^{t_3} \mathcal{H}(t, t_1)\rangle \mathcal{H}(t_1, t_2)\mathcal{H}(t_2, t_3)\mathcal{H}(t_3, t_4) \tag{30.15}$$

$$\times\langle b(t_1)b(t_2)b(t_3)b(t_4)\rangle \mathcal{H}(t_4, t')\mathrm{d}t_1\mathrm{d}t_2\mathrm{d}t_3\mathrm{d}t_4 + \cdots. \tag{30.16}$$

For a Gaussian process we have $\langle b(t_1)b(t_2)b(t_3)b(t_4)\rangle = \langle b(t_1)b(t_2)\rangle\langle b(t_3)b(t_4)\rangle + \langle b(t_1)b(t_3)\rangle\langle b(t_2)b(t_4)\rangle + \langle b(t_1)b(t_4)\rangle\langle b(t_3)b(t_2)\rangle \equiv q(t_1 - t_2)q(t_3 - t_4) + q(t_1 - t_3)$ $q(t_2 - t_4) + q(t_1 - t_4)q(t_3 - t_2)$ with the definitions introduced before for time stationary processes. We can write (30.16) as

$$\begin{aligned}\langle G(t, t')\rangle = \mathcal{H}(t, t') &+ (-i)^2 \int_{t'}^{t} \int_{t'}^{t_1} \mathcal{H}(t, t_1)\mathcal{H}(t_1, t_2)\mathcal{H}(t_2, t')q(t_1 - t_2)\mathrm{d}t_1\mathrm{d}t_2 \\ &+ (-i)^4 \int_{t'}^{t}\int_{t'}^{t_1}\int_{t'}^{t_2}\int_{t'}^{t_3} \mathcal{H}(t, t_1)\rangle\mathcal{H}(t_1, t_2)\mathcal{H}(t_2, t_3)\mathcal{H}(t_3, t_4)\mathcal{H}(t_4, t') \\ &\times [q(t_1 - t_2)q(t_3 - t_4) + q(t_1 - t_3)q(t_2 - t_4) + q(t_1 - t_4)q(t_3 - t_2)] \\ &\mathrm{d}t_1\mathrm{d}t_2\mathrm{d}t_3\mathrm{d}t_4 + \cdots.\end{aligned} \tag{30.17}$$

The two times t_1 and t_2 are connected by the correlation $q(t_1 - t_2)$ and similarly for other time-combinations. We therefore introduce a new symbol as

$(-i)^2 g(t_1 - t_2)$:

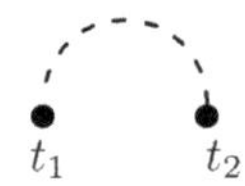

(30.18)

We can understand this symbol by writing

$(-i)^2 \langle b(t_1)\, b(t_2)\rangle =$

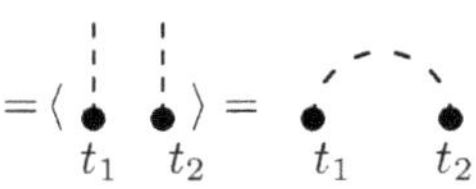

(30.19)

We also define a short hand notation of the *exact* average propagator as

$\langle G(t, t')\rangle =$

(30.20)

with a heavy line to be distinguished from $\mathcal{H}(t, t')$ in (30.12). We can now express (30.17) by diagrams in the form

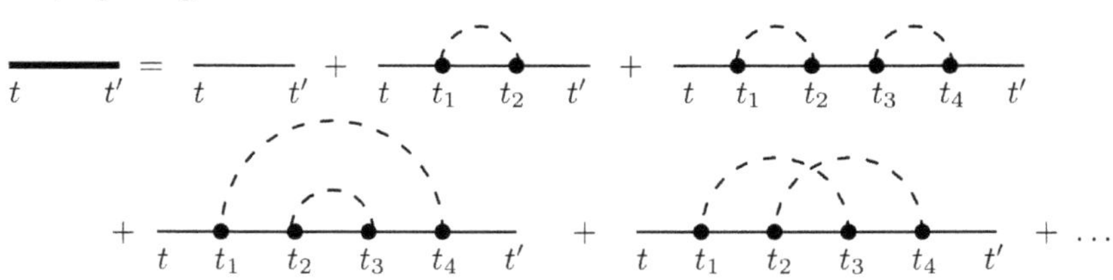

(30.21)

The elements of the second and third terms are often called 'bubbles' or 'speed bumps', and terms like the fourth one will be a 'tower of bubbles', the fifth one 'crossed bubbles'. In either representation, analytical expression as (30.16) or diagrams, the result is exact, assuming the series are converging.

The forms given in (30.21) are not unique (Krommes 2002): we can give an equivalent topology of for instance the last term as shown by

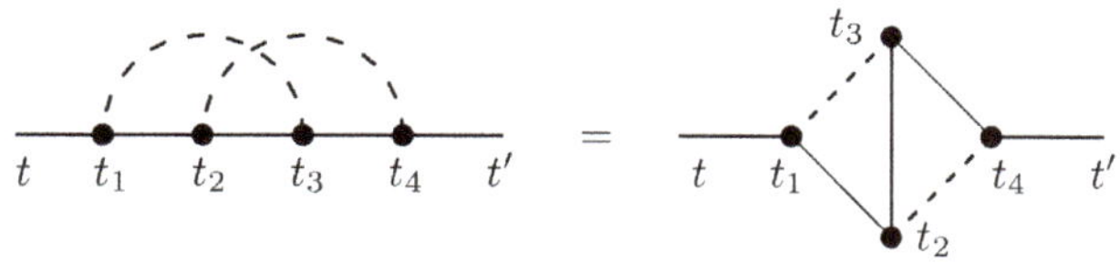

(30.22)

30.1 Approximate solution by the second order series method

As is so often the case when we are dealing with infinite series, the simplest approach is to truncate the series by throwing away all terms higher than some prescribed level. We can for instance keep only the two lowest order terms In a diagram representation this corresponds to

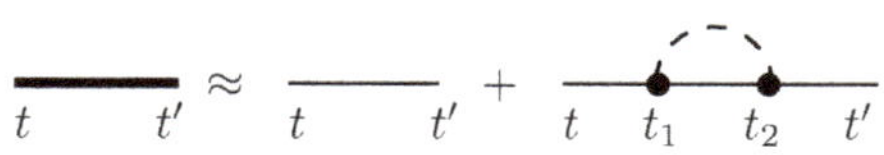

(30.23)

or if you like

$$\langle G(t, t')\rangle \approx \mathcal{H}(t, t') - \int_{t'}^{t}\int_{t'}^{t_1} q(t_1 - t_2)\mathrm{d}t_1\mathrm{d}t_2. \tag{30.24}$$

With the simple model $\langle b(t)b(t')\rangle = b_0^2$ introduced before we easily solve the integrals and find

$$\langle G(t, t')\rangle \approx 1 - \frac{1}{2}b_0^2(t - t')^2, \tag{30.25}$$

for $t > t'$. This is the result of what is called the second order series method. It is easily seen that this result corresponds to the first terms in a series expansion of the exact result (30.3), see also figure 30.1.

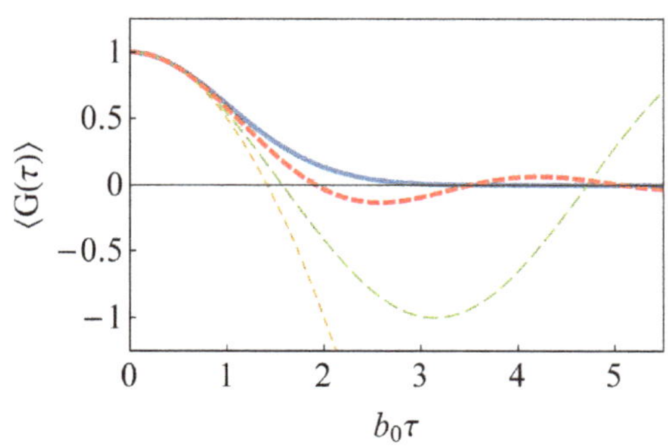

Figure 30.1. Comparison of the exact average propagator $\langle G(\tau)\rangle$ (full heavy line) with $\tau \equiv t - t'$, and approximate solutions obtained by (1) second order series approximation (parabola), (2) quasi-normal approximation (cos oscillation) and (3) the direct interaction approximation (DIA), all with dashed lines. See the text for details.

30.2 Approximate solution by the quasi-normal approximation

We may approach the solution for (30.1) along a somewhat different route and rewrite the expression by a time-integration to give

$$G(t, t') = 1 + (-i) \int_{t'}^{t_1} b(t_1) G(t_1, t') \mathrm{d}t_1 \tag{30.26}$$

for $t > 0$. We then insert this expression on the right-hand side of (30.1) to give

$$\frac{\partial}{\partial t} G(t, t') = -ib(t)\left(1 + (-i) \int_{t'}^{t_1} b(t_1) G(t_1, t') \mathrm{d}t_1\right). \tag{30.27}$$

Yes, this seems a strange thing to do, but wait and see!

The result (30.27) is now averaged over the ensemble to give

$$\frac{\partial}{\partial t}\langle G(t, t')\rangle = (-i)^2 \int_{t'}^{t} \langle b(t) b(t_1) G(t_1, t')\rangle \mathrm{d}t_1. \tag{30.28}$$

So far, this is an exact result, based on (30.1) with the given assumptions on the statistical properties of $b(t)$. The problematic part is the correlation $\langle b(t)b(t_1)G(t_1, t')\rangle$. The quasi-normal approximation consists of postulating $\langle b(t)b(t_1)G(t_1, t')\rangle \approx \langle b(t)b(t_1)\rangle\langle G(t_1, t')\rangle \equiv q(t - t_1)\langle G(t_1, t')\rangle$. If $b(t_1)$ and $G(t_1, t')$ were statistically independent, this would be an exact relation. Note that we met this idea of replacing an average of a product by product of averages already when discussing Corrsin's hypothesis in section 25.4.1. With the quasi-normal approximation we obtain a closed equation for $\langle G(t_1, t')\rangle$ in the form

$$\frac{\partial}{\partial t}\langle G(t, t')\rangle_{QN} = (-i)^2 \int_{t'}^{t} q(t - t_1)\langle G(t_1, t')\rangle_{QN} \mathrm{d}t_1, \tag{30.29}$$

where we added a subscript QN as a reminder of the quasi-normal approximation. The expression (30.29) can be expressed in a different way by a time-integration that gives

$$\begin{aligned}\langle G(t, t')\rangle_{QN} = {}& \mathcal{H}(t, t') \\ & + (-i)^2 \int_{t'}^{t} \int_{t'}^{t_1} \mathcal{H}(t, t_1) q(t_1 - t_2) \mathcal{H}(t_1, t_2) \\ & \langle G(t_2, t')\rangle_{QN} \mathrm{d}t_1 \mathrm{d}t_2.\end{aligned} \tag{30.30}$$

This result is easily written in terms of diagrams in the form

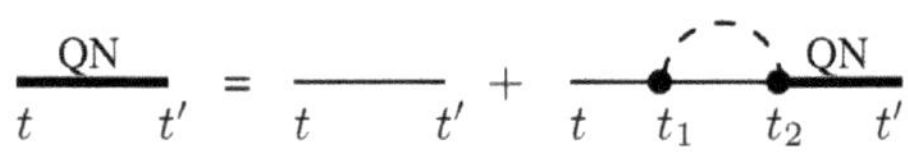

(30.31)

We see that the quasi-normal approximation corresponds to 'renormalizing' the second order series approximation by replacing the last propagator on the right-hand side by the quasi-normal propagator itself. In general, the term renormalization is used for the process of summing expansions to make them 'more' uniformly valid (Nayfeh 1973). The idea seems to originate from works of Rayleigh (1917) and later rediscovered by others. A detailed summary and reference list is given by Nayfeh (1973).

It is now easy to iterate (30.31) when we write it in terms of diagrams

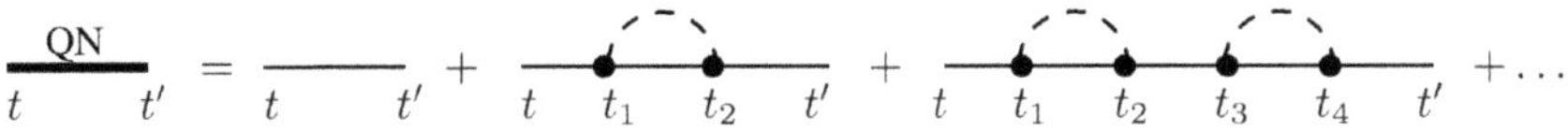

(30.32)

The details are left to the reader as an exercise. By comparing this diagram with the exact result (30.21) we realize that the quasi-normal expression corresponds to omitting all 'towers' and all 'crossed bubbles'!

For the simple case with $\langle b(t)b(t')\rangle = b_0^2$ we can solve (30.30) analytically and find with little effort the equation

$$\frac{\partial^2}{\partial t^2}\langle G(t, t')\rangle_{QN} = -b_0^2\langle G(t, t')\rangle_{QN},$$

with the simple solution $\langle G(t, t')\rangle_{QN} = \cos(b_0(t - t'))$ satisfying $\langle G(t, t)\rangle_{QN} = 1$, see also figure 30.1. The quasi-normal result is better than the one obtained by second order series method, but cannot be considered adequate, except for small times, where it is OK and a little better than the results from the series expansion.

30.3 Approximate solution by the DIA

Rather than going into details concerning the DIA we here illustrate some manipulations of diagrams and show how a result corresponding to the DIA is found as a special case (Wyld 1961, Orszag 1970, Kono and Ichikawa 1973, Krommes 2002). We take the quasi-normal approximation as a reference case, and *with the exception of the first bubble* denote all its constituent diagrams as 'reducible' in the sense that each of the bubbles can be flattened (or 'contracted' (Kraichnan 1961) in Kraichnan's terminology) individually without this affecting any other bubble (this would not be so for crossing bubbles, these cannot be flattened). We can write this symbolically as

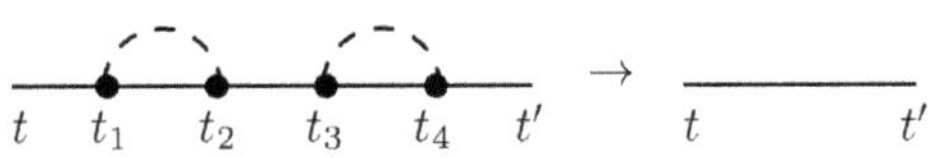

(30.33)

We will now use this definition in a more general sense. At first sight a tower of bubbles does not seem reducible, but with a little care we can fix that too

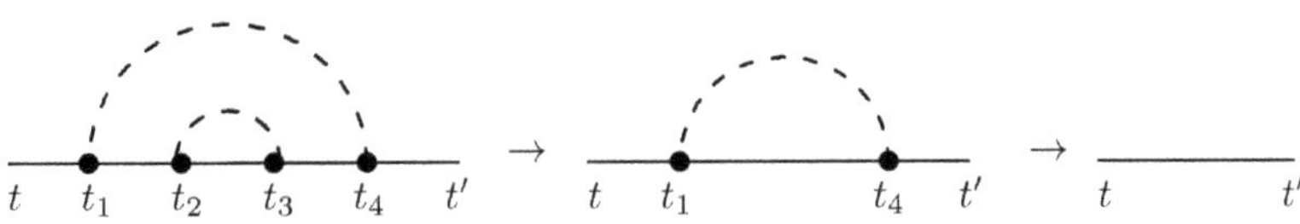

(30.34)

The first, third and fourth diagrams on the right-hand side of (30.21) can thus be flattened. Concerning the first bare propagator in any sum of diagrams (e.g. the first one on the right-hand side of (30.21)) there is nothing to be done: it is what *has* to remain if we set $b = 0$. The first bare propagator that connects t with t_1 in any of the diagrams there is nothing do be done either: this is the transit time before any interaction process initiates. Concerning the rest we now proceed with a bold prescription: flatten all that can be flattened and replace (i.e. 'renormalize') all remaining propagators by (30.20). By doing this we obtain

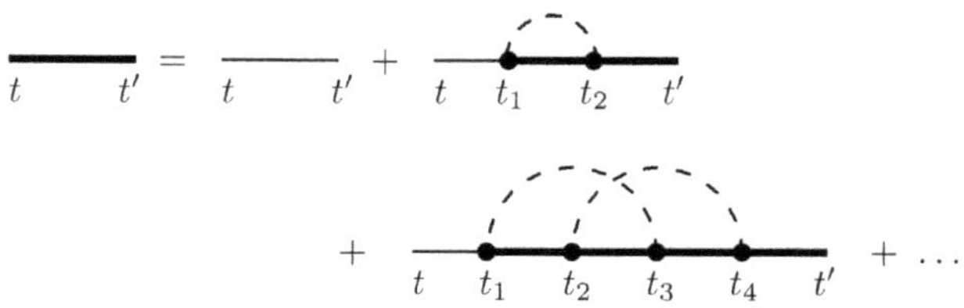

(30.35)

Amazingly enough this turns out to a different representation of the exact result! This is most easily shown by iteration, the same way as we did it by the Neumann series. One by one, we recover all the diagrams entering (30.21). By now the reader knows how to manipulate diagrams and can demonstrate this as an exercise.

The DIA result is found by keeping only the first two terms in (30.35) to give

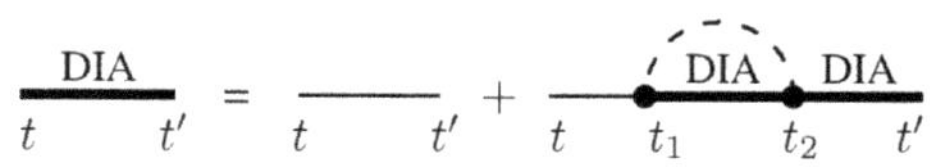

(30.36)

In this sense the DIA result can be understood as a generalization of the quasi-normal approximation, see (30.31) for comparison. It is easily demonstrated that (30.36) sums all bubbles and towers from the complete result (30.21), but omits crossed bubbles.

The DIA can also be applied for kinetic plasma phenomena as described by the Vlasov equation. Nice and clear dispositions are presented by Orszag (1969) and Kono and Ichikawa (1973) on this problem, and these can in a general sense also have value for understanding the renormalization procedure outlined here.

We can write (30.36) in equation form as

$$\langle G(t-t')\rangle = \mathcal{H}(t,t') + (-i)^2 \int_{t'}^{t}\int_{t'}^{t_1} \mathcal{H}(t,t_1)\langle G(t_1-t_2)\rangle q(t_1-t_2) \langle G(t_2-t')\rangle \mathrm{d}t_1\mathrm{d}t_2, \tag{30.37}$$

or after differentiation with respect to time t for $t > t'$ as

$$\frac{\partial}{\partial t}\langle G(t-t')\rangle = (-i)^2 \int_{t'}^{t} q(t-t_1)\langle G(t-t_1)\rangle\langle G(t_1-t')\rangle \mathrm{d}t_1, \tag{30.38}$$

and obtain a solution (Leslie 1973) for the special case with $\langle b(t)b(t')\rangle \equiv q(t-t') = b_0^2$. This is a bit tricky. It is best done by Laplace transforming (30.38) with the initial condition $\langle G(0)\rangle = 1$ to give

$$s\langle G(s)\rangle + b_0^2 \int_0^{\infty} e^{-st} \int_0^{t} \langle G(t-t_1)\rangle\langle G(t_1)\rangle \mathrm{d}t_1\mathrm{d}t = 1,$$

in terms of the complex transform variable s. The double integral can be evaluated (or you might simply look up the result in a textbook on the Laplace transform) and we find

$$s\langle G(s)\rangle + b_0^2\langle G(s)\rangle^2 = 1.$$

From this we have

$$\langle G(s)\rangle = -\frac{1}{2}\frac{s}{b_0^2} \pm \frac{1}{2}\sqrt{\frac{s^2}{b_0^4} + \frac{4}{b_0^2}}.$$

By tables of inverse Laplace transforms we then have the DIA result for this problem as

$$\langle G(t - t')\rangle = \frac{J_1(2b_0(t - t')}{b_0(t - t')}, \qquad (30.39)$$

as shown in figure 30.1. Although not exact, the DIA result gives a noticeably better approximation to the exact result than either of the two other approximations we tried. Higher order corrections can improve the accuracy (Mond and Knorr 1980, Knorr and Mond 1980).

The DIA (Kraichnan 1959) was one of the first field theoretical approaches to turbulence (Yakhot and Orszag 1986). The first results for the inertial subrange gave a spectrum $S(k) \sim k^{-3/2}$ in contradiction with the at that time accepted Kolmogorov–Oubokhov spectrum $S(k) \sim k^{-5/3}$. The problem with this original Eulerian formulation was that it did not distinguish between dynamic and kinematic interactions between eddies of widely separated length scales. As also discussed in section 25.2, small eddies are convected by large eddies in a purely kinematic way which should not lead to energy redistribution between different scales. The spurious effect of large-scale convection on small scales has been removed from the DIA by use of a Lagrangian description of the flow. This Lagrangian history direct interaction approximation (LHDIA) (Leslie 1973) gives the Kolmogorov $k^{-5/3}$-energy spectrum with the Kolmogorov constant $C_k = 1.77$ which is not too different from the generally accepted value. When applied to the problem of turbulent diffusion of a single particle (Kraichnan 1966), see also section 25.1, the results of the LHDIA did not perform very well. This was manifested (Yakhot and Orszag 1986) by a comparison between the calculated Prandtl number of ~0.14 and the currently accepted value of 0.7–0.9. This might seem surprising since the LHDIA is a Lagrangian analysis, that should be particularly suited for the problem. Significant improvements were obtained by renormalization group methods (Forster *et al* 1977, Yakhot and Orszag 1986).

References

Al'tshul' L M and Karpman V I 1966 Theory of nonlinear oscillations in a collisionless plasma *Soviet Phys. JETP-USSR* **22** 361–9

Al'tshul' L M and Karpman V I 1966 Russian original *J. Exptl. Theoret. Phys. (USSR)* **49** 515–28

Forster D, Nelson D R and Stephen M J 1977 Large-distance and long-time properties of a randomly stirred fluid *Phys. Rev.* A **16** 732–49

Kaiser D 2005 Physics and Feynman's diagrams *Am. Sci.* **93** 156–65

Knorr G and Mond M 1980 Iterative approach to strong turbulence theory *Phys. Fluids* **23** 1301–5

Kono M and Ichikawa Y H 1973 Renormalization of the wave-particle interaction in weakly turbulent plasmas *Prog. Theor. Phys.* **49** 754–63

Kraichnan R H 1959 The structure of isotropic turbulence at very high Reynolds numbers *J. Fluid Mech.* **5** 497–543

Kraichnan R H 1961 Dynamics of nonlinear stochastic systems *J. Math. Phys.* **2** 124–48

Kraichnan R H 1966 Isotropic turbulence and inertial-range structure *Phys. Fluids* **9** 1728–52

Krommes J A 2002 Fundamental statistical descriptions of plasma turbulence in magnetic fields *Phys. Reports* **360** 1–352

Leslie D C 1973 *Developments in the Theory of Turbulence* (Oxford: Oxford University Press)

Martin P C, Siggia E D and Rose H A 1973 Statistical dynamics of classical systems *Phys. Rev.* A **8** 423–37

Mattuck R D 1967 *A Guide to Feynman Diagrams in the Many-Body Problem* (London: McGraw-Hill)

Mond M and Knorr G 1980 Higher corrections to the direct interaction approximation in turbulence theory *Phys. Fluids* **23** 1306–10

Nayfeh A H 1973 *Perturbation Methods* (New York: Wiley)

Orszag S A 1969 *Turbulence in Fluids and Plasmas* (Brooklyn: Polytechnic Press) chapter 4 'Stochastic acceleration by strong electric fields'

Orszag S A 1970 Analytical theories of turbulence *J. Fluid Mech.* **41** 363–86

Rayleigh L 1917 On the reflection of light from a regularly stratified medium *Proc. R. Soc. London. Series A, Containing Papers of a Mathematical and Physical Character* **93** 565–77

Wyld H W 1961 Formulation of the theory of turbulence in an incompressible fluid *Annals Phys.* **14** 143–65

Yakhot V and Orszag S A 1986 Renormalization-group analysis of turbulence *Phys. Rev. Lett.* **57** 1722–4

Yeh K C and Liu C H 1972 *Theory of Ionospheric Waves, of International Geophysics Series* vol 17 (New York: Academic)

Appendix A

List of symbols used in the analysis of electrostatic drift waves

The literature on drift-waves, in particular, uses a large number of symbols that are more or less standard. They are summarized in the following. It might be a good idea to take a copy and have it next to you when reading the part on linear drift waves.

$c = 1/\sqrt{\varepsilon_0 \mu_0}$	Speed of light in vacuum, $c = 2.9979 \times 10^8$ m s^{-1}
e	Elementary charge, taken positive, $e = 1.602 \times 10^{-19}$ C
$\beta \equiv n\kappa T/(B^2/2\mu_0) = 2C_s^2/U_A^2$	Ratio of thermal and magnetic pressures
C_s	Ion acoustic velocity
Ω_{ci}	Ion cyclotron frequency
ω_{ce}	Electron cyclotron frequency
Ω_{pi}	Ion plasma frequency
ω_{pe}	Electron plasma frequency
$u_A \equiv B/\sqrt{nM\mu_0} \equiv c\Omega_{ci}/\Omega_{pi}$	Alfvén velocity
$u_{De} = -(\kappa T_e/eB)\mathrm{d}\ln n_0(x)/\mathrm{dx}$	Electron diamagnetic drift
$u_{Di} = (\kappa T_i/eB)\mathrm{d}\ln n_0(x)/\mathrm{dx}$	Ion diamagnetic drift at constant temperature
$\chi \equiv e\phi/\kappa T_e$	Normalized electrostatic potential
$\omega^* \equiv k_y u_{De} \equiv -k_y(\kappa T_e/eB)(n_0'/n_0) = -k_y(\kappa T_e/m\omega_{ce})(n_0'/n_0)$	Drift frequency
$\theta \equiv T_i/T_e$	Ion-electron temperature ratio
$\omega_i \equiv k_y u_{Di} = -\theta\omega^*$	Ion drift frequency
$\sigma_\parallel = (k_z/k_y)^2(\omega_{ce}\tau_{ei})\Omega_{ci}$	Reduced conductivity parameter
$\sigma_\perp = \frac{1}{4}\theta k_y^2 a_i^2/\tau_{ii} = \frac{1}{4}\theta b\nu_{ii}$	*B*-perpendicular conductivity parameter
$a_i \equiv \sqrt{\kappa T_e/M\Omega_{ci}^2}$	Ion Larmor radius obtained by the *electron* temperature

doi:10.1088/978-0-7503-1251-6ch31

$r_L \equiv \sqrt{2\kappa T_i/M\Omega_{ci}^2}$	Average ion Larmor radius
$b \equiv k_y^2 a_i^2 = k_y^2 r_L^2/2\theta$	Finite Larmor radius expansion parameter
$\eta \equiv n_1/n_0$	Normalized fluctuating plasma density
$\nu_{ei} = ne^4\pi \ln\Lambda/((4\pi\varepsilon_0)^2(\kappa T_e)^{3/2}\sqrt{m})$	Electron–ion collision frequency
$\xi = m\nu_{ei}/ne^2 = \nu_{ei}/(\omega_{pe}^2\varepsilon_0)$	Plasma resistivity
$\ln\Lambda \equiv \ln(n\lambda_D^3)$	The Spitzer logarithm
$\tau_{ii} = (32\pi M^2\varepsilon_0^2/ne^4 \ln\Lambda)(\kappa T_i/M)^{3/2}$	Ion-ion collision time
$\mu_m = \frac{1}{2}mU_\perp^2/B$	Magnetic moment of a particle with mass m and a velocity $U_\perp$ in a direction perpendicular to a magnetic field **B**.

Appendix B

Chemistry of the ionosphere

In the following some basic results are summarized concerning the chemistry and composition of the Earth's ionosphere. The given definitions of altitude ranges are approximative, and may differ depending on the literature being consulted (Brekke 2013).

The steady state ionosphere, day-time or night-time, is a balance between ionization and losses of charged particles, primarily by recombination. The most important recombination processes are:

- radiative recombination: $e + X^+ \rightarrow X + hf$ (airglow and proton aurora) with f being the photon frequency and h is Planck's constant.
- dissociative recombination: $e + XY^+ \rightarrow X + Y$.

The most important dissociative recombination processes in the atmosphere are

$$
\begin{aligned}
\mathrm{O}_2^+ + e &\rightarrow \mathrm{O} + \mathrm{O} \\
\mathrm{N}_2^+ + e &\rightarrow \mathrm{N} + \mathrm{N}.
\end{aligned}
$$

The recombination rates depend on the densities and on the recombination coefficient that is different for different processes and needs, in general, to be found empirically.

In the following we discuss individual regions of the Earth's ionosphere.

B.1 F-region, 150–400 km altitude

The most important source of charged particles is photo-ionization of oxygen atoms, O, and nitrogen molecules, N_2

$$
\begin{aligned}
\mathrm{O} + hf &\rightarrow \mathrm{O}^+ + e \\
\mathrm{N}_2 + hf &\rightarrow \mathrm{N}_2^+ + e,
\end{aligned}
$$

doi:10.1088/978-0-7503-1251-6ch32

where hf is the photon energy. We have, however, the processes

$$O^+ + O_2 \rightarrow O_2^+ + O$$
$$O^+ + N_2 \rightarrow NO^+ + N$$
$$N_2^+ + O \rightarrow NO^+ + N,$$

implying that the most important ions are O^+, NO^+ and O_2^+ in this region. Ions are lost by recombination via the processes

$$O_2^+ + e \rightarrow O^* + O^*$$
$$NO^+ + e \rightarrow N^* + O^*,$$

where * denotes an excited state. The processes are *dissociative recombinations.* We also have, for instance

$$O^+ + e \rightarrow O + hf,$$

but these and similar processes are relatively slow, as compared to the dissociative recombinations.

We can now write the equation of continuity for O^+ as

$$\frac{d}{dt} n_{O^+} = q - \gamma_1 n_{O^+} n_{O_2} - \gamma_2 n_{O^+} n_{N_2}, \tag{B.1}$$

where q again denotes the photo-ionization, and γ_1 and γ_2 represent the respective recombination coefficients. Similarly we have for O_2^+ and NO^+

$$\frac{d}{dt} n_{O_2^+} = \gamma_1 n_{O^+} n_{O_2} - \alpha_1 n_e n_{O_2^+} \tag{B.2}$$

$$\frac{d}{dt} n_{NO^+} = \gamma_2 n_{O^+} n_{N_2} - \alpha_2 n_e n_{NO^+}, \tag{B.3}$$

where we note that neither O_2^+ nor NO^+ are produced directly by photo-ionization in any significant quantity, but are the result of processes which give the *loss* of O^+ ions. We introduced again n_e as the local electron density. We now use that the overall charge neutrality implies $n_e = n_{O^+} + n_{O_2^+} + n_{NO^+}$ to a good accuracy. By addition of the relations (B.1)–(B.3) we find

$$\frac{d}{dt} n_e = q - \alpha_1 n_e n_{O_2^+} - \alpha_2 n_e n_{NO^+}, \tag{B.4}$$

This relation has little value as it stands, since it involves two unknowns, and in general time varying quantities $n_{O_2^+}$ and n_{NO^+}. We can however, identify two limiting simple cases; one where $n_{O^+} \gg n_{NO^+}$ as well as $n_{O^+} \gg n_{O_2^+}$, and another where $n_{O^+} \ll n_{NO^+}$ as well as $n_{O^+} \ll n_{O_2^+}$. In the first case, which is a valid approximation for altitudes above, say, 200 km, we have $n_e \approx n_{O^+}$, which by (B.1) implies

$$\frac{d}{dt} n_e = q - n_e\left(\gamma_1 n_{O_2} + \gamma_2 n_{N_2}\right). \tag{B.5}$$

In the other limit, relevant for the lower parts of the ionospheric F-region, we find $n_e \approx n_{NO^+} + n_{O_2^+}$. To the present approximation we can also assume $\alpha_1 \approx \alpha_2 \equiv \alpha$ giving by (B.4)

$$\frac{d}{dt} n_e = q - \alpha n_e^2. \tag{B.6}$$

The important difference between (B.5) and (B.6) is evidently that the loss term for electrons is proportional to n_e in the first case and n_e^2 in the second case. The standard Chapman ionosphere assumes one ion species only, and it is then possible to argue for a loss term proportional to the product of electron density and the density of an ion species.

B.2 E-region, 95–150 km altitude

In the ionospheric E-region we have O_2 and N_2 ionized by ultraviolet radiation and also some contributions from X-rays, giving O_2^+ and N_2^+. However, the N_2^+ ions are rapidly lost by charge exchange processes

$$N_2^+ + O_2 \rightarrow O_2^+ + N_2$$
$$N_2^+ + O \rightarrow NO^+ + N,$$

so that O_2^+ and NO^+ are the dominating ion species in this region. The recombinations are also dissociative in this case. The equation of continuity for the electron density obtains its standard form

$$\frac{d}{dt} n_e = q - \alpha n_e^2,$$

where now q and α assume the values appropriate for the E-region. In general, it is the E-region which is best described by the Chapman model for the ionosphere.

In the E-region we also have small concentrations of metal ions, originating mainly from meteors.

While q depends on altitude (i.e. the number density of particles available for ionization), the recombination coefficient in the ionosphere can as an order of magnitude be taken as a constant, $\alpha \sim 10^{-8}$ cm^3 s^{-1}, but a detailed and accurate analysis can become detailed and cumbersome (Bowhill 1961), the results being sensitive to the ion composition, in particular.

B.3 D-region, 50–95 km altitude

The plasma production in the D-region is modest, since most of the radiation from the Sun is depleted at higher altitudes. The strong H-Lyman-α line (with wavelengths $\approx$ 121.5 nm) is sufficiently intense to penetrate into the D-region, and ionizes parts of the modest NO concentration there. Below 70 km altitude we have the most significant contributions from energetic cosmic rays. The D-region also has significant concentrations of negative ion components, and the recombination processes are in general complicated.

References

Bowhill S 1961 The effective recombination coefficient of an ionosphere containing a mixture of ions *J. Atmos. Terrestrial Phys.* **20** 19–30

Brekke A 2013 *Physics of the Upper Polar Atmosphere* 2nd edn (Berlin: Springer)

Appendix C

Collisional cross-sections

C.1 Cross-sections in general

Interactions between particles are at times best described in terms of *cross-sections*. In general, such cross-sections can be defined for almost all sorts of interactions, but as an introduction we use here a model for a beam of particles interacting with a medium consisting of an arrangement of diluted scatterers. Assume that we take a thin slab with thickness dz and area A of this scattering medium and an incident beam of, for instance, electrons with flux density $\Gamma \equiv n_e U$, where the electron velocity is U and their density is n_e. The beam direction is perpendicular to the slab in consideration. See figure C.1 for an illustration of the problem. The basic interaction is first assumed to be a scattering of electrons by the particles distributed in the medium. If an electron hits a particle it is scattered out of the beam and is assumed lost, otherwise it is collected on the other side. The slab thickness is so small that we can ignore the possibility of two particles 'shadowing' each other. The entire area of scatterers seen from the direction of the electron beam is therefore $N\sigma$, where σ is the projected area of just one scatterer and N is the number of scatterers in the slab with the given thickness and area. Given the density, n_s, of scatterers we have $N = n_s A \mathrm{d}z$. There will be statistical variations in N, but the relative fluctuations around the average value $\langle N \rangle$ will be small if the volume is large, i.e. if A is large, so we might as well treat N as a deterministic quantity.

The relative change in flux is equal to the relative fraction of the entire area covered by the scatterers, in other words

$$\frac{\Gamma - (\Gamma + \mathrm{d}\Gamma)}{\Gamma} = \frac{N\sigma}{A} = \frac{n_s \sigma A \mathrm{d}z}{A} = n_s \sigma \mathrm{d}z, \tag{C.1}$$

or

$$\mathrm{d}\Gamma = -\Gamma n_s \sigma \mathrm{d}z. \tag{C.2}$$

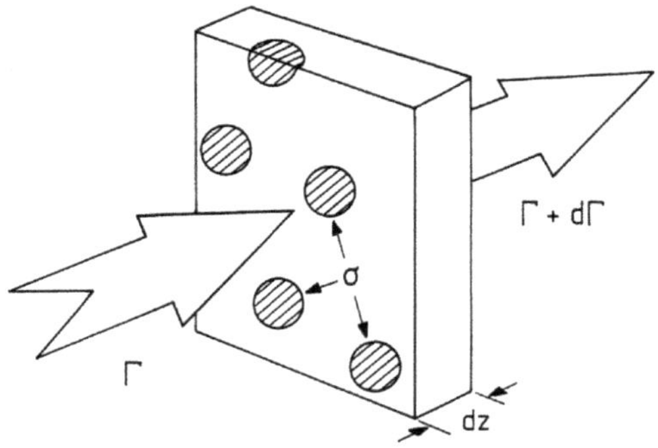

Figure C.1. Illustration of the geometry for flux depletion by scattering from scatterers with cross-section σ, being uniformly distributed in a statistical sense. The large arrows symbolize a spatially homogeneous flux, incident in the direction perpendicular to the surface of the reference volume shown.

Provided n_s is constant, independent of the position z, we easily find the solution

$$\Gamma = \Gamma_0 \mathrm{e}^{-\sigma n_s z} \equiv \Gamma_0 \mathrm{e}^{-z/\ell_c} \tag{C.3}$$

where γ_0 is the incoming flux at the position $z = 0$ and we defined a characteristic length $\ell_c \equiv 1/(n_s\sigma)$ which is the distance where the incoming flux is decreased by a factor e. It is natural to interpret ℓ_c as a *mean free path* for the collisions. For a particle with velocity U, we can obtain the collision frequency as $\nu = U/\ell_c = Un_s\sigma$. This is a dimensionally correct choice.

It may be timely here to add a few words on something which may appear as a technicality: we introduced ν as a collision frequency before, and evidently by its definition, we can interpret $1/\nu$ as the average time between two 'events', where we might leave the definition of what we actually mean by the interaction/scattering to later. We can define another time, namely the one between two *encounters*, opening up the possibility that two particles can meet without affecting each other, i.e. no collision occurs. The time for this latter case is, on average, given by L/U, where L is the average distance between particles, and again U is the velocity of our reference particle. Evidently, we expect $L/U \leqslant \ell_c/U$. Often, we find collision frequencies for scattering processes to be related to U/L.

If the distribution of scattering particles is inhomogeneous, we have an explicit spatial dependence of $n_s = n_s(\mathbf{r})$, and the situation becomes a bit more complicated. For the simple case where the dependence is on z only, $n_s = n_s(z)$, we find

$$\frac{\mathrm{d}}{\mathrm{d}z} \ln\left(\frac{\Gamma}{\Gamma_0}\right) = -\sigma n_s(z), \tag{C.4}$$

giving

$$\Gamma = \Gamma_0 \mathrm{e}^{-\sigma \int_0^z n_s(z')\mathrm{d}z'}, \tag{C.5}$$

where it is natural to assume σ to be constant, and take it outside the integration sign. The constant Γ_0 is determined by requiring $\Gamma = \Gamma_0$ for $z = 0$.

The concepts discussed in this appendix can be generalized considerably; we need not deal with an incoming beam of particles being scattered, they might as well be absorbed, for instance. In this case we deal with an *absorption cross-section*. The basic point is that we have something coming in, and that this something is changed somewhat when it propagates through the slab of the medium it interacts with. The ideas are readily generalized to incoming radiation, etc. Note that we *can* have amplification; more may come out than what we send in! Measurements of cross-section for various processes constitute a ‘cornerstone’ in experimental physics. The reader has to consult the (vast) literature for relevant information, but we might mention some useful works concerning electron collisions by Pack and Phelps (1961), Frost and Phelps (1964), Itikawa (1973), with a more general reference being Lieberman and Lichtenberg (1994). The mobilities of ions in a background of neutral gas is discussed by, for instance, Ellis *et al* (1976). Useful tables for ionospheric conditions can be found in Gurevich (1978).

The incoming flux is generally constituted by particles with a given velocity, and it will generally be found that the cross-sections depend on the velocity, with the exception of ‘billiard ball’-like scattering. In general, the velocity variation can be quite complicated, but a few idealized reference cases can be mentioned:

1) The hard-sphere, or ‘billiard ball’, collisions already discussed. Here $\sigma(U) =$ const. and consequently the collision frequency varies as $\nu(U) \propto U$.
2) For interactions given by polarization forces, for so called ‘Maxwellian molecules’ (Trubnikov 1965), we find for a wide velocity range that $\sigma \propto 1/U$ implying $\nu(U) \propto$ const. This process is often used for illustrations.
3) For Coulomb interactions, we find that $\sigma \propto 1/U^4$ implying $\nu(U) \propto 1/U^3$.

In general we have (Bekefi 1966) that for an inverse power law dependence of the interaction force, $F \propto 1/r^s$, we find, by a dimensional analysis, the velocity dependence of the collision frequency to be $\nu(U) \propto U^{(s-5)/(s-1)}$. For instance, for Maxwellian molecules we have $s = 5$, for Coulomb interactions $s = 2$.

Note that the discussion of cross-section and the collision frequencies derived from it, refers to explicit processes. One of the most important applications is the question of momentum relaxation, and we note here that particles can very well collide without giving up *all* of their momentum. As a ‘rule of thumb’ you can assume that for isotropic collisions, where the particles leave the interaction with a velocity vector pointing in any spherical direction with equal probability, the memory for momentum loss is the same as a collision time. If it is not so, then it takes several collisions before the original momentum is lost. Momentum losses are a bit special by involving both the magnitude and the direction of a vector.

A standard unit for cross-sections is given in terms of the Bohr radius, $a_0 = 0.53 \cdot 10^{-10}$ m, giving $\sigma_0 \equiv \pi a_0^2 = 0.88 \cdot 10^{-21}$ m^2 for the corresponding geometrical cross-section. The unit often used for reaction cross-sections is *barn*, being equal to 10^{-28} m^2. The name, coined by scientists with a certain sense of humor, is derived from the proverbial phrase ‘side of a barn’, i.e. something easy to hit. Values of cross-sections for a given nucleus, atom or molecule depend on the energy of the nature of the bombarding particle and the kind of reaction being considered.

Historically, the interest for nuclear cross-sections was accentuated in relation to nuclear reactions in fission reactors, cross-sections related to properties of neutrons in particular. Boron, for example, when bombarded by neutrons traveling 10^6 cm s^{-1} (or 22 500 miles per hour) has a cross-section for the neutron-capture reaction of about 120 barns. The cross-section increases to about 1 200 barns for neutrons traveling at 10^5 cm s^{-1}. These large cross-sections mean that boron is a good absorber of neutrons. The metal zirconium, on the other hand, having an absorption cross-section of only 0.18 barn for the low-energy neutrons that cause fission in nuclear reactors, is rather transparent to neutrons and is used to clad reactor fuel rods. The reactor fuel uranium-235 absorbs neutrons both by fission and by capture. The fission cross-section is 580 barns, whereas the total cross-section for neutron absorption (including the capture cross-section) is 680 barns.

C.2 Mean free path

We can try a somewhat different approach for discussing collisions and collisional mean free paths. As an aide for the discussion we take figure C.2, showing by the full line in a) the orbit of a small particle undergoing collisions. The scattering particles are indicated by solid circles. Note that the illustration does not assume billiard ball collisions, i.e. the orbit does not touch the scattering particles. It is implicitly assumed that the duration of the collisional interaction is small compared to the time between collisions. In figure C.2(*b*) we surround the particle trajectory by a tube having the cross-section to be the scattering cross-section $\sigma \approx \pi d^2$. Then in figure C.2(*c*) this tube is straightened out. The volume of the tube is $\sigma\sum_j^N \ell_j$ where ℓ_j is the length of the free path between scatter number j and number $j+1$. We assume that there are N scatterings in the record available. If the density of scattering particles is n_s, then we can estimate the number of scattering particles in the tube as $\sigma n_s \sum_j^N \ell_j$. This number is varying from one realization of the process to another, but

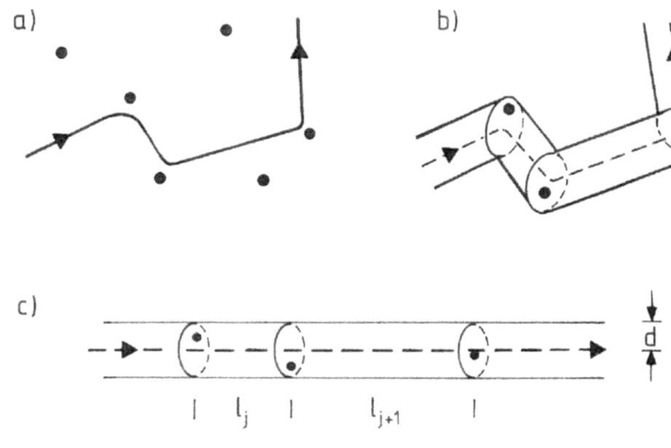

Figure C.2. Illustration of the orbit of a particle undergoing collisions. The scattering particles are illustrated by solid circles. The figure should be imagined as three-dimensional. Figures (*a*) and (*b*) are in principle identical, only in (*b*) we introduced a 'tube' with cross-section being the scattering cross-section, where its radius is *d*. In (*c*) this tube is 'straightened out'.

if we assume that N is large and the tube is long then the relative scatter of the number of particles in the tube is modest, and we can again take this number to be deterministic. By construction, each and every one of these particles has given rise to a collision, i.e. 'bendings' in the orbit of the scattered particle. The average mean free path is then the length of the total path divided by the number of such bendings

$$\ell_c = \frac{\sum_j^N \ell_j}{\sigma n_s \sum_j^N \ell_j} = \frac{1}{n_s \sigma}.$$

So far the arguments do not involve the velocities of the scattering particles or the one *being* scattered. These velocities are important for estimating the time between scatters. If the scattering particles are heavy and the scattered particle light and fast (as electrons scattered by ions, for instance), we can assume the electron energy to remain constant: the scattering process only changes the *direction* of propagation. The time between two collisions is then $\tau_j = \ell_j / U$ where U is the particle velocity, assumed constant in this time interval. If we let a light particle collide with very heavy ones, we can assume the energy to remain constant (at least for a long time) and assume that only the direction of propagation is changed. The average time between collisions is then ℓ_c / U and a collision frequency is $\nu_c = U/\ell_c$. If the collisional cross-section is velocity dependent (and this is generally the case), we will have to average over the actual velocity distribution $f(U)$ to obtain the average collision frequency (Gurevich 1978).

- **Exercise:** Consider a gas consisting of identical spherical particles with diameter $2d$, and density n, with mutual collisions of the type 1) discussed in section C.1. Demonstrate that the collisional mean free path is

  $$\ell_c = \frac{1}{4\pi d^2 n}.$$

  Introducing the density as the number of particles divided by the volume, demonstrate that

  $$\ell_c = \frac{RT}{4\pi d^2 N} \frac{1}{p},$$

  where p is the pressure. You could also use $p = n\kappa T$. For constant temperature, T, we find ℓ_c to be inversely proportional to the pressure if we vary the volume.

  As posed here, the problem assumes that the particle *being scattered* is moving, while the scatterers are at rest. Consider the effect of statistically moving scatterers, using qualitative arguments to compare particles with head on and overtaking collisions.
- **Exercise:** Assume that an incoming particle on figure C.1 can be scattered by *two* different types of particles, with different cross-sections. Assume that both types of particles are uniformly distributed in the medium, and that the

scattering processes are statistically independent. Demonstrate that the collision frequencies for the two scattering processes are *additive*, i.e. if the impurities of type 1 alone give a collision frequency ν_1 and those of type 2 give ν_2, then we have $\nu_{1+2} = \nu_1 + \nu_2$, when they are both present. Why is it important for the argument that the processes are statistically *independent*?

C.3 A statistical collision model

Basically, we have to realize that collisions form a statistical process, and we have to decide on an appropriate model. The collision between a charged particle and a neutral atom depends in general on their relative velocities in a complicated manner. In order to retain the model on a tractable level it will here be assumed that the neutrals are so heavy that they might be considered almost immobile, at least compared with the moving free electron. The simplest and almost universally used model is based on the assumption that the probability of a collision in a small time interval, dt, is proportional to the length of that time interval, where ν is a constant of proportionality. Let the probability of a collision in a short time interval $\{t;\, t + \Delta t\}$ be

$$P(\text{one collision in } \Delta t) = \nu \Delta t, \tag{C.6}$$

with ν being a constant, independent of the velocity of the electron (Pécseli 2000). Effectively it is thus assumed that the collisional cross-section is inversely proportional to the relative particle velocity. Although this is a non-trivial restriction, it is not unphysical, and is a good approximation to a number of relevant processes. In effect it is assumed that the collisional cross-section is proportional to the interaction time, i.e. the time an electron spends in the vicinity of the neutral; not an unreasonable assumption. These are not hard-sphere, or 'billiard-ball' collisions.

By assumption, the electron (or ion) leaves the neutral with a velocity component w, which is entirely independent of the one with which it arrived. There is no persistence of velocities. The constant ν is therefore independent of the 'history' of the electron, i.e. on times for foregoing collisions. This is a simple example of a Markov process, which is frequently encountered in practical applications. With the present model (C.6) the probability of a free flight starting at $t = 0$ and persisting at a time t can be calculated as follows:

Consider a time interval of length $t + \Delta t$ broken up into two subintervals, one of length t and one of Δt. Since the probability of a collision in Δt is independent of the prehistory it follows that $P(0, t + \Delta t) = P(0, t)P(0, \Delta t)$, where for brevity we introduced the notation $P(0, t + \Delta t) = P(\text{no collision in the interval } t + \Delta t)$, etc. Evidently $P(0, \Delta t) = 1 - \nu \Delta t$, with the probability of two or more collisions in the time interval Δt being negligible when Δt is small. Consequently $P(0, t + \Delta t) = (1 - \nu \Delta t)\Delta t P(0, t)$, giving

$$\frac{P(0, t + \Delta t) - P(0, t)}{\Delta t} = -\nu P(0, t).$$

In the limit $\Delta t \to 0$ this difference equation becomes

$$\frac{\mathrm{d}P(0,\,t)}{\mathrm{d}t} = -\nu P(0,\,t), \tag{C.7}$$

giving the result

$$P(0,\,t) = \mathrm{e}^{-\nu t}, \tag{C.8}$$

with the condition $P(0, t = 0) = 1$ from (C.6), since $P(0, \Delta t) = 1 - \nu\Delta t$, as before.

The average time between collisions can be found by noting that the probability of a free flight beginning at t_1 to be pertaining at $t_1 + t$ is $\exp(-\nu t)$. The probability that a free flight beginning at t_1 is terminated by a collision in the time interval $[t_1 + t;\, t_1 + t + \mathrm{d}t]$ is then $\exp(-\nu t)\nu\,\mathrm{d}t$. The average time between collisions is consequently given by

$$\int_0^\infty t\exp(-\nu t)\nu\,\mathrm{d}t = \frac{1}{\nu}.$$

The average collision frequency is thus ν. The argument can be applied when 'looking backwards' in time as well. Then $\exp(-\nu t)\nu\,\mathrm{d}t$ is the probability that an ongoing free flight at t_1 started with a collision in a narrow time interval $\mathrm{d}t$ at $t_1 - t$.

Assume now that the charged particles as well as the neutral background are embedded in a spatially uniform time varying deterministic electric field, with intensity $E(t) = \Re\{E_0\mathrm{e}^{-i\omega t}\}$. The equation of motion between collisions is now $m\,\mathrm{d}U(t)/\mathrm{d}t = -e\Re\{E_0\mathrm{e}^{-i\omega t}\}$. The electron velocity between collisions is given by

$$\begin{aligned} U(t) &= U(t_1) + \Re\left\{\frac{eE_0}{i\omega m}(\mathrm{e}^{-i\omega t} - \mathrm{e}^{-i\omega t_1})\right\} \\ &= U(t_1) + \frac{eE_0}{m}\Re\left\{\frac{(1 - \mathrm{e}^{-i\omega(t_1 - t)})\mathrm{e}^{-i\omega t}}{i\omega}\right\}, \end{aligned} \tag{C.9}$$

where it is assumed that the electron starts out with a velocity $U(t_1)$ after a collision at $t = t_1$. We use the capital letter U to denote the velocity of an individual electron. It is assumed implicitly that the electron starts out with a random velocity direction uniformly distributed in angle over 4π, so that the average is vanishing, $\langle U(t_1)\rangle = 0$.

The expression (C.9) remains valid until the next collision. The second term is statistically varying since the time t_1 is statistically distributed. The average value of the second term is different from zero. We can split the time interval into two parts; from t going backwards in time until $t_1 + d\Delta t$, and then a small time interval $d\Delta t$. In the first of these intervals we have no collisions, and then one collision in the last small one. The statistical distribution of $\Delta t \equiv t - t_1$ is given by the product of two probabilities, i.e. $\exp(-\nu\Delta t)\nu d\Delta t$, where Δt denotes the time since the previous collision. The average value of $U(t)$ is readily obtained as

$$\langle U(t)\rangle = -\frac{eE_0}{m}\Re\left\{\frac{\mathrm{e}^{-i\omega t}}{\nu - i\omega}\right\}, \tag{C.10}$$

using $\int_0^\infty [1 - \exp(-i\omega\tau)]\exp(-\nu\tau)\nu \, \mathrm{d}\tau = i\omega/(\nu + i\omega)$. The result is evidently valid also for DC-conditions, where $\omega = 0$. Details of the foregoing discussions are given for instance by Boström (1964) or Pécseli (2000).

- **Exercise:** We obtained in (C.8) the probability of *no* collisions in the interval $\{0; t\}$. Demonstrate that the probability of K collisions within the interval is given by the Poisson distribution

$$P(K, t) = \frac{(\nu t)^K \mathrm{e}^{-\nu t}}{K!}.$$

Show that the average number of collisions in a time interval $\mathcal{T}$ is

$$E(K) = \sum_{K=0}^{\infty} K \frac{(\nu \mathcal{T})^K \mathrm{e}^{-\nu \mathcal{T}}}{K!} = \nu \mathcal{T}. \tag{C.11}$$

Evidently, this average need not be an integer number, even though K *is* integer. The results (C.11), is consistent with the average collision frequency being ν.

C.3.1 Analytical collision model

A model equation can be proposed in the form

$$\frac{\mathrm{d}}{\mathrm{d}t} U(t) = \frac{q}{M} E_0 \mathrm{e}^{-i\omega t} - \nu \, U(t), \tag{C.12}$$

for describing collisions between neutrals and charged particles, subject to an oscillating electric field. The relation is written in the rest frame of the neutrals. We obtain trivially for this case

$$U(t) = i \frac{q}{M} E_0 \frac{\mathrm{e}^{-i\omega t}}{\omega + i\nu}.$$

The model (C.12) will be used often, and for collisional processes of other types. The physical argument in its derivation is found by noting its similarity with, for instance, a simple oscillating system subject to a restoring force and a friction. Taking, for instance, an oscillator with displacement Δx we have the equation $M\mathrm{d}^2\Delta x/\mathrm{d}t^2 = -K\Delta x - \nu \, \mathrm{d}\Delta x/\mathrm{d}t$, with M being the mass, $K\Delta x$ being the restoring force of a spring, and ν representing a friction with a supporting surface. Here, evidently, the friction plays the role of the collision frequency in (C.12).

By inspection it is evident that the result (C.10) is formally contained in the analytical solution of the model deterministic equation (C.12). Taking the real part, we find precisely the result for the *average* velocity obtained in (C.10). If we are not interested in the fine details in the particle velocity, we might simply identify the solution of (C.12) with the average particle velocity. The simple model (C.12) will be used as representative for the underlying statistical process, also for more complicated space-time varying fields. It is important, however, that the model is based on a force, which can be considered constant in space, at least locally. This implies, that we should use (C.12) with care in the case where the spatial field variations are on a scale comparable to, or smaller than, the collisional mean free paths.

C.3.2 Ohm's law

Consider for simplicity an unmagnetized plasma where the ions can be assumed to constitute an immobile background of positive charge density. The electron motion is (again) described by the equation

$$m\frac{\mathrm{d}}{\mathrm{d}t}\mathbf{U} = -e\mathbf{E} - m\nu\mathbf{U}, \tag{C.13}$$

with ν representing the appropriate collision frequency. Physically, ν^{-1} measures the time it takes an electron to 'forget' its original momentum. This can be seen by setting $\mathbf{E} = 0$ and solving the resulting expression (C.13) to give $\mathbf{U} = \mathbf{U}_0 \exp(-\nu t)$, where $m\mathbf{U}_0$ it the electron momentum at $t = 0$.

If we ignore electron inertia, i.e. the left-hand side of (C.13), we find trivially

$$\mathbf{U} = -\frac{e}{m\nu}\mathbf{E} \equiv -\mu\mathbf{E},$$

where we introduce the *electron mobility* $\mu \equiv e/(m\nu)$. We now introduce the current density $\mathbf{J} = -en\mathbf{U}$, in terms of the electron density n, and have $\mathbf{J} = -en\mu\mathbf{E}$. If the current density is constant over a plasma cross-section $\mathcal{A}$, we can introduce the net current in that cross-section as $\mathbf{I} = \mathbf{J}\mathcal{A}$. If the electric field is constant in that same cross-section and in addition also over a length $\mathcal{L}$, we have $\Phi = \mathcal{L}E$ in terms of a voltage difference Φ between the two end-points of $\mathcal{L}$, so that

$$I = \frac{\Phi}{R}, \qquad \text{with} \qquad R \equiv \frac{m\nu\mathcal{L}}{e^2 n\mathcal{A}}.$$

This is Ohm's law in its simplest form, with R being the resistance in the 'circuit'. With a constant R, the relation is seemingly based on a constant mobility assumption, with inertia being ignored. The electron mass enters through the collision term.

We generalize the previous results to the case where they describe the DC-conductivities and resistivities of a magnetized plasma where the dominant collision process is between a neutral component and charged particles. The basic phenomenological model becomes

$$M\frac{\mathrm{d}}{\mathrm{d}t}\mathbf{U} = q\mathbf{E} + q\mathbf{U} \times \mathbf{B} - M\nu\mathbf{U}, \tag{C.14}$$

where $\mathbf{U}(t)$ is the velocity of the selected charged particle moving in prescribed electric and magnetic fields, where ν is a collision frequency. The particle positions is $\mathbf{R}(t)$, with $\mathbf{U}(t) = \mathrm{d}\mathbf{R}/\mathrm{d}t$. In (C.14) we can let q and M representing the electron as well as the ion charges and masses, respectively. In the present phenomenological model we can take ν to represent a momentum loss of particles per time unit, without going into detail in discussing the underlying physical mechanisms. A great simplification can be achieved here by assuming that ν is a constant, irrespective of the relative particle velocities (Pécseli 2000). This implies that the collisional mean free path $\ell_c \sim u/\nu$ is assumed to decrease for decreasing velocity, giving the implicit assumption that the collisional cross-section σ decreases with increasing frequency, since $\ell_c \sim 1/(n\sigma)$. This assumption is not unphysical, but evidently somewhat restrictive (Trubnikov 1965, Bekefi 1966).

For time stationary conditions, we can obviously set the time derivative to zero in (C.14). It is readily seen that as an approximation we can also ignore $\mathrm{d}\mathbf{U}/\mathrm{d}t$ when the time variations of the electric field are very slow on the time scale given by $1/\nu$. Consequently, we solve first the simpler equation

$$q\mathbf{E} + q\mathbf{U} \times \mathbf{B} - M\nu\mathbf{U} = 0. \tag{C.15}$$

First we take the cross-product $\times\mathbf{B}$ 'from the right', use $(\mathbf{U} \times \mathbf{B}) \times \mathbf{B} = -\mathbf{U}_\perp B^2$, and find $q\mathbf{E} \times \mathbf{B} - q\mathbf{U}_\perp B^2 + M\nu\mathbf{U} \times \mathbf{B} = 0$. Insertion of $\mathbf{U} \times \mathbf{B}$ from (C.15) gives

$$q\mathbf{E} \times \mathbf{B} - q\mathbf{U}_\perp B^2 + \nu \frac{M}{q}\left(q\mathbf{E}_\perp - M\nu\mathbf{U}_\perp\right) = 0,$$

and we can express the velocity component in the direction $\perp\mathbf{B}$. The component *along* $\mathbf{B}$ is easily obtained as $U_{||} = E_{||}q/(M\nu)$. Note that we require $\nu \neq 0$ for this solution to exist with $E_{||} \neq 0$. *If* we have $\nu = 0$, we would have a steady acceleration of the particle along $\mathbf{B}$, and the steadily increasing particle velocity would be violating the basic assumption of time stationarity.

We can write the $\mathbf{B}$-perpendicular component of the particle velocity as

$$\mathbf{U}_\perp = \frac{q\mathbf{E} \times \mathbf{B} + M\nu\mathbf{E}_\perp}{qB^2 + M^2\nu^2/q} = \left(\frac{q}{M}\right)^2 \frac{\mathbf{E} \times \mathbf{B} + (M/q)\nu\mathbf{E}_\perp}{\Omega_c^2 + \nu^2}, \tag{C.16}$$

or by introducing the definition

$$\mathbf{U}_\perp \equiv \beta\frac{\mathbf{E} \times \mathbf{B}}{B^2} + \alpha\,\frac{\mathbf{E}_\perp}{B}, \tag{C.17}$$

with

$$\alpha \equiv \frac{\nu M}{qB + M^2\nu^2/(qB)} \equiv \frac{\nu/\Omega_c}{1 + \nu^2/\Omega_c^2} \quad \text{and} \quad \beta \equiv \frac{1}{1 + \nu^2/\Omega_c^2}, \tag{C.18}$$

in terms of the appropriate cyclotron frequency $\Omega_c \equiv qB/M$. Note that α has the sign of qB. Since the electric field has a component $||\mathbf{B}$, we wrote the $\mathbf{B}$-perpendicular component explicitly as $\mathbf{E}_\perp$. We have $0 \leqslant |\alpha| < 1/2$ and $0 < \beta < 1$, respectively.

The expressions can be used for any particle species specified by the appropriate mass M and charge q, and cyclotron frequencies Ω_c, noting though that we have to interpret these quantities with the correct sign of the charges and magnetic fields entering!

One could argue that the electric field component $\mathbf{E}_\perp$ could be made to vanish by a 'smart' change of reference (Feynman *et al* 1963, Pécseli 2012), moving with velocity $\mathbf{U}_* = -\mathbf{E}_\perp \times \mathbf{B}/B^2$. In this case, however, the neutral component is moving with velocity $-\mathbf{U}_*$ in this new frame of reference, and the last term in (C.15) has to be modified to $M\nu(\mathbf{U} - \mathbf{U}_*)$, so that nothing is gained (as to be expected!).

To derive the current, $\mathbf{J}$ induced in the plasma by the imposed electric field $\mathbf{E}$, we note that in general the ions as well as the electrons contribute, as given by the appropriate index. *Along* the magnetic field we expect the electron contribution to the current to be dominant, because of their small mass, but in the direction $\perp\mathbf{B}$ we expect that the ratios ν/Ω_c can vary significantly, and for some neutral densities the ion and for other cases the electron contribution can dominate the current. Generally, we have the plasma current given as $\mathbf{J} \equiv e(n_i\mathbf{U}_i - n_e\mathbf{U}_e)$. We ignore the possibility of slight deviations from local neutrality and set the densities of the two components equal, i.e. $n_e = n_i \equiv n$, and have

$$\mathbf{J}_\perp \equiv \sigma_P \mathbf{E}_\perp + \sigma_H \mathbf{E} \times \hat{\mathbf{b}} \tag{C.19}$$

$$J_\parallel \equiv \sigma_\parallel E_\parallel, \tag{C.20}$$

where $\hat{\mathbf{b}}$ is a unit vector along $\mathbf{B}$, and

$$\sigma_P \equiv \frac{en}{B}(\alpha_i - \alpha_e) \qquad \text{and} \qquad \sigma_H \equiv \frac{en}{B}(\beta_i - \beta_e)$$

are the 'Pedersen' and 'Hall'-conductivities, respectively. Sometimes we see a presentation with a change in sign in σ_H; this is because some authors consider it preferable to have the cross-product between the electric field and the unit vector $\hat{\mathbf{b}}$ the other way around! Note that α_e and α_i have opposite signs, ensuring here that $\sigma_P \geqslant 0$ for all plasma parameters. With $M \gg m$ we have $\beta_i - \beta_e < 0$, generally.

For the motion *along* the magnetic field we introduce $\sigma_\parallel \equiv e^2 n[1/(M\nu_i) + 1/(m\nu_e)] \approx e^2 n/(m\nu_e)$ to give $J_\parallel = \sigma_\parallel E_\parallel$.

- **Exercise:** Demonstrate that the total current can be expressed in tensor notation as

  $$\mathbf{J} = \boldsymbol{\sigma} \cdot \mathbf{E},$$

  with the conductivity tensor $\boldsymbol{\sigma}$ given as

  $$\boldsymbol{\sigma} \equiv \begin{pmatrix} \sigma_P & \sigma_H & 0 \\ -\sigma_H & \sigma_P & 0 \\ 0 & 0 & \sigma_\parallel \end{pmatrix}. \tag{C.21}$$

  This formulation is particularly useful when a transformation of the frame of reference is needed.

To understand the altitude variations of σ_P and σ_H in the ionosphere, we first note that both these quantities are proportional to the plasma density $n_e \approx n_i \equiv n$, which can be assumed to follow a Chapman model (Pécseli 2012), at least as an approximation. The neutral density is on the other hand, to an approximation, following an exponential profile, implying that ν_e as well as ν_i have the same functional altitude variation (but of course with different numerical values).

As a heuristic argument we can assume that the cross-sections, σ_c, for ion–neutral and electron–neutral collisions are approximately the same, so the mean free paths for collisions, $\ell_c = 1/(n\sigma_c)$ are also approximately the same for both species. (Note that *this* σ_c is *not* a conductivity!) The collision frequency is then $\nu \approx u_{th}/\ell_c$ with $u_{th} = \sqrt{\kappa T/m}$ being the appropriate thermal velocity. Consequently, $\Omega_{ci}/\nu_i \approx (eB/M)\ell_c\sqrt{M/\kappa T_i}$ and $\omega_{ce}/\nu_e \approx (eB/m)\ell_c\sqrt{m/\kappa T_e}$. *If* at some altitude we have $\Omega_{ci} \approx \nu_i$. Provided $T_e \sim T_i$, we then find $\omega_{ce}/\nu_e \approx \sqrt{M/m}\sqrt{T_i/T_e} \gg 1$ as stated. The foregoing discussion ignored the variations of cross-sections with velocity. These effects will in general strengthen the arguments, since cross-sections usually decrease at large velocities. Consequently, it is possible to have $\omega_{ce}/\nu_e \gg 1$ and $\Omega_{ci}/\nu_i \ll 1$ also in cases where $T_e \gg T_i$.

For all practical purposes, we can assume the magnetic field to be constant for relevant altitudes, and find that there is an altitude where $\omega_{ce} \approx \nu_e$, and a different, somewhat higher, altitude where $\Omega_{ci} \approx \nu_i$. The corresponding α_e and α_i have their maxima at these two altitudes, respectively, while β_e and β_i are steadily decreasing with altitude. For high altitudes above, say, 150 km both $\beta_e \approx 1$ and $\beta_i \approx 1$, implying that $\sigma_H \approx 0$. Similarly, for altitudes below 60 km or so, we have $\beta_e \approx 0$ and $\beta_i \approx 0$, implying that $\sigma_H \approx 0$ also here.

As a numerical example we have in the day time ionospheric E-region at an altitude of 110 km over northern Scandinavia: $\nu_i \approx 600\ \mathrm{s}^{-1}$ and $\nu_e \approx 1.5 \cdot 10^4\ \mathrm{s}^{-1}$. The temperatures are $T_i \approx 200$ K and $T_e \approx 400$ K. The magnetic fields are $B \approx 50\ \mu$T giving $\Omega_{ci} \approx 180$ rad/s, corresponding to an average ion mass of 31 AMU. For the same conditions $\omega_{ce} = 8.8 \cdot 10^6$ rad/s, giving $\Omega_{ci}/\nu_i \approx 0.3 < 1$ and $\omega_{ce}/\nu_e \approx 600 \gg 1$. Around 70 km altitudes we have $\omega_{ce} \approx \nu_e$ and then $\Omega_{ci} \ll \nu_i$. At such high collision frequencies, the ions are not magnetized at all, since they never manage to make a full turn in the cyclotron orbit, and as far as they are concerned, the magnetic field could as well have been absent.

Concerning electron collisionality in the ionosphere it seems generally accepted that electron–neutral and electron–ion collision frequencies are equal at an altitude of approximately 200 km. The conditions change with day/night-time conditions.

- **Exercise:** Demonstrate that at large altitudes we have $\sigma_H/\sigma_P \approx \nu_i/\Omega_{ci} \ll 1$. Give an expression for the angle θ between $\mathbf{E}_\perp$ and $\mathbf{J}_\perp$ in terms of σ_H and σ_P, and demonstrate that for high altitudes, the current is almost entirely a Pedersen current.

For large altitudes, where the neutral density is small, we have a large mobility for the charged particles in the direction $\perp\mathbf{B}$, but since this mobility is virtually the same for both species, the net current will be small in that direction. If we let the magnetic field be vertical, as appropriate for the polar regions, the altitude variation of σ_H and σ_P will imply that the direction of the current density vector $\mathbf{j}$ with respect to the imposed electric field is varying with altitude; large σ_H and small σ_P gives a current predominantly in the $\mathbf{E} \times \mathbf{B}$-direction, and vice versa.

Note that the $\sigma_\parallel$-conductivity is, in general, much larger than both σ_H and σ_P. If we, quite formally, consider the externally imposed electric field as being due to a generator with finite inner resistivity, we can easily imagine cases where the ionospheric plasma in effect 'short circuits' this generator along the magnetic field lines, and therefore it will only be the $\perp\mathbf{B}$ component of the electric field which in such cases will reach a significant magnitude in the plasma.

C.3.3 Time varying electric fields

The foregoing discussions were restricted to DC-electric fields. It is, however, evident that the analysis can be generalized to time varying, but spatially constant, electric fields quite easily. We can thus consider any externally applied forcing, $e\mathbf{E}(t)$, and make a Fourier expansion. Taking one Fourier component $\exp(-i\omega t)$, characterized by an angular frequency ω, we can the write (C.14) as

$$-i\omega M\mathbf{U} = q\mathbf{E} + q\mathbf{U}\times\mathbf{B} - M\nu\mathbf{U},$$

which can be written as

$$q\mathbf{E} + q\mathbf{U}\times\mathbf{B} - M(\nu - i\omega)\mathbf{U} = 0. \tag{C.22}$$

Obviously, we can make the trivial replacement $\nu \to (\nu - i\omega)$ in the previous results (C.18) and obtain for instance

$$\alpha = \frac{\nu\Omega_c - i\omega\Omega_c}{\Omega_c^2 + (\nu - i\omega)^2}$$

and

$$\beta = \frac{\Omega_c^2}{\Omega_c^2 + (\nu - i\omega)^2}.$$

From these quantities we can trivially obtain the elements of the frequency varying conductivity tensor. For finite frequencies, we will in general have that the velocity and the external electric fields are out of phase.

The foregoing results also apply to the collisionless case, and in the limit $\nu \to 0$ we find

$$\alpha = -i\frac{\omega\Omega_c}{\Omega_c^2 - \omega^2} \qquad \text{and} \qquad \beta = \frac{\Omega_c^2}{\Omega_c^2 - \omega^2}.$$

This limit evidently applies to the case where $\omega \gg \nu$. Note the resonance as $\omega \to \Omega_c$. The sign of Ω_c follows the sign of the particle charge q here as well.

For the motion along magnetic field lines we can ignore $\mathbf{B}$ altogether, and in this case we have the simple result $U_{\|} = E_{\|}q/(m(\nu - i\omega))$.

References

Bekefi G 1966 *Radiation Processes in Plasmas* (New York: Wiley)

Boström R 1964 A model for the Auroral electrojet *J. Geophys. Res.* **69** 4983–99

Ellis H W, Pal R Y, McDaniel E W, Mason E A and Viehland L A 1976 Transport properties of gaseous ions over a wide energy range *At. Data Nucl. Data Tables* **17** 177–210

Feynman R P, Leighton R B and Sands M 1963 *The Feynman Lectures on Physics* (New York: Addison-Wesley Publishing Co.)

Frost L S and Phelps A V 1964 Momentum-transfer cross sections for slow electrons in He, Ar, Kr, and Xe from transport coefficients *Phys. Rev.* **136** A1538–45

Gurevich A V 1978 *Nonlinear Phenomena in the Ionosphere* Vol 10 of *Physics and Chemistry in Space* (New York: Springer)

Itikawa Y 1973 Effective collision frequency of electrons in gases *Phys. Fluids* **16** 831–5

Lieberman M A and Lichtenberg A J 1994 *Principles of Plasma Discharges and Materials Processing* 2nd edn (New York: Wiley)

Pack J L and Phelps A V 1961 Drift velocities of slow electrons in Helium, Neon, Argon, Hydrogen, and Nitrogen *Phys. Rev.* **121** 798–806

Pécseli H L 2000 *Fluctuations in Physical Systems* (Cambridge, UK: Cambridge University Press)

Pécseli H L 2012 *Waves and Oscillations in Plasmas* (London: Taylor & Francis)

Trubnikov B A 1965 Particle interactions in fully ionized plasmas ed M A Leontovich *Reviews of Plasma Physics* vol 1 (New York: Consultants Bureau) pp 105–204

Appendix D

Negative temperatures

In standard applications of classical statistical mechanics the idea of negative temperatures might seem to lead to paradoxes and divergences. It might be worthwhile to give a simple illustrative summary demonstrating the properties of realizable physical systems which can logically be assigned a negative temperature (Marvan 1966, Braun *et al* 2013). Such systems consisting of simple particle spins are to some extent easier to analyze than a collection of line vortices.

Standard and intuitive objections to the concept of negative temperatures are often based on models derived from ideal gases, and indeed it will not be possible to give a meaningful definition of negative temperatures in that context. We heat a gas by supplying energy: the more energy we add, the warmer the gas becomes. On the other hand we will not be able to cool the gas below the absolute zero. Note, however, that although energy and temperature are indeed related, the relation is different for different physical systems. Consider for instance a simple system with N-spins having only two discrete states, one negative and one positive assigned the values +1 and −1 without loss of generality, see also figure D.1. If the system is subject to an externally imposed homogeneous magnetic field we will argue that the lowest energy state is one where all spins take the value −1. If we now supply a certain minimum energy to the system we might flip one of the spins to +1, see figure D.1(*b*). Adding more energy we might flip one more, etc. Eventually, we will, however, find that all spins are in the state +1 and no more energy can be accommodated in the system. It could be argued that the lowest state is assigned the temperature $T = 0$, while the final state has $T = T_{\max}$. This definition is, however, in variance with another intuitive interpretation of temperatures, namely that particles in hot systems are more randomly distributed than in cold systems. The classical measure of randomness in statistical mechanics is entropy. Some textbooks, Pathria (1996) for instance, actually use entropy as a basis for discussing statistical mechanics. Randomness is a concept we associate with lack of predictability, and it is indeed this property that is used when applying the ideas to communication theory

doi:10.1088/978-0-7503-1251-6ch34

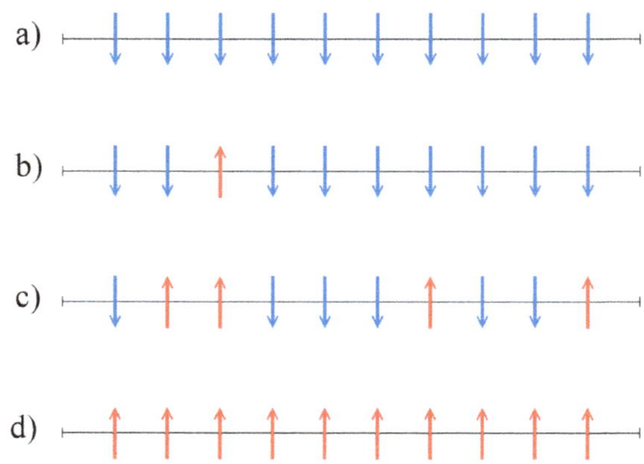

Figure D.1. Illustration of a simple spin system with two states, ±1 with the lowest energy state being one where all spins are −1 as in a). The number of possible micro-states is $w = N!/(m!(N - m)!)$, when m is the number spins with +1. For the case shown we have $N = 10$.

when introducing the 'Shannon entropy'. To sort out these different concepts it might be an advantage to give a short summary.

Disorder is a manifestation of the largeness of the number of micro-states the system can have with equal probability. The larger the choice of micro-states, the lesser the degree of predictability or the level of order in the system. Complete order prevails when and only when the system has no choice but to be in a unique state: this, in turn, corresponds to the state of vanishing entropy (Pathria 1996). Taking our example in figure D.1, we find that we have only one possible state for all spins +1 or −1. All spins with the exception one being −1 have N possible micro-states, where N is the number of spins. With all spins with the exception two being −1 have $N(N - 1)$ possible micro-states. We will denote the number of ways a micro-state can be realized by the letter w, the thermodynamic weight. For m positive spins we have in our case $w = N!/(m!(N - m)!)$ possible micro-states, where the maximum value is reached for $m = N/2$, or its closest integer value. In place of the thermodynamic weight, we frequently use the entropy, $S \equiv k \ln w$, where k is a universal constant independent of the substance considered.

We have generally that temperature is a measure of change of entropy corresponding to a unit change of energy, ΔU. The example illustrated in figure D.1 was used to demonstrate how the disorder changed by adding energy to the system, but also how the system became ordered again, when we exceeded the maximum number of available micro-states. We can also argue that temperature is a measure change of disorder in a system resulting from a unit change of energy, $\Delta S/\Delta U = 1/T$. Again for the example in figure D.1 we have $S = k \ln(N!/(m!(N - m)!))$, i.e. $1/T = \Delta S/\Delta m = k \ln(((m + 1)!(N - m - 1)!)/(m!(N - m)!))$ with $\Delta m = 1$ for the flipping on one spin in figure D.1. The result is illustrated in figure D.2 for $N = 20$.

In general it can be argued that negative temperatures are associated with systems in which the higher energy levels are more densely occupied than the lower levels. Matter in maser or laser states are associated with negative temperatures.

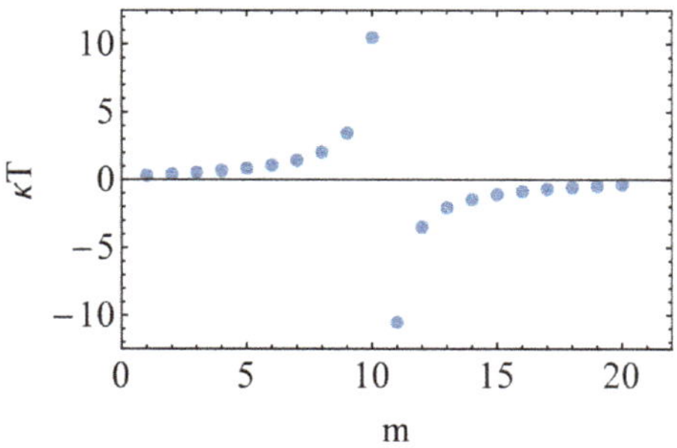

Figure D.2. Temperature of a spin system as illustrated in figure D.1, here with $N = 20$. Starting from the minimum state, as in figure D.1(*a*) and adding energy to flip the spins one by one, we see the temperature increasing, but after reaching the state with the maximum number of micro-states we obtain negative temperatures. Figures like this were for a time used to argue that negative temperatures were larger than infinity since they were associated with a larger energy!

- **Example:** Consider a generator or a source for X that can give only the symbols $+1$ and -1. Assume that they occur with the probabilities $P(+1) = p$ and $P(-1) = 1 - p \equiv q$, respectively. We define the *information* associated with the relevant value for x_i as $I(x_i) = \log(1/P(x_i)) = -\log P(x_i)$, measured in 'bit'. In this context the logarithm assumes the base 2. The basic idea is that if a symbol has a very large probability of occurrence then we do not get much information by encountering it, and vice versa for symbols with small probability of occurrence.

  The information generated on average is defined as

$$H(X) \equiv \sum_{i=1}^{2} P(x_i)\, I(x_i) = -\sum_{i=1}^{2} P(x_i) \log P(x_i) \qquad \text{bit/symbol.}$$

  This average information is also called the entropy of the source, because the expression resembles the one for entropy in statistical mechanics. For the present case we find $H = -p \log p - q \log q$. The entropy of the source will in general be maximal when all symbols appear with same probability, in this case $p = 1/2$, see also figure D.3.

- **Example:** The reader will hopefully allow a slight digression. Like negative temperatures, negative numbers were also at a time considered with some suspicion. Many mathematicians/scientists simply discarded them as being unphysical. In particular John Wallis (1616–1703) interpreted them as not only less than zero but also greater than infinity. He considered a series of ratios like $24/4 = 6$, $24/3 = 8$, $24/2 = 12$, etc. Each of the ratios giving 6, 8, 12 are finite and smaller than infinity. But as the denominators get smaller and smaller the value of the ratio eventually diverges. Wallis argued that as the denominators became even smaller, say -1 the ratio became $24/(-1) = -24$ and common logic would then imply that the negative

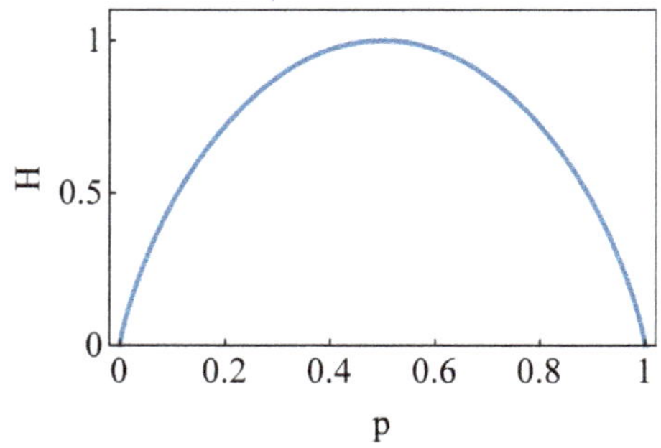

Figure D.3. $H = -p \log p - q \log q$ shown as a function of p, where $q = 1 - p$.

numbers were even larger than infinity, implying that the numbers are to be ordered in some cyclic fashion!

References

Braun S, Ronzheimer J P, Schreiber M, Hodgman S S, Rom T, Bloch I and Schneider U 2013 Negative absolute temperature for motional degrees of freedom *Science* **339** 52–5

Marvan M 1966 *Negative absolute temperatures* number 4 *in* 'Physics Paperbacks' (London: Iliffe)

Pathria R K 1996 *Statistical Mechanics* 2nd edn (Oxford: Butterworth-Heinemann)

IOP Publishing

Appendix E

Dimensional analysis

It has long been known that dimensional analysis provides more than just a means for carrying out a basic check of some analytical results. The basic observation is that expressions like exp(1 meter) or cos(1 meter) are meaningless, while $\exp(\ell/\ell_0)$ or $\cos(\ell/\ell_0)$ with $\ell_0 = 1$ meter are perfectly meaningful. (Logarithm functions are a bit special in this respect: while we can argue that a change in unit from $\ell_0 = 1$ meter to $\ell_0 = 1$ inch changes the functional variation of $\exp(\ell/\ell_0)$, we find that logarithms like $\ln(\ell/\ell_0)$ or $\log(\ell/\ell_0)$ are only *shifted* but not re-scaled by a change of variable.) If a problem contains a characteristic length scale, we expect that the solution of the problem depends in a systematic way on this scale, with the same argument for time scales.

Basically, it is argued (Buckingham 1914) that *any* physical result can be written in the form

$$f(Q_1, Q_2, \dots, Q_n) = 0, \tag{E.1}$$

where $Q_1, Q_2, \dots, Q_n$, etc denotes the physical variables of the problem in addition to time, position, etc. Using these variables we can obtain a dimensionless form

$$\pi \equiv Q_1^{\ell_1} \cdot Q_2^{\ell_2}, \dots Q_n^{\ell_n},$$

where the integers $\ell_1, \ell_2, \dots, \ell_n$ are chosen to make π dimensionless. As Professor McNamara told us at a Culham summer school many years ago: 'Theoretical physics is a study of the relation between dimensionless variables and only engineers need to know how big things are'!

We imagine a space spanned by the respective physical dimensions relevant for the problem. The dimension vector of any $Q_k = A_1^{b_1}, A_2^{b_2}, \dots A_n^{b_n}$, is denoted $\mathbf{b}_k = (b_1, b_2, \dots, b_n)$, with the As denoting the relevant units such as *length*, *mass*,

doi:10.1088/978-0-7503-1251-6ch35

etc. All b_js are integer. The dimension vector of $Q_i^{\ell_i}$ is then simply $\ell_i \mathbf{b}_i$. The condition for π being dimensionless is therefore simply

$$\sum_{i=1}^{m} \ell_i \mathbf{b}_i = 0.$$

We can introduce an $m \times n$-dimensional matrix as $(\mathbf{b}_1, \mathbf{b}_2, \ldots, \mathbf{b}_m)$, where m is the number of relevant dimensions, and n is the number of variables. It is then argued that the variables can be made dimensionless, and (E.1) written as an equivalent relation

$$g(\pi_1, \pi_2, \ldots, \pi_{n-r}) = 0, \tag{E.2}$$

where now $\pi_1, \pi_2, \ldots \pi_{n-r}$ etc are all independent, dimensionless variables, where r is the rank of the dimension matrix. This so called Π-theorem was first proven by Buckingham (1914), but has been elaborated later by many others (Bridgman 1931, Kurth 1972, Gibbings 1982, Jensen 2013). A short historical summary is given by Bolster *et al* (2011). The result of a correct analysis is invariant with respect to a change in base units. The statement can best be illustrated by some examples, where we first make a short summary of some basic definitions concerning matrices and sets of linear equations.

- **Example:** When A is an arbitrary matrix (not necessarily square), the determinant of any sub-matrix of A is called a *minor* of A.

  In a square matrix $\mathsf{A} = \{A\}_{i,k}$, the co-factor $a_{i,k}$ of the element $\{A\}_{i,k}$ is the particular minor obtained by deleting the i-th row and the k-th column, multiplied by $(-1)^{i+k}$. We have the identities

$$\sum_{j=1}^{n} A_{i,j} a_{k,j} = \begin{cases} \det A & \text{for } i = k \\ 0 & \text{for } i \neq k \end{cases}$$
$$\sum_{j=1}^{n} A_{j,i} a_{j,k} = \begin{cases} \det A & \text{for } i = k \\ 0 & \text{for } i \neq k \end{cases}.$$

  with det$A$ denoting the determinant of A. We also have

$$\det A = A_{\ell,\ell} a_{\ell,\ell} - \sum_{i,k=2}^{n} A_{i,\ell} A_{\ell,k} a_{\ell,\ell;i,k}$$

  where $a_{\ell,\ell;i,k}$ is the cofactor of $A_{i,k}$ in the matrix $A_{\ell,\ell}$.

  The *rank* of a matrix A (not necessarily square) is the greatest integer r such that A contains at least one minor of order of r, which is not zero (Cramér 1946). If all minors of A are zero, then A is a zero matrix, and in this case $r = 0$. When $\mathsf{A} = A_{m,n}$, the rank r is at most equal to the smaller of the numbers m and n.

  Let the rows and columns of A be considered as vectors. If A is of rank r, it is possible to find r linearly independent rows of A, while any $r + 1$ rows will be linearly dependent. The same is true for columns.

If $\mathsf{A}_1, \mathsf{A}_2, \ldots, \mathsf{A}_p$ are of ranks $r_1, r_2, \ldots, r_p$, then the rank of the sum $\mathsf{A}_1 + \mathsf{A}_2 + \cdots + \mathsf{A}_p$ is *at most* equal to the sum $r_1 + r_2 + \cdots + r_p$, while the rank of the product $\mathsf{A}_1 \cdot \mathsf{A}_2 \cdots \mathsf{A}_p$ is at most equal to the smaller of the ranks, $r_1, r_2, \ldots, r_p$.

If a square matrix $\mathsf{A} = \{A\}_{n,n}$ is such that $A \neq 0$, then A is of rank n. Such a matrix is said to be *non-singular*, while a square matrix with $A = 0$ is of rank $r < n$, and is called a *singular* matrix. If an arbitrary matrix B is multiplied (pre- or post-) by a non-singular matrix A, then the product has the same rank as B.

If A is symmetric and of rank r, there is at least one *principal* minor of order r in A which is non-zero. Hence, in particular the rank of a diagonal matrix is equal to the number of diagonal elements which are different from zero (Cramér 1946).

- **Example:** Consider an ideal classical pendulum. We expect the expression for its frequency to depend on various physical parameters entering the problem, namely its mass, m, and length, ℓ, and the gravitational acceleration g, while we wish to describe the displacement of the pendulum by an angle, which is a function of time. We assume the maximum angular displacement to be Θ. We write T for *time*, L for *length* and M for *mass*, and have the dimension matrix being

	m	ℓ	Θ	g	ω
T	0	0	0	−2	−1
L	0	1	0	1	0
M	1	0	0	0	0

where we use the short hand notation $[g] = \{-2, 1, 0\}$, i.e. the dimension of acceleration is *length*^{1} *time*$^{-2}$ *mass*^{0}. Evidently, we have five columns and three rows in the matrix, and the rank of the dimension matrix is $r = 2$ in this case. We have Θ to be dimensionless from the beginning, so we must be able to form one more dimensionless quantity. We write from left to right in the variable on the top in the dimension matrix

$$M^{\alpha_1} \times L^{\alpha_2} \times 1 \times \left(\frac{L}{T^2}\right)^{\alpha_3} \times \left(\frac{1}{T}\right)^{\alpha_4}$$

and determine the exponents α_j in such a way that the exponents of *mass*, of *time* and of *length* are each equal to zero. Evidently this requires $\alpha_1 = 0$, $\alpha_4 = -2\alpha_3$ and $\alpha_2 = -\alpha_3$.

The two dimensionless combinations we can make are then Θ and $\omega^2\ell/g$. Consequently, we can write

$$G\left(\Theta, \frac{\omega^2\ell}{g}\right) = 0.$$

We can solve this relation, at least formally, to give

$$\omega^2 = \frac{g}{\ell}\,\mathcal{G}(\Theta).$$

In view of this expression, we can reformulate the Π-theorem to state that any physical quantity can be written as a dimensionally correct constant multiplied with a dimensionless function of one (or, in general, more) dimensionless quantities.

In the small amplitude limit we expect that the angular frequency of the pendulum is independent of the maximum displacement, implying $\mathcal{G}(\Theta) \approx$ const. there, since it does not make physical sense for the result to depend on the *sign* of Θ, so presumably for small Θ we can approximate $\mathcal{G} \approx$ const. $+ \Theta^2$. This gives the (well known) result $\omega = \sqrt{g/\ell}$, apart from a numerical constant, which cannot be determined by the present dimensional arguments, i.e. we do not know *a priori what* constant $\mathcal{G}(\Theta)$ approaches for small Θ.

- **Example:** Consider fluctuations in voltage, V, across a resistor R in thermal equilibrium at temperature T. The fluctuating voltage may depend on frequency f, while Boltzmann's constant κ enters as a dimensional constant of nature, just as Planck's constant h. We know, however, that Boltzmann's constant will appear only as a coefficient for the temperature, and write κT from the outset. We are interested in the power spectrum for the fluctuations in voltage within a narrow frequency range Δf. Then $V^2/\Delta f$ is interpreted as the quantity which after integration with respect to frequency from zero to infinity, gives V^2. We can write the dimension matrix as

	κT	R	h	f	$V^2/\Delta f$
T	− 2	− 3	− 1	− 1	− 5
L	2	2	2	0	4
M	1	1	1	0	2
A	0	− 2	0	0	− 2

where A stands for *Ampere*. The second and third rows are proportional. The rank of the matrix is found to be 2. Omitting the second redundant row, we again write from left to right in the variable on the top in the dimension matrix

$$\left(\frac{M}{T^2}\right)^{\alpha_1} \times \left(\frac{M}{T^3A^2}\right)^{\alpha_2} \times \left(\frac{M}{T}\right)^{\alpha_3} \times \left(\frac{1}{T}\right)^{\alpha_4} \times \left(\frac{M^2}{T^5A^2}\right)^{\alpha_5}$$

and determine the exponents α_j in such a way that the exponents of *mass*, *time*, *length* and of *current* are each equal to zero. We introduce $\alpha_1 \equiv \alpha$ and $\alpha_5 \equiv \beta$ as parameters. After simple calculations we find $\alpha_2 = -\beta$, $\alpha_3 = -\alpha - \beta$, and $\alpha_4 = -\alpha - 2\beta$, implying that we have dimensionless quantities

$$\left(\frac{\kappa T}{hf}\right)^{\alpha} \quad \text{and} \quad \left(\frac{V^2}{Rhf\,\Delta f}\right)^{\beta}.$$

Evidently, we can have also $\kappa TRf/V^2$ as a dimensionless quantity, but given the two foregoing ones, it is not *independent*. With the two independent dimensionless quantities we can write

$$g\left(\frac{\kappa T}{hf}, \frac{V^2}{Rhf\,\Delta f}\right) = 0, \tag{E.3}$$

which we formally solve to give

$$V^2 = Rhf\,\mathcal{G}\left(\frac{\kappa T}{hf}\right)\Delta f. \tag{E.4}$$

In the classical limit $h \to 0$, we must have an expression independent of Planck's constant h. The only way this can be achieved is for $\mathcal{G}(x) \sim x$, which then gives

$$V^2 \sim R\kappa T\,\Delta f,$$

which is the well known Nyquist (1928) result. Evidently, we have made a serious simplification by assuming R to be constant, independent of frequency (Pécseli 2000), but at least for moderate frequencies the result applies.

It requires a much more detailed analysis to give the general functional form of $\mathcal{G}$. One finds the expression for the power spectrum of voltage fluctuations

$$V^2(f, ;f + \mathrm{d}f) = 4R\frac{hf}{\mathrm{e}^{hf/(\kappa T)} - 1}\mathrm{d}f,$$

where a zero point energy $\frac{1}{2}hf$ can be argued, but this is of little importance for practical applications.

In concluding this example we might add a few comments on the relation between power spectra and second order structure functions, when studied by dimensional analysis. Assume that we a fluctuating quantity X, and consider its correlation function R_X or structure function, which then has the dimension $[X^2]$. Given some parameters, we may for instance have found

$$R_X = Bg(t/\tau_o, t/\tau_1, \ldots, a, b, \ldots),$$

where the τ_js are some characteristic times for the problem, $a, b, \ldots$ are dimensionless parameters, and $[B] = [X^2]$. We can now obtain the power spectrum by Fourier transforming the correlation function (the Wiener–Khinchine theorem), and write

$$S(\omega, \ldots, a, b, \ldots) = \int Bg(t/\tau_o, t/\tau_1, \ldots, a, b, \ldots)\cos(\omega t)\mathrm{d}t.$$

We introduce a new variable $t\omega \equiv \gamma$ and write

$$S(\omega, \ldots, a, b, \ldots) = \frac{B}{\omega}\int f(\gamma/(\omega\tau_0), \gamma/(\omega\tau_1), \ldots, a, b, \ldots)\cos(\gamma)\mathrm{d}\gamma$$
$$\equiv \frac{B}{\omega}G(\omega\tau_0, \omega\tau_1, \ldots, a, b, \ldots).$$

Applying this result we could also obtain (E.3) or (E.4), without the somewhat clumsy introduction of Δf.

- **Example:** Take the equation of motion for an incompressible fluid

$$\frac{\partial}{\partial t}u_\ell + u_j\frac{\partial}{\partial r_j}u_\ell = \frac{1}{\rho}\frac{\partial}{\partial r_j}\sigma_{\ell,j} \tag{E.5}$$

and $\partial u_j/\partial r_j = 0$ for incompressibility, with the summation convention understood. Here $\sigma_{\ell,j}$ is the stress tensor. If the fluid is *Newtonian*, the stress tensor is given as $\sigma_{\ell,j} = -p\delta_{\ell,j} + 2\mu\tau_{\ell,j}$, where p is the hydrostatic pressure, μ the dynamic viscosity (taken to be constant) while $\tau_{\ell,j}$ is the rate of strain given as

$$\tau_{\ell,j} = \frac{1}{2}\left(\frac{\partial}{\partial r_j}u_\ell + \frac{\partial}{\partial r_\ell}u_j\right), \tag{E.6}$$

which inserted into (E.5) gives the Navier–Stokes equation for an incompressible fluid, here written in terms of velocity vectors and differential operators

$$\frac{\partial}{\partial t}\mathbf{u} + \mathbf{u}\cdot\nabla\mathbf{u} = -\frac{1}{\rho}\nabla p + \nu\nabla^2\mathbf{u}, \tag{E.7}$$

with $\nu \equiv \mu/\rho$ being the *kinematic viscosity*.

Assume that the geometry of the problem is properly characterized by a length scale $\mathcal{L}$, and a characteristic velocity U_0. Basically, these assumptions apply to Reynolds' classical experiments in tubes using different cross-sections and different flow velocities. The dimension matrix for this case becomes

	p	ν	ρ	u	r	t	$\mathcal{L}$	U_0
T	−2	−1	0	−1	0	1	0	−1
L	−1	2	−3	1	1	0	1	1
M	1	0	1	0	0	0	0	0

We can obtain the following dimensionless quantities $r' \equiv r/\mathcal{L}$, $u' \equiv u/U_0$, $p' \equiv p/(\rho U_0^2)$, $t' \equiv tU_0/\mathcal{L}$ and finally *Reynolds number* $R_e \equiv U_0\mathcal{L}/\nu$. In terms of these new variables we can rewrite Navier–Stokes equation in the form

$$\frac{\partial}{\partial t'}\mathbf{u}' + \mathbf{u}'\cdot\nabla'\mathbf{u}' = -\nabla' p' + \frac{1}{R_e}\nabla'^2\mathbf{u}'. \tag{E.8}$$

The interesting point is now that if we have results from *one* experiment, we automatically have the results from another self-similar one, constructed in such a way that Reynolds' number is the same. Many model experiments for ship-building are based on this observation. Note that the *nature* of the equations is fundamentally different when R_e is finite, and when it is infinite! In the former case we have a second order differential equation, in the latter, a first order differential equation in the spatial variables.

- **Example:** Consider the simple linear wave equation

$$\left(\frac{\partial^2}{\partial t^2} - C_s^2\nabla^2\right)n = 0,$$

describing the propagation of linear sound waves. This model has no length scale, implying that the space-time evolution of a scalar quantity n is controlled by the length scale $\mathcal{L}$ inherent in the initial condition. All we have is a characteristic velocity C_s as a natural parameter for the problem. A characteristic time for the problem will be $\mathcal{L}/C_s$.

- **Example:** A harmonically varying signal will propagate along a telephone cable according to the expression

$$I(x, t) = I_0 \mathrm{e}^{-\beta x} \cos\left(\omega t - \frac{\omega}{\vartheta}x\right),$$

where β (the spatial damping constant) and ϑ (the phase velocity) will depend on ω in general. For telephone cables with negligible leak-currents to the surroundings and large self-inductions it is found, however, that β and ϑ are independent of ω and depend only on the self-induction ℓ, the resistivity r, and the capacitance c of the cable, all taken per unit length. Determine by dimensional reasoning the variation of β and ϑ with varying ℓ, r, and c.

First we note that voltage differences U have dimension of energy per charge, i.e. $[U] = (M \times L^2 \times T^{-2})/(A \times T) = M \times L^2 \times T^{-3} \times A^{-1}$ where A again denotes the dimension of current. We express the dimensions of ℓ, r, and c in a self-evident notation as

$$\begin{aligned}[\ell] &= L^{-1}[U/(\mathrm{d}I/\mathrm{d}t)] \\ &= L^{-1}(M \times L^2 \times T^{-3} \times A^{-1})/(A \times T^{-1}) \\ &= M \times L \times T^{-2} \times A^{-2}\end{aligned}$$

$$\begin{aligned}[r] &= L^{-1}[U/I] \\ &= L^{-1}(M \times L^2 \times T^{-3} \times A^{-1})/A \\ &= M \times L \times T^{-3} \times A^{-2}\end{aligned}$$

$$[c] = L^{-1}[Q/U]$$
$$= L^{-1}(A \times T)/(M \times L^2 \times T^{-3} \times A^{-1})$$
$$= M^{-1} \times L^{-3} \times T^4 \times A^2.$$

We require that β and ϑ can be expressed as some combination of ℓ, r, and c, i.e. as, for instance

$$\vartheta \sim \ell^{a_1} r^{a_2} c^{a_3}.$$

Since β has the dimension L^{-1} we determine for this case the exponents a_1, a_2, a_3 by requiring $a_1 + a_2 - a_3 = 0$, $a_1 + a_2 - 3a_3 = -1$, $-2a_1 - 3a_2 + 4a_3 = 0$, and $-2a_1 - 2a_2 + 2a_3 = 0$. We find $a_1 = -1/2$, $a_2 = 1$, and $a_3 = 1/2$, so that

$$\beta = C_\beta \sqrt{\frac{r^2 c}{\ell}},$$

where C_β is some numerical constant unaccounted for by the dimensional analysis. Similarly, we introduce exponents b_1, b_2, b_3 for ϑ. The dimension of ϑ is $L \times T^{-1}$. We have the relations $b_1 + b_2 - b_3 = 0$, $b_1 + b_2 - 3b_3 = 1$, $-2b_1 - 2b_2 + 2b_3 = 0$, and $-2b_1 - 3b_2 + 4b_3 = -1$. The solution is $b_1 = -1/2$, $b_2 = 0$, $b_3 = -1/2$ giving

$$\vartheta = C_\vartheta \sqrt{\frac{1}{c\,\ell}},$$

with a new constant C_ϑ.

- **Example:** Consider gravity waves on the surface of deep oceans. The ocean is assumed to be so deep that the depth of the water does not enter the parameters. We are then left with the gravitational constant g, the mass density of water ρ, the phase velocity $\vartheta \equiv \omega/k$ and the wavenumber k. Given these parameters, we can write the dimension matrix as

	ϑ	k	g	ρ
T	-1	0	-2	0
L	1	-1	1	-3
M	0	0	0	1

To construct a phase velocity with dimensions $L \times T^{-1}$ we note that ρ cannot enter since there is no way to cancel the mass M. Left with k and g we can construct

$$\vartheta = C_\vartheta \sqrt{\frac{g}{k}},$$

with a numerical constant C_ϑ. If we prefer the dispersion relation in the classical sense, the result would be $\omega = C_\vartheta \sqrt{g\,k}$. The independence on mass density ρ is physically due to the fact that all masses are subject to the same gravitational acceleration, recalling the Galileo experiment (which, incidentally, may never have been conducted by him personally).

- **Example:** Consider a beam of light passing through the gravitational field of a spherical star. Without any understanding of the underlying physics we can argue that the result must depend on the distance R to the center of the star at closest approach and the mass $M_\odot$ of the star (Bolster *et al* 2011). The speed of light c must enter, just as the gravitational constant G. We could argue from the outset that $M_\odot$ and G will enter only by the product $M_\odot G$. The deflection angle Θ is a dimensionless quantity. The effects of refraction in the star's atmosphere are ignored, although in practice this is a rather naive assumption. We can write the dimension matrix as

	c	G	R	$M_\odot$	Θ
T	−1	−2	0	0	0
L	1	3	1	0	0
M	0	−1	0	1	0

The result, to lowest order (Kurth 1972, Bolster *et al* 2011), becomes

$$\Theta = C_\theta \frac{M_\odot G}{c^2 R}, \tag{E.9}$$

with an unknown numerical constant C_θ. Newton considered this problem, and arrived at the result (E.9) with $C_\theta = 2$. General relativity predicts $C_\theta = 4$. Early experiments measuring the deflections of light glancing the Sun (measured at Solar eclipse) forgot the effects of refraction in the atmosphere but arrived at surprisingly good agreement with (E.9). Later more accurate measurements show agreement with the predictions of general relativity within 0.02%.

This book is using MKSA-units; meter (m) for length, kilogram (kg) for mass, second (s) for time, and Ampere (A) for current. There is an additional, or supplementary unit for temperature, °K(elvin), usually the ° is omitted, though. At times it is argued that angles are a supplementary unit also, but these can be obtained as a length by measuring the length along the arc of a unit circle giving $180° \to \pi$, where 'radians' are preferred without reference to the unit radius circle. Luminous intensity is measured in candela (cd). If you stay within the MKSA-system the result of a correct calculation will always have the correct physical dimension. We do not always trust our ability to do correct calculations and it is wise to check, from time to time, whether results are dimensionally correct. In such cases one has to express other units in MKSA-units, for instance charge which is measured in C(oulomb). Now this is easy since current is charge per seconds by definition, giving Coulomb = Ampere ·second. In other cases the calculation might take a little more labor, so to save the reader some time, the following table is given, in terms of F(arad) for capacity, H(enry) for inductance, J(oule) for energy, N(ewton) for force, Ω (Ohm) for resistance, S(iemens) for conductance, Pa(scal) for pressure, T(esla) for magnetic flux density, V(olts) for voltage, W(att) for power, and Wb (Weber) for magnetic flux. The notation [a] will here denote the dimension of the quantity a.

Quantity	Unit	MKSA-unit
Boltzmann's constant, [κ]	J/K	$kg \cdot m^2 \cdot K^{-1} \cdot s^{-2}$
Capacitance	$F = A \cdot s \cdot V^{-1}$	$(A \cdot s^2)^2 \cdot kg^{-1} \cdot m^{-2}$
Charge	C	A·s
Conductance	$S = \Omega^{-1}$	$A^2 \cdot s^3 \cdot kg^{-1} \cdot m^{-2}$
Conductivity	$S \cdot m^{-1}$	$A^2 \cdot s^3 \cdot kg^{-1} \cdot m^{-3}$
Electric field	$V \cdot m^{-1}$	$kg \cdot m \cdot A^{-1} \cdot s^{-3}$
Energy	$J = N \cdot m$	$kg \cdot m^2 \cdot s^{-2}$
Force	N	$kg \cdot m \cdot s^{-2}$
Gravitational constant, [G]	$N \cdot m^2 \cdot kg^{-2}$	$m^3 \cdot kg^{-1} \cdot s^{-2}$
Heat capacity	$J \cdot kg^{-1} \cdot K^{-1}$	$m^2 \cdot s^{-2} \cdot K^{-1}$
Induction	$H = V \cdot s \cdot A^{-1}$	$kg \cdot m^2 \cdot A^{-2} \cdot s^{-2}$
Magnetic flux	$Wb = V \cdot s$	$kg \cdot m^2 \cdot A^{-1} \cdot s^{-2}$
Magnetic flux density	$T = Wb \cdot m^{-2}$	$kg \cdot A^{-1} \cdot s^{-2}$
Permeability of free space, [μ_0]	$H \cdot m^{-1}$	$kg \cdot m \cdot (A \cdot s)^{-2}$
Permittivity of free space, [ε_0]	$F \cdot m^{-1}$	$(A \cdot s^2)^2 \cdot kg^{-1} \cdot m^{-3}$
Impedance of the 'aether', [$\sqrt{\mu_0/\varepsilon_0}$]	Ω	$kg \cdot m^2 \cdot A^{-2} \cdot s^{-3}$
Planck's constant, [h]	$J \cdot s$	$kg \cdot m^2 \cdot s^{-1}$
Power	$W = J \cdot s^{-1} = V \cdot A$	$kg \cdot m^2 \cdot s^{-3}$
Pressure	$Pa = N \cdot m^{-2}$	$kg \cdot m^{-1} \cdot s^{-2}$
Resistance	$\Omega = V \cdot A^{-1}$	$kg \cdot m^2 \cdot A^{-2} \cdot s^{-3}$
Resistivity	$\Omega \cdot m$	$kg \cdot m^3 \cdot A^{-2} \cdot s^{-3}$
Viscosity (dynamic), [μ]		$kg \cdot m^{-1} \cdot s^{-1}$
Viscosity (kinematic), [$\nu = \mu \cdot \rho^{-1}$]		$m^2 \cdot s^{-1}$
Voltage	$V = W \cdot A^{-1} = J \cdot C^{-1}$	$kg \cdot m^2 \cdot A^{-1} \cdot s^{-3}$

References

Bolster D, Hershberger R E and Donnelly R J 2011 Dynamic similarity, the dimensionless science *Phys. Today* **64** 42–7

Bridgman P W 1931 *Dimensional Analysis* (New Haven, CT: Yale University Press)

Buckingham E 1914 On physically similar systems; illustrations of the use of dimensional equations *Phys. Rev.* **4** 345–76

Cramér H 1946 *Mathematical Methods of Statistics* (Princeton, NJ: Princeton University Press)

Gibbings J C 1982 A logic of dimensional analysis *J Phys. A: Math. Gen.* **15** 1991–2002

Jensen J H 2013 Introducing fluid dynamics using dimensional analysis *American J. Phys.* **81** 688–94

Kurth R 1972 *Dimensional Analysis and Group Theory in Astrophysics* (Oxford: Pergamon Press)

Nyquist H 1928 Thermal agitation of electric charge in conductors *Phys. Rev.* **32** 110–3

Pécseli H L 2000 *Fluctuations in Physical Systems* (Cambridge: Cambridge University Press)

Appendix F

Summation convention

We introduce here the *Einstein summation convention*. Introducing a short hand notation for the differential operator as $\nabla \equiv \{\partial/\partial x, \partial/\partial y, \partial/\partial z\} \equiv \{\partial_1, \partial_2, \partial_3\}$, or simply ∂_j with $j = 1, 2, 3$, we can write, for instance, the relation for incompressible flows $\nabla \cdot (\mathbf{u} \cdot \nabla\mathbf{u}) = \nabla \cdot ((1/\rho)\nabla p)$ as

$$\sum_i \sum_j \partial_i\left(u_i \partial_j u_j\right) = \frac{1}{\rho_0} \sum_j \partial_j^2 p, \tag{F.1}$$

where we assume for simplicity that $\rho = \rho_0$ is constant. Note that we can safely write ∂_j for $\partial/\partial r_j$, because there will not be any confusion with the partial time differentiation, which will be written as ∂_t.

Expressions like (F.1) are rather cumbersome because of all the summation signs. The Einstein summation convention says that any repeated index is to be summed over! This will hold also for the square of the operator ∂_j on the right hand side of (F.1); writing out the implied multiplication, it is in reality also a summed index. By the Einstein convention, the summation signs become redundant. With this in mind we can simply write (F.1) as

$$\begin{aligned} \partial_i\left(u_i \partial_j u_j\right) &= \left(\partial_j u_i\right)\left(\partial_i u_j\right) + u_i \partial_j \partial_i u_j \\ &= \left(\partial_j u_i\right)\left(\partial_i u_j\right) = \frac{1}{\rho_0} \partial_j^2 p, \end{aligned} \tag{F.2}$$

where we used that the incompressibility condition $\nabla \cdot \mathbf{u} = 0$ becomes $\partial_j u_j = 0$ with this elegant notation. For the product $\mathsf{K}_{ij}\mathsf{S}_{ji}$ we usually write $\mathsf{K} : \mathsf{S}$, which is the trace of the tensorial product $\mathsf{K} \cdot \mathsf{S}$. We can therefore formally express (F.2) as

$$\nabla^2 p = \rho_0 \nabla\nabla : \mathbf{uu} \tag{F.3}$$

doi:10.1088/978-0-7503-1251-6ch36

having the form of a Poisson equation, expressing pressure when the flow field **u** is known. For the case $\rho \neq$ const., we can still set up a differential equation for the pressure, only not as simple as (F.3).

By contraction of a tensor we mean the sum of the diagonal elements, or the *trace* of the tensor. With the Einstein convention, this is simply written as K_{jj}, for instance. There is a problem here, though: *if* you want to refer to one, but any one, of the diagonal elements of an $N \times N$ tensor M, you would write M_{jj}. This would, however, immediately be interpreted as $\sum_j^N \mathsf{M}_{jj}$, which is something else. Maybe one should invent a convention that only repeated *Latin* indices should be summed, and not repeated *Greek* indices? There may be a road to fame here!

- **Example:** The *scalar product* of two vectors $\mathbf{a} \cdot 0\mathbf{b}$ is written simply as $a_i b_i$ by the summation convention. In order to have a neat expression for the *vector* product we introduce a *permutation symbol* as

$$\epsilon_{ijk} = \begin{cases} 0 & \text{if any two of } i, j \text{ and } k \text{ are the same} \\ 1 & \text{if } i, k \text{ is an even permutation of } 1, 2, 3 \\ -1 & \text{if } ijk \text{ is an odd permutation of } 1, 2, 3 \end{cases}$$

  In terms of ϵ_{ijk} we have $\mathbf{a} \times b = \epsilon_{ijk} a_i b_j \hat{\mathbf{n}}_{(k)}$, with $\hat{\mathbf{n}}_{(1)} = \{1, 0, 0\}$, $\hat{\mathbf{n}}_{(2)} = \{0, 1, 0\}$, and $\hat{\mathbf{n}}_{(3)} = \{0, 0, 1\}$. In particular we have $a \cdot \mathbf{b} \times \mathbf{c} = \epsilon_{ijk} a_i b_j c_k$.

- **Exercise:** Demonstrate that $\epsilon_{ijk}\epsilon_{klm} = \delta_{il}\delta_{jm} - \delta_{im}\delta_{jl}$ in terms of the Kronecker δ.

- **Exercise:** Demonstrate that the condition for the three vectors **a**, **b**, and **c** to be co-planar can be written as $\epsilon_{ijk} a_i b_j c_k = 0$ with the summation convention.

At times you find the summation convention extended in the sense that any repeated variable is supposed to be integrated. This convention is not used so much, though.

Finally we might mention a *differentiation notation*, which does not seem to be much in use any longer. Here, differentiation is indicated by a subscript with a 'comma', for instance as

$$\frac{\partial}{\partial x_j} u_i \equiv u_{i,j}$$

or more generally

$$\frac{\partial}{\partial x_j} (*)_i \equiv (\overset{*}{i})_{,j}.$$

With this notation, for instance the continuity equation becomes

$$\frac{\partial}{\partial t} n + (n\, u_i)_{,i} = 0,$$

implying also the summation convention. Similarly, the simple momentum equation becomes

$$\frac{\partial}{\partial t}(n\, u_i) + (n\, u_i\, u_j)_{,i} = -p_{,j}\,,$$

etc.

In the opinion of the author, this ‘comma’ notation is potentially confusing!

Appendix G

Physical constants

Speed of light in vacuum	$c = 2.9979 \times 10^{8}$ m s^{-1}
Permittivity of free space	$\varepsilon_0 = 8.8542 \times 10^{-12}$ F m^{-1}
Permeability of free space	$\mu_0 = 4\pi \times 10^{-7}$ H m^{-1}
Impedance of the 'aether'	$\sqrt{\mu_0/\varepsilon_0} = 376.7301$ ohm
Gravitational constant	$G = 6.67408 \times 10^{-11}$ m^3 kg^{-1} s^{-2}
Elementary charge	$e = 1.6022 \times 10^{-19}$ C
Electron mass	$m = 9.1095 \times 10^{-31}$ kg
Proton mass	$M = 1.6726 \times 10^{-27}$ kg
Atomic mass unit (^{12}C $\times \frac{1}{12}$)	amu $= 1.6604 \times 10^{-27}$ kg
Boltzmann constant	$\kappa = 1.3807 \times 10^{-23}$ J K^{-1}
Planck constant	$h = 6.6256 \times 10^{-34}$ J s
	$\hbar \equiv h/2\pi = 1.0546 \times 10^{-37}$ J s
Bohr radius, $4\pi\varepsilon_0\hbar^2/e^2m$	$a_0 = 5.2918 \times 10^{-11}$ m
Atomic cross section	$\pi a_0^2 = 8.7974 \times 10^{-21}$ m^2
Classical electron radius	$e^2/mc^2 = 2.82 \times 10^{-15}$ m
Avogadro number (molecules/mol)	$N_A = 6.0220 \times 10^{23}$
Standard pressure (760 Torr = 1 atm)	$p_0 = 1.0133 \times 10^{5}$ Pa
Ideal gas mol-volume	$\mathcal{V}_{\rm mol} = 0.022414$ m^2 mol^{-1}
Loschmidt number (ideal gas density at STP)	$n_L = 2.6868 \times 10^{25}$ m^{-3}
Absolute zero temperature	$T_0 = -273.15$ °C

doi:10.1088/978-0-7503-1251-6ch37

Approximate expressions:

Electron plasma frequency, assuming densities inserted in particles m^{-3}	$f_{pe} \approx 9\sqrt{n}$ Hz $\omega_{pe} \approx 57\sqrt{n}$ rad s^{-1}
Electron cyclotron frequency, assuming magnetic fields inserted in Tesla	$f_{ce} \approx 2.8 \times 10^{10}\, B$ Hz $\omega_{ce} \approx 17.6 \times 10^{10}\, B$ rad s^{-1}
Proton cyclotron frequency, assuming magnetic fields inserted in Tesla	$f_{ci} \approx 1.5 \times 10^{7}\, B$ Hz $\Omega_{ci} \approx 9.4 \times 10^{7}\, B$ rad s^{-1}

For order of magnitude estimates, the following approximations (accurate to within few %) may come in handy

$$\sqrt{2\pi} \simeq 2.5 \qquad \pi^2 \simeq 10 \qquad e^3 \simeq 20 \qquad 2^{10} \simeq 10^3$$

It is interesting to know that 1 year is approximately $\pi \times 10^7$ s.

IOP Publishing

Low Frequency Waves and Turbulence in Magnetized Laboratory Plasmas and in the Ionosphere

Appendix H

Useful vector relations

A scalar is denoted by italic Roman letters, vectors by bold-face capital Roman letters and tensors by capital bold-face Greek letters. $\mathbf{r}$ is the position vector and $\mathbf{1}$ is the unit tensor and d is the dimensionality of the problem.

$$\mathbf{A}\cdot(\mathbf{B}\times\mathbf{C}) = \mathbf{C}\cdot(\mathbf{A}\times\mathbf{B}) = \mathbf{B}\cdot(\mathbf{C}\times\mathbf{A})$$
$$\mathbf{A}\times(\mathbf{B}\times\mathbf{C}) = \mathbf{A}\cdot\mathbf{CB} - \mathbf{A}\cdot\mathbf{BC}, \qquad (\mathbf{A}\times\mathbf{B})\times\mathbf{C} = \mathbf{A}\cdot\mathbf{CB} - \mathbf{B}\cdot\mathbf{CA}$$
$$\nabla(ab) = a\nabla b + b\nabla a$$
$$\nabla\cdot(a\mathbf{A}) = a\nabla\cdot\mathbf{A} + \mathbf{A}\cdot\nabla a$$
$$\nabla\times(a\mathbf{A}) = a\nabla\times\mathbf{A} + (\nabla a)\times\mathbf{A}$$
$$\nabla(\mathbf{A}\cdot\mathbf{B}) = (\nabla\mathbf{A})\cdot\mathbf{B} + \mathbf{A}\cdot\nabla\mathbf{B}$$
$$= \mathbf{A}\times(\nabla\times\mathbf{B}) + \mathbf{B}\times(\nabla\times\mathbf{A}) + (\mathbf{A}\cdot\nabla)\mathbf{B} + (\mathbf{B}\cdot\nabla)\mathbf{A}$$
$$\mathbf{A}\times(\nabla\times\mathbf{B}) = (\nabla\mathbf{B})\cdot\mathbf{A} - \mathbf{A}\cdot\nabla\mathbf{B}, \qquad (\mathbf{A}\times\nabla)\times\mathbf{B} = (\nabla\mathbf{B})\cdot\mathbf{A} - \mathbf{A}\nabla\cdot\mathbf{B}$$
$$\mathbf{A}\times(\nabla\times\mathbf{A}) = \frac{1}{2}\nabla(\mathbf{A}\cdot\mathbf{A}) - (\mathbf{A}\cdot\nabla)\mathbf{A}$$
$$\nabla\times(\nabla\times\mathbf{A}) = \nabla(\nabla\cdot\mathbf{A}) - \nabla^2\mathbf{A}, \quad \text{or} \quad \nabla^2\mathbf{A} = \nabla(\nabla\cdot\mathbf{A}) - \nabla\times(\nabla\times\mathbf{A})$$
$$\nabla\times\nabla a = 0$$
$$\nabla\cdot\nabla\times\mathbf{A} = 0$$
$$\nabla\cdot(\mathbf{A}\times\mathbf{B}) = (\nabla\times A)\cdot\mathbf{B} - \mathbf{A}\cdot(\nabla\times\mathbf{B})$$
$$\nabla\cdot(\mathbf{AB}) = (\nabla\cdot\mathbf{A})\mathbf{B} + (\mathbf{A}\cdot\nabla)\mathbf{B}$$
$$\nabla\times(\mathbf{AB}) = (\nabla\times\mathbf{A})\mathbf{B} - (\mathbf{A}\times\nabla)\mathbf{B}$$
$$\nabla\times(\mathbf{A}\times\mathbf{B}) = \nabla\cdot(\mathbf{BA} - \mathbf{AB}) = \mathbf{A}(\nabla\cdot\mathbf{B}) - \mathbf{B}(\nabla\cdot\mathbf{A}) + (\mathbf{B}\cdot\nabla)\mathbf{A} - (\mathbf{A}\cdot\nabla)\mathbf{B}$$
$$\nabla\cdot(a\Pi) = (\nabla a)\cdot\Pi + a\nabla\cdot\Pi$$
$$\nabla\times\mathbf{r} = 0 \qquad \nabla\cdot\mathbf{r} = d \qquad \nabla\mathbf{r} = \mathbf{1}$$

- Gauss theorem for a spatially varying vector field $\mathbf{A}$:

$$\iiint_{\text{Vol}} \nabla\cdot\mathbf{A}\,\mathrm{d}V = \iint_{\text{Surf}} \mathbf{A}\cdot\hat{\mathbf{n}}\mathrm{d}\sigma,$$

where the integrations are over a volume 'Vol', bounded by a closed surface 'Surf' with a normal unit vector $\hat{\mathbf{n}}$, where $\mathrm{d}V$ is a volume element and $\mathrm{d}\sigma$ a surface element.

Introducing an arbitrary constant vector $\mathbf{Z}$ so that $\mathbf{A} \to \mathbf{A} \times \mathbf{Z}$ it can be shown that

$$\iiint_{\mathrm{Vol}} \nabla \times \mathbf{A}\, \mathrm{d}V = \iint_{\mathrm{Surf}} \hat{\mathbf{n}} \times \mathbf{A}\, \mathrm{d}\sigma.$$

Taking $\mathbf{A} \to a\mathbf{Z}$ it can be shown that

$$\iiint_{\mathrm{Vol}} \nabla a\, \mathrm{d}V = \iint_{\mathrm{Surf}} a\, \hat{\mathbf{n}}\, \mathrm{d}\sigma.$$

Taking $\mathbf{A} \to a\nabla b$ we find Green's first theorem

$$\iiint_{\mathrm{Vol}} \nabla \cdot (a\nabla b)\, \mathrm{d}V = \iiint_{\mathrm{Vol}} \left(\nabla a \cdot \nabla b + a\nabla^2 b\right)\mathrm{d}V = \iint_{\mathrm{Surf}} \hat{\mathbf{n}} \cdot a\nabla b\, \mathrm{d}\sigma.$$

Taking $\mathbf{A} \to a\nabla b - b\nabla a$ we find Green's second theorem

$$\iiint_{\mathrm{Vol}} \left(a\nabla^2 b - b\nabla^2 a\right)\mathrm{d}V = \iint_{\mathrm{Surf}} (a\nabla b - b\nabla a) \cdot \hat{\mathbf{n}}\, \mathrm{d}\sigma.$$

- Stokes' theorem for a spatially varying vector field $\mathbf{A}$:

$$\iint (\nabla \times \mathbf{A}) \cdot \hat{\mathbf{n}}\, \mathrm{d}\sigma = \oint \mathbf{A} \cdot \mathrm{d}\boldsymbol{\ell},$$

where $\mathrm{d}\boldsymbol{\ell}$ is a vectorial line element along a closed integration curve.

With an arbitrary constant vector $\mathbf{Z}$ so that $\mathbf{A} \to \mathbf{A} \times \mathbf{Z}$ it can be shown that

$$\iint_{\mathrm{Surf}} (\hat{\mathbf{n}} \times \nabla) \times \mathbf{A}\, \mathrm{d}\sigma = \oint \mathrm{d}\boldsymbol{\ell} \times \mathbf{A}.$$

Taking $\mathbf{A} \to a\mathbf{Z}$ it can be shown that

$$\iint_{\mathrm{Surf}} \hat{\mathbf{n}} \times \nabla a\, \mathrm{d}V = \oint a\, \mathrm{d}\boldsymbol{\ell}.$$

All theorems above assume that there are no discontinuities in the integration volumes.

Appendix I

Differential operators in cylindrical geometry

Vectors are denoted $\mathbf{U} = \{U_r, U_\theta, U_z\}$, with their components in cylindrical geometry. Unit vectors are indicated by a 'hat'.

$$\nabla\phi = \frac{\partial}{\partial r}\phi\hat{\mathbf{r}} + \frac{1}{r}\frac{\partial}{\partial\theta}\phi\hat{\boldsymbol{\theta}} + \frac{\partial}{\partial z}\phi\hat{\mathbf{z}}$$

$$\nabla\cdot\mathbf{U} = \frac{1}{r}\frac{\partial}{\partial r}(rU_r) + \frac{1}{r}\frac{\partial}{\partial\theta}U_\theta + \frac{\partial}{\partial z}U_z$$

$$\nabla\times\mathbf{U} = \left(\frac{1}{r}\frac{\partial}{\partial\theta}\,U_z - \frac{\partial}{\partial z}\,U_\theta\right)\hat{\mathbf{r}} + \left(\frac{\partial}{\partial z}\,U_r - \frac{\partial}{\partial r}U_z\right)\hat{\boldsymbol{\theta}} + \left(\frac{1}{r}\frac{\partial}{\partial r}(rU_\theta) - \frac{1}{r}\frac{\partial}{\partial\theta}\,U_r\right)\hat{\mathbf{z}}$$

$$\nabla^2\phi = \frac{1}{r}\frac{\partial}{\partial r}\left(r\frac{\partial}{\partial r}\phi\right) + \frac{1}{r^2}\frac{\partial^2}{\partial\theta^2}\phi + \frac{\partial^2}{\partial z^2}\phi \equiv \frac{\partial^2}{\partial r^2}\phi + \frac{1}{r}\frac{\partial}{\partial r}\phi + \frac{1}{r^2}\frac{\partial^2}{\partial\theta^2}\phi + \frac{\partial^2}{\partial z^2}\phi$$

doi:10.1088/978-0-7503-1251-6ch39

Appendix J

Derivations of some special results

The calculations of section 29.3 are rather lengthy and messy, and were omitted from the discussion there. For completeness these calculations are reproduced here. Frankly speaking, the results as such are not so interesting nor important, but the assumptions made on the way are significant and the methods are valuable (Beran 1968).

First we form

$$\begin{aligned}\langle J(1)\rangle &= X_1(12)\langle\Psi(2)\rangle + \frac{1}{2}X_2(123)\langle\Psi(2)\Psi(3)\rangle \\ &= X_1(12)\langle\Psi(2)\rangle + \frac{1}{2}X_2(123)C(23) + \frac{1}{2}X_2(123)\langle\Psi(2)\rangle\langle\Psi(3)\rangle,\end{aligned} \tag{J.1}$$

where we introduced the correlation function. Soon we will assume that $\langle\Psi(j)\rangle = 0$ for all j, but note that this does not mean that $\delta\langle\Psi(j)\rangle = 0$. We now obtain

$$\begin{aligned}\frac{\delta\langle J(1)\rangle}{\delta\langle\Psi(1')\rangle} &\approx X_1(12)\frac{\delta\langle\Psi(2)\rangle}{\delta\langle\Psi(1')\rangle} \\ &\quad + \frac{1}{2}X_2(123)\langle\Psi(2)\rangle\frac{\delta\langle\Psi(3)\rangle}{\delta\langle\Psi(1')\rangle} + \frac{1}{2}X_2(123)\frac{\delta\langle\Psi(2)\rangle}{\delta\langle\Psi(1')\rangle}\langle\Psi(3)\rangle,\end{aligned}$$

where we ignored $\delta C/\delta\langle\Psi\rangle$ to begin with. We note that Rule 1 from section 29.1 can be used in three places to give δ-functions, so we have

$$\frac{\delta\langle J(1)\rangle}{\delta\langle\Psi(1')\rangle} \approx X_1(11') + \frac{1}{2}X_2(121')\langle\Psi(1')\rangle + \frac{1}{2}X_2(11'3)\langle\Psi(3)\rangle.$$

We now assume the symmetry $X_2(123) = X_2(132)$, which is somewhat restrictive, but not unphysical. This assumption allows us to write $X_2(11'3)\langle\Psi(3)\rangle = X_2(131')\langle\Psi(3)\rangle = X_2(121')\langle\Psi(2)\rangle$, recalling that the 'middle variable' is merely a

doi:10.1088/978-0-7503-1251-6ch40

dummy integration-variable, making the last two terms in $\delta\langle J(1)\rangle/\delta\langle\Psi(1')\rangle$ identical, giving

$$\frac{\delta\langle J(1)\rangle}{\delta\langle\Psi(1')\rangle} \approx X_1(11') + X_2(121')\langle\Psi(2)\rangle.$$

By this we obtain the approximation

$$\begin{aligned}\Delta J(1) &\approx X_1(12)\Psi(2) + \frac{1}{2}X_2(123)\Psi(2)\Psi(3) \\ &\quad - (X_1(12) + X_2(123)\langle\Psi(3)\rangle)\Psi(2) \\ &\approx \frac{1}{2}X_2(123)\Psi(2)\Psi(3) - X_2(123)\langle\Psi(3)\rangle\Psi(2),\end{aligned}$$

where we retained only the nonlinear physics, making several changes of variables along the way. With the assumption of $\langle\Psi(j)\rangle = 0$ for all j in the sense discussed before, we have the simpler result

$$\Delta J(1) \approx \frac{1}{2}X_2(123)\Psi(2)\Psi(3).$$

With this approximation at hand we can express the correlation as

$$\begin{aligned}C(11') &\approx R(12)R(1'2')(\langle\Delta J(2)\Delta J(2')\rangle - \langle\Delta J(2)\rangle\langle\Delta J(2')\rangle) \\ &= \frac{1}{4}R(12)R(1'2')X_2(234)X_2(2'3'4')[\langle\Psi(3)\Psi(4)\Psi(3')\Psi(4')\rangle \\ &\quad - \langle\Psi(3)\Psi(4)\rangle\,\langle\Psi(3')\Psi(4')\rangle].\end{aligned}$$

At this point we make an important approximation by using the quasi-normal model to give

$$\begin{aligned}&\langle\Psi(3)\Psi(4)\Psi(3')\Psi(4')\rangle \approx \\ &\langle\Psi(3)\Psi(4)\rangle\langle\Psi(3')\Psi(4')\rangle + \langle\Psi(3)\Psi(3')\rangle\langle\Psi(4)\Psi(4')\rangle + \langle\Psi(3)\Psi(4')\rangle\langle\Psi(3')\Psi(4)\rangle.\end{aligned}$$

With two terms canceling we now reduce the correlation function to

$$\begin{aligned}C(11') &\approx \frac{1}{4}R(12)R(1'2')X_2(234)X_2(2'3'4') \\ &\quad \times [C(33')C(44') + C(34')C(3'4)] \\ &= \frac{1}{2}R(12)R(1'2')X_2(234)X_2(2'3'4')C(33')C(44') \\ &\equiv R(12)R(1'2')S(22'),\end{aligned} \tag{J.2}$$

where we used the fact that the last two terms in the angular brackets of the top line are the same (see this by a variable change), and that $X_2(2'3'4') = X_2(2'4'3')$. A 'thermal noise correlator' $S(22')$ was introduced for simplicity. The result (J.2) agrees with a result from DuBois and Espedal (1978). The role of the quasi-normal approximation in relation to the direct interaction approximation is clarified best by diagram methods, as in chapter 30.

We also need an expression for $\delta\langle J\rangle/\delta\langle\psi\rangle$. Ignoring a term which does not contribute since we take $\langle\psi\rangle = 0$ eventually, we have by (J.1)

$$\frac{\delta\langle J(1)\rangle}{\delta\langle\Psi(2)\rangle} \approx X_1(12')\frac{\delta\langle\Psi(2')\rangle}{\delta\langle\Psi(2)\rangle} + \frac{1}{2}X_2(12'3)\frac{\delta C(2'3)}{\delta\langle\Psi(2)\rangle}$$
$$= X_1(12') + \frac{1}{2}X_2(12'3)\frac{\delta C(2'3)}{\delta\langle\Psi(2)\rangle}.$$

By use of the expression for $C(11')$ we find

$$\frac{\delta C(11')}{\delta\langle\Psi(3)\rangle} = \frac{\delta R(12)}{\delta\langle\Psi(2)\rangle}R(1'2')S(22') + R(12)\frac{\delta R(1'2')}{\delta\langle\Psi(2)\rangle}S(22')$$
$$+ R(12)R(1'2')\frac{\delta S(22')}{\delta\langle\Psi(2)\rangle},$$

where the last term is ignored in the following since it is of higher order. For the two other terms we need $\delta R/\delta\langle\Psi\rangle$. We take again the result (19.10). Taking its functional derivative we obtain

$$\frac{\partial}{\partial t_1}\delta(t_1 - t_2)\frac{\delta R(21')}{\delta\langle\Psi(3)\rangle} - \left(\frac{\delta}{\delta\langle\Psi(3)\rangle}\frac{\delta\langle J(1)\rangle}{\delta\langle\Psi(2)\rangle}\right)R(21') - \frac{\delta\langle J(1)\rangle}{\delta\langle\Psi(2)\rangle}\frac{\delta R(21')}{\delta\langle\Psi(3)\rangle} = 0.$$

This result can be expressed in a form we have seen before in (29.10), i.e. as

$$\left(\frac{\partial}{\partial t_1}\delta(t_1 - t_2) - \frac{\delta\langle J(1)\rangle}{\delta\langle\Psi(2)\rangle}\right)\frac{\delta R(21')}{\delta\langle\Psi(3)\rangle} = R(21')\frac{\delta}{\delta\langle\Psi(3)\rangle}\frac{\delta\langle J(1)\rangle}{\delta\langle\Psi(2)\rangle}. \tag{J.3}$$

We approximate the term on the right hand side as follows

$$\frac{\delta}{\delta\langle\Psi(3)\rangle}\frac{\delta}{\delta\langle\Psi(2)\rangle}\langle J(1)\rangle = \frac{\delta}{\delta\langle\Psi(3)\rangle}\left(\frac{\delta}{\delta\langle\Psi(2)\rangle}\frac{1}{2}X_2(12'3')\langle\Psi(2')\rangle\langle\Psi(3')\rangle\right)$$
$$= \frac{\delta}{\delta\langle\Psi(3)\rangle}X_2(12'3')\langle\Psi(3')\rangle\delta(2 - 2')$$
$$= \frac{\delta}{\delta\langle\Psi(3)\rangle}X_2(123')\langle\Psi(3')\rangle$$
$$= X_2(123')\delta(3 - 3')$$
$$= X_2(123),$$

where we have ignored a higher order term in

$$X_2\frac{\delta}{\delta\langle\Psi\rangle}\frac{\delta\langle C\rangle}{\delta\langle\Psi\rangle}.$$

With the present approximation we have (J.3) in the form

$$\left(\frac{\partial}{\partial t_1}\delta(t_1 - t_2) - \frac{\delta\langle J(1)\rangle}{\delta\langle\Psi(2)\rangle}\right)\frac{\delta R(21')}{\delta\langle\Psi(3)\rangle} = X_2(123)R(21').$$

From (29.10) we know that the Green's function for this equation is $R(21)$, so we can write

$$\frac{\delta\langle J(1)\rangle}{\delta\langle\Psi(2)\rangle} = R(21)X_2(12'3)R(2'1').$$

Using this result, we obtain

$$\frac{\delta C(11')}{\delta\langle\Psi(3)\rangle} = R(11'')X_2(1''2''3)R(2''2)R(1'2')S(22') + R(12)R(1'1')X_2(1''2''3)R(2''2')S(22').$$

On the other hand we have $R(2''2)R(1'2')S(22') = C(2''1')$ while $R(12)R(2''2')S(22') = C(12'')$. We can write

$$\frac{\delta C(11')}{\delta\langle\Psi(3)\rangle} = R(11'')X_2(1''2''3)C(2''1') + R(1'1'')X_2(1''2''3)C(12'').$$

Finally, we are in a position to obtain the expression

$$\frac{\delta\langle J(1)\rangle}{\delta\langle\Psi(3)\rangle} = X_1(12) + \frac{1}{2}X_2(12'3)\frac{\delta C(2'3)}{\delta\langle\Psi(3)\rangle}$$
$$= X_1(12) + \frac{1}{2}X_2(12'3)[R(2'1'')X_2(1''2''2)C(2''3) + R(31'')X_2(1''2''2)C(2'2'')].$$

By a simple change in variables we find that the two terms in the parenthesis on the right hand side are the same so we have the simpler result

$$\frac{\delta\langle J(1)\rangle}{\delta\langle\Psi(3)\rangle} = X_1(12) + X_2(12'3)X_2(1''2''2)R(2'1'')C(2''3).$$

The response equation (29.10) can finally be written in the form

$$\frac{\partial}{\partial t_1}\delta(1-2)R(21') = \delta(1-1') + X_1(12)R(21') + X_2(12'3)X_2(1''2''2)R(2'1'')C(2''3)R(21'),$$

which can be rewritten as

$$\frac{\partial}{\partial t_1}R(11') = \delta(1-1') + X_1(12)R(21') + X_2(123)X_2(1''2'3')R(21'')R(3'1')C(2'3).$$

This is the result quoted in (29.16). This equation is one of the two constituents of the direct interaction approximation used in section 29.3. Dwight had the energy to go through all these calculations!

References

Beran M J 1968 *Statistical Continuum Theories* Vol 9 *of Monographs in Statistical Physics and Thermodynamics* (New York: Interscience)

DuBois D F and Espedal M 1978 Direct interaction approximation and plasma turbulence theory *Plasma Phys.* **20** 1209–39

www.ingramcontent.com/pod-product-compliance
Ingram Content Group UK Ltd.
Pitfield, Milton Keynes, MK11 3LW, UK
UKHW051911100926
8556IPUK00005B/845